Springer Series in Computational Mathematics

44

For further volumes:
http://www.springer.com/series/797

Daniele Boffi • Franco Brezzi • Michel Fortin

Mixed Finite Element Methods and Applications

Springer

Daniele Boffi
Dipartimento di Matematica "F. Casorati"
University of Pavia
Pavia
Italy

Franco Brezzi
IUSS (Istituto Universitario di Studi Superiori)
Pavia
Italy

Michel Fortin
Département de mathématiques et de statistique
Université Laval
Québec
Canada

ISSN 0179-3632
ISBN 978-3-642-36518-8 ISBN 978-3-642-36519-5 (eBook)
DOI 10.1007/978-3-642-36519-5
Springer Heidelberg New York Dordrecht London

Library of Congress Control Number: 2013940257

Mathematics Subject Classification (2010): 6502, 65N15

Printed on acid-free paper

Springer is part of Springer Science+Business Media (www.springer.com)

Preface

About 10 years ago, Mixed and Hybrid Finite Element Methods by F. Brezzi and M. Fortin went out of print and we were asked to allow a second printing. The world had evolved and we thought that a revision was due and that some topics had to be added to the book. For this task, D. Boffi joined the team and we began to write the improved version. It turned out that this meant doubling the number of pages and essentially producing a new book.

We hope that the result is now a better, self-contained, presentation of the underlying issues, either from linear algebra or from functional analysis. The presentation of the basic results should now be accessible to readers which are not familiar with functional analysis, although willing to invest some effort in understanding mathematical issues.

The scope of finite element approximations was extended to $H(\underline{\text{curl}};\Omega)$ and the three-dimensional cases are now fully covered. Tensor elements were also considered for elasticity problems. The approximation of eigenvalue problems has been included as well.

Moreover, new applications have been introduced: mixed elasticity and electromagnetism. New results have been added to already treated applications such as the Stokes problem or mixed formulations of elliptic problems. Even so, some topics have been merely addressed. This is, for example, the case of a posteriori estimators, Discontinuous Galerkin methods and new developments on virtual elements which would have required a long development in an already (too?) long book. Indeed, each of these topics could be the subject of a whole book. The analysis of mixed methods is also relevant to many applications such as mortar methods or contact problems which were also reduced to a few remarks. This does not mean that these are not important. We had to stop somewhere. Indeed, we took a long time to do so.

We thus hope that this book will provide a good starting point for all those interested in mixed (and related) finite element methods.

Pavia, Italy
Québec, Canada

D. Boffi and F. Brezzi
M. Fortin

Contents

Chapter 1
Variational Formulations and Finite Element Methods

Although we shall not define in this chapter mixed and hybrid (or other non-standard) finite element methods in a very precise way, we would like to situate them in a sufficiently clear setting. As we shall see, boundaries between different methods are sometimes rather fuzzy. This will not be a real drawback if we nevertheless know how to apply correctly the principles underlying their analysis.

After having briefly recalled some basic facts about classical methods, we shall present a few model problems. The study of these problems will be the kernel of this book. Thereafter, we rapidly recall basic principles of *duality theory* as this will be our starting point to introduce mixed methods. *Domain decomposition* methods (allied to duality) will lead us to hybrid methods. Then we shall briefly discuss modified variational formulations that can be used to obtain better stability properties for the discretised versions.

1.1 Classical Methods

We recall here, in a very simplified way, some facts about optimisation methods and the classical finite element method. Such an introduction cannot be complete and does not want to be. We refer the reader to [146] or [334], among others, where standard finite element methods are clearly exposed. We also refer to [167] where an exhaustive analysis of many of our model problems can be found.

Let us consider a very common situation where the solution of a physical problem minimises some functional (usually an "energy functional"), in a "well chosen" space of admissible functions V that we take for the moment as a Hilbert space:

$$\inf_{v\in V} J(v). \tag{1.1.1}$$

D. Boffi et al., *Mixed Finite Element Methods and Applications*, Springer Series in Computational Mathematics 44, DOI 10.1007/978-3-642-36519-5_1,

If the functional $J(\cdot)$ is differentiable (cf. [184] for instance) the minimum (whenever it exists) will be characterised by *a variational equation*

$$\langle J'(u), v\rangle_{V'\times V} = 0, \quad \forall v \in V, \tag{1.1.2}$$

where $\langle\cdot,\cdot\rangle_{V'\times V}$ denotes duality between V and its topological dual V', the derivative $J'(u)$ at point u being considered as a linear form on V.

The classical Ritz's method to approximate the solution of (1.1.1) consists in choosing a finite dimensional subspace V_m of V, and then looking for $u_m \in V_m$ solution of the problem

$$\inf_{v_m \in V_m} J(v_m), \tag{1.1.3}$$

or, differentiating,

$$\langle J'(u_m), v_m\rangle_{V'\times V} = 0, \quad \forall v_m \in V_m. \tag{1.1.4}$$

Let us consider, to fix ideas, a quadratic functional

$$J(v) := \frac{1}{2}a(v,v) - L(v), \tag{1.1.5}$$

where $a(\cdot,\cdot)$ is a bilinear form on V, which we suppose continuous and symmetric, and $L(\cdot)$ a linear form on V. The variational equation (1.1.2) can then be written as

$$a(u,v) = L(v) \quad \forall v \in V, \tag{1.1.6}$$

while the discrete problem (1.1.4) becomes

$$a(u_m, v_m) = L(v_m), \quad \forall\, v_m \in V_m, \ \ u_m \in V_m. \tag{1.1.7}$$

If a basis $w_1, w_2, \ldots, w_m$ of V_m is chosen, and if we write

$$u_m = \sum_{i=1}^{m} \alpha_i w_i, \tag{1.1.8}$$

problem (1.1.7) is reduced to the solution of the linear system

$$\sum_{i=1}^{m} a_{ij}\alpha_i = b_j, \ \ 1 \le j \le m, \tag{1.1.9}$$

where we set

$$a_{ij} := a(w_i, w_j), \ \ b_j := L(w_j). \tag{1.1.10}$$

This formulation can be extended to the case where the bilinear form $a(\cdot,\cdot)$ *is not symmetric* and where problem (1.1.7) no longer corresponds to a minimisation problem. This is then usually called a Galerkin method. Let us recall that problems of type (1.1.7) will have a unique solution if, in particular, the bilinear form $a(\cdot,\cdot)$ is *coercive*, that is if there exists a positive real number α such that for all v in V

$$a(v,v) \geq \alpha \|v\|_V^2. \tag{1.1.11}$$

The above described methodology is very general and classical. We can consider the finite element method as a special case in the following sense.

The finite element method is a general technique to build finite dimensional subspaces of a Hilbert space V in order to apply the Ritz-Galerkin method to a variational problem.

This technique is based on a few simple ideas. The fundamental one is the *partition of the domain Ω* in which the problem is posed, into a set of "simple" sub-domains, called elements. These elements are usually triangles, quadrilaterals, tetrahedra, etc. A space V *of functions defined on Ω is then approximated* by "simple" functions, defined on each sub-domain with suitable matching conditions at interfaces. Simple functions are usually polynomials or functions obtained from polynomials by a change of variables.

This, of course, a very summarised way of defining finite elements and this is surely not the best way to understand it from the computational point of view. We shall come back to this in Chap. 2 with a much more workable approach.

The point that we want to emphasise here is the following. *A finite element method can only be considered in relation with a variational principle and a functional space. Changing the variational principle and the space in which it is posed leads to a different finite element approximation (even if the solution for the continuous problems can remain the same).*

In the remaining of this Chapter, we shall see how different variational formulations can be built for the same physical problem. Each of these formulations will lead to a new setting for finite element approximations. *The common point of the methods analysed in this book is that they are founded on a variational principle expressing an equilibrium* (saddle point) *condition* rather than on a minimisation principle. We shall now try to see, on some examples, how such equilibrium principles can be built.

1.2 Model Problems and Elementary Properties of Some Functional Spaces

The aim of this section is to introduce some notation and to present five model problems that will underlie almost all cases analysed in this book. They will be the Dirichlet problem for Laplace's equation, the linear elasticity problem, Stokes' problem, a fourth-order problem modelling the deflection of a thin clamped plate, and, finally, the time-harmonic Maxwell system. These problems are closely interrelated and methods to analyse them will also be.

We shall present, in this section, the most classical variational formulation of these problems. The following sections will lead us to less standard forms.

We shall assume, in our exposition, that the problems are posed in a domain Ω of $\mathbb{R}^n$, with a sufficiently smooth boundary $\partial\Omega = \Gamma$ (for instance a Lipschitz continuous boundary). In practice $n = 2$ or 3 and we shall present most of our examples in a two-dimensional setting for the sake of simplicity.

In the problems considered here, working in $\mathbb{R}^2$ rather than in $\mathbb{R}^3$ is not really restrictive and extensions are generally straightforward. (This is however not always the case for numerical methods.) Let us first recall some definitions. We shall constantly use Sobolev spaces [3, 281, 309]. They are based on

$$L^2(\Omega) := \left\{ v \Big| \int_\Omega |v|^2 \, dx = \|v\|^2_{L^2(\Omega)} < +\infty \right\}, \tag{1.2.1}$$

the space of square integrable functions on Ω. To be precise, instead of *functions* we should actually say *classes of measurable functions*, meaning that a class is made of functions that differ from each other only on a subset of Ω of zero Lebesgue measure. Having said this once, we shall keep calling them simply *functions*. We then define in general, for m integer ≥ 0,

$$H^m(\Omega) := \{ v \,|\, D^\alpha v \in L^2(\Omega), \quad \forall |\alpha| \leq m \}, \tag{1.2.2}$$

where

$$D^\alpha v := \frac{\partial^{|\alpha|} v}{\partial x_1^{\alpha_1} \cdots \partial x_n^{\alpha_n}}, \qquad |\alpha| = \alpha_1 + \cdots + \alpha_n,$$

these derivatives being taken *in the sense of distributions*. On this space, we shall use the semi-norms

$$|v|^2_{k,\Omega} := \sum_{|\alpha|=k} |D^\alpha v|^2_{L^2(\Omega)}, \qquad k = 0, 1, \ldots, m, \tag{1.2.3}$$

and the norm

$$\|v\|^2_{m,\Omega} := \sum_{k \leq m} |v|^2_{k,\Omega}. \tag{1.2.4}$$

Remark 1.2.1. The norm in (1.2.4) is definitely *the weirdest* aspect of the whole theory of Sobolev spaces. Indeed, one should take a *typical length* ℓ of the problem (as for instance the diameter of Ω) and use, instead of (1.2.4):

$$\|v\|^2_{m,\Omega} := \sum_{k \leq m} \ell^{2k} |v|^2_{k,\Omega}, \tag{1.2.5}$$

avoiding, in this way, to sum objects with different physical dimensions. The expression (1.2.4), which is by far *the most widely used in all the international literature* assumes implicitly that the problem has been adimensionalised, something that one is not always willing to do. Needless to say, (1.2.4) has a lot of advantages, and we are often going to use it. Nevertheless, we felt compelled to give at least a minor warning to our readers. □

The space $L^2(\Omega)$ is then $H^0(\Omega)$ and we shall usually write $\|v\|_{0,\Omega}$, to denote its norm $\|v\|_{L^2(\Omega)}$. Let us denote as usual by $\mathfrak{D}(\Omega)$ the space of infinitely differentiable functions having a compact support in Ω. We denote by $H_0^m(\Omega)$ the completion of $\mathfrak{D}(\Omega)$ for the topology defined by the norm (1.2.4). If the boundary is smooth enough (e.g. Lipschitz continuous boundary) this simple definition will coincide, without troublesome pathologies, with the more common

$$H_0^m(\Omega) := \{v |\; v \in H^m(\Omega) \text{ s. t. } v = \frac{\partial v}{\partial n} = \cdots = \frac{\partial^{m-1} v}{\partial n^{m-1}} = 0 \text{ on } \Gamma\}, \tag{1.2.6}$$

where $\underline{n}$ is the normal direction to $\Gamma = \partial\Omega$. The drawback in the definition (1.2.6) is however the difficulty in giving sense to the value of v (and, if $m > 1$, to its derivatives) on the boundary of Ω. We shall shortly give a hint on how this could be made precise for the most common cases $m = 1$ and $m = 2$.

Indeed, among the spaces introduced so far, the most commonly used, apart from $L^2(\Omega)$, will be $H^1(\Omega)$, $H_0^1(\Omega)$, $H^2(\Omega)$ and $H_0^2(\Omega)$.

If the boundary $\partial\Omega$ is sufficiently smooth (and, again, Lipschitz continuity will be enough), one can show that there exists a linear and continuous operator γ_0 : $H^1(\Omega) \to L^2(\Gamma)$ such that $\gamma_0 v$ *coincides with the restriction of* v *to* Γ *whenever* v *is smooth* (say, to fix the ideas, for every $v \in C^1(\bar{\Omega})$). It seems then natural to call $\gamma_0 v$ "the trace of v on Γ" and denote it by $v|_\Gamma$ even if v is a general function in $H^1(\Omega)$ that might not be in $C^1(\bar{\Omega})$.

A deeper analysis shows that by taking all the traces of all the functions of $H^1(\Omega)$, one does not obtain the whole space $L^2(\Gamma)$ but only a subspace of it. Further investigations show that such a subspace contains $H^1(\Gamma)$ as a proper subset. Hence we have,

$$H^1(\Gamma) \subset \gamma_0(H^1(\Omega)) \subset L^2(\Gamma) \equiv H^0(\Gamma), \tag{1.2.7}$$

where every inclusion is strict. It is finally recognised that the space $\gamma_0(H^1(\Omega))$ belongs to a family of spaces $H^s(\Gamma)$ (defined for all $s \in \mathbb{R}$, that we are not going to detail here) and corresponds exactly to the value $s = \frac{1}{2}$. Hence we have

$$H^{\frac{1}{2}}(\Gamma) := \gamma_0(H^1(\Omega)), \tag{1.2.8}$$

with

$$\|g\|_{H^{\frac{1}{2}}(\Gamma)} := \inf_{\substack{v \in H^1(\Omega) \\ \gamma_0 v = g}} \|v\|_{H^1(\Omega)}. \tag{1.2.9}$$

In a similar way we could see that the traces of functions in $H^2(\Omega)$ belong to a space $H^s(\Gamma)$ for $s = \frac{3}{2}$. We may therefore set

$$H^{\frac{3}{2}}(\Gamma) := \gamma_0(H^2(\Omega)), \tag{1.2.10}$$

$$\|g\|_{H^{\frac{3}{2}}(\Gamma)} := \inf_{\substack{v\in H^2(\Omega)\\ \gamma_0 v=g}} \|v\|_{H^2(\Omega)}. \tag{1.2.11}$$

This can be generalised to the traces of higher order derivatives. For instance, if the boundary Γ is Lipschitz continuous, one can define a linear continuous operator $\gamma_1 : H^2(\Omega) \to L^2(\Omega)$ such that $\gamma_1 v$ coincides with the trace of the normal derivative of v whenever v is smooth (say, $v \in C^2(\bar{\Omega})$). Proceeding as before we could then define, for $v \in H^2(\Omega)$, the *trace of the normal derivative*:

$$\frac{\partial v}{\partial n}\Big|_\Gamma \equiv \gamma_1 v.$$

If the boundary Γ is smooth enough (but, here, Lipschitz continuity will *not* be enough: one needs at least C^1) one gets

$$\gamma_1(H^2(\Omega)) = H^{\frac{1}{2}}(\Gamma) \qquad \text{(for } \partial\Omega \text{ smoother: not for a polygon).}$$

Note however that for less regular domains this will no longer be true: for instance, if Ω is a polygon we have that $\gamma_1 v$ belongs to a more complicated space that, roughly speaking, is made of functions whose restriction to each edge e belongs to $H^{\frac{1}{2}}(e)$. We shall not discuss in a more precise way trace theorems on Sobolev spaces of fractional order. (The reader may refer to the authors quoted above.) Intuitively, Sobolev spaces of fractional order can be considered as having regularity properties that are intermediate between the properties of the neighbouring integer order spaces and they can indeed be defined as *interpolation spaces*. Taking this as granted, we then have

$$H_0^1(\Omega) := \{v|\ v \in H^1(\Omega),\ v|_\Gamma := 0\}, \tag{1.2.12}$$

$$H_0^2(\Omega) = \left\{v|\ v \in H^2(\Omega),\ v|_\Gamma = 0,\ \frac{\partial v}{\partial n}\Big|_\Gamma = 0\right\}. \tag{1.2.13}$$

For $v \in H_0^1(\Omega)$, we have the *Poincaré inequality*

$$\|v\|_{0,\Omega} \le C(\Omega)|v|_{1,\Omega}, \tag{1.2.14}$$

and the semi-norm $|\cdot|_{1,\Omega}$ is therefore a norm on $H_0^1(\Omega)$, equivalent to $\|\cdot\|_{1,\Omega}$. We shall also need to consider functions that vanish on a part of the boundary; suppose that $\Gamma = D \cup N$ is a "reasonable" partition of Γ into disjoint parts; then we can define

$$H_{0,D}^1(\Omega) := \{v|\ v \in H^1(\Omega),\ v|_D = 0\}, \tag{1.2.15}$$

and one has $H_0^1(\Omega) \subset H_{0,D}^1(\Omega) \subset H^1(\Omega)$.

When considering vector-valued functions, additional spaces will be useful. In particular we shall use

$$H(\mathrm{div};\Omega) := \{\underline{v} \in L^2(\Omega)^n \,|\, \mathrm{div}\,\underline{v} \in L^2(\Omega)\}, \tag{1.2.16}$$

$$H(\underline{\mathrm{curl}};\Omega) := \{\underline{v} \in L^2(\Omega)^n \,|\, \underline{\mathrm{curl}}\,\underline{v} \in L^2(\Omega)^d\}, \tag{1.2.17}$$

where in (1.2.17) we have $d = 1$ when $n = 2$ and $d = 3$ when $n = 3$. The divergence and $\underline{\mathrm{curl}}$ operators are defined as usual; particular care has to be taken for the definition of the two-dimensional $\underline{\mathrm{curl}}$ operator which in this case is understood as $\underline{\mathrm{curl}}\,\underline{v} = \mathrm{div}(\underline{v}^{\perp})$, where $\underline{v}^{\perp}$ is the rotation of $\underline{v}$ by an angle of $\pi/2$. It can be shown that functions in $H(\mathrm{div};\Omega)$ (resp. $H(\underline{\mathrm{curl}};\Omega)$) admit traces of the normal (resp. tangential) component on Γ; namely, there exists a linear and continuous operator $\gamma_n : H(\mathrm{div};\Omega) \to H^{-1/2}(\Gamma)$ such that $\gamma_n \underline{v} =$ trace of $\underline{v} \cdot \underline{n}$ for every smooth $\underline{v}$, where $\underline{n}$ is the outward normal unit vector, and a similar property holds for the traces of the tangential components of vectors in $H(\underline{\mathrm{curl}};\Omega)$. More details on $H(\mathrm{div};\Omega)$ and $H(\underline{\mathrm{curl}};\Omega)$ will be given in Chap. 2. Spaces including homogeneous boundary conditions are denoted as follows

$$H_0(\mathrm{div};\Omega) := \{\underline{v} \mid \underline{v} \in H(\mathrm{div};\Omega),\ \underline{v}\cdot\underline{n} = 0 \text{ on } \Gamma\}, \tag{1.2.18}$$

$$H_0(\underline{\mathrm{curl}};\Omega) := \{\underline{v} \mid \underline{v} \in H(\underline{\mathrm{curl}};\Omega),\ \underline{v}\times\underline{n} = 0 \text{ on } \Gamma\}. \tag{1.2.19}$$

We shall come back in Chap. 2 to the properties of these spaces; the above definitions are sufficient to allow us to present some examples.

Example 1.2.1 (Boundary value problems for the Laplace equation). This is a very classical case that in fact led to the definition of Sobolev spaces. For f given in $L^2(\Omega)$, let us consider the following minimisation problem on $H_0^1(\Omega)$:

$$\inf_{q\in H_0^1(\Omega)} \Big(\frac{1}{2}\int_\Omega |\,\underline{\mathrm{grad}}\,q|^2\,dx - \int_\Omega fq\,dx\Big), \tag{1.2.20}$$

where $|\,\underline{\mathrm{grad}}\,q|^2 = |\frac{\partial q}{\partial x_1}|^2 + |\frac{\partial q}{\partial x_2}|^2 = \underline{\mathrm{grad}}\,q \cdot \underline{\mathrm{grad}}\,q$. One shows easily (cf. [141, 281, 309] for instance) that this problem has a unique solution p, characterised by: $p \in H_0^1(\Omega)$ and

$$\int_\Omega \underline{\mathrm{grad}}\,p \cdot \underline{\mathrm{grad}}\,q\,dx = \int_\Omega fq\,dx, \quad \forall q \in H_0^1(\Omega). \tag{1.2.21}$$

This is of the form (1.1.6), by setting

$$a(p,q) = \int_\Omega \underline{\mathrm{grad}}\,p \cdot \underline{\mathrm{grad}}\,q\,dx$$

This solution p satisfies, in the sense of distributions,

$$\begin{cases} -\Delta p = f \text{ in } \Omega, \\ p\big|_{\Gamma} = 0, \end{cases} \tag{1.2.22}$$

which is a standard Dirichlet problem for the Laplace operator Δ, commonly called *Poisson problem.* If $H_0^1(\Omega)$ were replaced by $H_{0,D}^1(\Omega)$ one would get instead of (1.2.22) a mixed type problem

$$\begin{cases} -\Delta p = f \text{ in } \Omega, \\ p = 0 \text{ on } \Gamma_D, \\ \dfrac{\partial p}{\partial n} = 0 \text{ on } \Gamma_N. \end{cases} \tag{1.2.23}$$

We thus have Dirichlet boundary conditions on Γ_D and Neumann conditions on Γ_N. In particular for $\Gamma_N = \Gamma$, we get a Neumann problem. It must be noted that minimising (1.2.20) on $H^1(\Omega)$ instead of $H_0^1(\Omega)$ will define p up to an additive constant and requires the compatibility condition

$$\int_{\Omega} f\, dx = 0,$$

which can be seen to be necessary from (1.2.21) by taking $q \equiv 1$ in Ω.

If we denote by $H^{-\frac{1}{2}}(\Gamma)$ the dual space of $H^{\frac{1}{2}}(\Gamma)$, and we take $g \in H^{-\frac{1}{2}}(\Gamma)$, we can consider the functional

$$\frac{1}{2}\int_{\Omega} |\,\underline{\mathrm{grad}}\, q|^2\, dx - \int_{\Omega} fq\, dx + \langle g, q\rangle, \tag{1.2.24}$$

where the bracket $\langle\cdot,\cdot\rangle$ denotes the duality between $H^{-\frac{1}{2}}(\Gamma)$ and $H^{\frac{1}{2}}(\Gamma)$. We shall sometimes write *formally* $\int_{\Gamma} gq\, ds$ instead of $\langle g, v\rangle$. Minimising (1.2.24) on $H_{0,D}^1(\Omega)$ leads to the problem

$$\begin{cases} -\Delta p = f \text{ in } \Omega, \\ p = 0 \text{ on } \Gamma_D, \\ \dfrac{\partial p}{\partial n} = g \text{ on } \Gamma_N. \end{cases} \tag{1.2.25}$$

When $\Gamma_D = \emptyset$ the solution is defined up to an additive constant and we must choose f and g such that

$$\int_{\Omega} f\, dx - \int_{\Gamma} g\, ds = 0.$$

These problems are among the most classical of mathematical physics and we do not have to emphasise their importance. In the following chapters we shall need to use *regularity results* for the problems introduced above. We have supposed up to now $f \in L^2(\Omega)$. For the Poisson problem (1.2.22) we could have assumed f to belong to a weaker space, namely $f \in H^{-1}(\Omega) = (H_0^1(\Omega))'$, and nevertheless obtained $p \in H_0^1(\Omega)$. On the other hand if f is taken in $L^2(\Omega)$ and Ω is convex one can prove [309] that $p \in H^2(\Omega)$ and that

$$\|p\|_{2,\Omega} \leq c\|f\|_{0,\Omega}, \tag{1.2.26}$$

where c is a constant depending only on Ω. Regularity results are essential to many approximations results and are fundamental to obtain error estimates. We refer the reader to [233] for the delicate questions of the regularity of the general problem (1.2.23) in a domain with corners. □

Example 1.2.2 (Linear elasticity). We want to determine the displacement $\underline{u} = \{u_1, u_2\}$ of an elastic material under the action of some external forces. We suppose the displacement to be small and the material to be isotropic and homogeneous [146, 295]. The domain Ω is the initial configuration of the body. To set our problem, we must introduce some notation from continuum mechanics. First we define the linearised strain tensor $\underline{\underline{\varepsilon}}(\underline{u})$ by

$$\underline{\underline{\varepsilon}}(\underline{u}) := \frac{1}{2}\Big(\frac{\partial u_i}{\partial x_j} + \frac{\partial u_j}{\partial x_i}\Big). \tag{1.2.27}$$

The trace, $\mathrm{tr}(\underline{\underline{\varepsilon}})$, of this tensor is nothing but the divergence of the displacement field

$$\mathrm{tr}(\underline{\underline{\varepsilon}})(\underline{v}) = \mathrm{div}\ \underline{v}. \tag{1.2.28}$$

We shall also use the *deviatoric* $\underline{\underline{\varepsilon}}^D$ of the tensor $\underline{\underline{\varepsilon}}$ that is

$$\underline{\underline{\varepsilon}}^D := \underline{\underline{\varepsilon}} - \frac{1}{n}\,\mathrm{tr}(\underline{\underline{\varepsilon}})\underline{\underline{\delta}}, \tag{1.2.29}$$

where $\underline{\underline{\delta}}$ is the standard Kronecker tensor and n is the space dimension. The deviatoric is evidently built to have $\mathrm{tr}(\underline{\underline{\varepsilon}}^D) = 0$. Let then Γ_0 be a part of Γ on which we assume $\underline{u} = 0$. We also assume the existence in Ω of a distributed force $\underline{f}$ (e.g. gravity) and on Γ_1 of a traction $\underline{g}$ that is decomposed into a normal part g_n and a tangential part g_t. We denote by $\underline{n}$ and $\underline{t}$ the normal and tangential unit vectors to Γ. Let us denote

$$|\underline{\underline{\varepsilon}}|^2 = \underline{\underline{\varepsilon}} : \underline{\underline{\varepsilon}} := \sum_{i,j} \varepsilon_{ij}^2 \tag{1.2.30}$$

and let us consider in $V := (H^1_{0,\Gamma_0}(\Omega))^2$ the minimisation problem

$$\inf_{\underline{v}\in V}\Big\{\int_\Omega \frac{1}{2}(\lambda|\,\mathrm{div}\ \underline{v}|^2 + 2\mu|\underline{\underline{\varepsilon}}(v)|^2)\,dx - \int_\Omega \underline{f}\cdot\underline{v}\,dx \\ - \int_{\Gamma_1} g_n\underline{v}\cdot\underline{n}\,ds - \int_{\Gamma_1} g_t\,\underline{v}\cdot\underline{t}\,ds\Big\}. \tag{1.2.31}$$

Constants λ and μ, the Lamé coefficients, depend on the physical properties of the material considered. The solution $\underline{u}$ of this problem is then characterised by

$$\begin{aligned} 2\mu \int_\Omega \underline{\underline{\varepsilon}}(\underline{u}) : \underline{\underline{\varepsilon}}(\underline{v})\, dx + \lambda \int_\Omega \operatorname{div} \underline{u} \ \operatorname{div} \underline{v}\, dx \\ = \int_\Omega \underline{f} \cdot \underline{v}\, dx + \int_{\Omega_1} g_n \underline{v} \cdot \underline{n}\, ds + \int_{\Gamma_2} g_t \underline{v} \cdot \underline{t}\, ds, \quad \forall \underline{v} \in V, \end{aligned} \tag{1.2.32}$$

which is still clearly of the form (1.1.6). We now use the classical integration by parts formula

$$\int_\Omega \underline{\underline{m}} : \underline{\underline{\varepsilon}}(\underline{v})\, dx = -\int_\Omega (\operatorname{div} \underline{\underline{m}}) \cdot \underline{v}\, dx + \int_\Gamma m_{nn} \underline{v} \cdot \underline{n} ds + \int_\Gamma m_{nt} \underline{v} \cdot \underline{t}\, ds \tag{1.2.33}$$

where $\underline{\underline{m}}$ is a (smooth enough) tensor, and m_{nn} and m_{nt} denote the normal and tangential parts of the traction vector m_n, i.e.

$$\begin{cases} m_{nn} := \sum_{i,j} m_{ij}\, n_i\, n_j = \sum_i \Big\{ \sum_j m_{ij} n_j \Big\} n_i = \sum_i (\underline{m}_n)_i\, n_i, \\ m_{nt} := \sum_{i,j} m_{ij} t_i\, n_j = \sum_i \Big\{ \sum_j m_{ij} n_j \Big\} t_i = \sum_i (\underline{m}_n)_i\, t_i. \end{cases} \tag{1.2.34}$$

Equation (1.2.32) can now be interpreted as

$$\begin{cases} -(2\mu\, \underline{\operatorname{div}}\, \underline{\underline{\varepsilon}}(\underline{u}) + \lambda\, \underline{\operatorname{grad}} \operatorname{div} \underline{u}) = \underline{f} \text{ in } \Omega, \\ \underline{u}\big|_{\Gamma_0} = 0, \\ 2\mu\varepsilon_{nn} + \lambda \ \operatorname{div}\ \underline{u} = g_n, \\ 2\mu\varepsilon_{nt} = g_t \text{ on } \Gamma_1. \end{cases} \tag{1.2.35}$$

Let us now introduce the stress tensor $\underline{\underline{\sigma}} := \underline{\underline{s}}^D + p\underline{\underline{\delta}}$ and the constitutive law

$$\begin{cases} \underline{\underline{s}}^D := 2\mu \underline{\underline{\varepsilon}}^D(\underline{u}), \\ p := 2(\lambda + \mu) \ \operatorname{div}\ \underline{u}, \end{cases} \tag{1.2.36}$$

relating stresses to displacements. It is now clear that the first equation of (1.2.35) expresses the equilibrium condition of continuum mechanics,

$$\underline{\operatorname{div}}\, \underline{\underline{\sigma}} + \underline{f} = 0. \tag{1.2.37}$$

In applications, the constitutive law (1.2.36) will vary depending on the type of materials and will sometimes take very non linear forms. In this case, the expression of the energy functional (1.2.31) will change accordingly. Moreover, large displacements will require a much more complex treatment. Nevertheless, the problem described above remains valuable as a model for more complicated situations.

The case of an incompressible material is especially important. It leads to the same equations as in the study of viscous incompressible flows. □

Example 1.2.3 (Stokes' problem for viscous incompressible flow.). We now consider a low velocity flow of a viscous incompressible fluid in a domain Ω. We denote by $\underline{u}$ the velocity field and by $\underline{\underline{\varepsilon}}(\underline{u})$ the (linearised) strain rate tensor defined in the same way as in (1.2.27). We thus consider the minimisation problem with the same notation and the same space V as in Example 1.2.2, but now with the incompressibility condition $\operatorname{div}\underline{v} = 0$, that is,

$$\inf_{\substack{\underline{v}\in V\\ \operatorname{div}\underline{v}=0}} \mu\int_\Omega |\underline{\underline{\varepsilon}}(\underline{v})|^2 - \int_\Omega \underline{f}\cdot\underline{v}\,dx + \int_{\Gamma_1} g_n\,\underline{v}\cdot\underline{n}\,ds + \int_{\Gamma_1} g_t\,\underline{v}\cdot\underline{t}\,ds. \quad (1.2.38)$$

As we shall see later, problem (1.2.31) can be considered, when λ is large, as an approximation (by a "penalty method") of problem (1.2.38). Indeed when λ is large the second constitutive relation of (1.2.36) forces, in some sense, $\operatorname{div}\underline{v}$ to be zero. In the limit for $\lambda = +\infty$ (1.2.36) becomes meaningless: we shall see in Sect. 1.3 that pressure can then be introduced as a Lagrange multiplier associated with the constraint $\operatorname{div}\underline{u} = 0$. □

We now present a fourth-order problem. It is again, from the physical point of view, an elasticity problem but in a special modelling.

Example 1.2.4 (Deflection of a thin clamped plate). We consider here the problem of a thin clamped plate deflected under a distributed load f. The physical model will be described in Chap. 10. We also refer to [147, 148] and [149] for more details on plate problems. Under reasonable assumptions (and setting, for simplicity, some physical constants equal to 1), one obtains that the vertical deflection ψ is solution of the minimisation problem

$$\inf_{\varphi\in H_0^2(\Omega)} \frac{1}{2}\int_\Omega |\Delta\varphi|^2\,dx - \int_\Omega f\varphi\,dx. \quad (1.2.39)$$

The unique solution ψ is characterised by

$$\int_\Omega \Delta\psi\,\Delta\varphi\,dx = \int_\Omega f\varphi\,dx, \quad \forall\varphi\in H_0^2(\Omega), \quad (1.2.40)$$

and is the solution of the boundary value problem

$$\begin{cases} \Delta^2 \psi = f, \\ \psi|_\Gamma = 0, \\ \dfrac{\partial \psi}{\partial n}|_\Gamma = 0. \end{cases} \tag{1.2.41}$$

For these boundary conditions (representing a clamped plate) (1.2.39) is equivalent to

$$\inf_{\varphi \in H_0^2(\Omega)} \left\{ \frac{1}{2} \int_\Omega \left[\left\{ \frac{\partial^2 \varphi}{\partial x_1^2} \right\}^2 + 2 \left\{ \frac{\partial^2 \varphi}{\partial x_1 \partial x_2} \right\}^2 + \left\{ \frac{\partial^2 v \varphi}{\partial x_2^2} \right\}^2 \right] dx - \int_\Omega f \varphi \, dx \right\}, \tag{1.2.42}$$

which is, in general, more physically sound. These two mathematically equivalent forms can lead to different numerical methods. It must also be noted that *natural boundary conditions* (those arising from integration by parts) will not be the same if (1.2.39) and (1.2.42) are minimised on a space larger than $H_0^2(\Omega)$, so that the equivalence only holds for plates that are *clamped* all over the boundary. Actually the true potential energy of the plate (that is, the true functional which has to be minimised) is given, for a clamped plate, by

$$\begin{aligned} J(\varphi) := \frac{Et^3}{12(1-\nu^2)} \int_\Omega \Big\{ \nu |\Delta \varphi|^2 + (1-\nu) \Big[\Big(\frac{\partial^2 \varphi}{\partial x_1^2} \Big)^2 + 2 \Big(\frac{\partial^2 \varphi}{\partial x_1 \partial x_2} \Big)^2 \\ + \Big(\frac{\partial^2 \varphi}{\partial x_2^2} \Big)^2 \Big] \Big\} \, dx - \int_\Omega f \varphi \, dx, \end{aligned} \tag{1.2.43}$$

where E is Young's modulus, ν is the Poisson's coefficient and t is the thickness of the plate. In particular E and ν can be expressed in terms of the Lamé coefficients λ, μ in the following way

$$E := \frac{\mu(3\lambda + 2\mu)}{\lambda + \mu}, \qquad \nu := \frac{\lambda}{2(\lambda + \mu)}. \tag{1.2.44}$$

We also recall that the Stokes problem (1.2.38) can also be expressed as a biharmonic problem by the introduction of a stream function ψ such that

$$\underline{u} = \left\{ \frac{\partial \psi}{\partial x_2}, -\frac{\partial \psi}{\partial x_1} \right\}. \tag{1.2.45}$$

We shall come back to this point in Sect. 1.3. □

Example 1.2.5 (The time-harmonic Maxwell system). Maxwell's equations represent one of the most elegant and concise ways to state the fundamentals of electricity and magnetism. The classical electromagnetic field is described by the four vectors $\mathcal{E}$, $\mathcal{D}$, $\mathcal{H}$, and $\mathcal{B}$ which are functions of the position $\underline{x} \in \mathbb{R}^3$ and of the time $t \in \mathbb{R}$. The vectors $\mathcal{E}$ and $\mathcal{H}$ are referred to as the electric and magnetic field, while $\mathcal{D}$ and $\mathcal{B}$ are the electric and magnetic displacements, respectively.

The Faraday law of induction states

$$\int_{\partial\Sigma} \mathcal{E} \cdot ds = -\frac{d}{dt} \int_{\Sigma} \mathcal{B} \cdot \underline{n}, \tag{1.2.46}$$

for any closed orientable surface Σ in $\mathbb{R}^2$ with normal $\underline{n}$; namely, the circulation of the electric field equals the negative of the rate of change of the magnetic flux.

The Ampère law says

$$\int_{\partial\Sigma} \mathcal{H} \cdot ds = \frac{d}{dt} \int_{\Sigma} \mathcal{D} \cdot \underline{n} + \int_{\Sigma} \mathcal{J} \cdot \underline{n}, \tag{1.2.47}$$

with the same notation as above and where $\mathcal{J}$ denotes the current density vector.

From Eqs. (1.2.46) and (1.2.47) it can be noticed that the fields $\mathcal{E}$, $\mathcal{H}$, $\mathcal{B}$ and $\mathcal{D}$ have a different nature. Indeed, the first two are integral 1-forms, while the latter two are integral 2-forms in the spirit of [247] (Definition 1). This remark is of fundamental importance for the design of finite element schemes.

The differential forms of (1.2.46) and (1.2.47) read

$$\begin{aligned} &\frac{\partial \mathcal{B}}{\partial t} + \underline{\mathrm{curl}}\, \mathcal{E} = 0, \\ &\frac{\partial \mathcal{D}}{\partial t} - \underline{\mathrm{curl}}\, \mathcal{H} = -\mathcal{J}, \end{aligned} \tag{1.2.48}$$

which are usually referred to as Maxwell's equations, together with the two Gauss Laws

$$\begin{aligned} &\mathrm{div}\, \mathcal{D} = \rho, \\ &\mathrm{div}\, \mathcal{B} = 0, \end{aligned} \tag{1.2.49}$$

where ρ denotes the charge density function. It is clear that in (1.2.48) and (1.2.49) the quantities $\mathcal{J}$ and ρ cannot be taken independently: taking the divergence of the second equation in (1.2.48) and comparing with the time derivative of the first equation in (1.2.49) we have indeed:

$$\frac{\partial \rho}{\partial t} + \mathrm{div}\, \mathcal{J} = 0, \tag{1.2.50}$$

which could be seen as a *compatibility condition* when $\mathcal{J}$ and ρ are considered as *given data*.

The time-harmonic Maxwell system is considered, for instance, when the Fourier transform in time is used or when the propagation of electromagnetic waves at a given frequency is studied. Then, given a frequency ω, we consider the *ansatz*:

$$\begin{aligned}
\mathcal{E}(\underline{x},t) &= \Re\left(e^{-i\omega t}E(\underline{x})\right),\\
\mathcal{D}(\underline{x},t) &= \Re\left(e^{-i\omega t}D(\underline{x})\right),\\
\mathcal{H}(\underline{x},t) &= \Re\left(e^{-i\omega t}H(\underline{x})\right),\\
\mathcal{B}(\underline{x},t) &= \Re\left(e^{-i\omega t}B(\underline{x})\right),
\end{aligned} \tag{1.2.51}$$

where $\Re$ denotes the *real part*. We define also

$$\begin{aligned}
\mathcal{J}(\underline{x},t) &= \Re\left(e^{-i\omega t}J(\underline{x})\right),\\
\rho(\underline{x},t) &= \Re\left(e^{-i\omega t}r(\underline{x})\right).
\end{aligned} \tag{1.2.52}$$

Standard constitutive equations for linear media read

$$D = \underline{\underline{\varepsilon}}E, \qquad B = \underline{\underline{\mu}}H, \tag{1.2.53}$$

where $\underline{\underline{\varepsilon}}$ and $\underline{\underline{\mu}}$ denote the electric permittivity and the magnetic permeability, respectively. For general inhomogeneous, anisotropic materials $\underline{\underline{\varepsilon}}$ and $\underline{\underline{\mu}}$ are 3×3 positive definite matrix functions.

Inserting constitutive relations (1.2.53) into (1.2.48) and (1.2.49), and considering the time-harmonic assumptions (1.2.51) and (1.2.52), we get the time harmonic Maxwell equations

$$\begin{aligned}
&\underline{\mathrm{curl}}\, E - i\omega\underline{\underline{\mu}}H = 0,\\
&\mathrm{div}(\underline{\underline{\varepsilon}}E) = r,\\
&\underline{\mathrm{curl}}\, H + i\omega\underline{\underline{\varepsilon}}E = J,\\
&\mathrm{div}(\underline{\underline{\mu}}H) = 0.
\end{aligned} \tag{1.2.54}$$

It is a standard procedure to eliminate one variable and to write (1.2.54) as a second order system. Eliminating for instance the field H, we get

$$\underline{\mathrm{curl}}(\underline{\underline{\mu}}^{-1}\,\underline{\mathrm{curl}}\, E) - \omega^2\underline{\underline{\varepsilon}}E = F, \tag{1.2.55}$$

where F is given by $i\omega J$, together with the divergence condition (which follows from the equation)

$$-\,\omega^2\,\mathrm{div}(\underline{\underline{\varepsilon}}E) = \mathrm{div}\, F. \tag{1.2.56}$$

Equation (1.2.55) is usually equipped with suitable boundary conditions. The simplest one is the perfect conducting boundary condition which reads:

$$E \times \underline{n} = 0, \tag{1.2.57}$$

where $\underline{n}$ is the outward unit vector. We shall discuss variational formulations of the time harmonic Maxwell equations in Chap. 11. □

The examples presented above are among the most fundamental of mathematical physics and engineering problems. A good understanding of their properties will enable to extend the results obtained to more complex situations.

1.2.1 Eigenvalue Problems

The above examples can also be associated to eigenvalue problems. It is worth making them explicit as we shall be concerned later with alternate formulations and their numerical properties. In the case of Example 1.2.1, restricting ourselves to the case of Dirichlet's conditions, we have

$$\begin{cases} -\Delta p = \lambda p \text{ in } \Omega, \\ p\big|_{\Gamma} = 0. \end{cases} \tag{1.2.58}$$

The solutions of this problem describe, for instance, the vibrational modes of a membrane. It can be written in a more precise way as,

$$\int_{\Omega} \underline{\text{grad}}\, p \cdot \underline{\text{grad}}\, q\, dx = \lambda \int_{\Omega} p\, q\, dx, \quad \forall q \in H_0^1(\Omega). \tag{1.2.59}$$

It is classical [172, 382] that this problem has an infinite countable set of solutions $(p_k, \lambda_k, k \in N)$ with $\lambda_k \to \infty$ as k increases. The key to this result is the compact inclusion of $H_0^1(\Omega)$ into $L^2(\Omega)$.

In the same way, the eigenvalue problem associated with the elasticity problem of Example 1.2.2 is fundamental for the study of vibrations in elastic structures. The problem is then, restricting ourselves again to Dirichlet's conditions,

$$\begin{cases} -(2\mu\, \underline{\text{div}}\, \underline{\underline{\varepsilon}}(\underline{u}) + \lambda\, \underline{\text{grad}}\, \text{div}\, \underline{u}) - \tilde{\lambda} \underline{u} \text{ in } \Omega, \\ \underline{u}\big|_{\Gamma_0} = 0. \end{cases} \tag{1.2.60}$$

We have denoted the eigenvalue as $\tilde{\lambda}$ to distinguish it from the Lamé coefficients. The variational form is then,

$$2\mu \int_{\Omega} \underline{\underline{\varepsilon}}(\underline{u}) : \underline{\underline{\varepsilon}}(\underline{v})\, dx + \lambda \int_{\Omega} \text{div}\, \underline{u}\ \text{div}\, \underline{v}\, dx$$
$$= \tilde{\lambda} \int_{\Omega} \underline{u} \cdot \underline{v}\, dx\, ds, \qquad \forall \underline{v} \in V. \tag{1.2.61}$$

Here again we have an infinite set of eigenvalues. Introducing in this problem the constraint $\operatorname{div} \underline{u} = 0$, we would get an eigenvalue problem for the Stokes problem of Example 1.2.3, or to a problem of incompressible elasticity as we shall see later. Eigenvalue problems related to Maxwell's equations will be discussed in Chap. 11.

1.3 Duality Methods

1.3.1 Generalities

Up to now, we have introduced equations that can be written as minimisation problems of some functionals in properly chosen functional spaces. This is the most classical way of setting these problems. Finite element approximations, based on the formulations described above, are routinely used in commercial codes. Various reasons justified the introduction, for these same problems and many other ones, of different variational formulations and therefore different finite element approximations. This was done at the beginning by many engineers. The reader may refer, for example to [321, 335, 336].

The first reason can be the presence in the variational formulation of a constraint, such as the condition $\operatorname{div} \underline{u} = 0$ in problem (1.2.38). As we shall see, it is difficult (and not necessary) to build finite element approximations satisfying exactly this constraint. It will be more efficient to modify the variational formulation and to introduce pressure.

A second reason may lie in the physical "importance" of the variables appearing in the problem. In elasticity problems, for example, it is often more useful to compute accurately stresses rather than displacements. In the standard formulation, stresses can be recovered from the displacements by (1.2.36) or some other similar law. Their computation requires the derivatives of the displacement field $\underline{u}$. From a numerical point of view, differentiating implies a loss of precision. It is therefore appealing to look for a formulation in which constraints are readily accessible.

A third reason comes from the difficulties arising in the discretisation of spaces of regular functions such as $H_0^2(\Omega)$ appearing in Example 1.2.4. Approximating this space by a finite element method implies ensuring continuity of the derivatives at interfaces between elements. This is possible but more cumbersome than approximating, say, $H^1(\Omega)$ or $H_0^1(\Omega)$. A variational formulation enabling to decompose a fourth-order problem into a system of second order problems permits to avoid building complicated elements, at the price of introducing some other difficulties.

Finally, a last reason could be to look for a *weaker variational formulation* corresponding better in some cases to available data (e.g. punctual loads) for which standard formulations may become meaningless due to a lack of regularity of the solution.

We must also point out that the "non-standard" formulations which we shall now describe have been initially introduced by engineers for one or some of the reasons discussed above. We quote in this respect, but in a totally non exhaustive way, [210, 243, 245, 320, 370]. On the other hand, very powerful tools for the transformation of variational problems can be found in convex analysis and duality theory [38,50,184, 338]. It is neither possible nor desirable to develop here duality theory and we shall restrict ourselves to the most basic facts. The fundamental idea of duality theory is that one can represent a convex function by the family of its tangent affine functions. This is indeed the principle of the classical Legendre transformation. More precisely, let us define for a given convex function $G(v)$, defined from a space V to $\mathbb{R}$, the conjugate function $G^*(v^*)$ on the dual space V' of V by

$$G^*(v^*) := \sup_{v \in V} \langle v, v^* \rangle_{V \times V'} - G(v). \tag{1.3.1}$$

Note that when $V = \mathbb{R}$, $G^*(v^*)$ is the intercept with the v axis of the tangent to G of slope v^*. The important point for what follows is that one can build $G(v)$ from $G^*(v^*)$ by the following formula, symmetrical to (1.3.1)

$$G(v) = \sup_{v^* \in V'} \langle v^*, v \rangle_{V' \times V} - G^*(v^*). \tag{1.3.2}$$

Given then a problem of the form

$$\inf_{v \in V} F(v) + G(v), \tag{1.3.3}$$

we can use (1.3.2) to obtain

$$\inf_{v \in V} \left\{ F(v) + \sup_{v^* \in V'} \langle v^*, v \rangle_{V' \times V} - G^*(v^*) \right\} \tag{1.3.4}$$

that is, the saddle point problem

$$\inf_{v \in V} \sup_{v^* \in V'} F(v) + \langle v^*, v \rangle_{V' \times V} - G^*(v^*). \tag{1.3.5}$$

Under simple regularity assumptions, one can also consider the *dual problem*

$$\sup_{v^* \in V'} \left\{ \inf_{v \in V} F(v) + \langle v^*, v \rangle_{V' \times V} - G^*(v^*) \right\}. \tag{1.3.6}$$

To fix ideas, it is worth considering a special and important case where we have as in (1.1.2), for $f \in V'$,

$$F(v) := \frac{1}{2} a(u, v) - \langle f, v \rangle. \tag{1.3.7}$$

We then introduce another Hilbert space Q and an operator B from V into Q' defined by a continuous bilinear form on $V \times Q$,

$$\langle Bv, q \rangle = b(v, q). \tag{1.3.8}$$

We then want to solve for $g \in Q'$ the constrained problem

$$\inf_{Bv=g} J(v). \tag{1.3.9}$$

This constrained problem can be written as an unconstrained problem, introducing the characteristic function $\delta(\cdot|\{0\})$ defined on Q' by

$$\delta(q|\{g\}) := \begin{cases} 0 & \text{if } v = g, \\ +\infty & \text{otherwise.} \end{cases} \tag{1.3.10}$$

We can then write (1.3.9) as

$$\inf_{v \in V} F(v) + \delta(Bv|\{g\}), \tag{1.3.11}$$

which can be readily transformed into the saddle-point problem

$$\inf_{v \in V} \sup_{q \in Q} \frac{1}{2} a(u, v) - b(v, q) - \langle f, v\rangle_{V' \times V} + \langle g, q\rangle_{Q' \times Q}, \tag{1.3.12}$$

for which the optimality system is

$$\begin{cases} a(u, v) + b(v, p) = \langle f, v\rangle_{V' \times V}, & \forall v \in V, \\ b(u, q) = \langle g, q\rangle_{Q' \times Q}, & \forall q \in Q, \end{cases} \tag{1.3.13}$$

or in operator form, denoting A the operator from V into V' defined by $a(\cdot, \cdot)$,

$$\begin{cases} Au + B^t p = f, \\ Bu = g. \end{cases} \tag{1.3.14}$$

Remark 1.3.1. Problem (1.3.12) has the general form

$$\inf_{v \in V} \sup_{q \in Q} L(v, q), \tag{1.3.15}$$

where $L(v, q)$ is a convex-concave functional on $V \times Q$. If one first eliminates q by computing

$$J(v) = \sup_{q \in Q} L(v, q),$$

one falls back on the original problem, the *primal problem*. Reversing the order of operations, (this cannot always be done, but no problems arise in the examples we present) and eliminating v from $L(v, q)$ by defining

$$D(q) := \inf_{v} L(v, q) \tag{1.3.16}$$

leads to the dual problem

$$\sup_{q\in Q} D(q). \tag{1.3.17}$$

□

The dual problem in Q corresponding to (1.3.12) would take the form

$$\inf_{q} \frac{1}{2}\langle A^{-1}B^t q, B^t q\rangle_{V\times V'} - \langle A^{-1}f, B^t q\rangle_{V\times V'} + \langle g, q\rangle_{Q'\times Q}, \tag{1.3.18}$$

or in operator form,

$$BA^{-1}B^t p = BA^{-1}f - g. \tag{1.3.19}$$

We now apply this idea to the previous examples. This form of problem and its variants will be central to our study and we shall proceed to introduce some examples.

1.3.2 Examples for Symmetric Problems

Example 1.3.1 (Introduction of pressure in Stokes' problem.). Let us consider problem (1.2.38) in which, to make the presentation easier, we take $\Gamma_0 = \Gamma$ that is pure Dirichlet conditions on the boundary. This constrained problem can be written as an unconstrained problem, introducing as above the characteristic function $\delta(\cdot|\{0\})$ defined on $L^2(\Omega)$ by

$$\delta(v|\{0\}) := \begin{cases} 0 & \text{if } v = 0, \\ +\infty & \text{otherwise.} \end{cases} \tag{1.3.20}$$

It is thus a pure change of notations to write instead of (1.2.38)

$$\inf_{v\in V} \mu \int_\Omega |\underline{\underline{\varepsilon}}(\underline{v})|^2\,dx - \int_\Omega \underline{f}\cdot\underline{v}\,dx + \delta(\operatorname{div}\underline{v}|\{0\}), \tag{1.3.21}$$

where $V = (H_0^1(\Omega))^2$. On the other hand, one clearly has,

$$\delta(\operatorname{div}\underline{u}|\{0\}) = \sup_{q\in L^2(\Omega)} \int_\Omega q \operatorname{div}\underline{u}\,dx, \quad \forall \underline{u} \in (H_0^1(\Omega))^2, \tag{1.3.22}$$

and the minimisation problem (1.3.21) can be transformed into the saddle point problem,

$$\inf_{v\in V}\sup_{q\in L^2(\Omega)} \mu \int_\Omega |\underline{\underline{\varepsilon}}(\underline{v})|^2\,dx - \int_\Omega \underline{f}\cdot\underline{v}\,dx - \int_\Omega q \operatorname{div}\underline{v}\,dx. \tag{1.3.23}$$

This apparently simple trick has in reality completely changed the nature of the problem. We now have to find a pair $(\underline{u}, p)$ solution of the variational system

$$\begin{cases} 2\mu \displaystyle\int_\Omega \underline{\underline{\varepsilon}}(\underline{u}) : \underline{\underline{\varepsilon}}(\underline{v})\, dx - \int_\Omega \underline{f} \cdot \underline{v}\, dx - \int_\Omega p \text{ div } \underline{v}\, dx = 0, \quad \forall \underline{v} \in V, \\ \displaystyle\int_\Omega q \text{ div } \underline{u}\, dx = 0, \quad \forall q \in L^2(\Omega). \end{cases} \tag{1.3.24}$$

The second equation of (1.3.24) evidently expresses the condition div $\underline{u} = 0$. In order to use (1.3.24), we shall have to show the existence of a saddle point $(\underline{u}, p)$, and in particular the existence of the Lagrange multiplier p. This will be done in Chap. 4. The variational system (1.3.24) can be interpreted in the form

$$\begin{cases} -2\mu A\underline{u} + \underline{\text{grad}}\, p = \underline{f}, \\ \text{div } \underline{u} = 0, \\ \underline{u}|_\Gamma = 0, \end{cases} \tag{1.3.25}$$

where we used the operator $A\underline{u} = \text{div}\, \underline{\underline{\varepsilon}}(\underline{u})$

$$A\underline{u} = \begin{pmatrix} \dfrac{\partial^2 u_1}{\partial x_1^2} + \dfrac{\partial}{\partial x_2}\dfrac{1}{2}\left\{\dfrac{\partial u_1}{\partial x_2} + \dfrac{\partial u_2}{\partial x_1}\right\} \\ \\ \dfrac{\partial^2 u_2}{\partial x_2^2} + \dfrac{\partial}{\partial x_1}\dfrac{1}{2}\left\{\dfrac{\partial u_1}{\partial x_2} + \dfrac{\partial u_2}{\partial x_1}\right\} \end{pmatrix}. \tag{1.3.26}$$

Under the *divergence-free condition* div $\underline{u} = 0$, this can also be written as

$$\begin{cases} -\mu \Delta \underline{u} + \underline{\text{grad}}\, p = \underline{f}, \\ \text{div } \underline{u} = 0, \end{cases} \tag{1.3.27}$$

which is the classical form of the Stokes problem. □

Example 1.3.2 (Dual problem for the Stokes problem). In the case of the Stokes problem, the dual problem can be expressed, as we shall see, in many equivalent ways. In order to find it we must, q being given, find the minimum in respect to $\underline{v}$ of $L(\underline{v}, q) = \mu \int_\Omega |\underline{\underline{\varepsilon}}(\underline{v})|^2\, dx - \int_\Omega \underline{f} \cdot \underline{v}\, dx - \int_\Omega q \text{ div } \underline{v}\, dx$. The minimum, that we denote by $\underline{u}_q$, is characterised by

$$2\mu \int_\Omega \underline{\underline{\varepsilon}}(\underline{u}_q) : \underline{\underline{\varepsilon}}(\underline{v})\, dx - \int_\Omega \underline{f} \cdot \underline{v}\, dx - \int_\Omega q \,\text{div}\, \underline{v}\, dx = 0, \quad \forall \underline{v} \in V. \tag{1.3.28}$$

Taking $\underline{v} = \underline{u}_q$ this gives,

$$2\mu \int_\Omega |\underline{\underline{\varepsilon}}(\underline{u}_q)|^2 \, dx - \int_\Omega \underline{f} \cdot \underline{u}_q \, dx - \int_\Omega q \operatorname{div} \underline{u}_q \, dx = 0. \tag{1.3.29}$$

Using (1.3.29) to evaluate $L(\underline{u}_q, q)$, the dual problem can be written as an optimal control problem,

$$\sup_{q \in L^2(\Omega)} -\mu \int_\Omega |\underline{\underline{\varepsilon}}(\underline{u}_q)|^2 \, dx, \tag{1.3.30}$$

where $\underline{u}_q$ is the solution of

$$\begin{cases} -2\mu \, A\underline{u}_q + \underline{\operatorname{grad}}\, q = \underline{f}, \\ \underline{u}_q|_\Gamma = 0. \end{cases} \tag{1.3.31}$$

Denoting by G the Green operator defining the solution of (1.3.31), that is

$$\underline{u}_q = G(\underline{f} - \underline{\operatorname{grad}}\, q) \tag{1.3.32}$$

and using (1.3.32) in (1.3.30), one can get from (1.3.29)

$$\inf_q \int_\Omega \underline{\operatorname{grad}}\, q \cdot G(\underline{\operatorname{grad}}\, q) \, dx - \int_\Omega G(\underline{f}) \cdot \underline{\operatorname{grad}}\, q \, dx. \tag{1.3.33}$$

One notices that this dual problem is a problem in $\underline{\operatorname{grad}}\, q$. It is well-known that the solution p is defined (for Dirichlet conditions on $\underline{u}$) only up to an additive constant. One can interpret (1.3.33) as the equation,

$$\operatorname{div}(G \, \underline{\operatorname{grad}}\, q) = \operatorname{div}(G \underline{f}). \tag{1.3.34}$$

If one defines on $V' := (H^{-1}(\Omega))^2$ the norm

$$\|\underline{f}\|_G^2 := \langle G\underline{f}, \underline{f} \rangle_{V \times V'}, \tag{1.3.35}$$

problem (1.3.33) can be written as a least-squares problem

$$\inf_{q \in L^2(\Omega)} \frac{1}{2} \| \underline{\operatorname{grad}}\, q - \underline{f} \|_G^2. \tag{1.3.36}$$

□

The presence of a Green operator makes this dual problem difficult to handle directly. It is however implicitly the basis of some numerical solution procedures [205, 368]. We shall meet below other dual problems that will have a large direct importance and that will be handled as such.

Example 1.3.3 (A duality method for nearly incompressible elasticity). We already noted in Examples 1.2.2 and 1.2.3 that the linear elasticity problem and the Stokes problem are very similar when a nearly incompressible material is considered. We now develop this analogy in the framework of Example 1.3.1. The starting point will be the obvious result,

$$\frac{\lambda}{2}\int_\Omega |\operatorname{div}\underline{v}|^2\,dx = \sup_{q\in L^2(\Omega)} \int_\Omega q\operatorname{div}\underline{v}\,dx - \frac{1}{2\lambda}\int_\Omega |q|^2\,dx. \tag{1.3.37}$$

Substituting (1.3.37) into (1.2.31) we get, by the same methods as in the previous examples, the problem

$$\begin{aligned}\inf_{\underline{v}\in V}\sup_{q\in L^2(\Omega)} \mu\int_\Omega |\underline{\underline{\varepsilon}}(\underline{v})|^2\,dx + \int_\Omega q\operatorname{div}\underline{v}\,dx - \frac{1}{2\lambda}\int_\Omega |q|^2\,dx \\ - \int_\Omega \underline{f}\cdot\underline{v}\,dx - \int_{\Gamma_1} g_n\,\underline{v}\cdot\underline{n}\,ds - \int_{\Gamma_1} g_t\,\underline{v}\cdot\underline{t}\,ds.\end{aligned} \tag{1.3.38}$$

The solution $(\underline{u}, p)$ of problem (1.3.38) is characterised by the system,

$$\begin{cases} 2\mu\displaystyle\int_\Omega \underline{\underline{\varepsilon}}(\underline{u}) : \underline{\underline{\varepsilon}}(\underline{v})\,dx + \int_\Omega p\operatorname{div}\underline{v}\,dx \\ \qquad = \displaystyle\int_\Omega \underline{f}\cdot\underline{v}\,dx + \int_{\Gamma_1} (g_n\underline{v}\cdot\underline{n} + g_t\underline{v}\cdot\underline{t})\,ds, \quad \forall\underline{v}\in V, \\ \displaystyle\int_\Omega q\operatorname{div}\underline{u}\,dx - \frac{1}{\lambda}\int_\Omega pq\,dx = 0, \quad \forall q\in L^2(\Omega). \end{cases} \tag{1.3.39}$$

This can be summarised by saying that we transformed our original problem into a system by introducing the auxiliary variable $p = \lambda \operatorname{div}\underline{u}$. It must be noted that this also makes our minimisation problem become a saddle point problem. We shall see in Chap. 8, that this apparently *tautological change has implications in the building of numerical approximations to* (1.2.31) *that remain valid when* λ *is large.* □

Example 1.3.4 (Dualisation of the Poisson problem). The result that we shall get here can be obtained by many methods. Techniques of convex analysis permit one to extend what appears to be a trick to much more complex situations. However it will be sufficient for our purpose to follow the simple development below. Let us then consider the Dirichlet problem,

$$\inf_{q\in H_0^1(\Omega)} \frac{1}{2}\int_\Omega |\underline{\operatorname{grad}}\,q|^2\,dx - \int_\Omega fq\,dx. \tag{1.3.40}$$

In many applications $\underline{\operatorname{grad}}\,p$ rather than p is the interesting variable. For instance in thermo-diffusion problems, $\underline{\operatorname{grad}}\,p$ will be the heat flux which is (very often) more important to know than the temperature p. What we now do is essentially to introduce the auxiliary variable $\underline{u} = \underline{\mathrm{grad}}\, p$ to transform our problem into a system. To do so, we use the same trick as in Example 1.3.3 and write

$$\frac{1}{2}\int_{\Omega} |\ \underline{\mathrm{grad}}\, q|^2\, dx = \sup_{\underline{v}\in(L^2(\Omega))^2} \int_{\Omega} \underline{v}\cdot\ \underline{\mathrm{grad}}\, q\, dx - \frac{1}{2}\int_{\Omega} |\underline{v}|^2\, dx, \tag{1.3.41}$$

which we use in (1.3.40) to get the saddle point problem,

$$\inf_{q\in Q}\sup_{\underline{v}\in V} -\frac{1}{2}\int_{\Omega} |\underline{v}|^2\, dx - \int_{\Omega} fq\, dx + \int_{\Omega} \underline{v}\cdot\ \underline{\mathrm{grad}}\, q\, dx, \tag{1.3.42}$$

where $Q := H_0^1(\Omega)$ and $V := (L^2(\Omega))^2$. The saddle point $(p, \underline{u})$ is characterised by

$$\begin{cases} \displaystyle\int_{\Omega} \underline{u}\cdot\underline{v}\, dx - \int_{\Omega} \underline{v}\cdot\ \underline{\mathrm{grad}}\, p\, dx = 0 \quad \forall \underline{v}\in V, \\ \displaystyle\int_{\Omega} \underline{u}\cdot\ \underline{\mathrm{grad}}\, q\, dx = \int_{\Omega} fq\, dx, \quad \forall q\in Q, \end{cases} \tag{1.3.43}$$

and this can be read as

$$\begin{cases} \underline{u} = \ \underline{\mathrm{grad}}\, p,\ \ p\in H_0^1(\Omega), \\ \mathrm{div}\ \underline{u} + f = 0, \end{cases} \tag{1.3.44}$$

which is evidently equivalent to a standard Dirichlet problem for Laplace operator.

The *dual problem* is readily made explicit. Writing it as a minimisation problem by changing the sign of the objective functional, we have

$$\inf_{\underline{v}\in Z_f} \frac{1}{2}\int_{\Omega} |\underline{v}|^2\, dx, \quad \text{where } Z_f := \{\underline{v}\in(L^2(\Omega))^2 |\ \mathrm{div}\ \underline{v} + f = 0\}. \tag{1.3.45}$$

This is the classical *complementary energy principle.* □

We now want to get a weaker form of this problem. In order to do so, we recall a functional space already introduced in (1.2.16):

$$H(\mathrm{div};\Omega) := \{\underline{v}|\ \underline{v}\in(L^2(\Omega))^2,\ \mathrm{div}\ \underline{v}\in L^2(\Omega)\}, \tag{1.3.46}$$

and we consider on it the following norm

$$\|\underline{v}\|^2_{H(\mathrm{div};\Omega)} := \|\underline{v}\|^2_{0,\Omega} + \|\ \mathrm{div}\ \underline{v}\|^2_{0,\Omega}, \tag{1.3.47}$$

that makes it a Hilbert space. As we have already said, vectors of $H(\mathrm{div};\Omega)$ admit a well defined *normal trace on* $\Gamma := \partial\Omega$. This normal trace $\underline{v}\cdot\underline{n}$, lies in $H^{-1/2}(\Gamma)$ and one has the following "integration by parts" formula,

$$\int_{\Omega} \underline{v} \cdot \underline{\operatorname{grad}}\, q \, dx + \int_{\Omega} \operatorname{div} \underline{v}\, q \, dx = \langle q, \underline{v} \cdot \underline{n} \rangle_{H^{\frac{1}{2}}(\Gamma) \times H^{-\frac{1}{2}}(\Gamma)}, \tag{1.3.48}$$

for any $\underline{v} \in H(\operatorname{div}; \Omega)$ and any $q \in H^1(\Omega)$. We shall often write formally $\int_{\Gamma} q \underline{v} \cdot \underline{n}\, ds$ instead of the duality product $\langle q, \ \underline{v} \cdot \underline{n} \rangle$.

Remark 1.3.2. The norm (1.3.47) is surely as *weird* as many other Sobolev norms. See Remark 1.2.1. The *proper way* to write it would be to take a characteristic length ℓ of the problem (for instance, the diameter of Ω), and consider instead

$$\|\underline{v}\|^2_{H(\operatorname{div};\Omega)} := \|\underline{v}\|^2_{0,\Omega} + \ell^2 \| \operatorname{div} \underline{v}\|^2_{0,\Omega}. \tag{1.3.49}$$

We do not do it here, nor, in general, throughout the book, in order to have simpler formulae to deal with. But, nevertheless, we consider it as *not healthy*. □

Example 1.3.5 (Weak form of the dual Poisson problem). If we take $f \in L^2(\Omega)$, problem (1.3.45) which is a constrained problem can be changed into a saddle point problem, as in Example 1.3.1, by introducing a Lagrange multiplier $p \in L^2(\Omega)$, that is, as $\underline{v}$ now belongs to $H(\operatorname{div}; \Omega)$,

$$\inf_{\underline{v} \in H(\operatorname{div};\Omega)} \ \sup_{q \in L^2(\Omega)} \frac{1}{2} \int_{\Omega} |\underline{v}|^2 \, dx + \int_{\Omega} f q \, dx + \int_{\Omega} q \ \operatorname{div} \underline{v} \, dx. \tag{1.3.50}$$

The functional spaces employed precisely enable us to write every term in (1.3.50) without ambiguity. We now look for a saddle point $(p, \underline{u})$ satisfying the variational system,

$$\begin{cases} \displaystyle\int_{\Omega} \underline{u} \cdot \underline{v} \, dx + \int_{\Omega} p \ \operatorname{div} \underline{v} \, dx = 0 & \forall \underline{v} \in H(\operatorname{div}; \Omega), \\[2ex] \displaystyle\int_{\Omega} (\operatorname{div} \underline{u} + f) q \, dx = 0 & \forall q \in L^2(\Omega). \end{cases} \tag{1.3.51}$$

Using (1.3.48) with $\underline{v} \cdot \underline{n}|_{\Gamma} = 0$, we obtain from (1.3.51)

$$\underline{u} = \underline{\operatorname{grad}}\, p. \tag{1.3.52}$$

Now $p \in L^2(\Omega)$ and $\underline{\operatorname{grad}}\, p = \underline{u} \in (L^2(\Omega))^2$ imply that $p \in H^1(\Omega)$ and it is justified to consider its trace. Using again (1.3.48) with a general $\underline{v}$ shows that $p|_{\Gamma} = 0$. The solution of our "weaker" problem is then the solution of the standard problem. However the discretisations of problem (1.3.51) will be quite different from those used for the standard formulation. □

Remark 1.3.3. The previous formulation enables us to write directly in a variational form a non-homogeneous Dirichlet problem. Indeed the solution $(p, \underline{u})$ of the saddle point problem with $g \in H^{\frac{1}{2}}(\Gamma)$,

$$\inf_{\underline{v}} \sup_{q} \frac{1}{2} \int_{\Omega} |\underline{v}|^2 \, dx + \int_{\Omega} (\text{div } \underline{v} + f) q \, dx - \int_{\Gamma} g \underline{v} \cdot \underline{n} \, ds, \tag{1.3.53}$$

leads to $\underline{u} = \underline{\text{grad}}\, p$, $\text{div } \underline{u} + f = 0$, $p|_{\Gamma} = g$.

On the other hand, Neumann conditions become *essential* conditions that have to be incorporated into the construction of $\underline{u}$, that is in the choice of the functional space. □

Example 1.3.6 (Dualisation for the linear elasticity problem). We now want to extend the previous results to the case of the *linear elasticity problem.* We shall thus get a second way to dualise problem (1.2.31). It is a general fact that there is no unique way to use duality techniques.

The lines of the development are the same as for the Poisson problem and we shall avoid to write the details. We just point out that now $\underline{u}$ and $\underline{v}$ play the role of p and q, while $\underline{\underline{\sigma}}$ and $\underline{\underline{\tau}}$ play the role of $\underline{u}$ and $\underline{v}$, respectively.

We start by introducing the space

$$\underline{\underline{H}}(\underline{\text{div}}; \Omega)_s := \{\underline{\underline{\tau}} |\ \tau_{ij} \in L^2(\Omega),\ \tau_{ij} = \tau_{ji},\ \underline{\text{div}}\, \underline{\underline{\tau}} \in (L^2(\Omega))^2\}, \tag{1.3.54}$$

where $\underline{\text{div}}\, \underline{\underline{\tau}}$ is the *vector* $\frac{\partial}{\partial x_1} \tau_{i1} + \frac{\partial}{\partial x_2} \tau_{i2}$. On this space we use the norm,

$$\|\underline{\underline{\tau}}\|^2_{\underline{\underline{H}}(\underline{\text{div}};\Omega)_s} := \sum_{i,j} \int_{\Omega} |\tau_{ij}|^2 \, dx + \| \underline{\text{div}}\, \underline{\underline{\tau}} \|^2_{(L^2(\Omega))^2}, \tag{1.3.55}$$

which makes it a Hilbert space.

One can then define, as for $H(\text{div}; \Omega)$, the vector $\underline{\tau}_n \in (H^{-\frac{1}{2}}(\Gamma))^2$

$$(\underline{\tau}_n)_i := \sum_j \tau_{ij}\, n_j \tag{1.3.56}$$

and we shall mostly use the normal and tangential components, τ_{nn} and τ_{nt}, of this vector, as defined in (1.2.34). We then have the following "integration by parts" formula,

$$\int_{\Omega} \underline{\underline{\tau}} : \underline{\underline{\varepsilon}}(\underline{v}) \, dx + \int_{\Omega} \underline{\text{div}}\, \underline{\underline{\tau}} \cdot \underline{v} \, dx = \langle \underline{\tau}_n, \underline{v} \rangle = \langle \tau_{nn}, \underline{v} \cdot \underline{n} \rangle + \langle \tau_{nt}, \underline{v} \cdot \underline{t} \rangle, \tag{1.3.57}$$

which is valid for any $\underline{\underline{\tau}}$ and $\underline{v}$ smooth enough. We have denoted $\langle \cdot, \cdot \rangle$ the duality between $H^{-\frac{1}{2}}(\Gamma)$ and $H^{\frac{1}{2}}(\Gamma)$ and shall often write the formal expression $\int_{\Gamma} \tau_{nn}\, \underline{v} \cdot \underline{n} \, ds + \int_{\Gamma} \tau_{nt}\, \underline{v} \cdot \underline{t} \, ds$.

We can now write our dual formulation for the linear elasticity problem. Following the same line as for the Poisson problem, we now write

$$\mu \int_{\Omega} |\underline{\underline{\varepsilon}}(\underline{v})|^2 \, dx + \frac{\lambda}{2} \int_{\Omega} |\operatorname{div} \underline{v}|^2 \, dx = \sup_{\underline{\underline{\tau}} \in \underline{\underline{H}}(\underline{\operatorname{div}};\Omega)_s} \int_{\Omega} \underline{\underline{\tau}}^D : \underline{\underline{\varepsilon}}^D \, dx$$
$$+ \int_{\Omega} tr\, \underline{\underline{\tau}}\, tr\, \underline{\underline{\varepsilon}} \, dx - \frac{1}{4\mu} \int_{\Omega} |\underline{\underline{\tau}}^D|^2 \, dx - \frac{1}{2(\lambda+\mu)} \int_{\Omega} (tr\, \underline{\underline{\tau}})^2 \, dx, \tag{1.3.58}$$

which leads us to the saddle point problem in $\underline{\underline{H}}(\underline{\operatorname{div}};\Omega) \times (L^2(\Omega))^2$,

$$\inf_{\underline{\underline{\tau}}} \sup_{\underline{v}} \frac{1}{2(\lambda+\mu)} \int_{\Omega} |tr\, \underline{\underline{\tau}}|^2 \, dx + \frac{1}{4\mu} \int_{\Omega} |\underline{\underline{\tau}}^D|^2 \, dx + \int_{\Omega} (\underline{\operatorname{div}}\, \underline{\underline{\tau}} + \underline{f}) \, \underline{v} \, dx. \tag{1.3.59}$$

The solution $(\underline{\underline{\sigma}}, \underline{u})$ of this saddle point problem is characterised by the system

$$\begin{cases} \underline{\operatorname{div}}\, \underline{\underline{\sigma}} + \underline{f} = 0, \\ tr\, \underline{\underline{\sigma}} = (\lambda+\mu)\, tr\, \underline{\underline{\varepsilon}}(\underline{u}), \\ \underline{\underline{\sigma}}^D = 2\mu\, \underline{\underline{\varepsilon}}^D(\underline{u}), \end{cases} \tag{1.3.60}$$

which are the equilibrium condition (1.2.37) and the constitutive relations (1.2.36). The dual problem then consists in minimising the *complementary energy*

$$\inf_{\underline{\underline{\tau}}} \frac{1}{4\mu} \int_{\Omega} |\underline{\underline{\tau}}^D|^2 \, dx + \frac{1}{2(\lambda+\mu)} \int_{\Omega} |tr\, \underline{\underline{\tau}}|^2 \, dx, \tag{1.3.61}$$

under the constraint $\underline{\operatorname{div}}\, \underline{\underline{\tau}} + \underline{f} = 0$. Both the *mixed formulation* (1.3.59) and the dual formulation (1.3.61) are used in practice. They lead to different although similar approximations. □

We now consider the thin plate problem of Example 1.2.4 to introduce a mixed formulation due to [152] and [298].

Example 1.3.7 (Decomposition of a biharmonic problem). Again using the same technique as in the dualisation of the Dirichlet problem in Example 1.3.4, it is a simple exercise to transform problem (1.2.39) into the saddle point problem

$$\inf_{\mu \in L^2(\Omega)} \sup_{\varphi \in H_0^2(\Omega)} \frac{1}{2} \int_{\Omega} |\mu|^2 \, dx + \int_{\Omega} \mu \, \Delta\varphi \, dx + \int_{\Omega} f\varphi \, dx, \tag{1.3.62}$$

and to get the dual problem,

$$\inf_{\mu \in M} \frac{1}{2} \int_{\Omega} |\mu|^2 \, dx, \tag{1.3.63}$$

where $M := \{\mu \in L^2(\Omega),\ \Delta\mu + f = 0\}$. Integrating by parts the term $\int_{\Omega} \mu \, \Delta\varphi \, dx$, we get, as in Example 1.3.5, a weaker formulation

$$\inf_{\mu \in L^2(\Omega)} \sup_{\varphi \in H_0^2(\Omega)} \frac{1}{2}\int_\Omega |\mu|^2\,dx - \int_\Omega \underline{\mathrm{grad}}\,\mu \cdot \underline{\mathrm{grad}}\,\varphi\,dx + \int_\Omega f\varphi\,dx. \quad (1.3.64)$$

Assume that (1.3.64) has a saddle point (ω, ψ) with $\omega \in H^1(\Omega)$. Then (ω, ψ) is characterised by the variational system

$$\begin{cases} \displaystyle\int_\Omega \omega\mu\,dx - \int_\Omega \underline{\mathrm{grad}}\,\mu \cdot \underline{\mathrm{grad}}\,\psi\,dx = 0 & \forall \mu \in H^1(\Omega), \\ \displaystyle\int_\Omega \underline{\mathrm{grad}}\,\omega \cdot \underline{\mathrm{grad}}\,\varphi\,dx = \int_\Omega f\varphi\,dx & \forall \varphi \in H_0^1(\Omega). \end{cases} \quad (1.3.65)$$

It is not difficult to see that the two equations of (1.3.65) imply

$$\begin{cases} -\Delta\psi = \omega \quad \text{and} \quad \dfrac{\partial\psi}{\partial n}\Big|_\Gamma = 0, \\ -\Delta\omega = f. \end{cases} \quad (1.3.66)$$

As we already have $\psi|_\Gamma = 0$ (since $\psi \in H_0^1(\Omega)$), we have in (1.3.66) too many boundary conditions on ψ and none on ω. The system however has a solution (ω, ψ) (provided Ω and f are smooth enough) such that the solution of the Dirichlet problem in ψ also satisfies (through the choice of the right-hand side) the extra Neumann condition. □

Example 1.3.8 (Decomposition of the plate bending problem). We now consider the plate bending problem (1.2.43). In order to make the dual problem easier to introduce, we first write the energy functional in the form

$$\frac{1}{2}\left(\frac{Et^3}{12(1-\nu^2)}\right)\int_\Omega \mathfrak{M}(\underline{\underline{D}}_2\varphi) : \underline{\underline{D}}_2\varphi\,dx - \int_\Omega f\varphi\,dx, \quad (1.3.67)$$

where the operator $\underline{\underline{D}}_2$ is defined by

$$(\underline{\underline{D}}_2\varphi)_{ij} := \frac{\partial^2\varphi}{\partial x_i \partial x_j}, \quad 1 \le i,\ j \le 2, \quad (1.3.68)$$

and the operator $\mathfrak{M}$ by

$$\mathfrak{M}(\underline{\underline{\tau}}) := \begin{pmatrix} \tau_{11} + \nu\tau_{22} & (1-\nu)\tau_{12} \\ (1-\nu)\tau_{12} & \nu\tau_{11} + \tau_{22} \end{pmatrix}, \quad (1.3.69)$$

for any symmetric tensor $\underline{\underline{\tau}}$. Using the same kind of analysis as in the previous examples, we then get the saddle point problem

$$\inf_{\underline{\underline{\tau}} \in (L^2(\Omega))^4_s} \sup_{\varphi \in H^2_0(\Omega)} \frac{1}{2}\Big(\frac{12(1-\nu^2)}{Et^3}\Big) \int_\Omega \mathfrak{M}^{-1}(\underline{\underline{\tau}}) : \underline{\underline{\tau}}\, dx + \int_\Omega \underline{\underline{\tau}} : \underline{\underline{D}}_2 \varphi\, dx - \int_\Omega f\varphi\, dx, \quad (1.3.70)$$

where $(L^2(\Omega))^4_s$ is the space of square integrable 2×2 symmetric tensors. We introduce, as dual variables, the bending moments, obtained from the second derivatives of the primal solution ψ by

$$\underline{\underline{\sigma}} := -\frac{Et^3}{12(1-\nu^2)} \mathfrak{M}(\underline{\underline{D}}_2 u), \quad (1.3.71)$$

or explicitly

$$\begin{cases} \sigma_{11} = -\dfrac{Et^3}{12(1-\nu^2)}\Big(\dfrac{\partial^2\psi}{\partial x_1^2} + \nu \dfrac{\partial^2\psi}{\partial x_2^2}\Big), \\ \sigma_{22} = -\dfrac{Et^3}{12(1-\nu^2)}\Big(\nu\dfrac{\partial^2\psi}{\partial x_1^2} + \dfrac{\partial^2\psi}{\partial x_2^2}\Big), \\ \sigma_{12} = -\dfrac{Et^3}{12(1+\nu)} \dfrac{\partial^2\psi}{\partial x_1 \partial x_2}. \end{cases} \quad (1.3.72)$$

The dual problem can then be written as

$$\inf_{\underline{\underline{\tau}}} \frac{1}{2}\Big(\frac{12}{Et^3}\Big) \int_\Omega [(\tau_{11} + \tau_{22})^2 + 2(1+\nu)(\tau_{12}^2 - \tau_{11}\tau_{22})]\, dx, \quad (1.3.73)$$

under the constraint

$$D_2^* \underline{\underline{\tau}} = f. \quad (1.3.74)$$

In (1.3.74) we denoted by D_2^* the transpose of the operator $\underline{\underline{D}}_2$ so that

$$D_2^* \underline{\underline{\tau}} = \frac{\partial^2 \tau_{11}}{\partial x_1^2} + 2\frac{\partial^2 \tau_{12}}{\partial x_1 \partial x_2} + \frac{\partial^2 \tau_{22}}{\partial x_2^2}. \quad (1.3.75)$$

It is possible, as in the previous case, to integrate by parts the expressions (1.3.74) and to obtain formulations in different functional spaces. We shall see an example of such a procedure in Sect. 10.2. □

1.3.3 Duality Methods for Non Symmetric Bilinear Forms

In all previous examples, our variational formulations were based on a minimisation problem for a functional and we were led to introduce a genuine saddle point problem. Even if this classical framework is suitable for a first presentation, it is not the sole possible, and the techniques developed before can also be applied to problems which are not optimisation problems. Let us consider for instance in $H_0^1(\Omega)$ a *continuous* and *coercive* bilinear form $a(p,q)$. If we do not require $a(\cdot,\cdot)$ to be symmetric, the variational problem

$$a(p,q) = \int_\Omega fq\,dx, \quad \forall q \in H_0^1(\Omega), \tag{1.3.76}$$

has for $f \in L^2(\Omega)$ a unique solution $p \in H_0^1(\Omega)$ but does not correspond to the minimisation of any functional. To fix ideas, let us suppose that $a(p,q)$ can be written as

$$a(p,q) = m(\underline{\operatorname{grad}}\, p,\ \underline{\operatorname{grad}}\, q) = \int_\Omega M(\underline{\operatorname{grad}}\, p)\cdot \underline{\operatorname{grad}}\, q\,dx \tag{1.3.77}$$

where $m(\cdot,\cdot)$ is a continuous bilinear form on $(L^2(\Omega))^2$, which, of course, is non symmetric, and M is the associated linear operator from $(L^2(\Omega))^2$ into $(L^2(\Omega))^2$. We can now introduce the auxiliary variable

$$\underline{u} = M(\underline{\operatorname{grad}}\, p) \tag{1.3.78}$$

and write problem (1.3.67) in the form

$$\begin{cases} \displaystyle\int_\Omega \underline{u}\cdot \underline{\operatorname{grad}}\, q\,dx = \int_\Omega fq\,dx, \\ \displaystyle\int_\Omega M^{-1}\underline{u}\cdot\underline{v}\,dx = \int_\Omega \underline{v}\cdot \underline{\operatorname{grad}}\, p\,dx. \end{cases} \tag{1.3.79}$$

This can be integrated by parts to yield, as in Example 1.3.5, for $\underline{u}$ in $H(\operatorname{div};\Omega)$ and p in $L^2(\Omega)$:

$$\begin{cases} \displaystyle\int_\Omega \operatorname{div}\underline{u}\, q\,dx + \int_\Omega fq\,dx = 0, \quad \forall q \in L^2(\Omega), \\ \displaystyle\int_\Omega M^{-1}\underline{u}\cdot\underline{v}\,dx + \int_\Omega p\ \operatorname{div}\underline{v}\,dx = 0, \quad \forall \underline{v} \in H(\operatorname{div};\Omega). \end{cases} \tag{1.3.80}$$

We shall thus consider in Chaps. 3 and 4 problems such as (1.3.80) without making reference to a saddle point problem. The same remark would apply to the methods of the following section. □

1.3.4 Mixed Eigenvalue Problems

We have considered earlier in Sect. 1.2.1 some eigenvalue problems associated with our examples. We shall now rapidly consider their counterpart in mixed form. Let us thus come back to problem (1.3.13). It is now possible to consider three distinct eigenvalue problems.

1. The primal eigenvalue problem,

$$\begin{cases} a(u,v) + b(v,p) = \lambda(u,v)_V, & \forall v \in V, \\ b(u,q) = 0, \quad \forall q \in Q. \end{cases} \tag{1.3.81}$$

2. The dual eigenvalue problem,

$$\begin{cases} a(u,v) + b(v,p) = 0, & \forall v \in V, \\ b(u,q) = -\lambda(p,q)_Q, & \forall q \in Q. \end{cases} \tag{1.3.82}$$

3. The global eigenvalue problem,

$$\begin{cases} a(u,v) + b(v,p) = \lambda(u,v)_V, & \forall v \in V, \\ b(u,q) = -\lambda(p,q)_Q, & \forall q \in Q. \end{cases} \tag{1.3.83}$$

In practice, the interesting cases, from the physical point of view, will be either the primal or the dual problem.

Example 1.3.9 (Eigenvalue problem for incompressible elasticity and the Stokes problem). This case is the simplest instance of a primal eigenvalue problem of type (1.3.81). We consider the eigenvalue problem corresponding to (1.3.39) in the limiting case of λ infinitely large.

$$\begin{cases} 2\mu \displaystyle\int_\Omega \underline{\underline{\varepsilon}}(\underline{u}) : \underline{\underline{\varepsilon}}(\underline{v})\, dx + \int_\Omega p \ \operatorname{div} \underline{v}\, dx = \lambda \int_\Omega \underline{u} \cdot \underline{v}\, dx, & \forall \underline{v} \in V, \\ \displaystyle\int_\Omega q \ \operatorname{div} \underline{u} = 0, \quad \forall q \in L^2(\Omega). \end{cases} \tag{1.3.84}$$

This problem is the equivalent of (1.2.60) in the incompressible case. The Lagrange multiplier p ensures the incompressibility of the eigenmodes. □

Example 1.3.10 (Eigenvalue problem for the mixed Poisson problem). This is the simplest example of a dual problem of type (1.3.82). We consider the eigenvalue problem associated with the saddle-point problem of (1.3.51).

$$\begin{cases} \displaystyle\int_\Omega \underline{u} \cdot \underline{v}\, dx + \int_\Omega p \ \operatorname{div} \underline{v}\, dx = 0, \ \forall \underline{v} \in H(\operatorname{div};\Omega), \\ \displaystyle\int_\Omega \operatorname{div} \underline{u} q\, dx = -\lambda \int_\Omega p\, q\, dx, \quad \forall q \in L^2(\Omega). \end{cases} \tag{1.3.85}$$

This corresponds to the standard eigenvalue problem of (1.2.58). The Lagrange multiplier p is now the important variable. □

As we shall see later in Chap. 6, the approximation of those two kinds of problems will need specific assumptions.

1.4 Domain Decomposition Methods, Hybrid Methods

We have shown in Sect. 1.3 that duality techniques enable us to obtain alternate variational formulations for some problems. The method that we shall now describe will yield a new family of variational principles that can be more or less grouped under the name of hybrid methods. The common point between the examples that follow is that in all cases the variational principle will depend explicitly, independently of any discretisation, on a partition of the domain Ω into sub-domains. To clarify some of the facts that will appear later, we first recall a very classical result.

Example 1.4.1 (A transmission problem). We consider the very classical case in which a domain Ω is split into two sub-domains Ω_1 and Ω_2 by a smooth enough internal boundary S (Fig. 1.1). We consider the case of a Dirichlet problem with variable coefficients $a_1(x)$, $a_2(x)$, defined respectively in Ω_1 and Ω_2 and being discontinuous on S. This classically leads to the variational problem: *find* $p \in H_0^1(\Omega)$ *such that*

$$\int_{\Omega_1} a_1(x)\ \underline{\mathrm{grad}}\, p \cdot \underline{\mathrm{grad}}\, q\, dx + \int_{\Omega_2} a_2(x)\ \underline{\mathrm{grad}}\, p \cdot \underline{\mathrm{grad}}\, q\, dx$$
$$= \int_{\Omega} f q\, dx, \quad \forall q \in H_0^1(\Omega). \quad (1.4.1)$$

We would like to decouple the above problem into two problems, one in each Ω_i, and add suitable continuity conditions at the interface S. For this we recall the following classical result.

Proposition 1.4.1. *Assume that Ω is a domain in $\mathbb{R}^2$ with a Lipschitz continuous boundary, and let S be a Lipschitz continuous curve that splits Ω in the two sub-domains Ω_1 and Ω_2. Let moreover $a(x)$ be a piecewise smooth function, and denote by $a_i(x)$ $(i = 1, 2)$ the restriction of $a(x)$ to Ω_i. Let $f \in L^2(\Omega)$ and let p be solution of the problem* (1.4.1) *Then, setting $p_1 = p|_{\Omega_1}$ and $p_2 = p|_{\Omega_2}$, it is equivalent to say that p is solution of the problem*

$$\begin{cases} -\,\mathrm{div}\,(a_1(x)\ \underline{\mathrm{grad}}\, p_1) = f \text{ in } \Omega_1, \\ -\,\mathrm{div}\,(a_2(x)\ \underline{\mathrm{grad}}\, p_2) = f \text{ in } \Omega_2, \\ p_1|_{\Gamma\cap\partial\Omega_1} = 0,\ p_2|_{\Gamma\cap\partial\Omega_2} = 0, \end{cases} \quad (1.4.2)$$

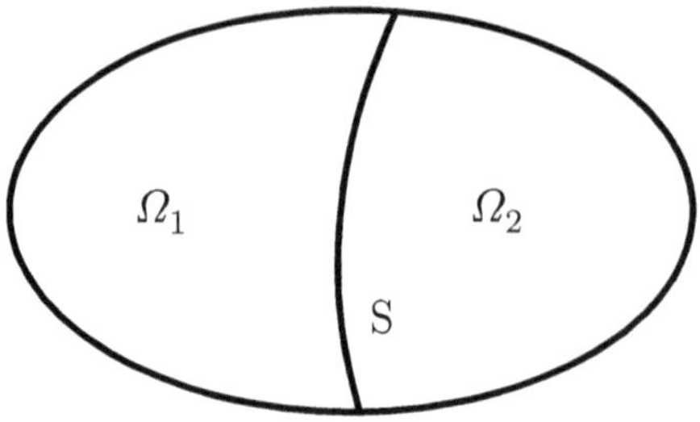

Fig. 1.1 The decomposed domain

$$p_1 = p_2 \text{ on } S,\ a_1 \frac{\partial p_1}{\partial n_1} + a_2 \frac{\partial p_2}{\partial n_2} = 0 \text{ on } S, \tag{1.4.3}$$

where n_1 and n_2 are the exterior normals to Ω_1 and Ω_2 (respectively) on S. □

□

An important special case is $a_1(x) = a_2(x) = 1$. In that case, problem (1.4.1) is obviously equivalent to

$$\begin{cases} -\Delta p = f, \\ u|_\Gamma = 0 \end{cases} \tag{1.4.4}$$

and conditions (1.4.2) and (1.4.3) can be written as

$$\begin{cases} -\Delta p_1 = f \text{ in } \Omega_1, \\ -\Delta p_2 = f \text{ in } \Omega_2, \\ p_1|_{\Gamma\cap\partial\Omega_1} = 0,\ p_2|_{\Gamma\cap\partial\Omega_2} = 0, \end{cases} \tag{1.4.5}$$

and

$$\begin{cases} p_1 = p_2 \text{ on } S, \\ \dfrac{\partial p_1}{\partial n_1} + \dfrac{\partial p_2}{\partial n_2} = 0 \text{ on } S. \end{cases} \tag{1.4.6}$$

Example 1.4.2 (A domain decomposition method for the Dirichlet problem). What we really want to do is to consider a general partition of Ω

$$\bar{\Omega} := \bigcup_{i=1}^{N} \bar{K}_i. \tag{1.4.7}$$

We now write the classical Dirichlet functional of Example 1.2.1 in the following way. First, we write the Dirichlet functional as

$$J(q) := \sum_{i=1}^{N} \Big\{ \frac{1}{2} \int_{K_i} |\ \underline{\text{grad}}\, q|^2\, dx - \int_{K_i} fq\, dx \Big\}. \tag{1.4.8}$$

Then, introducing the functional space

$$X(\Omega) := \{q|\ q|_{K_i} \in H^1(K_i)\} \approx \prod_{i=1}^{N} H^1(K_i), \tag{1.4.9}$$

we can extend $J(q)$ on $X(\Omega)$. Moreover, $H_0^1(\Omega)$ is a closed subspace of $X(\Omega)$ and we may consider "$q \in H_0^1(\Omega)$" as a linear constraint on $q \in X(\Omega)$. This constraint states that on $e_{ij} = \partial K_i \cap \partial K_j$ we must have (in $H^{\frac{1}{2}}(e_{ij})$) $p_i = p_j$, where $p_\ell = p|_{K_\ell}$.

We shall therefore, following a now familiar procedure, impose this constraint through a Lagrange multiplier properly chosen in $H^{-\frac{1}{2}}(e_{ij})$. As we shall see in Chap. 2, it will be more convenient to introduce $\underline{v} \in H(\mathrm{div};\Omega)$ and to use the normal trace of $\underline{v}$ on ∂K_i as a multiplier. This leads us to the saddle point problem

$$\inf_{p\in X(\Omega)} \sup_{\underline{v}\in H(\mathrm{div};\Omega)} \sum_{i=1}^{N} \left\{\frac{1}{2}\int_{K_i} |\ \underline{\mathrm{grad}}\, q|^2\, dx - \int_{\partial K_i} \underline{v}\cdot\underline{n}_i\, q\, ds - \int_{K_i} fq\, dx\right\}, \tag{1.4.10}$$

for which we have the following optimality conditions: for $i = 1, \dots, N$, find $p_i \in H^1(K_i)$ such that,

$$\int_{K_i} \underline{\mathrm{grad}}\, p_i \cdot \underline{\mathrm{grad}}\, q_i\, dx = \int_{K_i} fq_i\, dx + \int_{\partial K_i} \underline{u}\cdot\underline{n}\, q_i\, ds, \quad \forall q_i \in H^1(K_i), \tag{1.4.11}$$

$$\sum_{i=1}^{N} \int_{\partial K_i} \underline{v}\cdot\underline{n}_i\ p_i\, ds = 0, \quad \forall \underline{v} \in H(\mathrm{div};\Omega). \tag{1.4.12}$$

Condition (1.4.12) expresses continuity of p at interfaces e_{ij} and condition $p|_\Gamma = 0$. Condition (1.4.11) shows that each p_i is solution in K_i of a Neumann problem

$$\begin{cases} -\Delta p_i = f \text{ in } K_i, \\ \dfrac{\partial p_i}{\partial n_i} = \underline{u}\cdot\underline{n}_i \text{ on } \partial K_i. \end{cases} \tag{1.4.13}$$

Solving this problem obviously requires a compatibility condition (take $q_i = 1$ in (1.4.11))

$$\int_{\partial K_i} \underline{u}\cdot\underline{n}_i\, ds + \int_{K_i} f\, dx = 0 \tag{1.4.14}$$

on every sub-domain K_i. This condition can also be written as

$$\int_{K_i} (\mathrm{div}\, \underline{u} + f)\, dx = 0. \tag{1.4.15}$$

From (1.4.13) we have that the multiplier $\underline{u} \cdot \underline{n}$ can be seen as the normal derivative of p. Indeed, when equilibrium is attained, we have on interfaces $\frac{\partial p_i}{\partial n_i} = \underline{u} \cdot \underline{n}_i = -\underline{u} \cdot \underline{n}_j = \frac{-\partial p_j}{\partial n_j}$ and $p_i = p_j$. A suitable lifting of $\underline{u}$ in each K_i in order to have div $\underline{u} + f = 0$ can always be done because of (1.4.14) and (1.4.15). □

Example 1.4.3 (Dual problem of the domain decomposition method). We now consider the dual problem of the above saddle point formulation. It will be, as it can be expected, very similar to the dual problem introduced in Sect. 1.3 for the Poisson problem. Let us first remark that taking the *infimum* on the constant part of $q \in X(\Omega)$ on each K_i leads to the *constraint* (1.4.15) on $\underline{u}$. It is therefore possible to suppose div $\underline{u} + f = 0$ as this can be attained by modifications to $\underline{u}$ that are internal to K_i (that is not modifying $\underline{u} \cdot \underline{n}_i$) and are transparent to formulation (1.4.10). Writing

$$\int_{\partial K_i} \underline{v} \cdot \underline{n}_i \, q \, ds = \int_{K_i} \operatorname{div} \underline{u} \, q \, dx + \int_{K_i} \underline{v} \cdot \underline{\operatorname{grad}}\, q \, dx, \tag{1.4.16}$$

one may write from (1.4.10)

$$\sup_{\operatorname{div}(\underline{v})+f=0} \ \inf_{q_i \in H^1(K_i)/\mathbb{R}} \sum_{i=1}^{N} \Big\{ \frac{1}{2} \int_{K_i} |\, \underline{\operatorname{grad}}\, q_i|^2 \, dx - \int_{K_i} \underline{v} \cdot \underline{\operatorname{grad}}\, q_i \, dx \Big\}. \tag{1.4.17}$$

From (1.4.17) we evidently get, setting $\underline{u}_i = \underline{u}|_{K_i}$,

$$\underline{\operatorname{grad}}\, p_i = P(\underline{u}_i), \tag{1.4.18}$$

where P is the projection operator in $(L^2(K_i))^2$ on $\underline{\operatorname{grad}}(H^1(K_i))$. We shall indeed prove in Chap. 2 that one has

$$(L^2(\Omega))^2 = \{\underline{\operatorname{grad}}\, H^1(\Omega)\} \oplus \underline{\operatorname{curl}}\, H^1_0(\Omega)\}. \tag{1.4.19}$$

From this we can eliminate q_i and write the dual problem,

$$\sup_{\substack{\underline{v} \in H(\operatorname{div};\Omega) \\ \operatorname{div}(\underline{v})+f=0}} -\frac{1}{2} \sum_{i=1}^{N} \int_{K_i} |P(\underline{v}_i)|^2 \, dx. \tag{1.4.20}$$

We are therefore back to a variant of (1.3.45). Indeed, (1.3.45) shows that the projection operator P in (1.4.20) is unnecessary. □

Remark 1.4.1. One could obtain a variant of the above dual problem, without constraint (1.4.15) by using a "least-squares" solution of (1.4.13) whenever (1.4.14) does not hold. This could be done, for instance by solving on K_i, in a weak formulation that we shall not describe,

$$\begin{cases} \Delta^2 p_i = \Delta f \text{ in } K_i, \\ \dfrac{\partial}{\partial n_i} \Delta p_i = \dfrac{\partial f}{\partial n_i} \text{ on } \partial K_i, \\ \dfrac{\partial p_i}{\partial n_i} = \underline{v} \cdot \underline{n}_i \text{ on } \partial K_i, \end{cases} \tag{1.4.21}$$

for which a solution always exists, defined up to an additive constant. Such a procedure could be useful for algorithmic purposes for (1.4.21) is a local simple problem even if it is a fourth-order problem. □

Example 1.4.4 (Dual hybrid methods). We now consider the dual problem (1.3.45), that is the complementary energy principle, that we now pose in $H(\text{div}; \Omega)$,

$$\inf_{\substack{\underline{v} \in H(\text{div};\Omega) \\ \text{div}\,(\underline{v}) + f = 0}} \frac{1}{2} \int_{\Omega} |\underline{v}|^2 \, dx. \tag{1.4.22}$$

We can apply the domain decomposition principle to such a problem by introducing

$$Y(\Omega) := \{\underline{v} |\ \underline{v}|_{K_i} \in H(\text{div}; K_i)\} \approx \prod_{i=1}^{N} H(\text{div}; K_i). \tag{1.4.23}$$

As we shall see in Chap. 2, $H(\text{div}; \Omega)$ is now a closed subspace of $Y(\Omega)$ characterised by

$$\sum_{i=1}^{N} \int_{\partial K_i} (\underline{v} \cdot \underline{n}_i)\, q \, ds = 0, \quad \forall q \in H_0^1(\Omega). \tag{1.4.24}$$

We can then transform (1.4.22) into the saddle point problem

$$\inf_{v \in Y(\Omega)} \sup_{q \in H_0^1(\Omega)} \sum_{i=1}^{N} \left\{ \frac{1}{2} \int_{K_i} |\underline{v}_i|^2 \, dx + \int_{\partial K_i} \underline{v}_i \cdot \underline{n}_i \, q \, ds \right\} \tag{1.4.25}$$

under the local constraint

$$\text{div}\, \underline{v}_i + f = 0 \text{ on } K_i. \tag{1.4.26}$$

An advantage of this formulation is that it is easy to find $\underline{v}_i$ satisfying (1.4.26). We shall meet discretisation methods, based on such a principle, under the name of *dual hybrid methods* for the treatment of almost any example considered in this book: Dirichlet problems, elasticity problems, fourth-order problems, etc. □

Example 1.4.5 (The Hellan-Hermann-Johnson method in elasticity). This is an example in which a domain decomposition is introduced, not by dualising a continuity condition but by defining a variational formulation able to bypass this continuity by approximating weak derivatives. This formulation will not be developed but is amenable to the techniques of the book. We shall first present formal results and postpone a precise presentation of the functional framework. Our starting point will be the saddle point problem (1.3.59) and its optimality conditions (1.3.60) that we write, in variational form (with functional spaces to be defined), as

$$\frac{1}{\mu}\int_\Omega \underline{\underline{\sigma}}^D : \underline{\underline{\tau}}^D\,dx + \frac{1}{2(\lambda+\mu)}\int_\Omega tr\,\underline{\underline{\sigma}}\,tr\,\underline{\underline{\tau}}\,dx + \int_\Omega \underline{\underline{\varepsilon}}(\underline{u}) : \underline{\underline{\tau}}\,dx = 0, \quad \forall \underline{\underline{\tau}} \in \underline{\underline{H}}(\underline{\mathrm{div}};\Omega)_s, \tag{1.4.27}$$

$$\int_\Omega \underline{\underline{\varepsilon}}(\underline{v}) : \underline{\underline{\sigma}}\,dx + \int_\Omega \underline{f}\cdot\underline{v}\,dx = 0 \qquad \forall \underline{v} \in (H_0^1(\Omega))^2. \tag{1.4.28}$$

These conditions make sense for a space of $\underline{\underline{\sigma}}$ chosen so that $\underline{\mathrm{div}}\,\underline{\underline{\sigma}}$ is well defined, which implies, as we have seen, continuity of $\underline{\sigma}_n$ at interfaces. On the other hand $\underline{v}$ can be taken as completely discontinuous on these same interfaces. What we now try to do is to split continuity conditions between $\underline{\sigma}\cdot\underline{n}$ and $\underline{v}$. Let us consider indeed the well-known integration by parts formula,

$$\int_\Omega \underline{\mathrm{div}}\,\underline{\underline{\tau}}\cdot\underline{v}\,dx + \int_\Omega \underline{\underline{\tau}} : \underline{\underline{\varepsilon}}(\underline{v})\,dx = \int_{\partial\Omega} \tau_{nn}\,\underline{v}\cdot\underline{n}\,ds + \int_{\partial\Omega} \tau_{nt}\,\underline{v}\cdot\underline{t}\,ds. \tag{1.4.29}$$

Whenever $\underline{v}$ is a smooth (let us say $H^1(\Omega)$) vector, and τ_{nt} is continuous, we thus have

$$\int_\Omega \underline{\underline{\varepsilon}}(\underline{v}) : \underline{\underline{\sigma}}\,dx = -\sum_{i=1}^{N}\left\{\int_{K_i} \underline{\mathrm{div}}\,\underline{\underline{\sigma}}\cdot\underline{v}\,dx + \int_{\partial K_i} \sigma_{nn}\,\underline{v}\cdot\underline{n}\,ds\right\}, \tag{1.4.30}$$

so that we can rewrite (1.4.27) and (1.4.28) in the following form,

$$\frac{1}{\mu}\int_\Omega \underline{\underline{\sigma}}^D : \underline{\underline{\tau}}^D\,dx + \frac{1}{2(\lambda+\mu)}\int_\Omega tr\,\underline{\underline{\sigma}}\,tr\,\underline{\underline{\tau}}\,dx + \sum_{i=1}^{N}\left\{\int_{K_i} \underline{\mathrm{div}}\,\underline{\underline{\tau}}\cdot\underline{u}\,dx - \int_{\partial K_i} \tau_{nn}\,\underline{u}\cdot\underline{n}\,ds\right\} = 0 \quad \forall \underline{\underline{\tau}}, \tag{1.4.31}$$

$$\sum_{i=1}^{N}\left\{\int_{K_i} \underline{\mathrm{div}}\,\underline{\underline{\sigma}}\cdot\underline{v}\,dx - \int_{\partial K_i} \sigma_{nn}\,\underline{v}\cdot\underline{n}\,ds\right\} + \int_\Omega \underline{f}\cdot\underline{v}\,dx = 0 \quad \forall \underline{v}. \tag{1.4.32}$$

Formally, this is well defined for $\underline{\underline{\sigma}}$ chosen with σ_{nt} continuous at interfaces while $\underline{u}\cdot\underline{n}$ is continuous. Then the term

$$\sum_{i=1}^{N}\Big\{\int_{\partial K_i} \sigma_{nn}\, \underline{v}\cdot\underline{n}\, ds\Big\} \tag{1.4.33}$$

depends on the *jump* of σ_{nn} on ∂K_i and (1.4.32) can be read as $\underline{\mathrm{div}}\,\underline{\underline{\sigma}} + \underline{f} = 0$ in the sense of distributions.

Up to now we considered a purely formal problem. Giving a good framework to (1.4.31), and (1.4.32) is a task that requires some care. The presence of traces, appearing explicitly in the variational formulation, leads to deal with spaces $H^{\frac{1}{2}}(\partial K_i)$ and $H^{-\frac{1}{2}}(\partial K_i)$ and to subtle considerations about the behaviour of functions in these rather pathological spaces. Let us define

$$\Sigma := \prod_{K_i}(H^1(K_i))^4_s = \{\underline{\underline{\sigma}}|\ \sigma_{ij}|_{K_i} \in H^1(K_i),\ \sigma_{ij} = \sigma_{ji}\}. \tag{1.4.34}$$

This is a space of smooth tensors and we can consider σ_{nt} on each interface $e_{ij} = \partial K_i \cap \partial K_j$, (cf. Chap. 2). We have $\sigma_{nt} \in H^{\frac{1}{2}}(e_{ij})$ but we do not have $\sigma_{nt} \in H^{\frac{1}{2}}(\partial K)$ for this would require some continuity at vertices which cannot in general take place due to the change of direction of $\underline{n}$ and $\underline{t}$. We can nevertheless consider in Σ tensor functions $\underline{\underline{\sigma}}$ such that σ_{nt} is continuous on e_{ij}. To make (1.4.33) meaningful, we now have to choose $\underline{v}$ with $\underline{v}\cdot\underline{n}$ continuous on e_{ij}. We have already seen that for $\underline{v}$ in $H(\mathrm{div}; K_i)$ we can define $\underline{v}\cdot\underline{n}$ in $H^{-\frac{1}{2}}(\partial K_i)$. Unfortunately it is not possible to restrict $\underline{v}\cdot\underline{n}|_{e_{ij}}$ and get a result in $H^{-\frac{1}{2}}(e_{ij})$: something is lost at corners. In reality we only need an "infinitesimal" amount of extra smoothness and this will lead us to look for $\underline{v}$ in $(L^p(\Omega))^2 \cap H(\mathrm{div};\Omega)$ for $p > 2$. This will cause some problems in applying the theory of Chap. 4 and existence of a solution will have to be deduced through special considerations. □

1.5 Modified Variational Formulations

We shall present in this section modified variational principles associated to saddle-point problems or more general mixed methods. We shall distinguish between augmented formulations and perturbed formulations. In the following, *augmented formulations* will correspond to a modification of the variational formulation of the **continuous** problem. This will be done so that the solution is not changed (under perhaps some regularity conditions). The discretised versions will however in general be different. On the other hand, perturbed formulations will be meaningful only on a discrete problem. The rationale behind the introduction of these modified formulations is that they will have, for some discretisations, a better behaviour than the original one, in particular with respect to stability issues.

1.5.1 Augmented Formulations

As an example, let us consider the Galerkin least-squares methods introduced by Hughes and Franca [256]. To fix ideas, let us consider the simple cases of the saddle point formulation of the Poisson problem of Sect. 1.3,

$$\inf_{\underline{v}\in H(\mathrm{div};\Omega)} \sup_{q\in L^2(\Omega)} \frac{1}{2}\int_\Omega |\underline{v}|^2\,dx + \int_\Omega fq\,dx + \int_\Omega q \ \ \mathrm{div}\,\underline{v}\,dx, \tag{1.5.1}$$

for which the Euler equations are

$$\begin{cases} \displaystyle\int_\Omega \underline{u}\cdot\underline{v}\,dx + \int_\Omega p \ \ \mathrm{div}\,\underline{v}\,dx = 0 & \forall \underline{v}\in H(\mathrm{div};\Omega),\\ \displaystyle\int_\Omega (\mathrm{div}\,\underline{u} + f)q\,dx = 0 & \forall q\in L^2(\Omega). \end{cases} \tag{1.5.2}$$

To better understand stability issues, it will be convenient to write (1.5.2) using an antisymmetric bilinear form, obtained by subtracting the two equations

$$\begin{aligned} E((p,\underline{u}),(q,\underline{v})) :&= \int_\Omega \underline{u}\cdot\underline{v}\,dx + \int_\Omega p \ \ \mathrm{div}\,\underline{v}\,dx - \int_\Omega q \ \ \mathrm{div}\,\underline{u}\,dx \\ &= \int_\Omega fq\,dx, \qquad \forall \underline{v}\in H(\mathrm{div};\Omega), \forall q\in L^2(\Omega). \end{aligned} \tag{1.5.3}$$

One sees that

$$E((p,\underline{u}),(p,\underline{u})) = \int_\Omega |\underline{u}|^2\,dx, \tag{1.5.4}$$

so that our bilinear form is non-negative, but not coercive.

The approximation of (1.5.1) thus requires the special constructions that will be described in Chaps. 3–5.

Remark 1.5.1. **A basic philosophical issue:** It is important, at this point, to underline a basic philosophical issue: Assume that you are given a bilinear form (say, E) on $\mathcal{H}\times\mathcal{H}$, where $\mathcal{H}$ is a Hilbert space. There are several ways that could be used to prove that *E induces an isomorphism from $\mathcal{H}$ to its dual space $\mathcal{H}'$*. We mean by that that there are several properties that will imply such a result. For instance, if for simplicity $\mathcal{H} \equiv \mathbb{R}^n$, you can show that the associated matrix M_E has a determinant that is different from zero. Otherwise, you can show that the associated homogeneous system has only the (trivial) zero solution. Alternatively, you can show that the associated non-homogeneous system has at least one solution for every right-hand side, or you can show that there exists a constant c such that for every pair (X,F) that satisfies $M_E X = F$ you have $\|X\| \le c\,\|F\|$. All these properties are indeed *equivalent* to each other (in finite dimension). As a last possibility, you can show that E is *coercive*:

$$E(X,X) \geq \alpha \|X\|^2 \qquad \forall X \in \mathcal{H}. \tag{1.5.5}$$

This last condition (some sort of Cinderella among the other equivalent step-sisters) is not *necessary and sufficient* (as all the others are), but only *sufficient*. Indeed, there are zillions of non-singular matrices that are not coercive. Assume now that you want to play an additional game. You would like to consider *proper subspaces* $\tilde{\mathcal{H}}$ of $\mathbb{R}^n$ (say, to fix ideas, $\mathbb{R}^m$ with $m < n$) and the *restriction* $\tilde{E}$ of the bilinear form E to $\tilde{\mathcal{H}} \times \tilde{\mathcal{H}}$. We ask whether the bilinear form $\tilde{E}$ induces an isomorphism from $\tilde{\mathcal{H}}$ to its dual $\tilde{\mathcal{H}}'$. In fact, not a single one of the above necessary and sufficient conditions will be *automatically inherited* by $\tilde{E}$. Actually, as they are all equivalent, if one does not the others cannot do either. Surely there are zillions of non-singular matrices M whose first entry $M_{1,1}$ is equal to 0. Then you take $\tilde{\mathcal{H}}$ equal to $\mathbb{R}^1$ (using the first component of every $X \in \mathcal{H} \equiv \mathbb{R}^n$) and you are done: $\tilde{E}$ is 0.

But Cinderella *survives*: if E is coercive on $\mathcal{H}$, then $\tilde{E}$ is coercive on $\tilde{\mathcal{H}}$, *for every subspace* $\tilde{\mathcal{H}} \subseteq \mathcal{H}$ and the same value of α that makes (1.5.5) true will make

$$\tilde{E}(\tilde{X},\tilde{X}) \geq \alpha \|\tilde{X}\|^2 \qquad \forall \tilde{X} \in \tilde{\mathcal{H}}, \tag{1.5.6}$$

hold true as well. Hence, Cinderella becomes princess and superstar, and everybody, for every problem, would like to have a coercive bilinear form. If it is not, one would struggle to change the problem into an equivalent one, whose associated bilinear form is *coercive*. This, in short, is the essence of many stabilisation techniques. □

We have already seen in Sect. 1.3 that the solution of the "weaker" problem (1.5.2) is in fact the solution of the standard problem $-\Delta p = f$, written as the system

$$\begin{cases} \underline{u} - \underline{\mathrm{grad}}\, p = 0, & p \in H_0^1(\Omega), \\ \mathrm{div}\, \underline{u} + f = 0, & \underline{u} \in H(\mathrm{div};\Omega). \end{cases} \tag{1.5.7}$$

Starting from this system, we can also consider the other formulation

$$\inf_{\underline{v}\in(L^2(\Omega))^2} \ \sup_{q\in H_0^1(\Omega)} \ \frac{1}{2}\int_\Omega |\underline{v}|^2\,dx + \int_\Omega fq\,dx - \int_\Omega \underline{\mathrm{grad}}\, q \cdot \underline{v}\,dx, \tag{1.5.8}$$

for which the Euler equations are now

$$\begin{cases} \displaystyle\int_\Omega \underline{u}\cdot\underline{v}\,dx - \int_\Omega \underline{\mathrm{grad}}\, p\cdot\underline{v}\,dx = 0, & \forall \underline{v} \in (L^2(\Omega))^2, \\ \displaystyle -\int_\Omega \underline{u}\cdot\,\underline{\mathrm{grad}}\, q\,dx + \int_\Omega fq\,dx = 0, & \forall q \in H_0^1(\Omega). \end{cases} \tag{1.5.9}$$

Note that, comparing to (1.5.2), we had to change the regularity requirements on q in order to make the functional meaningful. One checks that the solution of the continuous Poisson problem, which belongs to $H_0^1(\Omega)$, is a solution of our new saddle point problem. However, it is now clear that the discrete problems will now have to employ finite element approximations of $H_0^1(\Omega)$.

It is clear that we can always add (or subtract) the square of one of the equations (1.5.7) to the Lagrangian of (1.5.1) or (1.5.8) without changing the min-max point. For instance, we can add to (1.5.1) the square of the second equation of (1.5.7) to obtain

$$\inf_{\underline{v}\in H(\mathrm{div};\Omega)} \sup_{q\in L^2(\Omega)} \left\{ \frac{1}{2}\int_\Omega |\underline{v}|^2\,dx + \int_\Omega fq\,dx + \int_\Omega q \;\mathrm{div}\,\underline{v}\,dx \right.$$
$$\left. + \frac{\kappa_1}{2}\int_\Omega (\mathrm{div}\,\underline{v}+f)^2\,dx \right\}, \quad (1.5.10)$$

where $\kappa_1 \geq 0$ is arbitrary. We write again the Euler equations using an antisymmetric bilinear form,

$$\begin{aligned} E((p,\underline{u}),(q,\underline{v})) &= \int_\Omega \underline{u}\cdot\underline{v}\,dx + \kappa_1\int_\Omega \mathrm{div}\,\underline{u}\;\mathrm{div}\,\underline{v}\,dx \\ &\quad + \int_\Omega p\;\mathrm{div}\,\underline{v}\,dx - \int_\Omega q\;\mathrm{div}\,\underline{u}\,dx \\ &= \int_\Omega fq\,dx - \kappa_1\int_\Omega f\;\mathrm{div}\,\underline{v}\,dx, \\ &\qquad \forall \underline{v}\in H(\mathrm{div};\Omega), \forall q\in L^2(\Omega). \end{aligned} \quad (1.5.11)$$

It is clear that the solution of problem (1.5.10) is exactly the same as that of problem (1.5.1). Approximate solutions might however be different and some choices of elements will be stable for (1.5.10) but not for (1.5.1). Indeed, we now have

$$\begin{aligned} E((q,\underline{v}),(q,\underline{v})) &= \int_\Omega |\underline{v}|^2\,dx + \kappa_1\int_\Omega |\;\mathrm{div}\,\underline{v}|^2\,dx \\ &\geq \min(1,\kappa_1)\|\underline{v}\|^2_{H(\mathrm{div};\Omega)}, \end{aligned} \quad (1.5.12)$$

which is not yet the coercivity property (1.5.5), but is much better than (1.5.4), as now at least the full norm of $\underline{v}$ is under control. Indeed, as we shall see, this enables the construction of otherwise impossible approximations such as in [124].

Remark 1.5.2. Terms like $\min(1,\kappa_1)$ appearing in (1.5.12) are also *weird*. Their presence has still to do with the weird choice of the norm (1.3.47) in $H(\mathrm{div};\Omega)$. Had we chosen the more reasonable (but, alas!, almost never used in the literature) definition (1.3.49), we would, instead, have reached a $\min(1,\kappa_1/\ell)$ which looks *much more healthy*, as κ_1 is clearly a length. □

Another reasonable possibility is available: one might take (1.5.8) and subtract from it the square of the first equation of (1.5.7), to get

$$\inf_{\underline{v}\in(L^2(\Omega))^2}\ \sup_{q\in H_0^1(\Omega)} \frac{1}{2}\int_\Omega |\underline{v}|^2\,dx + \int_\Omega fq\,dx - \int_\Omega \underline{\mathrm{grad}}\,q\,\underline{v}\,dx \\ - \frac{\kappa_2}{2}\int_\Omega |\underline{v} - \underline{\mathrm{grad}}\,q|^2 dx \quad (1.5.13)$$

and the Euler equations of this Lagrangian yield

$$E((p,\underline{u}),(q,\underline{v})) = (1-\kappa_2)\int_\Omega \underline{u}\cdot\underline{v}\,dx - (1-\kappa_2)\int_\Omega \underline{\mathrm{grad}}\,p\cdot\underline{v}\,dx \\ + (1-\kappa_2)\int_\Omega \underline{u}\cdot\underline{\mathrm{grad}}\,q\,dx + \kappa_2\int_\Omega \underline{\mathrm{grad}}\,p\cdot\underline{\mathrm{grad}}\,q\,dx \\ = \int_\Omega fq\,dx \qquad \forall\underline{v}\in(L^2(\Omega))^2, \forall q\in H_0^1(\Omega). \quad (1.5.14)$$

This equation is a convex combination of equation (1.5.3) and of a standard formulation of the Poisson problem. Although this may seem silly at first sight, such methods have been used in [208] to stabilise formulations in which the discrete space V_h for the approximation of $\underline{u}$ was too small with respect to the space spanned by $\underline{\mathrm{grad}}\,q_h$, resulting in a failure of the *inf-sup* condition. The reason to employ such a strange approximation was that the first equation was much more complex than $\underline{u} - \underline{\mathrm{grad}}\,p = 0$ but rather of the form $\underline{u} - \underline{\mathrm{grad}}\,p = F(\underline{u}, p)$ where the function $F(\underline{u}, p)$ contained terms prescribing low order elements in order to be manageable. Unsurprisingly, we now have

$$E((q,\underline{v}),(q,\underline{v})) = (1-\kappa_2)\int_\Omega |\underline{v}|^2\,dx + \kappa_2\int_\Omega |\,\underline{\mathrm{grad}}\,q|^2\,dx, \quad (1.5.15)$$

and we have coercivity on $(L^2(\Omega))^2 \times H_0^1(\Omega)$ for $0 < \kappa_2 < 1$.

Now, if one is really eager for stability, one could consider a "super-stabilised" formulation

$$\inf_{\underline{v}\in H(\mathrm{div};\Omega)}\ \sup_{q\in H_0^1(\Omega)} \frac{1}{2}\int_\Omega |\underline{v}|^2\,dx - \int_\Omega \underline{v}\cdot\underline{\mathrm{grad}}\,q\,dx + \int_\Omega fq\,dx \\ + \frac{\kappa_2}{2}\int_\Omega |\underline{v} - \underline{\mathrm{grad}}\,q|^2 dx + \frac{\kappa_1}{2}\int_\Omega (\mathrm{div}\,\underline{v} + f)^2\,dx \\ + \frac{\kappa_3}{2}\| - \Delta q - f\|_*^2, \quad (1.5.16)$$

where $\|\cdot\|_*$ is the norm in $H^{-1}(\Omega)$, which we can define from the corresponding scalar product:

$$\langle p', q'\rangle_* := \langle(-\Delta)^{-1}p', q'\rangle_{H_0^1(\Omega)\times H^{-1}(\Omega)}. \quad (1.5.17)$$

Taking the Euler equation and using

$$\langle -\Delta p - f, -\Delta q\rangle_* = \langle -\Delta p - f, q\rangle_{H^{-1}(\Omega)\times H_0^1(\Omega)} = \int_\Omega \underline{\text{grad}}\, p \cdot \underline{\text{grad}}\, q \, dx - \int_\Omega f q \, dx,$$

one obtains

$$\begin{aligned}
&(1+\kappa_2)\int_\Omega \underline{u}\cdot\underline{v}\,dx + \kappa_1\int_\Omega \text{div}\,\underline{u}\;\text{div}\,\underline{v}\,dx - (1+\kappa_2)\int_\Omega \underline{\text{grad}}\,p\cdot\underline{v}\,dx \\
&= -\kappa_1\int_\Omega f\;\text{div}\,\underline{v}\,dx, \qquad \forall \underline{v}\in H(\text{div};\Omega) \\
&-(1+\kappa_2)\int_\Omega \underline{u}\cdot\,\underline{\text{grad}}\,q\,dx + (\kappa_2+\kappa_3)\int_\Omega \underline{\text{grad}}\,p\cdot\,\underline{\text{grad}}\,q\,dx \\
&= (\kappa_3 - 1)\int_\Omega fq\,dx, \qquad \forall q\in H_0^1(\Omega).
\end{aligned} \tag{1.5.18}$$

Remark 1.5.3. If we take $\kappa_3 = 1,\ \kappa_2 = -1/2,\ \kappa_1 = 1/2$, the above system (1.5.18) reduces to

$$\begin{aligned}
&\int_\Omega (\underline{u} - \underline{\text{grad}}\,p)\cdot\underline{v}\,dx + \int_\Omega (\text{div}\,\underline{u} + f)\;\text{div}\,\underline{v}\,dx = 0, \quad \forall \underline{v}\in H(\text{div};\Omega) \\
&-\int_\Omega (\underline{u} - \underline{\text{grad}}\,p)\cdot\,\underline{\text{grad}}\,q\,dx = 0, \qquad \forall q\in H_0^1(\Omega)
\end{aligned} \tag{1.5.19}$$

which are the Euler equations of the problem

$$\inf_{\underline{v}\in H(\text{div};\Omega)}\ \inf_{q\in H_0^1(\Omega)} \frac{\kappa_2}{2}\int_\Omega |\underline{v} - \underline{\text{grad}}\,q|^2 dx + \frac{\kappa_1}{2}\int_\Omega (\text{div}\,\underline{v} + f)^2\,dx, \tag{1.5.20}$$

which is the least squares method introduced by Bramble et al. [108] and has given rise to a vast literature. □

Let us now consider the question of the stability of (1.5.18). This can be done in two ways. In the first one we subtract the two equations to get an antisymmetric bilinear form, $E_a((p,\underline{u}),(q,\underline{v}))$. Then, we clearly have

$$\begin{aligned}
E_a((q,\underline{v}),(q,\underline{v})) = (1+\kappa_2)\int_\Omega |\underline{v}|^2\,dx + \kappa_1\int_\Omega |\,\text{div}\,\underline{v}|^2\,dx \\
-(\kappa_2+\kappa_3)\int_\Omega |\,\underline{\text{grad}}\,q|^2\,dx.
\end{aligned} \tag{1.5.21}$$

We thus get coercivity if $\kappa_1 > 0,\ (1+\kappa_2) > 0,\ (\kappa_2+\kappa_3) < 0$. It is easy to see that these conditions imply $\kappa_3 < 1$ and that, this being given, one can take $\kappa_2 = -\frac{\kappa_3+1}{2}$.

On the other hand, one could want to consider a symmetrical bilinear form by adding the two equations instead of subtracting them. One then has

$$E_s((q,\underline{v}),(q,\underline{v})) = (1+\kappa_2)\int_\Omega |\underline{v}|^2\,dx + \kappa_1\int_\Omega |\operatorname{div}\underline{v}|^2\,dx$$
$$-2(1+\kappa_2)\int_\Omega \underline{\operatorname{grad}}\,q\cdot\underline{v}\,dx + (\kappa_2+\kappa_3)\int_\Omega |\underline{\operatorname{grad}}\,q|^2\,dx. \quad (1.5.22)$$

It is not difficult to check that (1.5.22) yields coercivity if

$$\kappa_1 > 0,\ \ \kappa_2 > -1,\ \ \kappa_2+\kappa_3 > 0,$$

and finally

$$(\kappa_2+\kappa_3)\min\{\kappa_1, 1+\kappa_2\} > (1+\kappa_2)^2.$$

Now, if $\min\{\kappa_1, 1+\kappa_2\} = 1+\kappa_2$, the last condition implies $\kappa_2+\kappa_3 < 1+\kappa_2$, that is $\kappa_3 < 1$.

Remark 1.5.4. As in Remark 1.5.2, instead of the expression $\min\{\kappa_1, 1+\kappa_2\}$, we should actually have $\min\{\kappa_1/\ell, 1+\kappa_2\}$ where ℓ is a characteristic length of the problem. □

Remark 1.5.5. We thus see that the least-squares formulation of Remark 1.5.3, which is obtained with $\kappa_3 = 1$, is a difficult case which cannot be studied by simple arguments and is therefore not a simple way to obtain a stable method. □

Remark 1.5.6. We presented here an example of a quite general idea which was introduced in [256] and [213]. Other examples of these ideas will be developed in the following sections of this chapter or in the following chapters. □

Remark 1.5.7. In the example presented above, Euler equations (1.5.2) were a system of first order equations. This is not the case of the Stokes problem

$$\inf_{\underline{v}\in(H_0^1(\Omega))^2}\ \sup_{q\in L^2(\Omega)}\ \mu\int_\Omega |\underline{\underline{\dot\varepsilon}}(\underline{v})|^2dx - \int_\Omega \underline{f}\cdot\underline{v}\,dx - \int_\Omega q\,\operatorname{div}\underline{v}\,dx, \quad (1.5.23)$$

for which one of the equations in strong form contains second derivatives. Applying the same procedure would lead to a fourth order problem in the variable $\underline{u}$ which would lead to undesirable complications. Indeed the analogue of (1.5.13) would here be obtained from (1.5.23) as follows:

$$\inf_{\underline{v}\in(H_0^1(\Omega))^2}\ \sup_{q\in L^2(\Omega)}\ \mu\int_\Omega |\underline{\underline{\dot\varepsilon}}(\underline{v})|^2dx - \int_\Omega \underline{f}\cdot\underline{v}\,dx$$
$$-\int_\Omega q\,\operatorname{div}\underline{v}\,dx - \beta\int_\Omega |A\underline{u} + \underline{\operatorname{grad}}\,p - \underline{f}|^2dx, \quad (1.5.24)$$

which would force us to use a very regular approximation for the variable $\underline{u}$ or to rather consider the formulation

$$\inf_{\underline{v}\in(H_0^1(\Omega))^2}\ \sup_{q\in L^2(\Omega)} \mu\int_\Omega |\underline{\underline{\dot{\varepsilon}}}(\underline{v})|^2 dx - \int_\Omega \underline{f}\cdot\underline{v}\, dx - \int_\Omega q \operatorname{div}\underline{v}\, dx - \beta\|A\underline{u} + \underline{\operatorname{grad}}\, p - \underline{f}\|_*^2, \quad (1.5.25)$$

where $\|\cdot\|_*$ denotes the norm in $H^{-1}(\Omega)$. If we take the derivative of (1.5.25) with respect to $\underline{v}$, we could write as in (1.5.17),

$$\langle A\underline{u} + \underline{\operatorname{grad}}\, p - \underline{f}, A\underline{v}\rangle_* = \langle A\underline{u} + \underline{\operatorname{grad}}\, p - \underline{f}, \underline{v}\rangle_{V'\times V}, \quad (1.5.26)$$

but the derivative with respect to p will yield a term of the form

$$\langle A\underline{u} + \underline{\operatorname{grad}}\, p - \underline{f}, A\underline{v}\rangle_*,$$

which is not readily computable and which will require some special handling when considering discrete approximations. The typical solution, when dealing with a decomposition of Ω into elements K and piecewise smooth functions $\underline{u}$, $\underline{v}$, p and q, is to consider terms of the form

$$\sum_K \beta_K(\operatorname{diam}(K))^2 \int_K |A\underline{u} + \underline{\operatorname{grad}}\, p - \underline{f}|^2 dx. \quad (1.5.27)$$

We shall consider solutions to this question in Chap. 8 where Stokes' problem will be studied in detail. □

Remark 1.5.8. If one looks carefully at the Euler equations of (1.5.14), one sees that we could be in some trouble for $\beta = 1$ as the first equation disappears and we obviously loose control over p. This justified, in [179], the introduction of another variant of the general idea developed above. The formulation cannot in this case be written as a modified Lagrangian but is rather derived from the antisymmetric bilinear form (1.5.3). Indeed, let us write,

$$\begin{aligned} E((u,\underline{p}),(v,\underline{q})) &= \int_\Omega \underline{p}\cdot\underline{q}\, dx + \int_\Omega \underline{p}\cdot \underline{\operatorname{grad}}\, v\, dx - \int_\Omega \underline{q}\cdot \underline{\operatorname{grad}}\, u\, dx \\ &\quad + \beta\int_\Omega (\underline{p} - \underline{\operatorname{grad}}\, u)\cdot(\underline{q} - \underline{\operatorname{grad}}\, v) dx \\ &= -\int_\Omega f v\, dx, \qquad \forall(\underline{q}, v) \in (L^2(\Omega))^2\times H_0^1(\Omega). \end{aligned} \quad (1.5.28)$$

This formulation cannot be obtained from a Lagrangian. It can easily be seen that it remains valid for $\beta > 0$ arbitrary. Indeed (1.5.28) can also be written as

$$(1+\beta)\int_\Omega (\underline{p}- \underline{\mathrm{grad}}\, u)\cdot\underline{q}\, dx = 0, \qquad \forall\underline{q} \in (L^2(\Omega))^2,$$

$$\beta\int_\Omega \underline{\mathrm{grad}}\, u\cdot \underline{\mathrm{grad}}\, v\, dx+(1-\beta)\int_\Omega \underline{p}\cdot \underline{\mathrm{grad}}\, v\, dx-\int_\Omega f v\, dx=0, \quad \forall v\in H_0^1(\Omega), \tag{1.5.29}$$

and this is equivalent to (1.5.7) for any $\beta \geq 0$ with the control of $\underline{p}$ remaining untouched. Moreover, we can easily see that, as long as $\beta > 0$, there exists a $\delta > 0$ such that

$$E((v,\underline{p}),(v,\underline{q})) =\int_\Omega |\underline{q}|^2\, dx + \beta\int_\Omega |\underline{q}- \underline{\mathrm{grad}}\, v|^2\, dx > \delta\int_\Omega |\underline{q}|^2\, dx + \int_\Omega |\, \underline{\mathrm{grad}}\, v|^2\, dx. \tag{1.5.30}$$

We shall see later on how this technique can indeed be included in the same general framework as the previous one. □

1.5.2 Perturbed Formulations

We can also consider another type of modified variational formulation which is introduced in the *discretised* problems and in which, hopefully, the additional term vanishes when the mesh is refined and the numerical solution converges to the solution of the original problem. As an example, we consider the Stokes problem presented above in (1.5.23). We now suppose that we have a subspace V_h of $(H_0^1(\Omega))^2$ and Q_h of $L^2(\Omega)$ and we try to solve

$$\inf_{\underline{v_h}\in V_h}\sup_{q_h\in Q_h} \mu\int_\Omega |\underline{\underline{\varepsilon}}(\underline{v_h})|^2 dx - \int_\Omega \underline{f}\cdot \underline{v_h}\, dx - \int_\Omega q_h\, \mathrm{div}\, \underline{v_h}\, dx. \tag{1.5.31}$$

Let us suppose that the couple $V_h \times Q_h$ does not satisfy the stability conditions developed on Chap. 8 but that there exists a subspace Q_h^S of Q_h such that the couple $V_h \times Q_h^S$ is stable. If for some reason we want to use Q_h instead of Q_h^S we could retrieve stability by working with the problem

$$\inf_{\underline{v_h}\in V_h}\sup_{q_h\in Q_h} \mu\int_\Omega |\underline{\underline{\varepsilon}}(\underline{v_h})|^2 dx - \int_\Omega \underline{f}\cdot \underline{v_h}\, dx - \int_\Omega q_h\, \mathrm{div}\, \underline{v_h}\, dx - \frac{\alpha}{2}\int_\Omega |q_h - P q_h|^2\, dx, \tag{1.5.32}$$

where Pq_h stands for the projection of q_h on Q_h^S, which we suppose to be (easily) computable. This is now a stable formulation which, for μ_1 large, yields almost exactly the solution in $V_h \times Q_h^S$. It must be noted that the modified problem is meaningless for the continuous problem so that we have a different way of stabilising. However, the difference sometimes disappears when augmented versions are discretised.

1.6 Bibliographical Remarks

The purpose of this chapter was to present examples which will be used later as a standing ground for our development. It was not possible in such a context to consider every case. We already referred the reader to [166, 167] or [277] where the mathematical analysis of the problems selected here (and of many others problems) can be found in a unified setting. Advanced presentation of elasticity problems can be found in [295] or in [147]. For the Navier-Stokes equations, among the huge amount of literature, one may consult [278, 350, 363] or [159], and for electromagnetic problems the classical [273]. We also refer to more engineering-oriented presentations such as [53, 254, 271] and [387]. In particular, non-linear problems and their treatment are described in these references.

Chapter 2
Function Spaces and Finite Element Approximations

In this chapter we present function spaces and suitable finite element approximations of them, which we shall use in order to apply the abstract theory of the previous chapters to problems of practical interest.

We do not aim at a general presentation of the subject of a vast literature, but we present the basic properties of the spaces we are going to use in the sequel of this book.

In particular, we consider standard results about the finite element approximation of Sobolev spaces and about approximations of $H(\mathrm{div};\Omega)$ and $H(\underline{\mathrm{curl}};\Omega)$. The results of Sect. 2.1 are technical and may be skipped by a reader interested mostly in numerical results.

Mainly for historical reasons, we present the finite element approximation of the spaces $H^1(\Omega)$, $H(\mathrm{div};\Omega)$, and $H(\underline{\mathrm{curl}};\Omega)$ separately, although they could be seen altogether in the framework of de Rham diagram. We shall briefly comment on it in Sect. 2.1.4.

2.1 Properties of the Spaces $H^m(\Omega)$, $H(\mathrm{div};\Omega)$, and $H(\underline{\mathrm{curl}};\Omega)$

2.1.1 Basic Properties

Some of the results of this section have been already anticipated in Chap. 1. Here, we present them in a unified setting.

- **Sobolev spaces $H^m(\Omega)$**. Given an integer number $m \geq 1$, standard Sobolev spaces read

$$H^m(\Omega) := \{v \mid v \in L^2(\Omega),\ D^\alpha v \in L^2(\Omega),\ |\alpha| \leq m\}, \qquad (2.1.1)$$

D. Boffi et al., *Mixed Finite Element Methods and Applications*, Springer Series in Computational Mathematics 44, DOI 10.1007/978-3-642-36519-5_2,

where

$$D^\alpha v := \frac{\partial^{|\alpha|} v}{\partial x_1^{\alpha_1} \partial x_2^{\alpha_2} \dots \partial x_n^{\alpha_n}}, \qquad |\alpha| := \alpha_1 + \alpha_2 + \cdots + \alpha_n. \tag{2.1.2}$$

We shall consider the standard norm

$$\|v\|_{m,\Omega}^2 := \sum_{|\alpha| \le m} \int |D^\alpha v|^2 \, dx, \tag{2.1.3}$$

associated with the usual inner product and the semi-norm

$$|v|_{m,\Omega}^2 := \sum_{|\alpha| = m} \int |D^\alpha v|^2 \, dx. \tag{2.1.4}$$

The most important of these spaces will be for us $H^1(\Omega)$ (and some of its subspaces) and for fourth-order problems $H^2(\Omega)$.

For the study of Sobolev spaces, we refer the reader to [281, 309], and [3]. It is well-known that if $\Gamma = \partial\Omega$ is smooth enough (for instance Lipschitz continuous), it is possible to define the trace $\gamma u = u|_\Gamma$ of $u \in H^1(\Omega)$ on the boundary Γ. The traces of functions in $H^1(\Omega)$ span a Hilbert space, denoted by $H^{\frac{1}{2}}(\Gamma)$, that is a proper dense subspace of $L^2(\Gamma)$. The mapping,

$$\gamma : \; H^1(\Omega) \to H^{\frac{1}{2}}(\Gamma), \tag{2.1.5}$$

is surjective and possesses a continuous lifting. The norm

$$\|\gamma v\|_{\frac{1}{2},\Gamma} := \inf_{\substack{w \in H^1(\Omega) \\ \gamma w = v}} \|w\|_{1,\Omega} \tag{2.1.6}$$

is then equivalent to more standard norms on $H^1(\Omega)$ as defined in [281]. Then we can write

$$\|v\|_{\frac{1}{2},\Gamma} = \|\bar{v}\|_{1,\Omega}, \tag{2.1.7}$$

where $\bar{v}$ is the unique solution in $H^1(\Omega)$ of the Dirichlet problem,

$$\begin{cases} -\Delta \bar{v} + \bar{v} = 0, \\ \bar{v}|_\Gamma = v. \end{cases} \tag{2.1.8}$$

We shall denote by $H^{-\frac{1}{2}}(\Gamma)$ the dual space of $H^{\frac{1}{2}}(\Gamma)$ with the dual norm,

$$\|v^*\|_{-\frac{1}{2},\Gamma} := \sup_{v \in H^{1/2}(\Gamma)} \frac{\langle v, v^* \rangle}{\|v\|_{\frac{1}{2},\Gamma}}, \tag{2.1.9}$$

where the bracket $\langle\cdot,\cdot\rangle$ denotes duality between $H^{-1/2}(\Gamma)$ and $H^{1/2}(\Gamma)$. It is easily checked that one has

$$\|v^*\|_{-\frac{1}{2},\Gamma} = \|\bar{v}^*\|_{1,\Omega}, \tag{2.1.10}$$

where $\bar{v}^*$ is the solution of the variational Neumann problem,

$$\int_\Omega \underline{\text{grad}}\,\bar{v}^* \cdot \underline{\text{grad}}\,v\,dx + \int_\Omega \bar{v}^* v\,dx = \langle v^*, v\rangle,\ \forall v \in H^1(\Omega). \tag{2.1.11}$$

Remark 2.1.1. We shall sometimes write formally $\int_\Gamma v^* v\,d\sigma$ instead of $\langle v^*, v\rangle$, to denote duality between $H^{\frac{1}{2}}(\Gamma)$ and $H^{-\frac{1}{2}}(\Gamma)$. □

We can define in the same way a trace operator γ on $H^2(\Omega)$. It is now possible to define $v|_\Gamma$ in a space denoted $H^{\frac{3}{2}}(\Gamma)$ but also traces of $\underline{\text{grad}}\,v|_\Gamma \in H^{\frac{1}{2}}(\Gamma)^n$ and thus the trace of the normal derivative $\frac{\partial v}{\partial n}$. We then define

$$H_0^1(\Omega) := \{v \mid v \in H^1(\Omega),\ v|_\Gamma = 0\}, \tag{2.1.12}$$

$$H_0^2(\Omega) := \{v \mid v \in H^2(\Omega),\ v|_\Gamma = 0,\ \frac{\partial v}{\partial n}|_\Gamma = 0\}. \tag{2.1.13}$$

Remark 2.1.2. Spaces $H^{1/2}(\Gamma)$ and $H^{3/2}(\Gamma)$ are particular cases of generalized Sobolev spaces $H^s(\cdot)$ for $s \in \mathbb{R}^+$. The reader should be aware that handling Sobolev spaces $H^s(\cdot)$ where s = integer $+1/2$ requires some caution [281]. In the case of $H^{1/2}(\Gamma)$ it is important to recall some facts. Let Γ_0 be a part of Γ; then $\phi \in H^{1/2}(\Gamma_0)$ cannot be extended by zero outside Γ_0 to a function in $H^{1/2}(\Gamma)$ (even if paradoxically $\mathcal{D}(\Gamma_0)$ is dense in $H^{1/2}(\Gamma_0)$). Dually, if $\Gamma = \Gamma_0 \cup \Gamma_1$, one does not get the whole of $H^{-1/2}(\Gamma)$ by patching functions of $H^{-1/2}(\Gamma_0)$ and $H^{-1/2}(\Gamma_1)$. Unfortunately, spaces $H^{1/2}(\partial K)$ and $H^{-1/2}(\partial K)$ with K an element of a partition of Ω are met very often in the analysis of hybrid and mixed methods and one must be very careful in handling them. □

- **The space $\boldsymbol{H(\text{div};\Omega)}$.** Having considered standard Sobolev spaces, we now present some properties of a space specially adapted to the study of mixed and hybrid methods.

  The mathematical analysis of mixed methods will use constantly

$$H(\text{div};\Omega) := \{\underline{q} \mid \underline{q} \in (L^2(\Omega))^n,\ \text{div}\,\underline{q} \in L^2(\Omega)\} \tag{2.1.14}$$

  with the norm

$$\|\underline{q}\|^2_{\text{div},\Omega} := |\underline{q}|^2_{0,\Omega} + |\,\text{div}\,\underline{q}|^2_{0,\Omega}. \tag{2.1.15}$$

  It is then possible to define $\underline{q}\cdot\underline{n}|_\Gamma$, the normal trace of $\underline{q}$ on Γ.

Lemma 2.1.1. *For $\underline{q} \in H(\text{div},\Omega)$, we can define $\underline{q}\cdot\underline{n}|_\Gamma \in H^{-\frac{1}{2}}(\Gamma)$ and we have Green's formula,*

$$\int_{\Omega} \operatorname{div} \underline{q}\, v\, dx + \int_{\Omega} \underline{q} \cdot \underline{\operatorname{grad}}\, v\, dx = \langle \underline{q} \cdot \underline{n}, v \rangle, \;\; \forall v \in H^1(\Omega). \tag{2.1.16}$$

Proof. For $\underline{q} \in (\mathcal{D}(\bar{\Omega}))^n$ and $v \in \mathcal{D}(\bar{\Omega})$, we have the standard Green's formula

$$\int_{\Gamma} \underline{q} \cdot \underline{n}\, v\, d\sigma = \int_{\Omega} \operatorname{div} \underline{q}\, v\, dx + \int_{\Omega} \underline{q} \cdot \underline{\operatorname{grad}}\, v\, dx, \tag{2.1.17}$$

and therefore

$$\left| \int_{\Gamma} \underline{q} \cdot \underline{n}\, v\, d\sigma \right| \leq \|\underline{q}\|_{\operatorname{div},\Omega} \|v\|_{1,\Omega}. \tag{2.1.18}$$

Moreover, the expression $\int_{\Omega} \operatorname{div} \underline{q}\, v\, dx + \int_{\Omega} \underline{q} \cdot \underline{\operatorname{grad}}\, v\, dx$ depends only on the trace $v|_{\Gamma} \in H^{\frac{1}{2}}(\Gamma)$. The result follows by density of $\mathcal{D}(\bar{\Omega})$ and $(\mathcal{D}(\bar{\Omega}))^n$ in $H^1(\Omega)$ and $H(\operatorname{div}; \Omega)$, respectively. □

The operator defined above also satisfies a surjectivity property.

Lemma 2.1.2. *The trace operator* $\underline{q} \in H(\operatorname{div}; \Omega) \rightarrow \underline{q} \cdot \underline{n}|_{\Gamma} \in H^{-\frac{1}{2}}(\Gamma)$ *is surjective.*

Proof. Let $g \in H^{-\frac{1}{2}}(\Gamma)$ be given. Then, solving in $H^1(\Omega)$

$$\int_{\Omega} \underline{\operatorname{grad}}\, \phi \cdot \underline{\operatorname{grad}}\, v\, dx + \int_{\Omega} \phi v\, dx = \langle g, v \rangle, \;\; \forall v \in H^1(\Omega), \tag{2.1.19}$$

and making $\underline{q} = \underline{\operatorname{grad}}\, \phi$ implies $\underline{q} \cdot \underline{n}|_{\Gamma} = g$. □

Let us now suppose a partition $(\Gamma = D \cup N)$ of the boundary Γ. We define

$$H^1_{0,D}(\Omega) := \{v \mid v \in H^1(\Omega), \;\; v|_D = 0\}. \tag{2.1.20}$$

In particular, we have $H^1_{0,D}(\Omega) = H^1_0(\Omega)$ if $D = \Gamma$ and $H^1_{0,D}(\Omega) = H^1(\Omega)$ if $D = \emptyset$. We shall also need the space

$$H_{0,N}(\operatorname{div}; \Omega) := \{\underline{q} \mid \underline{q} \in H(\operatorname{div} \Omega), \;\; \langle \underline{q} \cdot \underline{n}, v \rangle = 0, \;\; \forall v \in H^1_{0,D}(\Omega)\}. \tag{2.1.21}$$

Remark 2.1.3. This space contains functions of $H(\operatorname{div}; \Omega)$ whose normal traces vanish on N. For reasons related to pathological properties of $H^{\frac{1}{2}}(D)$ and $H^{-\frac{1}{2}}(N)$, it is necessary to use definition (2.1.21) and not an expression such as $\underline{q} \cdot \underline{n}|_N = 0$ in $H^{-\frac{1}{2}}(N)$. □

In particular, we denote $H_0(\operatorname{div}; \Omega) = H_{0,N}(\operatorname{div}; \Omega)$ when $N = \Gamma$. Finally, another important subspace of $H(\operatorname{div}; \Omega)$ will be

$$H^0(\operatorname{div}; \Omega) := \{\underline{q} \mid \underline{q} \in H(\operatorname{div}; \Omega), \;\; \operatorname{div}\, \underline{q} = 0\}, \tag{2.1.22}$$

from which we deduce the following result.

Lemma 2.1.3. *The normal trace operator* $\underline{q} \to \underline{q}\cdot\underline{n}|_\Gamma$ *is surjective from* $H^0(\text{div};\Omega)$ *onto* $\{\mu^* \mid \mu^* \in H^{-\frac{1}{2}}(\Gamma),\ \langle\mu^*, 1\rangle = 0\}$.

Proof. By Green's formula (2.1.14), we have $\langle\underline{q}\cdot\underline{n}, 1\rangle = 0$ if $\underline{q} \in N^0(\text{div};\Omega)$. Reciprocally, if $g \in H^{-\frac{1}{2}}(\Gamma)$ is given with $\langle g, 1\rangle = 0$, we can solve in $H^1(\Omega)/\mathbb{R}$ the Neumann problem

$$\int_\Omega \underline{\text{grad}}\,\phi \cdot \underline{\text{grad}}\,v\,dx = \langle g, \phi\rangle, \quad \forall\phi \in H^1(\Omega), \tag{2.1.23}$$

and taking $\underline{q} = \underline{\text{grad}}\,\phi$ yields $\underline{q}\cdot\underline{n} = g$. □

Remark 2.1.4. In applications, D will be the part of Γ where Dirichlet's conditions are given, and N the part with Neumann's conditions. □

- **The space $H(\underline{\text{curl}};\Omega)$.** To conclude this section, we consider the space $H(\underline{\text{curl}};\Omega)$ which will be used, in particular, for the approximation of problems arising from electro-magnetics in Chap. 11.

  For $\Omega \in \mathbb{R}^3$, we define

$$H(\underline{\text{curl}};\Omega) := \{\underline{\chi} \mid \underline{\chi} \in (L^2(\Omega))^3,\ \underline{\text{curl}}\,\underline{\chi} \in (L^2(\Omega))^3\}, \tag{2.1.24}$$

  where the curl operator is as usual defined as

$$\underline{\text{curl}}\,\underline{\chi} = \underline{\nabla}\wedge\underline{\chi} := \det\begin{pmatrix} i & j & k \\ \frac{\partial}{\partial x_1} & \frac{\partial}{\partial x_2} & \frac{\partial}{\partial x_3} \\ \chi_1 & \chi_2 & \chi_3 \end{pmatrix} \tag{2.1.25}$$

  and where we are using the standard norm

$$\|\underline{\chi}\|^2_{\text{curl},\Omega} := |\underline{\chi}|^2_{0,\Omega} + |\,\underline{\text{curl}}\,\underline{\chi}|^2_{0,\Omega}. \tag{2.1.26}$$

Remark 2.1.5. To complete these definitions, we must discuss the definition of $H(\text{curl};\Omega)$ when Ω belongs to $\mathbb{R}^2$. First of all, we recall the possible definitions of the curl operator in this setting. Given a vector $\underline{u}(x_1, x_2) = (u_1, u_2)$, we can evaluate $\text{curl}(u_1, u_2, 0)$, which is a vector oriented in the direction of x_3. This suggests the definition

$$\text{curl}\,\underline{u} := \frac{\partial u_2}{\partial x_1} - \frac{\partial u_1}{\partial x_2}. \tag{2.1.27}$$

On the other hand, given a scalar function $\phi(x_1, x_2)$, the vector $\underline{\text{curl}}(0, 0, \phi)$ is perpendicular to the direction of x_3 and drives to the following definition

$$\underline{\text{curl}}\,\phi := \left(\frac{\partial\phi}{\partial x_1}, -\frac{\partial\phi}{\partial x_2}\right). \tag{2.1.28}$$

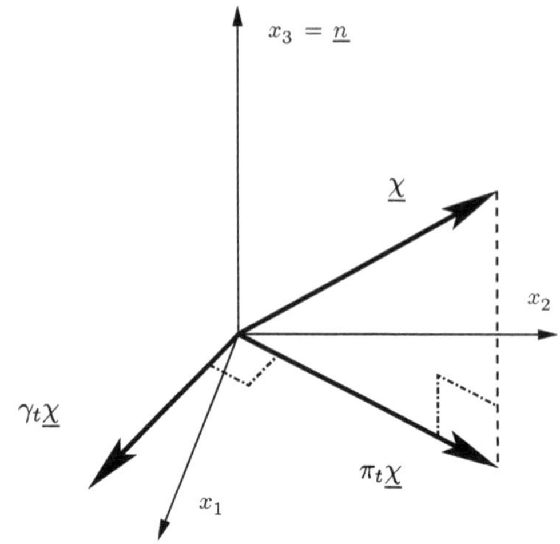

Fig. 2.1 $H(\underline{\text{curl}};\Omega)$ traces on the flat surface (x_1, x_2)

It is clear that the operators curl and $\underline{\text{curl}}$ introduced above are equivalent to the operators div and $\underline{\text{grad}}$, respectively, after rotation of the vectors by a right angle. More precisely, we have

$$\text{curl}\,\underline{u} = -\,\text{div}(\underline{u}^{\perp}), \tag{2.1.29}$$

$$\underline{\text{curl}}\,\phi = -(\underline{\text{grad}}\,\phi)^{\perp}, \tag{2.1.30}$$

with the standard notation $(v_1, v_2)^{\perp} = (-v_2, v_1)$. It turns out that *in two dimensions* the space $H(\text{curl};\Omega)$ is isomorphic to $H(\text{div};\Omega)$. This fact has important consequences for the construction of approximations of $H(\text{curl};\Omega)$; namely, any finite element space which is a good approximation of $H(\text{div};\Omega)$ can be turned into a good approximation of $H(\text{curl};\Omega)$ just by rotating the vectors by a right angle and vice versa. □

We are now going to state results concerning traces of $H(\underline{\text{curl}};\Omega)$ in three dimensions, in analogy of what has been done in the previous subsection for $H(\text{div};\Omega)$. Traces of $H(\underline{\text{curl}};\Omega)$ have been the object of a recent and active research by several authors [6,133–135,143,223,317] and it turns out that the theory is not as straightforward as in the case of $H(\text{div};\Omega)$; in particular we shall see that different trace definitions can be considered.

In order to get the reader acquainted with the topics which we are going to present, let us start with an easy but significant example.

Example 2.1.1 (Traces of $H(\underline{\text{curl}};\Omega)$ on a flat surface). Let $\Omega \subset \mathbb{R}^3$ be the half space $x_3 < 0$ and let Γ be the plane $x_3 = 0$. The aim of this example is to consider the basic situation of a flat boundary, i.e., when the tangent plane at any point of Γ coincides with the surface Γ itself. Given the vector $\underline{\chi} = (\chi_1, \chi_2, \chi_3)^T \in H(\underline{\text{curl}};\Omega)$, we investigate the possible definitions of tangential component of $\underline{\chi}$

along Γ (see Fig. 2.1). As usual, we denote by $\underline{n} = (0,0,1)^T$ the unit vector pointing in the outward normal direction of Γ with respect to Ω.

We are going to give a meaning to the trace of the tangential component of $\underline{\chi}$ along Γ. Whenever $\underline{\chi} \in C^0(\overline{\Omega})$, it makes sense to consider the vector $\gamma_t \underline{\chi} = (\underline{\chi} \wedge \underline{n})_{|\Gamma}$ which, by definition, is orthogonal to $\underline{n}$ and hence is aligned with the tangent plane $x_3 = 0$. More precisely, it can be easily checked that

$$\gamma_t \underline{\chi} = \begin{pmatrix} \chi_2 \\ -\chi_1 \\ 0 \end{pmatrix}_{|\Gamma} . \tag{2.1.31}$$

On the other hand, we can also consider the projection of $\underline{\chi}$ along the plane Γ, that is the vector $\pi_t \underline{\chi} = \underline{n} \wedge (\underline{\chi} \wedge \underline{n})_{|\Gamma}$, which is again orthogonal to $\underline{n}$ and can be evaluated explicitly as follows:

$$\pi_t \underline{\chi} = \begin{pmatrix} \chi_1 \\ \chi_2 \\ 0 \end{pmatrix}_{|\Gamma} . \tag{2.1.32}$$

It is clear that in this simple example we have that $\gamma_t \underline{\chi}$ is orthogonal to $\pi_t \underline{\chi}$. Moreover, let us denote by div_Γ and $\underline{\mathrm{curl}}_\Gamma$ the divergence and the curl operators in the (x_1, x_2) plane (see Remark 2.1.5 for the definition of $\underline{\mathrm{curl}}$ in two space dimensions and the relation between operators); then, the following relationships are easy to derive

$$\mathrm{div}_\Gamma\, \gamma_t \underline{\chi} = \underline{\mathrm{curl}}_\Gamma\, \pi_t \underline{\chi} = (\underline{\mathrm{curl}}\, \underline{\chi} \cdot \underline{n})_{|\Gamma} . \tag{2.1.33}$$

Although it should be clear from the context, we remark that in (2.1.33) we are formally applying surface operators div_Γ and $\underline{\mathrm{curl}}_\Gamma$ to three dimensional vectors $\gamma_t \underline{\chi}$ and $\pi_t \underline{\chi}$. On the other hand, vectors $\gamma_t \underline{\chi}$ and $\pi_t \underline{\chi}$ are orthogonal to the direction of x_3 so that we can identify them with two dimensional vectors in the tangent space $x_3 = 0$. We shall implicitly make use of this standard abuse of notation in the sequel of this chapter. □

Let us now come back to the general picture and see how the quantities considered in the previous example extend to general domains and boundaries. The following integration by parts, which is valid for smooth functions, constitutes the starting point for the analysis

$$\int_\Omega \underline{\chi} \cdot \underline{\mathrm{curl}}\, \underline{\phi}\, dx - \int_\Omega \underline{\mathrm{curl}}\, \underline{\chi} \cdot \underline{\phi}\, dx = \int_\Gamma (\underline{\chi} \wedge \underline{n}) \cdot \underline{\phi}\, ds. \tag{2.1.34}$$

Green's formula (2.1.34) allows us to define the trace $\gamma_t \underline{\chi}$ of a function $\underline{\chi} \in H(\underline{\mathrm{curl}};\Omega)$ by extending the classical tangential trace $(\underline{\chi} \wedge \underline{n})_{|\Gamma}$.

The following lemma is the analogue of Lemma 2.1.1 and can be proved by a similar technique.

Lemma 2.1.4. *For $\underline{\chi} \in H(\underline{curl};\Omega)$, we can define the tangential trace $\gamma_t\underline{\chi} = (\underline{\chi} \wedge \underline{n})_{|\Gamma} \in (H^{-1/2}(\Gamma))^3$ and we have Green's formula*

$$\int_\Omega \underline{\chi} \cdot \underline{curl}\,\underline{\phi}\,dx - \int_\Omega \underline{curl}\,\underline{\chi} \cdot \underline{\phi}\,dx = \langle \gamma_t\underline{\chi}, \underline{\phi}\rangle, \quad \forall\underline{\phi} \in (H^1(\Omega))^3. \qquad (2.1.35)$$

In particular, it is clear from our construction that, for smooth functions, the tangential trace $\gamma_t\underline{\chi}$ is equal to $(\underline{\chi} \wedge \underline{n})_{|\Gamma}$ in the classical sense.

Let us now consider the range of the trace operator γ_t. The linear and continuous operator γ_t cannot be surjective onto $(H^{-1/2}(\Gamma))^3$ since it has vanishing component in the direction of $\underline{n}$. Moreover, as we already observed in the simplified situation of Example 2.1.1, we remark that for smooth functions it makes sense to evaluate the surface divergence of $\gamma_t\underline{\chi}$ along Γ and that the following identity holds true

$$\mathrm{div}_\Gamma(\gamma_t\underline{\chi}) = (\underline{\mathrm{curl}}\,\underline{\chi} \cdot \underline{n})_{|\Gamma}. \qquad (2.1.36)$$

Indeed, the range of γ_t is usually denoted by $H^{-1/2}(\mathrm{div};\Gamma)$ and its dual space by $H^{-1/2}(\underline{\mathrm{curl}};\Gamma)$. In the case of smooth domains, it turns out that we have the following identities

$$\begin{aligned} H^{-1/2}(\mathrm{div};\Gamma) &= \{\underline{q} \in (H^{-1/2}(\Gamma))^3 | \underline{q}\cdot\underline{n} = 0 \text{ a.e. on } \Gamma,\ \mathrm{div}_\Gamma\,\underline{q} \in H^{-1/2}(\Gamma)\} \\ H^{-1/2}(\underline{\mathrm{curl}};\Gamma) &= \{\underline{q} \in (H^{-1/2}(\Gamma))^3 | \underline{q}\cdot\underline{n} = 0 \text{ a.e. on } \Gamma,\ \underline{\mathrm{curl}}_\Gamma\,\underline{q} \in H^{-1/2}(\Gamma)\}, \end{aligned} \qquad (2.1.37)$$

where the surface operators div_Γ and $\underline{\mathrm{curl}}_\Gamma$ are applied, by abuse of notation, to the (two-dimensional) tangential component of $\underline{q}$.

In the case of non smooth domains, several insidious aspects have to be taken into account. In general, computational domains are polyhedra and a polyhedron is not a smooth domain. In particular, one major issue is hidden in (2.1.37) and is related to the regularity of the normal vector $\underline{n}$. If Γ is a polyhedral surface, then $\underline{n}$ jumps across the edges of Γ and the product $\underline{q}\cdot\underline{n}$ is not well defined for $\underline{q} \in (H^{-1/2}(\Gamma))^3$ since $\underline{n}$ is not a multiplier for $(H^{1/2}(\Gamma))^3$ where jumps are not allowed. We refer the interested reader to [136] for the general picture and for more details on this issue.

Remark 2.1.6. In the most general situation, the definition of div_Γ and $\underline{\mathrm{curl}}_\Gamma$ is not trivial. Even for smooth domains, we need suitable definitions of differential operators on curved surfaces. This is a typical task of differential geometry and can be performed by means of covariant derivatives. We refer the interested reader, for instance, to [168] for a thorough introduction to this subject. On the other hand, if the considered finite elements have flat faces (as it is the case for usual tetrahedra), surface differential operators reduce to standard two-dimensional ones in a local coordinate system on the face and covariant derivatives along the face are plain directional derivatives. The situation is not trivial (and not completely understood yet for what concerns the construction of good finite element spaces) in the case of non-flat faces (as it usually occurs for isoparametric elements or general non-affine hexahedral decompositions). We shall detail some issues for non-affine finite elements in Sect. 2.2.4. □

Looking again at Example 2.1.1, another definition which arises when dealing with the space $H(\underline{\text{curl}};\Omega)$ is the projection of $\underline{\chi}$ along the tangent plane of Γ, also known as the tangential trace $\pi_t\,\underline{\chi}$, which can be defined as follows for smooth functions:

$$\pi_t\,\underline{\chi} = \underline{n} \wedge (\underline{\chi} \wedge \underline{n})_{|\Gamma}. \tag{2.1.38}$$

The following Theorem is a consequence of the results of [134–136] and states the well-posedness of the trace operator π_t together with the link between γ_t and π_t.

Theorem 2.1.1. *The trace operator π_t can be extended to an operator from $H(\underline{curl};\Omega)$ to $H^{-1/2}(\underline{curl};\Gamma)$ and the following Green's formula holds*

$$\int_\Omega \underline{curl}\,\underline{\chi}\cdot\underline{\phi}\,dx - \int_\Omega \underline{\chi}\cdot\underline{curl}\,\underline{\phi}\,dx = \langle \pi_t\underline{\phi}, \gamma_t\underline{\chi}\rangle, \quad \forall \underline{\chi},\underline{\phi} \in H(\underline{curl};\Omega), \tag{2.1.39}$$

where the brackets denote the duality pairing between $H^{-1/2}(\text{div};\Gamma)$ and $H^{-1/2}(\underline{curl};\Gamma)$.

If the boundary of Ω is split into two parts $\Gamma = D \cup N$, then we can consider the space

$$H_{0,N}(\underline{\text{curl}};\Omega) := \{\underline{\chi} \mid \underline{\chi} \in H(\underline{\text{curl}};\Omega), \langle \gamma_t\underline{\chi},\underline{\phi}\rangle = 0,\ \forall \underline{\phi} \in (H^1_{0,D}(\Omega))^3\}, \tag{2.1.40}$$

in analogy to what we have done in the previous subsection. When $D = \emptyset$ (and hence $N = \Gamma$), we shall make use of the notation $H_0(\underline{\text{curl}};\Omega) = H_{0,\Gamma}(\underline{\text{curl}};\Omega)$.

2.1.2 Properties Relative to a Partition of Ω

This section presents a short introduction to properties of some functional spaces. We refer to [331, 366] for more details.

Partitioning Ω into sub-domains is an essential feature of both standard and non-standard methods. Continuity properties at interfaces between sub-domains are an essential part in the definition of a finite element approximation. Moreover, we shall introduce here some notations that will be used throughout the book.

Let $\Omega = \bigcup_{r=1}^m K_r$ be partitioned into a family of sub-domains. In practice, K_r will be a triangle or a quadrilateral (resp., a tetrahedron or a hexahedron in three dimensions) and we shall call it *element*. We shall denote by $\mathcal{T}_h$ a partition of Ω into elements.

The edges of elements will be denoted by e_i ($i = 1,2,3$ or $i = 1,2,3,4$) in the two-dimensional case. For three-dimensional elements, unless differently stated, we denote again the faces of the elements by e_i ($1 \le i \le 4$ or $1 \le i \le 6$). We also denote by

$$e_{ij} = \partial K_i \cap \partial K_j, \tag{2.1.41}$$

the interface between element K_i and K_j and

$$\mathcal{E}_h = \bigcup_{ij} e_{ij} \bigcup \Gamma_h = \bigcup_K \partial K, \tag{2.1.42}$$

where Γ_h is the set of boundary edges or faces. We only deal with *compatible* meshes in the sense that the intersection between two elements is a common face, side or vertex. The situation when a mesh contains hanging nodes is out of the aims of this work.

Remark 2.1.7. The index h will of course be related to mesh size, that is to the size of elements. With an abuse of notation, we shall also use the symbol h for denoting the *maximum diameter* of the *elements of the decomposition.* □

We introduce the functional spaces

$$X(\Omega) := \{v \mid v \in L^2(\Omega),\ v|_{K_i} \in H^1(K_i),\ \forall i\} = \prod_r H^1(K_r), \tag{2.1.43}$$

with the norm

$$\|v\|^2_{X(\Omega)} := \sum_r \|v\|^2_{1,K_r}, \tag{2.1.44}$$

$$Y(\Omega) := \{\underline{q} \mid \underline{q} \in L^2(\Omega), \underline{q}\,|_{K_i} \in H(\mathrm{div}; K_i),\ \forall i\} = \prod_r H(\mathrm{div}; K_r), \tag{2.1.45}$$

with the norm

$$\|\underline{q}\|^2_{Y(\Omega)} := \sum_r \|\underline{q}\|^2_{\mathrm{div},\Omega} \tag{2.1.46}$$

and

$$W(\Omega) := \{\underline{\chi} \mid \underline{\chi} \in L^2(\Omega),\ \underline{\chi}\,|_{K_i} \in H(\mathrm{c\underline{url}}; K_i),\ \forall i\} = \prod_r H(\mathrm{c\underline{url}}; K_r), \tag{2.1.47}$$

with the norm

$$\|\underline{\chi}\|^2_{W(\Omega)} := \sum_r \|\underline{\chi}\|^2_{\mathrm{c\underline{url}},\Omega}. \tag{2.1.48}$$

We shall now characterize $H^1_{0,D}(\Omega)$, $H_{0,N}(\mathrm{div}; \Omega)$, and $H_{0,N}(\mathrm{c\underline{url}}; \Omega)$ as subspaces of $X(\Omega)$, $Y(\Omega)$, and $W(\Omega)$ respectively. Let us first remark that for $v \in H^1(\Omega)$ and $\underline{q} \in H(\mathrm{div}; \Omega)$, we have, denoting $\underline{n}_r$ the normal to $\Gamma_r = \partial K_r$,

$$\sum_r \langle \underline{q} \cdot \underline{n}_r, v\rangle_{\Gamma_r} = \langle \underline{q} \cdot \underline{n}, v\rangle_{\Gamma}, \tag{2.1.49}$$

where $\langle\cdot,\cdot\rangle$ denotes duality between $H^{\frac{1}{2}}(\Gamma_r)$ and $H^{-\frac{1}{2}}(\Gamma_r)$. Indeed we can decompose the Green formula as

$$\langle \underline{q}\cdot\underline{n}, v\rangle_\Gamma = \sum_r \left\{ \int_{K_r} \text{div}\,\underline{q}\; v\,dx + \int_{K_r} \underline{q}\cdot\underline{\text{grad}}\,v\,dx \right\} \tag{2.1.50}$$

and apply it inside each element. A similar splitting holds for the functions $\underline{\chi} \in H(\underline{\text{curl}};\Omega)$ and $\underline{\phi} \in H^1(\Omega)$ when we consider the tangential trace $\gamma_t\underline{\chi}$, namely

$$\sum_r \langle \gamma_{t_r}\underline{\chi}, \underline{\phi}\rangle_{\Gamma_r} = \langle \gamma_t\underline{\chi}, \underline{\phi}\rangle_\Gamma. \tag{2.1.51}$$

We can now state the following proposition.

Proposition 2.1.1. $H^1_{0,D}(\Omega) = \{v | v \in X(\Omega), \sum_r \langle \underline{q}\cdot\underline{n}_r, v\rangle = 0, \forall \underline{q} \in H_{0,N}(\text{div};\Omega)\}$.

Proof. It is clear by definition that if $v \in H^1_{0,D}(\Omega)$ we have by (2.1.49) that $\sum_r \langle \underline{q}\cdot\underline{n}_r, v\rangle = 0$, $\forall \underline{q} \in H_{0,N}(\text{div};\Omega)$. Let us consider the reciprocal. Using Green's formula, we get

$$\int_\Omega v\;\text{div}\,\underline{q}\,dx = -\sum_r \int_{K_r} \underline{\text{grad}}\,v\cdot\underline{q}\,dx,\ \ \forall \underline{q} \in H_{0,N}(\text{div};\Omega). \tag{2.1.52}$$

This implies for all $\underline{q}$, for instance $\underline{q} \in (\mathcal{D}(\Omega))^n$,

$$|\int v\;\text{div}\,\underline{q}\,dx| \le \Big(\sum_r |v|^2_{1,K_r}\Big)^{\frac{1}{2}}\ \|\underline{q}\|_{0,\Omega}, \tag{2.1.53}$$

and therefore $\underline{\text{grad}}\,v \in (L^2(\Omega))^n$, thus $v \in H^1(\Omega)$. We then have $\langle \underline{q}\cdot\underline{n}, v\rangle = 0$, $\forall \underline{q} \in H_{0,N}(\text{div};\Omega)$, so that $v \in H^1_{0,D}(\Omega)$. □

The same kind of proof would yield the following analogous results for $H(\text{div};\Omega)$ and $H(\underline{\text{curl}};\Omega)$.

Proposition 2.1.2. $H_{0,N}(\text{div};\Omega) = \{\underline{q} \mid \underline{q} \in Y(\Omega),\ \sum_r \langle \underline{q}\cdot\underline{n}_r, v\rangle = 0,\ \forall v \in H^1_{0,D}(\Omega)\}$. □

Proposition 2.1.3. $H_{0,N}(\underline{curl};\Omega) = \{\underline{\chi} \mid \underline{\chi} \in W(\Omega),\ \sum_r \langle \gamma_t\underline{\chi}, \underline{\phi}\rangle = 0,\ \forall \underline{\phi} \in H^1_{0,D}(\Omega)\}$. □

The last results state that functions of $Y(\Omega)$ (resp. $W(\Omega)$) belong to $H(\text{div};\Omega)$ (resp. $H(\underline{\text{curl}};\Omega)$) if and only if their normal (resp. tangential) traces are "continuous" at the interfaces. This will be an essential point for finite element approximations.

2.1.3 Properties Relative to a Change of Variables

The use of a *reference element*, and therefore of coordinate changes, is an essential ingredient of finite element methods, whether for convergence studies or for practical implementation. We must therefore study the effect of a change of variables on our function spaces. We refer to [147, 148] for a more complete presentation.

Let $\hat{K} \subset \mathbb{R}^n$. We denote by $\partial\hat{K}$ its boundary, by $\underline{\hat{n}}$ the outward oriented normal, by $d\hat{x}$ the Lebesgue measure on $\hat{K}$ and by $d\hat{\sigma}$ the superficial measure induced by it on $\partial\hat{K}$.

Let F be a smooth (at least C^1) mapping from $\mathbb{R}^n$ into $\mathbb{R}^n$. We define $K = F(\hat{K})$. We suppose that the Jacobian matrix $DF(\hat{x})$ is invertible for any $\hat{x}$ and that F is globally invertible on K. We then have

$$DF^{-1}(x) = (DF(\hat{x}))^{-1}. \tag{2.1.54}$$

An important case is $F(\hat{x}) = x_0 + B\hat{x}$, that is if F is an affine mapping. Then $DF(\hat{x}) = B$ is a constant matrix. We denote

$$\|DF\|_\infty := \sup_{\hat{x}\in K}\left(\sup_{\xi\in\mathbb{R}^n}\frac{|DF(\hat{x})\xi|}{|\xi|_{\mathbb{R}^n}}\right), \tag{2.1.55}$$

the norm in $L^\infty(\hat{K})$ of function $\hat{x} \to \|DF(\hat{x})\|$, that is, matrix norm of $DF(\hat{x})$. In the same way, we have

$$\|DF^{-1}\|_\infty := \sup_{x\in K}\left(\sup_{\xi\in\mathbb{R}^n}\frac{|(DF^{-1}(x))\xi|}{|\xi|_{\mathbb{R}^n}}\right). \tag{2.1.56}$$

We write

$$J(\hat{x}) := |\det DF(\hat{x})| \tag{2.1.57}$$

and, for $\hat{x} \in \partial\hat{K}$,

$$J_{\underline{\hat{n}}}(\hat{x}) := J(\hat{x})\ \|(DF^{-1})^t\ \underline{\hat{n}}\|_{\mathbb{R}^n}. \tag{2.1.58}$$

- **Sobolev spaces $\boldsymbol{H^s(\Omega)}$.** If $\hat{v}(\hat{x})$ is a function on $\hat{K}$, we define $v(x)$ on K by

$$v := \hat{v} \circ F^{-1}, \tag{2.1.59}$$

and we denote this by $v = \mathcal{F}(\hat{v})$. We then have the classical formulas,

$$\underline{\mathrm{grad}}\, v = (DF^{-1})^t\ \underline{\mathrm{grad}}\,\hat{v} \circ F^{-1} = \mathcal{F}((DF^{-1})^t\ \underline{\mathrm{grad}}\,\hat{v}) \tag{2.1.60}$$

and

$$\int_K \mathcal{F}(\hat{v})\, dx = \int_{\hat{K}} \hat{v}\ J\ d\hat{x}, \tag{2.1.61}$$

$$\int_{\partial K} \mathcal{F}(\hat{v})\, d\sigma = \int_{\partial\hat{K}} \hat{v}\ J_{\underline{\hat{n}}}\, d\hat{\sigma}. \tag{2.1.62}$$

From this, it is immediate to deduce the following lemma.

Lemma 2.1.5. *The mapping $\mathcal{F}$ is an isomorphism from $L^2(\hat{K})$ onto $L^2(K)$ and from $H^1(\hat{K})$ onto $H^1(K)$, satisfying,*

$$|v|_{0,K} \leq \left(\sup_{\hat{x}} J(\hat{x})\right)^{1/2} |\hat{v}|_{0,\hat{K}}, \tag{2.1.63}$$

$$|\hat{v}|_{0,\hat{K}} \leq \left(\inf_{\hat{x}}\ J(\hat{x})\right)^{-1/2} |v|_{0,K}, \tag{2.1.64}$$

$$|v|_{1,K} \leq \left(\sup_{\hat{x}} J(\hat{x})\right)^{1/2} \|DF^{-1}\|_\infty\ |\hat{v}|_{0,\hat{K}}, \tag{2.1.65}$$

$$|\hat{v}|_{1,\hat{K}} \leq \left(\inf_{\hat{x}}\ J(\hat{x})\right)^{-1/2} \|DF\|_\infty\ |v|_{1,K}. \tag{2.1.66}$$

Remark 2.1.8. If F is an affine mapping, we also have [146]

$$|v|_{m,K} \leq c\ (\det B)^{\frac{1}{2}}\ \|B^{-1}\|^m\ |\hat{v}|_{m,\hat{K}} \tag{2.1.67}$$

and similarly,

$$|\hat{v}|_{m,\hat{K}} \leq c\ (\det B)^{-\frac{1}{2}}\ \|B\|^m\ |v|_{m,K}, \tag{2.1.68}$$

where the constant c depends only on m and on the space dimension n. □

In the general case, one must use Leibnitz's formula and the final result is much more complex. For some comments in this directions, see Sect. 2.2.4.

- **The space H(div; Ω)**. When building approximations of $H(\text{div};\Omega)$ in Sect. 2.3, we shall be led to use the normal component of vectors as degrees of freedom. This is pretty natural according to Proposition 2.1.2. The above transformation obviously does not preserve normal components. It does neither map $H(\text{div};\hat{K})$ into $H(\text{div};K)$. To overcome this problem, we have to introduce a special (contravariant) transformation known as *Piola's transformation*.

Let, as before, $DF(\hat{x})$ be the Jacobian matrix of the transformation $F(\hat{x})$. We consider, for $\underline{\hat{q}} \in (L^2(\hat{K}))^n$, the mapping,

$$\mathcal{G}(\underline{\hat{q}})(x) := \frac{1}{J(\hat{x})} DF(\hat{x})\ \underline{\hat{q}}(\hat{x}),\ x = F(\hat{x}). \tag{2.1.69}$$

It is then elementary to check that one has (in $\mathbb{R}^2$, but the result holds for $\mathbb{R}^n$)

$$\begin{pmatrix} \frac{\partial q_1}{\partial x} & \frac{\partial q_1}{\partial y} \\ \frac{\partial q_2}{\partial x} & \frac{\partial q_2}{\partial y} \end{pmatrix} = \frac{1}{J}(DF) \begin{pmatrix} \frac{\partial \hat{q}_1}{\partial \hat{x}} & \frac{\partial \hat{q}_1}{\partial \hat{y}} \\ \frac{\partial \hat{q}_2}{\partial \hat{x}} & \frac{\partial \hat{q}_2}{\partial \hat{y}} \end{pmatrix} (DF^{-1}). \tag{2.1.70}$$

As the trace of a matrix is invariant by a change of variables, we have

$$\operatorname{div} \underline{q} = \frac{1}{J} \operatorname{div} \underline{\hat{q}}. \tag{2.1.71}$$

More generally, we have [366]

Lemma 2.1.6. *Let $v = \mathcal{F}(\hat{v})$ and $\underline{q} = \mathcal{G}(\underline{\hat{q}})$, then*

$$\int_K \underline{q} \cdot \underline{\text{grad}}\, v \, dx = \int_{\hat{K}} \underline{\hat{q}} \cdot \underline{\text{grad}}\, \hat{v} \, d\hat{x}, \tag{2.1.72}$$

$$\int_K v \cdot \operatorname{div} \underline{q} \, dx = \int_{\hat{K}} \underline{\hat{v}} \, \operatorname{div} \hat{q} \, d\hat{x}, \tag{2.1.73}$$

$$\int_{\partial K} \underline{q} \cdot \underline{n} v \, d\sigma = \int_{\partial \hat{K}} \underline{\hat{q}} \cdot \underline{\hat{n}} \, \hat{v} \, d\hat{\sigma}. \tag{2.1.74}$$

We refer to [366] and [331] for the proof of this result and most of the following ones.

From (2.1.74), we see that $\mathcal{G}$ preserves the normal trace in $H^{-\frac{1}{2}}$ and enables us to define subspaces of $H(\text{div}; K)$ through the reference element $\hat{K}$. More precisely we have,

Lemma 2.1.7. *The mapping $\mathcal{G}$ is an isomorphism of $H(\text{div}; \hat{K})$ onto $H(\text{div}; K)$ and of $H^0(\text{div}; \hat{K})$ onto $H^0(\text{div}; K)$. Moreover we have:*

$$|\underline{q}|_{0,K} \le \left(\inf_{\hat{x}} J(\hat{x})\right)^{-\frac{1}{2}} \|DF\|_\infty \, |\underline{q}|_{0,\hat{K}}, \tag{2.1.75}$$

$$|\underline{\hat{q}}|_{0,\hat{K}} \le \left(\sup_{\hat{x}} J(\hat{x})\right)^{\frac{1}{2}} \|DF^{-1}\|_\infty \, |\underline{q}|_{0,K}, \tag{2.1.76}$$

$$|\operatorname{div} \underline{q}|_{0,K} \le \left(\inf_{\hat{x}} J(\hat{x})\right)^{-\frac{1}{2}} |\operatorname{div} \underline{\hat{q}}|_{0,\hat{K}}, \tag{2.1.77}$$

$$|\operatorname{div} \underline{\hat{q}}|_{0,\hat{K}} \le \left(\sup_{\hat{x}} J(\hat{x})\right)^{\frac{1}{2}} |\operatorname{div} \underline{q}|_{0,K}. \tag{2.1.78}$$

It is also possible to obtain relations between $|\underline{q}|_{m,K}$ and $|\underline{\hat{q}}|_{m,\hat{K}}$ or between $|\operatorname{div} \underline{q}|_{m,K}$ and $|\operatorname{div} \underline{\hat{q}}|_{m,\hat{K}}$. We refer to [366] for details. The next lemma deals with the case where F is an affine transformation and $\underline{q} \in H^m(\text{div}; \Omega)$, where

$$H^m(\text{div};\Omega) := \{\underline{q} \mid \underline{q} \in (H^m(\Omega))^n,\ \text{div}\,\underline{q} \in H^m(\Omega)\}. \tag{2.1.79}$$

Lemma 2.1.8. *If the mapping F is affine and if $\underline{q} \in H^m(\text{div};\Omega)$, the following estimates hold, with $B = DF$,*

$$|\underline{q}|_{m,K} \le (det\ B)^{-\frac{1}{2}} \|B^{-1}\|^m\ \|B\|\ |\hat{\underline{q}}|_{m,\hat{K}}, \tag{2.1.80}$$

$$|\,\text{div}\,\underline{q}|_{m,K} \le (det\ B)^{-\frac{1}{2}} \|B^{-1}\|^m\ |\,\text{div}\,\hat{\underline{q}}|_{m,\hat{K}}. \tag{2.1.81}$$

The reverse inequalities also hold by a simple exchange of roles between K and $\hat{K}$. Such results are of course essential in the proofs of error estimates. The Piola transformation can be extended to tensor-valued functions with similar properties (cf. for instance [127, 295] or [146]).

- **The space $H(\underline{\text{curl}};\Omega)$**. According to Remark 2.1.5, in this subsection we restrict the discussion to the three-dimensional situation where $\hat{K} \subset \mathbb{R}^3$ since the two-dimensional case can be easily deduced from the result of the previous subsection about approximations of $H(\text{div};\Omega)$. In general, when dealing with reference elements, $\hat{K}$ will be a simple geometric entity like a cube or a tetrahedron. In particular, it will contain vertices, edges and faces. We shall denote by $\hat{e}$ and $\hat{f}$ generic edges and faces, respectively, and by e and f their corresponding images under the action of F. We shall then refer to an edge e or a face f of K. It is clear that, for general isomorphisms F, an edge of K might be curved and a face of K might not be contained in a plane. On the other hand, if F is a trilinear map, then edges of K will be straight and if, moreover, F is affine, then faces of K will be flat.

  In order to deal with the approximation of $H(\underline{\text{curl}};\Omega)$, we shall make use of the following (covariant) transformation:

$$\mathcal{H}(\hat{\underline{\chi}})(x) := [DF(\hat{x})]^{-T}\,\hat{\underline{\chi}}(\hat{x}), \quad x = F(\hat{x}). \tag{2.1.82}$$

  We notice that formula (2.1.82) is formally identical to (2.1.60); i.e., we have chosen to transform vectors of $H(\underline{\text{curl}};\Omega)$ like gradients. In particular, one of the main features of transformation (2.1.82) is that it preserves the tangential components in a sense that will soon be made clear.

  Before stating the general results, let us go back to the setting of Example 2.1.1.

Example 2.1.2 (Case when Γ is flat.). We recall that, in our example, $\Omega \subset \mathbb{R}^3$ is the half space $x_3 < 0$ and that Γ is the plane $x_3 = 0$. Let us consider a linear mapping $F : \mathbb{R}^3 \to \mathbb{R}^3$ defined as follows

$$F(\hat{x}_1, \hat{x}_2, \hat{x}_3) = (\alpha\hat{x}_1 + \beta\hat{x}_2, \gamma\hat{x}_1 + \delta\hat{x}_2, \hat{x}_3)^T, \tag{2.1.83}$$

so that

$$DF = \begin{pmatrix} \alpha & \beta & 0 \\ \gamma & \delta & 0 \\ 0 & 0 & 1 \end{pmatrix}. \tag{2.1.84}$$

Basically, we are considering a mapping that corresponds to a motion of Ω only in the direction of x_1 and x_2. In particular, the restriction of F to Γ maps linearly the plane Γ into itself. According to the notation introduced before, we shall denote by $\hat{\Omega}$ the reference domain and by Ω the transformed domain; analogously, Γ is the image of the reference boundary $\hat{\Gamma}$. Let us consider a vector field $\hat{\underline{\chi}} \in H(\underline{\mathrm{curl}}; \hat{\Omega})$; we are interested in investigating how the traces of $\hat{\underline{\chi}}$ transform under the effects of the covariant map (2.1.82).

It is immediate to evaluate the determinant of DF, given by the constant $J = \alpha\delta - \beta\gamma$, and the matrix $[DF]^{-T}$ given by

$$[DF]^{-T} = \frac{1}{J}\begin{pmatrix} \delta & -\gamma & 0 \\ -\beta & \alpha & 0 \\ 0 & 0 & J \end{pmatrix}. \tag{2.1.85}$$

Hence, transformation (2.1.82) reads

$$\underline{\chi} = \begin{pmatrix} \chi_1 \\ \chi_2 \\ \chi_3 \end{pmatrix} = \mathcal{H}(\hat{\underline{\chi}}) = \frac{1}{J}\begin{pmatrix} \delta\hat{\chi}_1 - \gamma\hat{\chi}_2 \\ \alpha\hat{\chi}_2 - \beta\hat{\chi}_1 \\ J\hat{\chi}_3 \end{pmatrix}. \tag{2.1.86}$$

The first important result which we are going to check is that the trace $\hat{\gamma}_t\hat{\underline{\chi}}$ transforms like vectors in $H(\mathrm{div}; \hat{\Gamma})$. More precisely, let $\mathcal{G}_\Gamma$ denote the Piola transformation defined in (2.1.69) related to the plane Γ, i.e.,

$$\mathcal{G}_\Gamma(\hat{\underline{q}}) := \frac{1}{J}\begin{pmatrix} \alpha & \beta \\ \gamma & \delta \end{pmatrix}\hat{\underline{q}}. \tag{2.1.87}$$

From (2.1.31) we obtain

$$\mathcal{G}_\Gamma(\hat{\gamma}_t\hat{\underline{\chi}}) = \frac{1}{J}\begin{pmatrix} \alpha & \beta \\ \gamma & \delta \end{pmatrix}\begin{pmatrix} \hat{\chi}_2 \\ -\hat{\chi}_1 \end{pmatrix} = \frac{1}{J}\begin{pmatrix} \alpha\hat{\chi}_2 - \beta\hat{\chi}_1 \\ \gamma\hat{\chi}_2 - \delta\hat{\chi}_1 \end{pmatrix}. \tag{2.1.88}$$

On the other hand, comparing (2.1.31), (2.1.86), and (2.1.88), we easily get

$$\mathcal{G}_\Gamma(\hat{\gamma}_t\hat{\underline{\chi}}) = \gamma_t(\mathcal{H}(\hat{\underline{\chi}})) = \gamma_t\underline{\chi}. \tag{2.1.89}$$

In a similar way, we can find how the trace $\hat{\pi}_t\hat{\underline{\chi}}$ transforms when $\hat{\underline{\chi}}$ is mapped like in (2.1.82). It turns out that $\hat{\pi}_t\hat{\underline{\chi}}$ transforms like vectors in $H(\underline{\mathrm{curl}}; \hat{\Gamma})$. In this case, we can make use of the mapping $\mathcal{H}_\Gamma$, defined as

$$\mathcal{H}_\Gamma\left(\begin{pmatrix}\chi_1\\ \chi_2\end{pmatrix}\right) := \frac{1}{J}\begin{pmatrix}\delta & -\gamma\\ -\beta & \alpha\end{pmatrix}\begin{pmatrix}\chi_1\\ \chi_2\end{pmatrix}, \tag{2.1.90}$$

in order to obtain the result

$$\mathcal{H}_\Gamma(\hat{\pi}_t\underline{\hat{\chi}}) = \pi_t(\mathcal{H}(\underline{\hat{\chi}})) = \pi_t\underline{\chi}. \tag{2.1.91}$$

□

Let us now come back to the general situation. For properties of the covariant transformation (2.1.82) and for the proofs of most of the results presented in the sequel of this subsection, we refer the interested reader to [222], to the abstract theory of differential forms presented in [248], to the book [302], and to the comprehensive review [33].

We shall make use, in particular, of the following relationship between the curl of $\underline{\chi}$ and the curl of $\underline{\hat{\chi}}$

$$\underline{\text{curl}}\,\underline{\chi}(x) = \frac{1}{J(\hat{x})}DF(\hat{x})\,\underline{\text{curl}}\,\underline{\hat{\chi}}(\hat{x}), \quad x = F(\hat{x}), \tag{2.1.92}$$

so that we have $\underline{\text{curl}}\,\underline{\chi} = \mathcal{G}(\underline{\text{curl}}\,\underline{\hat{\chi}})$. Formula (2.1.92) is a consequence of the more general transformation rule (see [310, Theorem 2] and [223])

$$\underline{\underline{\text{curl}}}\,\underline{\chi}(x) = [DF]^{-T}\,\underline{\underline{\text{curl}}}\,\underline{\hat{\chi}}(\hat{x})[DF]^{-1}, \quad x = F(\hat{x}), \tag{2.1.93}$$

where the tensor is defined by $\underline{\underline{\text{curl}}}\,\underline{\chi} = \left(\frac{\partial\chi_j}{\partial x_i} - \frac{\partial\chi_i}{\partial x_j}\right)_{i,j}$.

Before stating results which relate quantities evaluated on the reference element to corresponding quantities evaluated on the actual element (in the spirit of Lemma 2.1.6), we need to recall how tangent and normal vectors behave under the action of the mapping F. Namely, if $\underline{\hat{n}}$ is a normal vector at a given point of $\partial\hat{K}$, then

$$\underline{n}(x) = \frac{[DF]^{-T}\cdot\underline{\hat{n}}(\hat{x})}{\|[DF]^{-T}\cdot\underline{\hat{n}}(\hat{x})\|} \tag{2.1.94}$$

is the corresponding normal unit vector on ∂K. If $\underline{\hat{t}}$ is a tangent unit vector, then

$$\underline{t}(x) = \frac{DF\cdot\underline{\hat{t}}(\hat{x})}{\|DF\cdot\underline{\hat{t}}(\hat{x})\|} \tag{2.1.95}$$

is the corresponding tangent unit vector on ∂K. Moreover, if $\underline{\hat{t}}$ is tangent to an edge of $\hat{K}$, then $\underline{t}$ defined in (2.1.95) is tangent to the corresponding edge of K.

Lemma 2.1.9. *Let $\underline{\chi} := \mathcal{H}(\underline{\hat{\chi}})$. Let additional functions on K be constructed from reference functions as follows: $v := \mathcal{F}(\hat{v})$, $\underline{q} := \mathcal{G}(\underline{\hat{q}})$, and $\underline{\tau} := \mathcal{H}(\underline{\hat{\tau}})$. Then*

$$\int_e \underline{\chi} \cdot \underline{t} v \, ds = \int_{\hat{e}} \hat{\underline{\chi}} \cdot \hat{\underline{t}} \hat{v} \, d\hat{s}, \tag{2.1.96}$$

$$\int_f \pi_t \underline{\chi} \cdot \gamma_t \underline{\tau} \, d\sigma = \int_{\hat{f}} \hat{\pi}_t \hat{\underline{\chi}} \cdot \hat{\gamma}_t \hat{\underline{\tau}} \, d\hat{\sigma}, \tag{2.1.97}$$

$$\int_K \underline{\chi} \cdot \underline{q} \, dx = \int_{\hat{K}} \hat{\underline{\chi}} \cdot \hat{\underline{q}} \, d\hat{x}. \tag{2.1.98}$$

Remark 2.1.9. Formula (2.1.97) is worth some comments. For simplicity, let us consider the case where f is flat. From Example 2.1.2, it follows that, when $\underline{\tau} := \mathcal{H}(\hat{\underline{\tau}})$, then $(\gamma_t \underline{\tau})|_f$ is related to $(\gamma_t \hat{\underline{\tau}})|_{\hat{f}}$ by means of the two-dimensional Piola transform (2.1.69) on the face f. If we call $\mathcal{G}_f$ such a transform, then we have $\gamma_t \underline{\tau} = \mathcal{G}_f(\hat{\gamma}_t \hat{\underline{\tau}})$. On the other hand, in an analogous way we have that $\pi_t \underline{\chi} = H_f(\hat{\pi}_t \hat{\underline{\chi}})$ on the face f. Then, Eq. (2.1.97) might be rephrased in terms of functions defined on the face f as follows: let $\hat{\underline{\phi}}$ and $\hat{\underline{\psi}}$ be vector fields on $\hat{f}$, then

$$\int_{\hat{f}} \hat{\underline{\psi}} \cdot \hat{\underline{\phi}} \, d\hat{\sigma} = \int_f H_f(\hat{\underline{\psi}}) \cdot \mathcal{G}_f(\hat{\underline{\phi}}) \, d\sigma. \tag{2.1.99}$$

□

Finally, the following analogue of Lemma 2.1.7 holds true.

Lemma 2.1.10. *The mapping $\mathcal{H}$ is an isomorphism of $H(\underline{curl}; \hat{K})$ onto $H(\underline{curl}; K)$. Moreover, we have:*

$$\|\underline{\chi}\|_{0,K} \leq \left(\sup_{\hat{x}} J(\hat{x})\right)^{\frac{1}{2}} \|DF^{-1}\|_\infty \|\hat{\underline{\chi}}\|_{0,\hat{K}}, \tag{2.1.100}$$

$$\|\hat{\underline{\chi}}\|_{0,\hat{K}} \leq \left(\inf_{x} J(x)\right)^{-\frac{1}{2}} \|DF\|_\infty \|\underline{\chi}\|_{0,K}, \tag{2.1.101}$$

$$\|\operatorname{\underline{curl}} \underline{\chi}\|_{0,K} \leq \left(\inf_{\hat{x}} J(\hat{x})\right)^{-\frac{1}{2}} \|DF\|_\infty \|\operatorname{\underline{curl}} \hat{\underline{\chi}}\|_{0,\hat{K}}, \tag{2.1.102}$$

$$\|\operatorname{\underline{curl}} \hat{\underline{\chi}}\|_{0,\hat{K}} \leq \left(\sup_{x} J(x)\right)^{\frac{1}{2}} \|DF^{-1}\|_\infty \|\operatorname{\underline{curl}} \underline{\chi}\|_{0,K}. \tag{2.1.103}$$

2.1.4 De Rham Diagram

The topics introduced in this chapter can be presented in a general unified approach by means of tools of exterior algebra. The De Rham complex, in particular, has been rediscovered recently as a very convenient tool in order to provide a general setting for handling function spaces and their finite element approximations. In this framework, the commuting diagram property which will be presented throughout this chapter can be seen as particular cases of a more general picture. Even though it is out of the aims of this book to give a detailed presentation of this topic, we would like to provide the reader with a short introduction to this subject, which proves to be a very useful technique not only for the analysis of known finite elements, but also for designing new ones.

The importance of the de Rham complex in the analysis of finite element schemes has been detected independently by several authors in different fields. The commuting diagram properties for finite elements in $H(\text{div};\Omega)$ (see [177, 178]) has been the driving force for an active research in the approximation of second order elliptic problems in mixed form. Bossavit [102] was the first one who used the full form of the de Rham complex for the approximation of problems arising from electromagnetism. His pioneer idea has been exploited by several authors (see, for instance, [75, 170, 247, 248]). The general framework has been designed in [31] and successfully applied to the construction of new finite element spaces for the elasticity equations (see [32]). References [33, 34] present an excellent review on the state of the art of this active research field.

Here we recall the de Rham complex related with the spaces considered in this chapter. Let us suppose that $\Omega \subset \mathbb{R}^3$ is a simply connected domain, then the following sequence is exact

$$\mathbb{R} \hookrightarrow H^1(\Omega) \xrightarrow{\underline{\text{grad}}} H(\underline{\text{curl}};\Omega) \xrightarrow{\underline{\text{curl}}} H(\text{div};\Omega) \xrightarrow{\text{div}} L^2(\Omega) \to 0. \tag{2.1.104}$$

We shall present in the sequel some discrete variants of (2.1.104), which will be useful to understand interconnections between different approximations.

2.2 Finite Element Approximations of $H^1(\Omega)$ and $H^2(\Omega)$

This section will be mainly devoted to the approximation of $H^1(\Omega)$ and its subspace of the form $H^1_{0,D}(\Omega)$. We shall moreover sketch some results concerning the approximation of $H^2(\Omega)$. Standard approximations of Sobolev spaces can be subdivided into two classes: conforming and nonconforming methods. Even though nonconforming methods will be studied in the context of hybrid finite element methods, their importance makes it useful to introduce them here. We refer to [41, 146] or [334] for a detailed presentation of the following results.

2.2.1 *Conforming Methods*

Conforming methods are the most natural finite element methods. They yield *internal approximations* in the sense that they enable us to build finite dimensional subspaces of the function space that we want to approximate.

Given a partition of the domain Ω into polygonal or polyhedral elements, a conforming approximation of $H^1(\Omega)$ is a space of *continuous functions* defined by a finite number of parameters (or degrees of freedom).

The last condition is *usually* met by using a space of piecewise polynomial functions or functions obtained from polynomials by a change of variables like (using the notation of Sect. 2.1.3)

$$v_h|_K = \hat{v} \circ F^{-1}, \tag{2.2.1}$$

where $K = F(\hat{K})$ and $\hat{v}$ is a polynomial function on $\hat{K}$. Continuity is obtained by a clever choice of degrees of freedom.

Remark 2.2.1. For triangular elements, it is usual and convenient to use piecewise polynomial functions on K. For quadrilaterals, it is essential to use (2.2.1). It must then be noted that $v_h|_K$ is not in general a polynomial on K. This will be the case only for affine transformations. More comments on this issue will be discussed in Sect. 2.2.4. □

To give a more precise definition of our finite element approximations, we shall need a few definitions. Let us define on an element K

$$P_k(K) := \text{ the space of polynomials of degree } \leq \text{k}. \tag{2.2.2}$$

The dimension of $P_k(K)$ is $\frac{1}{2}(k+1)(k+2)$ for $n = 2$ and, for $n = 3$, it is $\frac{1}{6}(k+1)(k+2)(k+3)$. For a rectangular element, it will be convenient to define (for $n = 2$)

$$P_{k_1,k_2}(K) := \Big\{ p(x_1, x_2) \mid p(x_1, x_2) = \sum_{\substack{i \leq k_1 \\ j \leq k_2}} a_{ij}\, x_1^i\, x_2^j \Big\} \tag{2.2.3}$$

the space of polynomials of degree $\leq k_1$ in x_1 and $\leq k_2$ in x_2. In the same way, we can define on a rectangular hexahedron $P_{k_1,k_2,k_3}(K)$ for $n = 3$. The dimension of these spaces is $(k_1+1)(k_2+1)$ and $(k_1+1)(k_2+1)(k_3+1)$, respectively. We then define

$$Q_k(k) := \begin{cases} P_{k,k}(K), & \text{for } n = 2, \\ P_{k,k,k}(K), & \text{for } n = 3. \end{cases} \tag{2.2.4}$$

We shall also need polynomial spaces on the edges (or faces) of the elements. Using the notations of Sect. 2.1.2, we define

$$R_k(\partial K) := \{\phi \mid \phi \in L^2(\partial K),\ \phi|_{e_i} \in P_k(e_i),\ \forall e_i\}, \tag{2.2.5}$$

$$T_k(\partial K) := \{\phi \mid \phi \in R_k(\partial K) \cap C^0(\partial K)\}. \tag{2.2.6}$$

We define a subspace of P_k, which could have traces of a lower degree on the boundary of K

$$P_k^r := \{p \mid p \in P_k|_{\partial K},\ p \in T_r(\partial K),\ r \le k\}. \tag{2.2.7}$$

Functions of $R_k(\partial K)$ are polynomials of degree $\le k$ on each side (or face) of K. They do not have to be continuous at vertices. The dimensions of $R_k(\partial K)$ and $T_k(\partial K)$ are respectively for $k \ge 1$:

— $3(k+1)$ and $3k$ for triangles,
— $4(k+1)$ and $4k$ for quadrilaterals,
— $2(k+1)(k+2)$ and $2(k^2+1)$ for tetrahedra.

For hexahedra, it will *usually* be more convenient to consider functions in $Q_k(e_i)$ in the definition of $R_k(\partial K)$ and $T_k(\partial K)$.

In order to define a finite element, following [146], we need to specify three things.

— The geometry: we choose a reference element $\hat{K}$, a change of variables $F(\hat{x})$ and we set $K = F(\hat{K})$.
— A set $\hat{P}$ of polynomials on $\hat{K}$. For $\hat{p} \in \hat{P}$ we define on K, $p = \hat{p} \circ F^{-1}$.
— A set of degrees of freedom $\hat{\Sigma}$, that is, a set of linear forms $\{\hat{\ell}_i\}_{1 \le i \le \dim \hat{P}}$ on $\hat{P}$. We say that this set is unisolvent when these linear forms are linearly independent, i.e. the knowledge of $\hat{\ell}_i(\hat{p})$ for all i completely defines $\hat{p}$.

A finite element is of *Lagrange type* if its degrees of freedom are *point values*, that is, if one is given a set $\{\hat{a}_i\}_{1 \le i \le \dim \hat{P}}$ of points in $\hat{K}$ and one defines

$$\hat{\ell}_i(\hat{p}) = \hat{p}(\hat{a}_i),\ 1 \le i \le \dim \hat{P}. \tag{2.2.8}$$

For the approximation of $H^1(\Omega)$, Lagrange type elements will be sufficient but approximating $H^2(\Omega)$ requires Hermite type elements, that is degrees of freedom involving derivatives.

Remark 2.2.2. The reader should be aware that not any choice of points will yield a unisolvent set of degrees of freedom. Moreover, the points have to be chosen in order to ensure inter-element continuity. □

Example 2.2.1 (Affine Finite Elements). This is the most classical family of finite elements. The reference element is the triangle $\hat{K}$ of Fig. 2.2 and we use the affine transformation

$$F(\hat{x}) = x_0 + B\hat{x}. \tag{2.2.9}$$

The element K is still a triangle and it is not degenerate provided $\det(B) \ne 0$. We now take $\hat{P} = P_k(\hat{K})$ and choose an appropriate set of degrees of freedom. The standard choices for $k \le 3$ are presented in Fig. 2.3.

It can be easily observed that this choice of points ensures continuity at interfaces. □

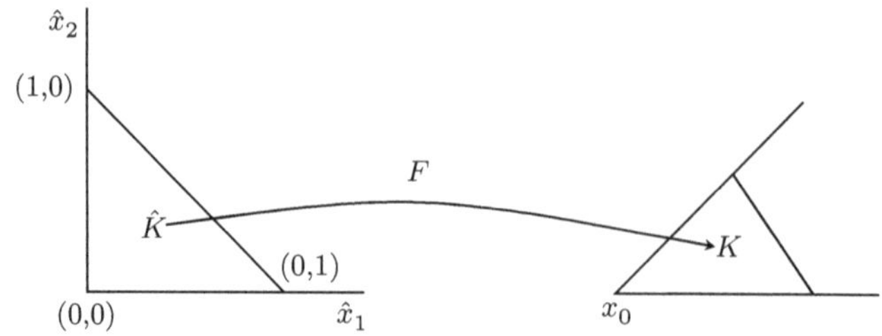

Fig. 2.2 An affine tranformation

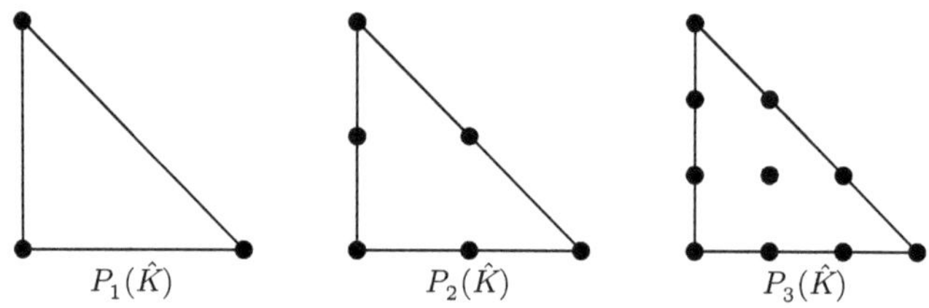

Fig. 2.3 Standard conforming elements

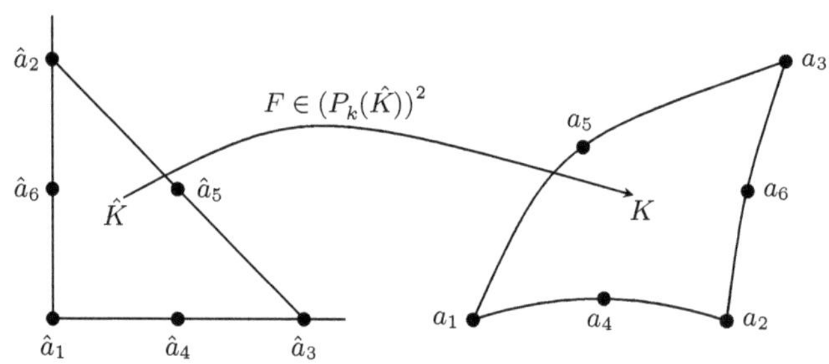

Fig. 2.4 Isoparametric triangle of degree 2

Example 2.2.2 (Isoparametric triangular elements). We use the same reference element and the same set $\hat{P}$ as in the previous example. We now take the transformation $F(\hat{x})$ such that each of its components F_i belongs to $P_k(\hat{K})$. For $k = 1$, nothing is changed but for $k \geq 2$, the element K now has curved boundaries. We present the case $k = 2$ in Fig. 2.4.

Using such curved triangles enables us to obtain a better approximation of curved boundaries. It must be noted that the curvature of boundaries introduces additional terms in the approximation error and the curved elements should be used only when they are really necessary [152] or [146]. □

Example 2.2.3 (Isoparametric quadrilateral elements). This is also a very classical family of finite elements. The reference element is the square $\hat{K} =]0, 1[\times]0, 1[$. We take $\hat{P} = Q_k(\hat{K})$ and a transformation F with each component in $Q_k(\hat{K})$.

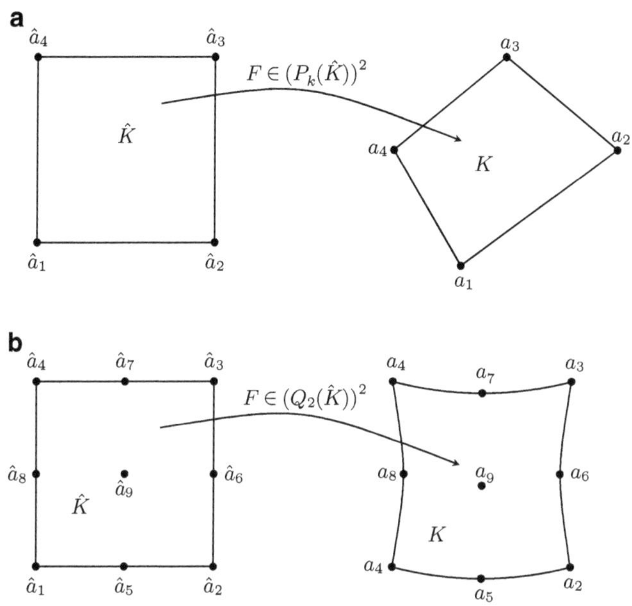

Fig. 2.5 (**a**) The Q1 isoparametric element. (**b**) The Q2 isoparametric element

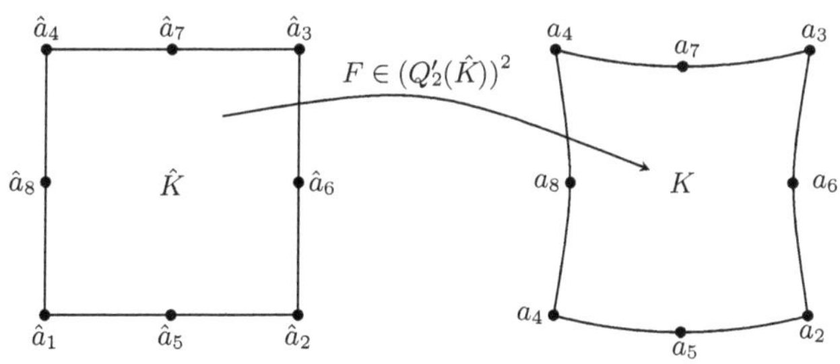

Fig. 2.6 Serendipity element

We present the standard choice of degrees of freedom for $k \leq 2$ in Fig. 2.5. It must be noted that we need $F \in (Q_1(\hat{K}))^2$ to define a general straight-sided quadrilateral.

Finally we recall that it is possible to eliminate internal nodes to get the so-called serendipity finite elements. For instance, if we take

$$\begin{aligned} \hat{P} = Q_2'(\hat{K}) : &= \{\hat{p} \mid \hat{p} \in Q_2(\hat{K}),\ 4\hat{p}(\hat{a}_9) + \sum_{i=1}^{4} \hat{p}(\hat{a}_i) - 2\sum_{i=5}^{8} \hat{p}(\hat{a}_i) = 0\} \\ &= P_3(\hat{K}) \cap Q_2(\hat{K}) \end{aligned} \tag{2.2.10}$$

we obtain the element of Fig. 2.6.

Fig. 2.7 (a) P_3 triangle. (b) Argyris' triangle

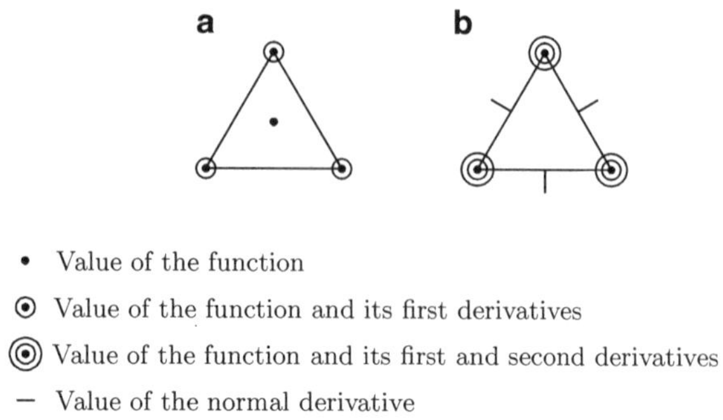

As before, the degrees of freedom have been chosen in order to ensure continuity between elements. The use of serendipity elements should be avoided on distorted (non-affine) meshes. More details on quadrilateral elements will be given in Sect. 2.2.4. □

Example 2.2.4 (Hermite type elements). Approximating $H^2(\Omega)$ will require continuity of derivatives at inter-element boundaries and leads to the introduction of elements in which values of the derivatives are used as degrees of freedom. The simplest Hermite type element is the P_3 triangle of Fig. 2.7a.

Here, the degrees of freedom are values of the function and its derivatives at vertices plus the function value at the barycentre. This element *does not* enable us to build an approximation of $H^2(\Omega)$. To do so, one must use Argyris' triangle (see Fig. 2.7.b) where polynomials of degree 5 are used. (Composite elements can also been used.) For quadrilaterals, the analogues are easily built. The difficulty of building approximations of $H^2(\Omega)$ by standard methods was one of the major reasons for the introduction of various kinds of mixed or hybrid methods for plate problems (cf. Sects. 10.2 or 10.3). □

The remaining part of this section is devoted to the analysis of the approximation of a given function v by the finite element spaces just described or similar ones. We shall not give proofs for which we refer to [146, 154, 358].

For a general set of degrees of freedom $\{\hat{\ell}_i\}$ on $\hat{K}$, we define the *interpolate* $\hat{r}_h\hat{v}$ of v by

$$\hat{\ell}_i(\hat{r}_h\hat{v}) := M(\hat{v}), \ 1 \leq i \leq \dim \hat{P}. \tag{2.2.11}$$

The operator M must be a well-defined continuous form. When the linear forms $\hat{\ell}_i$ are defined by (2.2.5), it is natural to set

$$(\hat{r}_h\hat{v})(\hat{a}_i) = \hat{v}(\hat{a}_i). \tag{2.2.12}$$

This definition makes sense only when $\hat{v}$ is a *continuous function* which is not the case when $v \in H^1(\Omega)$. For Lagrange type elements in $\mathbb{R}^2$ or $\mathbb{R}^3$, $\hat{v} \in H^2(\hat{K})$ is a sufficient condition for (2.2.12) to be justified and $\hat{r}_h\hat{v}$ is just the

Lagrange interpolate, in the classical sense, of $\hat{v}$. For $v \in H^1(\Omega)$, [154] *has defined a continuous interpolate* $\hat{r}_h$ *using averages of* u *instead of point values.* This also implies a more elaborate use of reference elements. In particular, the operator $\hat{r}_h \hat{v}$ is no longer defined on one single element. In fact the nodal values of $\hat{r}_h \hat{v}$ depend on the value of $\hat{v}$ on the adjacent elements through an averaging process.

Once $\hat{r}_h \hat{v}$ is defined, we can define on K

$$r_h v := (\hat{r}_h (v \circ F)) \circ F^{-1} = (\hat{r}_h \hat{v}) \circ F^{-1}. \tag{2.2.13}$$

We rapidly recall a few classical results. We refer the reader to [146] for a detailed presentation. We first consider the case of *affine elements*, assuming first r_h to be defined by a usual interpolate (2.2.12).

Proposition 2.2.1. *If the mapping* F *is affine, that is* $F(\hat{x}) = x_0 + B\hat{x}$, *and if* $r_h p_k = p_k$ *for any* $p_k \in P_k(K)$, *we have for* $v \in H^s(\Omega)$, $m \leq s$, $1 < s \leq k+1$,

$$|v - r_h v|_{m,K} \leq c \, \|B^{-1}\|^m \, \|B\|^s \, |v|_{s,K}. \tag{2.2.14}$$

The proof uses (2.1.67), its reciprocal, and the classical results stated below.

Lemma 2.2.1. $|\cdot|_{k+1,\Omega}$ *is a norm on* $H^{k+1}(\Omega)/P_k(\Omega)$, *equivalent to the standard quotient norm.*

From this lemma, the following classical result can be deduced.

Lemma 2.2.2 (Bramble-Hilbert's lemma). *Let* L *be a continuous linear form on* $H^{k+1}(\Omega)$ *such that* $L(p_k) = 0$ *for any* $p_k \in P_k(\Omega)$, *then there exists a constant* c *(depending on* L *and* Ω*) such that one has*

$$|L(v)| \leq c \, |v|_{k+1,\Omega}. \tag{2.2.15}$$

Results similar to (2.2.14), although more complex, can be obtained for general isoparametric elements [146, 151, 152]. Let then h_K be the diameter of K. Provided some classical conditions on the *shape of elements* forbidding degeneracy are fulfilled [146], relation (2.2.14) can be converted into a relation involving a power of h_K. For affine elements, one defines for instance

$$\sigma_K := \frac{h_K}{\rho_K}, \tag{2.2.16}$$

where ρ_K is the diameter of the largest inscribed disk (or sphere) in K.

We shall, in the following, always assume that the interpolation operator r_h is defined by the method of [154], that is, by a local projection instead of a point-wise interpolate. This allows us to get rid of the condition $s > 1$ of Proposition 2.2.1. To state this result, we define

$$\triangle K := \{K' | \bar{K}' \cap \bar{K} \neq \emptyset\}, \tag{2.2.17}$$

$$h_{\triangle K} := \sup_{K' \in \triangle K} h_{K'}, \tag{2.2.18}$$

$$\sigma_{\triangle K} := \sup_{K' \in \triangle K} \sigma_{K'}. \tag{2.2.19}$$

We then have the following proposition [154].

Proposition 2.2.2. *If the mapping F is affine and if $r_h p_k = p_k$ for any $p_k \in P_k(K)$, then there is a constant, depending on k and σ, such that for $m \leq s$, $1 \leq s \leq k+1$,*

$$|v - r_h v|_{m,K} \leq c\sigma_{\triangle K} h_{\triangle K}^{s-m} |v|_{s,\triangle K}. \tag{2.2.20}$$

□

We then say that *a family of triangulations $(\mathcal{T}_h)_{h \geq 0}$ is regular* if

$$\sigma_K < \sigma, \ \forall K \in \mathcal{T}_h, \ \forall h. \tag{2.2.21}$$

For the geometrical meaning of this condition, we refer to [146]. We may recall however that (2.2.21) can be written as a condition on angles excluding degenerate elements. For general curved elements, there is also a condition on the curvature of the sides.

We then have the following approximation result.

Corollary 2.2.1. *If $(\mathcal{T}_h)_{h \geq 0}$ is regular family of affine partitions, there exists a constant c depending on k and σ such that*

$$|v - r_h v|_{1,K} \leq c h^k |v|_{k+1,\triangle K}. \tag{2.2.22}$$

For more general partitions including general isoparametric elements, the result is qualitatively the same: we have an $O(h^k)$ approximation provided the family of partitions is regular in a sense to be specified (see Sect. 2.2.4 for more details).

We also refer the reader to [260] where some degenerate cases are analysed.

From the elements described above, we can build approximations of $H^1(\Omega)$ and $H^2(\Omega)$. The idea is of course to use functions whose restriction to an element belongs to a set of polynomial (or image S of polynomial) functions. Let $S_k(K)$ be a subspace of $P_k(K)$. We define, for a partition $\mathcal{T}_h$ of Ω,

$$\mathcal{L}^s(S_k, \mathcal{T}_h) := \{v \mid v \in H^s(\Omega), \ v|_K \in S_k(K)\}. \tag{2.2.23}$$

Remark 2.2.3. Since the elements of $\mathcal{L}^s(S_k, \mathcal{T}_h)$ are piecewise polynomials, we have $\mathcal{L}^s(S_k, \mathcal{T}_h) \subset C^{s-1}(\bar{\Omega})$ although $H^s(\Omega) \not\subset C^{s-1}(\bar{\Omega})$. □

We shall reduce this notation when no confusion is to be feared and write

$$\mathcal{L}^s(S_k) \tag{2.2.24}$$

when no ambiguity is possible as to $\mathcal{T}_h$ and still more compactly

$$\mathcal{L}_k^s = \mathcal{L}^s(P_k, \mathcal{T}_h), \tag{2.2.25}$$

when $\mathcal{T}_h$ is built from triangles and $S_k = P_k$, the space of polynomials of degree $\leq$k. In the same way, we shall write

$$\mathcal{L}_{[k]}^s = \mathcal{L}^s(Q_k, \mathcal{T}_h), \tag{2.2.26}$$

when $\mathcal{T}_h$ is built from quadrilaterals.

Remark 2.2.4 (Bubble functions). We shall often use in our constructions *bubble functions.* We consider them here in the case of H^1. For an element K, a bubble function is a function vanishing on ∂K. Thus, we say that S_k is a set of bubble functions if $S_k \subset H_0^1(K)$. We then denote

$$B(S_k) = \mathcal{L}^1(S_k, \mathcal{T}_h) = \mathcal{L}^0(S_k, \mathcal{T}_h) \tag{2.2.27}$$

and we shall use the compact notation,

$$\begin{cases} B_k = B(P_k \cap H_0^1(K)), \\ B_{[k]} = B(Q_k \cap H_0^1(K)), \end{cases} \tag{2.2.28}$$

when no ambiguity will be possible. Spaces of bubble functions will be used to build enriched spaces. For instance, the space $\mathcal{L}_2^1 \oplus B_3$ will be useful in Chap. 8 for the approximation of Stokes' problem. This space could also be written as $\mathcal{L}^1(P_3^2)$. □

When approximating a standard elliptic problem, the finite element spaces introduced up to now can be used directly in the variational formulation of the problem and error estimates follow from interpolation error estimates [146]. In many cases, however, nonconforming methods have proved to yield accurate (and sometimes easier to handle) approximations.

2.2.2 Explicit Basis Functions on Triangles and Tetrahedra

In the case of affine elements, it is often possible to define explicitly the basis functions associated to a choice of degrees of freedom. This is done using the following classical result.

Lemma 2.2.3. *Let K be an affine element of dimension k with vertices $\underline{x}_i$, $(1 \leq i \leq k)$. There exists a set $\lambda_i(\underline{x})$, $(1 \leq i \leq k)$ of linear functions on K, called barycentric coordinates, satisfying*

$$\lambda_i(\underline{x}_j) = \delta_{ij}$$

and

$$\sum_i \lambda_i(\underline{x}) = 1.$$

- It is then immediate that the barycentric coordinates are the basis functions of the affine $P_1(K)$ element.
- For the $P_2(K)$ element of Fig. 2.3 the basis function associated to the vertex $\underline{x}_i$ is $\lambda_i(2\lambda_i - 1)/2$ while for the node at midpoint of the edge $\underline{t}_{ij}$ between $\underline{x}_i$ and $\underline{x}_j$, the basis function is $4\lambda_i\lambda_j$.

  The bubbles of lower degree are on affine elements

$$\begin{cases} b_{3,K} = \lambda_1\lambda_2\lambda_3 & \text{in the two dimensional case ,} \\ b_{4,K} = \lambda_1\lambda_2\lambda_3\lambda_4 & \text{in the three-dimensional case.} \end{cases} \tag{2.2.29}$$

  We shall also define nonconforming bubbles in (2.2.39).
  Barycentric coordinates will also be employed in Sect. 2.6.

2.2.3 Nonconforming Methods

We shall meet later nonconforming methods when studying hybrid finite element methods. In many cases, it will however be more convenient to see them in the framework of *external approximations.*

Let us consider a variational problem (with $f \in V'$),

$$a(u, v) = \langle f, v\rangle_{V'\times V}, \ \forall v \in V, \ u \in V, \tag{2.2.30}$$

where V is some Hilbert space and $a(u, v)$ is a bilinear (coercive) form on $V \times V$.

Suppose that we can find a larger space $S \supset V$ such that there exists a canonical extension $\tilde{a}(\cdot, \cdot)$ of a to $S \times S$, satisfying

$$\tilde{a}(u, v) = a(u, v), \ \forall u, \ v \in V. \tag{2.2.31}$$

Moreover, let $V_h \subset S$ be a family of finite-dimensional subspaces of S such that, given $v_h \in V_h$,

$$v = \lim_{h\to 0} v_h \Rightarrow v \in V. \tag{2.2.32}$$

When (2.2.32) is satisfied, V_h is said to be an external approximation to V. Assuming that f can be extended to an element $\tilde{f}$ in S', we can now approximate problem (2.2.30) by: find $u_h \in V_h$ such that

$$\tilde{a}(u_h, v_h) = \langle \tilde{f}, v_h\rangle_{S'\times S}, \ \forall v_h \in V_h. \tag{2.2.33}$$

Using standard coerciveness and continuity assumptions, one gets from (2.2.30) and (2.2.33) a result known as Strang's lemma [146, 358].

$$\begin{aligned}\|u-u_h\|_S &\leq c \inf_{v_h \in V_h} \|u-v_h\|_S + \sup_{v_h \in V_h} \frac{|\tilde{a}(u,v_h) - \langle \tilde{f}, v_h \rangle|}{\|v_h\|_S} \\ &= c \inf_{v_h \in V_h} \|v-v_h\| + E_h(u,v_h). \end{aligned} \tag{2.2.34}$$

The last term can be seen as a *consistency* term: it measures how well the exact solution satisfies the discrete equation. This term vanishes when $V_h \subset V$ and we get the standard result for the conforming case.

In classical situations we have $V = H^1(\Omega)$ or $V = H^2(\Omega)$ (or one of their subspaces). Introducing a partition of the domain into m sub-domains K_r, and assuming $V = H^1(\Omega)$, we take $S = X(\Omega)$ as defined in Sect. 2.1.3. Any bilinear form of the type

$$\int_\Omega a(x) \,\underline{\text{grad}}\, u \cdot \underline{\text{grad}}\, v \, dx \tag{2.2.35}$$

can immediately be extended to $X(\Omega)$ by writing

$$\tilde{a}(u,v) = \sum_{r=1}^{m} \int_{K_r} a(x) \,\underline{\text{grad}}\, u \cdot \underline{\text{grad}}\, v \, dx. \tag{2.2.36}$$

We now want to find a subspace of $X(\Omega)$ approximating $H^1(\Omega)$ such that error estimates obtained from (2.2.27) are "optimal". Optimality is here relative to the degree of the polynomials from which the approximation is built: we would like to get $O(h^k)$ estimates when using polynomials of degree k. We are thus led to study the second term in the right-hand side of (2.2.34). We shall make this analysis later in the context of hybrid finite element methods; we shall therefore merely state the result which is quite classical [142, 165, 211, 259], which was discovered on empirical grounds, known as the *Céa patch test*: *the moments up to degree $k-1$ of u_h on any interface of the partition must be continuous, that is,*

$$\int_e u_h p_{k-1} \, dx, \ \forall p_{k-1} \in P_{k-1}(e) \tag{2.2.37}$$

is continuous across any interface e between two adjacent elements.

A more general form was given by Lascaux and Lesaint [275]. It states that the consistency term $E_h(u, v_h)$ must vanish whenever $u \in P_r(\Omega)$. For plate problems, this implies a condition similar to (2.2.37) for u_h and its derivatives. To fix ideas we recall a few classical examples.

In conformity to notation (2.2.23), we denote by $\mathcal{L}^{1,NC}(S_k, T_h)$ a nonconforming approximation of $H^1(\Omega)$ built from functions of $S_k(K)$. We shall simplify this notation whenever possible as in the following example.

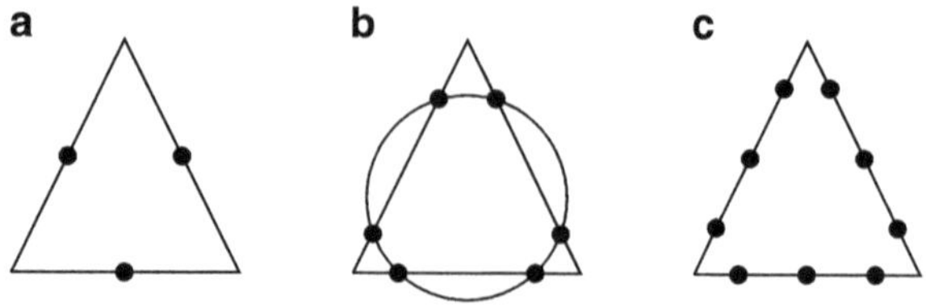

Fig. 2.8 Continuity points for nonconforming elements. **(a)** $k = 1$. **(b)** $k = 2$. **(c)** $k = 3$

Example 2.2.5 (Nonconforming elements on the triangle). Let us consider a partition of Ω into straight-sided triangles and an approximation

$$\mathcal{L}_k^{1,NC} := \{v_h \mid v_h \in L^2(\Omega),\ v_h|_K \in P_k(K),\ \forall K \in \mathcal{T}_h,$$
$$\sum_K \int_{\partial K} v_h \phi \, ds = 0,\ \forall \phi \in R_k(\partial K)\}. \quad (2.2.38)$$

It is then easy to see that the patch test implies that the functions of $\mathcal{L}_k^{1,NC}$ should be continuous at the k Gauss-Legendre points on every side of the triangles (see Fig. 2.8).

For k odd, those points, with internal points for $k \geq 3$, can be used as degrees of freedom, but for k even it is not so and the values at these points are not independent. We shall come back to this point in Sect. 7.1.4, in particular in Example 7.1.4. The trouble is that the six Gaussian points of the $k = 2$ case lie on an ellipse centered at the barycentre. This ellipse is easily expressed, in terms of barycentric coordinates by defining the *nonconforming bubble*, an ellipsoid taking value one at the barycentre and vanishing at the Gaussian points of the edges

$$b_{NC}(K) = 1 - \frac{4}{3}(\lambda_1^2 + \lambda_2^2 + \lambda_3^2). \quad (2.2.39)$$

It was however shown in [209] that this element can nevertheless be used in a very simple way. This was extended to the three-dimensional case in [200]. In this case the nonconforming bubble becomes

$$b_{NC}(K) = 1 - 2(\lambda_1^2 + \lambda_2^2 + \lambda_3^2 + \lambda_4^2). \quad (2.2.40)$$

It must be noted that, in three-dimensional nonconforming elements, the patch test implies, in general, no point continuity. However, there exists a six-point quadrature formula on the triangle, exact for polynomials of degree 3, of which all points lie on the ellipse defined by (2.2.40) and this ensures the patch-test. □

Nonconforming approximations of $H^2(\Omega)$, [147, 148], have been widely used because of the difficulty to obtain conforming elements. We refer the reader to [275] where several examples are given. We shall however use in Chaps. 8 and 10 the following nonconforming approximation of $H^2(\Omega)$.

Example 2.2.6 (Morley's triangle). In plate problems, where an approximation of $H^2(\Omega)$ is needed, an important nonconforming element is Morley's triangle (Fig. 2.9).

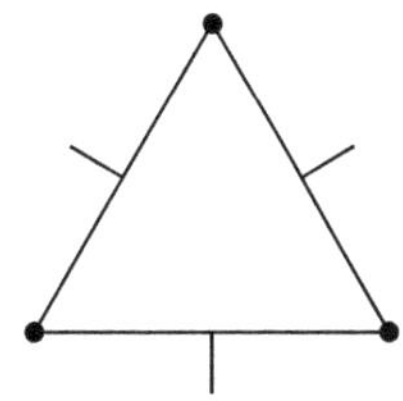

Fig. 2.9 Morley's triangle

The functions v_h are supposed to be in $P_2(K)$ for every K. The degrees of freedom are point values at the vertices of the triangle and normal derivatives at mid-side points. It can be shown by the method of [275] that this provides a consistent approximation that will converge as $O(h)$ in a *discrete* $H^2(\Omega)$-norm. We shall denote by $\mathcal{L}_2^{2,NC}$ the approximation of $H^2(\Omega)$ built from such elements. □

Example 2.2.7 (Nonconforming elements on the rectangle). We consider a partition of Ω into straight-sided quadrilaterals, and an approximation of $H^1(\Omega)$ defined by

$$\mathcal{L}_{[k]}^{1,NC} := \{u_h \mid u_h \in L^2(\Omega),\ u_h|_{K_r} = \hat{u}_h \circ F_r^{-1},\ \hat{u}_h \in Q_k(\hat{K}) \text{ and (2.2.37) holds}\}. \tag{2.2.41}$$

Here again the patch test implies continuity at the Gauss-Legendre points of the interfaces. It is never possible to use these points as degrees of freedom. For $k = 1$, the function $(\hat{x} - \frac{1}{2})(\hat{y} - \frac{1}{2}) \in Q(\hat{K})$ vanishes at the four Gauss-Legendre points of the sides that are indeed midpoints in this case. For $k = 2$, the points lie on an ellipse, and so on. It is however possible to extend the method of [209] to these cases. Another example of a nonconforming quadrilateral element is presented in [330].

□

The above examples are in no way exhaustive: many other nonconforming approximations can be built and some are indeed effectively used [244, 279]. As we shall see in the sequel, nonconforming methods are strongly related, and often equivalent, to hybrid methods or mixed methods. We think it is preferable to delay further examples until they are met in a proper context.

2.2.4 Quadrilateral Finite Elements on Non Affine Meshes

We present here some general results about finite elements on non affine meshes. A mesh $\mathcal{T}_h$ is called *affine* when for all elements $K \in \mathcal{T}_h$ the mapping F_K from the reference element $\hat{K}$ to K is an affine function. If this is the case, the Jacobian matrix DF_K is constant and this property has important consequences for the theoretical analysis.

On the other hand, non affine meshes are met very frequently in real world applications. They may occur in the approximation of general domains, when isoparametric elements are used (see, for instance, Example 2.2.2). Moreover, non affine meshes are generated when quadrilateral or hexahedral elements are used. The fact that the Jacobian matrix DF_K may not be constant not only makes the analysis of non affine elements more difficult, but may also lead to substantial degeneracy of their approximation properties.

The aim of this section is to recall recent results which hold for the quadrilateral finite element approximation of scalar functions (see [20]) and of vector-valued functions in $H(\mathrm{div};\Omega)$ (see [21]).

The possible sub-optimality of some quadrilateral finite elements has been observed by several authors, often as a result of numerical experiments; a non exhaustive list of relevant references is [270, 297, 330, 384, 386].

In order to understand where the trouble come from, let us make some comments on Corollary 2.2.1 in the case of non affine partitions. We consider, for instance, the reference triangle $\hat{K}$ and its image $K = F_K(\hat{K})$. Let us take a smooth function $\hat{v} : \hat{K} \to \mathbb{R}$, the corresponding mapped function $v = \hat{v} \circ F_K^{-1} : K \to \mathbb{R}$, and its linear interpolant $r_h v : K \to \mathbb{R}$. When F_K is affine, then K is a triangle with straight sides, the Jacobian matrix $DF(\hat{x}) = B$ is constant and Corollary 2.2.1 reads

$$|v - r_h v|_{1,K} \le c h_K |v|_{2,K}. \tag{2.2.42}$$

A fundamental ingredient for such an estimate is the scaling (2.1.68), which in this particular situation reads

$$|\hat{v}|_{2,\hat{K}} \le c|\det(B)|^{-1/2}\|B\|^2|v|_{2,K} \le c|\det(B)|^{-1/2}h_K^2|v|_{2,K}. \tag{2.2.43}$$

When the Jacobian matrix is not constant, we cannot use (2.1.68) and the chain rule gives

$$|\hat{v}|_{2,\hat{K}} \le c\left|\inf_{\hat{x}}(J(\hat{x}))\right|^{-1/2}\left(\|DF\|^2_{L^\infty(\hat{K})}|v|_{2,K} + \|D^2F\|_{L^\infty(\hat{K})}|v|_{1,K}\right). \tag{2.2.44}$$

The term $\|DF\|^2_{L^\infty(\hat{K})}$ in the right hand side is typically $O(h_K^2)$, while the norm of the Hessian matrix D^2F might be a lower order term. This fact is clearly a potential source of trouble for optimal order approximation.

Remark 2.2.5. Besides the situations presented in Sects. 2.2.5 and 2.5.5, there are other cases which could be studied but for which the analysis is not yet completed. Three dimensional $H(\mathrm{div},\Omega)$ and $H(\underline{\mathrm{curl}};\Omega)$ approximations on general hexahedral meshes, for instance, do not have a complete analysis yet (see, for instance, [19, 191]), while it is known that standard finite elements are suboptimal in several situations [64, 307]. □

2.2.5 Quadrilateral Approximation of Scalar Functions

Necessary and sufficient conditions for optimal order approximation by quadrilateral finite elements have been investigated in [20]. The theory applies to finite elements defined on a reference square element $\hat{K}$ and mapped to the actual quadrilateral element K by the standard transformation (2.1.59). The generic mapping F_K is bilinear in each component, so that K is a quadrilateral with straight sides.

Before stating the main results, let us recall a measure of shape regularity for quadrilateral meshes.

Definition 2.2.1. Let $\{\mathcal{T}_h\}$ be a family of partitions of *convex* quadrilaterals. For each K, consider the four triangles obtained by the possible choices of three vertices from the vertices of K and denote by ρ_K the smallest diameter of the circles inscribed in the four triangles. Define $\sigma_K = h_K/\rho_K$, where h_K is as usual the diameter of K. The **family of partitions** $\{\mathcal{T}_h\}$ is said **shape-regular** if

$$\sigma_K \leq C \quad \forall K,$$

uniformly in h.

The shape regularity is equivalent to a uniform bound on the ratio of any two sides of the elements and also to a bound away from 0 and π for the element angles.

Given a smooth function $u : \Omega \to \mathbb{R}$ and a finite element space family V_h, we are interested in the following optimal approximation properties (k is the polynomial degree of the reference finite element space and $\underline{\mathrm{grad}}_h$ is the element-by-element gradient):

$$\inf_{v_h \in V_h} \|u - v_h\|_{0,\Omega} = O(h^{k+1}), \tag{2.2.45}$$

$$\inf_{v_h \in V_h} \|\, \underline{\mathrm{grad}}_h (u - v_h)\|_{0,\Omega} = O(h^k). \tag{2.2.46}$$

It is well known (see, for instance, [146, 223]) that a sufficient condition for (2.2.46) to hold is that *the reference finite element space contains* $Q_k(\hat{K})$, the space of polynomials of degree less than or equal to k in each variable, separately. In this case, the following estimates are known to hold

$$\inf_{v_h \in V_h} \|u - v_h\|_{0,\Omega} \leq ch^{k+1} |u|_{k+1,\Omega}, \tag{2.2.47}$$

$$\inf_{v_h \in V_h} \|\, \underline{\mathrm{grad}}_h (u - v_h)\|_{0,\Omega} \leq ch^{k+1} |\, \underline{\mathrm{grad}}\, u|_{k+1,\Omega}. \tag{2.2.48}$$

Indeed, *this is also a necessary condition*. This result has been proved in [20] by exhibiting a very simple (and far from pathological) counterexample: the domain is a square, the mesh sequence, sketched in Fig. 2.10, is made of self similar trapezoids and the function to be approximated is as smooth as possible (a polynomial of degree k).

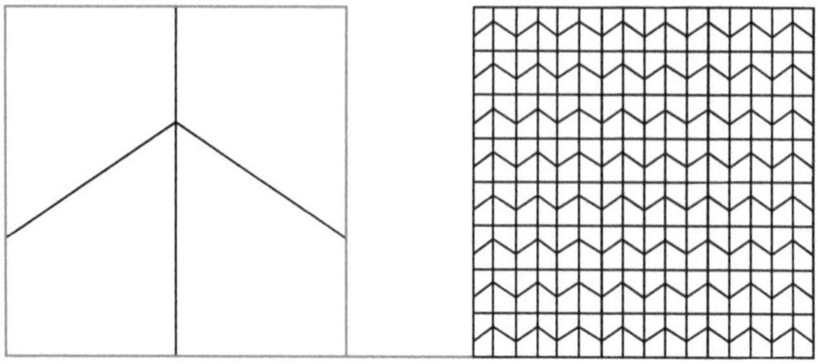

Fig. 2.10 Distorted quadrilateral mesh based on a trapezoid macro-element

Remark 2.2.6. The presented result has important consequences for some commonly used finite element space families. In particular, serendipity elements, obtained from standard Q_k elements by eliminating some internal degrees of freedom, cannot achieve optimal approximation order (2.2.46) on general quadrilateral meshes. □

Remark 2.2.7. The presented results extend to three space dimensions in a straightforward way [296]. □

Remark 2.2.8. The degeneracy of the approximation properties is related to the mesh distortion. In particular, on asymptotically affine partitions, the optimal approximation order is achieved. We refer the interested reader to [20] for the definition of an asymptotically affine mesh and for the proof of this result. □

2.2.6 Non Polynomial Approximations

2.2.6.1 Spaces $\mathcal{L}_k^s(\mathcal{E}_h)$

In the applications involving hybrid methods, it will be useful to consider approximation spaces built from functions that have a polynomial trace on ∂K but which are not necessarily polynomials inside K. These spaces will be useful whenever only the trace is computationally important: they can be thought of as defined only on $\mathcal{E}_h = \bigcup_K \partial K$ (cf. (2.1.42)). We thus define for $s \geq 1$

$$\mathcal{L}_h^s(\mathcal{E}_h) := \{v \mid v \in H^s(\Omega)\ v|_{\partial K} \in T_k(\partial K),\ \forall K\} \tag{2.2.49}$$

and for $s = 0$

$$\mathcal{L}_k^0(\mathcal{E}_h) := \{v \mid v \in L^2(\mathcal{E}_h),\ v|_{\partial K} \in R_k(\partial K),\ \forall K\}. \tag{2.2.50}$$

For $s \geq 1$, functions of $\mathcal{L}_k^s(\mathcal{E}_h)$ are evidently approximations of $H^1(\Omega)$ of optimal order with respect to k. It is also possible to get error estimates on the traces.

Virtual element methods. The spaces $\mathcal{L}_k^s(\mathcal{E}_h)$ can easily be defined on polygons of arbitrary shape, but (as we already pointed out) they can be used only on the boundary of elements, and, even restricted to a single element, are *infinite dimensional*. To reach a finite dimensional version, one has to *prolongate* them inside the elements. This can be done in several ways (see e.g. [367, 377] or [306] and the reference therein).

Recently a variant of these methods has been introduced, called *Virtual Elements*, in which the extension is made as the solution of a partial differential equation, but the construction is such that one does not need to know the solution of this PDE in order to compute the stiffness matrix.

Let us see a simple example, just to give the flavour of the idea. On an element K (a polygon of a practically arbitrary shape), we define $VM_1(K)$ as the space of functions that are linear on each edge of K and harmonic inside. The dimension of such a space is clearly equal to the number of vertices of K. If we want to compute

$$a^K(v, p) := \int_K \underline{\mathrm{grad}}\, v \cdot \underline{\mathrm{grad}}\, p \, dx, \tag{2.2.51}$$

where v is generic in $VM_1(K)$ and p is a polynomial of degree ≤ 1, we have

$$a^K(u, p) \equiv \int_K \underline{\mathrm{grad}}\, v \cdot \underline{\mathrm{grad}}\, p \, dx = \int_{\partial K} v \, \frac{\partial p}{\partial n} \, ds, \tag{2.2.52}$$

which is easily computable from the knowledge of v on ∂K. This allows to compute a projection operator (that we denote by Π_1^∇) from $VM_1(K)$ to $P_1(K)$ by

$$\int_{\partial K} (v - \Pi_1^\nabla v)\, ds = 0 \ \text{ and } \ a^K(v - \Pi_1^\nabla v, q) = 0 \, \forall q \in P_1(K). \tag{2.2.53}$$

Then we can take as *approximate local stiffness matrix* a_h^K the following expression:

$$a_h^K(u, v) := a^K(\Pi_1^\nabla u, \Pi_1^\nabla v) + S^K(u - \Pi_1^\nabla u, v - \Pi_1^\nabla v), \tag{2.2.54}$$

where S^K is any bilinear form acting on the vertex values and scaling like 1 (for instance, for a polygon with, say, five vertices, the usual scalar product in $\mathbb{R}^5$ will do). This will provide an optimal error bound (see [57]).

This can be extended to higher orders to improve the accuracy. Taking $k = 2$, for simplicity, we now define $VM_2(K)$ as the space of functions that are quadratic on each edge of K and whose Laplacian, inside K, is a constant (in general: a polynomial in $P_{k-2}(K)$). The dimension of such a space is clearly equal to twice the number of vertices of K plus one. As degrees of freedom we take the values at the vertices, the values at the midpoints of the edges, and the mean value on K. Now, if we want to compute

$$a^K(v,p) := \int_K \underline{\mathrm{grad}}\, v \cdot \underline{\mathrm{grad}}\, p \, dx, \tag{2.2.55}$$

where v is generic in $VM_2(K)$ and p is a polynomial of degree ≤ 2, we have

$$\int_K \underline{\mathrm{grad}}\, v \cdot \underline{\mathrm{grad}}\, p \, dx = -\int_K v\, \Delta p \, dx + \int_{\partial K} v \frac{\partial p}{\partial n} \, ds, \tag{2.2.56}$$

which is easily computable, as we know v on ∂K and its average on K. This allows to compute a projection operator (that we now denote by Π_2^∇) from $VM_2(K)$ to $P_2(K)$ by

$$\int_K (v - \Pi_2^\nabla v) \, ds = 0 \ \text{ and } \ a^K(v - \Pi_2^\nabla v, q) = 0 \ \forall q \in P_2(K). \tag{2.2.57}$$

Following (2.2.54) we can then take as *approximate local stiffness matrix* a_h^K the following one:

$$a_h^K(u,v) := a^K(\Pi_2^\nabla u, \Pi_2^\nabla v) + S^K(u - \Pi_2^\nabla u, v - \Pi_2^\nabla v), \tag{2.2.58}$$

where again S^K is any bilinear form acting on the vertex and inside values and scaling like 1 (for instance, for a polygon with, say, five vertices, the usual scalar product in $\mathbb{R}^{11}$ will do). We still refer to [57] for more details.

We observe that another possibility would be to consider the space $\widetilde{VM}_1$ of functions that are linear on each edge, whose Laplacian is constant, and such that

$$\int_K (v - \Pi_1^\nabla v) \, dx = 0, \tag{2.2.59}$$

where Π_1^∇ is still defined by (2.2.53) using only the boundary values of v. The advantage of $\widetilde{VM}_1$ over VM_1 is that in $\widetilde{VM}_1$ we can compute exactly the mean value of a function using only its boundary values; this can be useful, for instance, to define the Virtual Element spaces on a polygonal *face* of a polyhedron, in particular, if thereafter we want to use a formula like (2.2.56). For more information in this direction see [4].

2.2.7 Scaling Arguments

We shall briefly recall here the basic idea of the scaling arguments of [180]. We shall do it on a very simple example, but it will be clear how the idea applies to more general cases. Assume that we want to prove the following *inverse inequality* for elements $v_h \in \mathcal{L}_k^s$: there exists a constant c depending only on k and on the minimum angle θ_0 in $\mathcal{T}_h$ such that, on every element K, we have

$$|v_h|_{1,K} \leq c h_K^{-1} \, |v_h|_{0,K}. \tag{2.2.60}$$

We construct first a new element $\hat{K}$ such that the mapping $F: \hat{K} \to K$ is simply given by

$$\underline{x} = h_K \hat{\underline{x}} + \underline{b} \tag{2.2.61}$$

and K has a vertex at the origin. Formulas (2.1.67) and (2.1.68) then simply become (in two dimensions)

$$|\hat{v}|_{m,\hat{K}} = h_K^{m-1} |v|_{m,K} \tag{2.2.62}$$

and we easily get

$$|v_h|_{1,K} = |\hat{v}|_{1,\hat{K}} \leq c(k,\hat{K}) \, |\hat{v}|_{0,\hat{K}} \leq c(k,\hat{K}) \, h_K^{-1} \, |v|_{0,K}. \tag{2.2.63}$$

Now we remark that $c(k,\hat{K})$ actually depends *continuously* on the shape of $\hat{K}$ (a similar argument was already used in [127]). In particular, if one considers the family K_{θ_0} of all the triangles having diameter $= 1$, one vertex at the origin and a minimum angle $\geq \theta_0$, one easily gets

$$\sup_{\hat{K} \in K_{\theta_0}} c(k,\hat{K}) \leq c(k,\theta_0) \tag{2.2.64}$$

by compactness [180]. Hence from (2.2.63) and (2.2.64) we get

$$|v_h|_{1,K} \leq c(k,\theta_0) \, h_K^{-1} \, |v|_{0,K}, \tag{2.2.65}$$

that is (2.2.60).

Note that, in this particular case, it would have been equally easy (or even easier) to derive directly (2.2.60) by using (2.1.67) and (2.1.68) and a fixed $\hat{K}$ = unit triangle. However, (2.2.62) is easier to use and the continuity argument (2.2.64) is always essentially the same in many different applications, so that using the scaling (2.2.61) actually results in a simplification. For instance, one can get by this method the inequality

$$\int_{\partial K} |v_h| \, d\sigma = h_K \int_{\partial \hat{K}} |\hat{v}_h| \, d\hat{\sigma} \leq c(k,\theta_0) \, h_K \, |\hat{v}_h|_{o,\hat{K}} = c(k,\theta_o) \, |v_h|_{0,K}. \tag{2.2.66}$$

In the same way, one can guess, for instance, that one has

$$\|\partial v_h / \partial n\|_{L^\infty(\partial K)} \leq c(k,\theta_0) \, h_k^{-2} \, |v_h|_{0,K}, \tag{2.2.67}$$

because both sides scale like h_K^{-1} in the transformation (2.2.61) and the inequality holds on a fixed element of size = 1. However, note that an inequality of the type

$$\|v_h\|_{L^\infty(\partial K)} \leq c(k, \theta_0)\, |v_h|_{1,K} \tag{2.2.68}$$

is still hopeless (take $v_h = 1$!) unless we specify, for instance, that v_h has zero mean value in K.

2.3 Simplicial Approximations of $H(\text{div}; \Omega)$ and $H(\underline{\text{curl}}; \Omega)$

Although this section and the following one are important by themselves (we shall use $H(\text{div}; \Omega)$ or $H(\underline{\text{curl}}; \Omega)$ in many applications throughout this book), its importance also lies in its value as a model. The techniques introduced for the approximation of $H(\text{div}; \Omega)$ can indeed be applied to other situations and similar constructions have been employed in the discretisation of the Hellan-Hermann-Johnson mixed formulation for which we refer to [318] and [319]. The approximations that we shall present derive from the original work of [366], and [331] later generalized and extended to the three-dimensional case by Nédélec [310]. We shall also use the results of [118, 120], and [311] for the definition of elements that contain (for simplicial elements) the elements of [310] and [331]. In the case of rectangles, we introduce a general element containing the elements of [331], the elements of [120] and the ones of [119], thus clarifying the relation between those two. As the simplicial case is simple and more intuitive, we shall first consider it in detail. Quadrilateral and hexahedral elements will be treated afterwards.

2.3.1 Simplicial Approximations of $H(\text{div}; \Omega)$

In this section, the element K will be either a triangle ($n = 2$) or a tetrahedron ($n = 3$) and we will suppose that we have a mesh $\mathcal{T}_h$ built from such elements. We denote by e_i ($i = 1, 2, 3$ or $i = 1, 2, 3, 4$) the sides (or the faces) of K. We then start from the general space of piecewise polynomial vectors. It will be convenient to denote

$$\underline{P}_k(K) := (P_k(K))^n. \tag{2.3.1}$$

In analogy with (2.2.23), we want to build approximations of $H(\text{div}; \Omega)$ of the form

$$\mathcal{L}^{\text{div}}(\underline{S}_k, \mathcal{T}_h) := \{\underline{p} \in H(\text{div}; \Omega) : \underline{p}|_K \in \underline{S}_k(K)\}, \tag{2.3.2}$$

which will evidently imply continuity of the normal traces. We shall thus proceed to build suitable subspaces of to ensure this continuity. We first define, using (2.2.5)

$$\underline{P}_k^{n,s}(K) := \{\underline{p} \in \underline{P}_k(K) : \underline{p} \cdot \underline{n} \in R_s(\partial K)\} \tag{2.3.3}$$

and

$$\underline{P}^{n,s}_{k+\underline{x}k}(K) := \{\underline{p} \in (\underline{P}_k(K) + \underline{x}P_k(K)) : \underline{p}\cdot\underline{n} \in R_s(\partial K)\}, \tag{2.3.4}$$

where $\underline{x} = (x_1, x_2, \ldots, x_n)$. For these spaces, we shall write (2.3.2) in a compact way:

$$\mathcal{L}^{\mathrm{div}}_{k,s}(\mathcal{T}_h) = \mathcal{L}^{\mathrm{div}}(\underline{P}^{n,s}_k, \mathcal{T}_h) \tag{2.3.5}$$

$$\mathcal{L}^{\mathrm{div}}_{\underline{k}+\underline{x}k,s}(\mathcal{T}_h) = \mathcal{L}^{\mathrm{div}}(\underline{P}^{n,s}_{k+\underline{x}k}, \mathcal{T}_h) \tag{2.3.6}$$

or even $\mathcal{L}^{\mathrm{div}}_{k,s}$ and $\mathcal{L}^{\mathrm{div}}_{\underline{k}+\underline{x}k,s}$ when there will be no ambiguity as to the choice of $\mathcal{T}_h$. The case $s = k$ is the most natural and widely used. Taking $s \le k-1$ defines what we shall call *reduced* spaces. We introduce special names for a few classical cases. For $s = k$, the space $\underline{P}^{n,k}_k(K)$ was introduced in [120] (for $n = 2$) and [118] (for $n = 3$). We shall thus call it the Brezzi-Douglas-Marini space and write,

$$\mathcal{BDM}_k(K) := \underline{P}^{n,k}_k(K). \tag{2.3.7}$$

The dimension of $\mathcal{BDM}_k(K)$ is

$$\dim \mathcal{BDM}_k(K) = \begin{cases} (k+1)+(k+2), & \text{for } n = 2, \\ \frac{1}{2}(k+1)(k+2)(k+3), & \text{for } n = 3. \end{cases} \tag{2.3.8}$$

The space $\underline{P}^{n,k}_{k+\underline{x}k}$ was defined in [310] following [331]. We shall call it the Raviart-Thomas space and write

$$\mathcal{RT}_k(K) := \underline{P}^{n,k}_{k+\underline{x}k}. \tag{2.3.9}$$

Finally the reduced case $\underline{P}^{n,k-1}_k$ was considered in [119] and we shall write

$$\mathcal{BDFM}_k(K) := \underline{P}^{n,k-1}_k(K). \tag{2.3.10}$$

Remark 2.3.1. The original work of [331] used an expression equivalent to (2.3.4) with $s = k$ on the *reference element* $\hat{K}$ and defined $\mathcal{RT}_k(K)$ by the change of variable $\mathcal{G}$ of (2.1.69). It must be noted that this definition is not equivalent to the definition of $\mathcal{RT}_k(K)$ given above: it depends on the orientation of space. For simplicial elements, definition (2.3.4) is more natural and easier to handle. □

For the simplicial case, we thus have the following inclusions between the spaces just defined:

$$\mathcal{RT}_{k-1} \subset \mathcal{BDFM}_k \subset \mathcal{BDM}_k \subset \mathcal{RT}_k \subset \mathcal{BDFM}_{k+1} \subset \mathcal{BDM}_{k+1} \subset \mathcal{RT}_{k+1}. \tag{2.3.11}$$

We now have to define suitable degrees of freedom. For $\underline{q} \in \underline{P}^{n,s}_k(K)$, we evidently have $\mathrm{div}\,\underline{q} \in P_{k-1}(K)$. Moreover, the normal trace $\underline{q}\cdot\underline{n}$ on ∂K belongs to $R_s(\partial K)$. In order to build from $\underline{P}_k^{n,s}(K)$ an approximation of $H(\mathrm{div};\Omega)$, it will be necessary to ensure continuity of $\underline{q}\cdot\underline{n}$ at the interfaces. This will be made possible by the choice of appropriate degrees of freedom.

Proposition 2.3.1. *For $k \geq 1$ and for any $\underline{q} \in \underline{P}_k^{n,s}(K)$, the following relations imply $\underline{q} = 0$*

$$\int_{\partial K} \underline{q}\cdot\underline{n}\; p_s\, ds = 0, \qquad \forall\, p_s \in R_s(\partial K), \tag{2.3.12}$$

$$\int_K \underline{q}\cdot\underline{\mathrm{grad}}\; p_{k-1}\, dx = 0, \qquad p_{k-1} \in P_{k-1}(K), \tag{2.3.13}$$

$$\int_K \underline{q}\cdot\underline{p}_k\, dx = 0, \qquad \forall\, \underline{p}_k \in \underline{\mathbb{H}}_k(K), \tag{2.3.14}$$

where

$$\underline{\mathbb{H}}_k(K) := \{\underline{q}_k \mid \underline{q}_k \in \underline{P}_k(K),\ \mathrm{div}\,\underline{q}_k = 0,\ \underline{q}_k\cdot\underline{n}|_{\partial K} = 0\}. \tag{2.3.15}$$

Indeed, it is easy to check that (2.3.12) and (2.3.13) are equivalent to $\underline{q} \in \underline{\mathbb{H}}_k(K)$ as (2.3.12) implies $\underline{q}_k\cdot\underline{n}|_{\partial K} = 0$. Moreover,

$$\int_K \mathrm{div}\,\underline{q}\; p_{k-1}\, dx = -\int_K \underline{q}\cdot\underline{\mathrm{grad}}\, p_{k-1}\, dx + \int_{\partial K} \underline{q}\cdot\underline{n}\; p_{k-1}\, ds. \tag{2.3.16}$$

Thus (2.3.12) and (2.3.13) imply $\mathrm{div}\,\underline{q} = 0$. Reciprocally, it is trivial that (2.3.12) and (2.3.13) hold for $\underline{q}_k \in \underline{\mathbb{H}}_k(K)$. □

To prove that (2.3.12)–(2.3.14) can be used to define degrees of freedom for $\underline{P}_k^{n,s}(K)$ by choosing bases for $R_s(\partial K)$, $P_{k-1}(K)$, and $\underline{\mathbb{H}}_k(K)$, there remains to check that the set obtained from (2.3.12) and (2.3.13) is linearly independent. This is the object of the next lemma.

Lemma 2.3.1. *Let $g \in R_s(\partial K)$ and $f \in P_{k-1}(K)$ be such that*

$$\int_{\partial K} g\underline{q}\cdot\underline{n}\, d\sigma + \int_K \underline{q}\cdot\underline{\mathrm{grad}}\, f\, dx = 0,\ \forall \underline{q} \in \underline{P}_k^{n,s}(K). \tag{2.3.17}$$

Then, $g = 0$ and f = constant.

Proof. Using the change of variables (2.1.69) and Lemma 2.1.6, it is sufficient to prove the result on the reference element (see Fig. 2.11). We give the construction for $n = 3$ as the case $n = 2$ is a simple restriction of it. One first uses in (2.3.17)

$$q_1 = x\frac{\partial f}{\partial x}\lambda_4,\ q_2 = y\frac{\partial f}{\partial y}\lambda_4,\ q_3 = z\frac{\partial f}{\partial z}\lambda_4, \tag{2.3.18}$$

where λ_4 is the fourth barycentric coordinate, that is $\lambda_4 = 1 - x - y - z$.

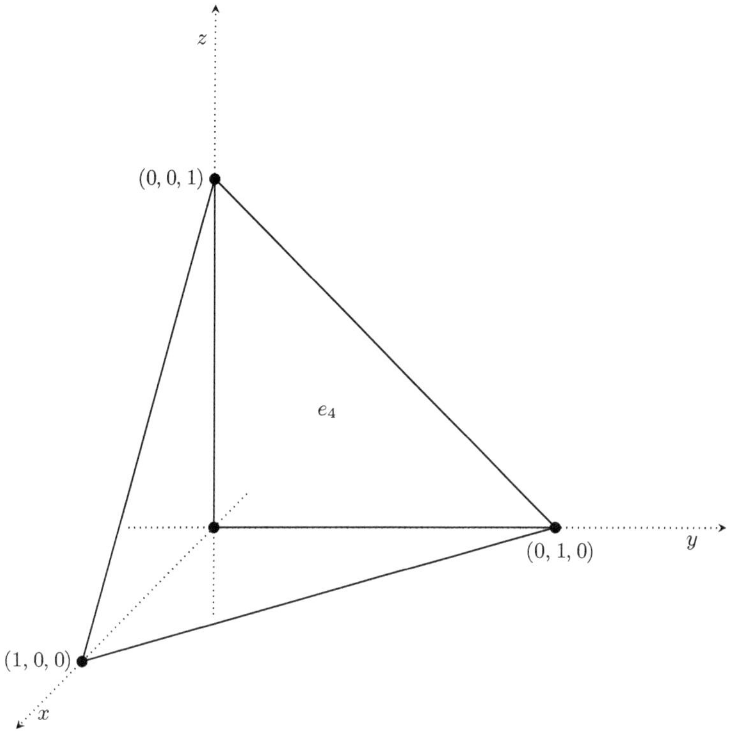

Fig. 2.11 The reference element

Then $\underline{q} \in \underline{P}_k(K)$ and $\underline{q} \cdot \underline{n}|_{\partial K} = 0$ and we get from (2.3.17)

$$\int_K \Big[x\Big(\frac{\partial f}{\partial x}\Big)^2 + y\Big(\frac{\partial f}{\partial y}\Big)^2 + z\Big(\frac{\partial f}{\partial z}\Big)^2\Big]\lambda_4 = 0, \tag{2.3.19}$$

which implies $\underline{\text{grad}}\, f = 0$ since all terms in the integral are positive. We now take $q_1 = xp_{s-1},\ q_2 = q_3 = 0$. From this we obtain $\int_{e_4} xg\ \ p_{s-1}\ \ ds = 0$. In the same way we get $\int_{e_4} yg\ \ p_{s-1}\ \ ds = \int_{e_4} zg\ \ p_{s-1}\ \ ds = 0$ and, as $x + y + z = 1$ on e_4, $\int_{e_4} g\ \ p_{s-1}\ \ ds = 0$. All these conditions imply $g|_{e_4} = 0$. Finally, we take $q_i = g|_{e_i} = g_i$ and (2.3.17) implies $\sum_{i=1}^{3} \int_{e_i} (g_i)^2\, ds = 0$, hence $g = 0$. □

Let us now count the number of conditions thus induced for $\mathcal{BDM}_k(K)$:

$$\dim R_k(\partial K) + \dim P_{k-1}(K) - 1 = \begin{cases} \frac{1}{2}k^2 + 7k + 4 & \text{for } n = 2, \\ \frac{1}{6}k^3 + 15k^2 + 38k + 18 & \text{for } n = 3. \end{cases} \tag{2.3.20}$$

From this we can deduce, by standard arguments of linear algebra,

$$\dim \underline{\mathbb{H}}_k = \begin{cases} \frac{1}{2}k(k-1) = \dim P_{k-2}(K) \text{ for } n=2,\ k \geq 2, \\ \frac{1}{6}3k^3 - 3k - \frac{1}{6}(k-2)(k-1)k = \begin{cases} \dim [(P_{k-2})^3] - \dim (P_{k-3}), \\ \qquad n=3,\ k \geq 3, \\ \dim [(P_{k-2})^3],\ n=3,\ k=2. \end{cases} \end{cases} \tag{2.3.21}$$

In the two-dimensional case, the space $\underline{\mathbb{H}}_k(K)$ can easily be characterized.

Lemma 2.3.2. *For $n = 2$, we have,*

$$\underline{\mathbb{H}}_k(K) = \{\underline{p}_k \mid \underline{p}_k = \underline{\mathrm{curl}}(b_K\, p_{k-2}),\ p_{k-2} \in P_{k-2}(K)\}, \tag{2.3.22}$$

where $b_K = \lambda_1\lambda_2\lambda_3 \in B_3(K)$ is the bubble function on K.

Proof. Any $\underline{p}_k \in \underline{\mathbb{H}}_k(K)$ is the curl of a polynomial of degree $k+1$. A simple count of degrees of freedom concludes the proof. □

In the three-dimensional case, the construction of $\underline{\mathbb{H}}_k(K)$ is less direct. It is still true that $\underline{p}_k \in \underline{\mathbb{H}}_k(K)$ implies that $\underline{p}_k$ is the $\underline{\mathrm{curl}}$ of a vector function polynomial of degree $k+1$. To characterise $\underline{\mathbb{H}}_k(K)$, we need the polynomial spaces that will be introduced in the next section for the approximation of $H(\underline{\mathrm{curl}};\Omega)$.

The next result shows, in particular, that the internal degrees of freedom coming from (2.3.13) and (2.3.14) can be replaced by a term involving the space $\mathcal{N}_{k-2}(K)$ which will be introduced in (2.3.37) for the approximation of $H(\underline{\mathrm{curl}};\Omega)$. The reader is referred to Sect. 2.3.2 for more details.

Proposition 2.3.2. *For $k \geq 1$ and for any $\underline{q} \in \underline{P}_k^{n,s}(K)$, the following relations imply $\underline{q} = 0$,*

$$\int_{\partial K} \underline{q}\cdot\underline{n}\, p_s\, ds = 0, \qquad \forall\, p_s \in R_s(\partial K), \tag{2.3.23}$$

$$\int_K \underline{q}\cdot\underline{w}_{k-2}\, dx = 0, \qquad \forall \underline{w}_{k-2} \in \mathcal{N}_{k-2}(K), \tag{2.3.24}$$

where $\mathcal{N}_{k-2}(K)$ is defined in (2.3.37).

Proof. Space $\mathcal{N}_{k-2}(K)$ contains the gradients of all polynomials of degree $k-1$ so that condition (2.3.24) is stronger than (2.3.13). We thus get $\underline{q} \in \underline{\mathbb{H}}_k(K)$. By Proposition 2.3.4, a polynomial of degree k which is divergence-free is the curl of $\underline{\phi} \in \mathcal{N}_k^+(K)$ and $\underline{q}\cdot\underline{n} = 0$ implies that we must take the tangential trace $\pi_t\underline{\phi}$ to vanish on ∂K. From Proposition 2.3.5 it follows that

$$\int_K \underline{\phi}\cdot\underline{p}_{k-2} dx = 0, \qquad \forall \underline{p}_{k-2} \in \underline{P}_{k-2} \tag{2.3.25}$$

which implies $\underline{\phi} = 0$. On the other hand, (2.3.24) implies

$$\int_K \underline{\mathrm{curl}}\,\underline{\phi}\cdot\underline{w}_{k-2}\,dx = \int_K \underline{\phi}\cdot\underline{\mathrm{curl}}\,\underline{w}_{k-2}\,dx = 0, \qquad \forall \underline{w}_{k-2}\in\mathcal{N}_{k-2}(K) \quad (2.3.26)$$

and $\underline{\mathrm{curl}}\,\mathcal{N}_{k-2}(K)$ spans $\mathcal{BDM}^0_{k-2}(K)$ (see Proposition 2.3.4). From Lemma 2.3.3, the complement of $\mathcal{BDM}^0_{k-2}(K)$ is made of gradients and the result follows. □

One must also say that the degrees of freedom described above have mainly *a theoretical importance*, for instance in building a B-compatible interpolation operator for proving the *inf-sup* condition. In practice, as we shall see in the applications of Chap. 7, any basis of $\underline{P}_k$ will be convenient and standard degrees of freedom can be used.

We can now consider the case of $\mathcal{RT}_k(K)$ as defined in (2.3.9). It can easily be checked that the dimension of $\mathcal{RT}_k(K)$ is given by

$$\dim \mathcal{RT}_k(K) = \begin{cases} (k+1)(k+3) & \text{for } n=2,\\ \frac{1}{2}(k+1)(k+2)(k+4) & \text{for } n=3, \end{cases} \quad (2.3.27)$$

and that only the part of $\underline{x}\,P_k(K)$ involving homogeneous polynomials of degree k is important. We now prove some basic results about $\mathcal{RT}_k(K)$ spaces. These spaces have indeed been tailor designed in order to satisfy the properties which we now state in the following proposition.

Proposition 2.3.3. *For any n-simplicial element K we have for* $\underline{q}\in\mathcal{RT}_k(K)$,

$$\begin{cases} \mathrm{div}\,\underline{q}\in P_k(K),\\ \underline{q}\cdot\underline{n}|_{\partial K}\in R_k(\partial K). \end{cases} \quad (2.3.28)$$

Moreover, the divergence operator is surjective from $\mathcal{RT}_k(K)$ *onto* $P_k(K)$.

Proof. $\underline{q}\in\mathcal{RT}_k(K)$ can be written as $\underline{q} = \underline{q}_0 + \underline{x}\,p_k$ with $\underline{q}_0\in\underline{P}_k(K)$. It is then clear that $\mathrm{div}\,\underline{q}$ is a polynomial of degree k. This proves the results about $\mathrm{div}\,\underline{q}$. On the other hand, let $\underline{n} = \{n_1,n_2\}$ be the normal to a side (we consider the two-dimensional case for simplicity)

$$\underline{q}\cdot\underline{n} = \underline{q_0}\cdot\underline{n} + p_k(x_1n_1 + x_2n_2). \quad (2.3.29)$$

Along a side, $x_1n_1 + x_2n_2$ is constant, so that $\underline{q}\cdot\underline{n}$ is a polynomial of degree k. The same argument works in $\mathbb{R}^n$. To end the proof, we observe that in $\mathbb{R}^n$

$$\int_K \mathrm{div}(\underline{x}\,p)p\,dx = \frac{n}{2}\int_K |p|^2\,dx + \frac{1}{2}\int_{\partial K}(\underline{x}\cdot\underline{n})|p|^2\,ds, \quad (2.3.30)$$

so that $\mathrm{div}(\underline{x}\,p_k) = 0$ implies $p_k = 0$. Hence, $\mathrm{div}(\underline{x}\,P_k)$ has the same dimension as P_k. □

Proposition 2.3.4. *For $k \geq 0$ and for any $\underline{q} \in \mathcal{RT}_k(K)$, the following relations imply $\underline{q} = 0$*

$$\int_{\partial K} \underline{q} \cdot \underline{n}\ p_k\ ds = 0,\ \forall p_k \in R_k(\partial K), \tag{2.3.31}$$

$$\int_k \underline{q} \cdot \underline{p}_{k-1}\ dx = 0,\ \forall \underline{p}_{k-1} \in (P_{k-1}(K))^n. \tag{2.3.32}$$

This is a variant of Proposition 2.3.1 and the proof is left as an exercise.

Let us now define

$$\mathcal{RT}_k^0(K) := \{\underline{q} \mid \underline{q} \in \mathcal{RT}_k(K), \text{div } \underline{q} = 0\}. \tag{2.3.33}$$

We can define in the same way $\mathcal{BDM}_k^0(K)$ and $\mathcal{BDFM}_k^0(K)$. From (2.3.4), we can easily deduce the following result.

Corollary 2.3.1. $\mathcal{RT}_k^0(K) \subset (P_k(K))^n$.

Therefore $\mathcal{RT}_k^0(K) = \mathcal{BDM}_k^0(K)$ while $\mathcal{BDFM}_k^0(K) = \mathcal{RT}_{k+1}^0$ contains the same divergence-free vectors.

Corollary 2.3.2.

- *For $n = 2$, any $\underline{q}_0 \in \mathcal{RT}_k^0(K) = \mathcal{BDM}^0(K)$ is the curl of a stream-function $\psi_{k+1} \in P_{k+1}(K)/\mathbb{R}$.*

  The dimension of $\mathcal{RT}_k^0(K)$ is equal to dim $(P_{k+1}(K) - 1) = \frac{1}{2}(k+1)(k+4)$.
- *For $n = 3$, from Lemma 2.3.4, any $\underline{q}_0 \in \mathcal{RT}_k^0(K)$ is the curl of a vector $\underline{\psi} \in \mathcal{N}_k^+(K) = \mathcal{N}_k(K)/\ \underline{\text{grad}}\ P_{k+1} = \underline{P}_{k+1}/\ \underline{\text{grad}}\ P_{k+2}$.*

Finally, one can obtain a complement of $\mathcal{RT}_k^0(K)$ by the following construction.

Lemma 2.3.3. *Any $\underline{p}_k \in \underline{P}_k$ can be written as the sum*

$$\underline{p}_k = \underline{p}^0 + \underline{\text{grad}}(b_{nc}(K) p_{k-1}) \tag{2.3.34}$$

with $\underline{p}^0 \in \mathcal{BDM}^0(K)$ where the nonconforming bubble is defined by (2.2.39).

Proof. It is clear that $b_{nc}(K) P_{k-1}$ vanishes on the ellipse $b_{nc} = 0$ and hence contains no harmonic functions. □

Before considering the case of $H(\underline{\text{curl}}; \Omega)$ we present a few examples.

Example 2.3.1 (The spaces $\mathcal{RT}_0$, $\mathcal{RT}_1$, $\mathcal{BDM}_1$). From the results above, we know that $\mathcal{RT}_0$ is a space of dimension 3 containing polynomials of the form

$$\begin{cases} q_1(x, y) = a + cx, \\ q_2(x, y) = b + cy. \end{cases} \tag{2.3.35}$$

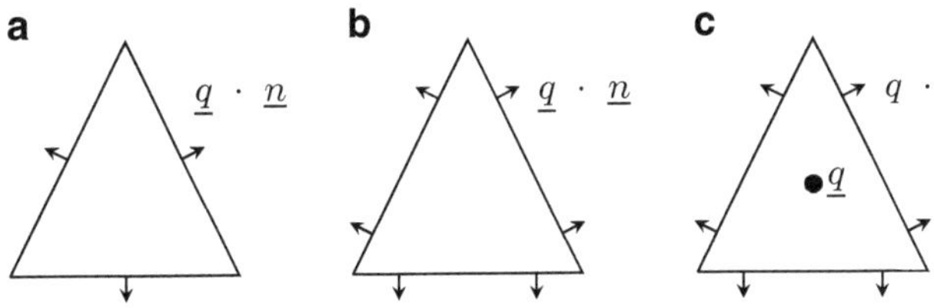

Fig. 2.12 (**a**) RT_0. (**b**) BDM_1. (**c**) RT_1

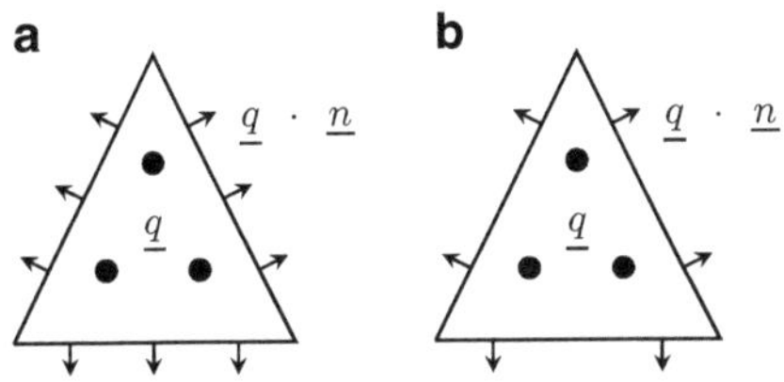

Fig. 2.13 (**a**) BDM_2. (**b**) $BDFM_2$

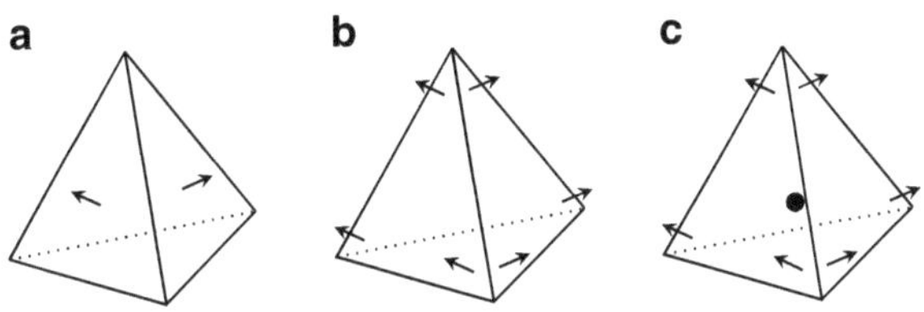

Fig. 2.14 (**a**) RT_0. (**b**) BDM_1. (**c**) RT_1

We can specify it by the three normal components of $\underline{q}$ on ∂K as sketched in Fig. 2.12. Space $\mathcal{BDM}_1$ is of dimension 6 and $\mathcal{RT}_1$ is of dimension 8. It must be noted that div $\mathcal{BDM}_1 = \text{div}\,\mathcal{RT}_0 = P_0$. In the same way, $\mathcal{BDM}_1$ is the subset of $\mathcal{RT}_1$ such that $\text{div}\,\underline{q} \in P_0$ instead of P_1. The same concepts can be extended to $\mathcal{BDM}_2$ and $\mathcal{BDFM}_2$ as shown in Fig. 2.13. □

Example 2.3.2 (Three-dimensional elements: $\mathcal{RT}_0$, $\mathcal{BDM}_1$, $\mathcal{RT}_1$). The simplest cases of three-dimensional elements are depicted in Fig. 2.14. □

Remark 2.3.2 (Two-dimensional approximations of $H(\text{curl};\Omega)$). Due to Remark 2.1.5, in the two dimensional case, approximations of $H(\text{curl};\Omega)$ can be derived directly from the approximations of $H(\text{div};\Omega)$ which we just presented since in two dimensions $H(\text{curl};\Omega)$ is isomorphic to $H(\text{div};\Omega)$ through a rotation by the angle $\pi/2$. Vector fields approximating $H(\text{div};\Omega)$ will provide an approximation of $H(\text{curl};\Omega)$ after a suitable rotation. □

2.3.2 Simplicial Approximation of $H(\underline{\text{curl}}; \Omega)$

As stated in Remark 2.3.2, two dimensional approximations of $H(\underline{\text{curl}}; \Omega)$ can be derived directly from those of $H(\text{div}; \Omega)$. For this reason, we restrict our presentation to the three-dimensional case. In the following, K will therefore be a tetrahedron. The problem of defining a subspace of $\underline{P}_k(K)$ which would lead naturally to the continuity of the tangential trace is more complex than for $H(\text{div}; \Omega)$. To understand this, we may consider the trace π_t defined in (2.1.32). From (2.1.37), this trace is in $H^{-\frac{1}{2}}(\underline{\text{curl}}; \partial K)$ which will imply continuity along the edges of K. This will also imply in the definition of degrees of freedom the appearance of the duality pairing with $H^{-\frac{1}{2}}(\text{div}; \partial K)$. Finite element spaces approximating $H(\underline{\text{curl}}; \Omega)$ have therefore often been referred to as *edge* finite elements, in analogy to *face* finite elements approximating $H(\text{div}; \Omega)$ and to *nodal* finite element approximations of $H^1(\Omega)$ as degrees of freedom will be associated to edges (see, in particular, Example 2.3.3), faces, and vertices, respectively.

Edge elements are also known as Nédélec elements since they have been originally presented in [310, 311]. Important contributions to the analysis can be found in [8, 102, 170, 223], in [248, 302] and in the references therein. In the simplicial case, a comprehensive list can be deduced from the presentation of [33] (see, in particular, Table 5.2) where they have been discussed in a more general setting.

Remark 2.3.3. As we did for approximations of $H(\text{div}, \Omega)$, we shall define, for $\underline{S}_k \subset \underline{P}_k$,

$$\mathcal{L}^{\text{curl}}(\underline{S}_k, \mathcal{T}_h) := \{\underline{p} \mid \underline{p} \in H(\underline{\text{curl}}; \Omega), \quad \underline{p}|_K \in \underline{S}_k(K)\}, \tag{2.3.36}$$

which will imply continuity of the tangential traces as defined in the previous sections. We now proceed to present classical ways of building some suitable $\underline{S}_k$.

□

The most popular choice, which has been introduced by Nédélec [310] and further analysed in [222] is often referred to as first kind Nédélec family.

Let $K \subset \mathbb{R}^3$ be a tetrahedron and let e_i, $i = 1, \ldots, 6$, denote its edges and f_i, $i = 1, \ldots, 4$, its faces. Given an integer $k \geq 0$, we define

$$\mathcal{N}_k(K) := (P_k(K))^3 \oplus [\underline{x} \wedge (P_k^h(K))^3], \tag{2.3.37}$$

where $\underline{x} = (x, y, z)$ and P_k^h denotes the space of homogeneous polynomials of degree k.

It is worth observing that definition (2.3.37) might also be given in the following equivalent way. Let us define first

$$\mathcal{S}^k := \{\underline{\chi} \in (P_k(K))^3 | \underline{\chi} \cdot \underline{x} = 0\}; \tag{2.3.38}$$

then we have $\mathcal{N}_k(K) := (P_k(K))^3 \oplus \mathcal{S}^{k+1}$.

Proposition 2.3.5. *For $k \geq 0$ and for any $\underline{\chi} \in \mathcal{N}_k(K)$, the degrees of freedom of $\mathcal{N}_k(K)$ can be defined by*

$$\int_{e_i} \underline{\chi} \cdot \underline{t}\, p_k \, ds \qquad \forall p_k \in P_k(e_i),\ \forall e_i \tag{2.3.39}$$

$$\int_{f_j} \pi_t \underline{\chi} \cdot \underline{\phi}_{k-1} \, d\sigma \qquad \forall \underline{\phi}_{k-1} \in (P_{k-1}(f_j))^2,\ \forall f_j \tag{2.3.40}$$

$$\int_K \underline{\chi} \cdot \underline{p}_{k-2} \, dx \qquad \forall \underline{p}_{k-2} \in (P_{k-2}(K))^3. \tag{2.3.41}$$

For the proof of the previous proposition, we refer the interested reader to [222]. In particular, let us check that the number of conditions involved with (2.3.39)–(2.3.41) is the same as the dimension of the space $\mathcal{N}_k(K)$. Indeed, (2.3.39) imposes $k+1$ conditions on each edge, (2.3.40) gives $k(k+1)/2$ conditions in each of the two components of the tangential component on each face, and (2.3.41) adds $(k-1)k(k+1)/6$ more conditions in each of the three components of the vector field. As it can be easily seen, the sum of the three contributions is equal to

$$\dim(\mathcal{N}_k(K)) = (k+1)(k+3)(k+4)/2. \tag{2.3.42}$$

Following (2.3.36), in analogy with what has been done for approximations of $H(\text{div};\Omega)$, we define

$$\mathcal{L}^{\text{curl}}_{\underline{k}+\underline{x}\wedge\underline{k}}(\mathcal{T}_h) := \mathcal{L}^{\text{curl}}(\mathcal{N}_k, \mathcal{T}_h). \tag{2.3.43}$$

Remark 2.3.4. It would be possible to introduce, if needed, reduced spaces where the tangential trace is of a lower degree. We can, for example, introduce a reduced space, $\mathcal{N}_k^r(K)$, which is the subspace of $\mathcal{N}_k(K)$ where the degree of the trace on the faces of K is lowered by one. We thus have for $k \geq 1$ the following degrees of freedom

$$\int_{e_i} \underline{\chi} \cdot \underline{t}\, p_{k-1} \, ds \qquad \forall p_{k-1} \in P_{k-1}(e_i),\ \forall e_i \tag{2.3.44}$$

$$\int_{f_j} \pi_t \underline{\chi} \cdot \underline{\phi}_{k-2} \, d\sigma \qquad \forall \underline{\phi}_{k-2} \in (P_{k-2}(f_j))^2,\ \forall f_j \tag{2.3.45}$$

$$\int_K \underline{\chi} \cdot \underline{p}_{k-2} \, dx \qquad \forall \underline{p}_{k-2} \in (P_{k-2}(K))^3. \tag{2.3.46}$$

□

Proposition 2.1.3 states that in order to construct an approximation of $H(\underline{\text{curl}};\Omega)$, given $\underline{\chi} \in \mathcal{N}_k(k)$, we need to ensure continuity of the tangential component $\gamma_t \underline{\chi}$ across elements. The following proposition implies that the degrees of freedom (2.3.39) and (2.3.40) guarantee such continuity.

Proposition 2.3.6. *Assume that* $\underline{\chi} \in \mathcal{N}_k(K)$ *is such that* (2.3.39) *and* (2.3.40) *vanish on a given face* $f \subset K$ *and on all edges* e *of* f. *Then* $\underline{\chi} \wedge \underline{n} = 0$ *on* f. □

Another possible choice for constructing approximations of $H(\underline{\text{curl}}; \Omega)$ has been introduced in [311] and is often referred to as the second kind Nédélec family. On a tetrahedron K, for $k \geq 1$, we consider the full polynomial space

$$\mathcal{NC}_k(K) := \underline{P}_k. \tag{2.3.47}$$

This is the same set of polynomials which we used to define $\mathcal{BDM}_k$. However, we must now build a set of degrees of freedom which would ensure the continuity of the tangential trace. Given $\underline{\chi} \in \mathcal{NC}_k(K)$, we introduce the following moments:

$$\int_{e_i} \underline{\chi} \cdot \underline{t} p_k \, ds, \qquad \forall p_k \in P_k(e_i), \ \forall e_i \tag{2.3.48}$$

$$\int_{f_j} \pi_t \underline{\chi} \cdot \underline{\phi}_{k-2} \, d\sigma, \qquad \forall \underline{\phi}_{k-2} \in \mathcal{RT}_{k-2}(f_j), \ \forall f_j \tag{2.3.49}$$

$$\int_K \underline{\chi} \cdot \underline{q}_{k-3} \, dx, \qquad \forall \underline{q}_{k-3} \in \mathcal{RT}_{k-3}(K). \tag{2.3.50}$$

It is not difficult to check that the total number of degrees of freedom introduced in (2.3.48)–(2.3.50) is equal to

$$\dim(\mathcal{NC}_k(K)) = \frac{(k+1)(k+2)(k+3)}{2}. \tag{2.3.51}$$

Indeed, moments in (2.3.48)–(2.3.50) correspond respectively to $6(k+1)$, $4(k+1)(k-1)$, and $(k+1)(k-1)(k-2)/2$ conditions. Lemma 2.1.9 guarantees that the moments are compatible with the mapping $\mathcal{H}$ and, finally, we refer the interested reader to [311] for the proof of the unisolvence.

We shall denote as in (2.3.43)

$$\mathcal{L}_{\underline{k}}^{\text{curl}}(\mathcal{T}_h) := \mathcal{L}^{\text{curl}}(\mathcal{NC}_k, \mathcal{T}_h). \tag{2.3.52}$$

Remark 2.3.5. We shall say that the vectors in $\mathcal{N}_k(K), \mathcal{N}_{\underline{k}}^r(K)$ or $\mathcal{NC}_k(K)$ for which $\underline{\chi} \wedge \underline{n} = 0$ on all faces of K and which are thus defined by the degrees of freedom (2.3.41), (2.3.46), and (2.3.50) are $H(\text{curl})$-bubbles. It is thus clear that $\mathcal{N}_k(K)$ and $\mathcal{N}_{\underline{k}}^r(K)$ contain exactly the same bubbles while $\mathcal{NC}_k(K)$ has more. □

Remark 2.3.6. Another important property is that the spaces $\mathcal{N}_k(K)$, $\mathcal{N}_{\underline{k}}^r(K)$ and $\mathcal{NC}_k(K)$ are invariant under the action of the covariant transform $\mathcal{H}$ in case of affine mappings (as it can be checked directly). Moreover, from Lemma 2.1.9 if follows that the degrees of freedom introduced, for example, in (2.3.39)–(2.3.41) are compatible with respect to $\mathcal{H}$. □

Our last comment is about the characterization of the kernel of the curl operator in $\mathcal{N}_k(K), \mathcal{N}_k^r(K)$ and $\mathcal{NC}_k(K)$. Using a similar notation to what we had introduced in the previous section, we define

$$\mathcal{N}_k^0(K) := \{\underline{\chi} \in \mathcal{N}_k(K) | \,\underline{\text{curl}}\,\underline{\chi} = 0\}, \tag{2.3.53}$$

$$\mathcal{N}_k^{r0}(K) := \{\underline{\chi} \in \mathcal{N}_k^r(K) | \,\underline{\text{curl}}\,\underline{\chi} = 0\}, \tag{2.3.54}$$

$$\mathcal{NC}_k^0(K) := \{\underline{\chi} \in \mathcal{NC}_k(K) | \,\underline{\text{curl}}\,\underline{\chi} = 0\}. \tag{2.3.55}$$

One easily sees that

$$\mathcal{N}_k^0(K) := \underline{\text{grad}}\, P_{k+1}(K), \tag{2.3.56}$$

$$\mathcal{N}_k^{r0}(K) := \underline{\text{grad}}\, P_k(K) + \underline{\text{grad}}\, B_4(K) P_{k-3}^h(K), \tag{2.3.57}$$

$$\mathcal{NC}_k^0(K) := \underline{\text{grad}}\, P_{k+1}(K), \tag{2.3.58}$$

where $B_4(K)$ is the bubble defined in (2.2.28) and $P_k^h(K)$ denotes, as above, the space of homogeneous polynomials of degree k. Let us define

$$\mathcal{N}_k^+(K) := \mathcal{N}_k(K) / \,\underline{\text{grad}}\, P_{k+1}(K), \tag{2.3.59}$$

$$\mathcal{N}_k^{r+}(K) := \mathcal{N}_k^r / \mathcal{N}_k^{r0}(K), \tag{2.3.60}$$

$$\mathcal{NC}_k^+(K) := \mathcal{NC}_k / \,\underline{\text{grad}}\, P_{k+1}(K). \tag{2.3.61}$$

We then have $\mathcal{N}_k^+(K) = \mathcal{NC}_{k+1}^+(K)$ while $\mathcal{N}_k^{r+}(K)$ is smaller. The important point is that we have

$$\underline{\text{curl}}\,\mathcal{N}_k(K) = \underline{\text{curl}}\,\mathcal{NC}_{k+1}(K) = \underline{\text{curl}}\,\mathcal{N}_k^+(K) = \mathcal{BDM}_k^0(K) = \mathcal{RT}_K^0(K) \tag{2.3.62}$$

$$\underline{\text{curl}}\,\mathcal{N}_k^r(K) = \underline{\text{curl}}\,\mathcal{N}_k^{r+}(K) = \mathcal{BDFM}_k^0(K). \tag{2.3.63}$$

We can, for example, check the following result.

Lemma 2.3.4. *The* $\underline{\text{curl}}$ *operator is surjective from* $\mathcal{N}_k^+(K)$ *onto* $\mathcal{BDM}_k^0(K) = \mathcal{RT}_k^0(K)$.

Proof. A simple count shows that the dimensions of $\mathcal{N}_k^+(K)$ and the dimension of $\mathcal{BDM}_k^0(K)$ are equal. □

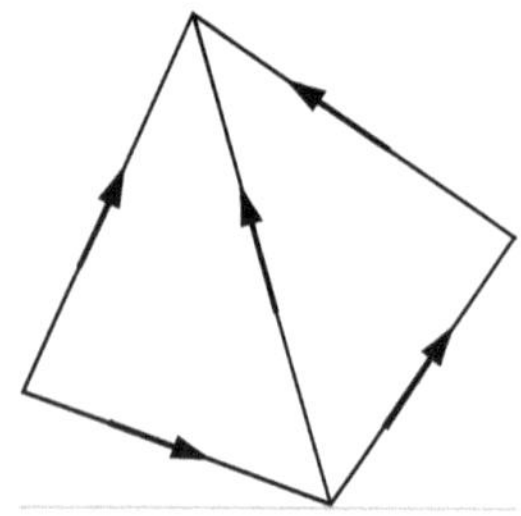

Fig. 2.15 Lowest order edge elements on a tetrahedron

This can be summarised in the exact sequences,

$$\mathcal{L}^1_{k+1} \xrightarrow{\underline{\text{grad}}} \mathcal{N}_k \xrightarrow{\underline{\text{curl}}} \mathcal{BDM}_k \xrightarrow{\text{div}} \mathcal{L}^0_k \tag{2.3.64}$$

$$\mathcal{L}^1_{k+1} \xrightarrow{\underline{\text{grad}}} \mathcal{N}^r_k \xrightarrow{\underline{\text{curl}}} \mathcal{BDFM}_k \xrightarrow{\text{div}} \mathcal{L}^0_k \tag{2.3.65}$$

This relation is part of a more general *commuting diagram property* which will be discussed in Sect. 2.1.4.

Example 2.3.3 (Lowest order edge elements). We conclude this section by explaining in more detail the case $k=0$. This is probably the most used edge finite element and it is also known as the Whitney element, since it has been used by Whitney in a different context [379]. The case $k=0$ is very particular, since the only meaningful degrees of freedom are those presented in (2.3.39) (and this is also a good explanation for the name *edge* elements). The space $\mathcal{N}_0$ is simply given by

$$\begin{aligned} \mathcal{N}_0 :&= (P_0(K))^3 \oplus [(\underline{x}) \wedge (P_0(K))^3] \\ &= \text{span}\left\{ \begin{pmatrix} 1 \\ 0 \\ 0 \end{pmatrix}, \begin{pmatrix} 0 \\ 1 \\ 0 \end{pmatrix}, \begin{pmatrix} 0 \\ 0 \\ 1 \end{pmatrix}, \begin{pmatrix} 0 \\ z \\ -y \end{pmatrix}, \begin{pmatrix} -z \\ 0 \\ x \end{pmatrix}, \begin{pmatrix} y \\ -x \\ 0 \end{pmatrix} \right\}. \end{aligned} \tag{2.3.66}$$

The moments (shown in Fig. 2.15), are given by the six degrees of freedom (2.3.39), that is

$$\int_e \underline{\chi} \cdot \underline{t}\, ds \tag{2.3.67}$$

and it can be checked that the quantity $\underline{\chi} \cdot \underline{t}$ is constant along the edges. □

2.4 Approximations of $H(\text{div}; K)$ on Rectangles and Cubes

We now consider the extension of the previous construction to rectangular elements. The general quadrilateral case is not straightforward; some considerations about it are presented in Sect. 2.2.4. In the present case, the use of a reference element is essential and *we shall build our spaces on* $\hat{K} =]-1,+1[^n$. Contrarily to the simplicial case, it will be simpler here to first introduce the approximations of Raviart and Thomas. The extension to the three-dimensional case can be found in [310].

2.4.1 Raviart-Thomas Elements on Rectangles and Cubes

We first consider a simple extension of the $\mathcal{RT}_k$ approximation introduced above for the simplicial case. Let us define

$$\mathcal{RT}_{[k]} := \begin{cases} P_{k+1,k} \times P_{k,k+1} & \text{for } n = 2, \\ P_{k+1,k,k} \times P_{k,k+1,k} \times P_{k,k,k+1} & \text{for } n = 3. \end{cases} \tag{2.4.1}$$

It is then easy to check that

$$\dim \mathcal{RT}_{[k]} = \begin{cases} 2(k+1)(k+2) & \text{for } n = 2, \\ 3(k+1)^2(k+2) & \text{for } n = 3. \end{cases} \tag{2.4.2}$$

These spaces have been defined in order to have

$$\text{div}\, \underline{q}_k \in Q_k. \tag{2.4.3}$$

Moreover, we have

$$\begin{cases} \underline{q} \cdot \underline{n}_k|_{e_i} \in P_k(e_i) \text{ on the edges for } n = 2, \\ \underline{q} \cdot \underline{n}_k|_{f_i} \in Q_k(e_i) \text{ on the faces for } n = 3. \end{cases} \tag{2.4.4}$$

Defining as in the simplicial case

$$RT^0_{[k]} := \{q \mid \underline{q} \in \mathcal{RT}_{[k]},\ \text{div}\, \underline{q} = 0\}, \tag{2.4.5}$$

we have the following result.

Lemma 2.4.1. *For $n = 2$, if $\underline{q} \in RT^0_{[k]}(\hat{K})$, there exists $\psi \in Q_{k+1}(\hat{K})$ such that $\underline{q} = \underline{curl}\, \psi$. The dimension of $RT^0_{[k]}(\hat{K})$ is $(k+1)(k+3)$.*

In order to choose an approximate set of degrees of freedom, we define

$$\Psi_k(K) := \begin{cases} P_{k-1,k}(K) \times P_{k,k-1}(K) & \text{for } n = 2, \\ P_{k-1,k,k}(K) \times P_{k,k-1,k}(K) \times P_{k,k,k-1}(K) & \text{for } n = 3. \end{cases} \tag{2.4.6}$$

Proposition 2.4.1. *For any $\underline{q} \in \mathcal{RT}_{[k]}(\hat{K})$, the relations*

$$\begin{aligned} \int_{e_i} \phi_i \underline{q} \cdot \underline{n}\, ds = 0 \quad & \forall \phi_i \in Q_k(e_i) \text{ for } n = 3 \\ & \forall \phi_i \in P_k(e_i) \text{ for } n = 2, \\ \int_{\hat{K}} \underline{\phi} \cdot \underline{q}\, dx = 0 \quad & \forall \underline{\phi} \in \Psi_k(\hat{K}) \end{aligned} \tag{2.4.7}$$

imply $\underline{q} = 0$.

For $n = 2$ the proof is analogous to the proof of Proposition 2.3.4. For $n = 3$ see [310]. Note that, for $n = 2$, the sides e_i are one-dimensional, so that actually $Q_k(e_i) = P_k(e_i)$ in (2.4.7).

The $\mathcal{RT}_{[k]}$ spaces just described are based on the idea that a finite element approximation on the rectangle should use a space of type Q_k. This is however by no means necessary in the present case.

2.4.2 Other Approximations of $H(\mathrm{div}; K)$ on Rectangles

In the following, we discuss rectangular finite element approximations of $H(\mathrm{div}; K)$, which are based on P_k polynomial spaces instead of Q_k. The original idea of the construction was introduced in [120] and a suitable modification was presented in [118]. Here we follow such approaches for $n = 2$. For the case when $n = 3$, we use the definitions given in [18], which are more natural and provide spaces which are independent of interchange of coordinate directions.

Let us define following [120] and [118], for $n = 2,\ k \geq 1$,

$$\begin{aligned} \mathcal{BDM}_{[k]} := \{\underline{q} \mid \underline{q} = \underline{p}_k(x, y) + r\, \underline{\mathrm{curl}}(x^{k+1} y) \\ + s\, \underline{\mathrm{curl}}(x y^{k+1}),\ \underline{p}_k \in (P_k)^2\}. \end{aligned} \tag{2.4.8}$$

These spaces have been carefully defined in order to have

$$\begin{cases} \mathrm{div}\, \underline{q} \in P_{k-1}(K), \\ \underline{q} \cdot \underline{n}|_{e_i} \in P_k(e_i). \end{cases} \tag{2.4.9}$$

It must be remarked that these last conditions are rather unusual for a rectangular approximation. We have by a simple count,

$$\dim \mathcal{BDM}_{[k]} = (k + 1)(k + 2) + 2 = k^2 + 3k + 4 \qquad (n = 2). \tag{2.4.10}$$

For the choice of degrees of freedom, we have the following proposition.

Proposition 2.4.2. *For $k \geq 1$, the following conditions imply $\underline{q} = 0$,*

$$\int_{e_i} \underline{q} \cdot \underline{n}\ p_k\, ds,\ \forall p_k \in P_k(e_i), \tag{2.4.11}$$

$$\int_{\hat{K}} \underline{q} \cdot \underline{p}_{k-2}\, dx = 0,\ \forall \underline{p}_{k-2} \in (P_{k-2})^n. \tag{2.4.12}$$

Proof. It is sufficient to prove that (2.4.11) implies $\underline{q} \in (P_k)^2$, that is, all terms introduced through $\underline{\text{curl}}$ vanish. Then, we have that if $\underline{q} \in (P_k)^2$, then $\underline{q}_0 \cdot \underline{n}|_{e_i} = 0$ implies $q_1 = (1 - x^2)p_{1,k-2}$ and $q_2 = (1 - y^2)p_{2,k-2}$, so that (2.4.12) implies $\underline{q} = 0$. Indeed, from (2.4.8) we have

$$q_1 = p_{1,k}(x, y) - rx^{k+1} + s(k + 1)xy^k$$
$$q_2 = p_{2,k}(x, y) + r(k + 1)x^k y - sy^{k+1}. \tag{2.4.13}$$

In order to have $q_1 = 0$ for $x = \pm 1$ and $q_2 = 0$ for $y = \pm 1$, we have that r and s must vanish, so that $\underline{q} \in (P_k)^2$. □

Remark 2.4.1. Definition (2.4.8) has been designed in order to keep div $\underline{q}$ in P_{k-1} by adding divergence-free functions to $(P_k)^n$ while providing terms with a normal component in $P_k(e_i)$ on each side or face e_i. □

We would now like to see what are the relations between $\mathcal{BDM}_{[k]}(K)$ and $\mathcal{RT}_{[k]}(K)$. First, one obviously has $\mathcal{BDM}_{[k]} \subset \mathcal{RT}_{[k]}$. However, the space obtained by restricting the normal component of $\mathcal{BDM}_{[k]}$ to belong to $P_{k-1}(e_i)$ on each side has no direct relation to $\mathcal{RT}_{[k-1]}$ and is a much smaller space (providing an approximation of the same accuracy). In order to get a pattern of inclusions, we define the space

$$S_{[k+1]} := \mathcal{RT}_{[k]} + \{\underline{\text{curl}}\, x^{k+2}y,\ \underline{\text{curl}}\, yx^{k+2}, \underline{\text{curl}}\, x^{k+2}, \underline{\text{curl}}\, y^{k+2}\}. \tag{2.4.14}$$

This space obviously contains $\mathcal{RT}_{[k]}$ but also contains $\mathcal{BDM}_{[k+1]}$.

We can also define the space $\mathcal{BDFM}_{[k+1]}$ by restricting the normal component of $\underline{q} \in \mathcal{BDM}_{[k+1]}$ to belong to $P_k(e_i)$ instead of $P_{k+1}(e_i)$ on each side [119].

It can easily be checked that $\mathcal{BDFM}_{[1]} = \mathcal{RT}_{[0]}$. To make things clear, let us consider a diagram in Fig. 2.16.

We can then summarize the previous facts in Fig. 2.17 in which arrows indexed by b represent a reduction in boundary degrees of freedom and arrows indexed by i represent a reduction in internal degrees of freedom. Space $\mathcal{RT}_{[0]}$ plays a special role in this set of spaces. It is the simplest possible space and it is related to the MAC space [240] that has been extensively used in fluid mechanical computations. It is clear from Fig. 2.17 that both $\mathcal{RT}_{[k]}$ and $\mathcal{BDFM}_{[k+1]}$ are a generalization of

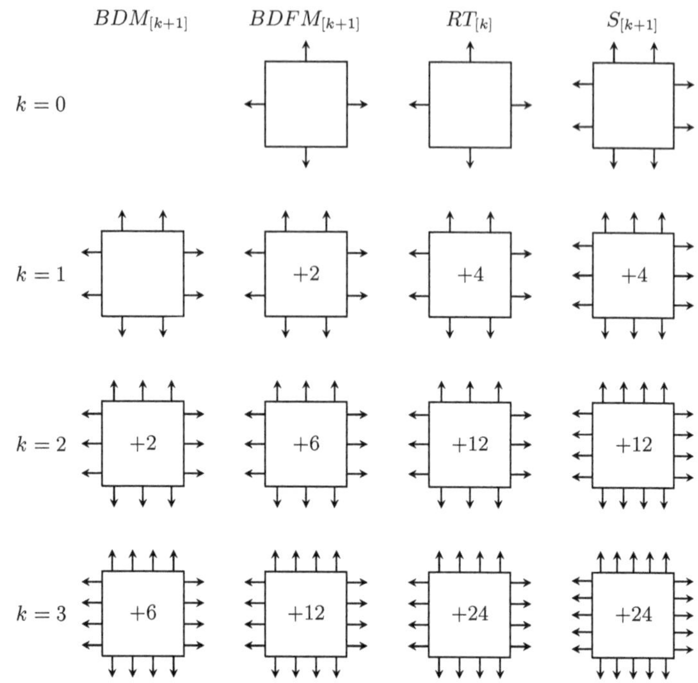

Fig. 2.16 Two dimensional approximations of $H(\text{div}; K)$

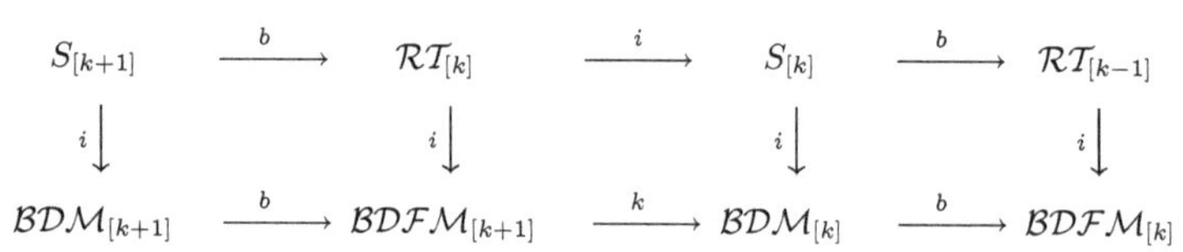

Fig. 2.17 Relations between elements approximating $H(\text{div}; K)$: operators b represent reduction in boundary degrees of freedom and i reduction in internal degrees of freedom

this space with the same order of accuracy. One uses $\mathcal{RT}_{[k]}$ whenever one wants $\text{div}\,\underline{q} \in Q_k$ and $\mathcal{BDFM}_{[k+1]}$ if $\text{div}\,\underline{q} \in P_k$ is sufficient. It is thus worth considering $\mathcal{BDFM}_{[k+1]}$ in more details. It is easy to check that

$$\mathcal{BDFM}_{[k+1]} = (P_{k+1})^2 \backslash \{0, x^{k+1}\} \backslash \{y^{k+1}, 0\}. \tag{2.4.15}$$

This shows that it is natural to move to $\mathcal{BDM}_{[k+1]}$ and get an extra order of accuracy whenever one is ready to pay for extra boundary nodes.

2.4.3 Other Approximations of $H(\text{div}; K)$ on cubes

To make our presentation complete, we now consider the extension of the elements of the previous section to the three-dimensional case. In [18] a general framework for designing finite element spaces on tensor product elements has been introduced. It turns out that this new construction is more natural than the original one. More precisely, we are considering the space denoted $\mathcal{S}_k \Lambda^2$ in this article: it turns out that, while in two-dimensions this space coincides with $\mathcal{BDM}_{[k]}$, in three-dimensions it provides a finite element with the same degrees of freedom as the original $\mathcal{BDM}_{[k]}$ space, but with no arbitrariness in the choice of the shape functions with respect to the order of the variables.

Following [18], we thus define, for $n = 3$, $k \geq 1$,

$$\begin{aligned}\mathcal{BDM}_{[k]} := (P_k(\hat{K}))^3 + \text{span}(\text{curl}\{&yz(w_2(x,z) - w_3(x,y)),\\ &zx(w_3(x,y) - w_1(y,z)),\\ &xy(w_1(y,z) - w_2(x,z))\}),\end{aligned} \tag{2.4.16}$$

where each w_i belongs to P_k. We have

$$\dim \mathcal{BDM}_{[k]}(K) = (k+1)(k^2 + 5k + 12)/2. \tag{2.4.17}$$

The following proposition is the natural extension of Proposition 2.4.2 and has been proved in a more general setting in [18, Theorem 3.6].

Proposition 2.4.3. *For $k \geq 1$, the following conditions imply $\underline{q} = 0$,*

$$\begin{aligned}&\int_{e_i} \underline{q} \cdot \underline{n}\; p_k\, ds, && \forall p_k \in P_k(e_i),\\ &\int_{\hat{K}} \underline{q} \cdot \underline{p}_{k-2}\, dx = 0, && \forall \underline{p}_{k-2} \in (P_{k-2})^n.\end{aligned} \tag{2.4.18}$$

We could also consider the three-dimensional $\mathcal{BDFM}_{[k]}$ case by restricting the degrees of the traces on the boundary. We leave this as an exercise for the reader.

2.4.4 Approximations of $H(\underline{\text{curl}}; K)$ on Cubes

Surprisingly enough, the construction of edge finite elements on cubes is less studied than the corresponding spaces on tetrahedrons. Namely, only one finite element family was basically known to provide a good approximation of problems involving the space $H(\underline{\text{curl}}; \Omega)$ before the recent paper [18]. We shall describe the associated space, also known as the Nédélec first kind space (see [310]). Given a cube K and an integer $k \geq 0$, we introduce the polynomial space

$$\mathcal{N}_{[k]} := P_{k,k+1,k+1}(K) \times P_{k+1,k,k+1}(K) \times P_{k+1,k+1,k}(K) \tag{2.4.19}$$

and the following degrees of freedom

$$\int_e \underline{\chi} \cdot \underline{t} p_k \, ds, \qquad \forall p_k \in P_k(e),\ \forall e \tag{2.4.20}$$

$$\int_f \pi_t \underline{\chi} \cdot \underline{\phi}_{k-1} \, d\sigma, \qquad \forall \underline{\phi}_{k-1} \in \mathcal{RT}_{[k-1]}(f),\ \forall f \tag{2.4.21}$$

$$\int_K \underline{\chi} \cdot \underline{q}_{k-1} \, dx, \qquad \forall \underline{q}_{k-1} \in \mathcal{RT}_{[k-1]}(K). \tag{2.4.22}$$

The number of degrees of freedom introduced in (2.4.20)–(2.4.22) is $12(k+1)$, $12k(k+1)$, and $3k^2(k+1)$, respectively, which sums up to

$$\dim \mathcal{N}_{[k]} = 3(k+1)(k+2)^2. \tag{2.4.23}$$

We refer the interested reader to [310] for the proof of unisolvence.

Remark 2.4.2. Sometimes people refer to second kind Nédélec finite elements on cubes as well. Although such an element has been introduced in [311], it should be noted that it does not seem to be a good choice for the approximation of problems arising from electromagnetism. See, for instance, [163] and [86]. □

A second discretisation of $H(\underline{\text{curl}}; \Omega)$ on cubes comes from the general framework of [18]. More precisely, the space $\mathcal{S}_k(\Lambda^1)$ can be defined as follows:

$$\begin{aligned} \mathcal{S}_k(\Lambda^1)(K) := (P_k(K))^3 + \text{span}(\{& yz(w_2(x,z) - w_3(x,y)), \\ & zx(w_3(x,y) - w_1(y,z)), \\ & xy(w_1(y,z) - w_2(x,z))\} \\ & + \underline{\text{grad}}\, s(x,y,z)), \end{aligned} \tag{2.4.24}$$

where each w_i belongs to P_k and s is a polynomial on K with *superlinear degree* at most $k+1$. The superlinear degree of a polynomial was defined in [18] as the ordinary degree ignoring variables which appear linearly. We have

$$\dim \mathcal{S}_k(\Lambda^1)(K) = (k+1)(k^2+5k+18)/2 \tag{2.4.25}$$

and a set of degrees of freedom is given by

$$\int_e \underline{\chi} \cdot \underline{t} p_k \, ds, \qquad \forall p_k \in P_k(e),\ \forall e \tag{2.4.26}$$

$$\int_f \pi_t \underline{\chi} \cdot \underline{\phi}_{k-2} \, d\sigma, \qquad \forall \underline{\phi}_{k-2} \in P_{k-2}(f),\ \forall f \tag{2.4.27}$$

$$\int_K \underline{\chi} \cdot \underline{q}_{k-4} \, dx, \qquad \forall \underline{q}_{k-4} \in P_{[k-4]}(K). \tag{2.4.28}$$

2.5 Interpolation Operator and Error Estimates

2.5.1 *Approximations of* $H(\mathrm{div}; K)$

Let now $\underline{q}$ be some function of $H(\mathrm{div}; K)$. Using for each of the spaces the degrees of freedom previously described, it is possible to define an interpolation operator $\rho_K \underline{q}$, *provided* $\underline{q}$ *is slightly smoother than merely belonging* to $H(\mathrm{div}; K)$. Indeed the degrees of freedom used always involve the moments of $\underline{q}$ on the faces (or sides) of an element. However, functions $p_k \in R_k(\partial K)$ do not belong to $H^{\frac{1}{2}}(\partial K)$ and it is not possible in general to compute expressions like $\int_{\partial K} \underline{q} \cdot \underline{n}\ p_k\ ds$ since $\underline{q} \cdot \underline{n}$ is only defined in $H^{-\frac{1}{2}}(\partial K)$.

However, it is easy to check that if $\underline{q}$ belongs to the space

$$W(K) := \{\underline{q} \in (L^s(K))^n \mid \mathrm{div}\ \underline{q} \in L^2 \in (\Omega)\} \tag{2.5.1}$$

(for a fixed $s > 2$), then such a construction is possible.

Remark 2.5.1. Readers less familiar with functional analysis will normally wonder why, given a triangle T and a function χ belonging to $H^{-1/2}(\partial T)$, even if we are allowed to take

$$\int_{\partial T} \chi$$

by interpreting it as a duality pairing

$$\int_{\partial T} \chi := \langle \chi, \varphi \rangle \text{ with } \varphi \equiv 1,$$

we cannot take the integral over an edge ℓ of ∂T. The typical answer, at this point, is: "Because the function identically equal to 1 on the whole boundary ∂T belongs to $H^{1/2}(\partial T)$, while the function that is equal to 1 on the edge ℓ and 0 on the rest of ∂T *does not* belong to $H^{1/2}(\partial T)$". The above answer is perfectly correct but, in general, leaves the person who asked the question *totally unhappy*. Let us see, therefore, some example which might help in shedding some more light.

Consider, to start with, the open circle

$$\Omega := \{(x, y)|\ x^2 + y^2 < (1/e)^2\},$$

(where $e = 2,718\ldots$ as usual), and the function

$$u(x, y) := \ln(|\ln(\sqrt{x^2 + y^2})|). \tag{2.5.2}$$

An easy computation (taking the derivatives of u and integrating their square) would show to everybody that $u \in H_0^1(\Omega)$. As a consequence, its restriction to the upper quarter

$$Q := \{(x, y)|\ x > 0,\ y > 0,\ x^2 + y^2 < (1/e)^2\},$$

will belong to $H^1(Q)$ and its trace on ∂Q will belong to $H^{1/2}(\partial Q)$. So far so good. Now we define χ as *the (anticlockwise) tangential derivative* on ∂Q of the trace of u. This will be a distribution over ∂Q, and it will belong to our puzzling space $H^{-1/2}(\partial Q)$. Now we have (finally!) in our hands an element of $H^{-1/2}(\partial Q)$ that is *irregular enough* to show some of the pathologies of the space. Let us see it.

To start with, the action of the distribution χ on a smooth function φ is easily described by

$$\langle \chi, \varphi \rangle := -\int_{\partial Q} u \frac{\partial \varphi}{\partial t} \tag{2.5.3}$$

where t is the anticlockwise tangent direction on ∂Q. Using the expression of u in (2.5.2) and taking the derivative, we easily see that in every open interval $]a, b[\subset]0, 1/e[$ of the x axis we have

$$\chi(x) = -\frac{1}{x \ln x},$$

while in every open interval $]a, b[\subset]0, 1/e[$ of the y axis we have

$$\chi(y) = \frac{1}{y \ln y}$$

and in every open subset of the curved part of ∂Q we have $\chi = 0$ (being the tangential derivative of the zero function).

The first (and now rather easy) fact that we can observe is that both

$$\int_0^{1/e} \chi(x)\mathrm{d}x$$

and

$$\int_0^{1/e} \chi(y)\mathrm{d}y$$

are diverging, so that neither of them can be properly defined. Hence, so to speak, *forget about taking the integral of an element of* $H^{-1/2}(\partial Q)$ *over* **a piece** *of* ∂Q.

One might still wonder, however, why we can take, instead, the integral *on the whole boundary* of the product of χ times a smooth enough function φ (including the function identically equal to 1). To do so, we parametrise the part of the boundary of Q where u does not vanish by a parameter $\sigma \in]-1/e, 1/e[$ such that

$$x(\sigma) = 0, y(\sigma) = |\sigma| \quad \text{if}\, \sigma \leq 0,$$
$$x(\sigma) = \sigma, y(\sigma) = 0 \quad \text{if}\, \sigma \geq 0.$$

Then, taking into account that $u \equiv 0$ on the remaining part of ∂S, (2.5.3) can be written as

$$\langle \psi, \varphi \rangle := \int_{-1/e}^{1/e} \ln(|\ln|\sigma||)\, \frac{\partial \varphi}{\partial \sigma} \mathrm{d}\sigma. \tag{2.5.4}$$

Moreover, integrating by parts, we easily see that in every open interval $]a, b[$ of the x axis included in $]-1/e, 1/e[$ and *not containing* 0 we have

$$\chi(x) = -\frac{1}{\sigma \ln|\sigma|}.$$

Hence (as we can see), it is still forbidden to compute, for instance, the integral

$$\int_0^{1/e} \chi(\sigma)\, \mathrm{d}\sigma = \int_0^{1/e} \frac{1}{\sigma \ln|\sigma|}\, \mathrm{d}\sigma = +\infty.$$

Let us see, however, what happens if we consider the integral

$$\int_{-1/e}^{1/e} \ln(|\ln|\sigma||)\, \frac{\partial \varphi}{\partial \sigma}\, \mathrm{d}\sigma. \tag{2.5.5}$$

As a first step, we introduce the *even* and *odd* parts of φ

$$\varphi_{even} := \frac{\varphi(\sigma) + \varphi(-\sigma)}{2} \qquad \varphi_{odd} := \frac{\varphi(\sigma) - \varphi(-\sigma)}{2}.$$

It is obvious that $\varphi = \varphi_{even} + \varphi_{odd}$, so that now (2.5.5) becomes

$$\int_{-1/e}^{1/e} \ln(|\ln|\sigma||)\, \frac{\partial(\varphi_{even} + \varphi_{odd})}{\partial \sigma}\, \mathrm{d}\sigma =$$
$$\int_{-1/e}^{1/e} \ln(|\ln|\sigma||)\, \frac{\partial \varphi_{even}}{\partial \sigma}\, \mathrm{d}\sigma + \int_{-1/e}^{1/e} \ln(|\ln|\sigma||)\, \frac{\partial \varphi_{odd}}{\partial \sigma}\, \mathrm{d}\sigma.$$

Clearly the derivative of φ_{even} is an odd function, and $\ln(|\ln|\sigma||)$ is even. Hence

$$\int_{-1/e}^{1/e} \ln(|\ln|\sigma||)\, \frac{\partial \varphi_{even}}{\partial \sigma}\, \mathrm{d}\sigma = 0,$$

and we only have to deal with

$$\int_{-1/e}^{1/e} \ln(|\ln|\sigma||)\, \frac{\partial \varphi_{odd}}{\partial \sigma}\, \mathrm{d}\sigma = 2 \int_0^{1/e} \ln(|\ln|\sigma||)\, \frac{\partial \varphi_{odd}}{\partial \sigma}\, \mathrm{d}\sigma. \tag{2.5.6}$$

Now the discussion becomes delicate, as the regularity of φ plays a crucial role. Roughly speaking, if φ is, say, in $H^1(\,]-1/e,1/e[\,)$, then $\varphi_{odd}(0) = 0$ and moreover:

$$|\varphi_{odd}(\sigma)| \le C|\sigma|^{1/2} \qquad \text{for } \sigma \to 0, \qquad \text{with } C = \|\varphi_{odd}\|_{H^1(\,]0,1/e[\,)}. \tag{2.5.7}$$

This allows the integration by parts in (2.5.6), yielding

$$-\int_0^{1/e} \frac{1}{\sigma \ln|\sigma|}\,\varphi_{odd}(\sigma)\,\mathrm{d}\sigma,$$

that can be easily seen to be convergent due to (2.5.7). On the other hand, if φ is simply in $H^{1/2}(\,]-1/e,1/e[\,)$, then φ_{odd} will "vanish at 0" only in a very weak sense, namely

$$\int_0^{1/e} \sigma^{-1}(\varphi_{odd}(\sigma))^2\,\mathrm{d}\sigma < +\infty.$$

That however will be enough to make the integral in (2.5.6) convergent. □

For the convenience of the reader we consider a few spaces introduced in this section and, for each of them, we shall define the corresponding operator ρ_K that we will always assume to be defined in $W(K)$ (see (2.5.1)).

Example 2.5.1 (Interpolation operator for $\underline{P}_k^{n,s}$). $\rho_K\colon W(K) \to \underline{P}_k^{n,s}$ is defined by

$$\begin{cases} \displaystyle\int_{\partial K} (\underline{q} - \rho_K \underline{q}) \cdot \underline{n}\; p_k\, ds = 0, \;\; \forall p_k \in R_k(\partial K), \\[2ex] \displaystyle\int_K (\underline{q} - \rho_K \underline{q}) \cdot \underline{w}_{k-2}\, dx = 0, \;\; \forall w_{k-2} \in \mathcal{N}_{k-2}(K), (k \ge 2). \end{cases} \tag{2.5.8}$$

□

Example 2.5.2 (Interpolation operator for $\mathcal{BDM}_{[k]}(K)$, case $n = 2$, $K =$ unit square). We recall that $\mathcal{BDM}_{[K]}(K) = (P_k(k))^2 \oplus \underline{\mathrm{curl}}(x^{k+1}y) \oplus \underline{\mathrm{curl}}(xy^{k+1})$, $(k \ge 1)$. $\rho_K : W(K) \to \mathcal{BDM}_{[k]}(K)$ is defined by

$$\begin{cases} \displaystyle\int_{\partial K} (\underline{q} - \rho_K \underline{q}) \cdot \underline{n}\; p_k\, d\sigma = 0, \quad \forall p_k \in R_k(\partial K), \\[2ex] \displaystyle\int_K (\underline{q} - \rho_K \underline{q}) \cdot \underline{p}_{k-2}\, dx = 0, \quad \forall \underline{p}_{k-2} \in (P_{k-2}(K))^2, \;(k \ge 2). \end{cases} \tag{2.5.9}$$

□

Example 2.5.3 (Interpolation operator for $\mathcal{RT}_k(K) = \underline{P}^{n,k}_{k+\underline{x}k}(K)$). $\rho_K : W(K) \to \mathcal{RT}_k(K)$ is defined by

$$\begin{cases} \displaystyle\int_{\partial K} (\underline{q} - \rho_K \underline{q}) \cdot \underline{n}\, p_k \, d\sigma = 0, & \forall p_k \in R_k(\partial K), \\ \displaystyle\int_K (\underline{q} - \rho_K \underline{q}) \cdot \underline{p}_{k-1} \, dx = 0, & \forall \underline{p}_{k-1} \in \underline{P}_{k-1}(K). \end{cases} \tag{2.5.10}$$

□

Note that for rectangular elements we used the unit square for K (or the unit cube for $n = 3$). For a general K, the spaces and the interpolation operators ρ_K have to be modified by means of the contravariant mapping $\mathcal{G}$ of (2.1.69). In particular, $\rho_K \underline{q} = \mathcal{G}(\rho_{\hat{K}} \underline{\hat{q}})$ where $\underline{\hat{q}} = \mathcal{G}^{-1}(\underline{q})$ and $\hat{K}$ is the unit square or the unit cube. As we already noted in Sect. 2.2.4, everything works in the case of affine elements while some complications may arise for general elements.

In the following, whenever it may be convenient, we will denote by the symbol $\underline{M}(K)$ anyone of the above approximations of $H(\text{div}; K)$. Since, as we shall see, the accuracy of these approximations in the L^2-norm is particularly relevant, we shall denote by $\underline{M}_k(K)$ anyone of the above spaces such that $\underline{P}_k(K) \subseteq \underline{M}_k(K)$ but $\underline{P}_{k+1}(K) \not\subset \underline{M}_k(K)$. Hence, in the following, $\underline{M}_k(K)$ might be, for example, one the following spaces: $\mathcal{BDM}_k(K)$, $\mathcal{BDM}_{[k]}(K)$, $\mathcal{RT}_k(K)$, $\mathcal{RT}_{[k]}(K)$, $\mathcal{BDFM}_{k+1}(K)$, $\mathcal{BDFM}_{[k+1]}(K)$.

Using Lemmas 2.1.7 and 2.1.8 and usual techniques [146] we have immediately the following result.

Proposition 2.5.1. *Let K be an affine element and ρ_K be the interpolation operator $W(K) \to \underline{M}_k(K)$. There exists a constant c depending only on k and on the shape of K, such that, for $1 \le m \le k+1$, for $s = 0$ or 1 and for any $\underline{q}$ in $(H^m(K))^n$, we have*

$$\|\underline{q} - \rho_K \underline{q}\|_{s,K} \le c h_K^{m-s} \, |\underline{q}|_{m,K}. \tag{2.5.11}$$

□

We now want to analyse the behaviour of the error in $H(\text{div}; K)$. We need to characterize the space of the divergences of the vectors in $\underline{M}_k(K)$. Let

$$D_k(K) := \text{div}(\underline{M}_k(K)). \tag{2.5.12}$$

For affine elements, we have, for example,

$$\text{div}(\mathcal{BDM}_k(K)) = \text{div}(\mathcal{BDM}_{[k]}(K)) \qquad = P_{k-1}(K), \tag{2.5.13}$$

$$\text{div}(\mathcal{BDFM}_{k+1}(K)) = \text{div}(\mathcal{BDFM}_{[k+1]}(K)) = P_k(K), \tag{2.5.14}$$

$$\text{div}(\mathcal{RT}_k(K)) = P_k, \tag{2.5.15}$$

$$\text{div}(\mathcal{RT}_{[k]}(K)) = \mathcal{F}(Q_k(K)), \tag{2.5.16}$$

where the definition of $\mathcal{F}$ is given immediately after (2.1.59) (note that Q_k is not invariant under affine transformations). The following result is of paramount importance in the study of these approximations.

Proposition 2.5.2. *Let K be an affine element and ρ_K the interpolation operator: $W(K) \rightarrow \underline{M}_k(K)$. Let moreover π_K be the L^2-projection on $D_k(K) = \mathrm{div}(\underline{M}_k(K))$. Then we have, for all $\underline{q} \in W(K)$,*

$$\mathrm{div}(\rho_K \underline{q}) = \pi_K \ \mathrm{div}\ \underline{q}. \tag{2.5.17}$$

Proof. Since $\mathrm{div}_K\, \underline{q} \in D_k(K)$ by definition, we only have to prove that

$$\int_K v\ \mathrm{div}(\rho_K \underline{q})\, dx = \int_K v\ \mathrm{div}\, \underline{q}\, dx,\ \forall v \in D_k(K). \tag{2.5.18}$$

Indeed,

$$\begin{aligned}\int_K v(\mathrm{div}\, \rho_K \underline{q} - \mathrm{div}\, \underline{q})\, dx &= \int_K (\underline{q} - \rho_K \underline{q}) \cdot \underline{\mathrm{grad}}\, v\, dx \\ &\quad - \int_K (\underline{q} - \rho_K \underline{q}) \cdot \underline{n}\ v\, dx,\end{aligned} \tag{2.5.19}$$

and it is easy to check that, for all the possible choices of ρ_K, the right-hand side of (2.5.19) vanishes. □

Remark 2.5.2. The statement of Proposition 2.5.2 can also be expressed as

$$\begin{array}{ccc} W(K) & \xrightarrow{\mathrm{div}} & L^2(K) \\ \rho_K \big\downarrow & & \pi_K \big\downarrow \\ \underline{M}_k(K) & \xrightarrow{\mathrm{div}} & D_k(K) \end{array} \tag{2.5.20}$$

and is often called the "commuting diagram property" (see [177, 178]). We shall comment more on this property in Sect. 2.5.6. □

From Proposition 2.5.2, using Lemmas 2.1.7 and 2.1.8 and usual techniques, we easily have the following result.

Proposition 2.5.3. *Let K be an affine element and ρ_K the interpolation operator: $W(K) \rightarrow \underline{M}_k(K)$. There exists a constant c depending only on k and on the shape of K such that, for $1 \le m \le \phi(\underline{M}_k)$, we have*

$$\| \mathrm{div}(\underline{q} - \rho_K \underline{q})\|_{0,K} \le c h_K^m\ |\mathrm{div}\, \underline{q}|_{m,K}, \tag{2.5.21}$$

where $\phi(\underline{M}_k) = k$ for $\mathcal{BDM}_k(K)$ or $\mathcal{BDM}_{[k]}$ and $\phi(\underline{M}_k) = k+1$ for the other choices.

Remark 2.5.3. Proposition 2.5.3 shows that choosing $\mathcal{RT}_{[k]}$, $\mathcal{BDFM}_{k+1}$ or $\mathcal{BDFM}_{[k+1]}$ leads to the same accuracy in $H(\mathrm{div}; K)$ as we have in $(L^2(K))^n$. This is not the case for $\mathcal{BDM}_k$ or $\mathcal{BDM}_{[k]}$ where the accuracy in $H(\mathrm{div}; K)$

is of one order less than the accuracy in $(L^2(K))^n$. However, as we shall see in Chap. 7, the commuting diagram property is so strong that this drawback can be circumvented. □

Remark 2.5.4. For non-affine elements, the situation is more complicated. In particular, we have now to define $D_k(K)$ and $\mathcal{F}(D_k(\hat{K}))$, where $\hat{K}$ is the reference element and $\mathcal{F}$ is defined in (2.1.59). On the other hand, div($\underline{M}_k(K)$) will be $\mathcal{F}(J^{-1}\operatorname{div}\underline{M}_k(\hat{K}))$. Hence it is clear that Proposition 2.5.2 will not hold any more. Moreover, Proposition 2.5.3 does not hold (at least for $\mathcal{RT}_{[k]}$-elements; see again [366]). More comments on these issues will be given in Sect. 2.5.5. □

2.5.2 *Approximation Spaces for* $H(\mathbf{div};\Omega)$

It is clear that the spaces defined in the previous sections can be used to define internal approximations of $H(\operatorname{div};\Omega)$. The choice of degrees of freedom has obviously been made in order to ensure continuity of $\underline{q}\cdot\underline{n}$ at interfaces of elements. We can then define, for each choice of $\underline{M}_k(K)$, the space

$$\underline{M}_k(\Omega,\mathcal{T}_h) := \{\underline{q} \mid \underline{q} \in H(\operatorname{div};\Omega),\ \underline{q}|_K \in \underline{M}_k(K)\}. \tag{2.5.22}$$

In a similar manner we have, in agreement with the notation (2.2.23),

$$\mathcal{L}^0(D_k,\mathcal{T}_h) := \{v \mid v \in L^2(\Omega),\ v|_K \in D_k(K)\}. \tag{2.5.23}$$

It is clear that for affine elements

$$\operatorname{div}\underline{M}_k(\Omega,\mathcal{T}_h) \subset \mathcal{L}^0(D_k,\mathcal{T}_h). \tag{2.5.24}$$

Moreover, we can now define a *global* interpolation operator from

$$W := H(\operatorname{div};\Omega) \cap L^s(\Omega)^n \tag{2.5.25}$$

(for a fixed $s > 2$) into $\underline{M}_k(\Omega;\mathcal{T}_h)$ by simply setting

$$\Pi_h\underline{q}|_K = \rho_K(\underline{q}|_K). \tag{2.5.26}$$

By defining $P_h :=$ projection on $\mathcal{L}^0(D_k,\mathcal{T}_h)$ we have therefore the following commuting diagram:

$$\begin{array}{ccc} W & \xrightarrow{\operatorname{div}} & L^2(\Omega) \\ {\scriptstyle\Pi_h}\big\downarrow & & {\scriptstyle P_h}\big\downarrow \\ \underline{M}_k(\Omega,\mathcal{T}_h) & \xrightarrow{\operatorname{div}} & \mathcal{L}^0(D_k,\mathcal{T}_h) \end{array} \tag{2.5.27}$$

This will imply in particular that

$$\operatorname{div} \underline{M}_k(\Omega; \mathcal{T}_h) = \mathcal{L}^0(D_k, \mathcal{T}_h). \tag{2.5.28}$$

Finally we have from Propositions 2.5.1 and 2.5.3 the following estimates for the interpolation operator Π_h.

Proposition 2.5.4. *Let $\mathcal{T}_h$ be a regular family of decompositions of Ω and let Π_h be defined as in* (2.5.26)*. Then there exists a constant c independent of h such that*

$$\|\underline{q} - \Pi_h \underline{q}\|_{0,\Omega} \le c h^m \, |\underline{q}|_{m,\Omega} \tag{2.5.29}$$

for $1 \le m \le k+1$. Moreover,

$$\| \operatorname{div}(\underline{q} - \Pi_h \underline{q})\|_{0,\Omega} \le c h^s \, | \operatorname{div} \underline{q}|_{s,\Omega}, \tag{2.5.30}$$

where $s \le k$ for $\mathcal{BDM}_k$ or $\mathcal{BDM}_{[k]}$ and $s \le k+1$ for the other choices of $\underline{M}_k$.

□

2.5.3 Approximations of $H(\underline{\mathrm{curl}}; \Omega)$

Now, we show how to use the definitions given in the previous subsection in order to construct conforming approximations of $H(\underline{\mathrm{curl}}; \Omega)$. First of all, we need to define an interpolation operator. We start with the description of the first case we discussed, namely the tetrahedral space $\mathcal{N}_k$.

We follow the theory developed in [8]; for the error estimate we refer to [248] and to the improved modification proposed in [85].

The main question for the definition of the interpolant concerns the regularity assumptions on the function to be interpolated. We have already seen that $H^1(\Omega)$ regularity does not allow for the existence of a *nodal* interpolant (since point-wise values are not defined in $H^1(\Omega)$) and that $H(\mathrm{div}; \Omega)$ regularity does not guarantee the existence of a *face* interpolant (essentially because it is not possible to evaluate the integral of $\underline{q} \cdot \underline{n}$ on a single face of K if $\underline{q} \cdot \underline{n}$ belongs only to $H^{-1/2}(\partial K)$). The case of the *edge* interpolant is more tricky than the previous ones since, according to the theory developed in the previous subsection, we have two families of degrees of freedom associated with the boundary of K: degrees of freedom defined in (2.3.39) (corresponding to the edges of K) and the ones defined in (2.3.40) (corresponding to the faces of K). In [8] it has been proved that if $\underline{\chi}$ belongs to the following space

$$X(K) := \{\underline{\chi} \mid \underline{\chi} \in (L^s(K))^3, \underline{\mathrm{curl}}\, \underline{\chi} \in (L^s(K))^3,\ (\underline{\chi} \wedge \underline{n})_{|\partial K} \in (L^s(K))^2\} \tag{2.5.31}$$

(for a fixed $s > 2$), then the moments defined in Proposition 2.3.5 make sense.

We are then in the position of defining the interpolation operator for the edge element spaces introduced in the previous subsection.

Case 2.5.1 (Case $n = 2$). As it has been explained at the beginning of Sect. 2.5.3 (see also Remark 2.1.5), two dimensional approximations of $H(\mathrm{curl};\Omega)$ can be obtained from corresponding approximations of $H(\mathrm{div};\Omega)$ through a rotation by a right angle.

Case 2.5.2 (Case $n = 3$, tetrahedral elements).

(i) $\mathcal{N}_k(K) := (P_k(K))^3 \oplus [\underline{x} \wedge (\tilde{P}_k(K))^3]$.
 $\sigma_K : X(K) \to \mathcal{N}_k(K)$ is defined by

$$\begin{cases} \displaystyle\int_e (\underline{\chi} - \sigma_K \underline{\chi}) \cdot \underline{t}\, p_k \, ds = 0 & \forall p_k \in P_k(e), \ \forall e \\ \displaystyle\int_f (\pi_t \underline{\chi} - \pi_t \sigma_K \underline{\chi}) \cdot \underline{\phi}_{k-1} \, d\sigma = 0 & \forall \underline{\phi}_{k-1} \in (P_{k-1}(f))^2, \ \forall f \\ \displaystyle\int_K (\underline{\chi} - \sigma_K \underline{\chi}) \cdot \underline{p}_{k-2} \, dx = 0 & \forall \underline{p}_{k-2} \in (P_{k-2}(K))^3. \end{cases} \tag{2.5.32}$$

(ii) $\mathcal{NC}_k(K) := (P_k(K))^3$.
 $\sigma_K : X(K) \to \mathcal{NC}_k(K)$ is defined by

$$\begin{cases} \displaystyle\int_e (\underline{\chi} - \sigma_K \underline{\chi}) \cdot \underline{t}\, p_k \, ds = 0 & \forall p_k \in P_k(e), \ \forall e \\ \displaystyle\int_f (\pi_t \underline{\chi} - \pi_t \sigma_K \underline{\chi}) \cdot \underline{\phi}_{k-1} \, d\sigma = 0 & \forall \underline{\phi}_{k-1} \in \mathcal{RT}_{k-1}(f), \ \forall f \\ \displaystyle\int_K (\underline{\chi} - \sigma_K \underline{\chi}) \cdot \underline{p}_{k-2} \, dx = 0 & \forall \underline{p}_{k-2} \in \mathcal{RT}_{k-2}(K). \end{cases} \tag{2.5.33}$$

Case 2.5.3 (Case $n = 3$, cubic elements).

(i) $\mathcal{N}_{[k]}(K) := P_{k,k+1,k+1}(K) \times P_{k+1,k,k+1}(K) \times P_{k+1,k+1,k}(K)$.
 $\sigma_K : X(K) \to \mathcal{N}_{[k]}(K)$ is defined by

$$\begin{cases} \displaystyle\int_e (\underline{\chi} - \sigma_K \underline{\chi}) \cdot \underline{t}\, p_k \, ds, & \forall p_k \in P_k(e), \ \forall e \\ \displaystyle\int_f (\pi_t \underline{\chi} - \pi_t \sigma_K \underline{\chi}) \cdot \underline{\phi}_{k-1} \, d\sigma, & \forall \underline{\phi}_{k-1} \in \mathcal{RT}_{[k-1]}(f), \ \forall f \\ \displaystyle\int_K (\underline{\chi} - \sigma_K \underline{\chi}) \cdot \underline{q}_{k-1} \, dx, & \forall \underline{q}_{k-1} \in \mathcal{RT}_{[k-1]}(K). \end{cases} \tag{2.5.34}$$

We now conclude this section with error estimates, in analogy to what has been done for $H(\mathrm{div};\Omega)$ at the end of Sect. 2.5. Let us denote by $N(K)$ any of the elements presented so far. More precisely, let $N_k(K)$ denote an approximation of order k of $H(\underline{\mathrm{curl}};\Omega)$ (i.e., $\mathcal{N}_k(K)$, $\mathcal{NC}_k(K)$, or $\mathcal{N}_{[k]}(K)$).

An important difference between the error estimates for edge elements with respect to the ones presented in Sect. 2.5 for face elements is that, in general, stronger regularity assumptions are required. The following result has been proved in [85].

Proposition 2.5.5. *Let K be an affine element and σ_K be the interpolation operator $X(K) \to N_k(K)$. Then there exists a constant c depending only on k and on the shape of K, such that, for $1 < m \leq k+1$, for any $\underline{\chi} \in (H^m(K))^3$, we have*

$$\|\underline{\chi} - \sigma_K \underline{\chi}\|_{0,K} \leq c h_K^m |\underline{\chi}|_{m,K}. \tag{2.5.35}$$

Moreover, for $1/2 < s \leq 1$, we have ($p > 2$)

$$\|\underline{\chi} - \sigma_K \underline{\chi}\|_{0,K} \leq c h_K^s (|\underline{\chi}|_{s,K} + \|\,\underline{\mathrm{curl}}\ \underline{\chi}\|_{L^p(K)}). \tag{2.5.36}$$

□

Let us now characterize the space

$$E_k(K) := \underline{\mathrm{curl}}(N_k(K)). \tag{2.5.37}$$

Taking into account (2.1.92), we have

$$\underline{\mathrm{curl}}(\mathcal{N}) \subset P_k(K), \tag{2.5.38}$$

$$\underline{\mathrm{curl}}(\mathcal{NC}) \subset P_{k-1}(K), \tag{2.5.39}$$

$$\underline{\mathrm{curl}}(\mathcal{N}_{[k]}) \subset \mathcal{G}(\mathcal{RT}_{[k]}(K)), \tag{2.5.40}$$

with $\mathcal{G}$ defined in (2.1.69). On the other hand, arguing as in Lemma III.5.11 of [223] it is possible to show that

$$\underline{\mathrm{curl}}(\mathcal{N}) = P_k(K) \cap \underline{\mathrm{curl}}(H(\underline{\mathrm{curl}}; K)), \tag{2.5.41}$$

$$\underline{\mathrm{curl}}(\mathcal{NC}) = P_{k-1}(K) \cap \underline{\mathrm{curl}}(H(\underline{\mathrm{curl}}; K)), \tag{2.5.42}$$

$$\underline{\mathrm{curl}}(\mathcal{N}_{[k]}) = \mathcal{G}(\mathcal{RT}_{[k]}(K) \cap \underline{\mathrm{curl}}(H(\underline{\mathrm{curl}}; K)). \tag{2.5.43}$$

Arguing as in Propositions 2.5.2 and 2.5.3, we can get an estimate for the interpolation error in the $H(\underline{\mathrm{curl}}; \Omega)$ norm. Indeed, the following *commuting diagram* holds true

$$\begin{array}{ccc} X(K) & \xrightarrow{\underline{\mathrm{curl}}} & W(K) \\ {\scriptstyle \sigma_K}\big\downarrow & & {\scriptstyle \rho_K}\big\downarrow \\ N_k(K) & \xrightarrow{\underline{\mathrm{curl}}} & E_k(K) \end{array} \tag{2.5.44}$$

from which the next result can be deduced.

Proposition 2.5.6. *Let K be an affine element and σ_K the interpolation operator: $X(K) \to N_k(K)$. Then there exists a constant c depending only on k and on the shape of K such that*

$$\| \underline{\mathrm{curl}}(\underline{\chi} - \sigma_K \underline{\chi}) \|_{0,K} \leq c h_K^m | \underline{\mathrm{curl}}\, \underline{\chi} |_{m,K} \tag{2.5.45}$$

for $1 \leq m \leq \phi_N(k)$, where $\phi_N(k) = k + 1$ for $\mathcal{N}$ or $\mathcal{N}_{[k]}$ and $\phi_N(k) = k$ for $\mathcal{NC}$. □

2.5.4 Approximation Spaces for $H(\underline{curl}; \Omega)$

Let Ω be a domain in $\mathbb{R}^3$ (the two-dimensional case can be dealt with according to Remark 2.1.5).

The spaces introduced in the previous sections can be used to define internal approximations of $H(\underline{\mathrm{curl}}; \Omega)$. The choice of degrees of freedom has been made in such a way that the continuity of the trace γ_t is enforced across the inter-element boundaries, which is the natural condition for being conforming in $H(\underline{\mathrm{curl}}; \Omega)$, according to Proposition 2.1.3.

In analogy to what has been done in Sect. 2.5.2, we can define, for each choice of $N_k(K)$, the spaces

$$N_k(\Omega, \mathcal{T}_h) := \{\underline{\chi} \mid \underline{\chi} \in H(\underline{\mathrm{curl}}; \Omega),\ \underline{\chi}|_K \in N_k(K)\} \tag{2.5.46}$$

and

$$\mathcal{L}^0(E_k, \mathcal{T}_h) := \{\underline{q} \mid \underline{q} \in (L^2(\Omega))^3,\ \underline{q}|_K \in E_k(K)\}. \tag{2.5.47}$$

Given $\varepsilon > 0$, let us consider the space

$$X := H(\underline{\mathrm{curl}}; \Omega) \cap (H^{1/2+\varepsilon}(\Omega))^3, \tag{2.5.48}$$

so that the *global* interpolation operator

$$\Sigma_h \underline{\chi}|_K := \sigma(\underline{\chi}|_K) \tag{2.5.49}$$

can be defined (see (2.5.31)).

Given the space $W_0 := \underline{\mathrm{curl}}(X)$, then we have the following commuting diagram

$$\begin{array}{ccc} X & \xrightarrow{\underline{\mathrm{curl}}} & W_0 \\ {\scriptstyle \Sigma_h}\downarrow & & {\scriptstyle \Pi_h}\downarrow \\ N_k(\Omega, \mathcal{T}_h) & \xrightarrow{\underline{\mathrm{curl}}} & \mathcal{L}^0(E_k, \mathcal{T}_h) \end{array} \tag{2.5.50}$$

where Π_h is the face interpolant defined in (2.5.26).

The final error estimates are summarized in the following result.

Proposition 2.5.7. *Let $\mathcal{T}_h$ be a regular family of decompositions of Ω and let Σ_h be defined as in* (2.5.49)*. Then there exists a constant c independent of h such that, for* $1 < m \leq k+1$,

$$\|\underline{\chi} - \Sigma_h \underline{\chi}\|_{0,\Omega} \leq ch^m |\underline{\chi}|_{m,\Omega}. \tag{2.5.51}$$

Moreover, for $1/2 < s \leq 1$ *and* $p > 2$,

$$\|\underline{\chi} - \Sigma_h \underline{\chi}\|_{0,\Omega} \leq ch^s \left(|\underline{\chi}|_{s,\Omega} + \| \underline{\operatorname{curl}}\, \underline{\chi}\|_{L^p(\Omega)} \right). \tag{2.5.52}$$

Finally,

$$\| \underline{\operatorname{curl}}(\underline{\chi} - \Sigma_h \underline{\chi})\|_{0,\Omega} \leq ch^t \| \underline{\operatorname{curl}}\, \underline{\chi}\|_{t,\Omega}, \tag{2.5.53}$$

where $t \leq k+1$ *for* $\mathcal{N}_k$ *or* $\mathcal{N}_{[k]}$ *and* $t \leq k$ *for* $\mathcal{NC}_k$. □

2.5.5 Quadrilateral and Hexahedral Approximation of Vector-Valued Functions in $H(\mathrm{div};\Omega)$ and $H(\underline{\mathrm{curl}};\Omega)$

The results of Sect. 2.2.4 extend to vector valued functions. We refer the interested reader to [21], where necessary and sufficient conditions have been presented for optimal order approximations of functions in $H(\mathrm{div};\Omega)$ in two dimensions. The general situation is more complicated and only partial results exists so far. The most general result can be found in [19] where sufficient conditions are investigated and sharper results are shown in [191] for particular three-dimensional situations.

The theory of [21] applies to the approximation by means of vector-valued discrete functions defined on a reference square element $\hat{K}$ and mapped to the actual quadrilateral element K via the Piola transformation (2.1.69).

Given a smooth function $\underline{q} : \Omega \to \mathbb{R}^2$, we say that a finite element space family $\{X_h\}$ is optimally convergent in $L^2(\Omega)$ if

$$\inf_{\underline{p_h} \in X_h} \|\underline{q} - \underline{p_h}\|_{0,\Omega} = O(h^{k+1}), \tag{2.5.54}$$

where k refers to the polynomial degree of the reference finite element space. Similarly, the finite element space family $\{X_h\}$ is optimally convergent in the $H(\mathrm{div};\Omega)$ semi-norm if

$$\inf_{\underline{p_h} \in X_h} \| \mathrm{div}_h(\underline{q} - \underline{p_h})\|_{0,\Omega} = O(h^{k+1}), \tag{2.5.55}$$

where div_h is the divergence operator evaluated element by element.

Let $\widetilde{\mathcal{RT}}_{[k]}(\hat{K})$ be the subspace of co-dimension one of $\mathcal{RT}_{[k]}(\hat{K})$, where the two highest order fields $(\hat{x}^{k+1}\hat{y}^k, 0)$ and $(0, \hat{x}^k\hat{y}^{k+1})$ are replaced by the single field $(\hat{x}^{k+1}\hat{y}^k, -\hat{x}^k\hat{y}^{k+1})$. Then, for shape-regular families of quadrilateral meshes, condition (2.5.54) is valid if and only if the reference space contains $\widetilde{\mathcal{RT}}_{[k]}(\hat{K})$. Moreover, if the reference space contains $\widetilde{\mathcal{RT}}_{[k]}(\hat{K})$, then the following estimate holds true:

$$\inf_{\underline{p}_h \in X_h} \|\underline{q} - \underline{p}_h\|_{0,\Omega} \le ch^{k+1} |\underline{q}|_{k+1,\Omega}. \tag{2.5.56}$$

Finally, let $\tilde{Q}_k(\hat{K})$ be the subspace of co-dimension one of $Q_k(\hat{K})$ obtained by eliminating the higher order term $\hat{x}^k \hat{y}^k$. Then, condition (2.5.55) holds true if and only if the divergence of the reference space contains $\tilde{Q}_{k+1}(\hat{K})$. In this case the following estimate holds true:

$$\inf_{\underline{p}_h \in X_h} \| \operatorname{div}(\underline{q} - \underline{p}_h)\|_{0,\Omega} \le ch^{k+1} |\operatorname{div} \underline{q}|_{k+1,\Omega}. \tag{2.5.57}$$

Remark 2.5.5. The reported results have dramatic consequences for the finite elements presented in Sect. 2.4. In particular, it turns out that none of the presented finite element families achieve optimal convergence in $H(\text{div};\Omega)$ on general quadrilateral meshes. Actually, $\mathcal{RT}_{[k]}$ is optimal in $L^2(\Omega)$ (since, of course, $\mathcal{RT}_{[k]}(\hat{K})$ contains $\widetilde{\mathcal{RT}}_{[k]}(\hat{K})$), but not in $H(\text{div};\Omega)$ (in particular, there is no convergence of the divergence for $k = 0$); while $\mathcal{BDM}_{[k]}$ and $\mathcal{BDFM}_{[k]}$ are suboptimal both in $L^2(\Omega)$ and $H(\text{div};\Omega)$. □

A possible cure to the pathologies outlined in Remark 2.5.5 has been presented in [21] where the family of spaces $\mathcal{ABF}$ is introduced. The basic idea is to add $H(\text{div})$-conforming bubbles to the $\mathcal{RT}$ spaces so that optimal convergence properties can be achieved.

The results of this section, as it has been already observed, apply to finite element spaces which are defined on the reference element and mapped to the actual element by means of standard transformations. The suboptimal approximation orders have been observed in practical computations (see [20–22, 91]).

However, other finite element definitions are possible for which the (negative) results of this section might not apply. An example of such definitions is the non-conforming quadrilateral element presented in [330] which is constructed locally on the physical element or the reduced integration technique (interpreted as a local projection technique) presented in [94], where optimal convergence in $H(\text{div};\Omega)$ for the $\mathcal{RT}_{[k]}$ family is recovered.

2.5.6 Discrete Exact Sequences

We have introduced in Sect. 2.1.4 the exact sequence (2.1.104). We now show how this translates to some of the finite element approximations introduced above. Let us make a particular choice of finite elements approximating the functional spaces involved with (2.1.104). We consider a simplicial decomposition of a simply connected domain Ω in $\mathbb{R}^3$ and use the following finite elements: we take

$\mathcal{L}^1_{k+1}$ (i.e., standard continuous piecewise polynomials of degree $k+1$) for the approximation of $H^1(\Omega)$, first kind Nédélec elements $\mathcal{N}_k$ for the approximation of $H(\underline{\text{curl}};\Omega)$, Raviart–Thomas elements $\mathcal{RT}_k$ for the approximation of $H(\text{div};\Omega)$, and discontinuous elements $\mathcal{L}^0_k$ for the approximation of $L^2(\Omega)$. Moreover, we consider the three sets of interpolation operators onto these spaces r_h, Σ_h, Π_h, respectively, and the L^2 projection P_h. With these particular choices, de Rham complex reads as follows:

$$\begin{array}{ccccccc}
C^\infty(\Omega) & \xrightarrow{\underline{\text{grad}}} & (C^\infty(\Omega))^3 & \xrightarrow{\underline{\text{curl}}} & (C^\infty(\Omega))^3 & \xrightarrow{\text{div}} & C^\infty(\Omega) \\
{\scriptstyle r_h}\downarrow & & {\scriptstyle \Sigma_h}\downarrow & & {\scriptstyle \Pi_h}\downarrow & & {\scriptstyle P_h}\downarrow \\
\mathcal{L}^1_{k+1} & \xrightarrow{\underline{\text{grad}}} & \mathcal{N}_k & \xrightarrow{\underline{\text{curl}}} & \mathcal{RT}_k & \xrightarrow{\text{div}} & \mathcal{L}^0_k
\end{array} \tag{2.5.58}$$

Diagram (2.5.58) has to be understood in the sense that the two lines are exact and the entire diagram commutes. Thus, for instance, we have that $\underline{\text{grad}}\, r_h v = \Sigma_h \,\underline{\text{grad}}\, v$ or $\Pi_h \,\underline{\text{curl}}\, \underline{\chi} = \underline{\text{curl}}\, \Sigma_h \underline{\chi}$. Some of these properties have been already recalled in this chapter (see, in particular, (2.5.27) and (2.5.50)) when analysing the finite element spaces. We refer the interested reader to [18,33] where the general results are stated and where it has been shown that several other finite element choices are possible for the diagram to commute. We could indeed consider, for example, instead of (2.5.58),

$$\begin{array}{ccccccc}
C^\infty(\Omega) & \xrightarrow{\underline{\text{grad}}} & (C^\infty(\Omega))^3 & \xrightarrow{\underline{\text{curl}}} & (C^\infty(\Omega))^3 & \xrightarrow{\text{div}} & C^\infty(\Omega) \\
{\scriptstyle r_h}\downarrow & & {\scriptstyle \Sigma_h}\downarrow & & {\scriptstyle \Pi_h}\downarrow & & {\scriptstyle P_h}\downarrow \\
\mathcal{L}^1_{k+2} & \xrightarrow{\underline{\text{grad}}} & \mathcal{NC}_{k+1} & \xrightarrow{\underline{\text{curl}}} & \mathcal{BDM}_{k+1} & \xrightarrow{\text{div}} & \mathcal{L}^0_k
\end{array} \tag{2.5.59}$$

Remark 2.5.6. The information contained in Sect. 2.1.4 and in this section are by far not exhaustive of the connections between exterior calculus and finite element analysis. This active research area, not only proves useful for the analysis and the understanding of existing finite elements, but is also of fundamental importance for the design of new schemes. The reader who needs an introduction into this fascinating field is referred to the seminal papers [33, 34]. A more recent and succinct overview of finite element spaces constructed in the language of differential forms can be found in [12].

2.6 Explicit Basis Functions for $H(\text{div}; K)$ and $H(\underline{\text{curl}}; K)$ on Triangles and Tetrahedra

Although it is obviously possible to make explicit the previously defined spaces on a reference element and to transfer them on an arbitrary element by the Piola transformation (2.1.69) or (2.1.82), it is worth mentioning that there exist, on simplicial elements, general formulas using barycentric coordinates which are therefore totally general. We shall first introduce some notation.

- In the two-dimensional case, $\underline{x}_i, \underline{x}_j, \underline{x}_k$ are the three vertices of a triangle and $\lambda_i, \lambda_j, \lambda_k$ the associated barycentric coordinates.
- In the three-dimensional case, $\underline{x}_i, \underline{x}_j, \underline{x}_k, \underline{x}_l$ are the four vertices of a tetrahedron and $\lambda_i, \lambda_j, \lambda_k, \lambda_l$ the associated barycentric coordinates.
- We shall denote by $\underline{t}_{ij} = \underline{x}_j - \underline{x}_i$ the edge connecting $\underline{x}_i$ and $\underline{x}_j$ and similarly for other pairs of indices. We shall write l_{ij} the length of this vector.
- In the three-dimensional case, we denote f_{ijk} the face of the tetrahedron defined by the vertices $\underline{x}_i$, $\underline{x}_j$, $\underline{x}_k$ and similarly for other indices.
- The height of the triangle from $\underline{x}_k$ to the opposite edge $\underline{t}_{ij}$ will be denoted by h_k.
- The height of the tetrahedron from $\underline{x}_l$ to the opposite face f_{ijk} will be denoted by h_l.
- In a triangle, the outward normal to an edge t_{ij} is denoted by $\underline{n}_{ij}$. In a tetrahedron the outward normal to a face f_{ijk} is denoted by $\underline{n}_{ijk}$.

The gradient of the barycentric coordinates is related to the normals. In the two-dimensional case we have

$$\underline{\text{grad}}\,\lambda_k = -\frac{1}{h_k}\underline{n}_{ij} \tag{2.6.1}$$

and the similar three-dimensional formula is

$$\underline{\text{grad}}\,\lambda_l = -\frac{1}{h_l}\underline{n}_{ijk}. \tag{2.6.2}$$

When building basis functions, we shall distinguish between those associated with edge or face degrees of freedom and 'bubble' basis functions associated to degrees of freedom internal to K. We shall make explicit the lower order cases.

2.6.1 Basis Functions for $H(\text{div}; K)$: The Two-Dimensional Case

A basis for $\mathcal{BDM}_1(K)$ (6 degrees of freedom). Let us define for an edge $\underline{t}_{ij}$, $\underline{x}_k$ being the opposite vertex,

$$\phi_{i,ij} := \frac{1}{h_k}\underline{t}_{ki}\lambda_i \tag{2.6.3}$$

and

$$\phi_{j,ij} := \frac{1}{h_k}\underline{t}_{kj}\lambda_j. \tag{2.6.4}$$

Let n_{ij} be the normal to $\underline{t}_{ij}$. We have

$$\phi_{i,ij} \cdot n_{ij} = \lambda_i \tag{2.6.5}$$

$$\phi_{j,ij} \cdot n_{ij} = \lambda_j. \tag{2.6.6}$$

Using all possible indices, we get six such functions, two for each edge. From (2.6.5) and (2.6.6), it is clear that we have thus obtained a basis for $\mathcal{BDM}_1(K)$. □

A basis for $\mathcal{RT}_0(K)$ (3 degrees of freedom). Summing (2.6.5) and (2.6.6), we obtain

$$\phi_{i,ij} \cdot n_{ij} + \phi_{j,ij} \cdot n_{ij} = 1. \tag{2.6.7}$$

Defining

$$\phi_{ij} := \phi_{i,ij} + \phi_{j,ij} \tag{2.6.8}$$

we therefore have a basis function for $\mathcal{RT}_0(K)$, associated with the edge $\underline{t}_{ij}$. It is easy to see that we have

$$\phi_{ij} = \frac{1}{h_k}(\underline{x} - \underline{x}_k). \tag{2.6.9}$$

which is a more standard way of writing this basis. □

A basis for $\mathcal{BDM}_2(K)$ (12 degrees of freedom). To make this construction, we need to increase the order of polynomials on the edges. We can indeed define on each edge,

$$\underline{t}_{m,ij} := (\underline{t}_{ki} + \underline{t}_{kj})/2 \tag{2.6.10}$$

and

$$\phi_{m,ij} := \frac{1}{h_k}\underline{t}_{m,ij}\lambda_i\lambda_j. \tag{2.6.11}$$

To get $\mathcal{BDM}_2(K)$, we add these three functions to those previously obtained in (2.6.3) and (2.6.4). This yields 3 degrees of freedom per edge.

To complete the construction, we also need to consider 'bubble' functions in the sense of $H(\text{div}; K)$. This means that their normal component must vanish on ∂K. They can be built in an easy way from the quadratic functions

$$b_{ij} := \underline{t}_{ij}\lambda_i\lambda_j. \tag{2.6.12}$$

We have three such expressions, one associated with each edge. With the previously defined 9, we obtain a basis for $\mathcal{BDFM}_2(K)$. □

A basis for $\mathcal{BDFM}_1(K)$ (9 degrees of freedom) and $\mathcal{RT}_1(K)$ (8 degrees of freedom). A basis for $\mathcal{BDFM}_1(K)$ is readily obtained by suppressing the function defined in (2.6.10) from the previous construction. To get a basis for $\mathcal{RT}_1(K)$ we have to suppress a combination of the bubbles. This is allowed because there exists such a combination which is divergence-free and which, in a sense, does not contribute. However, this construction shows that, in a way, the slightly richer space $\mathcal{BDFM}_1(K)$ is respecting better the symmetries of the triangle. □

A basis for $\mathcal{BDFM}_3(K)$ (20 degrees of freedom). We follow the same lines. We can build 4 degrees of freedom on each edge, using for example (2.6.3) and (2.6.4), and

$$\phi_{\frac{1}{3},ij} = \frac{1}{h_k}\left(\frac{1}{3}\underline{t}_{ki} + \frac{2}{3}\underline{t}_{kj}\right), \tag{2.6.13}$$

$$\phi_{\frac{2}{3},ij} = \frac{1}{h_k}\left(\frac{2}{3}\underline{t}_{ki} + \frac{1}{3}\underline{t}_{kj}\right). \tag{2.6.14}$$

Bubbles can be generated from (2.6.12) using the following expression

$$b_{\alpha_1,\alpha_2,\alpha_3,ij} := \underline{t}_{ij}\lambda_i^{\alpha_i}\lambda_j^{\alpha_j}\lambda_k^{\alpha_k}, \; (\alpha_i + \alpha_j + \alpha_k) = 3. \tag{2.6.15}$$

We remark that $b_{1,2,0,ij} + b_{2,1,0,ij} = b_{ij}$ so that we still have the bubbles of $\mathcal{BDM}_1(K)$. Moreover, we note that the three functions of the form $b_{1,1,1,ij}$ are not linearly independent as we have

$$\underline{t}_{ij} + \underline{t}_{jk} + \underline{t}_{ki} = 0. \tag{2.6.16}$$

We should then select two linear combinations of these three functions to get a basis. □

A basis for $\mathcal{BDFM}_2(K)$ (17 degrees of freedom) and $\mathcal{RT}_2(K)$ (15 degrees of freedom). To obtain $\mathcal{BDFM}_2(K)$, the easiest way is to consider on the edges the basis functions for $\mathcal{BDM}_2(K)$ and the eight bubbles of $\mathcal{BDM}_3(K)$. For $\mathcal{RT}_2(K)$, we must again eliminate the divergence-free combination of bubbles. □

2.6.2 *Basis Functions for $H(\mathrm{div};K)$: The Three-Dimensional Case*

We can now consider the three-dimensional case.

A basis for $\mathcal{BDM}_1(K)$ (12 degrees of freedom) and $\mathcal{RT}_0(K)$ (4 degrees of freedom). On the face f_{ijk} we have three basis functions

$$\phi_{i,ijk} := \frac{1}{h_l}\underline{t}_{li}\,\lambda_i \tag{2.6.17}$$

$$\phi_{j,ijk} := \frac{1}{h_l}\underline{t}_{lj}\,\lambda_j \tag{2.6.18}$$

$$\phi_{k,ijk} := \frac{1}{h_l}\underline{t}_{lk}\,\lambda_k. \tag{2.6.19}$$

Considering all four faces, we have obtained a basis for $\mathcal{BDM}_1(K)$. It is interesting to note that it employs the same edge-based basis functions as in the two-dimensional case. The sum

$$\phi_{ijk} := \phi_{i,ijk} + \phi_{j,ijk} + \phi_{k,ijk} = \frac{1}{h_l}(\underline{x} - \underline{x}_l) \tag{2.6.20}$$

is a basis function for $\mathcal{RT}_0(K)$. □

A basis for $\mathcal{BDFM}_1(K)$ (18 degrees of freedom) and $\mathcal{RT}_1(K)$ (4 degrees of freedom). As in the two-dimensional case, we can associate a 'bubble', function to each edge, using exactly the same formula (2.6.12) as in the two-dimensional case. We now have six such bubbles which we can use to build $\mathcal{BDFM}_1(K)$ by adding them to the basis of $\mathcal{BDM}_1(K)$. There exist three linear combinations of these bubbles which are divergence free and which can be eliminated to obtain $\mathcal{RT}_1(K)$. □

The reader should have now understood the mechanism and be able to move to higher degrees.

2.6.3 Basis Functions for $H(curl; K)$: The Two-Dimensional Case

As we have noted previously, in the two-dimensional case, the space $H(\underline{\text{curl}}; K)$ is essentially the same as $H(\text{div}; K)$. A basis for the discrete spaces can readily be obtained by a rotation of the vectors. However, we can write the basic construction in a slightly different way that will be more suitable for the extension to the three-dimensional case. Instead of (2.6.3) and (2.6.4), let us define for edge $\underline{t}_{ij}$

$$\psi_{i,ij} := l_{ij}\,\lambda_i\ \underline{\text{grad}}\,\lambda_j \tag{2.6.21}$$

and

$$\psi_{j,ij} := l_{ij}\,\lambda_j\ \underline{\text{grad}}\,\lambda_i. \tag{2.6.22}$$

Given that $\underline{\text{grad}}\,\lambda_i$ is a vector orthogonal to $\underline{t}_{kj}$ of length $1/h_i$ it is easy to check that those definitions correspond exactly to what we obtain when replacing in (2.6.3) and (2.6.4) the vectors $\underline{t}_{ki}$ and $\underline{t}_{kj}$ by their orthogonal. It is also easy to see that the tangential components of $\psi_{i,ij}$ and $\psi_{j,ij}$ along $\underline{t}_{ij}$ are respectively λ_i and $-\lambda_j$.

2.6.4 Basis Functions for $H(\underline{curl}; K)$: The Three-Dimensional Case

A basis for $\mathcal{NC}_1(K)$ (12 degrees of freedom) and $\mathcal{N}_0(K)$ (6 degrees of freedom). In order to define the lowest degree spaces, we define 12 basis functions associated to edges. The interesting fact is that we can still use on any edge $\underline{t}_{ij}$ the expressions

(2.6.21) and (2.6.22). On the other edges, the vectors obtained in this way are either zero because, for example, λ_i is zero or orthogonal to the edge. It is thus clear that we have obtained a basis for $\mathcal{NC}_1(K)$. Summing the two expressions (up to the orientation), we get the basis for $\mathcal{N}_0(K)$

$$\psi_{ij} := l_{ij}\,(\lambda_j \;\underline{\text{grad}}\,\lambda_i - \lambda_i \;\underline{\text{grad}}\,\lambda_j). \tag{2.6.23}$$

□

We can move to higher degree elements by using similar constructions.

***H*(c<u>url</u>; *K*)-Bubbles.** We shall be interested in the $H(\underline{\text{curl}}; K)$-bubbles which correspond to the internal degrees of freedom of $\mathcal{N}_k(K)$ and $\mathcal{NC}_k(K)$. We recall that from (2.3.46) and (2.3.50) these degrees of freedom are associated respectively to $(P_{k-2})^3$ and to $\mathcal{RT}_{k-2}(K) = (P_{k-2})^3 + \underline{x}\,P_{k-2}$. We thus suppose that $k \geq 2$ and we define on a tetrahedron, for the four faces defined by the choice of three barycentric coordinates $\lambda_i, \lambda_j, \lambda_k$,

$$\Phi_{ijk} = P_{k-2}(f_{ijk})\,\lambda_i\lambda_j\lambda_k\,\underline{n}_{ijk}. \tag{2.6.24}$$

For $k = 2$ this defines four functions which are associated with the four faces and which are the bubbles of $\mathcal{NC}_{k+1}(K)$. To obtain the bubbles of $\mathcal{N}_k(K)$ we must suppress the gradient of the standard bubble $\lambda_1\lambda_2\lambda_3\lambda_4$ which is

$$\lambda_1\lambda_2\lambda_3\,\underline{\text{grad}}\,\lambda_4 + \lambda_1\lambda_2\lambda_4\,\underline{\text{grad}}\,\lambda_3 + \lambda_1\lambda_3\lambda_4\,\underline{\text{grad}}\,\lambda_2 + \lambda_2\lambda_3\lambda_4\,\underline{\text{grad}}\,\lambda_1 \tag{2.6.25}$$

which is a combination of the four bubbles of (2.6.24) by (2.6.2).

For $k \geq 3$ we must add to the functions defined by (2.6.24) standard bubbles of the form

$$\underline{P}_{k-3}(K)\lambda_1\lambda_2\lambda_3\lambda_4 = \underline{P}_{k-3}(K)B_4(K), \tag{2.6.26}$$

which vanish totally on the boundary of K. This defines a basis for the bubbles of $\mathcal{NC}_k(K)$ and to obtain $\mathcal{N}_k(K)$ we need to remove the gradients of the standard bubbles. For $k = 3$, for example, we must remove the gradients of the four bubbles $\underline{P}_k(K)B_4(K)$.

2.7 Concluding Remarks

This chapter is evidently not a complete presentation of finite element approximation methods. It cannot be, unless it becomes a book by itself. Our aim was therefore to present examples of the most classical cases and to consider a construction for the less standard case $H(\text{div};\Omega)$ and $H(\underline{\text{curl}};\Omega)$. On the other hand, approximations of elasticity problems will also require special spaces. They will be described in due time. We however believe that the present chapter will then provide a sound basis for these developments.

Chapter 3
Algebraic Aspects of Saddle Point Problems

The examples of Chap. 1 clearly showed that several formulations typically lead to linear systems of the general form

$$\begin{pmatrix} A & B^T \\ B & 0 \end{pmatrix} \begin{pmatrix} \mathbf{x} \\ \mathbf{y} \end{pmatrix} = \begin{pmatrix} \mathbf{f} \\ \mathbf{g} \end{pmatrix}, \tag{3.0.1}$$

where A and B are linear differential operators from some functional space to another (which often is its dual space). The general abstract theory for systems of the type (3.0.1) in Hilbert spaces will be given in Chap. 4. As we shall see, it involves from time to time non-trivial results in functional analysis that can be difficult to understand for readers with a weaker mathematical background.

The purpose of this chapter is to present first the basic results of the general abstract theory in the much simpler context of *finite dimensional spaces*, where we can avoid all the subtleties of functional analysis. We shall therefore study systems of the form (3.0.1) where A and B are respectively an $n \times n$ matrix and an $m \times n$ matrix, while $\mathbf{x}$ and $\mathbf{f}$ are $n \times 1$ vectors and $\mathbf{y}$ and $\mathbf{g}$ are $m \times 1$ vectors.

It is clear that the present finite dimensional case will usually be reached after the discretisation of more general systems in abstract Hilbert spaces, so that we cannot be afraid of wasting our time in analysing it in detail. Moreover, many results that will be proved in the next chapter can be seen, formally, as simple extensions of the present algebraic version (although the proofs in the infinite dimensional case are often more tricky).

Hence, in a sense, the present chapter is dedicated to the readers that have a weaker background in mathematics, and in particular in functional analysis. We hope that, for them, a good grasp of the finite dimensional cases will be sufficient to understand *the results* (if not the proofs) that will be discussed in the next chapter.

In the study of linear systems of the type (3.0.1), our first need will be to express in proper form the conditions for their *solvability* in terms of the properties of the matrices A and B. By solvability we mean that, for every right-hand side $\mathbf{f}$ and $\mathbf{g}$,

D. Boffi et al., *Mixed Finite Element Methods and Applications*, Springer Series in Computational Mathematics 44, DOI 10.1007/978-3-642-36519-5_3,

the system (3.0.1) has a unique solution. It is well known that this property holds *if and only if* the $(n+m) \times (n+m)$ matrix

$$M = \begin{pmatrix} A & B^T \\ B & 0 \end{pmatrix} \tag{3.0.2}$$

is *non-singular*, i.e. if and only if its determinant is different from zero. We shall therefore give necessary and sufficient conditions on the sub-matrices A and B for producing a non-singular M.

In order to have a good numerical method, however, solvability is not enough. An additional property that we also require is *stability*. Let us see in more detail what we mean by that. For a solvable finite-dimensional linear system, we always have continuous dependence of the solution upon the data. This means that there exists a constant c such that for every set of vectors $\mathbf{x}, \mathbf{y}, \mathbf{f}, \mathbf{g}$ satisfying (3.0.1) we have

$$\|\mathbf{x}\| + \|\mathbf{y}\| \le c(\|\mathbf{f}\| + \|\mathbf{g}\|). \tag{3.0.3}$$

In turn, this property implies solvability. Indeed, if we assume that (3.0.3) holds for every set of vectors $\mathbf{x}, \mathbf{y}, \mathbf{f}, \mathbf{g}$ satisfying (3.0.1), then, whenever $\mathbf{f}$ and $\mathbf{g}$ are both zero, $\mathbf{x}$ and $\mathbf{y}$ must also be equal to zero. This is another way of saying that the homogeneous system has only the trivial solution, which implies that the determinant of the matrix (3.0.2) is different from zero, and hence the system is solvable.

However, formula (3.0.3) deserves another very important comment. Actually, we did not specify the norms adopted for $\mathbf{x}, \mathbf{y}, \mathbf{f}, \mathbf{g}$. We had the right to do so since, in finite dimension, all norms are equivalent. Hence, the change of one norm with another would only result in a change of the numerical value of the constant c, but it would not change the basic fact that such a constant exists. However, in dealing with linear systems resulting from the discretisation of a partial differential equation, we face a slightly different situation. In fact, if we want to analyse the behaviour of a given *method* when the mesh-size becomes smaller and smaller, we must ideally consider *a sequence* of linear systems whose dimension increases and approaches infinity when the mesh-size tends to zero. As it is well known (and it can also be easily verified), the constants involved in the equivalence of different norms depend on the dimension of the space. For instance, in $\mathbb{R}^n$, the two norms

$$\|\mathbf{x}\|_1 := \sum_{i=1}^{n} |x_i| \qquad \text{and} \qquad \|\mathbf{x}\|_2 := \left(\sum_{i=1}^{n} |x_i|^2\right)^{1/2} \tag{3.0.4}$$

are indeed equivalent, in the sense that there exist two positive constants c_1 and c_2 such that

$$c_1\|\mathbf{x}\|_2 \le \|\mathbf{x}\|_1 \le c_2\|\mathbf{x}\|_2 \tag{3.0.5}$$

for all $\mathbf{x}$ in $\mathbb{R}^n$. However, it can be rather easily checked that the *best* constants one can choose in (3.0.5) are

$$\|\mathbf{x}\|_2 \le \|\mathbf{x}\|_1 \le \sqrt{n}\|\mathbf{x}\|_2; \tag{3.0.6}$$

in particular, the first inequality becomes an equality, for instance, when x_1 is equal to 1 and all the other x_i's are zero, while the second inequality becomes an equality, for instance, when all the x_i are equal to 1.

When considering a discretisation method for a boundary value problem, which gives rise to a sequence of algebraic problems with increasing dimension, we have to take into account that n becomes unbounded. It is then most natural to ask the following question. *Is it possible, for a given choice of the sequence of matrices A and B and norms $\|\mathbf{x}\|$, $\|\mathbf{y}\|$, $\|\mathbf{f}\|$, and $\|\mathbf{g}\|$, to find a constant c independent of the mesh-size that makes (3.0.3) hold true for all mesh-sizes?* If this is true (with some additional relations between the matrices and the norms that will be made precise later on, in Sect. 3.4), we consider *the method* to be *stable*. We point out that, in this context, stability is a property of methods and not a property of linear systems.

However, in this preliminary chapter, we will not deal directly with boundary value problems and related methods. We will consider generic sequences of matrices A and B with the corresponding sequences of norms; then we will require A and B to satisfy suitable properties expressed in terms of constants (say, α and β) that will be assumed to be *the same constants for all the sequence*; finally, we will show that this gives rise to a constant c in (3.0.3) that depends only on α and β, and is therefore valid for all the linear systems of the sequence.

To read the present chapter, only a rudimentary background in linear algebra will be needed, but we hope that the basic ideas will still come out clear enough. The chapter is therefore mostly recommended for readers with a weak mathematical background. Some proofs, in particular in the last two sections, although simple, are somewhat lengthy. The readers with less mathematical inclination might skip them. On the other hand, the chapter could be considered as useless for people with a stronger mathematical formation. Indeed, essentially everything will be repeated, in the more general context of Hilbert spaces, in the next chapter. However, the examples and the counterexamples of the last two Sections might still have some interest, and at least a glance at them is recommended for everybody.

We summarise the outline of the chapter: we first (in Sect. 3.1) recall some elementary facts in linear algebra. The main goal for that is to fix the notation, and to refresh the memory for people with a low mathematical background. Then, in Sect. 3.2 we consider the unique solvability of problems of the type (3.0.1), and we describe necessary and sufficient conditions in terms of properties of matrices A and B. At this level, all norms are considered to be equivalent. Next, in Sect. 3.3 we extend part of the theory to matrices of the type

$$M = \begin{pmatrix} A & B^T \\ B & C \end{pmatrix}, \tag{3.0.7}$$

which is indeed *very generic*. However, we shall play the game that (3.0.7) is, in some sense, *a perturbation of* (3.0.2). Roughly speaking, we shall assume that A and B are such that, for $C = 0$, the matrix (3.0.7) is non-singular, and we look for conditions on C that would preserve this non-singularity. In that section as well, all norms will be considered as equivalent. In the following Sect. 3.4, we start dealing with *big matrices*, and for this we introduce different norms, together with the problem of *stability* of a sequence of problems for a given choice of the sequences of norms. As announced, our conditions will involve stability constants (to be precise: M_a, M_b, α, and β, that will be defined later on), depending on properties of matrices A and B, respectively. The dependence of the global stability constants upon M_a, M_b, α, and β (and in particular upon α and β) will be tracked down with care, and some simple examples will show the optimality of our results. Some additional results are presented in Sect. 3.5. Finally, the stability conditions for the perturbed problems of the type (3.0.7) will be considered in Sect. 3.6.

3.1 Notation, and Basic Results in Linear Algebra

3.1.1 Basic Definitions

Let r and s be positive integers, and $M : \mathbb{R}^r \to \mathbb{R}^s$ an $s \times r$ real matrix. We denote by M^T the **transposed matrix** of M, given by

$$M^T_{i,j} = M_{j,i} \qquad i = 1, \dots, r, \ j = 1, \dots, s. \tag{3.1.1}$$

It is clear that M^T is an $r \times s$ matrix, and therefore $M^T : \mathbb{R}^s \to \mathbb{R}^r$. It is also immediate to check that

$$(M^T)^T \equiv M. \tag{3.1.2}$$

If we have two matrices $M : \mathbb{R}^r \to \mathbb{R}^s$ and $N : \mathbb{R}^k \to \mathbb{R}^r$, the **product** MN of the two matrices will be the usual *rows times columns* one, namely

$$(M\,N)_{m,n} = \sum_{i=1}^{r} M_{m,i} N_{i,n} \qquad 1 \le m \le s, \ 1 \le n \le k. \tag{3.1.3}$$

Vectors in $\mathbb{R}^n$ will be considered as *columns*, that is as $n \times 1$ matrices. It is elementary to check that, in the above assumptions on N and M, we have

$$(M\,N)^T = N^T M^T \tag{3.1.4}$$

and (since the transposed of a 1×1 matrix is the matrix itself)

$$\mathbf{y}^T M\mathbf{x} \equiv \mathbf{x}^T M^T \mathbf{y} \qquad \forall \mathbf{x} \in \mathbb{R}^r, \ \forall \mathbf{y} \in \mathbb{R}^s. \tag{3.1.5}$$

Throughout this section, which is very elementary, we shall denote by $\mathbf{0}_r$ and $\mathbf{0}_s$ the zero vectors in $\mathbb{R}^r$ and in $\mathbb{R}^s$ respectively. This notation will be abandoned in the sequel, with only a few exceptions. Throughout the first three sections of this chapter, unless it is otherwise explicitly specified, the **norm** in $\mathbb{R}^r$, for every integer $r \geq 1$, will be the usual *Euclidean norm* defined by

$$\|\mathbf{x}\|^2 := \sum_{i=1}^{r} x_i^2 \equiv \mathbf{x}^T\mathbf{x}. \tag{3.1.6}$$

We define the **kernel** and the **Range** (or **image**) of M and M^T as follows:

$$\begin{aligned}
&(i)\ \mathrm{Ker}M := \{\mathbf{x} \in \mathbb{R}^r \text{ such that } M\mathbf{x} = \mathbf{0}_s\},\\
&(ii)\ \mathrm{Ker}M^T := \{\mathbf{y} \in \mathbb{R}^s \text{ such that } M^T\mathbf{y} = \mathbf{0}_r\},\\
&(iii)\ \mathrm{Im}M := \{\mathbf{y} \in \mathbb{R}^s \text{ such that } M\mathbf{x} = \mathbf{y} \text{ for some } \mathbf{x} \in \mathbb{R}^r\},\\
&(iv)\ \mathrm{Im}M^T := \{\mathbf{x} \in \mathbb{R}^r \text{ such that } M^T\mathbf{y} = \mathbf{x} \text{ for some } \mathbf{y} \in \mathbb{R}^s\}.
\end{aligned} \tag{3.1.7}$$

3.1.2 Subspaces

As usual, we shall say that Z is a subspace of $\mathbb{R}^r$ if $Z \subset \mathbb{R}^r$ and Z is itself a linear space.

Remark 3.1.1. We recall that a subset Z of a linear space $\mathbb{R}^r$ *is itself a linear space* (and hence is *a subspace*) if, for any two elements $\mathbf{z}_1$ and $\mathbf{z}_2$ in Z, their sum $\mathbf{z}_1 + \mathbf{z}_2$ also belongs to Z and moreover, for any $z \in Z$ and for any real number λ, the product $\lambda\mathbf{z}$ also belongs to Z. □

Remark 3.1.2. According to the previous definition, when, for instance, $r = 3$, any subspace Z of $\mathbb{R}^3$ has to be made of triplets. However, it is quite common to consider, say, $\mathbb{R}^2$ as a subspace of $\mathbb{R}^3$ by considering $(x_1, x_2)^T$ as identified with the triplet $(x_1, x_2, 0)^T$. This, strictly speaking, is not 100% correct. However, on some occasion, it might turn out to be convenient, as we are going to see immediately in the Example 3.1.1 here below. Therefore we will accept it sometimes, while being very careful with what we do. □

If Z is a linear subspace of $\mathbb{R}^r$, the image of the restriction of M to Z will be denoted by $M(Z)$. Hence,

$$M(Z) := \{\mathbf{y} \in \mathbb{R}^s \text{ such that } M\mathbf{z} = \mathbf{y} \text{ for some } \mathbf{z} \in Z\}. \tag{3.1.8}$$

It is clear that $M(\mathbb{R}^r) \equiv \mathrm{Im}M$.

Example 3.1.1. Assume that $r = 5$, $s = 2$, and consider the operator $M : \mathbb{R}^5 \to \mathbb{R}^2$ defined by

$$M = \begin{pmatrix} 1 & 0 & 0 & 0 & 0 \\ 0 & 1 & 0 & 0 & 0 \end{pmatrix}. \tag{3.1.9}$$

If Z is the subspace $Z := \{x_3 = x_4 = x_5 = 0\}$ (that is the space of quintuples of the type $(x_1, x_2, 0, 0, 0)^T$), the temptation to identify the restriction of M to Z with the matrix

$$M_Z = \begin{pmatrix} 1 & 0 \\ 0 & 1 \end{pmatrix} \tag{3.1.10}$$

is actually quite strong. If, instead of a 2×5 matrix, we had a 2×500 matrix, then the temptation would be much stronger (as well as the economy in using the form (3.1.10)). □

Definition 3.1.1. Let M be an $s \times r$ matrix. Let Z be a subspace of $\mathbb{R}^r$ and S a subspace of $\mathbb{R}^s$. We say that M **restricted to** Z is **injective** if

$$\forall \mathbf{z}^1 \in Z,\ \forall \mathbf{z}^2 \in Z \text{ we have: } \{M\mathbf{z}^1 = M\mathbf{z}^2\} \Rightarrow \{\mathbf{z}^1 = \mathbf{z}^2\}. \tag{3.1.11}$$

We say that M **from** Z **to** S is **surjective** if

$$\forall \mathbf{w} \in S\ \exists \mathbf{z} \in Z \text{ such that } M\mathbf{z} = \mathbf{w}. \tag{3.1.12}$$

It is easy to see that, if for instance $Z \equiv \mathbb{R}^r$, then M is injective if and only if $\mathrm{Ker}M = \mathbf{0}_r$. More generally, M restricted to Z is injective if and only if $\mathrm{Ker}M \cap Z = \mathbf{0}_r$. On the other hand, if $S \equiv \mathbb{R}^s$, then M is surjective if and only if $M(Z) = \mathbb{R}^s$. More generally, M is surjective from Z to S if and only if $M(Z) \supseteq S$.

From now on, if we say that an $s \times r$ matrix M is injective or surjective, without specifying the subspaces Z and S, we intend that $\mathrm{Ker}M = \mathbf{0}_r$ or $\mathrm{Im}M = \mathbb{R}^s$, respectively. In other words, by default we intend that $Z = \mathbb{R}^r$ and $S = \mathbb{R}^s$.

The **dimension** of a linear space will be denoted by *dim*. Hence, for instance, $dim(\mathbb{R}^r) = r$, and if Z is a subspace $\subseteq \mathbb{R}^r$, then $dim(Z) \le r$. Moreover,

$$Z \text{ subspace of } \mathbb{R}^r \text{ and } dim(Z) = r \quad \Rightarrow \quad Z \equiv \mathbb{R}^r. \tag{3.1.13}$$

The **rank** of M is defined as the dimension of its range:

$$rank(M) := dim(\mathrm{Im}M). \tag{3.1.14}$$

Example 3.1.2. In order to become familiar with the notation, it will be convenient to consider an elementary example, made by the family of matrices

$$M_\alpha = \begin{pmatrix} 0 & 0 & 1 & 0 & 0 \\ 0 & 0 & 0 & 1 & 0 \\ 0 & 0 & 0 & 0 & \alpha \end{pmatrix}, \tag{3.1.15}$$

where α is a real parameter. We have clearly $r = 5$ and $s = 3$. For our present purposes, only the cases $\alpha = 0$ and $\alpha = 1$ will be relevant. The transposed matrix will be

$$M_\alpha^T = \begin{pmatrix} 0 & 0 & 0 \\ 0 & 0 & 0 \\ 1 & 0 & 0 \\ 0 & 1 & 0 \\ 0 & 0 & \alpha \end{pmatrix}. \tag{3.1.16}$$

It is immediate to check that for $\alpha = 0$ we have:

$$\begin{aligned}
&\mathrm{Ker}M_0 = \{\mathbf{x} \in \mathbb{R}^5 \text{ s. t. } x_3 = x_4 = 0\} \ \mathit{dim}(\mathrm{Ker}M_0) = 3, \\
&\mathrm{Ker}M_0^T = \{\mathbf{y} \in \mathbb{R}^3 \text{ s. t. } y_1 = y_2 = 0\} \ \mathit{dim}(\mathrm{Ker}M_0^T) = 1, \\
&\mathrm{Im}M_0 = \{\mathbf{y} \in \mathbb{R}^3 \text{ s. t. } y_3 = 0\} \ \mathit{dim}(\mathrm{Im}M_0) = 2, \\
&\mathrm{Im}M_0^T = \{\mathbf{x} \in \mathbb{R}^5 \text{ s. t. } x_1 = x_2 = x_5 = 0\} \ \mathit{dim}(\mathrm{Im}M_0^T) = 2,
\end{aligned} \tag{3.1.17}$$

while for $\alpha = 1$, instead, we have

$$\begin{aligned}
&\mathrm{Ker}M_1 = \{\mathbf{x} \in \mathbb{R}^5 \text{ s. t. } x_3 = x_4 = x_5 = 0\} \ \mathit{dim}(\mathrm{Ker}M_1) = 2, \\
&\mathrm{Ker}M_1^T = \mathbf{0_3} \ \mathit{dim}(\mathrm{Ker}M_1^T) = 0, \\
&\mathrm{Im}M_1 = \mathbb{R}^3 \ \mathit{dim}(\mathrm{Im}M_1) = 3, \\
&\mathrm{Im}M_1^T = \{\mathbf{x} \in \mathbb{R}^5 \text{ s. t. } x_1 = x_2 = 0\} \ \mathit{dim}(\mathrm{Im}M_1^T) = 3.
\end{aligned} \tag{3.1.18}$$

In particular, M_1 is surjective from $\mathbb{R}^5$ to $\mathbb{R}^3$, and M_1^T is injective from $\mathbb{R}^3$ to $\mathbb{R}^5$. The same properties are not true for M_0 and M_0^T respectively. These simple cases might also be useful to check several other properties that will be discussed in the rest of the section. □

3.1.3 Orthogonal Subspaces

For a given linear subspace Z of $\mathbb{R}^r$, we define its **orthogonal subspace** $Z^\perp$ as follows

$$Z^\perp := \{\mathbf{x} \in \mathbb{R}^r \text{ such that } \mathbf{x}^T\mathbf{z} = 0 \ \forall \mathbf{z} \in Z\}. \tag{3.1.19}$$

It is not difficult (and quite intuitive) to check that

$$\mathit{dim}\,(Z^\perp) + \mathit{dim}\,(Z) = r, \tag{3.1.20}$$

and each $\mathbf{x}$ of $\mathbb{R}^r$ can be split *in a unique way* in its two components $\mathbf{x}_Z \in Z$ and $\mathbf{x}_\perp$

$$\mathbf{x} = \mathbf{x}_Z + \mathbf{x}_\perp. \tag{3.1.21}$$

We also have that

$$Z \cap Z^{\perp} = \mathbf{0}_r, \tag{3.1.22}$$

that

$$(Z^{\perp})^{\perp} \equiv Z \tag{3.1.23}$$

and that for two subspaces Z_1 and Z_2

$$Z_1 \subseteq Z_2 \;\Rightarrow Z_2^{\perp} \subseteq Z_1^{\perp}. \tag{3.1.24}$$

Example 3.1.3. For instance, with the notation of the previous example, if $Z = \mathrm{Ker}M_{\alpha}$, we have in $\mathbb{R}^5$: for $\alpha = 0$

$$\begin{aligned} &(\mathrm{Ker}M_0)^{\perp} = \{\mathbf{x} \in \mathbb{R}^5 \text{ such that } x_1 = x_2 = x_5 = 0\} \\ &dim((\mathrm{Ker}M_0)^{\perp}) = 2, \end{aligned} \tag{3.1.25}$$

and for $\alpha = 1$

$$\begin{aligned} &(\mathrm{Ker}M_1)^{\perp} = \{\mathbf{x} \in \mathbb{R}^5 \text{ such that } x_1 = x_2 = 0\} \\ &dim((\mathrm{Ker}M_1)^{\perp}) = 3. \end{aligned} \tag{3.1.26}$$

Always referring to the previous example, we have instead, in $\mathbb{R}^3$: for $\alpha = 0$

$$(\mathrm{Ker}M_0^T)^{\perp} = \{\mathbf{y} \in \mathbb{R}^3 \text{ such that } y_3 = 0\} \;\; dim((\mathrm{Ker}M_0^T)^{\perp}) = 2, \tag{3.1.27}$$

and for $\alpha = 1$

$$(\mathrm{Ker}M_1^T)^{\perp} = \{ \text{ the whole } \mathbb{R}^3\} \;\; dim((\mathrm{Ker}M_1^T)^{\perp}) = 3. \tag{3.1.28}$$

□

Remark 3.1.3. Note that the definition of the orthogonal subspace relies on the choice of the whole space. For instance, as we have already seen in Remark 3.1.2, it is quite common to accept that $\mathbb{R}^r \subset \mathbb{R}^{r+1}$ by identifying $(x_1, \ldots, x_r)$ with $(x_1, \ldots, x_r, 0)$. In this case, for $Z \subseteq \mathbb{R}^r$ we could consider Z both to be a subspace of $\mathbb{R}^r$ and a subspace of $\mathbb{R}^{r+1}$. Clearly, its orthogonal in $\mathbb{R}^r$ and its orthogonal in $\mathbb{R}^{r+1}$ will be different. We will try to be careful whenever this type of confusion can occur. □

3.1.4 Orthogonal Projections

The notion of orthogonal projection on a subspace will play an important role in the next Section. We recall it here, briefly.

For a given subspace Z, say, of $\mathbb{R}^r$, we introduce the **orthogonal projection** $\pi_Z\colon \mathbb{R}^r \to Z$ as follows. For a given $\mathbf{x} \in \mathbb{R}^r$, its orthogonal projection $\pi_Z\mathbf{x}$ is the minimiser in Z of the quantity $\|\mathbf{x} - \mathbf{z}\|$. Hence, we have

$$\pi_Z\mathbf{x} \in Z \quad \text{and} \quad \|\mathbf{x} - \pi_Z\mathbf{x}\| \le \|\mathbf{x} - \mathbf{z}\|, \quad \forall \mathbf{z} \in Z. \tag{3.1.29}$$

An alternative and equivalent way of writing (3.1.29) is

$$\pi_Z\mathbf{x} := \arg\min_{\mathbf{z}\in Z} \|\mathbf{z} - \mathbf{x}\|. \tag{3.1.30}$$

It is easy to see that such a minimiser exists, is unique and is the unique solution of

$$\pi_Z\mathbf{x} \in Z \qquad \text{and} \qquad \mathbf{z}^T\pi_Z\mathbf{x} = \mathbf{z}^T\mathbf{x}, \qquad \forall \mathbf{z} \in Z. \tag{3.1.31}$$

An obvious consequence of (3.1.31) is

$$\{\mathbf{x} \in Z^\perp\} \quad \Leftrightarrow \quad \{\pi_Z\mathbf{x} = \mathbf{0}_r\}. \tag{3.1.32}$$

Example 3.1.4. Always referring to the cases of Example 3.1.2, if, for instance, $Z = \mathrm{Ker}M_0$ and $\mathbf{x} = (1,2,3,4,5)^T$, then $\pi_Z\mathbf{x} = (1,2,0,0,5)^T$. □

It will also be convenient to associate to a subspace $Z \subseteq \mathbb{R}^r$ the **extension** operator E_Z, defined as the linear operator that to every $z \in Z$ associates the same z, thought as a member of $\mathbb{R}^r$. At first sight, this appears to be **obnoxiously redundant**. However, as we have seen in Remark 3.1.2, it is quite common, for instance, to identify $Z = \mathbb{R}^2$ as the subspace of $\mathbb{R}^3$ made by the triplets $(x_1, x_2, 0)^T$. Note that, if we consider

$$Z := \{(x_1, x_2, 0)^T\}, \tag{3.1.33}$$

then E_Z is just the *identity matrix*. If however we consider

$$Z := \{(x_1, x_2)^T\}, \tag{3.1.34}$$

then the operator E_Z would correspond to the matrix

$$E_Z = \begin{pmatrix} 1 & 0 \\ 0 & 1 \\ 0 & 0 \end{pmatrix}, \tag{3.1.35}$$

and its *transposed operator* would be

$$E_Z^T = \begin{pmatrix} 1 & 0 & 0 \\ 0 & 1 & 0 \end{pmatrix} \equiv \pi_Z. \tag{3.1.36}$$

Considering now the general case, we note that if we follow a notation of the type of (3.1.34), then the equality

$$E_Z^T \equiv \pi_Z, \tag{3.1.37}$$

in fact, holds for a general Z. Indeed, for every $\mathbf{z} \in Z$, we can consider the element $E_Z\mathbf{z}$ defined as $\mathbf{z} + \mathbf{0}_{Z^\perp}$ and for every $\mathbf{y} \in \mathbb{R}^r$, we can split it into its components on Z and on $Z^\perp$ and write $\mathbf{y} = \mathbf{y}_Z + \mathbf{y}_{Z^\perp}$, getting

$$\mathbf{y}^T E_Z\mathbf{z} = (\mathbf{y}_Z + \mathbf{y}_{Z^\perp})^T(\mathbf{z} + \mathbf{0}_{Z^\perp}) = (\mathbf{z} + \mathbf{0}_{Z^\perp})^T(\mathbf{y}_Z + \mathbf{y}_{Z^\perp}) = \mathbf{z}^T\pi_Z\mathbf{y}. \tag{3.1.38}$$

On the other hand, following a notation of the type (3.1.33), we would have (also in general)

$$E_Z \equiv E_Z^T \equiv \pi_Z \equiv \pi_Z^T. \tag{3.1.39}$$

3.1.5 Basic Results

We start by proving an easy but useful proposition.

Proposition 3.1.1. *Let M be an $s \times r$ matrix. Then, the restriction of M to* $(\mathrm{Ker}M)^\perp$ *is a one-to-one mapping between* $(\mathrm{Ker}M)^\perp$ *and* $\mathrm{Im}M$.

Proof. Let us see first that M, restricted to $(\mathrm{Ker}M)^\perp$, is injective: according to the definition (3.1.11), we have to prove that, if $\mathbf{z}^1$ and $\mathbf{z}^2$ belong to $(\mathrm{Ker}M)^\perp$, and $M\mathbf{z}^1 = M\mathbf{z}^2$, then we must have $\mathbf{z}^1 = \mathbf{z}^2$. Indeed, setting $\tilde{\mathbf{z}} := \mathbf{z}^1 - \mathbf{z}^2$ we have $M\tilde{\mathbf{z}} = 0$ and hence $\tilde{\mathbf{z}} \in \mathrm{Ker}M$. On the other hand, the vector $\tilde{\mathbf{z}}$, as the difference between two elements of $(\mathrm{Ker}M)^\perp$, must also be in $(\mathrm{Ker}M)^\perp$. Hence, $\tilde{\mathbf{z}}$ belongs, at the same time, to $\mathrm{Ker}M$ and to $(\mathrm{Ker}M)^\perp$. Due to (3.1.22), this implies $\tilde{\mathbf{z}} = \mathbf{0}_r$, that means $\mathbf{z}^1 = \mathbf{z}^2$, as we wanted.

Let us now see that M, as a mapping from $(\mathrm{Ker}M)^\perp$ to $\mathrm{Im}M$, is surjective. According to the definition (3.1.12), we have to prove that, for every element $\mathbf{w} \in \mathrm{Im}M$, there exists a $\mathbf{z} \in (\mathrm{Ker}M)^\perp$ such that $M\mathbf{z} = \mathbf{w}$. For this, let $\mathbf{w}$ be an element of $\mathrm{Im}M$. By definition, there exists an $\mathbf{x} \in \mathbb{R}^r$ such that $M\mathbf{x} = \mathbf{w}$. Split this $\mathbf{x}$ into its components along $\mathrm{Ker}M$ and $(\mathrm{Ker}M)^\perp$. Let $\mathbf{x} = \mathbf{x}_K + \mathbf{z}$ be the splitting, with $\mathbf{x}_K \in \mathrm{Ker}M$ and $\mathbf{z} \in (\mathrm{Ker}M)^\perp$. By definition of kernel, $M\mathbf{x}_K = 0$, so that $M\mathbf{z} = M\mathbf{x}_K + M\mathbf{z} = M\mathbf{x} = \mathbf{w}$, as we wanted. □

As immediate consequences, we have now the following properties.

Corollary 3.1.1. *Let M be an $s \times r$ matrix. Then, there exists a lifting L_M, linear from* $\mathrm{Im}M$ *to* $(\mathrm{Ker}M)^\perp$, *such that*

$$L_M M\mathbf{x} = \mathbf{x} \quad \forall \mathbf{x} \in (\mathrm{Ker}M)^\perp. \tag{3.1.40}$$

Moreover, there exists a $\mu > 0$ such that

$$\mu \| L_M \mathbf{y} \| \leq \| \mathbf{y} \| \quad \forall \mathbf{y} \in \mathrm{Im} M \text{ and } \quad \mu \| \mathbf{x} \| \leq \| M \mathbf{x} \| \quad \forall \mathbf{x} \in (\mathrm{Ker} M)^{\perp}. \tag{3.1.41}$$

Proof. The existence of L_M satisfying (3.1.40) is obvious. Since all linear operators are continuous in finite dimension, the two inequalities in (3.1.41) (that are actually, in this context, *the same* inequality) are also obvious. □

Remark 3.1.4. We point out that (3.1.40) easily implies (applying M to both sides) that

$$M L_M \mathbf{y} = \mathbf{y} \quad \forall\, \mathbf{y} \in \mathrm{Im} M. \tag{3.1.42}$$

We also point out that exchanging M with M^T in (3.1.41) we have that there exists a $\mu > 0$ such that

$$\mu \| \mathbf{y} \| \leq \| M^T \mathbf{y} \| \quad \forall \mathbf{y} \in (\mathrm{Ker} M^T)^{\perp}. \tag{3.1.43}$$

□

Remark 3.1.5. We used the same letter (μ) to denote the two constants that appear in (3.1.41) and (3.1.43). This was not by chance. Actually, as we shall see in a while (see e.g. Proposition 3.4.3), the two constants coincide, in the sense that if, for a certain value of μ, (3.1.41) is verified, then (3.1.43) is also verified, and vice-versa.

□

Corollary 3.1.2. *Let M be an $s \times r$ matrix. Then,*

$$dim((\mathrm{Ker} M)^{\perp}) = dim(\mathrm{Im} M), \tag{3.1.44}$$

$$dim(\mathrm{Im} M) + dim(\mathrm{Ker} M) = r, \tag{3.1.45}$$

$$dim((\mathrm{Ker} M^T)^{\perp}) = dim(\mathrm{Im} M^T), \tag{3.1.46}$$

and:

$$dim(\mathrm{Im} M^T) + dim(\mathrm{Ker} M^T) = s. \tag{3.1.47}$$

Proof. Equation (3.1.44) is an obvious consequence of Proposition 3.1.1. Equation (3.1.45) follows from (3.1.44) using (3.1.20). Then (3.1.46) and (3.1.47) follow by exchanging M and M^T in (3.1.44) and in (3.1.45).

Remark 3.1.6. Note that (3.1.44) is well in agreement with the previous examples: for $\alpha = 0$, we have from (3.1.17) that $dim(\mathrm{Im} M_0) = 2$ and from (3.1.25) that $dim((\mathrm{Ker} M_0)^{\perp}) = 2$, while for $\alpha = 1$, we have from (3.1.18) that $dim(\mathrm{Im} M_1) = 3$ and from (3.1.26) that $dim((\mathrm{Ker} M_1)^{\perp}) = 3$. The agreement of (3.1.46), (3.1.45) and (3.1.47) with the previous examples can be checked in a similar way. We leave it as an exercise. □

Moreover, the following property is very commonly used.

Corollary 3.1.3. *A square $r \times r$ matrix M is injective if and only if it is surjective.*

Proof. The proof follows immediately from (3.1.45). Indeed,

$$\begin{aligned} &M \text{ is injective} \Leftrightarrow \text{Ker}M = \{\mathbf{0}_r\} \Leftrightarrow dim(\text{Ker}M) = 0 \\ &\Leftrightarrow dim(\text{Im}M) = r \Leftrightarrow \text{Im}M = \mathbb{R}^r \Leftrightarrow M \text{ is surjective.} \end{aligned} \tag{3.1.48}$$

Remark 3.1.7. In different words, Corollary 3.1.3 says that, for a square $r \times r$ matrix M, the system

$$M\mathbf{x} = \mathbf{f} \tag{3.1.49}$$

has a unique solution for every right-hand side $\mathbf{f} \in \mathbb{R}^r$ (= surjectivity) if and only if the homogeneous system $M\mathbf{x} = \mathbf{0}_r$ has $\mathbf{x} = \mathbf{0}_r$ as a unique solution, that is if and only if

$$\{M\mathbf{x} = \mathbf{0}_r\} \Rightarrow \{\mathbf{x} = \mathbf{0}_r\} \tag{3.1.50}$$

(= injectivity). It can also be proved (although we are not going to do it here) that both properties are equivalent to say that *the determinant of the matrix M is different from zero.* □

In particular, we recall the following definition.

Definition 3.1.2. A square $r \times r$ matrix M is said to be **non-singular** if it is injective (or, which is the same, if it is surjective, or, which is again the same, if its determinant is different from zero).

It is well known that if M is a non-singular $r \times r$ matrix, then it has an *inverse matrix*, denoted by M^{-1} such that

$$M^{-1}M = M\,M^{-1} = \mathbb{I}_r \tag{3.1.51}$$

where $\mathbb{I}_r$ is the identity matrix in $\mathbb{R}^r$. It is easy to check that whenever M is non-singular, then M^T is also non-singular, and its inverse is given by $(M^T)^{-1} = (M^{-1})^T$. With a (quite common) abuse of notation, we will indicate it simply by M^{-T}, that is

$$M^{-T} = (M^T)^{-1} = (M^{-1})^T. \tag{3.1.52}$$

An important property is given by the following proposition.

Proposition 3.1.2. *Let M be an $s \times r$ matrix. Then,*

$$\text{Ker}M^T = (\text{Im}M)^{\perp}. \tag{3.1.53}$$

Proof. We start by proving that $\text{Ker}M^T \subseteq (\text{Im}M)^{\perp}$. Let $\mathbf{y} \in \mathbb{R}^s$ be in $\text{Ker}M^T$ (that is, $M^T\mathbf{y} = \mathbf{0}_r$). We want to prove that $\mathbf{y} \in (\text{Im}M)^{\perp}$, that is

$$\mathbf{y}^T(M\mathbf{x}) = 0 \quad \forall \mathbf{x} \in \mathbb{R}^r. \tag{3.1.54}$$

This, however, is immediate since

$$\mathbf{y}^T(M\mathbf{x}) = \mathbf{x}^T M^T \mathbf{y} = 0. \tag{3.1.55}$$

Now, we prove that $(\mathrm{Im}M)^{\perp} \subseteq \mathrm{Ker}M^T$. Let therefore $\mathbf{z} \in \mathbb{R}^s$ be in $(\mathrm{Im}M)^{\perp}$ (that is $\mathbf{z}^T M\mathbf{x} = 0$ for all $\mathbf{x} \in \mathbb{R}^r$). Then,

$$\mathbf{x}^T(M^T\mathbf{z}) = 0 \quad \forall \mathbf{x} \in \mathbb{R}^r, \tag{3.1.56}$$

implying that $M^T\mathbf{z} = \mathbf{0}_r$, that is, $\mathbf{z} \in \mathrm{Ker}M^T$.

□

We then have the following theorem.

Theorem 3.1.1. *Let M be an $s \times r$ matrix. Then:*

$$\mathrm{Ker}M^T = (\mathrm{Im}M)^{\perp}, \tag{3.1.57}$$

$$\mathrm{Im}M = (\mathrm{Ker}M^T)^{\perp}, \tag{3.1.58}$$

$$\mathrm{Ker}M = (\mathrm{Im}M^T)^{\perp}, \tag{3.1.59}$$

$$\mathrm{Im}M^T = (\mathrm{Ker}M)^{\perp}. \tag{3.1.60}$$

Proof. Property (3.1.57) has already been seen in (3.1.53). Property (3.1.58) follows from (3.1.53) and (3.1.23). Properties (3.1.59) and (3.1.60) then follow exchanging M and M^T. □

We note that from Theorem 3.1.1 we can easily deduce some useful properties:

$$\{\mathrm{Im}M \equiv \mathbb{R}^s\} \Leftrightarrow \{\mathrm{Ker}M^T = \mathbf{0}_s\}, \quad \mathrm{Im}\{M^T \equiv \mathbb{R}^r\} \Leftrightarrow \{\mathrm{Ker}M = \mathbf{0}_r\}. \tag{3.1.61}$$

All the above properties can also be easily checked on the example of matrices M_α in (3.1.15) and their transposed.

Remark 3.1.8. In spite of its immediate proof, Theorem 3.1.1 can be considered as the finite dimensional version of a very important theorem of functional analysis (that we shall see in the next chapter) which goes under the name of the *Banach Closed Range Theorem.* □

Collecting the results of Proposition 3.1.1, of Corollary 3.1.40 and of Theorem 3.1.1, we now have immediately the following result.

Corollary 3.1.4. *Let M be an $s \times r$ matrix. Then, setting $K := KerM$ and $H := KerM^T$, we have:*

$$M \textit{ is one-to-one from } K^{\perp} \textit{ to } ImM \equiv H^{\perp}, \tag{3.1.62}$$

$$M^T \textit{ is one-to-one from } H^{\perp} \textit{ to } ImM^T \equiv K^{\perp}, \tag{3.1.63}$$

$$\exists L_M : H^{\perp} \to K^{\perp} \textit{ such that } L_M(M\mathbf{x}) = \mathbf{x} \;\; \forall \mathbf{x} \in K^{\perp}, \tag{3.1.64}$$

$$\exists L_{M^T} : K^{\perp} \to H^{\perp} \textit{ such that } L_{M^T}(M^T\mathbf{y}) = \mathbf{y} \;\; \forall \mathbf{y} \in H^{\perp}, \tag{3.1.65}$$

$$(L_M)^T = L_{M^T}. \tag{3.1.66}$$

Example 3.1.5. Assume that the matrix M has the following form

$$M = \begin{pmatrix} \mu_1 & 0 & \cdot & 0 & 0 & 0 & 0 & 0 \\ 0 & \mu_2 & \cdot & 0 & 0 & 0 & 0 & 0 \\ \cdot & \cdot & \cdot & \cdot & \cdot & \cdot & \cdot & \cdot \\ 0 & 0 & \cdot & \mu_k & 0 & 0 & 0 & 0 \\ 0 & 0 & \cdot & 0 & 0 & 0 & 0 & 0 \\ 0 & 0 & \cdot & 0 & 0 & 0 & 0 & 0 \end{pmatrix}, \tag{3.1.67}$$

where k is the dimension of $K^\perp \equiv (\mathrm{Ker}M)^\perp$, which, due to (3.1.44) and (3.1.58) coincides with the dimension of $H^\perp \equiv (\mathrm{Ker}M^T)^\perp$. Here we have $r = k + 4$ and $s = k + 2$. We obviously have

$$M^T = \begin{pmatrix} \mu_1 & 0 & \cdot & 0 & 0 & 0 \\ 0 & \mu_2 & \cdot & 0 & 0 & 0 \\ \cdot & \cdot & \cdot & \cdot & \cdot & \cdot \\ 0 & 0 & \cdot & \mu_k & 0 & 0 \\ 0 & 0 & \cdot & 0 & 0 & 0 \\ 0 & 0 & \cdot & 0 & 0 & 0 \\ 0 & 0 & \cdot & 0 & 0 & 0 \\ 0 & 0 & \cdot & 0 & 0 & 0 \end{pmatrix}, \tag{3.1.68}$$

and

$$L_M = \begin{pmatrix} \mu_1^{-1} & 0 & \cdot & 0 & 0 & 0 \\ 0 & \mu_2^{-1} & \cdot & 0 & 0 & 0 \\ \cdot & \cdot & \cdot & \cdot & \cdot & \cdot \\ 0 & 0 & \cdot & \mu_k^{-1} & 0 & 0 \\ 0 & 0 & \cdot & 0 & 0 & 0 \\ 0 & 0 & \cdot & 0 & 0 & 0 \\ 0 & 0 & \cdot & 0 & 0 & 0 \\ 0 & 0 & \cdot & 0 & 0 & 0 \end{pmatrix} \qquad L_{M^T} = \begin{pmatrix} \mu_1^{-1} & 0 & \cdot & 0 & 0 & 0 & 0 & 0 \\ 0 & \mu_2^{-1} & \cdot & 0 & 0 & 0 & 0 & 0 \\ \cdot & \cdot & \cdot & \cdot & \cdot & \cdot & \cdot & \cdot \\ 0 & 0 & \cdot & \mu_k^{-1} & 0 & 0 & 0 & 0 \\ 0 & 0 & \cdot & 0 & 0 & 0 & 0 & 0 \\ 0 & 0 & \cdot & 0 & 0 & 0 & 0 & 0 \end{pmatrix}. \tag{3.1.69}$$

Remark 3.1.9. Although the form of the matrix M in Example 3.1.5 might appear very special, using the so-called *singular-value decomposition* (see e.g. [228]) for every $s \times r$ matrix B, we can always choose an orthonormal basis in $\mathbb{R}^r$ and an orthonormal basis in $\mathbb{R}^s$ that will transform the matrix B in the form (3.1.67). We shall come back to this later on. □

3.1.6 Restrictions of Operators

Assume that we have a subspace $Z \subseteq \mathbb{R}^r$ and an $s \times r$ matrix M. To M we can associate its restriction M_Z to Z defined as

$$M_Z\mathbf{z} = M(E_Z(\mathbf{z})) \quad \forall \mathbf{z} \in Z \qquad \text{that is } M_Z = M\, E_Z, \tag{3.1.70}$$

where E_Z, here and in all this chapter, is the *extension operator* as defined in Sect. 3.1.4.

If now S is a subspace of $\mathbb{R}^s$, we can consider **the operator M_{ZS}, from Z to S**, defined as

$$M_{ZS} = \pi_S\, M\, E_Z. \tag{3.1.71}$$

Clearly, the transposed operator $(M_{ZS})^T$ will be

$$(M_{ZS})^T = \pi_Z\, M^T\, E_S = (M^T)_{SZ}. \tag{3.1.72}$$

Remark 3.1.10. We point out that all the results that we have seen in the previous subsections (and in particular Theorem 3.1.1) still hold for operators like M_{ZS}, but we have to be careful in the interpretation of the orthogonal complement. In particular, we have

$$\text{Ker}M_{SZ}^T = (\text{Im}M_{ZS})^{\perp_S}, \tag{3.1.73}$$

$$\text{Im}M_{ZS} = (\text{Ker}M_{SZ}^T)^{\perp_S}, \tag{3.1.74}$$

$$\text{Ker}M_{ZS} = (\text{Im}M_{SZ}^T)^{\perp_Z}, \tag{3.1.75}$$

$$\text{Im}M_{SZ}^T = (\text{Ker}M_{ZS})^{\perp_Z}, \tag{3.1.76}$$

where, for three spaces $U \subseteq V \subseteq W$, the notation $U^{\perp_V}$ stands (rather obviously) for *the elements of V that are orthogonal to all the elements of U*. □

Example 3.1.6. In the same spirit, considering once more the matrix (3.1.15) (which describes a linear operator from $\mathbb{R}^5$ to $\mathbb{R}^3$), if the subspace $Z \subseteq \mathbb{R}^5$ is defined by $\{x_1 = x_4 = 0\}$, we can indeed either follow the example of (3.1.33) and consider Z as the set of quintuplets $(0, x_2, x_3, 0, x_5)^T$ and describe *the restriction of M to Z* again with the matrix (3.1.15). Otherwise, we can follow the example of (3.1.34), and consider Z as a set of triples $(x_2, x_3, x_5)^T$, and describe it with the matrix

$$M_Z = M\, E_Z = \begin{pmatrix} 0 & 1 & 0 \\ 0 & 0 & 1 \\ 0 & 0 & \alpha \end{pmatrix}. \tag{3.1.77}$$

So far there is no big difference, and the first option seems actually much cleaner. □

Example 3.1.7. Coming back to the Example 3.1.6 above, if we consider now the space $S \subset \mathbb{R}^3$, defined by $\{y_2 = 0\}$, and if we want to analyse the behaviour of M as an operator from Z to S, the first option would lead us to consider the matrix

$$M_{ZS}^{*}=\begin{pmatrix}0&0&1&0&0\\0&0&0&0&0\\0&0&0&0&\alpha\end{pmatrix},\tag{3.1.78}$$

while the second option would lead to the (simpler) matrix

$$M_{ZS}=\begin{pmatrix}0&1&0\\0&0&\alpha\end{pmatrix}.\tag{3.1.79}$$

Apparently, the advantage of M_{ZS} over M_{ZS}^{*} is just simplicity. However, if you want to apply the general results of the previous subsection (as e.g. (3.1.58)) to the operator "M from Z to S", you see that the use of the form (3.1.79) makes life much easier: for instance, for $\alpha \neq 0$, the image of M_{ZS} coincides with the whole S while the kernel of $(M_{ZS})^T$ is reduced to $\mathbf{0}$. On the other hand, for $\alpha = 0$, then the image of M_{ZS} will be made by the pairs $(y_1, 0)^T$ and the kernel of $(M_{ZS})^T$ is made by the pairs $(0, y_2)^T$ so that again $\mathrm{Im}M_{ZS}$ is orthogonal to $\mathrm{Ker}(M_{ZS})^T$, and so on. Looking carefully, you can also see everything using the form (3.1.78), but with a bigger effort. □

Remark 3.1.11. We must be careful when discussing the *kernel* and the *image* of operators restricted to subspaces. Indeed, in general, $\mathrm{Ker}M_{ZS}$ will not be a subspace of $\mathrm{Ker}M$, and $\mathrm{Im}M_{ZS}$ will not be a subspace of $\mathrm{Im}M$. Let us see some examples. Assume that we consider operators $\mathbb{R}^2 \to \mathbb{R}^2$. We start with

$$M=\begin{pmatrix}1&1\\1&1\end{pmatrix}.\tag{3.1.80}$$

Clearly, the kernel of M is given by $\mathrm{Ker}M = \{(x_1, x_2)^T \,|\, \text{with } x_1 = -x_2\}$ and the image by $\mathrm{Im}M = \{(y_1, y_2)^T \,|\, \text{with } y_1 = y_2\}$. If we take

$$Z := \{(x_1, x_2)^T \,|\, \text{with } x_1 = x_2\} \qquad S := \{(y_1, y_2)^T \,|\, \text{with } y_2 = 0\},$$

then $\mathrm{Ker}M_{ZS} = \{(0,0)^T\}$ and $\mathrm{Im}M_{ZS} := \{(y_1, y_2)^T \,|\, \text{with } y_2 = 0\}$ so that $\mathrm{Ker}M_{ZS} \subseteq \mathrm{Ker}M$ but $\mathrm{Im}M_{ZS} \not\subseteq \mathrm{Im}M$. If we take instead

$$M=\begin{pmatrix}1&-1\\0&1\end{pmatrix},\tag{3.1.81}$$

then $\mathrm{Ker}M = \{(0,0)^T\}$ and $\mathrm{Im}M := \mathbb{R}^2$. Choosing Z and S as before, we have now $\mathrm{Ker}M_{ZS} = \{(x_1, x_2)^T \,|\, \text{with } x_1 = x_2\}$ and $\mathrm{Im}M_{ZS} := \{(0,0)^T\}$ so that now $\mathrm{Im}M_{ZS} \subseteq \mathrm{Im}M$ but $\mathrm{Ker}M_{ZS} \not\subseteq \mathrm{Ker}M$. □

The following result deals with the possible inclusions of kernels and images of M_{ZS} and M and their transposed operators.

Proposition 3.1.3. *Let M be an $s \times r$ matrix, let Z be a subspace of $\mathbb{R}^r$, let S be a subspace of $\mathbb{R}^s$ and let finally $M_{ZS} \equiv \pi_S M E_Z$ be the restriction of M operating from Z to S. Finally, let M^T and M_{SZ}^T be the transposed operators of M and M_{ZS}, respectively. Then, the two following inclusions are equivalent*

$$\mathrm{Ker} M_{ZS} \subseteq \mathrm{Ker} M \tag{3.1.82}$$

$$\mathrm{Im}(\pi_Z M^T) \subseteq \mathrm{Im} M_{SZ}^T. \tag{3.1.83}$$

Moreover, exchanging the operators with their transposed, we obviously also have

$$\mathrm{Ker} M_{SZ}^T \subseteq \mathrm{Ker} M^T \quad \Leftrightarrow \quad \mathrm{Im}(\pi_S M) \subseteq \mathrm{Im} M_{ZS}. \tag{3.1.84}$$

Proof. We start by noting that (3.1.82) is equivalent to

$$\mathrm{Ker} M_{ZS} = Z \cap \mathrm{Ker} M. \tag{3.1.85}$$

On the other hand, from (3.1.74) we have that an element of Z belongs to $\mathrm{Im} M_{SZ}^T$ if and only if it is orthogonal to all $\mathbf{z} \in \mathrm{Ker} M_{ZS}$. Taking into account that the generic element of $\mathrm{Im}(\pi_Z M^T)$ is $\pi_Z M \mathbf{y}$ (with $\mathbf{y}$ generic in $\mathbb{R}^s$), and that obviously (by transposition) $\mathbf{z}^T \pi_Z M^T \mathbf{y} = \mathbf{y}^T M E_Z \mathbf{z}$, we deduce that (3.1.83) is equivalent to

$$\mathbf{y}^T M E_Z \mathbf{z} = 0 \quad \forall \mathbf{y} \in \mathbb{R}^s, \quad \forall \mathbf{z} \in \mathrm{Ker} M_{ZS}, \tag{3.1.86}$$

which in turn is clearly equivalent to (3.1.85). □

Remark 3.1.12. An equivalent way of looking at Proposition 3.1.3 is as follows. Using (3.1.24), we immediately have that (3.1.82) holds if and only if $(\mathrm{Ker} M)^\perp \subseteq (\mathrm{Ker} M_{ZS})^\perp$, where both the orthogonals are taken in $\mathbb{R}^r$. On the other hand, from (3.1.60) we have that $(\mathrm{Ker} M)^\perp = \mathrm{Im} M^T$ while an elementary argument using (3.1.76) gives that

$$(\mathrm{Ker} M_{ZS})^{\perp_{\mathbb{R}^r}} = (\mathrm{Ker} M_{ZS})^{\perp_Z} \cup Z^{\perp_{\mathbb{R}^r}} = \mathrm{Im} M_{SZ}^T \cup Z^\perp. \tag{3.1.87}$$

Hence, (3.1.82) is equivalent to

$$\mathrm{Im} M^T \subseteq \mathrm{Im} M_{SZ}^T \cup Z^\perp, \tag{3.1.88}$$

which is clearly equivalent to (3.1.83). □

Example 3.1.8. In the case of the matrix M of Example 3.1.5, we see that the matrix $M_{K^\perp H^\perp}$ would be

$$M_{K^{\perp}H^{\perp}} = \begin{pmatrix} \mu_1 & 0 & \cdot & 0 \\ 0 & \mu_2 & \cdot & 0 \\ \cdot & \cdot & \cdot & \cdot & \cdot \\ 0 & 0 & \cdot & \mu_k \end{pmatrix}, \tag{3.1.89}$$

showing its nice nature as a $k \times k$ non singular matrix. With this notation, $(L_M)_{K^{\perp}H^{\perp}}$ would just be the *inverse matrix*

$$L_{M_{K^{\perp}H^{\perp}}} = \begin{pmatrix} \mu_1^{-1} & 0 & \cdot & 0 \\ 0 & \mu_2^{-1} & \cdot & 0 \\ \cdot & \cdot & \cdot & \cdot & \cdot \\ 0 & 0 & \cdot & \mu_k^{-1} \end{pmatrix}. \tag{3.1.90}$$

□

3.2 Existence and Uniqueness of Solutions: The Solvability Problem

We go back to our general form (3.0.1), which we repeat here for the convenience of the reader:

$$A\mathbf{x} + B^T\mathbf{y} = \mathbf{f}, \tag{3.2.1}$$

$$B\mathbf{x} = \mathbf{g}. \tag{3.2.2}$$

We assume that $\mathbf{f}$ and $\mathbf{g}$ are given in $\mathbb{R}^n$ and $\mathbb{R}^m$ respectively (n and m being given integer numbers ≥ 1), and that $\mathbf{x}$ and $\mathbf{y}$ are also sought in $\mathbb{R}^n$ and $\mathbb{R}^m$, respectively. This implies that A must be a square matrix $n \times n$ and B a rectangular matrix $m \times n$.

An important role will be played by the kernels of the operators B and B^T. Hence, we set

$$K := \mathrm{Ker}B \qquad H := \mathrm{Ker}B^T. \tag{3.2.3}$$

An easy consequence of Theorem 3.1.1 that will be used quite often in the sequel is: *for all* $\mathbf{x} \in \mathbb{R}^n$ *and for all* $\mathbf{y} \in \mathbb{R}^m$,

$$\mathbf{x} \in \mathrm{Ker}B \quad \Rightarrow \quad \mathbf{x}^T B^T \mathbf{y} \equiv \mathbf{y}^T B\mathbf{x} = 0, \tag{3.2.4}$$

or equivalently, for $K = \mathrm{Ker}B$,

$$\pi_K B^T \mathbf{y} = 0 \qquad \forall\, \mathbf{y} \in \mathbb{R}^m. \tag{3.2.5}$$

3.2.1 A Preliminary Discussion

Our present aim is to give conditions on A and B in order that (3.2.1) and (3.2.2) have a unique solution.

Let us discuss first some heuristic ideas: according to Remark 3.1.7, the global matrix

$$M = \begin{pmatrix} A & B^T \\ B & 0 \end{pmatrix} \tag{3.2.6}$$

will be non-singular if and only if the corresponding homogeneous system

$$A\mathbf{x} + B^T\mathbf{y} = 0, \tag{3.2.7}$$

$$B\mathbf{x} = 0, \tag{3.2.8}$$

has the pair $\mathbf{x} = 0$ and $\mathbf{y} = 0$ as a unique solution. Hence, we start our discussion assuming that $\mathbf{f}$ and $\mathbf{g}$ are both zero. What do we know about $\mathbf{x}$? From the second equation (3.2.8), we see that

$$\mathbf{x} \in K = \mathrm{Ker} B. \tag{3.2.9}$$

Moreover, we can take the projection π_K of the first equation (3.2.7). We note that, using (3.2.5), we have $\pi_K B^T\mathbf{y} = 0$ so that the projection of the first equation onto the kernel K is

$$\pi_K A\mathbf{x} = 0. \tag{3.2.10}$$

We wonder whether

$$\{\mathbf{x} \in K \text{ and } \pi_K A\mathbf{x} = 0\} \Rightarrow \{\mathbf{x} = 0\}\,. \tag{3.2.11}$$

Actually, it depends on the matrix A and on K. Either it does or it doesn't. For the moment, we just set, with the notation of (3.1.71),

$$A_{KK} := \pi_K A E_K. \tag{3.2.12}$$

Coming back to the question (3.2.11), let us analyse the two cases.

- If the answer to (3.2.11) is *no*, then we *surely lose* (meaning that the matrix will indeed be singular). Why do we say that? This is subtle, but not really difficult. We claim that *if the answer is* no, *then there exists a non-zero solution of the homogeneous system.* Let us see why. If the answer to (3.2.11) is *no*, it means that there exists an $\mathbf{x}^* \neq 0$ such that both (3.2.9) and (3.2.10) hold. Now, using (3.1.32), we note that (3.2.10) implies

$$A\mathbf{x}^* \in K^{\perp}. \tag{3.2.13}$$

Moreover, we remember that $K = \mathrm{Ker}B$ and that, from (3.1.60), $(\mathrm{Ker}B)^{\perp} = \mathrm{Im}B^T$. Hence, from (3.2.13), we have $A\mathbf{x}^* \in \mathrm{Im}B^T$, and therefore there must exist a $\mathbf{y}^*$ such that

$$B^T\mathbf{y}^* = A\mathbf{x}^*. \tag{3.2.14}$$

This is why we lose: indeed, the pair $(\mathbf{x}^*, -\mathbf{y}^*)$ satisfies *both* equations $A\mathbf{x}^* + B^T(-\mathbf{y}^*) = 0$ and $B\mathbf{x}^* = 0$, and we have a *non-zero* solution of the homogeneous problem (3.2.7) and (3.2.8), since at least $\mathbf{x}^* \neq 0$.

- If instead the answer to (3.2.11) is *yes*, we can conclude that, for every pair $(\mathbf{x}, \mathbf{y})$ solving the homogeneous system (3.2.7) and (3.2.8), we must have $\mathbf{x} = 0$. However, we still have to work on $\mathbf{y}$. Once we know that $\mathbf{x} = 0$, the first equation (3.2.7) becomes

$$B^T\mathbf{y} = 0, \tag{3.2.15}$$

and we face a second *dilemma*: do we have

$$\{B^T\mathbf{y} = 0\} \Rightarrow \{\mathbf{y} = 0\}\ ? \tag{3.2.16}$$

Clearly, the answer depends on the matrix B^T. If it is injective, the answer to (3.2.16) will be *yes*, otherwise it will be *no*. Here, however, the situation is simpler: indeed, if the answer is *no*, it means that there exists a $\hat{\mathbf{y}} \neq 0$ such that $B^T\hat{\mathbf{y}} = 0$, and we lose again because the pair $(\mathbf{0}_n, \hat{\mathbf{y}})$ will clearly be a non-zero solution to the homogeneous system (3.2.7) and (3.2.8). If instead the answer to (3.2.16) is also *yes*, then we can conclude: every solution $(\mathbf{x}, \mathbf{y})$ of the homogeneous system (3.2.7) and (3.2.8) will necessarily be zero, and the matrix M will be non-singular.

In conclusion to our heuristic analysis, it seems that, in order to have a non-singular global matrix M, we need a "yes" for both questions (3.2.11) and (3.2.16). This indeed is what we are going to *prove*, in a more precise way, in the next subsection.

3.2.2 *The Necessary and Sufficient Condition*

We start with the basic result that provides necessary and sufficient conditions for solvability.

Theorem 3.2.1. *Let n and m be two integers ≥ 1. Let A and B be an $n \times n$ matrix and an $m \times n$ matrix, respectively. Let K be the kernel of B as in (3.2.3), and let A_{KK} be defined as in* (3.2.12)*. Then, the matrix*

$$M = \begin{pmatrix} A & B^T \\ B & 0 \end{pmatrix} \tag{3.2.17}$$

is non-singular if and only if the following two conditions are both satisfied:

$$A_{KK} : K \to K \text{ is surjective (or, equivalently, is injective),} \tag{3.2.18}$$

$$B : \mathbb{R}^n \to \mathbb{R}^m \text{ is surjective (or, equivalently, } B^T \text{ is injective).} \tag{3.2.19}$$

Proof. We start by noting that the equivalence claimed in (3.2.18) has been made clear in Proposition 3.1.2, while the equivalence claimed in (3.2.19) is an easy consequence of (3.1.61). We also note that, in some sense, the theorem has been proved already during the heuristic discussion above. However, here we re-start and give a more detailed proof.

To this aim, assume first that (3.2.17) is non-singular, that is to say that the system (3.2.1) has a unique solution for every right-hand side $(\mathbf{f}, \mathbf{g}) \in \mathbb{R}^n \times \mathbb{R}^m$. In particular, looking at (3.2.2) we see that it must have a solution for every $\mathbf{g} \in \mathbb{R}^m$, and hence $\mathrm{Im} B \equiv \mathbb{R}^m$ and (3.2.19) holds. Moreover, for every $\mathbf{f} = \mathbf{f}_K \in K$ the system

$$\begin{pmatrix} A & B^T \\ B & 0 \end{pmatrix} \begin{pmatrix} \mathbf{x} \\ \mathbf{y} \end{pmatrix} = \begin{pmatrix} \mathbf{f}_K \\ \mathbf{0}_m \end{pmatrix} \tag{3.2.20}$$

must have a solution. For every such solution, we clearly have $B\mathbf{x} = 0$, that is $\mathbf{x} \in K$. We also note that for every $\mathbf{y} \in \mathbb{R}^m$, from (3.2.5) we have that $\pi_K B^T \mathbf{y} = 0$. Hence, taking the projection π_K of the first equation of (3.2.20) yields:

$$\pi_K A\mathbf{x} = \mathbf{f}_K. \tag{3.2.21}$$

In other words, solving (3.2.20), we have that: for every $\mathbf{f} = \mathbf{f}_K \in K$, there exists an $\mathbf{x} \in K$ such that (3.2.21) holds. Hence, A_{KK} is surjective from K to K, and (3.2.18) holds.

Assume, conversely, that (3.2.18) and (3.2.19) hold. We want to show that the matrix (3.2.17) is non-singular. This will follow if we show that the homogeneous system (3.2.7) and (3.2.8) has $\mathbf{x} = 0$, $\mathbf{y} = 0$ as a unique solution. Indeed, from $B\mathbf{x} = 0$, we first get that $\mathbf{x} \in K$. Taking the projection π_K of the first equation (and noting again that $\pi_K B^T \mathbf{y} = 0$), we have then $\pi_K A\mathbf{x} = 0$. This, together with $\mathbf{x} \in K$, implies $\mathbf{x} = 0$ thanks to the injectivity in (3.2.18). Finally, the first equation now becomes $B^T \mathbf{y} = 0$, and this gives $\mathbf{y} = 0$ thanks to the injectivity in (3.2.19). □

Remark 3.2.1. It follows easily from (3.2.19), using for instance (3.1.45), that *a necessary condition* for the solvability is $n \geq m$. This was pretty obvious from the very beginning, but it could be a valuable first simple check for users that are truly illiterate from the mathematical point of view. □

Remark 3.2.2. We point out that a necessary and sufficient condition is somehow a delicate mathematical item: all possible necessary and sufficient conditions for a matrix to be non-singular are *mathematically* equivalent to each other, and all equivalent to the obvious *it is non-singular if and only if the determinant is different from zero* or even to the tautology *it is non-singular if and only if it is non-singular.*

It is only the commodity of usage that, in each context, makes a necessary and sufficient condition a useful instrument or a sterile mathematical exercise. In this respect, we may say that *it is not true, in practice*, that all necessary and sufficient conditions are equivalent. Moreover, it often happens that conditions that are *only necessary* or *only sufficient* are more useful, in practice, than the necessary and sufficient ones. This is what we shall discuss in the next subsection. □

Remark 3.2.3. We note that the result of Theorem 3.2.1 could have been obtained in a different, more algebraic way. As the result is particularly important, we report this alternative way as well, in the hope that two different points of view could provide a deeper understanding of the whole result.

For this, together with the kernel K of B, we now consider its orthogonal complement $K^{\perp}$ in $\mathbb{R}^n$ that we call J. Let n_K be the dimension of K and n_J the dimension of J. From (3.1.20) we have

$$n_K + n_J = n. \tag{3.2.22}$$

We now take a basis $\{\mathbf{x}_1^J, \ldots, \mathbf{x}_{n_J}^J\}$ in J and a basis $\{\mathbf{x}_1^K, \ldots, \mathbf{x}_{n_K}^K\}$ in K. It is clear that

$$\{\mathbf{x}_1^J, \ldots, \mathbf{x}_{n_J}^J, \mathbf{x}_1^K, \ldots, \mathbf{x}_{n_K}^K\} \tag{3.2.23}$$

will be a basis for $\mathbb{R}^n$. *With respect to this basis*, we can re-write the matrices A, B, and B^T as follows:

$$A = \begin{pmatrix} A_{JJ} & A_{JK} \\ A_{KJ} & A_{KK} \end{pmatrix} \qquad B = \begin{pmatrix} B_J & B_K \end{pmatrix} \qquad B^T = \begin{pmatrix} B_J^T \\ B_K^T \end{pmatrix}. \tag{3.2.24}$$

Now, from the definition (3.1.7) of K, we immediately have that $B_K = 0$ (that is the zero $m \times n_K$ matrix) so that $B_K^T = 0$ as well. Splitting $\mathbf{x}$ and $\mathbf{f}$ in their orthogonal components $\mathbf{x}_J$ and $\mathbf{x}_K$, and $\mathbf{f}_J$ and $\mathbf{f}_K$, respectively, we can now write the original system (3.2.1) as follows

$$\begin{pmatrix} A_{JJ} & A_{JK} & B_J^T \\ A_{KJ} & A_{KK} & 0 \\ B_J & 0 & 0 \end{pmatrix} \begin{pmatrix} \mathbf{x}_J \\ \mathbf{x}_K \\ \mathbf{y} \end{pmatrix} = \begin{pmatrix} \mathbf{f}_J \\ \mathbf{f}_K \\ \mathbf{g} \end{pmatrix}. \tag{3.2.25}$$

With a little additional work we can see that B_J is a non-singular square matrix if and only if B is surjective, and the result of Theorem (3.2.1) follows from the block-triangular structure of (3.2.25) since $A_{KK} \equiv \pi_K A$. □

3.2.3 Sufficient Conditions

The problem of checking whether (3.2.18) holds or not could be simplified or even avoided in some particular cases, as pointed out in the following corollaries to the basic Theorem 3.2.1. We recall that, in general, a square $r \times r$ matrix M is said to be **positive semi-definite** if

$$\mathbf{x}^T M\mathbf{x} \geq 0 \qquad \forall \mathbf{x} \in \mathbb{R}^r \tag{3.2.26}$$

and it is said to be **positive definite** if

$$\mathbf{x}^T M\mathbf{x} > 0 \qquad \forall \mathbf{x} \in \mathbb{R}^r \text{ with } \mathbf{x} \neq 0. \tag{3.2.27}$$

More generally, if Z is a subspace of $\mathbb{R}^r$, we say that M is **positive semi-definite on** Z if M_{ZZ} is positive semi-definite, that is

$$\forall \mathbf{x} \in Z \qquad \mathbf{x}^T M_{ZZ}\mathbf{x} \equiv \mathbf{x}^T M\mathbf{x} \geq 0, \tag{3.2.28}$$

and we say that M is **positive definite on** Z if M_{ZZ} is positive definite, that is

$$\forall \mathbf{x} \in Z \text{ with } \mathbf{x} \neq 0 \qquad \mathbf{x}^T M_{ZZ}\mathbf{x} \equiv \mathbf{x}^T M\mathbf{x} > 0. \tag{3.2.29}$$

We observe that a positive definite matrix is always non-singular, since (3.2.27) easily implies (3.1.50). Hence, in particular, if M is positive definite on a subspace Z, then M_{ZZ} will be non-singular $Z \to Z$. It is also obvious that if a matrix M is positive definite (or positive semi definite), then its restriction to every subspace Z will also be positive definite (resp. semi-definite).

From the above discussion, we have the following useful result.

Corollary 3.2.1. *Let A be an $n \times n$ matrix, and B an $m \times n$ matrix. If $B : \mathbb{R}^n \to \mathbb{R}^m$ is surjective and A is positive definite on the kernel K of B, then the matrix M in* (3.2.17) *is non-singular.*

The proof follows immediately from Theorem 3.2.1. The following corollary has more restrictive assumptions, but its use is even simpler.

Corollary 3.2.2. *Let A be an $n \times n$ positive definite matrix, and B an $m \times n$ matrix. If $B : \mathbb{R}^n \to \mathbb{R}^m$ is surjective then the matrix* (3.2.17) *is non-singular.*

Again, the proof is immediate. The advantage of Corollary 3.2.2 (when we can use it!) is that there is no need to characterise the kernel K, which, in some cases, can be a non-trivial task.

Among the various sufficient conditions, we could point out that if A_{KK} is an isomorphism from K to K, then the condition $\mathbf{g} \in \mathrm{Im}B$ will be *sufficient* to guarantee the *existence* of a solution for the system (3.2.1). We have in particular the following result.

Proposition 3.2.1. *Let n and m be two integers ≥ 1. Let A and B be an $n \times n$ matrix and an $m \times n$ matrix, respectively. Let K be the kernel of B as in (3.2.3), and let A_{KK} be defined as in* (3.2.12)*. Assume that A_{KK} is an isomorphism from K to K and that $\mathbf{g} \in ImB$. Then the system* (3.2.1) *has at least one solution. Moreover, if $(\mathbf{x}_1, \mathbf{y}_1)$ and $(\mathbf{x}_2, \mathbf{y}_2)$ are two solutions of* (3.2.1)*, then $\mathbf{x}_1 = \mathbf{x}_2$ and $(\mathbf{y}_1 - \mathbf{y}_2) \in H = \mathrm{Ker}B^T$.*

Proof. Indeed, if $\mathbf{g} \in \mathrm{Im}B$ then, by definition, there exists an $\mathbf{x}_g \in \mathbb{R}^n$ such that $B\mathbf{x}_g = \mathbf{g}$. Looking for $\mathbf{x}_0 \in K$, solution of the problem $A_{KK}\mathbf{x}_0 = \pi_K(\mathbf{f} - A\mathbf{x}_g)$, we can set $\mathbf{x} := \mathbf{x}_0 + \mathbf{x}_g$ and note that, projecting the first equation on K, we have $\pi_K(\mathbf{f} - A\mathbf{x}) = \mathbf{0}$, because $\pi_K\mathbf{f} - A_{KK}\mathbf{x}_0 - \pi_K A\mathbf{x}_g = \mathbf{0}$. In other words, $\mathbf{f} - A\mathbf{x} \in K^\perp$ which, thanks to (3.1.60), implies $\mathbf{f} - A\mathbf{x} \in \mathrm{Im}B^T$. Hence, there exists a $\mathbf{y} \in \mathbb{R}^m$ such that $B^T\mathbf{y} = \mathbf{f} - A\mathbf{x}$. It is immediate to check that $(\mathbf{x}, \mathbf{y})$ is a solution of (3.2.1). Assume now that $(\mathbf{x}_1, \mathbf{y}_1)$ and $(\mathbf{x}_2, \mathbf{y}_2)$ are two solutions of (3.2.1), and set $\mathbf{x}^* := \mathbf{x}_1 - \mathbf{x}_2$ and $\mathbf{y}^* := \mathbf{y}_1 - \mathbf{y}_2$. Clearly, $(\mathbf{x}^*, \mathbf{y}^*)$ is a solution of the homogeneous system (that is, (3.2.1) with $\mathbf{f} = \mathbf{0}$ and $\mathbf{g} = \mathbf{0}$). In particular we have, from the second equation, that $\mathbf{x}^* \in K$, and from the projection on K of the first equation we have $A_{KK}\mathbf{x}^* = \mathbf{0}$ and since A_{KK} is an isomorphism we have $\mathbf{x}^* = \mathbf{0}$. This implies $A\mathbf{x}^* = \mathbf{0}$ and, using again the first equation: $B^T\mathbf{y}^* = \mathbf{0}$ (that is $\mathbf{y}^* \in H$). □

Remark 3.2.4. In the framework of Proposition 3.2.1, the solution will never be unique, unless we have $H = \mathbf{0}_m$ (that however brings us back to Theorem 3.2.1). On the other hand, we could change the problem and look for $\mathbf{y}$ in $H^\perp$. This actually is *the* way to recover a well posed problem when B is not surjective. However, it obviously works only when $\mathbf{g} \in \mathrm{Im}B$. A particular case in which this would work systematically is whenever $\mathbf{g} \equiv \mathbf{0}$ (as it is often the case when the second equation expresses some incompressibility condition, or some sort of conservation property). □

3.2.4 Examples

Let us see now some examples and exercises. We start by emphasising that the part of A that *must* be non-singular is actually A_{KK}, and **not** A itself. Take for instance, for $n = 2$ and $m = 1$, the matrices

$$A = \begin{pmatrix} 1 & 1 \\ 1 & 0 \end{pmatrix} \qquad B = \begin{pmatrix} 1 & 0 \end{pmatrix} \qquad B^T = \begin{pmatrix} 1 \\ 0 \end{pmatrix}. \tag{3.2.30}$$

Then, the rank of B is 1 ($= m$), and (3.2.19) holds true. On the other hand we have that $K = \mathrm{Ker}B = \{\mathbf{x} \in \mathbb{R}^2 \text{ such that } x_1 = 0\}$. Hence, in this case, the *new basis* (3.2.23) coincides with the original one, and the matrices are in the form (3.2.24) already. It is then easy to check that A itself is non-singular, but $A_{KK} = (0)$ and hence (3.2.18) does not hold. Indeed, the whole matrix is

$$M = \begin{pmatrix} 1 & 1 & 1 \\ 1 & 0 & 0 \\ 1 & 0 & 0 \end{pmatrix} \tag{3.2.31}$$

which is clearly singular.

On the other hand, consider the choice

$$A = \begin{pmatrix} 0 & 0 \\ 0 & 1 \end{pmatrix} \qquad B = \begin{pmatrix} 1 & 0 \end{pmatrix} \qquad B^T = \begin{pmatrix} 1 \\ 0 \end{pmatrix}, \tag{3.2.32}$$

where A is singular. Since K is the same as before, the new coordinates (3.2.23) coincide again with the old ones, and we have easily that $A_{KK} = (1)$. This is clearly non-singular, so that (3.2.18) is now satisfied. Indeed, the whole matrix is now non-singular:

$$M = \begin{pmatrix} 0 & 0 & 1 \\ 0 & 1 & 0 \\ 1 & 0 & 0 \end{pmatrix}. \tag{3.2.33}$$

Along the same lines, referring to Corollary 3.2.2 we notice that it would not be enough to require that A is positive *semi*-definite (that is $\mathbf{x}^T A\mathbf{x} \geq 0$ for all $\mathbf{x} \in \mathbb{R}^n$). Indeed, for the choice

$$A = \begin{pmatrix} 1 & 0 \\ 0 & 0 \end{pmatrix} \qquad B = \begin{pmatrix} 1 & 0 \end{pmatrix} \qquad B^T = \begin{pmatrix} 1 \\ 0 \end{pmatrix}, \tag{3.2.34}$$

we have that A is positive semi-definite, we have that (3.2.19) is verified, but the whole matrix

$$M = \begin{pmatrix} 1 & 0 & 1 \\ 0 & 0 & 0 \\ 1 & 0 & 0 \end{pmatrix} \tag{3.2.35}$$

is clearly singular.

In many cases, however, it is not immediate to see, at first glance, what the matrix A_{KK} is. Consider for instance the case

$$A = \begin{pmatrix} a & b \\ c & d \end{pmatrix} \qquad B = \begin{pmatrix} 1 & -1 \end{pmatrix} \qquad B^T = \begin{pmatrix} 1 \\ -1 \end{pmatrix}. \tag{3.2.36}$$

We have in this case

$$K := \{\mathbf{x} \in \mathbb{R}^2 \text{ such that } x_1 - x_2 = 0\}. \tag{3.2.37}$$

Hence, K can be presented as the one-dimensional subset of $\mathbb{R}^2$ made of vectors of the type $(\alpha, \alpha)^T$ with $\alpha \in \mathbb{R}$. In its turn, J can now be presented as the one-dimensional subset of $\mathbb{R}^2$ made of vectors of the type $(\beta, -\beta)^T$ with $\beta \in \mathbb{R}$. In order to reach the form (3.2.24), we now have to express the matrix A in the new basis $\{\mathbf{x}_1^J, \ldots, \mathbf{x}_{n_J}^J, \mathbf{x}_1^K, \ldots, \mathbf{x}_{n_K}^K\}$ that is now simply $\{\mathbf{x}_1^J, \mathbf{x}_1^K\}$ with $\mathbf{x}_1^J = (1, -1)^T$ and $\mathbf{x}_1^K = (1, 1)^T$ (and if we want an orthonormal basis, we can take $\mathbf{x}_1^J = (1/\sqrt{2}, -1/\sqrt{2})^T$ and $\mathbf{x}_1^K = (1/\sqrt{2}, 1/\sqrt{2})^T$). After some classical computations, we can see that, *in this new basis*, the matrix A takes the form

$$\tilde{A} = \frac{1}{2}\begin{pmatrix} a-b-c+d & a+b-c-d \\ a-b+c-d & a+b+c+d \end{pmatrix}. \tag{3.2.38}$$

From (3.2.38) we have that A_{KK} is the 1×1 matrix $(\frac{1}{2}(a+b+c+d))$, which is non-singular if and only if $a+b+c+d \neq 0$.

Indeed, one can check easily (for instance, by computing the determinant) that the condition $a+b+c+d \neq 0$ is necessary and sufficient for the matrix

$$\begin{pmatrix} a & b & 1 \\ c & d & -1 \\ 1 & -1 & 0 \end{pmatrix} \tag{3.2.39}$$

to be non-singular. In cases like this (which are the majority), it would possibly be simpler to deal directly with the restriction of $\pi_K A$ to K, which is A_{KK} in the original variables. This would require to apply the (original) matrix A

$$\begin{pmatrix} a & b \\ c & d \end{pmatrix} \tag{3.2.40}$$

to the general vector (in the original coordinates) $\mathbf{x}_K = (\alpha, \alpha)^T$ in K, obtaining the vector

$$A\mathbf{x}_K = \begin{pmatrix} \alpha(a+b) \\ \alpha(c+d) \end{pmatrix}. \tag{3.2.41}$$

Then, we have to check whether the component of $A\mathbf{x}_K$ in K (that is $\pi_K A\mathbf{x}$) is different from zero. As K is one-dimensional, this amounts to take the scalar product

$$(\mathbf{x}_1^K)^T A\mathbf{x}_K = (1/\sqrt{2}, \quad 1/\sqrt{2})A\mathbf{x}_K = \frac{\alpha}{\sqrt{2}}(a+b+c+d), \tag{3.2.42}$$

and see if it is different from zero when α is different from zero. We clearly obtain again the condition $a+b+c+d \neq 0$.

We point out, however, that if, by chance, a and d are positive and $ad > bc$, then A will be positive definite on the whole $\mathbb{R}^2$, and we can have the solvability directly from Corollary 3.2.2 without any additional work.

3.2.5 Composite Matrices

Sometimes, the matrix A itself has a block structure of the type

$$\mathbb{A} = \begin{pmatrix} C & D^T \\ D & 0 \end{pmatrix}. \tag{3.2.43}$$

Then again, one has to be careful and require the non-singularity of $\mathbb{A}$ just on the kernel of B. In some cases, together with an A with the structure (3.2.43), we have a B with the structure $\mathbb{B} = (E \quad 0)$ or $\mathbb{B} = (0 \quad E)$, so that the whole matrix has the block structure

$$M = \begin{pmatrix} C & D^T & E^T \\ D & 0 & 0 \\ E & 0 & 0 \end{pmatrix} \quad \text{or} \quad M = \begin{pmatrix} C & D^T & 0 \\ D & 0 & E^T \\ 0 & E & 0 \end{pmatrix}, \tag{3.2.44}$$

respectively. In these cases, it can be a useful exercise to rewrite conditions (3.2.18) and (3.2.19) in terms of properties of the matrices C, D, and E.

To fix the ideas, let us assume that, in the *first case* of (3.2.44), C is an $r \times r$ matrix, D is an $s \times r$ matrix, and E a $k \times r$ matrix. We also assume that $r \geq s + k$, otherwise, according to Remark 3.2.1, the Matrix M will surely be singular. It is clear that we can directly use Theorem 3.2.1, with

$$\mathbb{A} := C \text{ with } n = r \quad \text{and } \mathbb{B} := \begin{pmatrix} D \\ E \end{pmatrix} \text{ with } m = s + k. \tag{3.2.45}$$

With a minor effort, one can recognise that

$$\mathbb{K} := \mathrm{Ker}\mathbb{B} = \mathrm{Ker}D \cap \mathrm{Ker}E \qquad \mathrm{Im}\mathbb{B} = \begin{pmatrix} \mathrm{Im}D \\ \mathrm{Im}E \end{pmatrix} \tag{3.2.46}$$

and that

$$\begin{aligned} &\left\{\mathrm{Ker}\mathbb{B}^T = \begin{pmatrix} \mathbf{0}_s \\ \mathbf{0}_k \end{pmatrix}\right\} \\ &\Leftrightarrow \left\{\{D^T\mathbf{y} + E^T\mathbf{z} = \mathbf{0}_r\} \Rightarrow \{\mathbf{y} = \mathbf{0}_s \text{ and } \mathbf{z} = \mathbf{0}_k\}\right\} \\ &\Leftrightarrow \left\{\mathrm{Im}D^T \cap \mathrm{Im}E^T = \mathbf{0}_r\right\}. \end{aligned} \tag{3.2.47}$$

Conditions (3.2.18) and (3.2.19), in terms of the matrices C, D, and E, are then

$$\begin{aligned} &\mathrm{Im}D^T \cap \mathrm{Im}E^T = \mathbf{0}_r, \\ &C_{\mathbb{K}\mathbb{K}} \text{ is non-singular } \mathbb{K} \to \mathbb{K} \quad \text{where } \mathbb{K} = \mathrm{Ker}D \cap \mathrm{Ker}E. \end{aligned} \tag{3.2.48}$$

It is not difficult to verify that conditions (3.2.48) are necessary and sufficient for the non-singularity of the whole matrix M.

To deal with the *second case* of (3.2.44), we assume instead that C is an $r \times r$ matrix, D is an $s \times r$ matrix, and E a $k \times s$ matrix. We also assume, this time, that $r + k \geq s \geq k$, otherwise, according to Remark 3.2.1, the Matrix M will surely be singular. Possibly the easiest way to apply Theorem 3.2.1 consists in performing first an exchange of rows and columns to reach the form

$$\begin{pmatrix} C & 0 & D^T \\ 0 & 0 & E \\ D & E^T & 0 \end{pmatrix}. \tag{3.2.49}$$

Then, we can take $n = r + k$ and $m = s$ with

$$\mathbb{A} = \begin{pmatrix} C & 0 \\ 0 & 0 \end{pmatrix} \qquad \mathbb{B} = \begin{pmatrix} D & E^T \end{pmatrix} \qquad \mathbb{B}^T = \begin{pmatrix} D^T \\ E \end{pmatrix}. \tag{3.2.50}$$

It is now immediate to see that

$$\mathrm{Ker}\mathbb{B}^T = \mathrm{Ker}D^T \cap \mathrm{Ker}E$$

so that condition (3.2.19) (that now becomes: $\mathrm{Ker}\mathbb{B}^T = \mathbf{0}_s = \mathbf{0}_m$) requires in this case that $\mathrm{Ker}D^T \cap \mathrm{Ker}E = \mathbf{0}_s$. Then, we have to look at the kernel of $\mathbb{B}$ and require the non-singularity of $\mathbb{A}$ on it. It is clear that the kernel of $\mathbb{B}$, in this case, is given by

$$\mathbb{K} = \{(\mathbf{x}, \mathbf{z}) \in \mathbb{R}^r \times \mathbb{R}^k \text{ such that } D\mathbf{x} + E^T\mathbf{z} = \mathbf{0}_s\}. \tag{3.2.51}$$

This includes all pairs of the form $(\mathbf{0}_r, \tilde{\mathbf{z}})$, with $\tilde{\mathbf{z}} \in \mathrm{Ker}E^T$. When we apply the matrix $\mathbb{A}$ to one of these vectors, we obviously obtain the zero vector. Hence, if we want the restriction of $\mathbb{A}$ to $\mathbb{K}$ to be non-singular, we must first require that these pairs are reduced to $(\mathbf{0}_r, \mathbf{0}_k)$, that is, we must require first that $\mathrm{Ker}E^T = \mathbf{0}_k$. However, $\mathbb{K}$ might also contain pairs $(\mathbf{x}, \mathbf{z})$ with $\mathbf{x} \neq \mathbf{0}_r$, provided $D\mathbf{x} \in \mathrm{Im}E^T$. This subset of $\mathbb{R}^r$ can be characterised, using also (3.1.60), as

$$\begin{aligned} \tilde{\mathbb{K}} &= \{\mathbf{x} \in \mathbb{R}^r \text{ such that } D\mathbf{x} = E^T\mathbf{z} \text{ for some } \mathbf{z} \in \mathbb{R}^k\} \\ &\equiv \{\mathbf{x} \in \mathbb{R}^r \text{ such that } \tilde{\mathbf{z}}^T D\mathbf{x} = 0 \quad \forall \tilde{\mathbf{z}} \in \mathrm{Ker}E\}. \end{aligned} \tag{3.2.52}$$

Hence, the conditions for the *second case* can be summarised in terms of the matrices C, D and E as:

$$\begin{aligned} &\mathrm{Ker}D^T \cap \mathrm{Ker}E = \mathbf{0}_s, \\ &\mathrm{Ker}E^T = \mathbf{0}_k, \\ &C_{\tilde{\mathbb{K}}\tilde{\mathbb{K}}} \text{ is non-singular } \tilde{\mathbb{K}} \to \tilde{\mathbb{K}} \quad \text{where } \tilde{\mathbb{K}} \text{ is given in (3.2.52).} \end{aligned} \tag{3.2.53}$$

Again, it is not difficult to verify that conditions (3.2.53) are necessary and sufficient for the non-singularity of the whole matrix.

There are obviously other equivalent ways to apply Theorem 3.2.1. For instance, we can, in both cases, consider directly $n = r + s$, $m = k$ and

$$\mathbb{A} = \begin{pmatrix} C & D^T \\ D & 0 \end{pmatrix} \qquad \mathbb{B} = (E \quad 0) \text{ or } \mathbb{B} = (0 \quad E). \tag{3.2.54}$$

As we are dealing with necessary and sufficient conditions, we would find exactly the same conditions as before, possibly with a longer argument.

In a similar way, one could treat the case when the space $\mathbb{R}^m \times \mathbb{R}^n$ is split into a bigger number of subspaces (four, five, etc.). We do not insist too much on these exercises.

3.3 The Solvability Problem for Perturbed Matrices

A different, more interesting variant arises when we consider the case of systems of the type

$$\begin{pmatrix} A & B^T \\ B & -C \end{pmatrix} \begin{pmatrix} \mathbf{x} \\ \mathbf{y} \end{pmatrix} = \begin{pmatrix} \mathbf{f} \\ \mathbf{g} \end{pmatrix}, \tag{3.3.1}$$

where again A and B are $n \times n$ and $m \times n$ matrices, respectively, and C is an $m \times m$ matrix. The name of the game here is to see C as a perturbation of the original problem (3.2.1). We shall therefore assume that matrices A and B satisfy (3.2.18) and (3.2.19), plus, possibly, some minor additional requirement, and we look for conditions on C in order to have the unique solvability of (3.3.1).

The minus sign in front of the matrix C is due to the fact that, in what follows, we are going to assume the perturbation, in some sense, to be *negative* (and hence C to be positive), in order to have existence and uniqueness results.

3.3.1 Preliminary Results

A first sufficient condition for solvability is quite obvious.

Proposition 3.3.1. *Assume that A and C are positive definite. Then problem* (3.3.1) *is uniquely solvable.* □

Indeed, it is easy to check that in this case the matrix

$$\begin{pmatrix} A & B^T \\ -B & C \end{pmatrix} \tag{3.3.2}$$

is positive definite.

Another more or less obvious sufficient condition is given in the following proposition.

Proposition 3.3.2. *Assume that* (3.2.18) *and* (3.2.19) *are satisfied. Then there exists an* $\varepsilon > 0$ *such that, for every* $m \times m$ *matrix* C *satisfying*

$$\|C\mathbf{y}\| \le \varepsilon \|\mathbf{y}\|, \quad \forall \mathbf{y} \in \mathbb{R}^m, \tag{3.3.3}$$

problem (3.3.1) *is uniquely solvable.* □

The proof is based on the following obvious fact: if the determinant of a matrix is different from zero, and if we perturb the matrix by a small enough quantity, the determinant will still be different from zero. We omit the mathematical details.

In the next subsection, we shall provide a theorem that is more interesting, and more relevant for the applications. In order to prove it, however, we are going to need the following elementary (and classical) lemma, that will also be useful in other occasions.

Lemma 3.3.1. *Assume that* A *is a symmetric* $n \times n$ *matrix satisfying*

$$\mathbf{x}^T A\mathbf{x} \ge 0, \quad \forall \mathbf{x} \in \mathbb{R}^n \tag{3.3.4}$$

(that is: A *is positive semi-definite). Then, for every* $\mathbf{x} \in \mathbb{R}^n$ *and for every* $\mathbf{z} \in \mathbb{R}^n$*, we have*

$$(\mathbf{z}^T A\mathbf{x})^2 \le (\mathbf{x}^T A\mathbf{x})\,(\mathbf{z}^T A\mathbf{z}), \tag{3.3.5}$$

and consequently, always for every $\mathbf{x} \in \mathbb{R}^n$*,*

$$\mathbf{x}^T A\mathbf{x} = 0 \quad \Rightarrow \quad A\mathbf{x} = 0. \tag{3.3.6}$$

Proof. Using (3.3.4), we easily have that, for any $\mathbf{z} \in \mathbb{R}^n$ and for any real number λ,

$$(\mathbf{x} + \lambda\mathbf{z})^T A(\mathbf{x} + \lambda\mathbf{z}) \ge 0. \tag{3.3.7}$$

Expanding (3.3.7) in powers of λ and using the symmetry of A, we have

$$\mathbf{x}^T A\mathbf{x} + 2\lambda\mathbf{z}^T A\mathbf{x} + \lambda^2\mathbf{z}^T A\mathbf{z} \ge 0, \tag{3.3.8}$$

implying that the equation (in the unknown λ) $\mathbf{x}^T A\mathbf{x} + 2\lambda\mathbf{z}^T A\mathbf{x} + \lambda^2\mathbf{z}^T A\mathbf{z} = 0$ cannot have distinct real roots, and therefore

$$\Delta \equiv (2\mathbf{z}^T A\mathbf{x})^2 - 4(\mathbf{x}^T A\mathbf{x})\,(\mathbf{z}^T A\mathbf{z}) \le 0, \tag{3.3.9}$$

which, divided by four, gives exactly (3.3.5). From this we see that $\mathbf{x}^T A\mathbf{x} = 0$ implies that $\mathbf{z}^T A\mathbf{x} = 0$ *for all* $\mathbf{z} \in \mathbb{R}^n$, and therefore $A\mathbf{x} = 0$. This is what is claimed in (3.3.6). □

3.3.2 Main Results

We are now ready to present the main theorem of this section.

Theorem 3.3.1. *Let A be an $n \times n$ matrix, B an $m \times n$ matrix and let C be an $m \times m$ matrix. Assume (as in the basic Theorem 3.2.1) that B^T is injective and A_{KK} is non-singular from K to K, where $K = \operatorname{Ker} B$. Assume further that A and C are positive semi-definite and that, moreover, A is symmetric. Then, problem* (3.3.1) *is uniquely solvable for every right-hand side* **f**, **g**.

Proof. The proof can be easily done by showing that the homogeneous version of (3.3.1) (that is when **f** and **g** are both equal to zero) has $\mathbf{x} = 0, \mathbf{y} = 0$ as the unique solution. For this, let $(\mathbf{x}, \mathbf{y})$ be the solution of the homogeneous system. Taking the scalar product of the first equation of (3.3.1) times **x**, we get

$$\mathbf{x}^T A\mathbf{x} + \mathbf{x}^T B^T \mathbf{y} = 0, \tag{3.3.10}$$

while, taking the scalar product of the second equation of (3.3.1) times y, we obtain

$$\mathbf{y}^T B\mathbf{x} - \mathbf{y}^T C\mathbf{y} = 0. \tag{3.3.11}$$

Subtracting (3.3.11) from (3.3.10), and using (3.1.5), we therefore have

$$\mathbf{x}^T A\mathbf{x} + \mathbf{y}^T C\mathbf{y} = 0. \tag{3.3.12}$$

Using the fact that A and C are positive semi-definite in (3.3.12), we then have

$$\mathbf{x}^T A\mathbf{x} = \mathbf{y}^T C\mathbf{y} = 0. \tag{3.3.13}$$

We can now use (3.3.13) and Lemma 3.3.1 to deduce that $A\mathbf{x} = 0$. Using this in the first equation, we obtain now $B^T\mathbf{y} = 0$ which, as B^T is injective, implies $\mathbf{y} = 0$. This, in turn, gives $C\mathbf{y} = 0$, so that, from the second equation, $B\mathbf{x} = 0$. Hence, **x** belongs to $\operatorname{Ker} B$. Having already $A\mathbf{x} = 0$, we deduce $A_{KK}\mathbf{x} = 0$, and since A_{KK} is non-singular $K \to K$, we conclude that **x** is also equal to zero. □

Remark 3.3.1. Looking at the proof of Theorem 3.3.1, we also see that we can trade the *symmetry* assumption on A with the condition that A is *positive definite on the whole* $\mathbb{R}^n$. Indeed, the symmetry was only used in Lemma 3.3.1 to show that $\mathbf{x}^T A\mathbf{x} = 0$ implies $A\mathbf{x} = 0$. If A is supposed to be positive definite, from $\mathbf{x}^T A\mathbf{x} = 0$ we have immediately $\mathbf{x} = 0$ and then $\mathbf{y} = 0$ as before. □

Theorem 3.3.1 has a counterpart, in which the symmetry assumption is shifted from A to C.

Theorem 3.3.2. *Let A be an $n \times n$ matrix, B an $m \times n$ matrix and let C be an $m \times m$ matrix. Assume (as in the basic Theorem 3.2.1) that B^T is injective and A_{KK} is non-singular from K to K, where $K = \operatorname{Ker} B$. Assume further that A and C are* *positive semi-definite and moreover that C is symmetric. Then, problem* (3.3.1) *is uniquely solvable for every right-hand side* **f**, **g**.

Proof. We proceed exactly as in the proof of Theorem 3.3.1. Let $(\mathbf{x}, \mathbf{y})$ be a solution of the homogeneous system. Taking the scalar products of the first equation times x, the scalar product of the second equation times y, and finally taking the difference, we reach again (3.3.12) and (3.3.13). This time, we apply Lemma 3.3.1 to the matrix C, obtaining $C\mathbf{y} = 0$. Then we can go back to Theorem 3.2.1 and, using (3.2.18) and (3.2.19), we obtain $\mathbf{x} = 0$ and $\mathbf{y} = 0$. □

Remark 3.3.2. The above results could be summarised as follows. Assume that A and B verify the assumptions of the basic Theorem 3.2.1, that is: B is surjective (or, equivalently, B^T is injective) and A_{KK} is non-singular from K to K, where K is the kernel of B. Then, problem (3.3.1) is uniquely solvable under the following assumptions:

- A and C are positive semi-definite and A is symmetric;
- A is positive definite and C is positive semi-definite;
- A and C are positive semi-definite and C is symmetric. □

3.3.3 Examples

In the following Examples, we shall discuss the *necessity* of the conditions that we have used so far. The form (3.3.1) is clearly too general to allow non-trivial necessary and sufficient conditions. We shall therefore discuss the possibility of finding more general, but still easy, sufficient conditions.

In the first example, we shall see that the symmetry assumptions in Theorem 3.3.1 or in Theorem 3.3.2 cannot be easily reduced. Indeed, if we consider the case

$$A = \begin{pmatrix} 1 & 0 & 0 \\ 0 & 1 & -1 \\ 0 & 1 & 0 \end{pmatrix} \quad B = \begin{pmatrix} 0 & 1 & 0 \\ 0 & 0 & 1 \end{pmatrix} \quad C = \begin{pmatrix} 0 & -1 \\ 1 & 1 \end{pmatrix}, \tag{3.3.14}$$

we see that A and C are positive semi-definite, B is surjective and A is non-singular when restricted to the Ker$B$ which in this case is $\{\mathbf{x} \in \mathbb{R}^3$ such that $x_2 = x_3 = 0\}$. Hence, all the assumptions of Theorem 3.3.1 are satisfied but the symmetry assumption (since *neither A nor C is symmetric*). It is easy to see that the whole matrix

$$M = \begin{pmatrix} 1 & 0 & 0 & 0 & 0 \\ 0 & 1 & -1 & 1 & 0 \\ 0 & 1 & 0 & 0 & 1 \\ 0 & 1 & 0 & 0 & 1 \\ 0 & 0 & 1 & -1 & -1 \end{pmatrix} \tag{3.3.15}$$

is singular, since the third and fourth rows are equal. Note that A is symmetric when restricted to Ker$B$, but this is not enough.

On the other hand, it is obvious that we cannot give up the assumption that A and C have, in some weak sense, the same sign, because the elementary choice

$$A = (1) \qquad B = (1) \qquad C = (-1) \tag{3.3.16}$$

gives rise to the singular matrix

$$M = \begin{pmatrix} 1 & 1 \\ 1 & 1 \end{pmatrix}. \tag{3.3.17}$$

Similarly, we cannot even accept that one of the two matrices, A or C, is indefinite: for instance, the choice

$$A = \begin{pmatrix} 0 & 1 \\ 1 & 1 \end{pmatrix} \qquad B = \begin{pmatrix} 1 & 0 \end{pmatrix} \qquad C = (1) \tag{3.3.18}$$

with C symmetric and positive definite and A symmetric but indefinite, produces the singular matrix

$$M = \begin{pmatrix} 0 & 1 & 1 \\ 1 & 1 & 0 \\ 1 & 0 & -1 \end{pmatrix}. \tag{3.3.19}$$

Hence, although the conditions discussed in Theorems 3.3.1 and 3.3.2 are clearly only sufficient and by no way necessary, it does not seem easy to write down more convenient ones.

3.4 Stability

We saw at the beginning of this Chapter that *solvability* will not be sufficient to provide a *good method* to discretise partial differential equations, and some *stability* (in a sense to be made precise) is actually needed.

Here, we suppose that we are actually given a *sequence* of problems with increasing dimensions. It is clear that this will be the case when we are going to consider discretisations of a given, say, partial differential equation, with a sequence of finer and finer meshes. Consider therefore for $k = 1, 2, \ldots$ the problems

$$\begin{pmatrix} A_k & B_k^T \\ B_k & 0 \end{pmatrix} \begin{pmatrix} \mathbf{x}_k \\ \mathbf{y}_k \end{pmatrix} = \begin{pmatrix} \mathbf{f}_k \\ \mathbf{g}_k \end{pmatrix}, \tag{3.4.1}$$

where A_k is an $n_k \times n_k$ matrix, B_k an $m_k \times n_k$ matrix, and the dimensions n_k and m_k tend to infinity when k goes to infinity. Roughly speaking, we can imagine that each value of k will correspond to a different decomposition, and when we will say that some constant is *independent of the decomposition*, we will actually mean that it does not depend on the index k in (3.4.1).

We are therefore interested in conditions that ensure not only the unique solvability of each problem (3.4.1), but also a stability estimate of the type (3.0.3):

$$\|\mathbf{x}_k\| + \|\mathbf{y}_k\| \leq c(\|\mathbf{f}_k\| + \|\mathbf{g}_k\|), \tag{3.4.2}$$

where the constant c *does not depend on k*. This requirement is obviously meaningless, unless we specify the norms that we intend to use. As anticipated at the beginning of this chapter, the choice of the norms, in this case, is not irrelevant: although they are all equivalent, the *constants* involved in the equivalence may (and, in general, do) depend on the dimensions, which we are assuming to be going to infinity.

On the other hand, if we want to use these abstract results in order to provide a priori error bounds for some realistic discretisation of a differential problem, we are not totally free in the choice of the norms.

In general, in the finite element context, the norms to be used will be the norms in some functional space, where the differential problem itself is set. Hence, in practice, we are going to have little choice.

For instance (anticipating some ideas from the following chapters), our unknown vector $\mathbf{x}$ could represent the *nodal values* of a piecewise linear continuous function defined on a domain Ω that has been decomposed into triangles T. This means that we have a one-to-one mapping from $\mathbb{R}^n$ to the space $\mathcal{L}_1^1$ of piecewise linear continuous functions on Ω, that associates to a vector $\mathbf{v}$ in $\mathbb{R}^n$ the function $\varphi_{\mathbf{v}}$ such that, at every node N_j of the decomposition ($j = 1, 2 \ldots n$), we have $\varphi_{\mathbf{v}}(N_j) = v_j$. In this case, a very natural choice of norm for $\mathbf{v}$ would be

$$\|\mathbf{v}\|_0 := \Big(\int_\Omega \varphi_{\mathbf{v}}^2 \, d\Omega\Big)^{1/2}, \tag{3.4.3}$$

or, alternatively,

$$\|\mathbf{v}\|_1 := \Big(\int_\Omega |\nabla\varphi_{\mathbf{v}}|^2 \, d\Omega\Big)^{1/2}, \tag{3.4.4}$$

representing, respectively, the L^2-norm and the H_0^1-norm of the corresponding function $\varphi_{\mathbf{v}}$ (if this function vanishes on boundary nodes). At the present level, however, we have no functional spaces yet (nor, for what matters, a differential problem). Hence, we are going to consider norms, or, rather, families of norms, that are defined independently of functional spaces and discretisation schemes. However, having that target in mind, we shall make assumptions that are somehow tailored for it. In the present section, we shall then reconsider several aspects that were discussed in Sect. 3.2 but, this time, introducing norms, and analysing the behaviour of the various constants in dependence of the chosen norms.

For the sake of simplicity, from now on we shall drop the index k unless it will really be necessary, and we will just **remember** that m, n, A, and B depend on k.

Remark 3.4.1. We point out that, as we have already seen, **stability** is not a concept that can be applied to a single discretised problem, but only to a **sequence** of discretised problems, or to a discretisation **method** (that in turn gives rise to sequences of discretised problems). □

Remark 3.4.2. **Important warning** In what follows, we will often consider the *infimum* or the *supremum* of quotients of the type

$$\frac{\ell(\boldsymbol{\xi})}{\|\boldsymbol{\xi}\|} \quad \text{or} \quad \frac{|\ell(\boldsymbol{\xi})|}{\|\boldsymbol{\xi}\|} \tag{3.4.5}$$

where $\ell(\boldsymbol{\xi})$ is a real number depending linearly on $\boldsymbol{\xi}$. It is clear that the quotients in (3.4.5) make no sense for $\boldsymbol{\xi} = \mathbf{0}$, so that the value $\boldsymbol{\xi} = \mathbf{0}$ should be discarded when taking the *infimum* or the *supremum*. On the other hand, due to the linearity of ℓ, it is clear that for every $\boldsymbol{\xi}_0 \neq \mathbf{0}$ the quotients in (3.4.5) take the same value over the ray $\boldsymbol{\xi} = \kappa\boldsymbol{\xi}_0$ when κ ranges over the positive real numbers. Hence, the limit of the quotients (3.4.5) for $\boldsymbol{\xi} \to \mathbf{0}$, in general, will not exist (we would have a different limit on every ray coming out of the origin), but the meaning of, say,

$$\sup_{\boldsymbol{\xi}} \frac{\ell(\boldsymbol{\xi})}{\|\boldsymbol{\xi}\|} \tag{3.4.6}$$

will not be "seriously ambiguous". Hence, for the sake of brevity, we shall write in these cases

$$\sup_{\boldsymbol{\xi}} \frac{\ell(\boldsymbol{\xi})}{\|\boldsymbol{\xi}\|} \quad \text{instead of} \quad \sup_{\boldsymbol{\xi}\neq\mathbf{0}} \frac{\ell(\boldsymbol{\xi})}{\|\boldsymbol{\xi}\|}. \tag{3.4.7}$$

□

3.4.1 Assumptions on the Norms

We denote by **X**, **Y**, **F**, **G**, respectively, the spaces of vectors **x**, **y**, **f**, **g**. Hence, we have

$$\mathbf{X} = \mathbb{R}^n, \quad \mathbf{Y} = \mathbb{R}^m, \quad \mathbf{F} = \mathbb{R}^n, \quad \mathbf{G} = \mathbb{R}^m. \tag{3.4.8}$$

Then, we assume that:

1. The spaces **X** and **Y** are equipped with norms $\|\cdot\|_X$ and $\|\cdot\|_Y$. For the sake of simplicity, we will assume that there exist two symmetric and positive definite matrices S_X (an $n \times n$ matrix) and S_Y (an $m \times m$ matrix) such that

$$\begin{aligned} \|\mathbf{x}\|_X^2 &= (S_X\,\mathbf{x})^T(S_X\,\mathbf{x}) \equiv \mathbf{x}^T S_X^T S_X \mathbf{x} \quad \forall\, \mathbf{x} \in \mathbf{X}, \\ \|\mathbf{y}\|_Y^2 &= (S_Y\,\mathbf{y})^T(S_Y\,\mathbf{y}) \equiv \mathbf{y}^T S_Y^T S_Y \mathbf{y} \quad \forall\, \mathbf{y} \in \mathbf{Y}. \end{aligned} \tag{3.4.9}$$

2. the spaces **F** and **G** are equipped with norms $\|\cdot\|_F$ and $\|\cdot\|_G$ defined as the **dual norms** of $\|\cdot\|_X$ and $\|\cdot\|_Y$, i.e.

$$\|\mathbf{f}\|_F := \sup_{\mathbf{x}\in\mathbf{X}} \frac{\mathbf{x}^T\mathbf{f}}{\|\mathbf{x}\|_X} \qquad \text{and} \qquad \|\mathbf{g}\|_G := \sup_{\mathbf{y}\in\mathbf{Y}} \frac{\mathbf{y}^T\mathbf{g}}{\|\mathbf{y}\|_Y}. \tag{3.4.10}$$

It is not difficult to check that

$$\begin{aligned} \|\mathbf{f}\|_F^2 &= (S_X^{-1}\mathbf{f})^T(S_X^{-1}\mathbf{f}) \equiv \mathbf{f}^T S_X^{-T} S_X^{-1}\mathbf{f} \;\; \forall\, \mathbf{f} \in \mathbf{F}, \\ \|\mathbf{g}\|_G^2 &= (S_Y^{-1}\mathbf{g})^T(S_Y^{-1}\mathbf{g}) \equiv \mathbf{g}^T S_Y^{-T} S_Y^{-1}\mathbf{g} \;\; \forall\, \mathbf{g} \in \mathbf{G}. \end{aligned} \tag{3.4.11}$$

3. Given the norms in **X**, **Y**, **F** and **G**, we can define the **induced norms** of the matrices A and B as follows

$$\|A\| := \sup_{\mathbf{x}\in\mathbf{X}} \frac{\|A\mathbf{x}\|_{\mathbf{F}}}{\|\mathbf{x}\|_{\mathbf{X}}} \qquad \|B\| := \sup_{\mathbf{x}\in\mathbf{X}} \frac{\|B\mathbf{x}\|_{\mathbf{G}}}{\|\mathbf{x}\|_{\mathbf{X}}}. \tag{3.4.12}$$

4. The norms of the transposed matrices A^T and B^T are obviously defined in the same way as in (3.4.12). Moreover, we have the following immediate result.

Proposition 3.4.1. *In the above assumptions, we have*

$$\|A\| = \|A^T\| \equiv \sup_{\mathbf{x}\in\mathbf{X}} \sup_{\mathbf{z}\in\mathbf{X}} \frac{\mathbf{z}^T A\mathbf{x}}{\|\mathbf{z}\|_{\mathbf{X}} \|\mathbf{x}\|_{\mathbf{X}}} \tag{3.4.13}$$

and

$$\|B\| = \|B^T\| \equiv \sup_{\mathbf{x}\in\mathbf{X}} \sup_{\mathbf{y}\in\mathbf{Y}} \frac{\mathbf{y}^T B\mathbf{x}}{\|\mathbf{y}\|_{\mathbf{Y}} \|\mathbf{x}\|_{\mathbf{X}}}. \tag{3.4.14}$$

□

The proof follows immediately from (3.1.5), which implies that $\mathbf{z}^T A\mathbf{x} = \mathbf{x}^T A^T \mathbf{z}$ and $\mathbf{y}^T B\mathbf{x} = \mathbf{x}^T B^T \mathbf{y}$.

5. We will assume that *there exist two constants M_a and M_b, independent of the mesh-size, such that*

$$\|A\| = \|A^T\| \le M_a \qquad \|B\| = \|B^T\| \le M_b. \tag{3.4.15}$$

Sometimes, for **K** a subspace of **X**, we will also use the norm

$$\|\mathbf{f}\|_{K'} := \sup_{\mathbf{x}\in\mathbf{K}} \frac{\mathbf{x}^T\mathbf{f}}{\|\mathbf{x}\|_X}. \tag{3.4.16}$$

The following very useful properties are immediate consequences of the above assumptions.

Proposition 3.4.2. *Assume that the properties* (3.4.8)–(3.4.15) *hold true. Then, for every* $\mathbf{x}$ *and* $\mathbf{f}$ *in* $\mathbb{R}^n$ *and for every* $\mathbf{y}$ *and* $\mathbf{g}$ *in* $\mathbb{R}^m$, *we have*

$$\mathbf{x}^T\mathbf{f} \le \|\mathbf{x}\|_X\,\|\mathbf{f}\|_F, \qquad \mathbf{y}^T\mathbf{g} \le \|\mathbf{y}\|_Y\,\|\mathbf{g}\|_G, \tag{3.4.17}$$

$$\|A\mathbf{x}\|_F \le M_a\|\mathbf{x}\|_X, \qquad \|A^T\mathbf{x}\|_F \le M_a\|\mathbf{x}\|_X, \tag{3.4.18}$$

$$\|B\mathbf{x}\|_G \le M_b\|\mathbf{x}\|_X, \qquad \|B^T\mathbf{y}\|_F \le M_b\|\mathbf{y}\|_Y, \tag{3.4.19}$$

$$\mathbf{x}^T A\mathbf{z} \le M_a\|\mathbf{x}\|_X\,\|\mathbf{z}\|_X, \qquad \mathbf{x}^T B^T\mathbf{y} \le M_b\|\mathbf{x}\|_X\,\|\mathbf{y}\|_Y, \tag{3.4.20}$$

and

$$\|\mathbf{f}\|_{K'} \le \|\mathbf{f}\|_F. \tag{3.4.21}$$

If moreover A *is symmetric and positive semi-definite, then* (3.4.18) *can be improved to*

$$\|A\mathbf{x}\|_F \le M_a^{1/2}(\mathbf{x}^T A\mathbf{x})^{1/2}. \tag{3.4.22}$$

Proof. The proof of (3.4.17) is immediate. For instance, the first inequality follows from the fact that for every fixed $\tilde{\mathbf{x}} \in \mathbf{X} \setminus \{0\}$ we obviously have

$$\frac{\tilde{\mathbf{x}}^T\mathbf{f}}{\|\tilde{\mathbf{x}}\|_X} \le \sup_{\mathbf{x}\in\mathbf{X}} \frac{\mathbf{x}^T\mathbf{f}}{\|\mathbf{x}\|_X} \equiv \|\mathbf{f}\|_F, \tag{3.4.23}$$

which multiplied by $\|\tilde{\mathbf{x}}\|_X$ gives $\tilde{\mathbf{x}}^T\mathbf{f} \le \|\tilde{\mathbf{x}}\|_X\,\|\mathbf{f}\|_F$. The second one can be proven in exactly the same way. The proof of (3.4.18) and (3.4.19) is also immediate, as is the proof of (3.4.21) (in the right-hand side we take the supremum over a bigger set). Let us see for instance the proof of (3.4.18) (as the proofs of the other two are identical): using first (3.4.12) and then (3.4.15), we have:

$$\|A\mathbf{x}\|_F \le \|A\|\,\|\mathbf{x}\|_{\mathbf{X}} \le M_a\,\|\mathbf{x}\|_{\mathbf{X}}. \tag{3.4.24}$$

Property (3.4.20) will then follow immediately from (3.4.18) and (3.4.19), and the proof of (3.4.21) is immediate. Finally, for the proof of (3.4.22), we can first use Lemma 3.3.1, which, for every $\mathbf{x}, \mathbf{z} \in \mathbf{X}$, gives

$$|\mathbf{z}^T A\mathbf{x}| \le (\mathbf{z}^T A\mathbf{z})^{1/2}(\mathbf{x}^T A\mathbf{x})^{1/2}. \tag{3.4.25}$$

Then, we use (3.4.10), (3.4.25), and (3.4.20) to get

$$\begin{aligned}\|A\mathbf{x}\|_F &= \sup_{\mathbf{z}\in\mathbf{X}} \frac{\mathbf{z}^T A\mathbf{x}}{\|\mathbf{z}\|_X} \le \sup_{\mathbf{z}\in\mathbf{X}} \frac{(\mathbf{z}^T A\mathbf{z})^{1/2}(\mathbf{x}^T A\mathbf{x})^{1/2}}{\|\mathbf{z}\|_X} \\ &\le \sup_{\mathbf{z}\in\mathbf{X}} \frac{(M_a\|\mathbf{z}\|_X^2)^{1/2}(\mathbf{x}^T A\mathbf{x})^{1/2}}{\|\mathbf{z}\|_X} = M_a^{1/2}(\mathbf{x}^T A\mathbf{x})^{1/2}.\end{aligned} \tag{3.4.26}$$

□

From now on, in this chapter, the Euclidean norm will be denoted by $\|\cdot\|_E$, that is

$$\|\mathbf{z}\|_E^2 := \mathbf{z}^T\mathbf{z}. \tag{3.4.27}$$

The following proposition is an elementary consequence of Corollary 3.1.4.

Proposition 3.4.3. *Let B be an $m \times n$ matrix, and set $K := Ker B$ (as usual) and $H := Ker B^T$. Then, there exists a positive constant $\tilde{\beta}$ such that*

$$\inf_{\mathbf{y}\in H^\perp}\sup_{\mathbf{x}\in K^\perp}\frac{\mathbf{x}^T B^T\mathbf{y}}{\|\mathbf{x}\|_X\,\|\mathbf{y}\|_Y} = \inf_{\mathbf{x}\in K^\perp}\sup_{\mathbf{y}\in H^\perp}\frac{\mathbf{y}^T B\mathbf{x}}{\|\mathbf{x}\|_X\,\|\mathbf{y}\|_Y} = \tilde{\beta} > 0. \tag{3.4.28}$$

Moreover, with the notation of Proposition 3.1.1, we have exactly

$$\frac{1}{\tilde{\beta}} \equiv \|L_B\| \equiv \|L_{B^T}\|. \tag{3.4.29}$$

Proof. Corollary (3.1.4) implies that B is one-to-one from $K^\perp$ to $H^\perp$ and B^T is one-to-one from $K^\perp$ to $H^\perp$. It is not difficult to see that $\tilde{\beta}$ in (3.4.28) is exactly the value of the norms of L_B and L_{B^T} (that are equal to each other). See also Examples 3.1.5 and 3.1.8.

We are now ready to introduce a *precise definition of stability*.

Definition of stability. *Given a numerical method that produces a sequence of matrices A and B when applied to a given sequence of meshes (with the mesh-size h going to zero), we choose norms $\|\cdot\|_X$ and $\|\cdot\|_Y$ that satisfy the continuity condition* (3.4.20), *and dual norms $\|\cdot\|_F$ and $\|\cdot\|_G$ according to* (3.4.10). *Then, we say that* ***the method is stable*** *if there exists a constant c,* ***independent of the mesh size****, such that for all vectors $\mathbf{x}, \mathbf{y}, \mathbf{f}, \mathbf{g}$ satisfying the general system* (3.2.1) *and* (3.2.2), *it holds*

$$\|\mathbf{x}\|_X + \|\mathbf{y}\|_Y \le c(\|\mathbf{f}\|_F + \|\mathbf{g}\|_G). \tag{3.4.30}$$

Remark 3.4.3. We recall (as we have also seen in Remark 3.1.7) that for a square matrix, we have unique solvability for every right-hand side if and only if the only solution of the homogeneous system is the zero solution. We note here that (3.4.30) implies that, whenever $\mathbf{f}$ and $\mathbf{g}$ are zero, the only possible solution of (3.2.1) and (3.2.2) is $\mathbf{x} = \mathbf{0}$ and $\mathbf{y} = \mathbf{0}$. Hence, we deduce that (3.4.30) implies the unique solvability of (3.2.1) and (3.2.2). This is the reason why, on several occasions in this section, we will state theorems that ensure the stability (3.4.30) without mentioning explicitly that we have unique solvability for every right-hand side $\mathbf{f}$ and $\mathbf{g}$. □

Having now a precise definition of stability, we can look for suitable assumptions on the matrices A and B that may provide the stability result (3.4.30). In Sect. 3.2, we started with the basic Theorem 3.2.1, giving the necessary and sufficient conditions for solvability, and then we discussed possible variants with stronger assumptions which gave only sufficient conditions but were easier to deal with. In the present section, we shall follow somehow the opposite path: we shall start with stronger assumptions (allowing an easier proof) and move progressively towards weaker assumptions.

In particular, as we did in the previous sections, we will consider essentially three possible situations, with three different levels of generality. In all three cases, we shall assume an *inf-sup* condition on the matrix B. On the other hand, for the matrix A, we shall consider the three cases: ellipticity on the whole space V, ellipticity only on the kernel K, and a non-singularity condition on A_{KK} of the type of (3.2.18).

Different assumptions on the *symmetry* of A will often affect the dependence of the final stability constants on the *inf-sup* and ellipticity constants.

As a first step, however, we shall discuss the basic assumption to be made on the matrix B (the *inf-sup* condition) that will be used in all the theorems of the Section. In several applications, checking whether the *inf-sup* condition holds or not will be *the main difficulty*. It is therefore necessary to try to have a good understanding of it.

3.4.2 *The* inf-sup *Condition for the Matrix* B*: An Elementary Discussion*

As we are going to see at the end of this subsection, with the definitions and the notation that we introduced in the previous part of this chapter, the so-called *inf-sup* condition can be expressed rather quickly.

However, as it is often one of the main difficulties (to check or to enforce) in many applications, we expect a certain number of readers to pick up the book and start reading this subsection first.

This, clearly, is not recommended, and, frankly speaking, cannot be done. Nevertheless, we tried, in the beginning of this subsection, to be softer than usual, rephrasing many concepts that were seen before, and (if not really restarting from scratch, that would be a total nonsense) to recover some concepts in a more heuristic way.

Let us start from one of its most common formulations.

Inf-sup condition on B. *There exists a positive constant β,* ***independent of the mesh-size*** *h, such that:*

$$\forall\, \mathbf{y} \in \mathbf{Y} \quad \exists\, \mathbf{x} \in \mathbf{X} \setminus \{\mathbf{0}\} \text{ such that } \mathbf{x}^T B^T \mathbf{y} \geq \beta \|\mathbf{x}\|_X \|\mathbf{y}\|_Y. \tag{3.4.31}$$

In order to understand it better, we start by rewriting condition (3.4.31) in different equivalent forms, which will also clarify the reason why it is called *inf-sup condition*.

Since, by assumption, $\mathbf{x}$ is different from zero, condition (3.4.31) can equivalently be written as:

$$\forall\, \mathbf{y} \in \mathbf{Y} \quad \exists\, \mathbf{x} \in \mathbf{X}\backslash\{\mathbf{0}\} \quad \text{such that} \quad \frac{\mathbf{x}^T B^T \mathbf{y}}{\|\mathbf{x}\|_X} \geq \beta \|\mathbf{y}\|_Y. \tag{3.4.32}$$

Given $\mathbf{y} \in \mathbf{Y}$, the most suitable $\mathbf{x} \in \mathbf{X}$ (for making the inequality in (3.4.32) hold) is clearly the one that makes the left-hand side of the inequality as big as possible. Hence, the best we can do is to take the *supremum* of the left-hand side when $\mathbf{x}$ varies among all possible $\mathbf{x} \in \mathbf{X}$ different from $\mathbf{0}$. Hence, recalling also the notation in (3.4.7), we may equivalently require that

$$\forall\, \mathbf{y} \in \mathbf{Y} \quad \sup_{\mathbf{x} \in \mathbf{X}} \frac{\mathbf{x}^T B^T \mathbf{y}}{\|\mathbf{x}\|_X} \geq \beta \|\mathbf{y}\|_Y. \tag{3.4.33}$$

In a sense, we got rid of the task of choosing $\mathbf{x}$. We observe that, making use of the notation of (3.4.10) for dual norms, we immediately have

$$\sup_{\mathbf{x} \in \mathbf{X}} \frac{\mathbf{x}^T B^T \mathbf{y}}{\|\mathbf{x}\|_X} \equiv \|B^T \mathbf{y}\|_F, \tag{3.4.34}$$

so that condition (3.4.33) could easily be rewritten as

$$\forall\, \mathbf{y} \in \mathbf{Y} \qquad \|B^T \mathbf{y}\|_F \geq \beta \|\mathbf{y}\|_Y. \tag{3.4.35}$$

We recall now that the usual condition required in the previous section for the matrix B (see (3.2.19)) was: *B is surjective or, equivalently, B^T is injective*. We also recall that the injectivity (3.1.11) could be written as

$$\{\|B^T \mathbf{y}\| = 0\} \;\Rightarrow\; \{\|\mathbf{y}\| = 0\}. \tag{3.4.36}$$

Looking back at the basic algebraic property (3.1.41) (that, in finite dimension, is always true), with $M = B^T$ we see that here we are first asking that the inequality holds for every $\mathbf{y} \in \mathbf{Y}$ (and not, as in (3.1.41), for every $\mathbf{y} \in (\mathrm{Ker} B^T)^\perp$). Hence, we require that, for every k in our sequence, $(\mathrm{Ker} B^T)^\perp = \{\mathbf{0}\}$. Moreover, we require that the constant μ that appears in (3.1.41) is uniformly bounded from below by a uniform constant β.

We also easily recognise that the *inf-sup* condition, in its equivalent form (3.4.35), easily implies (3.4.36). Hence, it can be seen as *a stronger form* of the plain injectivity (3.4.36), depending on the choice of the norms, and requiring *a uniform bound, β, independent of the mesh-sizes.*

However: why is it called *inf-sup* condition? We note that condition (3.4.35) still depends on $\mathbf{y}$. We also note that it clearly always holds for $\mathbf{y} = \mathbf{0}$, and therefore we can concentrate on the $\mathbf{y}$'s that are different from $\mathbf{0}$; in particular, for $\mathbf{y} \neq 0$, condition (3.4.35) can be also written as

$$\forall\, \mathbf{y} \in \mathbf{Y}\backslash\{\mathbf{0}\} \qquad \frac{\|B^T\mathbf{y}\|_F}{\|\mathbf{y}\|_Y} \geq \beta. \tag{3.4.37}$$

The worst possible **y** is therefore the one that makes the left-hand side of (3.4.37) as small as possible. If we want (3.4.37) to hold *for every* $\mathbf{y} \in \mathbf{Y} \setminus \{\mathbf{0}\}$, we might as well consider the worst case, looking directly at the *infimum* of the left-hand side of (3.4.37) among all possible **y**'s, requiring that

$$\inf_{\mathbf{y}\in\mathbf{Y}} \frac{\|B^T\mathbf{y}\|_F}{\|\mathbf{y}\|_Y} \geq \beta, \tag{3.4.38}$$

(still following the notation (3.4.7)) that is, recalling (3.4.34),

$$\inf_{\mathbf{y}\in\mathbf{Y}} \sup_{\mathbf{x}\in\mathbf{X}} \frac{\mathbf{x}^T B^T \mathbf{y}}{\|\mathbf{x}\|_X \,\|\mathbf{y}\|_Y} \geq \beta, \tag{3.4.39}$$

which is possibly the most used equivalent presentation of the assumption, and which gave it its name. The advantage of formulation (3.4.39) over the original formulation (3.4.31), if any, is that we got rid of the dependence on **y** and **x**. Indeed, condition (3.4.39) is now clearly a condition on *the matrix* B, on *the spaces* **X** and **Y** (together with their *norms*), as well as on the crucial *constant* β.

Remark 3.4.4. We point out once more that the *inf-sup* condition is *stronger* than the simple injectivity (3.4.36). Considering for simplicity the matrix

$$B_\theta := \begin{pmatrix} 1 & 0 & 0 \\ 0 & \theta & 0 \end{pmatrix} \tag{3.4.40}$$

and taking the Euclidean norm for all the spaces, we easily see that, for $0 < \theta < 1$,

$$\inf_{\mathbf{y}\in\mathbb{R}^2} \frac{\|B^T\mathbf{y}\|}{\|\mathbf{y}\|} = \inf_{\mathbf{y}\in\mathbb{R}^2} \frac{(y_1^2 + (\theta y_2)^2)^{1/2}}{(y_1^2 + y_2^2)^{1/2}} = \theta.$$

In a sequence of problems, sub-matrices as B_θ can appear, in crucial places, with smaller and smaller θ's. In these cases, for every single problem of the sequence, we shall have a positive infimum in (3.4.38), but there will **not** be a positive uniform β bounding them all from below. □

We collect the previous discussion in the following proposition.

Proposition 3.4.4. *Given a sequence of spaces* **X**, **Y**, *a sequence of matrices* A *and* B *and a single positive constant* β, *then the inf-sup condition* (3.4.31) *is equivalent to*

$$\beta\|\mathbf{y}\|_Y \leq \|B^T\mathbf{y}\|_F. \qquad \forall\, \mathbf{y} \in \mathbf{Y}. \tag{3.4.41}$$

Moreover, recalling Proposition 3.4.3, we have that the inf-sup condition (3.4.31) *is also equivalent to*

$$\exists L_B : \mathbf{G} \to \mathbf{X} \textit{ such that } BL_B\mathbf{g} = \mathbf{g}\ \forall\, \mathbf{g} \in \mathbf{G} \tag{3.4.42}$$

with

$$\beta \|L_B\mathbf{g}\|_X \le \|\mathbf{g}\|_G \qquad \forall\, \mathbf{g} \in \mathbf{G}. \tag{3.4.43}$$

Therefore, in particular, the inf-sup condition (3.4.31) *implies that all the matrices B in the sequence are surjective and all the matrices B^T are injective.* □

3.4.3 The inf-sup *Condition and the Singular Values*

Now we shall see that, using the definitions and the notation of the previous part of this chapter, the discussion of the previous subsection could be drastically shortened. However, first we recall some basic notion on the *singular value decomposition* (see e.g. [228]). Given an $m \times n$ matrix M, it is always possible to find an $n \times n$ unitary matrix U and an $m \times m$ unitary matrix V such that

$$M = V\,\Sigma\,U \tag{3.4.44}$$

where Σ is an $m \times n$ non-negative diagonal matrix. We recall that a rectangular matrix Σ is said to be a *non-negative diagonal matrix* if all its entries are non-negative and for all $i \neq j$ we have $\sigma_{ij} = 0$. On the other hand, an $r \times r$ matrix Λ is said to be a *unitary matrix* when the product $\Lambda^T\,\Lambda$ is equal to the identity $r \times r$ matrix $\mathbb{I}_r$. Note that this implies that $(\Lambda\mathbf{z})^T\Lambda\mathbf{z} = \mathbf{z}^T\mathbf{z}$ for all $\mathbf{z} \in \mathbb{R}^r$, so that Λ *does not change the Euclidean norm.*

In (3.4.44), the diagonal entries of Σ are known as the **singular values** of M. It can be shown that the non-zero singular values of M are the square roots of the non-zero eigenvalues of $M^T M$.

We now focus our attention on a fundamental example already considered in Sect. 3.1.

Example 3.4.1. Let us go back to the Example 3.1.5, and consider the matrix (that we now denote by Σ) given by

$$\Sigma = \begin{pmatrix} \mu_1 & 0 & \cdot & 0 & 0 & 0 & 0 & 0 \\ 0 & \mu_2 & \cdot & 0 & 0 & 0 & 0 & 0 \\ \cdot & \cdot & \cdot & \cdot & \cdot & \cdot & \cdot & \cdot \\ 0 & 0 & \cdot & \mu_k & 0 & 0 & 0 & 0 \\ 0 & 0 & \cdot & 0 & 0 & 0 & 0 & 0 \\ 0 & 0 & \cdot & 0 & 0 & 0 & 0 & 0 \end{pmatrix}, \tag{3.4.45}$$

where again k is the dimension of $(\mathrm{Ker}\Sigma)^{\perp}$, which coincides with the dimension of $(\mathrm{Ker}\Sigma^T)^{\perp}$. Here we have $n = k + 4$ and $m = k + 2$. Assuming that the singular values μ_i have been ordered in decreasing order, that is

$$\mu_1 \geq \mu_2 \geq \ldots . \geq \mu_{k-1} \geq \mu_k, \tag{3.4.46}$$

we clearly have (referring to Corollary 3.1.4)

$$\sup_{\boldsymbol{\xi} \in \mathbb{R}^n} \frac{\|\Sigma \boldsymbol{\xi}\|_E}{\|\boldsymbol{\xi}\|_E} \equiv \mu_1 \qquad \text{and} \quad \sup_{\boldsymbol{\eta} \in \mathrm{Im}\Sigma} \frac{\|L_{\Sigma} \boldsymbol{\eta}\|_E}{\|\boldsymbol{\eta}\|_E} \equiv \mu_k^{-1}, \tag{3.4.47}$$

which, using Proposition 3.4.3, gives immediately

$$\inf_{\boldsymbol{\eta} \in (\mathrm{Ker}\Sigma^T)^{\perp}} \sup_{\boldsymbol{\xi} \in (\mathrm{Ker}\Sigma)^{\perp}} \frac{\boldsymbol{\eta}^T \Sigma \boldsymbol{\xi}}{\|\boldsymbol{\xi}\|_E \|\boldsymbol{\eta}\|_E} =: \tilde{\beta}_{\Sigma} \equiv \mu_k. \tag{3.4.48}$$

Now, we remark that in (3.4.48) there would be no gain and no loss in taking the supremum for $\boldsymbol{\xi} \in \mathbb{R}^n$ rather than for $\boldsymbol{\xi} \in (\mathrm{Ker}\Sigma)^{\perp} \subseteq \mathbb{R}^n$. In general, taking the supremum on a bigger set will provide a bigger (or equal) supremum. Here, for $\boldsymbol{\xi} \in \mathrm{Ker}\Sigma$, the numerator in (3.4.48) (that is $\boldsymbol{\eta}^T \Sigma \boldsymbol{\xi}$) will always be zero and therefore the supremum will not change. Hence,

$$\inf_{\boldsymbol{\eta} \in (\mathrm{Ker}\Sigma^T)^{\perp}} \sup_{\boldsymbol{\xi} \in \mathbb{R}^n} \frac{\boldsymbol{\eta}^T \Sigma \boldsymbol{\xi}}{\|\boldsymbol{\xi}\|_E \|\boldsymbol{\eta}\|_E} = \mu_k. \tag{3.4.49}$$

□

Now, given an $m \times n$ matrix B, we set (recalling assumption (3.4.9))

$$M := S_Y \, B \, S_X, \qquad \text{so that} \quad B = S_Y^{-1} M \, S_X^{-1}. \tag{3.4.50}$$

Taking the singular value decomposition (3.4.44) for M will correspond to writing B as

$$B = S_Y \, V \, \Sigma \, U \, S_X. \tag{3.4.51}$$

It is not difficult to check that writing $\mathbf{x} = S_X^{-1} U^T \boldsymbol{\xi}$ and $\mathbf{y} = S_Y^{-1} V \boldsymbol{\eta}$ yields

$$\frac{\mathbf{y}^T B \mathbf{x}}{\|\mathbf{x}\|_X \, \|\mathbf{y}\|_Y} = \frac{\boldsymbol{\eta}^T \, V^T \, S_Y^{-1} \, S_Y \, V \Sigma \, U \, S_X \, S_X^{-1} \, U^T \boldsymbol{\xi}}{\|S_X \, S_X^{-1} U^T \boldsymbol{\xi}\|_E \, \|S_Y \, S_Y^{-1} V \boldsymbol{\eta}\|_E} = \frac{\boldsymbol{\eta}^T \Sigma \boldsymbol{\xi}}{\|\boldsymbol{\xi}\|_E \, \|\boldsymbol{\eta}\|_E} \tag{3.4.52}$$

where, in the last step, we used the definition of the norms (3.4.9) and the fact that U and V are unitary.

Noting that, as it can be easily checked, for $\mathbf{y} = S_Y^{-1} V \boldsymbol{\eta}$ and B given by (3.4.51) (so that $B^T = S_X \, U^T \, \Sigma^T \, V^T \, S_Y$), we have

$$\mathbf{y} \in \mathrm{Ker} B^T \quad \textbf{iff} \quad \eta \in \mathrm{Ker}\Sigma^T,$$

we conclude that

$$\inf_{\mathbf{y}\in(\mathrm{Ker}B^T)^{\perp}} \sup_{\mathbf{x}\in\mathbb{R}^n} \frac{\mathbf{y}^T B\mathbf{x}}{\|\mathbf{x}\|_X \, \|\mathbf{y}\|_Y} = \inf_{\eta\in(\mathrm{Ker}\Sigma^T)^{\perp}} \sup_{\xi\in\mathbb{R}^n} \frac{\eta^T \Sigma \xi}{\|\xi\|_E \, \|\eta\|_E} = \mu_k. \tag{3.4.53}$$

We collect the result in the following proposition.

Proposition 3.4.5. *Let B be an $m \times n$ matrix, let the norms in $\mathbf{X}$ and $\mathbf{Y}$ be defined as in* (3.4.9) *through the matrices S_X and S_Y, respectively, and let $\tilde{\beta}$ be defined as*

$$\inf_{\mathbf{y}\in H^{\perp}} \sup_{\mathbf{x}\in K^{\perp}} \frac{\mathbf{x}^T B^T \mathbf{y}}{\|\mathbf{x}\|_X \, \|\mathbf{y}\|_Y} \equiv \inf_{\mathbf{y}\in H^{\perp}} \sup_{\mathbf{x}\in\mathbf{X}} \frac{\mathbf{x}^T B^T \mathbf{y}}{\|\mathbf{x}\|_X \, \|\mathbf{y}\|_Y} =: \tilde{\beta}, \tag{3.4.54}$$

*where, as usual, $K := \mathrm{Ker}B$ and $H := \mathrm{Ker}B^T$. Then, $\tilde{\beta}$ coincides with **the smallest positive singular value** of the matrix $S_Y \, B \, S_X$. In particular, the inf-sup condition* (3.4.31) *is equivalent to say that "all the singular values of $S_Y \, B \, S_X$ are positive, and the smallest singular value $\tilde{\beta}$ is bounded from below by a fixed positive constant β, independent of the decomposition".* □

3.4.4 The Case of A Elliptic on the Whole Space

As we have seen when discussing solvability, the *inf-sup* condition alone cannot be sufficient for having stability for problems of the general form (3.2.1) and (3.2.2). In order to have sufficient conditions, we now introduce a further assumption on the matrix A. As discussed at the end of Sect. 3.4.1, we start considering a strong condition. More precisely, we make the following assumption.

Ellipticity condition. *There exists a positive constant α, independent of the mesh-size h, such that*

$$\alpha \|\mathbf{x}\|_X^2 \le \mathbf{x}^T A\mathbf{x} \qquad \forall\, \mathbf{x} \in \mathbf{X}. \tag{3.4.55}$$

We immediately note that, from (3.4.20) and (3.4.55), we easily deduce that

$$\alpha \le M_a. \tag{3.4.56}$$

We now have the following Theorem.

Theorem 3.4.1. *Let the assumptions* (3.4.8)–(3.4.15) *on spaces, norms and matrices be satisfied. Let $\mathbf{x}, \mathbf{y}, \mathbf{f}, \mathbf{g}$ satisfy the general system of equations* (3.2.1) *and* (3.2.2)*. Assume moreover that the inf-sup condition* (3.4.31) *and the ellipticity* (3.4.55) *are satisfied. Then, we have*

$$\|\mathbf{x}\|_X \leq \frac{1}{\alpha}\|\mathbf{f}\|_F + \frac{M_a}{\alpha\beta}\|\mathbf{g}\|_G, \tag{3.4.57}$$

$$\|\mathbf{y}\|_Y \leq \frac{2M_a}{\alpha\beta}\|\mathbf{f}\|_F + \frac{M_a^2}{\alpha\beta^2}\|\mathbf{g}\|_G. \tag{3.4.58}$$

Proof. We shall prove the result by splitting $\mathbf{x} = \mathbf{x}_f + \mathbf{x}_g$ and $\mathbf{y} = \mathbf{y}_f + \mathbf{y}_g$, defined as the solutions of

$$\begin{cases} A\mathbf{x}_f + B^T\mathbf{y}_f = \mathbf{f}, \\ B\mathbf{x}_f = 0, \end{cases} \tag{3.4.59}$$

and

$$\begin{cases} A\mathbf{x}_g + B^T\mathbf{y}_g = 0, \\ B\mathbf{x}_g = \mathbf{g}. \end{cases} \tag{3.4.60}$$

We proceed in several steps.

- *Step 1 – Estimate of* $\mathbf{x}_f$ *and* $A\mathbf{x}_f$
  We multiply the first equation of (3.4.59) to the left by $\mathbf{x}_f^T$ and we note that $\mathbf{x}_f^T B^T\mathbf{y}_f \equiv \mathbf{y}^T B\mathbf{x}_f = 0$ (by the second equation). Hence,

$$\mathbf{x}_f^T A\mathbf{x}_f = \mathbf{x}^T\mathbf{f} \tag{3.4.61}$$

  and, using the ellipticity condition (3.4.55), relation (3.4.61) and the first of the dual norm estimates (3.4.17), we have

$$\alpha\|\mathbf{x}_f\|_X^2 \leq \mathbf{x}_f^T A\mathbf{x}_f = \mathbf{x}^T\mathbf{f} \leq \|\mathbf{x}_f\|_X\,\|\mathbf{f}\|_F, \tag{3.4.62}$$

  giving immediately

$$\|\mathbf{x}_f\|_X \leq \frac{1}{\alpha}\|\mathbf{f}\|_F, \tag{3.4.63}$$

  and using (3.4.18),

$$\|A\mathbf{x}_f\|_F \leq \frac{M_a}{\alpha}\|\mathbf{f}\|_F. \tag{3.4.64}$$

- *Step 2 – Estimate of* $\mathbf{y}_f$
  Using the equivalent form of the inf-sup condition (3.4.41), we have

$$\beta\|\mathbf{y}_f\|_Y \leq \|B^T\mathbf{y}_f\|_F = \|\mathbf{f} - A\mathbf{x}_f\|_F. \tag{3.4.65}$$

  Then, using (3.4.65), (3.4.64) and (3.4.56), we obtain

$$\|\mathbf{y}_f\|_Y \leq \frac{1}{\beta}\|\mathbf{f} - A\mathbf{x}_f\|_F \leq \frac{1}{\beta}\left(1 + \frac{M_a}{\alpha}\right)\|\mathbf{f}\|_F \leq \frac{2M_a}{\alpha\beta}\|\mathbf{f}\|_F. \tag{3.4.66}$$

- *Step 3 – Estimate of* $\|\mathbf{x}_g\|_X^2$ *by* $\|\mathbf{y}_g\|_Y$
  We use the ellipticity (3.4.55), then the first equation of (3.4.60), then (3.1.5), then the second equation of (3.4.60), and finally the second of the dual norm estimates (3.4.17):

$$\alpha\|\mathbf{x}_g\|_X^2 \le \mathbf{x}_g^T A\mathbf{x}_g = -\mathbf{x}_g^T B^T\mathbf{y}_g \equiv -\mathbf{y}_g^T B\mathbf{x}_g = -\mathbf{y}_g^T\mathbf{g} \le \|\mathbf{y}_g\|_Y\|\mathbf{g}\|_G. \quad (3.4.67)$$

- *Step 4 – Estimate of* $\|\mathbf{y}_g\|_Y$ *by* $\|\mathbf{x}_g\|_X$
  Using again the inf-sup condition in the from (3.4.41), the first equation of (3.4.60) and the continuity property (3.4.18), we have

$$\beta\|\mathbf{y}_g\|_Y \le \|B^T\mathbf{y}_g\|_F = \|A\mathbf{x}_g\|_F \le M_a\|\mathbf{x}_g\|_X. \quad (3.4.68)$$

- *Step 5 – Estimate of* $\|\mathbf{x}_g\|_X$ *and* $\|\mathbf{y}_g\|_Y$
  We combine (3.4.67) and (3.4.68) to obtain

$$\alpha\|\mathbf{x}_g\|_X^2 \le \frac{M_a}{\beta}\|\mathbf{g}\|_G\|\mathbf{x}_g\|_X, \quad (3.4.69)$$

  which immediately implies

$$\|\mathbf{x}_g\|_X \le \frac{M_a}{\alpha\beta}\|\mathbf{g}\|_G. \quad (3.4.70)$$

  Using this in (3.4.68), we therefore have

$$\|\mathbf{y}_g\|_X \le \frac{M_a^2}{\alpha\beta^2}\|\mathbf{g}\|_G. \quad (3.4.71)$$

  The final estimate then follows by simply collecting the separate estimates (3.4.63), (3.4.66), (3.4.70) and (3.4.71).

□

Remark 3.4.5. In some applications (and in particular for the Stokes problem), the matrix A will always be symmetric and positive definite, essentially for all possible types of finite element discretisations, with an α easily bounded away from 0. In these cases, the only condition that we must check will be the *inf-sup* condition on B. This led some people to believe that the *inf-sup* condition for B is *the assumption* to be made for getting a good method when dealing with mixed formulations. This, however, is a superstition, based (as all superstitions) on a narrow horizon. We will see in Chap. 5, Sect. 5.2.4, some examples of discretisations of simple one-dimensional problems that illustrate this point. □

Remark 3.4.6. In some applications it might happen that the constants α and β either depend on h (and tend to zero as h tends to zero) or have a fixed value that is however very small. It is therefore important to keep track of the possible degeneracy of the constants in our estimates when α and/or β are very small. In particular, it is relevant to know whether our stability constants degenerate and tend to infinity, for example, as $1/\beta$ or $1/\beta^2$ or other powers of $1/\beta$ (and, similarly, of $1/\alpha$). In this respect, we point out that the behaviour indicated in (3.4.57) and (3.4.58) is optimal. This means that we cannot hope to find a better proof giving a better behaviour of the constants in terms of powers of $1/\alpha$ and $1/\beta$, as shown by the following example. Considering the system

$$\begin{pmatrix} 1 & -1 & b \\ 1 & a & 0 \\ b & 0 & 0 \end{pmatrix} \begin{pmatrix} x_1 \\ x_2 \\ y \end{pmatrix} = \begin{pmatrix} f_1 \\ f_2 \\ g \end{pmatrix} \qquad 0 < a, b \ll 1, \tag{3.4.72}$$

one easily obtains

$$x_1 = \frac{g}{b}, \quad x_2 = \frac{f_2}{a} - \frac{g}{ab}, \quad y = \frac{f_1}{b} + \frac{f_2}{ab} - \frac{(1+a)g}{ab^2}. \tag{3.4.73}$$

Since $\alpha = a$ and $\beta = b$, from (3.4.73) we deduce that the bounds of Theorem 3.4.1 cannot be improved. □

The dependence of the stability constants on α and β *can* however be improved if we add as a further assumption the symmetry of the matrix A. We have indeed the following result.

Theorem 3.4.2. *Let the assumptions* (3.4.8)–(3.4.15) *on spaces, norms and matrices be satisfied. Let* $\mathbf{x}, \mathbf{y}, \mathbf{f}, \mathbf{g}$ *satisfy the general system of equations* (3.2.1) *and* (3.2.2)*. Assume moreover that the inf-sup condition* (3.4.31) *and the ellipticity* (3.4.55) *are satisfied, and assume moreover that A is symmetric. Then, we have*

$$\|\mathbf{x}\|_X \leq \frac{1}{\alpha}\|\mathbf{f}\|_F + \frac{M_a^{1/2}}{\alpha^{1/2}\beta}\|\mathbf{g}\|_G, \tag{3.4.74}$$

$$\|\mathbf{y}\|_Y \leq \frac{2M_a^{1/2}}{\alpha^{1/2}\beta}\|\mathbf{f}\|_F + \frac{M_a}{\beta^2}\|\mathbf{g}\|_G. \tag{3.4.75}$$

Proof. The following proof mimics rather closely the path of the previous one. In particular, it is done again analysing separately the two problems: (3.4.59), for $\mathbf{g} = 0$, and (3.4.60) for $\mathbf{f} = 0$. However, instead of just indicating the differences between the two proofs, we prefer to report also the second one in detail.

- *Step 1 – Estimate of* $\mathbf{x}_f$ *and* $A\mathbf{x}_f$
  We multiply the first equation of (3.4.59) to the left by $\mathbf{x}_f^T$ and we note that $\mathbf{x}_f^T B^T \mathbf{y}_f \equiv \mathbf{y}^T B \mathbf{x}_f = 0$ (by the second equation). Hence,

$$\mathbf{x}_f^T A\mathbf{x}_f = \mathbf{x}^T\mathbf{f} \tag{3.4.76}$$

and, using the ellipticity condition (3.4.55), relation (3.4.76) and the first of the dual norm estimates (3.4.17), we have

$$\alpha\|\mathbf{x}_f\|_X^2 \le \mathbf{x}_f^T A\mathbf{x}_f = \mathbf{x}^T\mathbf{f} \le \|\mathbf{x}_f\|_X \|\mathbf{f}\|_F,$$

giving immediately

$$\|\mathbf{x}_f\|_X \le \frac{1}{\alpha}\|\mathbf{f}\|_F \tag{3.4.77}$$

as well as

$$\mathbf{x}_f^T A\mathbf{x}_f \le \frac{1}{\alpha}\|\mathbf{f}\|_F^2. \tag{3.4.78}$$

Therefore, using (3.4.22), we also get

$$\|A\mathbf{x}_f\|_F \le \frac{M_a^{1/2}}{\alpha^{1/2}}\|\mathbf{f}\|_F, \tag{3.4.79}$$

which improves estimate (3.4.64).

- *Step 2 – Estimate of* $\mathbf{y}_f$
  We now use the equivalent form of the inf-sup condition (3.4.41) with $\mathbf{y} = \mathbf{y}_f$. We have

$$\beta\|\mathbf{y}_f\|_Y \le \|B^T\mathbf{y}_f\|_F = \|\mathbf{f} - A\mathbf{x}_f\|_F. \tag{3.4.80}$$

  Then, using (3.4.80), (3.4.79) and (3.4.56), we obtain

$$\|\mathbf{y}_f\|_Y \le \frac{1}{\beta}\|\mathbf{f} - A\mathbf{x}_f\|_F \le \left(\frac{1}{\beta} + \frac{M_a^{1/2}}{\alpha^{1/2}\beta}\right)\|\mathbf{f}\|_F \le \frac{2M_a^{1/2}}{\alpha^{1/2}\beta}\|\mathbf{f}\|_F. \tag{3.4.81}$$

- *Step 3 – Estimate of* $\mathbf{x}_g^T A\mathbf{x}_g$ *by* $\|\mathbf{y}_g\|_Y$
  We multiply the first equation of (3.4.60) by $\mathbf{x}_g^T$. Using the second equation of (3.4.60) and the second of the dual norm estimates (3.4.17), we have

$$\mathbf{x}_g^T A\mathbf{x}_g = -\mathbf{x}_g^T B^T\mathbf{y}_g \equiv -\mathbf{y}_g^T B\mathbf{x}_g = -\mathbf{y}_g^T\mathbf{g} \le \|\mathbf{y}_g\|_Y \|\mathbf{g}\|_G. \tag{3.4.82}$$

- *Step 4 – Estimate of* $\|\mathbf{y}_g\|_Y$ *by* $(\mathbf{x}_g^T A\mathbf{x}_g)^{1/2}$
  Using now the inf-sup condition in the form (3.4.31) with $\mathbf{y} = \mathbf{y}_g$, we get that there exists an $\tilde{\mathbf{x}} \neq 0$ such that $\tilde{\mathbf{x}}^T B^T\mathbf{y}_g \ge \beta\|\tilde{\mathbf{x}}\|_X\|\mathbf{y}_g\|_Y$. This relation, the first equation of (3.4.60) and the continuity property (3.4.25), yield

$$\beta\|\tilde{\mathbf{x}}\|_X\|\mathbf{y}_g\|_Y \le \tilde{\mathbf{x}}^T B^T\mathbf{y}_g = -\tilde{\mathbf{x}}^T A\mathbf{x}_g \le M_a^{1/2}\|\tilde{\mathbf{x}}\|_X (\mathbf{x}_g^T A\mathbf{x}_g)^{1/2}, \tag{3.4.83}$$

  giving (as $\tilde{\mathbf{x}} \neq 0$):

$$\|\mathbf{y}_g\|_Y \le \frac{M_a^{1/2}}{\beta}(\mathbf{x}_g^T A\mathbf{x}_g)^{1/2}. \tag{3.4.84}$$

- *Step 5 – Estimate of* $\|\mathbf{x}_g\|_X$ *and* $\|\mathbf{y}_g\|_Y$
  We first combine (3.4.82) and (3.4.84) to obtain

$$\|\mathbf{y}_g\|_Y \leq \frac{M_a}{\beta^2}\|\mathbf{g}\|_G. \tag{3.4.85}$$

  Moreover, using the ellipticity assumption (3.4.55), then (3.4.82) and finally (3.4.85), we have

$$\alpha\|\mathbf{x}_g\|_X^2 \leq \mathbf{x}_g^T A\mathbf{x}_g \leq \|\mathbf{y}_g\|_Y\|\mathbf{g}\|_G \leq \frac{M_a}{\beta^2}\|\mathbf{g}\|_G^2,$$

  which can be rewritten as

$$\|\mathbf{x}_g\|_X \leq \frac{M_a^{1/2}}{\alpha^{1/2}\beta}\|\mathbf{g}\|_G. \tag{3.4.86}$$

  The final estimate follows then by simply collecting the separate estimates (3.4.77), (3.4.81), (3.4.86) and (3.4.85). □

Remark 3.4.7. We point out that the behaviour indicated in (3.4.74) and (3.4.75) is also optimal, in the sense that, as in the previous case, we cannot hope to find a better proof giving a better behaviour of the constants in terms of powers of $1/\alpha$ and $1/\beta$. Indeed, consider the system

$$\begin{pmatrix} 2 & \sqrt{a} & b \\ \sqrt{a} & a & 0 \\ b & 0 & 0 \end{pmatrix}\begin{pmatrix} x_1 \\ x_2 \\ y \end{pmatrix} = \begin{pmatrix} f_1 \\ f_2 \\ g \end{pmatrix} \qquad 0 < a, b \ll 1,$$

whose solution is

$$x_1 = \frac{g}{b}, \quad x_2 = \frac{f_2}{a} - \frac{g}{a^{1/2}b}, \quad y = \frac{f_1}{b} - \frac{f_2}{a^{1/2}b} - \frac{g}{b^2}. \tag{3.4.87}$$

Since the constants α and β are given by

$$\alpha = \frac{2 + a - \sqrt{a^2+4}}{2} = \frac{4a}{2\left(2 + a + \sqrt{a^2+4}\right)} \approx \frac{a}{2}$$

and

$$\beta = b,$$

we see from (3.4.87) that there are cases in which the actual stability constants behave exactly as predicted by the theory. □

3.4.5 The Case of A Elliptic on the Kernel of B

We now consider, together with the *inf-sup* condition on B, a condition on A that is weaker than the full ellipticity (3.4.55). In particular, we require the ellipticity of A to hold only in the kernel K of B.

More precisely, we make the following requirement.

***Elker* condition.** *There exists a positive constant α_0, independent of the mesh-size h, such that*

$$\alpha_0 \|\mathbf{x}\|_X^2 \le \mathbf{x}^T A\mathbf{x} \qquad \forall \mathbf{x} \in K, \tag{3.4.88}$$

where K is the kernel of B.

The above condition is often called *elker* since it requires the ellipticity on the kernel.

We remark, for future use, that from (3.4.20) and (3.4.88) we get

$$\alpha_0 \le M_a. \tag{3.4.89}$$

The following Theorem generalises Theorem 3.4.1.

Theorem 3.4.3. *Let the assumptions* (3.4.8)–(3.4.15) *on spaces, norms and matrices be satisfied. Let $\mathbf{x}, \mathbf{y}, \mathbf{f}, \mathbf{g}$ satisfy the general system of equations* (3.2.1) *and* (3.2.2)*. Assume moreover that the inf-sup* (3.4.31) *and the elker condition* (3.4.88) *are satisfied. Then, we have*

$$\|\mathbf{x}\|_X \le \frac{1}{\alpha_0}\|\mathbf{f}\|_F + \frac{2M_a}{\alpha_0\beta}\|\mathbf{g}\|_G, \tag{3.4.90}$$

$$\|\mathbf{y}\|_Y \le \frac{2M_a}{\alpha_0\beta}\|\mathbf{f}\|_F + \frac{2M_a^2}{\alpha_0\beta^2}\|\mathbf{g}\|_G. \tag{3.4.91}$$

Proof. We first set $\mathbf{x}_g := \mathbf{L}\mathbf{g}$ where $\mathbf{L}$ is the lifting operator defined by Proposition 3.4.4. We also point out the following estimates on $\mathbf{x}_g$: from the continuity of the lifting $\mathbf{L}$ (3.4.43) we have

$$\beta\|\mathbf{x}_g\|_X \le \|\mathbf{g}\|_G \tag{3.4.92}$$

and using (3.4.18) and (3.4.92) we obtain

$$\|A\mathbf{x}_g\|_F \le M_a\|\mathbf{x}_g\|_X \le \frac{M_a}{\beta}\|\mathbf{g}\|_G. \tag{3.4.93}$$

Then we set

$$\mathbf{x}_K := \mathbf{x} - \mathbf{x}_g = \mathbf{x} - \mathbf{L}\mathbf{g} \tag{3.4.94}$$

and we note that $\mathbf{x}_K \in K$. Moreover, $(\mathbf{x}_K, \mathbf{y})$ solves the linear system

$$\begin{cases} A\mathbf{x}_K + B^T\mathbf{y} = \mathbf{f} - A\mathbf{x}_g, \\ B\mathbf{x}_K = \mathbf{0}. \end{cases} \tag{3.4.95}$$

We can now proceed as in *Steps 1* and *2* of the proof of Theorem 3.4.1. We note that our weaker assumption *elker* (3.4.88) is sufficient for allowing the first step in (3.4.62). Proceeding as in the first part of *Step 1*, and using (3.4.93) at the end, we get

$$\|\mathbf{x}_K\|_X \le \frac{1}{\alpha_0}\|\mathbf{f} - A\mathbf{x}_g\|_F \le \frac{1}{\alpha_0}\left(\|\mathbf{f}\|_F + \frac{M_a}{\beta}\|\mathbf{g}\|_G\right). \tag{3.4.96}$$

This allows to reconstruct the estimate on $\mathbf{x}$:

$$\begin{aligned} \|\mathbf{x}\|_X = \|\mathbf{x}_K + \mathbf{x}_g\|_X &\le \frac{1}{\alpha_0}\|\mathbf{f}\|_F + \left(\frac{M_a}{\alpha_0\beta} + \frac{1}{\beta}\right)\|\mathbf{g}\|_G \\ &\le \frac{1}{\alpha_0}\|\mathbf{f}\|_F + \frac{2M_a}{\alpha_0\beta}\|\mathbf{g}\|_G, \end{aligned} \tag{3.4.97}$$

where we have used (3.4.89) in the last inequality. Combining (3.4.18) and (3.4.97), we also have

$$\|A\mathbf{x}\|_F \le M_a\|\mathbf{x}\|_X \le \frac{M_a}{\alpha_0}\|\mathbf{f}\|_F + \frac{2M_a^2}{\alpha_0\beta}\|\mathbf{g}\|_G. \tag{3.4.98}$$

Then, we proceed as in *Step 2* to obtain, as in (3.4.81),

$$\beta\|\mathbf{y}\|_Y \le \|\mathbf{f} - A\mathbf{x}\|_F \tag{3.4.99}$$

and, using the above estimate (3.4.98) on $A\mathbf{x}$ in (3.4.99), we obtain

$$\|\mathbf{y}\|_Y \le \left(\frac{1}{\beta} + \frac{M_a}{\alpha_0\beta}\right)\|\mathbf{f}\|_F + \frac{2M_a^2}{\alpha_0\beta^2}\|\mathbf{g}\|_G \le \frac{2M_a}{\alpha_0\beta}\|\mathbf{f}\|_F + \frac{2M_a^2}{\alpha_0\beta^2}\|\mathbf{g}\|_G, \tag{3.4.100}$$

and the proof is concluded. □

Remark 3.4.8. In the spirit of Remark 3.4.6, we note that the dependence of the stability constants on α_0 and β is optimal. Indeed, the dependence is the same as the one proved in Theorem 3.4.1 under stronger assumptions. Hence, the optimality is again shown by example (3.4.72), for which we have $\alpha_0 = a$ and $\beta = b$. It is interesting to note that, contrary to the result of Theorem (3.4.2), **adding the assumption that A is symmetric would not improve the bounds** (!). Indeed, considering the system

$$\begin{pmatrix} 1 & 1 & b \\ 1 & a & 0 \\ b & 0 & 0 \end{pmatrix} \begin{pmatrix} x_1 \\ x_2 \\ y \end{pmatrix} = \begin{pmatrix} f_1 \\ f_2 \\ g \end{pmatrix} \qquad 0 < a, b \ll 1, \tag{3.4.101}$$

one easily obtains

$$x_1 = \frac{g}{b}, \quad x_2 = \frac{f_2}{a} - \frac{g}{ab}, \quad y = \frac{f_1}{b} - \frac{f_2}{ab} + \frac{(1-a)g}{ab^2}. \tag{3.4.102}$$

Since $\alpha_0 = a$ and $\beta = b$, system (3.4.101) shows the same behaviour as the bounds of Theorem 3.4.3 (and not better), even though A is symmetric. □

In order to recover the better bounds found in Theorem 3.4.2, we have to assume that A, on top of satisfying the ellipticity in the kernel (3.4.88), is symmetric *and* positive semi-definite in the whole $\mathbb{R}^n$ (a property that the matrix A in (3.4.101) does not have for $a < 1$). This is because, in order to improve the bounds, one has to use (3.4.22) that requires A to be symmetric and positive semi-definite. We have indeed the following result, that we state without proof: indeed, we shall see in the next section that this result can be obtained as a particular case of a more general estimate (see Remark 3.6.4).

Theorem 3.4.4. *Let the assumptions* (3.4.8)–(3.4.15) *on spaces, norms and matrices be satisfied. Let* $\mathbf{x}, \mathbf{y}, \mathbf{f}, \mathbf{g}$ *satisfy the general system of equations* (3.2.1) *and* (3.2.2). *Assume that the inf-sup* (3.4.31) *and the elker condition* (3.4.88) *are satisfied, and assume moreover that A is symmetric and positive semi-definite on the whole space* $\mathbf{X}$. *Then, we have*

$$\|\mathbf{x}\|_X \leq \frac{1}{\alpha_0}\|\mathbf{f}\|_F + \frac{2M_a^{1/2}}{\alpha_0^{1/2}\beta}\|\mathbf{g}\|_G, \tag{3.4.103}$$

$$\|\mathbf{y}\|_Y \leq \frac{2M_a^{1/2}}{\alpha_0^{1/2}\beta}\|\mathbf{f}\|_F + \frac{M_a}{\beta^2}\|\mathbf{g}\|_G. \tag{3.4.104}$$

3.4.6 *The Case of A Satisfying an* inf-sup *on the Kernel of B*

As we have seen in the previous sections, the ellipticity in the kernel for the matrix A is not the weakest condition we can use. Indeed, in order to get necessary and sufficient conditions for solvability, we used the surjectivity of B (here replaced with the *inf-sup* condition on B) and the non-singularity of A_{KK} on the kernel K of B. Hence, it is clear that we still have room to improve the result of Theorem 3.4.3 by assuming on A some property weaker than (3.4.88). In particular we can assume

Inf-sup condition on A_{KK}: *There exists a positive constant* α_1, *independent of the mesh-size h, such that*

$$\inf_{\mathbf{x}\in K}\sup_{\mathbf{z}\in K}\frac{\mathbf{z}^T A\mathbf{x}}{\|\mathbf{z}\|_X\,\|\mathbf{x}\|_X} \geq \alpha_1. \tag{3.4.105}$$

We note that (3.4.105) can be equivalently written as

$$\alpha_1\|\mathbf{x}\|_X \leq \sup_{\mathbf{z}\in K}\frac{\mathbf{z}^T A\mathbf{x}}{\|\mathbf{z}\|_X} \qquad \forall\,\mathbf{x}\in K, \tag{3.4.106}$$

or

$$\alpha_1\|\mathbf{x}\|_X \leq \|A_{KK}\mathbf{x}\|_{K'} \qquad \forall\,\mathbf{x}\in K, \tag{3.4.107}$$

where we used the notation of (3.4.16).

We have then the following result.

Theorem 3.4.5. *Let the assumptions* (3.4.8)–(3.4.15) *on spaces, norms and matrices be satisfied. Let* $\mathbf{x},\mathbf{y},\mathbf{f},\mathbf{g}$ *satisfy the general system of equations* (3.2.1) *and* (3.2.2)*. Assume, moreover, that the inf-sup condition* (3.4.31) *on* B *and the bounding condition* (3.4.107) *on* A_{KK} *are satisfied. Then, we have*

$$\|\mathbf{x}\|_X \leq \frac{1}{\alpha_1}\|\mathbf{f}\|_F + \frac{2M_a}{\alpha_1\beta}\|\mathbf{g}\|_G, \tag{3.4.108}$$

$$\|\mathbf{y}\|_Y \leq \frac{2M_a}{\alpha_1\beta}\|\mathbf{f}\|_F + \frac{2M_a^2}{\alpha_1\beta^2}\|\mathbf{g}\|_G. \tag{3.4.109}$$

Proof. The proof is identical to that of Theorem 3.4.3. The only change is in the first inequality in (3.4.96). Using this time (3.4.107), and noting once more that from (3.2.5), we easily obtain

$$\alpha_1\|\mathbf{x}_K\|_X \leq \|A_{KK}\mathbf{x}_K\|_{K'} \leq \|\mathbf{f} - A_{KK}\mathbf{x}_g\|_{K'} \leq \|\mathbf{f}\|_F + \|A\mathbf{x}_g\|_F, \tag{3.4.110}$$

so that the first inequality of (3.4.96) still holds if we replace α_0 by α_1. The rest of the proof goes on unchanged. □

So far, for every type of bounding conditions on the matrix A (global ellipticity and ellipticity on K), we considered separately the special cases in which A had some additional property. In particular, after Theorem 3.4.1 (where A was assumed to be elliptic on the whole $\mathbf{X}$), we considered in Theorem 3.4.2 the case where A was also symmetric. Similarly, after Theorem 3.4.3 (where A was assumed to be elliptic on K), we considered in Theorem 3.4.4 the case where A was also symmetric and positive semi-definite on the whole $\mathbf{X}$. Now, after Theorem 3.4.5 (where A is supposed to satisfy the bounding condition (3.4.107) on K), we could ask ourselves what happens if we assume further that A is also symmetric and positive semi-definite on the whole $\mathbf{X}$. This, however, would bring us back to the case of Theorem 3.4.4, thanks to the following proposition.

Proposition 3.4.6. *Let A be an $n \times n$ matrix, and K a subspace of $\mathbb{R}^n$. Assume that A is symmetric, positive semi-definite, and verifies* (3.4.107) *on K. Then, A is elliptic on K.* □

Proof. Indeed, for $x \in K$, using (3.4.106) and then (3.3.1), we have

$$\alpha_1^2 \|\mathbf{x}\|_X^2 \le \sup_{\mathbf{z}\in K} \frac{(\mathbf{z}^T A\mathbf{x})^2}{\|\mathbf{z}\|_X^2} \le \sup_{\mathbf{z}\in K} \frac{\mathbf{x}^T A\mathbf{x}\,\mathbf{z}^T A\mathbf{z}}{\|\mathbf{z}\|_X^2} \le M_a \mathbf{x}^T A\mathbf{x}, \tag{3.4.111}$$

and the result follows with $\alpha_0 = \alpha_1^2/M_a$. □

3.5 Additional Results

In this section, we present some additional results concerning necessary conditions, modified problems and special cases.

3.5.1 Some Necessary Conditions

We see in this subsection that the above sufficient conditions for having existence and uniqueness of the solution, together with stability estimates, are indeed *necessary*.

Theorem 3.5.1. *Assume that there exists a constant C such that, for any quadruple $(\mathbf{x}, \mathbf{y}, \mathbf{f}, \mathbf{g})$ in $X \times Y \times F \times G$ solution of* (3.2.1) *and* (3.2.2)*, we have*

$$\|\mathbf{x}\|_X + \|\mathbf{y}\|_Y \le C(\|\mathbf{f}\|_F + \|\mathbf{g}\|_G). \tag{3.5.1}$$

Then, (3.4.107) *and* (3.4.31) *are verified with $\alpha_1 = \beta = 1/C$.*

Proof. For every $\mathbf{y} \in \mathbf{Y}$, it is easy to see that $(0, \mathbf{y}, B^T\mathbf{y}, 0)$ satisfies (3.2.1) and (3.2.2). Hence, (3.5.1) shows that the *inf-sup* condition (3.4.31) is satisfied in the equivalent form (3.4.41), with $\beta = 1/C$. Then, for every $\mathbf{x} \in K = \mathrm{Ker}B$, set $\mathbf{f} := \pi_K A\mathbf{x} \equiv A_{KK}\mathbf{x}$. Note that $\pi_K(\mathbf{f} - A\mathbf{x}) = \mathbf{0}$, and hence $\mathbf{f} - A\mathbf{x}$ belongs to $K^\perp$. From (3.1.60) we have that there exists a $\mathbf{y} \in \mathbf{Y}$ such that $B^T\mathbf{y} = \mathbf{f} - A\mathbf{x}$, and since $\mathbf{x} \in K$, we have that $(\mathbf{x}, \mathbf{y}, A\mathbf{x}, 0)$ satisfies (3.2.1) and (3.2.2). Hence, inequality (3.5.1) gives now (3.4.107) with $\alpha_1 = 1/C$. □

Remark 3.5.1. Note that an inequality like (3.5.1) implies that the problem (3.2.1) and (3.2.2) has been adimensionalised. This is not the case for the results of the previous section. See also Remark 3.6.6 at the end of this chapter. □

Theorem 3.5.1 dealt with the necessity of the assumptions in Theorem 3.4.5. The following result deals with the necessity of the assumptions in Theorem 3.4.4.

Theorem 3.5.2. *Let A be symmetric and positive semi-definite. Assume that there exists a constant C such that for any quadruple $(\mathbf{x}, \mathbf{y}, \mathbf{f}, \mathbf{g})$ in $X \times Y \times F \times G$ solution of* (3.2.1) *and* (3.2.2) *we have that the bound* (3.5.1) *holds. Then,* (3.4.88) *and* (3.4.31) *are verified with $\alpha_0 = 1/(C^2 M_a)$ (where M_a is the continuity constant of A defined in* (3.4.18)*) and $\beta = 1/C$, respectively.*

Proof. The result is an immediate consequence of Theorem 3.5.1 and Proposition 3.4.6 □

Remark 3.5.2. As we have seen in Theorem 3.2.1, the estimate (3.5.1) implies the *inf-sup* condition (3.2.19) and the non singularity of A_{KK} on the kernel K (3.2.18). The purpose of Theorems 3.5.1 and 3.5.2 is mainly to show that a uniform bound for C implies uniform bounds for the constants α_1 (or α_0) and β. □

3.5.2 The Case of B Not Surjective. Modification of the Problem

Here, we come back, somehow, to the case of Remark 3.2.1. To start with, we observe that, proceeding as in Remark 3.2.1 we, immediately have the following result.

Proposition 3.5.1. *Assume that A_{KK} satisfies* (3.4.105)*, and $\mathbf{g} \in \operatorname{Im}B$. Then, problem* (3.2.1) *and* (3.2.2) *has at least one solution $(\mathbf{x}, \mathbf{y})$. Moreover, $\mathbf{x}$ is uniquely determined and*

$$||\mathbf{x}||_X \le \frac{1}{\alpha_1}(||\mathbf{f}||_F + \frac{M_a}{\tilde{\beta}}\|\mathbf{g}\|_G) \tag{3.5.2}$$

where $\tilde{\beta}$ is defined in (3.4.28)*.* □

We note that (3.5.2) does not provide any estimate for the variable $\mathbf{y}$. This should be expected since in Proposition 3.5.1 we did not assume that the *inf-sup* condition (3.4.31) holds true. However, (3.4.28) will always hold so that for $\mathbf{g} \in \operatorname{Im}B$ we might consider the problem (3.2.1) and (3.2.2) in $\mathbf{X} \times H^{\perp}$ instead of $\mathbf{X} \times \mathbf{Y}$, keeping in H the same norm that we had in $\mathbf{Y}$. Hence, we can apply any of the previous theorems of this section (that is, one of the Theorems 3.4.1–3.4.5) and have an estimate in $\mathbf{X} \times H^{\perp}$ as a function of the norms of $\mathbf{f}$ and $\mathbf{g}$, of the constant α (or α_0, or α_1) and of the constant $\tilde{\beta}$ appearing in (3.4.28). For instance, applying Theorem 3.4.2, we have the following result.

Theorem 3.5.3. *Assume that the assumptions* (3.4.8)–(3.4.15) *on spaces, norms and matrices are satisfied. Let $\mathbf{x}, \mathbf{y}, \mathbf{f}, \mathbf{g}$ satisfy the general system of equations* (3.2.1) *and* (3.2.2)*, with $\mathbf{y} \in H^{\perp}$. Assume moreover that A is symmetric and satisfies* (3.4.55) *and that the constant $\tilde{\beta}$ is defined by* (3.4.28)*. Then, we have:*

$$\|\mathbf{x}\|_X \leq \frac{1}{\alpha}\|\mathbf{f}\|_F + \frac{M_a^{1/2}}{\alpha^{1/2}\tilde{\beta}}\|\mathbf{g}\|_G, \tag{3.5.3}$$

$$\|\mathbf{y}\|_{H^\perp} \leq \frac{2M_a^{1/2}}{\alpha^{1/2}\tilde{\beta}}\|\mathbf{f}\|_F + \frac{M_a}{\tilde{\beta}^2}\|\mathbf{g}\|_G. \tag{3.5.4}$$

3.5.3 Some Special Cases

In some applications, we shall encounter situations where the right-hand side has the special form $(\mathbf{f}, 0)$ or $(0, \mathbf{g})$. In fact, the proofs of the previous Theorems often used explicitly those special cases. We now consider them in more detail. For the sake of simplicity, we will restrict our attention to the case of A symmetric and positive semi-definite.

3.5.3.1 The case $(\mathbf{f}, 0)$

From Proposition 3.5.1, we have immediately the following particular case.

Proposition 3.5.2. *Assume that A satisfies* (3.4.105) *and* $\mathbf{g} = 0$. *Then, problem* (3.2.1) *and* (3.2.2) *has at least one solution* $(\mathbf{x}, \mathbf{y})$. *Moreover,* $\mathbf{x}$ *is uniquely determined by* $\mathbf{f}$ *and*

$$||\mathbf{x}||_X \leq \frac{||\mathbf{f}||_F}{\alpha_1}. \tag{3.5.5}$$

Finally, $\mathbf{y}$ *is unique up to an element in* $H \equiv \mathrm{Ker}B^T$ *and*

$$||\pi_{H^\perp}\mathbf{y}||_Y \leq \frac{M_a||\mathbf{f}||_F}{\alpha_1\tilde{\beta}}. \tag{3.5.6}$$

□

Conversely, we have that Theorem 3.5.2 has two correspondents in the $(\mathbf{f}, 0)$ case.

Proposition 3.5.3. *Assume that A is symmetric and positive semi-definite, and assume that there exists a constant $C > 0$ such that, for every quadruple* $(\mathbf{x}, \mathbf{y}, \mathbf{f}, 0) \in X \times Y \times F \times G$ *satisfying* (3.2.1) *and* (3.2.2), *one has*

$$||\mathbf{x}||_X \leq C||\mathbf{f}||_F. \tag{3.5.7}$$

Then, the discrete ellipticity on the kernel (3.4.88) *holds with* $\alpha_0 = 1/(C^2 M_a)$, M_a *being the continuity constant of A defined in* (3.4.18). □

Proof. The proof is identical to the first part of the proof of Theorem 3.5.1, using Proposition 3.4.6. □

Proposition 3.5.4. *Assume that A is symmetric and positive semi-definite, and assume that there exists a constant $C > 0$ such that, for every quadruple $(\mathbf{x}, \mathbf{y}, \mathbf{f}, 0) \in X \times Y \times F \times G$ satisfying* (3.2.1) *and* (3.2.2)*, one has*

$$||\mathbf{y}||_Y \leq C||\mathbf{f}||_F. \tag{3.5.8}$$

Then, the inf-sup condition (3.4.41) *holds with $\beta = 1/C$.* □

Proof. The proof is identical to the second part of the proof of Theorem 3.5.2. □

3.5.3.2 The case $(\mathbf{0}, \mathbf{g})$

We begin with a simple lemma.

Lemma 3.5.1. *Assume that A is symmetric and positive semi-definite, and let Z be a subspace of* $\mathbf{X}$. *Then, $Ker(A_{ZZ}) \subset Ker A$.*

Proof. If $\mathbf{z}$ is in the kernel of A_{ZZ}, we immediately have that

$$\mathbf{z}^T A\mathbf{z} = 0, \tag{3.5.9}$$

which, using (3.4.22), implies $A\mathbf{z} = 0$. □

We can now prove the following result.

Proposition 3.5.5. *Assume that A is symmetric and positive semi-definite and that the inf-sup condition* (3.4.31) *holds. Then, for every $\mathbf{g} \in G$ and $\mathbf{f} = 0$, problem* (3.2.1) *and* (3.2.2) *has at least one solution* $(\mathbf{x}, \mathbf{y})$*. Moreover,* $\mathbf{y}$ *is uniquely determined by* $\mathbf{g}$ *and we have the bound*

$$\|\mathbf{y}\|_Y \leq \frac{M_a}{\beta^2}\|\mathbf{g}\|_G. \tag{3.5.10}$$

□

Proof. Using Proposition 3.4.4, we have that, for every $\mathbf{g} \in G$, there exists at least one $\mathbf{x}_g \in X$ such that $B\mathbf{x}_g = \mathbf{g}$ and

$$\|\mathbf{x}_g\| \leq \frac{1}{\beta}\|\mathbf{g}\|_G. \tag{3.5.11}$$

Using Lemma 3.5.1 with $Z = K$, we see that $\mathrm{Ker}A_{KK} \subseteq \mathrm{Ker}A$. Then, using Proposition 3.1.3 with $r = s$ and $S = Z = K$, we have that $\pi_K \mathrm{Im}A \subseteq \mathrm{Im}A_{KK}$. Hence, the problem: *find* $\mathbf{x}_K \in K$ *such that*

$$A_{KK}\mathbf{x}_K = -\pi_K A\mathbf{x}_g \tag{3.5.12}$$

has at least one solution. Using (3.4.22), then using (3.5.12) (multiplied to the left by $\mathbf{x}_K$), and finally using the symmetry of A, one gets

$$\|A\mathbf{x}_K\|_F^2 \le M_a \mathbf{x}_K^T A\mathbf{x}_K = M_a \mathbf{x}_K^T A\mathbf{x}_g \le M_a \|\mathbf{x}_g\|_X \|A\mathbf{x}_K\|_F, \tag{3.5.13}$$

which, using (3.5.11), gives immediately

$$\|A\mathbf{x}_K\|_F \le \frac{M_a}{\beta}\|\mathbf{g}\|_G. \tag{3.5.14}$$

Note that (3.5.12) implies that $A(\mathbf{x}_K + \mathbf{x}_g) \in K^\perp$, so that by (3.1.60) there exists a $\mathbf{y} \in Y$ such that $B^T\mathbf{y} = -(A\mathbf{x}_K + A\mathbf{x}_g)$, and by (3.4.41), (3.5.11), and (3.5.14) we have

$$\|\mathbf{y}\|_Y \le \frac{1}{\beta}\|A(\mathbf{x}_K + \mathbf{x}_g)\|_F \le \frac{M_a}{\beta^2}\|\mathbf{g}\|_G. \tag{3.5.15}$$

Finally, observe that $(\mathbf{x}_g + \mathbf{x}_K, \mathbf{y})$ solves (3.2.1) and (3.2.2) with $(0, \mathbf{g})$ as right-hand side.

To see the uniqueness, assume that $(\mathbf{x}^i, \mathbf{y}^i)$ $(i = 1, 2)$ are two solutions. Clearly, $\pi_K A(\mathbf{x}^1 - \mathbf{x}^2) = \pi_K B^T(\mathbf{y}^2 - \mathbf{y}^1) = 0$ and hence $\mathbf{x}^1 - \mathbf{x}^2$ is in the kernel of A_{KK}. Using Lemma 3.5.1, we see that $A(\mathbf{x}^1 - \mathbf{x}^2) = 0$ so that, from the first equations, $B^T(\mathbf{y}^2 - \mathbf{y}^1) = 0$ and the *inf-sup* condition (3.4.31) implies $\mathbf{y}^1 = \mathbf{y}^2$. □

Proposition 3.5.6. *Assume that A is symmetric and positive semi-definite, and that there exists a constant $C > 0$ such that, for every quadruple $(\mathbf{x}, \mathbf{y}, 0, \mathbf{g}) \in X \times Y \times F \times G$ satisfying* (3.2.1) *and* (3.2.2)*, one has*

$$||\mathbf{y}||_Y \le C||\mathbf{g}||_G. \tag{3.5.16}$$

Then, the inf-sup condition (3.4.31) *holds. However, we cannot bound β in terms of the constant C appearing in* (3.5.16). □

Proof. Let us first remark that assumption (3.5.16) implies that B^T is injective, and this implies (3.4.31). In order to see that the value of β cannot be deduced in general, consider the case when $A = 0$, $X = Y$ and B is γ times the identity. Then, the *inf-sup* condition holds with $\beta = |\gamma|$ and (3.5.16) holds with $C = 0$. □

Proposition 3.5.7. *Assume that A is symmetric and positive semi-definite, and that there exists a constant $C > 0$ such that for every quadruple $(\mathbf{x}, \mathbf{y}, 0, \mathbf{g}) \in X \times Y \times F \times G$ satisfying* (3.2.1) *and* (3.2.2) *one has,*

$$||\mathbf{x}||_X + ||\mathbf{y}||_Y \le C||\mathbf{g}||_G, \tag{3.5.17}$$

then (3.2.1) *and* (3.2.2) *has a solution for any $\mathbf{f} \in F$ and $\mathbf{g} \in G$, and* (3.4.31) *holds with $\beta = 1/C$.* □

Proof. Clearly, (3.5.17) implies that (3.2.1) and (3.2.2) for $\mathbf{f} = 0$ and $\mathbf{g} = 0$ has only the zero solution. Hence, Corollary 3.1.3 implies the solvability of (3.2.1) and (3.2.2) for general $\mathbf{f}$ and $\mathbf{g}$, and then Theorem 3.2.1 gives us (3.2.18) and (3.2.19). Hence, we just have to deal with the estimate of β. Note that, now (as we already have the unique solvability), (3.5.17) ensures the existence of a lifting operator that associates to every $\mathbf{g} \in G$ the first component $\mathbf{x}$ of the unique solution of (3.2.1) and (3.2.2) with right-hand side $(0, \mathbf{g})$. Hence, the result follows from Proposition 3.4.4. □

3.5.4 Composite Matrices

In the previous section, we considered the case in which the matrix A has a block structure of the type

$$\mathbb{A} = \begin{pmatrix} C & D^T \\ D & 0 \end{pmatrix}, \tag{3.5.18}$$

and B has the structure $\mathbb{B} = (E \quad 0)$ or $\mathbb{B} = (0 \quad E)$, so that the whole matrix has the block structure

$$M = \begin{pmatrix} C & D^T & E^T \\ D & 0 & 0 \\ E & 0 & 0 \end{pmatrix} \tag{3.5.19}$$

or

$$M = \begin{pmatrix} C & D^T & 0 \\ D & 0 & E^T \\ 0 & E & 0 \end{pmatrix}, \tag{3.5.20}$$

respectively. We were also able to find necessary and sufficient conditions for the *solvability*, simply using in a reasonable way the conditions dictated by the basic Theorem 3.2.1.

Here, we would like to consider the associated *stability* properties. These again can be deduced from the general case. It is clear that we would need three sequences of spaces $\mathbf{X}$, $\mathbf{Y}$ and $\mathbf{Z}$, with norms that ensure the continuity of the quadratic forms associated with the matrices

- C (on $\mathbf{X} \times \mathbf{X}$),
- D (on $\mathbf{X} \times \mathbf{Y}$),
- E (on $\mathbf{X} \times \mathbf{Z}$ for (3.5.19) and $\mathbf{Y} \times \mathbf{Z}$ for (3.5.20)),

as we did in (3.4.20), together with dual norms as in (3.4.10). Then, we just have to change the non-singularity conditions into their corresponding *uniform bounds*.

For instance, in the case (3.5.19), we easily obtained the algebraic conditions (3.2.48), that we recall for convenience of the reader:

$$\begin{aligned} &\mathrm{Im} D^T \cap \mathrm{Im} E^T = \mathbf{0}_r, \\ &\pi_{\mathbb{K}} C \text{ is non-singular } \mathbb{K} \to \mathbb{K} \quad \text{where } \mathbb{K} = \mathrm{Ker} D \cap \mathrm{Ker} E. \end{aligned} \tag{3.5.21}$$

It is not difficult to verify that the corresponding stability conditions are:

$$\begin{aligned} &\inf_{(\mathbf{y},\mathbf{z})\in \mathbf{Y}\times\mathbf{Z}} \sup_{\mathbf{x}\in\mathbf{X}} \frac{\mathbf{x}^T D^T \mathbf{y} + \mathbf{x}^T E^T \mathbf{z}}{\|\mathbf{x}\|_{\mathbf{X}} (\|\mathbf{y}\|_{\mathbf{Y}} + \|\mathbf{z}\|_{\mathbf{Z}})} \geq \delta > 0, \\ &\inf_{\tilde{\mathbf{x}}\in\mathbb{K}} \sup_{\mathbf{x}\in\mathbb{K}} \frac{\mathbf{x}^T C \tilde{\mathbf{x}}}{\|\tilde{\mathbf{x}}\|_{\mathbf{X}} \|\mathbf{x}\|_{\mathbf{Y}}} \geq \alpha > 0, \quad \text{where } \mathbb{K} = \mathrm{Ker} D \cap \mathrm{Ker} E. \end{aligned} \tag{3.5.22}$$

Clearly, we could simplify the condition on C by requiring ellipticity on $\mathbb{K}$, or ellipticity on the whole $\mathbf{X}$.

For (3.5.20), we performed first an exchange of rows and columns, to reach the form

$$M = \begin{pmatrix} C & 0 & D^T \\ 0 & 0 & E \\ D & E^T & 0 \end{pmatrix},$$

and we found the following solvability conditions:

$$\begin{aligned} &\mathrm{Ker} D^T \cap \mathrm{Ker} E = 0, \\ &\mathrm{Ker} E^T = 0, \\ &\pi_{\tilde{\mathbb{K}}} C \text{ is non-singular } \tilde{\mathbb{K}} \to \tilde{\mathbb{K}}, \end{aligned} \tag{3.5.23}$$

where $\tilde{\mathbb{K}}$ (cfr. (3.2.52)) is given by

$$\tilde{\mathbb{K}} = \{\mathbf{x} \in \mathbf{X} \text{ such that } D\mathbf{x} \in (\mathrm{Ker}\, E)^{\perp}\}. \tag{3.5.24}$$

Again, it is not difficult to verify that the corresponding stability conditions are:

$$\begin{aligned} &\inf_{\mathbf{y}\in\mathbf{Y}} \sup_{(\mathbf{x},\mathbf{z})\in(\mathbf{X}\times\mathbf{Z})} \frac{\mathbf{y}^T D\mathbf{x} + \mathbf{y}^T E^T \mathbf{z}}{(\|\mathbf{x}\|_{\mathbf{X}} + \|\mathbf{z}\|_{\mathbf{Z}}) \|\mathbf{y}\|_{\mathbf{Y}}} \geq \delta > 0, \\ &\inf_{\mathbf{z}\in\mathbf{Z}} \sup_{\mathbf{y}\in\mathbf{Y}} \frac{\mathbf{y}^T E^T \mathbf{z}}{\|\mathbf{y}\|_{\mathbf{Y}} \|\mathbf{z}\|_{\mathbf{Z}}} \geq \eta > 0, \\ &\inf_{\tilde{\mathbf{x}}\in\tilde{\mathbb{K}}} \sup_{\mathbf{x}\in\tilde{\mathbb{K}}} \frac{\mathbf{x}^T C \tilde{\mathbf{x}}}{\|\tilde{\mathbf{x}}\|_{\mathbf{X}} \|\mathbf{x}\|_{\mathbf{Y}}} \geq \alpha > 0. \end{aligned} \tag{3.5.25}$$

Here too, the third condition could possibly by replaced by an ellipticity condition. Moreover, it is easy to see that, in order to get the first condition, it would be sufficient to assume that one of the two matrices D or E^T satisfies an *inf-sup* condition by itself. However, this would often be an assumption too strong and difficult to obtain in practice. As we did in the previous section, we do not insist on these matters, and we shall not analyse the optimal dependence of the stability constants from δ, η and α appearing in (3.5.22) and (3.5.25).

3.6 Stability of Perturbed Matrices

We shall now discuss the case of problems of the type (3.3.1) where an additional matrix C is present. We assume that we are given, for each $k \in \mathbb{N}$, an $m(k) \times m(k)$ matrix C_k. Together with the matrices A_k and B_k, this will give us a sequence of perturbed problems

$$\begin{pmatrix} A_k & B_k^T \\ B_k & -C_k \end{pmatrix} \begin{pmatrix} \mathbf{x}_k \\ \mathbf{y}_k \end{pmatrix} = \begin{pmatrix} \mathbf{f}_k \\ \mathbf{g}_k \end{pmatrix}. \tag{3.6.1}$$

As we did for the unperturbed case (3.4.1), we *drop the index* k, and we just **remember** that we are actually dealing with a sequence of problems instead of a single one.

As a first step, we have to extend our assumptions (3.4.20) on the continuity of matrices A and B, requiring the continuity of C as well. Hence, we assume that there exists a constant M_c, independent of k, such that

$$\forall\, \mathbf{z} \in \mathbf{Y},\ \forall\, \mathbf{y} \in \mathbf{Y} \quad \mathbf{z}^T C \mathbf{y} \le M_c\, \|\mathbf{z}\|_Y\, \|\mathbf{y}\|_Y. \tag{3.6.2}$$

We note that, as in (3.4.18) and (3.4.19), we now have for every $\mathbf{y} \in \mathbf{Y}$:

$$\|C\mathbf{y}\|_G \equiv \sup_{\mathbf{z} \in \mathbb{R}^m} \frac{\mathbf{z}^T C \mathbf{y}}{\|\mathbf{z}\|_Y} \le M_c \|\mathbf{y}\|_Y. \tag{3.6.3}$$

We would like to extend the results of the previous subsection to the perturbed problem (3.6.1).

3.6.1 The Basic Estimate

Following Theorem 3.3.1 we are going to assume that A is symmetric and non-singular on $K = \operatorname{Ker} B$. It will therefore be useful, in order to reach optimal estimates in an easier way, to use directly (3.4.88), that we repeat for the convenience of the reader

$$\alpha_0 \|\mathbf{x}\|_X^2 \le \mathbf{x}^T A \mathbf{x} \qquad \forall \mathbf{x} \in K, \tag{3.6.4}$$

instead of (3.4.105). For technical reasons, it will also be easier to deal separately with the case $\mathbf{f} = 0$ and the case $\mathbf{g} = 0$, as we did, for instance, in the proofs of Theorems 3.4.1 and 3.4.2. This time, however, it will be more convenient to split the results in two different lemmata, and join them afterwards. We start therefore with the following lemma.

Lemma 3.6.1. *Let the assumptions* (3.4.8)–(3.4.15) *and* (3.6.2) *on spaces, norms and matrices be satisfied. Assume that the inf-sup condition* (3.4.31) *and the ellipticity requirement* (3.6.4) *are satisfied, and assume moreover that A is symmetric, and A and C are positive semi-definite. Then, if* $\mathbf{x}$, $\mathbf{y}$, *and* $\mathbf{g}$ *satisfy*

$$\begin{cases} A\mathbf{x} + B^T\mathbf{y} = 0 \\ B\mathbf{x} - C\mathbf{y} = \mathbf{g}, \end{cases} \tag{3.6.5}$$

we have the estimate

$$\|\mathbf{x}\|_X \leq \frac{2M_a^{1/2}(\beta^2 + M_c M_a)}{\alpha_0^{1/2}\beta^3}\|\mathbf{g}\|_G, \tag{3.6.6}$$

$$\|\mathbf{y}\|_Y \leq \frac{M_a}{\beta^2}\|\mathbf{g}\|_G. \tag{3.6.7}$$

Proof. Using the *inf-sup* condition in the form (3.4.41) together with the first equation of (3.6.5), we obtain

$$\beta\|\mathbf{y}\|_Y \leq \|B^T\mathbf{y}\|_F = \|A\mathbf{x}\|_F. \tag{3.6.8}$$

Now, we take the scalar product of the first equation of (3.6.5) times $\mathbf{x}$, we take the scalar product of the second equation of (3.6.5) times $\mathbf{y}$, and we take the difference, obtaining

$$\mathbf{x}^T A\mathbf{x} + \mathbf{y}^T C\mathbf{y} = -\mathbf{y}^T\mathbf{g}. \tag{3.6.9}$$

Using (3.4.22), then Eq. (3.6.9) with the assumption that C is positive semi-definite, and finally (3.4.17), we have

$$\|A\mathbf{x}\|_F^2 \leq M_a\mathbf{x}^T A\mathbf{x} \leq -M_a\mathbf{y}^T\mathbf{g} \leq M_a\|\mathbf{y}\|_Y\,\|\mathbf{g}\|_G, \tag{3.6.10}$$

which, combined with (3.6.8), yields

$$\|A\mathbf{x}\|_F \leq \frac{M_a}{\beta}\|\mathbf{g}\|_G, \tag{3.6.11}$$

so that, using again (3.6.8),

$$\|\mathbf{y}\|_Y \leq \frac{M_a}{\beta^2}\|\mathbf{g}\|_G, \tag{3.6.12}$$

which proves (3.6.7). The proof now becomes similar to that of Theorem 3.4.2. Using Proposition 3.4.4, we set

$$\tilde{\mathbf{x}} := \mathbf{L}(\mathbf{g} + C\mathbf{y}) \tag{3.6.13}$$

so that, from (3.4.43),

$$B\tilde{\mathbf{x}} = \mathbf{g} + C\mathbf{y}, \tag{3.6.14}$$

together with

$$\beta\|\tilde{\mathbf{x}}\|_X \le \|\mathbf{g} + C\mathbf{y}\|_G \le (1 + \frac{M_c M_a}{\beta^2})\|\mathbf{g}\|_G, \tag{3.6.15}$$

where, in the last step, we used (3.6.3) and (3.6.12). From (3.6.15), we have then immediately

$$\|\tilde{\mathbf{x}}\|_X \le \frac{\beta^2 + M_c M_a}{\beta^3}\|\mathbf{g}\|_G. \tag{3.6.16}$$

Setting now

$$\mathbf{x}_K := \mathbf{x} - \tilde{\mathbf{x}}, \tag{3.6.17}$$

we have from (3.6.14) and the second equation of (3.6.5) that $\mathbf{x}_K \in K$ (the kernel of B). We then note that, from the first equation of (3.6.1):

$$\mathbf{x}_K^T A\mathbf{x} = -\mathbf{x}_K^T B^T\mathbf{y} = -\mathbf{y}^T B\mathbf{x}_K = 0. \tag{3.6.18}$$

Moreover, using (3.6.17), (3.6.18), and then (3.3.5), we have

$$\mathbf{x}_K^T A\mathbf{x}_K = -\mathbf{x}_K^T A\tilde{\mathbf{x}} \le (\mathbf{x}_K^T A\mathbf{x}_K)^{1/2}(\tilde{\mathbf{x}}^T A\tilde{\mathbf{x}})^{1/2}, \tag{3.6.19}$$

which easily gives

$$\mathbf{x}_K^T A\mathbf{x}_K \le \tilde{\mathbf{x}}^T A\tilde{\mathbf{x}}. \tag{3.6.20}$$

Hence, we can use (3.6.4) and (3.6.20) to obtain

$$\alpha_0\|\mathbf{x}_K\|_X^2 \le \mathbf{x}_K^T A\mathbf{x}_K \le \tilde{\mathbf{x}}^T A\tilde{\mathbf{x}}, \tag{3.6.21}$$

and finally from (3.6.21) and (3.4.20)

$$\|\mathbf{x}_K\|_X \le \left(\frac{M_a}{\alpha_0}\right)^{1/2}\|\tilde{\mathbf{x}}\|_X. \tag{3.6.22}$$

Finally, we can collect (3.6.17), (3.6.22) and (3.6.16) and have an estimate for $\mathbf{x}$:

$$\begin{aligned}\|\mathbf{x}\|_X &\le \|\mathbf{x}_K\|_X + \|\tilde{\mathbf{x}}\|_X \le (1 + (\frac{M_a}{\alpha})^{1/2})\|\tilde{\mathbf{x}}\|_X \\ &\le (1 + (\frac{M_a}{\alpha_0})^{1/2})\,\frac{\beta^2 + M_c M_a}{\beta^3}\|\mathbf{g}\|_G.\end{aligned} \tag{3.6.23}$$

Using (3.4.89) in (3.6.23), we obtain (3.6.6) and the proof is completed. □

Remark 3.6.1. The dependence of the constants in (3.6.6) and (3.6.7) on α_0 and β cannot be improved. Indeed, considering for instance (for $0 < a, b \ll 1$) the problem

$$\begin{pmatrix} 2a & \sqrt{a} & -\sqrt{a} & 0 & 0 \\ \sqrt{a} & 2 & 1 & b & 0 \\ -\sqrt{a} & 1 & 2 & 0 & b \\ 0 & b & 0 & 0 & 1 \\ 0 & 0 & b & -1 & 0 \end{pmatrix} \cdot \begin{pmatrix} x_1 \\ x_2 \\ x_3 \\ y_1 \\ y_2 \end{pmatrix} = \begin{pmatrix} 0 \\ 0 \\ 0 \\ -1 \\ -1 \end{pmatrix}, \tag{3.6.24}$$

we easily have, as unique solution,

$$x_1 = \frac{3}{b^3 a^{1/2}}, \quad x_2 = -\frac{3+b^2}{b^3}, \quad x_3 = \frac{3-b^2}{b^3}, \tag{3.6.25}$$

$$y_1 = \frac{3}{b^2}, \quad y_2 = \frac{3}{b^2}. \tag{3.6.26}$$

We can easily check that we have $\alpha_0 = 2a$ and $\beta = b$, and we verify the optimality of (3.6.12) and (3.6.23). □

We now consider the case where $\mathbf{g}$ is equal to zero and $\mathbf{f}$ is not.

Lemma 3.6.2. *Let the assumptions* (3.4.8)–(3.4.15) *and* (3.6.2) *on spaces, norms and matrices be satisfied. Assume that the inf-sup condition* (3.4.31) *and the ellipticity requirement* (3.6.4) *are satisfied, and assume moreover that A is symmetric, and A and C are positive semi-definite. Then, if* $\mathbf{x}$, $\mathbf{y}$, *and* $\mathbf{f}$ *satisfy*

$$\begin{cases} A\mathbf{x} + B^T\mathbf{y} = \mathbf{f} \\ B\mathbf{x} - C\mathbf{y} = 0, \end{cases} \tag{3.6.27}$$

we have the estimates

$$\|\mathbf{x}\|_X \le \frac{(\beta^2 + 2M_c M_a)^2 + 4(M_c M_a)^2}{\alpha_0 \beta^4} \|\mathbf{f}\|_F, \tag{3.6.28}$$

$$\|\mathbf{y}\|_Y \le \frac{2M_a^{1/2}(2M_c M_a + \beta^2)}{\alpha_0^{1/2}\beta^3} \|\mathbf{f}\|_F. \tag{3.6.29}$$

Proof. As in the previous lemma, we take the scalar product of the first equation of (3.6.27) with $\mathbf{x}$, then we take the scalar product of the second equation of (3.6.27) with $\mathbf{y}$, and we take the difference, obtaining

$$\mathbf{x}^T A\mathbf{x} + \mathbf{y}^T C\mathbf{y} = \mathbf{x}^T\mathbf{f}. \tag{3.6.30}$$

Using then (3.4.22), Eq. (3.6.30) with the assumption that C is positive semi-definite, and finally (3.4.17), we have

$$\|A\mathbf{x}\|_F^2 \le M_a \mathbf{x}^T A\mathbf{x} \le M_a \mathbf{x}^T \mathbf{f} \le M_a \|\mathbf{x}\|_X \, \|\mathbf{f}\|_F. \tag{3.6.31}$$

Next, we use the *inf-sup* condition in the form (3.4.41) to obtain, from the first equation of (3.6.27),

$$\begin{aligned} \beta\|\mathbf{y}\|_Y \le \|B^T\mathbf{y}\|_F &\equiv \|\mathbf{f} - A\mathbf{x}\|_F \\ &\le \|A\mathbf{x}\|_F + \|\mathbf{f}\|_F. \end{aligned} \tag{3.6.32}$$

We now consider, as we did before, the lifting operator $\mathbf{L}$ as defined in Proposition 3.4.4 and we set

$$\tilde{\mathbf{x}} := \mathbf{L}(C\mathbf{y}) \tag{3.6.33}$$

so that

$$B\tilde{\mathbf{x}} - C\mathbf{y} = 0. \tag{3.6.34}$$

Then, using (3.4.43) and (3.6.3),

$$\beta\|\tilde{\mathbf{x}}\|_X \le \|C\mathbf{y}\|_G \le M_c \, \|\mathbf{y}\|_Y. \tag{3.6.35}$$

We now set

$$\mathbf{x}_K := \mathbf{x} - \tilde{\mathbf{x}} \tag{3.6.36}$$

and we note that, clearly, $B\mathbf{x}_K = 0$, so that $\mathbf{x}_K \in K = \operatorname{Ker} B$. Our next (and most delicate) step will be to estimate $\mathbf{x}_K$ in terms of $\tilde{\mathbf{x}}$. We first note that, using (3.6.4),

$$\alpha_0\|\mathbf{x}_K\|_X^2 \le \mathbf{x}_K^T A\mathbf{x}_K, \tag{3.6.37}$$

which implies that

$$\|\mathbf{x}_K\|_X \le \Big(\frac{\mathbf{x}_K^T A\mathbf{x}_K}{\alpha_0}\Big)^{1/2}. \tag{3.6.38}$$

Then, we estimate $\mathbf{x}_K^T A\mathbf{x}_K$. We remember again that $\mathbf{x}_K^T B^T\mathbf{y} = 0$ (since $\mathbf{x}_K \in \operatorname{Ker} B$), so that, using (3.6.36) and the first equation of (3.6.27),

$$\mathbf{x}_K^T A\mathbf{x}_K = \mathbf{x}_K^T A\mathbf{x} - \mathbf{x}_K^T A\tilde{\mathbf{x}} = \mathbf{x}_K^T\mathbf{f} - \mathbf{x}_K^T A\tilde{\mathbf{x}}. \tag{3.6.39}$$

We now use (3.6.39) with (3.4.17) and (3.3.5), and then (3.6.38) to obtain

$$\begin{aligned} \mathbf{x}_K^T A\mathbf{x}_K &\le \|\mathbf{f}\|_F \, \|\mathbf{x}_K\|_X + (\mathbf{x}_K^T A\mathbf{x}_K)^{1/2}(\tilde{\mathbf{x}}^T A\tilde{\mathbf{x}})^{1/2} \\ &\le \|\mathbf{f}\|_F \Big(\frac{\mathbf{x}_K^T A\mathbf{x}_K}{\alpha_0}\Big)^{1/2} + (\mathbf{x}_K^T A\mathbf{x}_K)^{1/2}(\tilde{\mathbf{x}}^T A\tilde{\mathbf{x}})^{1/2} \\ &\le (\mathbf{x}_K^T A\mathbf{x}_K)^{1/2}\Big(\frac{1}{{\alpha_0}^{1/2}}\|\mathbf{f}\|_F + (\tilde{\mathbf{x}}^T A\tilde{\mathbf{x}})^{1/2}\Big), \end{aligned} \tag{3.6.40}$$

implying

$$(\mathbf{x}_K^T A \mathbf{x}_K)^{1/2} \leq \frac{1}{{\alpha_0}^{1/2}} \|\mathbf{f}\|_F + (\tilde{\mathbf{x}}^T A \tilde{\mathbf{x}})^{1/2}. \tag{3.6.41}$$

Inserting (3.6.41) into (3.6.38), and then using (3.4.20), we now have

$$\|\mathbf{x}_K\|_X \leq \frac{1}{\alpha_0} \|\mathbf{f}\|_F + \Big(\frac{\tilde{\mathbf{x}}^T A \tilde{\mathbf{x}}}{\alpha_0}\Big)^{1/2} \leq \frac{1}{\alpha_0} \|\mathbf{f}\|_F + \frac{{M_a}^{1/2}}{{\alpha_0}^{1/2}} \|\tilde{\mathbf{x}}\|_X. \tag{3.6.42}$$

We can now collect (3.6.36), (3.6.42) and (3.6.35) to obtain an estimate for $\mathbf{x}$

$$\begin{aligned} \|\mathbf{x}\|_X \leq \|\mathbf{x}_K\|_X + \|\tilde{\mathbf{x}}\|_X &\leq \frac{1}{\alpha_0} \|\mathbf{f}\|_F + \Big(\frac{M_c M_a^{1/2}}{{\alpha_0}^{1/2}\beta} + \frac{M_c}{\beta}\Big) \|\mathbf{y}\|_Y \\ &\leq \frac{1}{\alpha_0} \|\mathbf{f}\|_F + \frac{2 M_c M_a^{1/2}}{{\alpha_0}^{1/2}\beta} \|\mathbf{y}\|_Y. \end{aligned} \tag{3.6.43}$$

Now, we take the square of both sides of (3.6.32), we use $(a+b)^2 \leq 2(a^2+b^2)$, we insert (3.6.31) and finally (3.6.43):

$$\begin{aligned} \beta^2 \|\mathbf{y}\|_Y^2 &\leq 2\|A\mathbf{x}\|_F^2 + 2\|\mathbf{f}\|_F^2 \leq 2 M_a \|\mathbf{x}\|_X \, \|\mathbf{f}\|_F + 2\|\mathbf{f}\|_F^2 \\ &\leq 2\|\mathbf{f}\|_F \Big(\frac{2 M_c M_a^{3/2}}{{\alpha_0}^{1/2}\beta} \|\mathbf{y}\|_Y + \frac{M_a}{\alpha_0} \|\mathbf{f}\|_F\Big) + 2\|\mathbf{f}\|_F^2. \end{aligned} \tag{3.6.44}$$

We now use the fact that, for positive real numbers t, a, and b, if $t^2 \leq at + b$, then $t \leq a + \sqrt{b}$. Applied to (3.6.44), this gives

$$\|\mathbf{y}\|_Y \leq \frac{4 M_c M_a^{3/2}}{{\alpha_0}^{1/2}\beta^3} \|\mathbf{f}\|_F + \frac{(2M_a + 2\alpha_0)^{1/2}}{{\alpha_0}^{1/2}\beta} \|\mathbf{f}\|_F. \tag{3.6.45}$$

Using again the fact that $\alpha_0 \leq M_a$, we can rewrite (3.6.45) as

$$\|\mathbf{y}\|_Y \leq \frac{2 M_a^{1/2}(2 M_c M_a + \beta^2)}{{\alpha_0}^{1/2}\beta^3} \|\mathbf{f}\|_F. \tag{3.6.46}$$

Inserting (3.6.46) into (3.6.43), we obtain the corresponding estimate for $\mathbf{x}$:

$$\begin{aligned} \|\mathbf{x}\|_X &\leq \Big(\frac{8(M_c M_a)^2 + 4 M_c M_a \beta^2}{\alpha_0 \beta^4} + \frac{1}{\alpha_0}\Big) \|\mathbf{f}\|_F \\ &= \frac{(\beta^2 + 2 M_c M_a)^2 + 4(M_c M_a)^2}{\alpha_0 \beta^4} \|\mathbf{f}\|_F, \end{aligned} \tag{3.6.47}$$

which concludes the proof. □

Remark 3.6.2. The result (3.6.28) and (3.6.29) cannot be improved in its dependence from the constants α_0 and β. Indeed, if we consider, for $0 < a, b \ll 1$, the system

$$\begin{pmatrix} 2a & \sqrt{a} & -\sqrt{a} & 0 & 0 \\ \sqrt{a} & 2 & 1 & b & 0 \\ -\sqrt{a} & 1 & 2 & 0 & b \\ 0 & b & 0 & 0 & 1 \\ 0 & 0 & b & -1 & 0 \end{pmatrix} \cdot \begin{pmatrix} x_1 \\ x_2 \\ x_3 \\ y_1 \\ y_2 \end{pmatrix} = \begin{pmatrix} 2 \\ 0 \\ 0 \\ 0 \\ 0 \end{pmatrix}, \tag{3.6.48}$$

we easily have, as unique solution,

$$x_1 = \frac{3+b^4}{a\,b^4}, \quad x_2 = \frac{-3-b^2}{a^{1/2}\,b^4}, \quad x_3 = \frac{3-b^2}{a^{1/2}\,b^4}, \tag{3.6.49}$$

$$y_1 = \frac{3-b^2}{a^{1/2}\,b^3}, \quad y_2 = \frac{3+b^2}{a^{1/2}\,b^3}. \tag{3.6.50}$$

It is not difficult to check that $\alpha_0 = 2a$ and $\beta = b$. Hence, (3.6.49) and (3.6.50) shows the optimality of (3.6.28) and (3.6.29). □

We can now collect the results of the previous two lemmata.

Theorem 3.6.1. *Let the assumptions* (3.4.8)–(3.4.15) *and* (3.6.2) *on spaces, norms and matrices be satisfied. Assume that the inf-sup condition* (3.4.31) *and the ellipticity requirement* (3.6.4) *are satisfied, and assume moreover that A is symmetric and that A and C are positive semi-definite. Then, if* $\mathbf{x}$, $\mathbf{y}$, $\mathbf{f}$, *and* $\mathbf{g}$ *satisfy*

$$\begin{cases} A\mathbf{x} + B^T\mathbf{y} = \mathbf{f} \\ B\mathbf{x} - C\mathbf{y} = \mathbf{g}, \end{cases} \tag{3.6.51}$$

we have the estimates

$$\|\mathbf{x}\|_X \le \frac{(\beta^2 + 2M_cM_a)^2 + 4(M_cM_a)^2}{\alpha_0\beta^4}\|\mathbf{f}\|_F + \frac{2M_a^{1/2}(\beta^2 + M_cM_a)}{\alpha_0^{1/2}\beta^3}\|\mathbf{g}\|_G, \tag{3.6.52}$$

$$\|\mathbf{y}\|_Y \le \frac{2M_a^{1/2}(2M_cM_a + \beta^2)}{\alpha_0^{1/2}\beta^3}\|\mathbf{f}\|_F + \frac{M_a}{\beta^2}\|\mathbf{g}\|_G. \tag{3.6.53}$$

The proof easily follows by linearity.

3.6.2 The Symmetric Case for Perturbed Matrices

The dependence of the constants in (3.6.52) and (3.6.53) on α_0 and β improves noticeably if we assume that C is symmetric as well. As an example, we can consider the particular case (relevant in applications) of systems still having the structure (3.6.1), where C is a symmetric and positive definite matrix verifying

$$\gamma \|\mathbf{y}\|_Y^2 \leq \mathbf{y}^T C \mathbf{y} \leq M_c \|\mathbf{y}\|_Y^2 \quad \forall\, \mathbf{y} \in \mathbf{Y}. \tag{3.6.54}$$

We note that our assumption (3.6.54) easily implies that

$$\frac{1}{M_c}\|\mathbf{z}\|_G^2 \leq \mathbf{z}^T C^{-1}\mathbf{z} \leq \frac{1}{\gamma}\|\mathbf{z}\|_G^2 \quad \forall\, \mathbf{z} \in \mathbf{Y}. \tag{3.6.55}$$

From (3.6.54), we easily obtain as well that

$$\|\mathbf{y}\|_Y \leq \frac{1}{\gamma}\|C\mathbf{y}\|_G \quad \forall\, \mathbf{y} \in \mathbf{Y} \tag{3.6.56}$$

and from (3.6.55)

$$\|\mathbf{z}\|_G \leq M_c \|C^{-1}\mathbf{z}\|_Y \quad \forall\, \mathbf{z} \in \mathbf{Y}. \tag{3.6.57}$$

We are now ready to prove our improved estimates.

Remark 3.6.3. We shall prove in the next chapter, Sect. 4.3, additional related results (in the infinite dimensional case) which may be considered as more elegant, but for which we have no example showing optimality. □

Theorem 3.6.2. *Let the assumptions* (3.4.8)–(3.4.15) *and* (3.6.2) *on spaces, norms and matrices be satisfied. Assume that the inf-sup condition* (3.4.31) *and the ellipticity requirement* (3.6.4) *are satisfied, and assume moreover that A is symmetric and positive semi-definite and that C is* ***symmetric*** *and satisfies* (3.6.54)*. Then, if* $\mathbf{x}$, $\mathbf{y}$, $\mathbf{f}$, *and* $\mathbf{g}$ *satisfy* (3.6.51)*, we have the estimate*

$$\|\mathbf{x}\|_X \leq \frac{\beta^2 + 4M_c M_a}{\alpha_0 \beta^2}\|\mathbf{f}\|_F + \frac{2M_a^{1/2} M_c}{\alpha_0^{1/2}\gamma\beta}\|\mathbf{g}\|_G \tag{3.6.58}$$

and

$$\|\mathbf{y}\|_Y \leq \frac{2M_c M_a^{1/2}}{\gamma \alpha_0^{1/2}\beta}\|\mathbf{f}\|_F + \frac{2\, M_a(M_c+\gamma)}{M_a\gamma^2 + (M_c+\gamma)\beta^2}\|\mathbf{g}\|_G. \tag{3.6.59}$$

Proof. As we are already used to, we shall split the two cases $\mathbf{f} = 0$ and $\mathbf{g} = 0$, and then combine the estimates by linearity.

Let us consider first the case $\mathbf{f} = 0$, and assume that $\mathbf{x}$, $\mathbf{y}$, and $\mathbf{g}$ satisfy (3.6.5).

Following the notation of Lemma 3.6.1, we still have (3.6.11), (3.6.12) and (3.6.22). Our target is to improve (3.6.16), which is suboptimal in our (stronger) assumptions. For this we restart by taking once more the scalar product of the first equation of (3.6.5) times $\mathbf{x}$, getting

$$\mathbf{x}^T A\mathbf{x} + \mathbf{x}^T B^T \mathbf{y} = 0 \tag{3.6.60}$$

and we substitute $\mathbf{y} = C^{-1}(B\mathbf{x} - \mathbf{g})$. Recalling that A is positive semi-definite, we obtain

$$\mathbf{x}^T B^T C^{-1} B\mathbf{x} \le \mathbf{x}^T B^T C^{-1}\mathbf{g} = \mathbf{g}^T C^{-1} B\mathbf{x}. \tag{3.6.61}$$

Using (3.6.55) with $\mathbf{z} = B\mathbf{x}$, then (3.6.61), then (3.4.17), and finally (3.6.56) with $\mathbf{y} = C^{-1}B\mathbf{x}$, we have

$$\begin{aligned}\|B\mathbf{x}\|_G^2 &\le M_c(\mathbf{x}^T B^T C^{-1} B\mathbf{x}) \le M_c(\mathbf{g}^T C^{-1} B\mathbf{x})\\ &\le M_c\|\mathbf{g}\|_G\,\|C^{-1}B\mathbf{x}\|_Y \le \frac{M_c}{\gamma}\|\mathbf{g}\|_G\|B\mathbf{x}\|_G,\end{aligned} \tag{3.6.62}$$

which easily gives

$$\|B\mathbf{x}\|_G \le \frac{M_c}{\gamma}\|\mathbf{g}\|_G. \tag{3.6.63}$$

As in Lemma 3.6.1, we set again (see (3.6.13) and (3.6.14)) $\tilde{\mathbf{x}} := \mathbf{L}(\mathbf{g}+C\mathbf{y})$, getting $B\tilde{\mathbf{x}} = \mathbf{g} + C\mathbf{y} = B\mathbf{x}$. Using (3.4.43), we have therefore

$$\beta\|\tilde{\mathbf{x}}\|_X \le \|B\tilde{\mathbf{x}}\|_G = \|B\mathbf{x}\|_G \tag{3.6.64}$$

and combining (3.6.63) and (3.6.64), we obtain

$$\|\tilde{\mathbf{x}}\|_X \le \frac{M_c}{\gamma\beta}\|\mathbf{g}\|_G, \tag{3.6.65}$$

which is the required improvement of (3.6.16). We can now use this improved estimate in (3.6.22), and we obtain

$$\|\mathbf{x}_K\|_X \le \left(\frac{M_a}{\alpha}\right)^{1/2}\|\tilde{\mathbf{x}}\|_X \le \frac{M_c M_a^{1/2}}{\gamma\beta\alpha^{1/2}}\|\mathbf{g}\|_G. \tag{3.6.66}$$

We note at this point that we have another way to obtain an estimate for $\mathbf{y}$, apart from (3.6.12) that we keep from the previous analysis; actually, from (3.6.56) and the second equation of (3.6.5), and then (3.6.63):

$$\|\mathbf{y}\|_Y \le \frac{1}{\gamma}\|B\mathbf{x} - \mathbf{g}\|_G \le \left(\frac{1}{\gamma} + \frac{M_c}{\gamma^2}\right)\|\mathbf{g}\|_G = \frac{\gamma + M_c}{\gamma^2}\|\mathbf{g}\|_G. \tag{3.6.67}$$

With some manipulations, we see that (3.6.12) and (3.6.67) can be combined into

$$\|\mathbf{y}\|_Y \le \frac{2\,M_a(M_c+\gamma)}{M_a\gamma^2 + (M_c+\gamma)\beta^2}\|\mathbf{g}\|_G. \tag{3.6.68}$$

We collect the results for $\mathbf{f} = 0$:

$$\|\mathbf{x}\|_X \le \|\mathbf{x}_K\|_X + \|\tilde{\mathbf{x}}\|_X \le \left(\left(\frac{M_a}{\alpha_0}\right)^{1/2} + 1\right)\frac{M_c}{\gamma\beta}\|\mathbf{g}\|_G, \tag{3.6.69}$$

$$\|\mathbf{y}\|_Y \le \frac{2\,M_a(M_c+\gamma)}{M_a\gamma^2 + (M_c+\gamma)\beta^2}\|\mathbf{g}\|_G. \tag{3.6.70}$$

We also note that, using $\alpha_0 \le M_a$, the estimate (3.6.69) becomes

$$\|\mathbf{x}\|_X \le \frac{2M_a^{1/2}M_c}{\alpha_0^{1/2}\gamma\beta}\|\mathbf{g}\|_G. \tag{3.6.71}$$

We consider now the case in which $\mathbf{g} = 0$ and assume that $\mathbf{x}$, $\mathbf{y}$, and $\mathbf{f}$ satisfy (3.6.27). As before, we can keep part of the previous analysis, but we can improve it in several places. From the proof of Lemma 3.6.2, we keep the definition of $\tilde{\mathbf{x}}$ and $\mathbf{x}_K$, and the estimates (3.6.41) and (3.6.42). We now take the scalar product of the first equation of (3.6.27) times $\tilde{\mathbf{x}}$, and substitute $\mathbf{y} = C^{-1}B\mathbf{x}$:

$$\tilde{\mathbf{x}}^T A\mathbf{x} + \tilde{\mathbf{x}}^T B^T C^{-1}B\mathbf{x} = \tilde{\mathbf{x}}^T\mathbf{f}. \tag{3.6.72}$$

We now recall that $B\tilde{\mathbf{x}} = B\mathbf{x}$, and rewrite (3.6.72) as follows

$$\mathbf{x}^T B^T C^{-1}B\mathbf{x} = \tilde{\mathbf{x}}^T\mathbf{f} - \tilde{\mathbf{x}}^T A\mathbf{x}. \tag{3.6.73}$$

We now apply (3.6.55) with $\mathbf{z} = B\mathbf{x}$ and we use (3.6.73) to obtain:

$$\frac{1}{M_c}\|B\mathbf{x}\|_G^2 \le \mathbf{x}^T B^T C^{-1}B\mathbf{x} = \tilde{\mathbf{x}}^T\mathbf{f} - \tilde{\mathbf{x}}^T A\mathbf{x}.$$

We then use (3.4.17) and the estimate $\beta\|\tilde{\mathbf{x}}\|_G \le \|B\tilde{\mathbf{x}}\|_G = \|B\mathbf{x}\|_G$ as in (3.6.64) and we reach

$$\frac{1}{M_c}\|B\mathbf{x}\|_G^2 \le \frac{1}{\beta}\|\mathbf{f}\|_F\,\|B\mathbf{x}\|_G - \tilde{\mathbf{x}}^T A\mathbf{x}. \tag{3.6.74}$$

We leave (3.6.74) for a while, and we estimate $-\tilde{\mathbf{x}}^T A\mathbf{x}$. Using the fact that $\mathbf{x} = \tilde{\mathbf{x}} + \mathbf{x}_K$, then (3.3.5), then (3.6.41), and finally some little algebra, we have

$$\begin{aligned} -\tilde{\mathbf{x}}^T A\mathbf{x} &= -\tilde{\mathbf{x}}^T A\tilde{\mathbf{x}} - \tilde{\mathbf{x}}^T A\mathbf{x}_K \\ &\le -\tilde{\mathbf{x}}^T A\tilde{\mathbf{x}} + (\tilde{\mathbf{x}}^T A\tilde{\mathbf{x}})^{1/2}(\mathbf{x}_K^T A\mathbf{x}_K)^{1/2} \\ &\le -\tilde{\mathbf{x}}^T A\tilde{\mathbf{x}} + (\tilde{\mathbf{x}}^T A\tilde{\mathbf{x}})^{1/2}\left(\frac{1}{\alpha_0{}^{1/2}}\|\mathbf{f}\|_F + (\tilde{\mathbf{x}}^T A\tilde{\mathbf{x}})^{1/2}\right) \\ &= \frac{1}{\alpha_0{}^{1/2}}\|\mathbf{f}\|_F(\tilde{\mathbf{x}}^T A\tilde{\mathbf{x}})^{1/2}, \end{aligned} \tag{3.6.75}$$

which inserted in (3.6.74) gives

$$\frac{1}{M_c}\|B\mathbf{x}\|_G^2 \le \frac{1}{\alpha_0^{1/2}}\|\mathbf{f}\|_F(\tilde{\mathbf{x}}^T A\tilde{\mathbf{x}})^{1/2} + \frac{1}{\beta}\|\mathbf{f}\|_F\,\|B\mathbf{x}\|_G. \tag{3.6.76}$$

Using the continuity of A (3.4.18) and once more $\beta\|\tilde{\mathbf{x}}\|_X \le \|B\mathbf{x}\|_G$, inequality (3.6.76) gives:

$$\frac{1}{M_c}\|B\mathbf{x}\|_G^2 \le \frac{M_a^{1/2}}{\alpha_0^{1/2}\beta}\|\mathbf{f}\|_F\|B\mathbf{x}\|_G + \frac{1}{\beta}\|\mathbf{f}\|_F\,\|B\mathbf{x}\|_G, \tag{3.6.77}$$

so that we can divide both sides by $\|B\mathbf{x}\|_G$, obtaining

$$\frac{1}{M_c}\|B\mathbf{x}\|_G \le \frac{M_a^{1/2}}{\alpha_0^{1/2}\beta}\|\mathbf{f}\|_F + \frac{1}{\beta}\|\mathbf{f}\|_F \le \frac{M_a^{1/2}+\alpha_0^{1/2}}{\alpha_0^{1/2}\beta}\|\mathbf{f}\|_F, \tag{3.6.78}$$

which is the basis of our improved estimates. From (3.6.78), we first derive

$$\|\tilde{\mathbf{x}}\|_X \le \frac{1}{\beta}\|B\tilde{\mathbf{x}}\|_G \le \frac{M_c\,(M_a^{1/2}+\alpha_0^{1/2})}{\alpha_0^{1/2}\beta^2}\|\mathbf{f}\|_F, \tag{3.6.79}$$

and then we use it in (3.6.42)

$$\begin{aligned}
\|\mathbf{x}_K\|_X &\le \frac{1}{\alpha_0}\|\mathbf{f}\|_F + \frac{M_a^{1/2}}{\alpha_0^{1/2}}\|\tilde{\mathbf{x}}\|_X \\
&\le \Big(\frac{1}{\alpha_0} + \frac{M_a^{1/2}}{\alpha_0^{1/2}}\,\frac{M_c\,(M_a^{1/2}+\alpha_0^{1/2})}{\alpha_0^{1/2}\beta^2}\Big)\|\mathbf{f}\|_F \\
&\le \Big(\frac{1}{\alpha_0} + \frac{M_cM_a + M_c(M_a\alpha_0)^{1/2}}{\alpha_0\beta^2}\Big)\|\mathbf{f}\|_F.
\end{aligned} \tag{3.6.80}$$

From the second equation of (3.6.27), (3.6.56) and (3.6.78), we also derive our improved estimate for $\mathbf{y}$

$$\|\mathbf{y}\|_Y = \|C^{-1}B\mathbf{x}\|_Y \le \frac{1}{\gamma}\|B\mathbf{x}\|_G \le \frac{M_c}{\gamma}\,\frac{M_a^{1/2}+\alpha_0^{1/2}}{\alpha_0^{1/2}\beta}\|\mathbf{f}\|_F. \tag{3.6.81}$$

We collect the results for $\mathbf{g} = 0$, using the fact that $\alpha \le M_a$. From (3.6.79) and (3.6.80), we have the estimate on $\mathbf{x}$

$$\|\mathbf{x}\|_X \le \|\tilde{\mathbf{x}}\|_X + \|\mathbf{x}_K\|_X$$
$$\le \Big(\frac{M_c\,(M_a^{1/2}+\alpha_0^{1/2})}{\alpha_0^{1/2}\beta^2} + \frac{1}{\alpha_0} + \frac{M_cM_a + M_c(M_a\alpha_0)^{1/2}}{\alpha_0\beta^2}\Big)\|\mathbf{f}\|_F$$
$$\le \frac{\beta^2 + 4M_cM_a}{\alpha_0\beta^2}\|\mathbf{f}\|_F, \quad (3.6.82)$$

while, from (3.6.81), we have the estimate on $\mathbf{y}$

$$\|\mathbf{y}\|_Y \le \frac{2M_cM_a^{1/2}}{\gamma\alpha_0^{1/2}\beta}\|\mathbf{f}\|_F. \quad (3.6.83)$$

The final results can then be obtained collecting (3.6.70), (3.6.71), (3.6.82) and (3.6.83). □

Remark 3.6.4. We remark that in several applications we have $C = \varepsilon Identity$, so that $M_c = \gamma = \varepsilon$. In this case, the estimates (3.6.58) and (3.6.59) become

$$\|\mathbf{x}\|_X \le \frac{\beta^2 + 4\varepsilon M_a}{\alpha_0\beta^2}\|\mathbf{f}\|_F + \frac{2M_a^{1/2}}{\alpha_0^{1/2}\beta}\|\mathbf{g}\|_G \quad (3.6.84)$$

and

$$\|\mathbf{y}\|_Y \le \frac{2M_a^{1/2}}{\alpha_0^{1/2}\beta}\|\mathbf{f}\|_F + \frac{4M_a}{M_a\varepsilon + 2\beta^2}\|\mathbf{g}\|_G. \quad (3.6.85)$$

We also note that in the limit for $\varepsilon \to 0$ we recover the result of Theorem 3.4.4. □

Remark 3.6.5. We also point out that (3.6.84) and (3.6.85) are optimal, with respect to the dependency of the stability constants on the parameters α_0, β and ε. To see that, consider for $0 < a, b \ll 1$ the problem

$$\begin{pmatrix} 2a & \sqrt{a} & -\sqrt{a} & 0 & 0 \\ \sqrt{a} & 2 & 1 & b & 0 \\ -\sqrt{a} & 1 & 2 & 0 & b \\ 0 & b & 0 & -\varepsilon & 0 \\ 0 & 0 & b & 0 & -\varepsilon \end{pmatrix} \cdot \begin{pmatrix} x_1 \\ x_2 \\ x_3 \\ y_1 \\ y_2 \end{pmatrix} = \begin{pmatrix} 2f \\ 0 \\ 0 \\ 0 \\ 2g \end{pmatrix}, \quad (3.6.86)$$

whose solution is given by

$$x_1 = \frac{f(b^2+\varepsilon)}{ab^2} + \frac{g}{ba^{1/2}}, \qquad x_2 = -\frac{f\varepsilon}{a^{1/2}b^2} - \frac{3g\varepsilon}{b(3\varepsilon+b^2)},$$

$$x_3 = \frac{f\varepsilon}{a^{1/2}b^2} + \frac{g(3\varepsilon+2b^2)}{b(3\varepsilon+b^2)}, \tag{3.6.87}$$

$$y_1 = -\frac{f}{a^{1/2}b} - \frac{3g}{3\varepsilon+b^2}, \qquad y_2 = \frac{f}{a^{1/2}b} - \frac{3g}{3\varepsilon+b^2}. \tag{3.6.88}$$

Indeed, it is easy to recognise that $\alpha_0 = 2a$ and $\beta = b$, and hence the optimality of estimates (3.6.84) and (3.6.85). □

Remark 3.6.6. It is of some interest to check that all our estimates are "dimensionally correct". Indeed, denoting by $[a]$, $[b]$, $[c]$, $[x]$, $[y]$, $[f]$ and $[g]$, respectively, the physical dimensions of A, B, C, $\mathbf{x}$, $\mathbf{y}$, $\mathbf{f}$ and $\mathbf{g}$, we have that M_a and α have the same dimensions as A (and hence, in particular, M_a/α is a pure number). Similarly, M_c and γ have the same dimension as $[c]$. Moreover, from the two equations of our system, we have

$$[x] = \frac{[f]}{[a]} = \frac{[g]}{[b]}, \quad [y] = \frac{[f]}{[b]} = \frac{[a][g]}{[b^2]} = \frac{[g]}{[c]} \tag{3.6.89}$$

from which we easily deduce that $[a]\,[c]$ equals to $[b^2]$, so that for instance $M_a M_c/\beta^2$ is also a pure number. Taking this into account, we can verify that, in every stability inequality, an $[x]$ is bounded by a $\frac{1}{[a]}[f]$ times a pure number or by a $\frac{1}{[b]}[g]$ times a pure number, while a $[y]$ is bounded by a $\frac{1}{[b]}[f]$ times a pure number or by a $\frac{[a]}{[b^2]}[g]$ times a pure number. □

Chapter 4
Saddle Point Problems in Hilbert Spaces

In the first chapter of this book, we introduced a large number of saddle point problems or generalisations of such problems. In most cases, the question of existence and uniqueness of solutions was left aside. In the previous chapter, we considered the solvability of *finite dimensional* problems in mixed form, together with the stability of sequences of such problems. We now introduce an abstract frame that is sufficiently general to cover all our needs, from the problems of existence and uniqueness in infinite dimension to the stability of their Finite Element discretisations.

As a first step, we shall recall some basic definitions of Functional Analysis: Hilbert spaces, continuous functionals, bilinear forms, and linear operators associated with bilinear forms.

In Sect. 4.2, we discuss conditions that ensure existence and uniqueness for mixed formulations in Hilbert spaces. Several examples of mixed formulations related to Partial Differential equations will illustrate the theoretical results. Different stability estimates will then be provided for different sets of assumptions.

The last section (Sect. 4.2.2) will be devoted to the study of perturbed problems (whose algebraic aspects were discussed in Sect. 3.6 of the previous chapter).

We shall follow essentially the analysis of [112] and [122]. We also refer the reader to other presentations, as can be found in the books [41, 106, 222, 315, 337].

4.1 Reminders on Hilbert Spaces

In this section, we recall some basic notions on Hilbert spaces. Most readers, and in particular those with a better mathematical background, will already be familiar with all the contents of the section. For them, the aim of the section will just be to fix the notation. For other people with a weaker mathematical background, it could be useful to refresh some notions. On the other hand, we do not pretend to provide a complete mastering of Hilbert spaces to people that never heard of them before.

D. Boffi et al., *Mixed Finite Element Methods and Applications*, Springer Series in Computational Mathematics 44, DOI 10.1007/978-3-642-36519-5_4,

For these people, a superficial reading will be enough to convince them that things, in Hilbert spaces, are *very similar* to their counterparts in finite dimensional spaces.

4.1.1 Scalar Products, Norms, Completeness

We assume that the reader is familiar with the concept of *linear space over* $\mathbb{R}$. This, roughly speaking, means that you are allowed to *sum* two elements of the space, and to *multiply* each element of the space times a real number.

Let H_1 and H_2 be two linear spaces over $\mathbb{R}$. A map $a : H_1 \times H_2 \to \mathbb{R}$ is said to be **a bilinear form** on $H_1 \times H_2$ if, for every $u_1, v_1, w_1 \in H_1$, for every $u_2, v_2, w_2 \in H_2$ and for every $\lambda, \mu \in \mathbb{R}$, we have

$$\begin{aligned} a(\lambda u_1 + \mu v_1, w_2) &= \lambda a(u_1, w_2) + \mu a(v_1, w_2) \\ a(w_1, \lambda u_2 + \mu v_2) &= \lambda a(w_1, u_2) + \mu a(w_1, v_2). \end{aligned} \tag{4.1.1}$$

When both H_1 and H_2 coincide in a single linear space H, we shall often say that a is a bilinear form on H, meaning that it is a bilinear form on $H \times H$.

A bilinear form a on H is said to be **symmetric** if, for every $u, v \in H$, we have

$$a(u, v) = a(v, u). \tag{4.1.2}$$

A bilinear form s on H is said to be **a scalar product** if it is symmetric and if, moreover,

$$s(v, v) \geq 0 \quad \forall\, v \in H \qquad \text{and} \qquad s(v, v) = 0 \Rightarrow v = 0. \tag{4.1.3}$$

We assume that we have a scalar product given on $H \times H$, and from now on we shall write $(u, v)_H$ (or simply (u, v) when no confusion can occur) instead of $s(u, v)$. To a scalar product, we can always associate a **norm**

$$\|v\|_H := \Big((v, v)_H\Big)^{1/2} \qquad \forall\, v \in H. \tag{4.1.4}$$

Again, we shall simply write $\|v\|$ instead of $\|v\|_H$ when no confusion is likely to occur. It is interesting to note that the norm, as defined in (4.1.4), has the usual properties of the norms in finite dimension:

$$\begin{aligned} &\|\lambda v\| = |\lambda|\, \|v\| \qquad \forall\, v \in H,\ \forall\, \lambda \in \mathbb{R}, \\ &\|v\| \geq 0\ \forall\, v \in H \quad \text{and} \quad \|v\| = 0 \Rightarrow v = 0, \\ &\|v_1 + v_2\| \leq \|v_1\| + \|v_2\| \qquad \forall\, v_1, v_2 \in H. \end{aligned} \tag{4.1.5}$$

It is also worth noting that, even in infinite dimension, we have the Cauchy inequality

$$(u, v)_H \leq \|u\|_H \|v\|_H \qquad \forall\, u, v \in H, \tag{4.1.6}$$

whose proof can be easily done mimicking the proof of Lemma 3.3.1 of the previous chapter.

It is a *strong temptation* to start defining a norm first (as a mapping from H to $\mathbb{R}$ satisfying (4.1.5)), and then getting a scalar product out of it, for instance by

$$(u, v) := \left(\|u + v\|^2 - \|u - v\|^2\right)/4. \tag{4.1.7}$$

Smart, isn't it? But doomed. That would work *if and only if* the norm you started with satisfies the so called *parallelogram identity*:

$$\|v + u\|^2 + \|v - u\|^2 = 2(\|u\|^2 + \|v\|^2). \tag{4.1.8}$$

A norm that satisfies (4.1.5) *and* (4.1.8) is said to be **a pre-Hilbert norm**, and induces a scalar product associated to it through (4.1.7).

A linear space H with a norm $\|\cdot\|_H$ that satisfies (4.1.5) is called **a normed space**. If, on top of that, the norm satisfies the parallelogram identity (4.1.8), then we say that H is a **pre-Hilbert space**.

As soon as we have a norm (no matter if it is a pre-Hilbert norm or not), we can talk about **convergence** and **limits**. We say that the sequence $\{v_n\}$ of elements of H converges to $v \in H$ (or that v is the limit of v_n for $n \to +\infty$) if

$$\lim_{n \to +\infty} \|v_n - v\|_H = 0. \tag{4.1.9}$$

The *limit* in (4.1.9) is obviously the one of elementary calculus (dealing with sequences of real numbers). When the type of norm to be used cannot be confused, we will also write, more simply, $v_n \to v$.

Example 4.1.1. It is immediate to see that for every integer $k \geq 1$ the space $\mathbb{R}^k$ with the usual Euclidean norm (3.1.6) used in the previous chapter is a pre-Hilbert space. Indeed, the Euclidean norm does come from a scalar product, so that

$$\|\mathbf{x}+\mathbf{y}\|^2+\|\mathbf{x}-\mathbf{y}\|^2 = \|\mathbf{x}\|^2+\|\mathbf{y}\|^2+2\mathbf{x}^T\mathbf{y}+\|\mathbf{x}\|^2+\|\mathbf{y}\|^2-2\mathbf{x}^T\mathbf{y} = 2(\|\mathbf{x}\|^2+\|\mathbf{y}\|^2).$$

On the other hand, for instance $\mathbb{R}^2$ with the norm $\|\mathbf{x}\|_1 := |x_1| + |x_2|$, already seen in (3.0.4), is *not* a pre-Hilbert space, since the norm $\|\cdot\|_1$ does not satisfy (4.1.8): try it with $u = (1, 0)$ and $v = (0, 1)$. □

Once we have a norm in H, we can measure the *distance* of two elements u and v of H by $\|u-v\|$. Given a non-empty subset $T \subseteq H$, we can measure its *diameter* by

$$diam(T) := \sup_{u,v \in T} \|u - v\|. \tag{4.1.10}$$

Loosely speaking, the diameter of T is the "maximum" distance of any two elements in T. It is obvious that for two subsets S and T, if $T \subseteq S$, then $diam(T) \le diam(S)$ (if you increase the set of possible choices, the supremum cannot go down).

Now, for every sequence $\{v_n\}_{n\in\mathbb{N}}$ of elements of H, and for every integer $m \in \mathbb{N}$, we can consider its m-th *tail* T_m, defined as the set

$$T_m := \{v_m,\ v_{m+1},\ v_{m+2}\ldots\} \equiv \{v_n|\ n \ge m\}. \tag{4.1.11}$$

We clearly have $T_{m+1} \subseteq T_m$ for every $m \in \mathbb{N}$ (the farther you cut, the lesser is left in the tail). Hence, whatever the sequence $\{v_n\}$ from which you started, the sequence of real numbers $diam(T_m)$ that you get out of it is obviously *non increasing*: that is, $diam(T_{m+1}) \le diam(T_m)$ for every m. Hence, the sequence $diam(T_n)$ will always have a limit, which is ≥ 0. A sequence $\{v_n\}$ of elements of a normed space H is said to be **a Cauchy sequence** in H if the sequence of real numbers $\{diam(T_m)\}$ that you get out of it verifies

$$\lim_{m\to+\infty} diam(T_m) = 0. \tag{4.1.12}$$

Note that, in order to speak about Cauchy sequences, what you need is to be able to measure the distance of two objects. This is always possible if, as in our case, you have a norm. This is also possible in more general situations, but we are not interested in them here.

A normed linear space H is said to be **complete** if for every Cauchy sequence $\{v_n\}$ in H there exists an element $v \in H$ such that $v_n \to v$ in the sense of (4.1.9). In other words, a normed linear space is complete if every Cauchy sequence has a limit. We are almost done.

Definition 4.1.1. A **Banach space** is a normed linear space that is complete.

Definition 4.1.2. A **Hilbert space** is a pre-Hilbert space that is complete.

Note that we could have defined alternatively a Hilbert space as a Banach space whose norm satisfies the parallelogram identity (4.1.8). Hence, every Hilbert space is also a Banach space, but the converse is not true: in Hilbert spaces, you have a scalar product, and in Banach spaces that are not Hilbert spaces, you do not (and can not) have one.

Example 4.1.2. It is immediate to have, from elementary Calculus, that for every integer $k \ge 1$ the space $\mathbb{R}^k$, with the usual Euclidean scalar product and norm, is a Hilbert space. In particular, $\mathbb{R}$ itself is a Hilbert space if we take the usual product of two numbers as scalar product (and hence the absolute value as norm). We also saw in the previous chapter that, for instance in $\mathbb{R}^2$, the norm $\|\mathbf{x}\|_1 := |x_1| + |x_2|$ is *equivalent* to the Euclidean norm (in the sense of (3.0.5)). On the other hand, we have already seen in Example 4.1.1 that $\mathbb{R}^2$ with the norm $\|\cdot\|_1$ is not even a pre-Hilbert space, and hence it cannot be a Hilbert space although, still by elementary Calculus, it is easily seen to be a Banach space. Actually, it is not difficult to check that the property of being complete is not lost if you exchange your norm with an equivalent norm (while the property (4.1.8) might indeed be lost). □

Example 4.1.3. Regarding the functional spaces already used in the first chapter, we can see that if Ω is a bounded open domain, then $L^1(\Omega)$ (the space of Lebesgue integrable functions over Ω), with the norm

$$\|v\|_{L^1(\Omega)} := \int_\Omega |v(x)|\, dx \tag{4.1.13}$$

is a Banach space, but not a Hilbert space. Note that, as we did already in the first chapter (and as we are going to do all over this book), we used the term *functions* in lieu of the (more precise) *classes of measurable functions*.

Instead, the space $L^2(\Omega)$ (the space of Lebesgue *square* integrable functions over Ω), with the norm

$$\|v\|^2_{L^2(\Omega)} := \int_\Omega v^2(x)\, dx \tag{4.1.14}$$

is a Hilbert space, and the corresponding scalar product is given by

$$(u, v)_{L^2(\Omega)} := \int_\Omega u(x)v(x)\, dx. \tag{4.1.15}$$

Similarly, the space $H^1_0(\Omega)$ with the scalar product

$$(u, v)_{H^1_0(\Omega)} := \int_\Omega \underline{\mathrm{grad}}\, u(x) \cdot \underline{\mathrm{grad}}\, v(x)\, dx \tag{4.1.16}$$

is a Hilbert space. □

In the following discussion, we shall mostly use only Hilbert spaces. Hence, from now on, we shall mainly concentrate on them, although most of the concepts and results could be extended easily to Banach spaces.

4.1.2 Closed Subspaces and Dense Subspaces

Definition 4.1.3. A subset T of a Hilbert space H is said to be **closed** if, for every Cauchy sequence $\{v_n\}_{n\in\mathbb{N}}$ of elements of T, the limit v (which surely exists in H, since H is complete) belongs to T as well.

If T is a *linear subspace* of a Hilbert space H, and if T is closed, then we will say that T is **a closed subspace** of H. Then T itself will be a Hilbert space, with the same norm as H.

Example 4.1.4. For instance, in $L^2(\Omega)$, we can consider the subspace $L^2_0(\Omega)$ made of functions that have zero mean value in Ω. It is easy to see that it is a closed subspace (since the L^2-limit of functions with zero mean value has itself zero mean value). On the other hand, $C^0(\overline{\Omega})$ is a linear subspace of $L^2(\Omega)$, but it is not closed: for instance, for $\Omega =]-1, 1[$, the sequence $f_n(x) := \arctan(nx)$ converges in $L^2(\Omega)$ to $f_\infty(x) := (\pi/2)\, sign\,(x)$, which does not belong to $C^0(\overline{\Omega})$. □

Definition 4.1.4. Let H be a Hilbert space, and let Z be a subset of H. The **closure** of Z, that we denote by $\overline{Z}$, is the set of elements $v \in H$ such that there exists a sequence $\{z_n\}_{n\in\mathbb{N}}$ of elements of Z that converges to v.

We abviously have that Z is closed if and only if $\overline{Z} = Z$.

Another important concept regarding subspaces is that of a *dense subspace.*

Definition 4.1.5. A subset Z of a Hilbert space H is said to be **dense** if its closure $\overline{Z}$ coincides with the whole space H. If Z is also a linear subspace of H, then we say that it is **a dense subspace**.

In other words, Z is dense in H if for every element v of H there exists a sequence $\{z_n\}_{n\in\mathbb{N}}$ of elements of Z such that

$$\lim_{n\to+\infty} \|v - z_n\|_H = 0.$$

Example 4.1.5. It is not difficult to see that $Z := H_0^1(\Omega)$ is a dense subspace of $H = L^2(\Omega)$. It is also clear that Z is not a closed subspace of H: for instance, for $\Omega =]-1, 1[$, the sequence of functions defined by

$$z_n(x) := \min(1, n - n|x|) \equiv \begin{cases} 1 & \text{when } |x| \le 1 - 1/n \\ n(1-|x|) & \text{when } |x| > 1 - 1/n \end{cases}$$

verifies $z_n \in H_0^1(\Omega)$ for all n, and its limit in L^2 equals the constant 1, which is not in $H_0^1(\Omega)$ (as it does not vanish at the boundary). □

Note that a dense closed subspace of a Hilbert space H coincides necessarily with the whole space H. Hence, in general, we consider subspaces that are closed, but not dense, and subspaces that are dense, but not closed. These two categories of subspaces are both very important, and we cannot restrict our attention to just one of them. We point out however, from the very beginning, that closed subspaces are the ones that, loosely speaking, inherit most of the properties of subspaces of finite dimensional spaces. In particular, a finite dimensional subspace is *always closed* and is *never dense* (unless it coincides with the whole space).

4.1.3 *Orthogonality*

Some very useful instruments available in Hilbert spaces (and not in Banach spaces) are related to the concept of **orthogonality**. We say that two elements u and v of a Hilbert space H are orthogonal if $(u, v)_H = 0$. It is the same as in the finite dimensional case, with the only difference that there, the scalar product was denoted by $\mathbf{v}^T\mathbf{u}$. If Z is a linear subspace of a Hilbert space H, we can define its **orthogonal complement** $Z^{\perp}$

$$Z^{\perp} := \{w \in H \text{ such that } (w, z) = 0 \;\forall\, z \in Z\}. \tag{4.1.17}$$

It is not difficult to see that an orthogonal complement $Z^{\perp}$ is always closed (even when Z itself is not closed). As in (3.1.24), if Z_1 and Z_2 are both subspaces of a Hilbert space H, then

$$Z_1 \subseteq Z_2 \quad \Rightarrow \quad Z_2^{\perp} \subseteq Z_1^{\perp}. \tag{4.1.18}$$

Moreover, we have the following useful property.

Proposition 4.1.1. *Let H be a Hilbert space, let Z be a subspace of it, and let $\overline{Z}$ be its closure. Then,*

$$\overline{Z}^{\perp} = Z^{\perp}. \tag{4.1.19}$$

Proof. Since $Z \subseteq \overline{Z}$, we obviously have $\overline{Z}^{\perp} \subseteq Z^{\perp}$. On the other hand, let $w \in Z^{\perp}$. We want to see that $(\overline{z}, w)_H = 0$ for all $\overline{z} \in \overline{Z}$. Indeed, for every $\overline{z} \in \overline{Z}$, there exists a sequence $\{z_n\}_{n \in \mathbb{N}}$ of elements of Z that converges to $\overline{z}$. As $w \in Z^{\perp}$, we have $(z_n, w)_H = 0$ for all n. Hence,

$$(\overline{z}, w)_H = \lim_{n \to +\infty} (z_n, w) = 0. \tag{4.1.20}$$

□

Remark 4.1.1. Note that, as we had in Remark 3.1.3, the notion of orthogonal space depends heavily on the choice of the "whole space" H. Indeed, if H_1 and H_2 are Hilbert spaces, and Z is a subspace of H_1 and also a subspace of H_2, then the orthogonal of Z in H_1 will, in general, be different from the orthogonal of Z in H_2. This is rather obvious. However, the common notation (that we are using here) does not distinguish among the two (we should, for this, use something like $Z^{\perp_{H_1}}$ and $Z^{\perp_{H_2}}$, which would be tremendously ugly). As a consequence, one should be careful when confusion is possible. □

As we did in the finite dimensional case, if Z is a closed subspace, we can define the **projection operator** $\pi_Z : H \to Z$ defined for every $v \in H$ by

$$\pi_Z v \in Z \qquad \textit{and} \qquad (\pi_Z v - v) \in Z^{\perp}. \tag{4.1.21}$$

Compare with (3.1.31) to see that we are just extending the definitions given in the previous chapter for the finite dimensional case. As we had in the finite dimensional case, $\pi_Z v$ can be seen as the element in Z that minimises the distance from v, namely

$$\|\pi_Z v - v\|_H = \min_{z \in Z} \|z - v\|_H. \tag{4.1.22}$$

Remark 4.1.2. In the definition of π_Z, we assumed that Z was *a closed subspace* of H. Of the two properties (of being closed, and of being a subspace), the second is not very important. Indeed, it is easy to see that (4.1.22) can be used to define the projection mapping π_Z in more general cases, for instance when Z is not a subspace but simply a closed convex subset, as for instance a closed affine manifold (which, roughly speaking, is the translation of a closed subspace). On the other hand, closedness is more essential. To see what can happen when you remove it, assume that Z is a dense subspace. Then, for v in H but not in Z, the projection $\pi_Z v$ cannot be defined: indeed, we recall that, from Proposition 4.1.1, if Z is dense (and hence $\overline{Z} = H$), then $Z^\perp = H^\perp = \{0\}$. Hence, looking at the definition (4.1.21), if Z is dense the only $w \in H$ such that $(w - v) \in Z^\perp$ is $w = v$. However, such a w is not in Z, so that there is no element that we could choose as $\pi_Z v$ that satisfies *both* properties required in (4.1.21). Hence, $\pi_Z v$ does not exist. Note that the alternative definition (4.1.22) would not be of any help either. Actually, always for Z dense and $v \in H$ with $v \notin Z$, the *minimum* of $\|z - v\|$ for $z \in Z$ does not exist, and the *infimum* is equal to zero. □

It is easy to check that if H is a Hilbert space, and if Z is a closed linear subspace, then every element v of H can be split in a unique way into its two components in Z and in $Z^\perp$:

$$v = v_Z + v_{Z^\perp}, \tag{4.1.23}$$

just by setting $v_Z := \pi_Z v$.

Example 4.1.6. For instance, if $H := L^2(\Omega)$ and $Z := L^2_0(\Omega)$, then $Z^\perp$ is the (one-dimensional) space made of constant functions. The projection of $v \in L^2(\Omega)$ onto Z is given by

$$\pi_Z = v - \frac{1}{|\Omega|}\int_\Omega v(x)\,dx, \tag{4.1.24}$$

where $|\Omega|$ is the *Lebesgue measure* of Ω. □

We have, moreover, the following property.

Proposition 4.1.2. *Let H be a Hilbert space and Z a closed subspace of it. Then, either $Z \equiv H$ or $Z^\perp$ is* **not** *reduced to $\{0\}$.*

Proof. If Z does not coincide with H, then there exists a $v \in H$ such that $v \notin Z$. Hence, $\pi_Z v - v \neq 0$. As (4.1.21) also gives $\pi v - v \in Z^\perp$, the proof is concluded. □

We can now see the equivalent of (3.1.23) in general Hilbert spaces.

Proposition 4.1.3. *Let H be a Hilbert space, and $Z \subseteq H$ a subspace. Then,*

$$(Z^\perp)^\perp = Z \quad \textbf{iff} \quad Z \textit{ is closed.} \tag{4.1.25}$$

Proof. Indeed, if $Z \equiv (Z^\perp)^\perp$, then Z, being the orthogonal of something, is closed. To see the converse, we remark first that we always have the inclusion

$Z \subseteq (Z^{\perp})^{\perp}$. If Z is closed, suppose, by contradiction, that Z does not coincide with $(Z^{\perp})^{\perp}$. From Proposition 4.1.2 (applied with $H = (Z^{\perp})^{\perp}$), we should have a $v \in (Z^{\perp})^{\perp}$, with $v \neq 0$, that is orthogonal to all z in Z. As such, v will hence be also in $Z^{\perp}$. However, $Z^{\perp} \cap (Z^{\perp})^{\perp} = \{0\}$ and this contradicts the fact that $v \neq 0$. □

Moreover, we have the following additional property.

Proposition 4.1.4. *Let H be a Hilbert space, and let Z be a subspace of H. Then,*

$$Z^{\perp} = \{0\} \quad \textbf{iff} \quad Z \textit{ is dense.} \tag{4.1.26}$$

Proof. Assume first that Z is dense. Then, $\overline{Z} \equiv H$ and hence $\overline{Z}^{\perp} = \{0\}$, and the result follows from Proposition 4.1.1. Assume conversely that $Z^{\perp} = \{0\}$. Always from Proposition 4.1.1, we have now $\overline{Z}^{\perp} = \{0\}$. However, $\overline{Z}$ is closed, and hence, by Proposition 4.1.2 $\overline{Z} = H$, and Z is therefore dense. □

4.1.4 Continuous Linear Operators, Dual spaces, Polar Spaces

We can now recall several other important definitions.

Definition 4.1.6. Let V and W be Hilbert spaces, and let M be a linear mapping from V to W. We say that M is **bounded** or that it is **continuous** if there exists a constant μ^* such that

$$\|Mv\|_W \leq \mu^* \|v\|_V \qquad \forall\, v \in V. \tag{4.1.27}$$

Note that we have two different names for that (*bounded* and *continuous*) because the two definitions do not coincide if the operator is not *linear*. Actually, for a more general operator, (4.1.27) defines a bounded operator, while continuity can be taken as in the usual Calculus books: for every $v \in V$ and for every sequence v_n converging to v, we have that $M v_n$ converges to $M v$. Here, however, we only deal with linear operators, and the two concepts coincide.

Example 4.1.7. For instance, the operator $v \rightarrow \dfrac{\partial v}{\partial x_1}$ is continuous from $H^1_0(\Omega)$ to $L^2(\Omega)$. Similarly, if ϕ is a given (fixed) bounded function, then the mapping $v \rightarrow \phi\, v$ is continuous from $L^2(\Omega)$ into itself. □

The following definition is less common but very useful.

Definition 4.1.7. Let V and W be Hilbert spaces and let M be a linear mapping from V to W. We say that M is **bounding** if there exists a constant μ_* such that

$$\|Mv\|_W \geq \mu_* \|v\|_V \qquad \forall\, v \in V. \tag{4.1.28}$$

In other words, *bounding operators* are injective operators whose inverse is continuous.

The set of all linear continuous operators from a Hilbert space V into another Hilbert space W is also a linear space (after defining, in an obvious way, the *sum* of two operators or the *multiplication* of an operator times a real number). Such a space is usually denoted by $\mathcal{L}(V, W)$. In $\mathcal{L}(V, W)$, we can also introduce a norm:

$$\|M\|_{\mathcal{L}(V,W)} := \sup_{v \in V} \frac{\|M v\|_W}{\|v\|_V}. \tag{4.1.29}$$

When no confusion can occur, the norm in (4.1.29) is simply denoted by $\|M\|$. Hence, for instance, (4.1.29) implies

$$\|M v\|_W \leq \|M\| \, \|v\|_V \qquad \forall\, v \in V. \tag{4.1.30}$$

One can prove that (4.1.29) actually defines a norm, and that such a norm verifies (4.1.8), so that with this norm $\mathcal{L}(V, W)$ is itself a Hilbert space.

A remarkable result concerning linear continuous and *one-to-one* operators is the following one, due to Banach.

Theorem 4.1.1 (Banach Theorem). *Let V and W be Hilbert spaces and let $M \in \mathcal{L}(V, W)$ be a one to one mapping. Then, its inverse operator M^{-1}, from W to V, is also continuous.*

Proof. The proof can be found in any book of Functional Analysis. □

As we did for finite dimensional spaces, given a subspace $Z \subseteq V$, we can consider the **extension operator** $E_{Z \to V}$, from Z to V which to every $z \in Z$ associates the same z, thought as an element of V. If there is no risk of confusion, this will, more simply, be denoted by E_Z as we did in the previous chapter. Always in agreement with the finite dimensional case, given an operator $M \in \mathcal{L}(V, W)$, we can consider the **restriction** M_Z of M to Z, that could be defined as

$$M_Z\, z = M\, E_Z\, z \qquad \forall z \in Z. \tag{4.1.31}$$

Since for every $z \in Z \subseteq V$ we have obviously $M_Z\, z = M\, z$, in several occasions, the extension operator E_Z will not be explicitly written. In other cases, however, such notation will turn out to be very useful.

If we assume that Z is a closed subspace of V, that S is a closed subspace of W and $M \in \mathcal{L}(V, W)$, we can also consider its *restriction* M_{ZS}, defined as

$$M_{ZS}\, z = \pi_S\, M\, E_Z\, z \qquad \forall z \in Z. \tag{4.1.32}$$

It is easy to check that $M_{ZS} \in \mathcal{L}(Z, S)$. Conversely, given an operator L in $\mathcal{L}(Z, S)$, we can always consider its *extension* $\tilde{L} \in \mathcal{L}(V, W)$ defined by

$$\tilde{L}\, v = E_S\, L\, \pi_Z\, v \qquad \forall v \in V. \tag{4.1.33}$$

A particular case of linear operators, *of paramount importance*, is found when the *arrival space* is $\mathbb{R}$. In this case, linear operators $V \to \mathbb{R}$ are called **linear functionals** on V. The space of all linear *continuous* functionals on a Hilbert space V is called **the dual space** of V, and is usually denoted by V'. Hence, $V' \equiv \mathcal{L}(V, \mathbb{R})$. As a particular case of the previous situation, V' is itself a Hilbert space, and its norm (often called **the dual norm** of $\|\cdot\|_V$) is given by

$$\|f\|_{V'} := \sup_{v \in V} \frac{|f(v)|}{\|v\|_V}. \tag{4.1.34}$$

We easily recognise the definition of dual norms that were given in finite dimension. The value of f at v (denoted by $f(v)$ in (4.1.34)) is often denoted in a different way: either by ${}_{V'}\langle f, v\rangle_V$ or by $\langle f, v\rangle_{V'\times V}$, or simply $\langle f, v\rangle$ when no confusion can occur. It is not too difficult to check (although we shall not do it here) that, if V is a Hilbert space, then *the dual space of* V' (often called the *bi-dual* space), actually can be identified with V itself (see the Ritz representation Theorem (4.1.37) here below).

Example 4.1.8. For instance, in one dimension, it is easy to see that the mapping $\delta_0 : v \to v(0) \in \mathbb{R}$ is continuous from $H^1_0(]-1,1[)$ to $\mathbb{R}$: indeed,

$$v(0) = \int_{-1}^{0} v'(t)\, dt \le \Big(\int_{-1}^{0} 1^2\, dt\Big)^{1/2}\Big(\int_{-1}^{0} (v'(t))^2\, dt\Big)^{1/2}$$
$$\le 1\Big(\int_{-1}^{1} \big(v'(t)\big)^2\, dt\Big)^{1/2} = \|v\|_{H^1_0(]-1,1[)}. \tag{4.1.35}$$

Hence, δ_0 is an element of the dual space of $H^1_0(]-1,1[)$ (usually denoted by $H^{-1}(]-1,1[)$). Note that a similar result does not hold in dimension $d > 1$. Indeed, if Ω is the disk centred at the origin O and radius $1/\sqrt{e}$, a simple explicit computation shows that the function

$$v(x,y) := \log|\log(x^2+y^2)|$$

is indeed in $H^1_0(\Omega)$. Setting

$$v_n(x,y) := \min\{n, v(x,y)\},$$

it is not difficult to see that v_n converges to v in $H^1_0(\Omega)$. However, $v_n(0,0) = n$ so that the bound

$$v_n(0,0) \le C\,\|v_n\|_{H^1_0(\Omega)}$$

cannot be true with a constant C (no matter how big) independent of n, as the left-hand side tends to $+\infty$ and the right-hand side stays finite. Similarly, the estimate (4.1.35) becomes false if we try to replace, in the right-hand side, the H^1 norm with the L^2 norm: consider, for instance, the sequence of functions defined by

$$v_n(x) = \begin{cases} 1 \text{ for } x \text{ in } [-1/n, 1/n] \\ 0 \text{ for } x \text{ outside } [-1/n, 1/n]. \end{cases}$$

We have

$$\|v_n\|^2_{L^2]-1,1[} = \int_{-1}^{1} v_n^2(x)\,dx = \int_{-1/n}^{1/n} 1\,dx = \frac{2}{n} \to 0$$

while $v_n(0) = 1$ for all n. Hence, there is no constant C (independent of n) such that

$$v_n(0) \le C\,\|v_n\|_{L^2(]-1,1[)}. \qquad \square$$

Example 4.1.9. Let us also see an example of dual norm: let n be an integer (larger than 1) and consider in $\Omega =]0,\pi[$ the function $f_n(x) := \sin(nx)$. It is immediate to check that

$$\|f_n\|_{L^2(\Omega)} = \sqrt{\pi/2} \quad \text{and that} \quad \|f_n\|_{H^1_0(\Omega)} = n\sqrt{\pi/2}.$$

To f_n we can associate an element, that we still call f_n, of $H^{-1}(\Omega)$ (that is the dual space of $H^1_0(\Omega)$) as follows

$$\langle f_n, \varphi\rangle_{H^{-1}\times H^1_0} := \int_0^{\pi} \varphi(x) f_n(x)\,dx \quad \forall \varphi \in H^1_0(\Omega).$$

Let us compute the norm of f_n in $H^{-1}(\Omega)$. For every $\varphi \in H^1_0(\Omega)$ we have (integrating by parts):

$$\begin{aligned}\langle f_n, \varphi\rangle &= \int_0^{\pi} f_n(x)\varphi(x)\,dx = \int_0^{\pi} \sin(nx)\varphi(x)\,dx \\ &= \frac{1}{n}\int_0^{\pi} \cos(nx)\varphi'(x)\,dx \le \frac{1}{n}\|\cos(nx)\|_{L^2}\,\|\varphi\|_{H^1_0} = \frac{\sqrt{\pi/2}}{n}\|\varphi\|_{H^1_0},\end{aligned}$$

giving us, always for every $\varphi \in H^1_0(\Omega)$:

$$\frac{\langle f_n, \varphi\rangle}{\|\varphi\|_{H^1_0}} \le \frac{\sqrt{\pi/2}}{n}. \tag{4.1.36}$$

On the other hand, it is not difficult to see that, taking $\varphi \equiv \sin(nx)$, we get

$$\frac{\langle f_n, \varphi \rangle}{\|\varphi\|_{H_0^1}} = \frac{\pi/2}{n\sqrt{\pi/2}} = \frac{\sqrt{\pi/2}}{n},$$

showing that $(1/n)\sqrt{\pi/2}$ actually realises the *supremum* (over all possible φ's) of the left-hand side of (4.1.36). In conclusion, we have

$$\|f_n\|_{L^2(\Omega)} = \sqrt{\pi/2} \qquad \|f_n\|_{H_0^1(\Omega)} = n\sqrt{\pi/2} \qquad \|f_n\|_{H^{-1}(\Omega)} = \frac{\sqrt{\pi/2}}{n}.$$

This shows that the three norms $\|\cdot\|_{L^2}$, $\|\cdot\|_{H_0^1}$, and $\|\cdot\|_{H^{-1}}$ cannot be equivalent. This also shows that a *high frequency* function can have, at the same time, an L^2-norm (and a maximum norm) of the order of 1, a huge H^1-norm ($\simeq$ energy norm), and a tiny H^{-1}-norm. This will show up in the next chapter, when the use of finer and finer grids will allow the presence of highly oscillating piecewise linear (or piecewise polynomial) functions. □

While several properties that we saw and that we will see hold in a much more general setting (for instance, in all Banach spaces), the following theorem is, in a certain sense, characteristic of Hilbert spaces.

Theorem 4.1.2 (Ritz's Theorem). *Let H be a Hilbert space, and let R_H be the operator $H \to H'$ that to each $z \in H$ associates the functional $f_z = R_H z \in H'$ defined as*

$$\langle f_z, v \rangle_{H' \times H} = (z, v)_H \quad \forall v \in H. \tag{4.1.37}$$

Then, R_H is one to one, and $\|R_H\|_{\mathcal{L}(H,H')} = \|R_H^{-1}\|_{\mathcal{L}(H',H)} = 1$. Moreover, if we identify (as it is natural) H with $(H')'$, then $R_H^{-1} = R_{H'}$.

Proof. The proof can be found in every Functional Analysis textbook. □

Another result that we are going to use later on is the following theorem, that can be seen as a particular case of a more general result, known as the *Kato Theorem.*

Theorem 4.1.3 (Kato Theorem). *Let V and W be Hilbert spaces and let T_1 and T_2 be in $\mathcal{L}(V, W)$. If T_1 is bounding, then there exists an $\varepsilon_0 > 0$ such that for all $\varepsilon \in \mathbb{R}$ with $|\varepsilon| \le \varepsilon_0$ the perturbed operator $T_1 + \varepsilon T_2$ is also bounding, and we have moreover*

$$\|T_1^{-1} - (T_1 + \varepsilon T_2)^{-1}\|_{\mathcal{L}(W,V)} \le C\,|\varepsilon| \tag{4.1.38}$$

with C depending on ε_0 but independent of ε.

If Z is a subspace of a Hilbert space H, we can spot a special subset of H', usually called the **polar space** of Z, made of all functionals $f \in H'$ that vanish identically on Z. The polar space of Z is usually denoted by Z^0: hence, we have

$$Z^0 := \{f \in H' \text{ such that } \langle f, z \rangle_{H' \times H} = 0 \;\; \forall\, z \in Z\}. \tag{4.1.39}$$

It is clear that the definition of polar space of Z makes sense only when Z is considered as a subspace of another space (in this case, H). In particular, the polar space of $Z = \{0\}$ coincides with the whole H' while the polar space of $Z = H$ is reduced to the zero functional.

Remark 4.1.3. It is easy to check that a polar space is always closed. Indeed, roughly speaking, if $\langle f_n, z\rangle = 0$ for every n and for every z, and if $f_n \to f$ in H', then $\langle f, z\rangle = 0$ for all z. □

The concept of polar space is commonly used for general Banach spaces. In Hilbert spaces, however, it becomes particularly simple using the Ritz Theorem. Indeed, from (4.1.39), we immediately have

$$Z^0 \equiv R_H(Z^{\perp}). \tag{4.1.40}$$

From this and Proposition 4.1.3, we then have

$$\left(Z^0\right)^0 = Z \quad \textbf{iff} \quad Z \text{ is closed}, \tag{4.1.41}$$

and from (4.1.26),

$$Z^0 = \{0\} \quad \textbf{iff} \quad Z \text{ is dense}. \tag{4.1.42}$$

Remark 4.1.4. Property (4.1.42) is a particular case (or, if you want, the restriction to Hilbert spaces) of a fundamental theorem of Functional Analysis, known as the **Hahn-Banach Theorem**. □

Remark 4.1.5. As we had in Remark 4.1.1 for orthogonal spaces, if Z can be seen as a subspace of two different spaces H_1 and H_2, then the polar of Z in H_1' will be different from the polar of Z in H_2'. □

Similarly to (4.1.18), when Z_1 and Z_2 are subspaces of the same space H, then

$$Z_1 \subseteq Z_2 \quad \Rightarrow \quad Z_2^0 \subseteq Z_1^0. \tag{4.1.43}$$

4.1.5 Bilinear Forms and Associated Operators; Transposed Operators

Another important particular case is that of bilinear forms. Assume that V and Q are Hilbert spaces: we say that a bilinear form b from $V \times Q$ to $\mathbb{R}$ is **continuous** if there exists a constant μ_b such that

$$b(v,q) \le \mu_b \|v\|_V \, \|q\|_Q \qquad \forall\, v \in V,\ \forall\, q \in Q. \tag{4.1.44}$$

The **norm of the continuous bilinear form** $\|b\|_{\mathcal{L}(V\times Q,\mathbb{R})}$ is then defined as

$$\|b\|_{\mathcal{L}(V\times Q,\mathbb{R})} := \sup_{\substack{v\in V\\ q\in Q}} \frac{b(v,q)}{\|v\|_V\,\|q\|_Q}, \tag{4.1.45}$$

and it will be denoted simply by $\|b\|$ when no confusion can occur. Hence, from (4.1.44) and (4.1.45), we have

$$b(v,q) \le \|b\|\,\|v\|_V\,\|q\|_Q \qquad \forall\, v \in V,\ \forall\, q \in Q. \tag{4.1.46}$$

It is important to note that continuous bilinear forms on $V\times Q$ are strictly connected to linear continuous operators from V to Q': indeed, if b is a bilinear form on $V\times Q$, we can associate to it a linear operator B from V to Q', defined as

$$\langle Bv, q\rangle_{Q'\times Q} := b(v,q) \qquad \forall\, v \in V,\ \forall\, q \in Q. \tag{4.1.47}$$

Conversely, if B is a linear operator from V to Q', we can associate to it the bilinear form

$$b(v,q) := \langle Bv, q\rangle_{Q'\times Q} \qquad \forall\, v \in V,\ \forall\, q \in Q. \tag{4.1.48}$$

It is elementary to check that B is continuous (from V to Q') *if and only if* the associated bilinear form b is continuous from $V\times Q$ to $\mathbb{R}$. To $B : V \to Q'$ we can also associate another operator, that we call **transposed operator** $B^t : Q \to V'$, given by

$$\langle v, B^t q\rangle_{V\times V'} := \langle Bv, q\rangle_{Q'\times Q} = b(v,q). \tag{4.1.49}$$

Example 4.1.10. It is easy to see that if $V := \mathbb{R}^n$ and $Q := \mathbb{R}^m$, then the linear operators from V to $Q' \simeq Q$ are just $(m\times n)$ *matrices*. In particular, the *transposed operator* will simply be the *transposed matrix*. □

It is worth noting that the continuity of the three objects b, B, and B^t is just the same property. In particular we have

$$\|B\|_{\mathcal{L}(V,Q')} \equiv \|B^t\|_{\mathcal{L}(Q,V')} \equiv \|b\|_{\mathcal{B}(V\times Q,\mathbb{R})} \equiv \sup_{\substack{v\in V\\ q\in Q}} \frac{b(v,q)}{\|v\|_V\,\|q\|_Q}. \tag{4.1.50}$$

For a linear operator M from a Hilbert space V to another Hilbert space W, we can define the **kernel** and the **image** (or **range**) as we did in (3.1.7) for the finite dimensional case:

$$\begin{aligned} \mathrm{Ker}M &:= \{v \in V \text{ such that } Mv = 0\},\\ \mathrm{Im}M &:= \{w \in W \text{ such that } \exists\, v \in V \text{ with } Mv = w\}. \end{aligned} \tag{4.1.51}$$

Remark 4.1.6. Note that the *kernel* of a continuous operator M is always closed. Indeed, if $Mv_n = 0$ and $v_n \to v$ in V, the continuity of M will imply that $Mv = 0$. This is not true for the *image*. Referring to the case of Example 4.1.5, take $V = H^1_0(\Omega)$ and $W = L^2(\Omega)$, with $Mv = v$ for every $v \in V$. Clearly, M is continuous, but $\text{Im}M = V$ is not a closed subspace of W. This fact (that the image might not be closed) *puts pains thousandfold upon the* mathematicians (whether *Achaians* or not). However, as you will see, one can survive. □

We concentrate now our attention on the case of linear operators B from V to $W = Q'$, with their associated bilinear form b and transposed operator B^t, as in (4.1.49). In this case, we can see that $\text{Ker}B$ and $\text{Ker}B^t$ can be written, respectively, as

$$\begin{aligned}\text{Ker}B :&= \{v \in V \text{ such that } b(v,q) = 0\ \forall\, q \in Q\} \\ &= \{v \in V \text{ such that } \langle v, B^t q\rangle_{V\times V'} = 0\ \forall\, q \in Q\}\end{aligned} \tag{4.1.52}$$

and

$$\begin{aligned}\text{Ker}B^t :&= \{q \in Q \text{ such that } b(v,q) = 0\ \forall\, v \in V\} \\ &= \{q \in Q \text{ such that } \langle Bv, q\rangle_{Q'\times Q} = 0\ \forall\, v \in V\}.\end{aligned} \tag{4.1.53}$$

In finite dimensional problems (see Proposition 3.1.2), we did interpret (4.1.52) and (4.1.53) as

$$\text{Ker}B = (\text{Im}B^T)^{\perp} \text{ and } \text{Ker}B^T = (\text{Im}B)^{\perp} \tag{4.1.54}$$

respectively. This, however, cannot be done in the present infinite dimensional case, because, for instance, $\text{Im}B$ is not a subset of Q but a subset of Q' (the two spaces were *identified* in finite dimension without telling you anything; sorry for that!). We have, however introduced a special definition for that: the polar space (see (4.1.39)). Hence, we can interpret (4.1.52) and (4.1.53) as

$$\text{Ker}B = (\text{Im}B^t)^0 \quad \text{and} \quad \text{Ker}B^t = (\text{Im}B)^0 \tag{4.1.55}$$

respectively. In finite dimension, in Theorem 3.1.1, we also had $\text{Im}B^T = (\text{Ker}B)^{\perp}$ and $\text{Im}B = (\text{Ker}B^T)^{\perp}$. Here we might hope to have

$$(\text{Ker}B)^0 = \text{Im}B^t \quad \text{and} \quad (\text{Ker}B^t)^0 = \text{Im}B. \tag{4.1.56}$$

Actually, for instance, the equality

$$(\text{Ker}B)^0 = \text{Im}B^t \tag{4.1.57}$$

will follow easily from the second of (4.1.55) using (4.1.41) if we only knew that $\text{Im}B$ is closed. However, unfortunately, this is not always the case. On the other hand, if $\text{Im}B$ is not closed, then (4.1.57) is hopeless, as a polar space is always closed. Indeed, we can see that we have the following generalisation of Corollary 3.1.1 and Theorem 3.1.1 to the infinite dimensional case.

Theorem 4.1.4. *Let V and Q be Hilbert spaces, and B a linear continuous operator from V to Q' (that is: $B \in \mathcal{L}(V, Q')$). Then, the following three properties are equivalent:*

$$\mathrm{Im}B \textit{ is closed in } Q' \tag{4.1.58}$$

$$\mathrm{Im}B = (\mathrm{Ker}B^t)^0 \tag{4.1.59}$$

$\exists\, L_B \in \mathcal{L}(\mathrm{Im}B, (\mathrm{Ker}B)^\perp)$ *and* $\beta > 0$ *such that:*

$$B\, L_B\, g = g \quad \forall\, g \in \mathrm{Im}B \quad \textit{and} \quad \beta \|L_B\, g\|_V \le \|g\|_{Q'} \quad \forall\, g \in \mathrm{Im}B. \tag{4.1.60}$$

Proof. We already discussed the equivalence of (4.1.58) and (4.1.59). Moreover, if (4.1.58) holds, then B (or actually its restriction to $(\mathrm{Ker}B)^\perp$) becomes (with the same argument used in Proposition 3.1.1) a continuous one-to-one operator between the two Hilbert spaces $(\mathrm{Ker}B)^\perp$ and $\mathrm{Im}B$, and Theorem 4.1.1 gives us (4.1.60). Finally, (4.1.60) easily implies (4.1.58): if $g_n = B\, v_n$ is a Cauchy sequence in Q' then, using (4.1.60), we have that v_n (equal to $L\, g_n$) is a Cauchy sequence in V. Then, it converges to a $v \in V$, and the continuity of B implies that g_n converges to $B\, v$ in Q'. Hence, the limit of g_n is in $\mathrm{Im}B$. □

Exchanging B and B^t, we immediately have the equivalence of the three properties

$$\mathrm{Im}B^t \text{ is closed in } V' \tag{4.1.61}$$

$$\mathrm{Im}B^t = (\mathrm{Ker}B)^0 \tag{4.1.62}$$

$\exists\, L_{B^t} \in \mathcal{L}(\mathrm{Im}B^t, (\mathrm{Ker}B^t)^\perp)$ and $\beta > 0$ such that:

$$B^t\, L_{B^t}\, f = f \quad \forall\, f \in \mathrm{Im}B^t \quad \text{and} \quad \beta \|L_{B^t}\, f\|_Q \le \|f\|_{V'} \;\forall\, f \in \mathrm{Im}B^t. \tag{4.1.63}$$

What is somehow remarkable is that, actually, the two triplets of properties (4.1.58)–(4.1.60) and (4.1.61)–(4.1.63) are equivalent to each other. This actually follows easily from the following proposition.

Proposition 4.1.5. *Let V and Q be Hilbert spaces, and B a linear continuous operator from V to Q' (that is: $B \in \mathcal{L}(V, Q')$). Then, $\mathrm{Im}B$ is closed **iff** $\mathrm{Im}B^t$ is closed.*

Proof. In view of the above equivalences, we only need to prove that (4.1.58)–(4.1.60) imply (4.1.61). For this, consider $q \in (\mathrm{Ker}B^t)^\perp$ and set $g = R_Q q$ where R_Q is the Ritz operator $Q \to Q'$. Using (4.1.40) we have $g \in (\mathrm{Ker}B^t)^0$. Hence, using (4.1.59), we have $g \in \mathrm{Im}B$ so that $g = Bx$ for $x = L_B g$ and from (4.1.60): $\beta \|x\|_V \le \|g\|_{Q'} = \|q\|_Q$. Then, we have

$$\begin{aligned}\|q\|_Q^2 &= {}_{Q'}\langle R_Q q, q\rangle_Q = {}_{Q'}\langle g, q\rangle_Q = {}_{Q'}\langle Bx, q\rangle_Q \\ &= {}_V\langle x, B^t q\rangle_{V'} \le \|x\|_V\, \|B^t q\|_{V'} \le \frac{1}{\beta}\|q\|_Q\, \|B^t q\|_{V'}\end{aligned} \tag{4.1.64}$$

which easily gives

$$\beta \|q\|_Q \leq \|B^t q\|_{V'} \qquad \forall q \in (\mathrm{Ker} B^t)^{\perp} \tag{4.1.65}$$

which, in turn, proves that $\mathrm{Im}B^t$ is closed by the same argument used in the proof of Theorem 4.1.4 □

We can summarise the above results in the following theorem, that is a particular case of a more general (and important) theorem, also due to Banach, and mostly known as the *Closed Range Theorem*.

Theorem 4.1.5 (Banach Closed Range Theorem). *Let V and Q be Hilbert spaces and let B be a linear continuous operator from V to Q'. Set*

$$K := \mathrm{Ker} B \subset V \quad \textit{and} \quad H := \mathrm{Ker} B^t \subset Q. \tag{4.1.66}$$

Then, the following statements are equivalent:

- $\mathrm{Im}B$ *is closed in* Q',
- $\mathrm{Im}B^t$ *is closed in* V',
- $K^0 = \mathrm{Im}B^t$,
- $H^0 = \mathrm{Im}B$,
- $\exists\, L_B \in \mathcal{L}(\mathrm{Im}B, K^{\perp})$ *and* $\exists\, \beta > 0$ *such that* $B(L_B(g)) = g \quad \forall\, g \in \mathrm{Im}B$ *and moreover* $\beta \|L_B g\|_V \leq \|g\|_{Q'} \quad \forall\, g \in \mathrm{Im}B$,
- $\exists\, L_{B^t} \in \mathcal{L}(\mathrm{Im}B^t, H^{\perp})$ *and* $\exists\, \beta > 0$ *such that* $B^t(L_{B^t}(f)) = f \quad \forall\, f \in \mathrm{Im}B^t$ *and moreover* $\beta \|L_{B^t} f\|_Q \leq \|f\|_{V'} \quad \forall\, f \in \mathrm{Im}B^t$. □

In the following treatment, we shall often assume that B is surjective. Let us see what the Closed Range Theorem has to say in this case.

Corollary 4.1.1. *Let V and Q be Hilbert spaces, and let B be a linear continuous operator from V to Q'. Then, the following statements are equivalent:*

- $\mathrm{Im}B = Q'$,
- $\mathrm{Im}B^t$ *is closed and* B^t *is injective,*
- B^t *is bounding:* $\exists\, \beta > 0$ *s.t.* $\|B^t q\|_{V'} \geq \beta \|q\|_Q \quad \forall\, q \in Q$,
- $\exists\, L_B \in \mathcal{L}(Q', V)$ *such that* $B(L_B(g)) = g \quad \forall\, g \in Q'$, *with* $\|L_B\| = 1/\beta$.

The proof is immediate.

A useful consequence of Corollary 4.1.1 is the well known Lax-Milgram Lemma:

Theorem 4.1.6 (Lax-Milgram Lemma). *Let V be a Hilbert space, and let $a(\cdot,\cdot)$ be a bilinear continuous form on V. Assume that a is coercive, that is*

$$\exists\, \alpha > 0 \textit{ such that } a(v,v) \geq \alpha\, \|v\|_V^2,\ \forall\, v \in V. \tag{4.1.67}$$

Then, for every $f \in V'$, the problem: find $u \in V$ such that

$$a(u, v) = \langle f, v \rangle_{V' \times V} \quad \forall\, v \in V \tag{4.1.68}$$

has a unique solution.

Proof. Note that (4.1.68) is equivalent to $Au = f$, where $A \in \mathcal{L}(V, V')$ is the operator associated to the bilinear form a. We have to prove that A is injective and surjective. Condition (4.1.67) immediately implies that A is *bounding*:

$$\|Av\|_{V'} = \sup_{w \in V \setminus \{0\}} \frac{a(v, w)}{\|w\|_V} \geq \frac{a(v, v)}{\|v\|_V} \geq \alpha \|v\|_V. \tag{4.1.69}$$

Hence, A is injective. With an identical proof, we see that A^t is also bounding. Hence, A^t is injective and (due to Corollary 4.1.1) A is surjective. □

Remark 4.1.7. Roughly speaking, we can summarise the result of the Closed Range Theorem by saying that operators with a closed range have essentially all the well-known properties of operators in finite dimensional spaces (whose range is always trivially closed) that we have seen in the previous chapter. In particular, Corollary 4.1.1 is *exactly what we need* to extend the properties and the results of Sect. 3.4 to the infinite dimensional case. See in particular Proposition 3.4.4. □

4.1.6 Dual Spaces of Linear Subspaces

We have seen two (very different) types of subspaces: closed subspaces and dense subspaces. We shall see now that they also behave quite differently when we consider their *dual spaces*. Let us see the difference.

Assume first that Z is a *closed* subspace of a Hilbert space H. Then, we already pointed out that, using on Z *the same norm* that we already have on H, the space Z becomes itself a Hilbert space, and, as such, it will have a dual space Z' of its own. It is easy to see that Z' could be identified with a particular subset of H', made of all functionals $f \in H'$ that vanish identically on the orthogonal complement $Z^\perp$ of Z. Note that we already have a name for that space, that is $(Z^\perp)^0$. We have therefore, in a natural way,

$$Z' \equiv (Z^\perp)^0 \equiv R_H(Z) \subset H' \quad \text{and} \quad (Z^\perp)' \equiv Z^0 \equiv R_H(Z^\perp) \subset H'. \tag{4.1.70}$$

Hence, *the dual space Z' of a closed subspace $Z \subset H$ can be identified with a closed subspace of H'*. Once this identification is made, we can also consider the extension operator $E_{Z' \to H'}$ (that we shall often denote simply as $E_{Z'}$), and the projection operator $\pi_{Z'}$ from H' to Z'. Note that, for $\phi \in Z'$, the functional $E_{Z' \to H'}\phi$ can also be described as

$$_{H'}\langle E_{Z' \to H'}\phi, v \rangle_H := {}_{Z'}\langle \phi, \pi_Z v \rangle_Z \tag{4.1.71}$$

while for $\psi \in H'$ the functional $\pi_{Z'}\psi$ can be described as

$$ {}_{Z'}\langle \pi_{Z'}\psi, z\rangle_Z := {}_{H'}\langle \psi, E_{Z\to H} z\rangle_H . \tag{4.1.72} $$

In other terms

$$ (\pi_{Z'})^t \equiv (E_Z) \tag{4.1.73} $$

and

$$ (\pi_Z)^t \equiv (E_{Z'}). \tag{4.1.74} $$

Example 4.1.11. For instance, if $H = L^2(\Omega)$ and Z is the space of constant functions, it is not difficult to see that $Z^\perp = L^2_0(\Omega)$ (the space of functions having zero mean value). Now, the dual space of Z will be the space of functionals that can be written as

$$ q \to k \int_\Omega q\,dx \qquad k \in \mathbb{R} $$

(meaning that for each $k \in \mathbb{R}$ we have a different functional). On the other hand, the dual space of $Z^\perp$ will be the space of functionals that can be written as

$$ q \to \int_\Omega k\,q\,dx \qquad\qquad k \in Z^\perp $$

(meaning that for each $k \in Z^\perp$ we have a different functional). On the other hand, $(Z^\perp)'$ could also be identified with the subset of H' made of functionals that vanish identically on constant functions (that is, with the polar set of the space of constants, which is the polar set of Z, as in (4.1.70)). Using the Ritz operator R_H of Theorem 4.1.2, we could write $Z' = R_H(Z)$ and $(Z^\perp)' = R_H(Z^\perp)$. If, as is done almost every time, we identify $L^2(\Omega)$ with its dual space, then we could write $Z' = Z$ and $(Z^\perp)' = Z^\perp$. □

Let us consider now the case of a *dense* subspace $S \subset H$ of a Hilbert space H. If we take on S the same norm as on H, we cannot (in the present setting) consider its dual space, as S will not be closed (unless $S \equiv H$, a case without any interest). Hence, we assume that on S we take a *different norm*. More precisely, we assume that on S we are given another norm, $\|\cdot\|_S$, that makes S a Hilbert space. We assume, moreover, that this other norm is (up to a multiplicative constant) *bigger* than the $\|\cdot\|_H$ norm:

$$ \exists\, C_{SH} > 0 \text{ such that } \|s\|_H \le C_{SH}\|s\|_S \quad \forall\, s \in S. \tag{4.1.75} $$

In this case, we will say that *S is continuously embedded in H*. Indeed, (4.1.75) means exactly that the identity operator is continuous from S into H. There is a special symbol for that: instead of $S \subset H$, we write $S \hookrightarrow H$.

Example 4.1.12. If we take, as in Example 4.1.5, $S = H^1_0(\Omega)$ and $H = L^2(\Omega)$, then inequality (4.1.75) is just the Poincaré inequality. □

Now S, being a Hilbert space, has a dual space S'. Let us see the relationship between S' and H'. As $S \subset H$, for each element $g \in H'$, we can consider its restriction $g_{|S} \in S'$ defined by $\langle g_{|S}, s\rangle_{S'\times S} = \langle g, s\rangle_{H'\times H}$ for all $s \in S$. Indeed, from (4.1.75) we have easily for every $s \in S$

$$\langle g_{|S}, s\rangle_{S'\times S} \equiv \langle g, s\rangle_{H'\times H} \leq \|g\|_{H'} \|s\|_H \leq C_{SH} \|g\|_{H'} \|s\|_S, \qquad (4.1.76)$$

implying the continuity of $g_{|S} : S \to \mathbb{R}$, as well as the *continuity of the restriction operator*: namely,

$$\|g_{|S}\|_{S'} \leq C_{SH}\|g\|_{H'}. \qquad (4.1.77)$$

Using the Hahn-Banach theorem (here simplified to (4.1.42)), we see that we cannot have in H' two different g's having the same restriction to S: indeed, if g^1 and g^2 have the same restriction to S (that is, if $g^1_{|S} = g^2_{|S}$), then the difference $g^1 - g^2$ is in S^0, and hence it must be zero.

We can then summarise the above discussion by saying that: every $g \in H'$ has a restriction $g_{|S}$ in S' and the mapping $g \to g_{|S}$ from H' to S' is *injective*. This allows us to *identify* H' with a subset of S':

$$H' \subseteq S'. \qquad (4.1.78)$$

On the other hand, there are, in general, elements in S' that cannot be presented as the restriction of any $g \in H'$: indeed, S has a norm which is bigger than that of H, and g could be continuous from S to $\mathbb{R}$ and not from H to $\mathbb{R}$. As we have seen for instance in Example 3.1.6, for $I :=]-1, 1[$, taking $H := L^2(I)$ and $S := H^1_0(I)$, the mapping $v \to v(0)$ belongs to S' but *cannot be seen* as the restriction to S of an element of H'

In other words, (4.1.77) cannot be reversed. Hence, we have $H' \subset S'$, and using (4.1.77), we see that we actually have $H' \hookrightarrow S'$, and in general the inclusion is *strict*. On the other hand, one can also prove that H' is dense in S'. Moreover, out of the previous discussion, we easily have that

$$\langle g, s\rangle_{S'\times S} = \langle g, s\rangle_{H'\times H} \quad \text{whenever } g \in H' \textbf{ and } s \in S. \qquad (4.1.79)$$

Hence, if we have two Hilbert spaces S and H with $S \subset H$ and S *dense* in H, then

$$S \hookrightarrow H \quad \Rightarrow \quad H' \hookrightarrow S'. \qquad (4.1.80)$$

The difference between the two cases, (4.1.70) and (4.1.80), that might be surprising at first sight, is due to the fact that in the first case we used on S the same norm that we had on H, while in the second case we used a different, stronger norm.

Example 4.1.13. We have already seen the example of δ_0, which belongs to the dual space of $S := H^1_0(]-1, 1[)$ but not to the dual space of $H := L^2(]-1, 1[)$. Let us see another simple example. For a general domain Ω, taking always $S = H^1_0(\Omega)$ and $H = L^2(\Omega)$, and taking in H' the functional

$$v \to \int_\Omega v\, dx,$$

it is clear that its restriction to S leaves the functional (essentially) unchanged. On the other hand, for f fixed in $L^2(\Omega)$, the functional

$$v \to \int_\Omega f \frac{\partial v}{\partial x}\, dx,$$

linear and continuous on S, cannot be extended to a continuous functional on H.

□

4.1.7 *Identification of a Space with its Dual Space*

It is usually a strong temptation, when dealing with Hilbert spaces, to use the Ritz Theorem 4.1.2 to **identify a Hilbert space with its dual**. After all, this is what is done most of the times when dealing with finite dimensional spaces. However, when dealing with *functional spaces* (that is, spaces made of functions), it is **highly recommended** to limit such identification to L^2 with its dual (or of a closed subspace Z of L^2 with its dual Z'). Every other identification will be *calling for a total disaster*. Let us see why. Assume that in (4.1.80) we have $H = L^2(\Omega)$ and $S = H^1_0(\Omega)$. Identifying L^2 with its dual space, we would have $H \equiv H'$, and (4.1.80) will become

$$S \hookrightarrow H \equiv H' \hookrightarrow S'. \tag{4.1.81}$$

So far, so good. Everybody does that, and nobody suffers. Assume, however, that, in spite of all recommendations, you **also** identify S with S'. Then, in (4.1.81), you compress the four spaces $S \equiv H \equiv H' \equiv S'$ into one, identifying at the same time a function with itself **and** with its Laplacian. This is *the beginning of the end*. Now, the question that everybody asks (the first time one hears about that) is "What is so special with L^2?". It is a very good question. Actually, there is nothing special, *mathematically*, about it, apart from the fact that we are so used to identify a function $f \in L^2(\Omega)$ with the mapping (defined for $\varphi \in L^2(\Omega)$):

$$\varphi \to \int_\Omega f\,\varphi\, dx \tag{4.1.82}$$

that we do it all the time, without even realising it. In principle, we might as well identify a function $f \in H^1_0(\Omega)$ with the mapping (defined for φ in $H^1_0(\Omega)$):

$$\varphi \to \int_\Omega \underline{\text{grad}}\, f \cdot \underline{\text{grad}}\, \varphi\, dx \tag{4.1.83}$$

and don't use the identification (4.1.82). This will be mathematically correct but psychologically very, very difficult; and before the rooster crows, you will have used (4.1.82) three times. Hence, our advice is: *No matter whether the above discussion was clear or not, just avoid any identification of a functional space that is not L^2 (or a multiple copy of it, or, exceptionally, a closed subspace of it) with its dual space!* This, of course, unless you are very skilled in Functional Analysis. Although, if you are... why are you reading all this? □

4.1.8 Restrictions of Operators to Closed Subspaces

We shall now deal briefly with a situation that we will meet constantly in the following chapter. We have (as before) two Hilbert spaces V and Q, we have a linear continuous operator $B \in \mathcal{L}(V, Q')$, and we have two closed subspaces $Z \subset V$ and $S \subset Q$. In the applications of the next chapter, Z and S will typically be *finite dimensional spaces* (and hence automatically closed).

As we have seen, B (and its transposed operator B^t) can be associated to a bilinear form b defined on $V \times Q$. It is not difficult to see that, restricting the bilinear form to $Z \times S$, we have as associated operators

$$B_{ZS'} \equiv \pi_{S'} B E_Z \quad \text{and} \quad B^t_{SZ'} \equiv \pi_{Z'} B^t E_S \tag{4.1.84}$$

and obviously $(B_{ZS'})^t = B^t_{SZ'}$.

Remark 4.1.8. As we have already pointed out in Remark 3.1.11 of the previous chapter, in general, we cannot expect the kernel of $B_{ZS'}$ to be a subspace of the kernel of B, nor the image of $B_{ZS'}$ to be a subset of the image of B. The same is obviously true for the images and the kernels of $B^t_{SZ'}$ and B^t. □

Proposition 4.1.6. *Let V and Q be Hilbert spaces, let $B \in \mathcal{L}(V, Q')$, and let $Z \subset V$ and $S \subset Q$ be closed subspaces, with S finite dimensional. Then, the inclusion*

$$\mathrm{Ker} B^t_{SZ'} \subseteq \mathrm{Ker} B^t \tag{4.1.85}$$

holds **iff** *we have*

$$\pi_{S'}(\mathrm{Im} B) \subseteq \mathrm{Im} B_{ZS'}. \tag{4.1.86}$$

Proof. Assume first that (4.1.85) (that, to be precise, we should actually write as $E_S \mathrm{Ker} B^t_{SZ'} \subseteq \mathrm{Ker} B^t$) holds, and let $g = Bv \in \mathrm{Im} B$. As $\mathrm{Im} B_{ZS'}$ is closed (since S is finite dimensional), to show that $\pi_{S'} g \in \mathrm{Im} B_{ZS'}$, we just have to check that $\pi_{S'} g \in (\mathrm{Ker} B^t_{SZ'})^0$, that is,

$${}_{Q'}\langle g, q \rangle_Q = 0 \quad \forall q \in \mathrm{Ker} B^t_{SZ'}. \tag{4.1.87}$$

If the inclusion (4.1.85) is satisfied, then every $q \in \operatorname{Ker} B^t_{SZ'}$ will also be in $\operatorname{Ker} B^t$. However, for $q \in \operatorname{Ker} B^t$ we have

$$ {}_{Q'}\langle g, q\rangle_Q = {}_{Q'}\langle Bv, q\rangle_Q = {}_V\langle v, B^t q\rangle_{V'} = 0, \tag{4.1.88}$$

giving (4.1.87) and ending the first part of the proof.

Assume now that (4.1.86) holds, and let $q_s \in S$ be in $\operatorname{Ker} B^t_{SZ'}$, that is: $\pi_{Z'} B^t q_s = 0$. For such a q_s we have, for every $z \in Z$, that

$$\begin{aligned} &{}_S\langle q_s, \pi_{S'} Bz\rangle_{S'} \\ &= {}_Q\langle q_s, Bz\rangle_{Q'} = {}_{V'}\langle B^t q_s, z\rangle_V = {}_{Z'}\langle \pi_{Z'} B^t q_s, z\rangle_Z \\ &= 0, \end{aligned} \tag{4.1.89}$$

meaning that q_s is in the polar space of $\operatorname{Im} B_{ZS'}$. Inclusion (4.1.86) together with (4.1.43) implies then that q_s is in the polar space of $\pi_{S'} \operatorname{Im} B$, so that for all $v \in V$ we have ${}_S\langle q_s, \pi_{S'} Bv\rangle_{S'} = 0$, hence ${}_{V'}\langle B^t q_s, v\rangle_V = 0$ and therefore $q_s \in \operatorname{Ker} B^t$.

□

Remark 4.1.9. The assumption that S is finite dimensional, in Proposition 4.1.6, is clearly stronger than necessary. Indeed, looking at the proof, we see that for the first part we only need $\operatorname{Im} B_{ZS'}$ to be closed, while the second part does not even need that. However, as we said, we are going to use the result in the case of Z and S being finite dimensional, so that we did not struggle to minimise this type of assumptions. □

Exchanging the roles of B and B^t, we have, moreover, in the same assumptions of Proposition 4.1.6 (but requiring Z to be finite dimensional instead of S), that

$$\operatorname{Ker} B_{ZS'} \subseteq \operatorname{Ker} B \tag{4.1.90}$$

is equivalent to

$$\pi_{Z'}(\operatorname{Im} B^t) \subseteq \operatorname{Im} B^t_{SZ'}. \tag{4.1.91}$$

The case in which the subspaces Z and S are related to the kernels and images of a linear operator $B \in \mathcal{L}(V, Q')$ (and of its transposed) is obviously of special interest. In particular, we can present a corollary of the Closed Range Theorem 4.1.5 that will often be useful.

Corollary 4.1.2. *In the same assumptions of Theorem 4.1.5, if one of the six equivalent properties is satisfied, then $L_B \in \mathcal{L}(K^\perp, H^0)$ is the transposed operator of $L_{B^t} \in \mathcal{L}(H^\perp, K^0)$ and in particular,*

$$\|L_B\|_{\mathcal{L}(K^\perp, H^0)} = \|L_{B^t}\|_{\mathcal{L}(H^\perp, K^0)} =: \mu. \tag{4.1.92}$$

Moreover, setting $\beta := 1/\mu$ we have

$$\beta\|v\|_V \leq \|Bv\|_{V'} \qquad \forall\, v \in K^{\perp}, \tag{4.1.93}$$

and

$$\beta\|q\|_Q \leq \|B^t q\|_{V'} \qquad \forall\, q \in H^{\perp}. \tag{4.1.94}$$

Proof. If, say, $\mathrm{Im}B$ is closed, then B will be an isomorphism from $K^{\perp}$ to $\mathrm{Im}B$ which, however, coincides with H^0. Similarly, B^t will be an isomorphism from $H^{\perp}$ to $\mathrm{Im}B^t$ that coincides with K^0. Hence, L_B coincides with $(B_{K^{\perp}H^0})^{-1}$ and L_{B^t} coincides with $(B^t_{H^{\perp}K^0})^{-1}$. We also recall from (4.1.70) that

$$(K^{\perp})^0 = K' \quad K^0 = (K^{\perp})' \qquad (H^{\perp})^0 = H' \quad H^0 = (H^{\perp})' \tag{4.1.95}$$

so that it is immediate to see that L_{B^t} is the transposed operator of L_B. Now (4.1.92) will follow immediately from (4.1.50). Finally, (4.1.93) and (4.1.94) are now immediate since, for $v \in K^{\perp}$, we have $v = L_B(Bv)$ and for $q \in H^{\perp}$, we have $q = L_{B^t}(B^t q)$. □

4.1.9 *Quotient Spaces*

Assume that Q is a Hilbert space and let H be a closed subspace of Q. We also assume that H is a *proper* subspace, meaning that H does not coincide with Q. We consider then *the quotient space* $Q_{/H}$ defined as *the space whose elements are the equivalence classes induced by the equivalence relation:*

$$v_1 \cong v_2 \qquad \text{if and only if } (v_1 - v_2) \in H. \tag{4.1.96}$$

In other words, two elements are equivalent if their difference belongs to H. It is immediate to see that all the elements of H will then be equivalent to 0. In view of this definition, an element of $Q_{/H}$ will then be a subset of Q made by elements that are all equivalent to each other.

Example 4.1.14. For a bounded domain $\Omega \subset \mathbb{R}^d$, we take $Q := L^2(\Omega)$ and we consider the (one-dimensional) subspace H made of *constant functions*. Then, $Q_{/H}$ will be made of *classes of functions that differ from each other by a constant function.* □

Note that *if two classes C_1 and C_2 have an element v^* in common, then they must coincide*. Indeed, for every v_1 in C_1, we have $v_1 - v^* \in H$ and, for every $v_2 \in C_2$, we have $v_2 - v^* \in H$. As a consequence, for every $v_1 \in C_1$ and every $v_2 \in C_2$, we have $v_1 - v_2 = (v_1 - v^*) - (v_2 - v^*) \in H$ (as difference of two elements of H). This implies that for every $v_1 \in C_1$ and for every $v_2 \in C_2$, we have $v_1 \cong v_2$, which is to say that the two classes C_1 and C_2 coincide. We conclude that two *different* classes have no elements in common.

It is then easy to verify that *there is a one-to-one correspondence between* $Q_{/H}$ *and the orthogonal complement* $H^{\perp}$ *of* H *in* Q. Let us see it in more detail. Let q^* be an element of $H^{\perp}$: to it we associate the class C_{q^*} defined by

$$C_{q^*} := \{v \in Q | \; v \cong q^*\} \equiv \{v \in Q | \; v - q^* \in H\}. \tag{4.1.97}$$

It is clear that the mapping $q^* \to C_{q^*}$, from $H^{\perp}$ to $Q_{/H}$, is *injective*: indeed, assume that q^* and q^{**} are two elements in $H^{\perp}$ such that the two corresponding classes C_{q^*} and $C_{q^{**}}$ coincide. This implies that, say, $q^{**} \in C_{q^*}$, that is $q^{**} - q^* \in H$. Since $q^{**} - q^*$ must also belong to $H^{\perp}$ (as difference of two elements both in $H^{\perp}$), we conclude that $q^{**} = q^*$.

Let us see that the mapping $q^* \to C_{q^*}$ is also *surjective*: let therefore the class C^* be an element of $Q_{/H}$ and let $\bar{q} \in C^*$. The class C^* could then be characterised as

$$C^* := \{v \in Q | \; v \cong \bar{q}\} \equiv \{v \in Q | \; v - \bar{q} \in H\}. \tag{4.1.98}$$

At this point, it is not difficult to see that C^* is a closed convex subset of Q and hence (see (4.1.22) and Remark 4.1.2) we can define $q^*_{C^*}$ as the projection $\pi_C 0$ of 0 on C^* (that can also be seen as the element of C^* having minimum norm). It is then elementary to check that

$$(q^*_{C^*}, v) = 0 \qquad \forall\, v \in H, \tag{4.1.99}$$

implying that $q^*_{C^*} \in H^{\perp}$ and that, actually, $C^* \equiv C_{q^*_{C^*}}$. This also allows us to define *a norm* in $Q_{/H}$: for every $C \in Q_{/H}$, we define

$$\|C\|_{Q_{/H}} := \|\pi_C 0\|_Q \equiv \|q^*_C\|_Q. \tag{4.1.100}$$

Hence, if we prefer, we could choose in each class (= element of $Q_{/H}$) the unique element, in the class, which belongs to $H^{\perp}$, and identify $Q_{/H}$ with $H^{\perp}$.

Example 4.1.15. Let us go back to the case of Example 4.1.14 where $Q := L^2(\Omega)$ and H is the subspace made of *constant functions*. We recall that $Q_{/H}$ is made of classes of functions that differ from each other by a constant function. For every such class, we could always take one function q in the class, and describe the class as the set of all functions of the form $q + c$ with c constant. In doing so, we could however decide to choose as "representative" the unique function, in the class, that has zero mean value. This is the same as picking $q^* \in H^{\perp}$, since $H^{\perp}$ is clearly the subspace of Q made of functions having zero mean value. □

4.2 Existence and Uniqueness of Solutions

4.2.1 Mixed Formulations in Hilbert Spaces

From here to the end of this chapter, we will consistently remain in the same notational framework. As this framework will also include some assumptions, we summarise all these assumptions under the name of *Assumption* $\mathcal{AB}$.

Assumption $\mathcal{AB}$: *We are given two Hilbert spaces,* V *and* Q*, and two continuous bilinear forms:* $a(\cdot,\cdot)$ *on* $V\times V$ *and* $b(\cdot,\cdot)$ *on* $V\times Q$*. We denote by* A *and* B*, respectively, the linear continuous operators associated with them. We also set*

$$K := \mathrm{Ker}\,B \qquad \text{and} \qquad H := \mathrm{Ker}\,B^t. \tag{4.2.1}$$

We recall from the previous subsection that we have

$$|a(u,v)| \le \|a\|\;\|u\|_V\;\|v\|_V, \tag{4.2.2}$$

and that the two linear continuous operators $A : V \to V'$ and $A^t : V \to V'$ satisfy

$$\langle Au, v\rangle_{V'\times V} = \langle u, A^t v\rangle_{V\times V'} = a(u,v), \qquad \forall\, v\in V\;\forall\, u\in V. \tag{4.2.3}$$

Similarly,

$$|b(v,q)| \le \|b\|\;\|v\|_V\;\|q\|_Q, \tag{4.2.4}$$

and the two linear operators $B : V \to Q'$, and $B^t : Q \to V'$ satisfy

$$\langle Bv, q\rangle_{Q'\times Q} = \langle v, B^t q\rangle_{V\times V'} = b(v,q) \qquad \forall\, v\in V,\;\forall\, q\in Q. \tag{4.2.5}$$

We now consider our **basic problem**. *Given* $f\in V'$ *and* $g\in Q'$*, we want to find* $(u,p)\in V\times Q$ *solution of*

$$\begin{cases} a(u,v) + b(v,p) = \langle f, v\rangle_{V'\times V}, \;\forall\, v\in V,\\ b(u,q) = \langle g, q\rangle_{Q'\times Q}, \;\forall\, q\in Q. \end{cases} \tag{4.2.6}$$

Note that problem (4.2.6) can also be written as

$$\begin{cases} Au + B^t p = f \;\text{ in } V',\\ Bu = g \;\text{ in } Q', \end{cases} \tag{4.2.7}$$

and from now on we shall consider the formulations (4.2.6) and (4.2.7) to be the same, referring to one or the other according to the convenience of the moment. We now want to find conditions implying existence and possibly uniqueness of solutions to this problem.

Remark 4.2.1. If the bilinear form $a(\cdot,\cdot)$ is symmetric, the equations (4.2.6) are the optimality conditions of the minimisation problem

$$\inf_{Bv=g} \frac{1}{2}a(v,v) - \langle f, v\rangle_{V'\times V}. \tag{4.2.8}$$

The variable p is then the Lagrange multiplier associated with the constraint $Bu = g$, and the associated saddle point problem is

$$\inf_{v\in V}\sup_{q\in Q}\left\{\frac{1}{2}a(v,v) + b(v,q) - \langle f, v\rangle_{V'\times V} - \langle g, q\rangle_{Q'\times Q}\right\}. \tag{4.2.9}$$

This is the reason for the title of this chapter, in spite of the fact that we deal in fact with a more general case. □

Remark 4.2.2. The two equations in (4.2.6) can sometimes be written as a unique variational equation, setting

$$\mathcal{A}((u,p),(v,q)) = a(u,v) + b(v,p) - b(u,q) \qquad \forall (u,p),\ (v,q) \in V\times Q \tag{4.2.10}$$

and then requiring that

$$\mathcal{A}((u,p),(v,q)) = \langle f, v\rangle_{V'\times V} - \langle g, q\rangle_{Q'\times Q} \qquad \forall (v,q) \in V\times Q. \tag{4.2.11}$$

One can obviously go from (4.2.6) to (4.2.11), subtracting the two equations, and from (4.2.11) to (4.2.6) by considering separately the pairs $(v,0)$ and $(0,-q)$. □

It is clear from the second equation of (4.2.7) that, in order to have existence of a solution for every $g \in Q'$, we must have $\mathrm{Im}B = Q'$. Following the path of the previous chapter, we first consider a simpler case, in which we have *sufficient conditions* on a and b for having a unique solution.

Theorem 4.2.1. *Together with Assumption $\mathcal{AB}$, assume that $\mathrm{Im}B = Q'$ and that the bilinear form $a(\cdot,\cdot)$ is coercive on K, that is*

$$\exists\, \alpha_0 > 0 \text{ such that } a(v_0,v_0) \geq \alpha_0\, \|v_0\|_V^2,\ \forall\, v_0 \in K. \tag{4.2.12}$$

Then, for every $(f,g) \in V'\times Q'$, problem (4.2.6) *has a unique solution.*

Proof. Let us first prove the *existence* of a solution. From the surjectivity of B and Corollary 4.1.1, we have that there exists a lifting operator L_B such that $B(L_B g) = g$ for every $g \in Q'$. Setting $u_g := L_B g$, we therefore have $Bu_g = g$. We now consider the new unknown $u_0 := u - u_g$ and, in order to have $Bu = g$, we require $u_0 \in K$. For every $v_0 \in K$, we obviously have $b(v_0,q) = 0$ for every $q \in Q$, so that the first equation of (4.2.6) now implies

$$a(u_0,v_0) = \langle f, v_0\rangle_{V'\times V} - a(u_g,v_0),\ \forall v_0 \in K, \tag{4.2.13}$$

and the Lax-Milgram Lemma, using (4.2.12), ensures that we have a unique $u_0 \in \mathrm{Ker}B$ satisfying (4.2.13). Remark now that the functional

$$v \to \ell(v) := \langle f, v\rangle_{V'\times V} - a(u_g + u_0, v), \tag{4.2.14}$$

thanks to (4.2.13), vanishes identically for every $v \in K$. Hence, $\ell \in K^0$ (the polar space of K), which, due to Theorem 4.1.4, coincides with $\mathrm{Im}B^t$. Hence, ℓ is in the image of B^t, and there exists a $p \in Q$ such that $B^t p = \ell$. This means that

$$\langle B^t p, v\rangle_{V'\times V} = \langle \ell, v\rangle_{V'\times V} = \langle f, v\rangle_{V'\times V} - a(u_g + u_0, v) \tag{4.2.15}$$

for every $v \in V$, and since $u = u_g + u_0$, the first equation is satisfied. On the other hand, $Bu = Bu_g + Bu_0 = g$ and the second equation is also satisfied.

We now prove *uniqueness*. By linearity, assume that $f = 0$ and $g = 0$: then, $u \in K$. Testing the first equation on $v = u$ we get $a(u,u) = 0$ and then $u = 0$ from (4.2.12). Using $u = 0$ and $f = 0$ in the first equation of (4.2.7), we have then $B^t p = 0$, and from Corollary 4.1.1 we have $p = 0$. Then, problem (4.2.6) has a unique solution. □

Remark 4.2.3. The coercivity of $a(\cdot,\cdot)$ on K may hold while there is no coercivity on V. We have already seen examples of this situation in finite dimension and we shall see in the next chapters many other examples coming from partial differential equations. □

The result of Theorem 4.2.1 will be *the most commonly used* in our applications. However, as we had in the finite-dimensional case of the previous chapter, it is clear that it does not give a *necessary and sufficient* condition. To get it, we must weaken the coercivity condition (4.2.12). For this, we recall that K is a closed subspace of V, and hence it is itself a Hilbert space (with the same norm as V). As such, as we have seen, K has a dual space, that we denote by K'. Moreover, we note that, *restricting the bilinear form* $a(\cdot,\cdot)$ *to* K, we have two operators, which, according to the notation (4.1.84), we denote by $A_{KK'}$ and $A^t_{KK'}$, from K to K', given as in (4.2.3) by

$$\langle A_{KK'}u_0, v_0\rangle_{K'\times K} = \langle u_0, A^t_{KK'}v_0\rangle_{K\times K'} = a(u_0, v_0),\ \forall\, u_0, v_0 \in K. \tag{4.2.16}$$

We also recall that K' could be identified, through (4.1.70), to a subspace of Q', and precisely to $(K^\perp)^0$ (the polar space of $K^\perp$). Moreover, it is easy to check that

$$A_{KK'} = \pi_{K'} A\, E_K \tag{4.2.17}$$

coincides, in the finite dimensional case, with the operator that (identifying K and K') was denoted by A_{KK} in the previous chapter.

We are now ready to state and prove the following theorem, which is, from the theoretical point of view, the most relevant of this section. As we shall see, it generalises Theorem 4.2.1 and gives the required *necessary and sufficient conditions*.

Theorem 4.2.2. *Assume that $\mathcal{AB}$ holds, and let $A_{KK'}$ be defined as in* (4.2.16). *Then, problem* (4.2.6) *has a unique solution for every $(f,g) \in V' \times Q'$* ***if and only if*** *the two following conditions are satisfied:*

$$A_{KK'} \text{ is an isomorphism from } K \text{ to } K', \tag{4.2.18}$$

$$ImB = Q'. \tag{4.2.19}$$

Proof. Assume first that (4.2.18) and (4.2.19) are satisfied. The existence and uniqueness of the solution of (4.2.6) follow as in the proof of Theorem 4.2.1. The only difference is in the solution of (4.2.13), which in the present notation can be written as

$$A_{KK'}u_0 = \pi_{K'}f - \pi_{K'}Au_g. \tag{4.2.20}$$

Indeed, here we now have to use (4.2.18) (in order to get the existence of a solution u_0) instead of Lax-Milgram as we did there.

Assume conversely that the problem (4.2.6) has a unique solution for every $(f, g) \in V' \times Q'$. It is clear that, in particular for every $g \in Q'$, we can take $(0, g)$ as right-hand side in (4.2.6) and have a $u \in V$ such that $Bu = g$ (from the second equation of (4.2.7)). Hence, $\mathrm{Im}B = Q'$ and therefore (4.2.19) holds. To show that (4.2.18) also holds, we proceed as follows.

First, *for every* $\phi \in K'$, we take in (4.2.7) $f = E_{K' \to V'}\phi$ (as defined in (4.1.71)), and $g = 0$. By assumption, we have a unique solution (u_ϕ, p_ϕ), and we observe that $u_\phi \in K$ since $g = 0$. Testing the first equation of problem (4.2.6) on $v_0 \in K$, and using (4.1.71), we have

$$a(u_\phi, v_0) = \langle f_\phi, v_0 \rangle_{K' \times K} \equiv \langle \phi, v_0 \rangle_{K' \times K} \qquad \forall\, v_0 \in K. \tag{4.2.21}$$

This implies that $A_{KK'}u_\phi = \phi$, and hence that $A_{KK'}$ is *surjective*. Hence, we are left to show that $A_{KK'}$ is also *injective*. Assume, by contradiction, that we had $A_{KK'}w = 0$ for some $w \in K$ *different from zero*. Then, we would have $a(w, v_0) = 0$ for all $v_0 \in K$, implying that $Aw \in K^0$. Due to Corollary 4.1.1 (as we already saw that (4.2.19) holds), this would imply $Aw \in \mathrm{Im}B^t$ and we would have the existence of a $p_w \in Q$ such that $B^t p_w = Aw$. Then, the pair $(w, -p_w)$ (different from zero) would satisfy the homogeneous version of problem (4.2.6), and uniqueness would be lost. Hence, such a $w \neq 0$ cannot exist. This shows that $A_{KK'}$ must also be injective, and hence (4.2.18) holds. □

4.2.2 *Stability Constants and* inf-sup *Conditions*

In this subsection, we would like to express condition (4.2.19) and condition (4.2.18) in a different way, to emphasise the role of the stability constants.

Let us start from condition (4.2.19). According to Corollary 4.1.1 of the Closed Range Theorem, we already know that (4.2.19) holds if and only if the operator B^t is bounding, that is, if and only if there exists a constant $\beta > 0$ such that

$$\|B^t q\|_{V'} \geq \beta \|q\|_Q \qquad \forall\, q \in Q. \tag{4.2.22}$$

Always from the same corollary, we have that this is also equivalent to the existence of a lifting $L_B : Q' \to V$ of the operator B

$$B(L_B(g)) = g \qquad \forall\, g \in Q', \tag{4.2.23}$$

with its norm being bounded by:

$$\|L_B\|_{\mathcal{L}(Q',V)} \leq \frac{1}{\beta}, \tag{4.2.24}$$

where β is the same constant as in (4.2.22) and $\mathrm{Im}L_B = K^{\perp}$.

We now want to define, somehow, the *best possible constant that would fit in* (4.2.22). For this, we note that (4.2.22) is equivalent to

$$\inf_{q\in Q} \frac{\|B^t q\|_{V'}}{\|q\|_Q} \geq \beta, \tag{4.2.25}$$

which, recalling the definition of norm in a dual space (4.1.34) and (4.1.49), becomes

$$\inf_{q\in Q} \sup_{v\in V} \frac{b(v,q)}{\|v\|_V \, \|q\|_Q} \geq \beta, \tag{4.2.26}$$

which is possibly the most commonly used among the many equivalent formulations of assumption (4.2.19).

With similar arguments, we see that condition (4.2.18) is equivalent to saying that there exists an $\alpha_1 > 0$ such that

$$\begin{aligned} &\inf_{v_0\in K} \sup_{w_0\in K} \frac{a(v_0,w_0)}{\|v_0\|_V \, \|w_0\|_V} \geq \alpha_1 \\ &\inf_{w_0\in K} \sup_{v_0\in K} \frac{a(v_0,w_0)}{\|v_0\|_V \, \|w_0\|_V} \geq \alpha_1. \end{aligned} \tag{4.2.27}$$

Remark 4.2.4. Note that in (4.2.25), in (4.2.26), and in (4.2.27), as we did in the previous chapter and we shall do in the rest of the book, we assumed *implicitly* that for fractions of the type

$$\frac{\ell(v)}{\|v\|_V} \quad \text{or} \quad \frac{|\ell(v)|}{\|v\|_V}$$

where $\ell(\cdot)$ is a linear functional on a Banach space V, the supremum and the infimum are taken for $v \neq \{0\}$, and therefore we wrote the supremum (or infimum) for $v \in V$ rather than for $v \in V \setminus \{\mathbf{0}\}$ (as it would have been more correct, since these fractions do not make sense for $v = \{\mathbf{0}\}$). □

We now want to point out, for future use, the following extension of Lemma 3.3.1 of the previous chapter, that is an important ingredient in the proof of the present Theorem 4.2.3.

Lemma 4.2.1. *Let V be a Hilbert space and let $a(\cdot\,,\cdot)$ be a symmetric bilinear continuous form on V. Assume that*

$$a(v,v) \geq 0 \qquad \forall\, v \in V. \tag{4.2.28}$$

Then, we have

$$(a(v,w))^2 \leq a(v,v)\, a(w,w) \qquad \forall\, v,w \in V \tag{4.2.29}$$

and, for the associated operator A,

$$\|Av\|^2_{V'} \leq \|a\|\, a(v,v) \equiv \|A\|\langle Av,v\rangle \qquad \forall\, v \in V. \tag{4.2.30}$$

Apart from the different notation, the proof is identical to that of Lemma 3.3.1.

Moreover, under the assumptions of the previous lemma, we also have the following result.

Lemma 4.2.2. *Let V be a Hilbert space, and let $a(\cdot\,,\cdot)$ be a symmetric bilinear continuous form on V. Assume that*

$$a(v,v) \geq 0 \qquad \forall\, v \in V. \tag{4.2.31}$$

Then, (4.2.27) *implies ellipticity on the kernel* (4.2.12).

Proof. Indeed, from (4.2.27), we have for $v_0 \in K$, using (4.2.29),

$$\alpha_1^2\|v\|_V^2 \leq \sup_{w\in K}\frac{a(v,w)^2}{\|w\|_V^2} \leq \sup_{w\in K}\frac{a(v,v)a(w,w)}{\|w\|_V^2} \leq \|a\|a(v,v), \tag{4.2.32}$$

hence the result with $\alpha_0 = \alpha_1^2/\|a\|$. □

4.2.3 The Main Result

As we had in the previous chapter (in Theorems 3.4.1 and 3.4.2), we have here the following final result, that could be considered as *the main result of this chapter.*

Theorem 4.2.3. *Together with $\mathcal{AB}$, assume that there exist two positive constants α and β such that the inf-sup condition* (4.2.26) *on $b(\cdot,\cdot)$, and the double-inf-sup condition* (4.2.27) *on the restriction of $a(\cdot,\cdot)$ to K are satisfied. Then, for every $f \in V'$ and for every $g \in Q'$, problem* (4.2.6) *has a unique solution that is bounded by*

$$\|u\|_V \leq \frac{1}{\alpha_1}\|f\|_{V'} + \frac{2\|a\|}{\alpha_1\beta}\|g\|_{Q'}, \tag{4.2.33}$$

$$\|p\|_Q \leq \frac{2\|a\|}{\alpha_1\beta}\|f\|_{V'}F + \frac{2\|a\|^2}{\alpha_1\beta^2}\|g\|_{Q'}. \tag{4.2.34}$$

If, moreover, $a(\cdot,\cdot)$ *is symmetric and satisfies*

$$a(v, v) \geq 0 \qquad \forall\, v \in V, \tag{4.2.35}$$

then we have the improved estimates

$$\|u\|_V \leq \frac{1}{\alpha_0}\|f\|_{V'} + \frac{2\|a\|^{1/2}}{\alpha_0^{1/2}\beta}\|g\|_{Q'}, \tag{4.2.36}$$

$$\|p\|_Q \leq \frac{2\|a\|^{1/2}}{\alpha_0^{1/2}\beta}\|f\|_{V'} + \frac{\|a\|}{\beta^2}\|g\|_{Q'}, \tag{4.2.37}$$

where α_0 *is the constant appearing in* (4.2.12)*.*

The proof is identical to that given in the previous chapter for Theorems 3.4.1 and 3.4.2. This is indeed the gift of the Closed Range Theorem, which allows us to extend all the instruments that were used in finite dimension to the more general case of Hilbert spaces.

Remark 4.2.5. As we did in Theorem 3.5.2, we could restate Theorem 4.2.2 in terms of necessary and sufficient conditions. In the present context, this means that if the bounds (4.2.33) and (4.2.34) hold for all right-hand sides f and g, then (4.2.27) and (4.2.26) hold. Indeed, for an arbitrary $u_0 \in K$, let us define $f_0 \in K'$ by

$$\langle f_0, v_0 \rangle = a(u_0, v_0) \quad \forall\, v_0 \in K. \tag{4.2.38}$$

We then use the prolongation $E_{K'} f_0$ of f_0 to V', as in (4.1.71), and we take $g = 0$. We now have that $(u_0, 0)$ is solution of (4.2.6) with $f = f_0$, and by (4.2.33) we have

$$\|u\|_V \leq \frac{1}{\alpha_1}\|f_0\|_{V'} = \frac{1}{\alpha_1} \sup_{w_0 \in K} \frac{a(u_0, w_0)}{\|w_0\|}. \tag{4.2.39}$$

Similarly, taking $\langle f_p, v \rangle = b(v, p)$ and again $g = 0$, we have that $(0, p)$ is solution of (4.2.6) with $f = f_p$, and (4.2.34) implies (4.2.26). All this can be seen as the natural extension of Lemma 3.5.2 to the infinite dimensional case. □

If the bilinear form $a(\cdot,\cdot)$ is coercive on the whole space, we have immediately the following corollary (particularly useful for Stokes addicts that do not even want to know what a kernel is).

Corollary 4.2.1. *Let the assumptions* $\mathcal{AB}$ *hold. Suppose that there exist two positive constants* α *and* β *such that the inf-sup condition* (4.2.26) *on* $b(\cdot,\cdot)$*, and the global coercivity condition* (4.1.67) *on* $a(\cdot,\cdot)$ *are satisfied. Then, for every* $f \in V'$ *and for every* $g \in Q'$*, problem* (4.2.6) *has a unique solution that is bounded by*

$$\|u\|_V \leq \frac{1}{\alpha}\|f\|_{V'} + \frac{2\|a\|}{\alpha\beta}\|g\|_{Q'}, \tag{4.2.40}$$

$$\|p\|_Q \leq \frac{2\|a\|}{\alpha\beta}\|f\|_{V'} + \frac{2\|a\|^2}{\alpha\beta^2}\|g\|_{Q'}. \tag{4.2.41}$$

If, moreover, $a(\cdot,\cdot)$ *is symmetric, then we have the improved estimates*

$$\|u\|_V \leq \frac{1}{\alpha}\|f\|_{V'} + \frac{2\|a\|^{1/2}}{\alpha^{1/2}\beta}\|g\|_{Q'}, \tag{4.2.42}$$

$$\|p\|_Q \leq \frac{2\|a\|^{1/2}}{\alpha^{1/2}\beta}\|f\|_{V'} + \frac{\|a\|}{\beta^2}\|g\|_{Q'}. \tag{4.2.43}$$

4.2.4 The Case of $\mathrm{Im} B \neq Q'$

We now want to discuss briefly the case in which the *inf-sup* condition on the bilinear form b does not hold.

Essentially, if $\mathrm{Im}B$ does not coincide with Q', we can distinguish two cases. Either $\mathrm{Im}B$ is closed in Q', or it is not (everybody surely agrees with that).

If $\mathrm{Im}B$ is not closed, then we are in *deep trouble*. Generally speaking, *we should better look for a different formulation*.

If instead the image of B is closed, we survive rather easily. Let us analyse the situation. We first observe that in this case $H = \mathrm{Ker}B^t$ will be a closed subspace of Q that is *not* reduced to $\{0\}$. In this case, it is clear that problem (4.2.7) *cannot* have a unique solution for every $f \in V'$ and for every $g \in Q'$. To start with, if $\mathrm{Im}B \neq Q'$, and if $g \in Q'$ does not belong to $\mathrm{Im}B$, we cannot have a solution. Hence, the *existence* of the solution will *not always* hold. Moreover, if by chance we have $g \in \mathrm{Im}B$ and we have a solution (u, p), then for every $p^* \in H$ with $p^* \neq 0$, we easily have that $(u, p + p^*)$ is another, different solution. Hence, the *uniqueness* of the solution will *never* hold. Apparently, we are not so well off.

However, if we have $g \in \mathrm{Im}B$, then there is an easy way out. Indeed, we observe first that if A_{KK} is non-singular, we could proceed as we did in the finite dimensional case (see Proposition 3.2.1) and deduce that we still have at least one solution, whose first component is unique and whose second component is unique only up to an element of H. Moreover, we could note that $b(v, q) = 0$ for every $q \in H$. Hence, following what has been done in Remark 3.2.4 (for the finite-dimensional case), we can consider the restriction $\tilde{b}$ of b to $V \times H^{\perp}$ without loosing any information. However, this time, $\tilde{B}$ will be surjective from V to $(H^{\perp})'$. Indeed, using (4.1.70) we have that $(H^{\perp})' = H^0$ $(= (\mathrm{Ker}B^t)^0)$. On the other hand, from the Closed Range Theorem 4.1.5, we have $(\mathrm{Ker}B^t)^0 = \mathrm{Im}B$, and joining the two we get $(H^{\perp})' = \mathrm{Im}B$ and everything works.

Hence, *the theory developed so far in the case of B surjective applies to the case where* $\mathrm{Im}B$ *is closed and* $g \in \mathrm{Im}B$, *by just replacing* Q *with* $H^{\perp}$.

An alternative path (whose difference from the one above is mainly psychological) consists in replacing Q with the *quotient space*

$$\tilde{Q} := Q_{/H}. \tag{4.2.44}$$

We recall that the elements of $\tilde{Q}$ are subsets of elements of Q that differ from each other by an element of H. As we have seen in Sect. 4.1.9, $\tilde{Q}$ can be identified to $H^{\perp}$. Hence, as we said, the difference between using $V \times H^{\perp}$ and using $V \times Q_{/H}$ is mainly psychological. Nevertheless, some people seem to be in love with this second option and dislike the first. Just for them, we re-state one of our previous results in terms of the original space Q and the original bilinear form b in the following theorem, which is just Theorem 4.2.3 applied to $V \times Q_{/H}$, and stated in terms of V and Q.

Theorem 4.2.4. *Together with Assumption $\mathcal{AB}$, assume that ImB is closed and that the double-inf-sup condition* (4.2.27) *is satisfied. Then, for every $f \in V'$ and for every $g \in ImB$, problem* (4.2.6) *has a solution (u, p) where u is uniquely determined, and p is determined up to an element of H. Moreover, setting*

$$\tilde{\beta} := \inf_{q \in Q} \sup_{v \in V} \frac{b(v,q)}{\|v\|_V \, \|q\|_{Q/H}} \tag{4.2.45}$$

we have

$$\|u\|_V \leq \frac{1}{\alpha_1}\|f\|_{V'} + \frac{2\|a\|}{\alpha_1 \tilde{\beta}}\|g\|_{Q'}, \tag{4.2.46}$$

$$\|p\|_{Q/H} \leq \frac{2\|a\|}{\alpha_1 \tilde{\beta}}\|f\|_{V'} + \frac{2\|a\|^2}{\alpha_1 \tilde{\beta}^2}\|g\|_{Q'}. \tag{4.2.47}$$

If, moreover, $a(\cdot,\cdot)$ is symmetric and satisfies

$$a(v, v) \geq 0 \qquad \forall\, v \in V, \tag{4.2.48}$$

then we have the improved estimates

$$\|u\|_V \leq \frac{1}{\alpha_0}\|f\|_{V'} + \frac{2\|a\|^{1/2}}{\alpha_0^{1/2}\tilde{\beta}}\|g\|_{Q'}, \tag{4.2.49}$$

$$\|p\|_{Q/H} \leq \frac{2\|a\|^{1/2}}{\alpha_0^{1/2}\tilde{\beta}}\|f\|_{V'} + \frac{\|a\|}{\tilde{\beta}^2}\|g\|_{Q'}, \tag{4.2.50}$$

where again α_0 is the constant appearing in (4.2.12). □

Remark 4.2.6. We point out that the estimates (4.2.46) and (4.2.47), valid for every $f \in V'$ and for every $g \in \mathrm{Im}B$, imply in particular that, under the assumptions of Theorem 4.2.4, the image of the operator $\mathbb{M} : (u, p) \to (Au + B^t p, Bu)$ from $V \times Q$ to $V' \times Q'$ *is also closed.* □

Remark 4.2.7. Another type of generalisation was considered in [312] and [68]. They consider a problem of type (4.3.1) but employing two bilinear forms $b_1(\cdot,\cdot)$ and $b_2(\cdot,\cdot)$ on $V \times Q$, that is,

$$\begin{cases} a(u,v) + b_1(v,p) = \langle f, v\rangle_{V'\times V}, \ \forall\, v \in V \\ b_2(u,q) = \langle g, q\rangle_{Q'\times Q}, \ \forall\, q \in Q. \end{cases} \tag{4.2.51}$$

Conditions for existence of a solution are now that both $b_1(\cdot,\cdot)$ and $b_2(\cdot,\cdot)$ should satisfy an inf-sup condition of the type (4.2.45), and $a(u,v)$ should satisfy an invertibility condition from $\mathrm{Ker}B_2$ on $(\mathrm{Ker}B_1)'$, that is,

$$\inf_{u_0\in\mathrm{Ker}B_1} \sup_{v_0\in\mathrm{Ker}B_2} \frac{a(u_0,v_0)}{\|u_0\|\,\|v_0\|} \ge \alpha_1, \tag{4.2.52}$$

$$\inf_{v_0\in\mathrm{Ker}B_1} \sup_{u_0\in\mathrm{Ker}B_2} \frac{a(u_0,v_0)}{\|u_0\|\,\|v_0\|} \ge \alpha_1. \tag{4.2.53}$$

This condition is in general rather hard to check, and the ellipticity on the whole space V, when applicable, can bring a considerable relief.

For more details, we refer to [68]. □

Remark 4.2.8 (Special cases $(\mathbf{f},0)$ *and* $(0,\mathbf{g})$*).* We have considered these special cases in the Sect. 3.5.3 in the finite dimensional framework. In these cases, it is possible to obtain existence and stability results under weakened assumptions. We shall not make them explicit here. However, we refer to the proofs of Theorem 3.4.1 and the following ones in Sect. 3.4 where detailed proofs of related situations are presented. We just want to point out here that in the case $(\mathbf{f},0)$, the a priori estimates (e.g. (4.2.46)) on u do not depend on the inf-sup constant of B. Conversely, in the case $(0,\mathbf{g})$, for $a(\cdot,\cdot)$ symmetric and positive semi-definite, the estimates on p (e.g. (4.2.50)) do not depend on the constant α_0. □

4.2.5 Examples

To fix ideas, we shall apply the results just obtained to some of the examples introduced in Chap. 1.

Example 4.2.1 (Mixed formulation of the Poisson problem). We consider here the case of Example 1.3.5. Given f in $L^2(\Omega)$, we look for $\underline{u} \in H(\mathrm{div};\Omega) =: V$ *and* $p \in L^2(\Omega) =: Q$ such that:

$$\begin{cases} \displaystyle\int_\Omega \underline{u}\cdot\underline{v}\,dx + \int_\Omega p \ \mathrm{div}\ \underline{v}\,dx = 0, \ \forall\, \underline{v} \in H(\mathrm{div};\Omega), \\ \displaystyle\int_\Omega (\mathrm{div}\ \underline{u} + f) q\,dx = 0, \ \forall\, q \in L^2(\Omega). \end{cases} \tag{4.2.54}$$

Here we have

$$b(\underline{v}, q) = \int_\Omega \operatorname{div} \underline{v}\, q\, dx, \tag{4.2.55}$$

and B is the divergence operator from $H(\operatorname{div};\Omega)$ into $L^2(\Omega)$. It is not difficult to check that it is surjective: for instance, for every $g \in L^2(\Omega)$, consider the auxiliary problem: *find* $\psi \in H_0^1(\Omega)$ *such that* $\Delta\psi = g$. Its (traditional) variational formulation is

$$\int_\Omega \underline{\operatorname{grad}}\, \psi \cdot \underline{\operatorname{grad}}\, \phi\, dx = -\int_\Omega g\, \phi\, dx \qquad \forall\, \phi \in H_0^1(\Omega), \tag{4.2.56}$$

and it has a unique solution thanks to the Lax-Milgram lemma. Then, take $\underline{v}_g := \underline{\operatorname{grad}}\,\psi$ and you immediately have $\underline{v}_g \in H(\operatorname{div};\,\Omega)$ and $\operatorname{div} \underline{v}_g = g$ as wanted. The kernel of B is made of the vectors $\underline{v}_0 \in H(\operatorname{div};\Omega)$ such that $\operatorname{div} \underline{v}_0 = 0$. The bilinear form a is given by

$$a(\underline{u}, \underline{v}) = \int_\Omega \underline{u} \cdot \underline{v}\, dx, \tag{4.2.57}$$

while we remember that in (1.3.44) the norm in $H(\operatorname{div};\Omega)$ was defined as

$$\|\underline{v}\|^2_{H(\operatorname{div};\,\Omega)} := \|\underline{v}\|^2_{(L^2(\Omega))^2} + \|\operatorname{div} \underline{v}\|^2_{L^2(\Omega)}. \tag{4.2.58}$$

Hence, a is coercive on $\operatorname{Ker} B$ (although it is *not coercive* on $H(\operatorname{div};\Omega)$). Our abstract theory (in particular Theorem 4.2.1) applies immediately, and we have existence and uniqueness of the solution. □

Example 4.2.2 (The Stokes problem). Let us go back to Example 1.3.1. We take $V := (H_0^1(\Omega))^2$, $Q := L^2(\Omega)$, and, given $\underline{f} \in V'$, we look for $(\underline{u}, p) \in V \times Q$, solution of

$$\begin{cases} 2\mu \displaystyle\int_\Omega \underline{\underline{\varepsilon}}(\underline{u}) : \underline{\underline{\varepsilon}}(\underline{v})\, dx - \int_\Omega p\ \operatorname{div} \underline{v}\, dx = \int_\Omega \underline{v} \cdot \underline{f}\, dx,\ \forall\, v \in V, \\ \displaystyle\int_\Omega q\ \operatorname{div} \underline{u}\, dx = 0,\ \forall\, q \in Q. \end{cases} \tag{4.2.59}$$

Here, we have $g = 0$. Moreover, the bilinear form $a(u, v) = 2\mu \int_\Omega \underline{\underline{\varepsilon}}(\underline{u}) : \underline{\underline{\varepsilon}}(\underline{v})\, dx$ is coercive on V, due to the *Korn inequality* [183, 362]

$$\exists\, \kappa = \kappa(\Omega) > 0 \text{ s.t. } \|\underline{\underline{\varepsilon}}(\underline{v})\|^2_{(L^2(\Omega))^4} \geq \kappa |\underline{v}|^2_{1,\Omega} \qquad \forall \underline{v} \in (H_0^1(\Omega))^2. \tag{4.2.60}$$

On the other hand, we have

$$b(\underline{v}, q) = -\int_\Omega q\ \operatorname{div} \underline{v}\, dx \tag{4.2.61}$$

and B is the divergence operator from $(H_0^1(\Omega))^2$ into $L^2(\Omega)$. This time, the study of its image is much harder than in the previous example. Due to a non-trivial result by O. Ladyzhenskaya, we have [272, 362] that

$$\mathrm{Im}B = L_0^2(\Omega) = \{q \mid q \in L^2(\Omega), \int_\Omega q\, dx = 0\} \tag{4.2.62}$$

and that this subspace of $L^2(\Omega)$ is closed and has co-dimension one. In agreement with the Closed Range Theorem, $\mathrm{Ker}B^t$ has also dimension 1:

$$\mathrm{Ker}B^t = \mathrm{Ker}(-\underline{\mathrm{grad}}) = \{q \mid q \text{ is constant on } \Omega\}. \tag{4.2.63}$$

We are therefore in the case where B^t is not injective. As we did in the last subsection (see Sect. 4.2.4), we can easily survive by considering $\tilde{Q}$, defined as in (4.2.44), instead of Q. However, in this case, the space $\tilde{Q}$ (that is the space of *classes* of functions in $L^2(\Omega)$ that differ from each other by an additive constant) is often identified with the space $H^\perp$ of functions in $L^2(\Omega)$ having zero mean value, as discussed in Example 4.1.14. Actually, in practice, we simply take

$$Q := \left\{q | q \in L^2(\Omega), \int_\Omega q\, dx = 0\right\} \equiv L_0^2(\Omega) \tag{4.2.64}$$

and we can apply directly Theorem 4.2.1, which will give the existence and uniqueness of the velocity $\underline{u}$, together with the existence of a pressure p that is unique up to an additive constant.

The example of Stokes' problem is paradigmatic of the typical escape that is usually performed when $\mathrm{Im}B$ is closed but different from Q'. □

Example 4.2.3 (Domain decomposition for the Poisson problem). Referring to Example 1.4.2, we have to solve the following problem: find $(p, \underline{u})$ with $p \in X(\Omega) =: V$, $\underline{u} \in H(\mathrm{div}; \Omega) =: Q$, solution of

$$\begin{cases} \displaystyle\int_{K_i} \underline{\mathrm{grad}}\, p_i \cdot \underline{\mathrm{grad}}\, q_i\, dx - \int_{\partial K_i} \underline{u} \cdot \underline{n}_i\, q_i\, d\sigma = \int_{K_i} f\, q_i\, dx, \\ \qquad\qquad\qquad\qquad\qquad\qquad \forall\, q_i \in H^1(K_i),\ \forall\, K_i, \\ \displaystyle\sum_i \int_{\partial K_i} \underline{v} \cdot \underline{n}_i\, p_i\, d\sigma = 0, \forall \underline{v} \in H(\mathrm{div}; \Omega). \end{cases} \tag{4.2.65}$$

We thus have $b(q, \underline{v}) = -\sum_i \int_{\partial K_i} \underline{v} \cdot \underline{n}_i\, q_i\, d\sigma$, and the operator B, roughly speaking, associates to $q \in X(\Omega)$ its "DG jumps" $q_i \underline{n}_i + q_j \underline{n}_j$ on the interfaces $e_{ij} = \partial K_i \cap \partial K_j$. The kernel of B is nothing but $H_0^1(\Omega)$ and the problem corresponding to (4.2.65) is the standard Poisson problem. To prove the existence of u, we shall have to prove that $\mathrm{Im}B$ is closed in $(H(\mathrm{div}; \Omega))'$ and we shall have to characterise $\mathrm{Ker}B^t$. This will be done in Chap. 7. □

We shall of course come back to these problems when studying more precisely mixed and hybrid methods. Checking the closedness of $\mathrm{Im}B$, even if existence proofs can be obtained through other considerations, is a crucial step ensuring that we have a well-posed problem and that we are working with the right functional spaces. This last fact is essential to obtain "natural" error estimates.

We end this subsection with a few rather academic examples, just in order to see formulations that *do not* work (or present some sort of difficulty) and understand why.

We shall consider the problem (*very loosely* related to plate bending problems, as in Example 1.2.4, or to the Stokes problem in the so-called *streamline-vorticity* formulation):

$$\Delta^2 \psi = f \qquad \text{in } \Omega \tag{4.2.66}$$

on a reasonably smooth domain Ω (for instance, a convex polygon). We introduce $\omega := -\Delta\psi$, and we are going to consider *various boundary conditions, and different possible mixed formulations.*

- We start with the easiest choice of boundary conditions, that is

$$\psi = \omega = 0 \quad \text{on } \Gamma. \tag{4.2.67}$$

In this case, we can set $V \equiv Q := H^1_0(\Omega)$, and consider the formulation

$$\begin{cases} \displaystyle\int_\Omega \omega\, \mu\, dx - \int_\Omega \underline{\mathrm{grad}}\, \mu\ \underline{\mathrm{grad}}\, \psi\, dx = 0, & \forall\, \mu \in V, \\ \displaystyle -\int_\Omega \underline{\mathrm{grad}}\, \omega\ \underline{\mathrm{grad}}\, \varphi\, dx = -\langle f, \varphi\rangle, & \forall\, \varphi \in Q. \end{cases} \tag{4.2.68}$$

In this case, both the operators B and B^t coincide with the Laplace operator $\Delta : H^1_0(\Omega) \to H^{-1}(\Omega)$, which is an isomorphism. In particular, $\mathrm{Im}B = Q'$ and $\mathrm{Ker}B = \{0\}$, so that the ellipticity in the kernel (4.2.12) is also trivially satisfied. All is well and good. However, one could object that, with the boundary conditions (4.2.67), we are *almost cheating*. Indeed, the problem is equivalent to the cascade of sub-problems: $-\Delta\omega = f$ and $-\Delta\psi = \omega$ which are both well posed if we look for $\omega \in H^1_0$ and $\psi \in H^1_0$.
- We now consider the "clamped plate" boundary conditions

$$\psi = \frac{\partial \psi}{\partial n} = 0 \quad \text{on } \Gamma. \tag{4.2.69}$$

Setting $V := L^2(\Omega)$ and $Q := H^2_0(\Omega)$, it is immediate to see that (ω, ψ) satisfies the equations

$$\begin{cases} \displaystyle\int_\Omega \omega\, \mu\, dx + \int_\Omega \mu\, \Delta\psi\, dx = 0, & \forall\, \mu \in V, \\ \displaystyle\int_\Omega \omega\, \Delta\varphi\, dx = -\langle f, \varphi\rangle, & \forall\, \varphi \in Q. \end{cases} \tag{4.2.70}$$

Here, the operator B^t is just the Laplace operator from $H_0^2(\Omega)$ to $L^2(\Omega)$, and it is clearly injective and bounding, since $\|\varphi\|_{2,\Omega} \leq C\,\|\Delta\varphi\|_{0,\Omega}$ for some constant C. Hence, the image of B coincides with Q'. The kernel of B is a little more sophisticated. With some work, we discover that it can be characterised as

$$\mathrm{Ker} B = (L^2_{harm})^{\perp}, \tag{4.2.71}$$

where L^2_{harm} is the (closed) subspace of $L^2(\Omega)$ made of harmonic functions (that is, functions w such that $\Delta w = 0$ in the distributional sense). Indeed, for such functions, it is not difficult to see that $(w, \Delta\varphi) = 0$ for all $\varphi \in H_0^2(\Omega)$. In any case, we don't care too much about $\mathrm{Ker} B$, since the bilinear form a is coercive on the whole V. Our theory applies, and we are happy.

- Still with the "clamped plate" boundary conditions (4.2.69), if we are not willing to use spaces involving two derivatives (as $H_0^2(\Omega)$), we could take $V := H^1(\Omega)$ and $Q := H_0^1(\Omega)$. It is not difficult to see that (ω, ψ) solves

$$\begin{cases} \displaystyle\int_\Omega \omega\,\mu\,dx - \int_\Omega \underline{\mathrm{grad}}\,\mu\ \underline{\mathrm{grad}}\,\psi\,dx = 0, & \forall\,\mu \in V,\\ \displaystyle -\int_\Omega \underline{\mathrm{grad}}\,\omega\ \underline{\mathrm{grad}}\,\varphi\,dx = -\langle f,\varphi\rangle, & \forall\,\varphi \in Q. \end{cases} \tag{4.2.72}$$

This time, B is the Laplace operator from $H^1(\Omega)$ to $H^{-1}(\Omega)$ (dual space of H_0^1), which is clearly surjective. However, its kernel is made of the harmonic functions in $H^1(\Omega)$ and the bilinear form a (which in this case is just the L^2-inner product) cannot be coercive (in $H^1(\Omega)$), not even if you restrict it to the harmonic functions. If you are not convinced of that, consider in $\Omega :=]0,\pi[\times]0,1[$ the sequence of functions

$$\phi_k := \sin(kx)\,e^{ky}.$$

Clearly, $\Delta\phi_k = 0$ for all k. However, a simple computation shows that

$$\|\,\underline{\mathrm{grad}}\,\phi_k\|^2_{0,\Omega} = 2k^2\|\phi_k\|^2_{0,\Omega}$$

and you cannot bound $\|\phi_k\|^2_V$ (that is $\|\,\underline{\mathrm{grad}}\,\phi_k\|^2_{0,\Omega}$) with $a(\phi_k,\phi_k)$ (that is $\|\phi_k\|^2_{0,\Omega}$) uniformly in k. Hence, formulation (4.2.72) is not really healthy, as the ellipticity in the kernel fails. Indeed, if for instance the domain Ω is not convex, you are likely to have a problem without existence, as ψ usually will not be in $H^3(\Omega)$ and therefore ω might not be in $H^1(\Omega)$. We shall see in the following chapters that methods based on this formulation might exhibit a suboptimal rate of convergence.

- We now consider the boundary conditions

$$\frac{\partial\psi}{\partial n} = \frac{\partial\omega}{\partial n} = 0 \quad \text{on } \Gamma. \tag{4.2.73}$$

Here, we can take $V := L^2(\Omega)$ and $Q := {}_0H^2(\Omega)$ defined as

$$ {}_0H^2(\Omega) := \{\varphi \mid \varphi \in H^2(\Omega),\ \frac{\partial \varphi}{\partial n} = 0 \text{ on } \Gamma\} $$

and use the formulation (4.2.70). We see that B^t is again the Laplace operator, but this time ${}_0H^2(\Omega) \to (L^2(\Omega))' \equiv L^2(\Omega)$. If, for instance, the domain Ω is convex, then $H := \mathrm{Ker}B^t$ is the space of constants, and $\mathrm{Im}B^t$ will be the subset of $V' \equiv V = L^2(\Omega)$ made of functions with zero mean value. The kernel of B will also be the space of constants, and the image of B will be the polar set of $\mathrm{Ker}B^t$, made of those functionals that vanish on constants. We are in a situation similar to that faced for the Stokes problem in Example 4.2.2. Here, we can adjust everything by redefining Q as the subset of ${}_0H^2(\Omega)$ made of functions with zero mean value (that is, $Q = H^\perp$). The compatibility condition $\langle f, c\rangle = 0$ for every constant c will still have to be required, in order to have $f \in Q'$ (now $= (H^\perp)'$). Doing that, we have that a is elliptic on V and $\mathrm{Im}B = Q'$ (the new Q', of course), and everything will work.

- It is time to see a *really weird* case. Consider, to fix the ideas, the case of $\Omega :=]0,\pi[\times]0,1[$, and split its boundary into the bottom part $\Gamma_b :=]0,\pi[\times\{0\}$, the top part $\Gamma_t :=]0,\pi[\times\{1\}$, and the lateral part $\Gamma_\ell := \partial\Omega \setminus (\Gamma_b \cup \Gamma_t)$. Consider now the boundary conditions

$$ \psi = \frac{\partial \psi}{\partial n} = 0 \ \text{ on } \Gamma_b; \quad \omega = \frac{\partial \omega}{\partial n} = 0 \ \text{ on } \Gamma_t; \quad \frac{\partial \psi}{\partial n} = \frac{\partial \omega}{\partial n} = 0 \ \text{ on } \Gamma_\ell \tag{4.2.74} $$

and the spaces $V := L^2$ and $Q := \tilde{H}^2$ defined as

$$ \tilde{H}^2 := \{\varphi \mid \varphi \in H^2(\Omega),\ \varphi = \frac{\partial \varphi}{\partial n} = 0 \ \text{ on } \Gamma_b \text{ and } \frac{\partial \varphi}{\partial n} = 0 \text{ on } \Gamma_\ell\}. $$

It is clear that, if you have a solution (ω, ψ) of the problem, then it will satisfy

$$ \begin{cases} \displaystyle\int_\Omega \omega\,\mu\,dx + \int_\Omega \mu\,\Delta\psi\,dx = 0, & \forall\,\mu \in V, \\ \displaystyle\int_\Omega \omega\,\Delta\varphi\,dx = -\langle f, \varphi\rangle, & \forall\,\varphi \in Q. \end{cases} \tag{4.2.75} $$

This time, B^t will be the Laplace operator from $\tilde{H}^2$ to $L^2(\Omega)$. A non-trivial result of complex analysis (Cauchy-Kovalewskaya Theorem) ensures that $\mathrm{Ker}B^t = \{0\}$ (the boundary conditions at the bottom are enough to give you that). However, we can check that B^t *is not bounding*. To see that, consider the sequence

$$ \phi_k := \frac{1}{k^2}\cos(kx)(1 - \cosh(ky)). $$

It is clear that $\phi_k \in \tilde{H}^2$ for all k. A simple computation shows that $-\Delta\phi_k = \cos(kx)$, so that

$$\|\Delta\phi_k\|^2_{L^2(\Omega)} = \frac{\pi}{2}.$$

On the other hand, it is also simple to check that

$$\|\phi_k\|^2_{L^2(\Omega)} \simeq \frac{e^{2k}}{k^4}$$

goes to $+\infty$ for $k \to +\infty$, so that a uniform bound (in k) of the form

$$\|\Delta\varphi\|_{L^2(\Omega)} \geq \beta\|\varphi\|_{\tilde{H}^2} \qquad \forall\varphi \in \tilde{H}^2$$

is hopeless. Hence, $\mathrm{Im}B^t$ *is not closed* and therefore $\mathrm{Im}B$ is not closed either. Indeed, the problem is severely ill posed, and you cannot solve it in practice unless you add some sort of regularisation.

4.3 Existence and Uniqueness for Perturbed Problems

Some applications, in particular nearly incompressible materials (Sect. 8.13), will require a more general formulation than Problem (4.2.6). Although the first generalisation introduced will appear to be simple, we shall see that its analysis is rather more intricate.

4.3.1 Regular Perturbations

We assume that we are also given a continuous bilinear form $c(\cdot\,,\cdot)$ on $Q \times Q$, and we denote by C its associated operator $Q \to Q'$.

We now consider the following extension of problem (4.2.6): *given* $f \in V'$ *and* $g \in Q'$, *find* $u \in V$ *and* $p \in Q$ *such that*

$$\begin{cases} a(u,v) + b(v,p) = \langle f, v\rangle_{V'\times V}, & \forall\, v \in V, \\ b(u,q) - c(p,q) = \langle g, q\rangle_{Q'\times Q}. & \forall\, q \in Q. \end{cases} \tag{4.3.1}$$

Remark 4.3.1. Whenever $a(\cdot,\cdot)$ and $c(\cdot,\cdot)$ are symmetric, this problem corresponds to the saddle point problem

$$\inf_{v\in V}\sup_{q\in Q} \frac{1}{2}a(v,v) + b(v,q) - \frac{1}{2}c(q,q) - \langle f, v\rangle + \langle g, q\rangle$$

and it is no longer equivalent to a minimisation problem on u. □

Remark 4.3.2. As in Remark 4.2.2, the two equations in (4.3.1) can sometimes be written as a unique variational equation, setting

$$\mathcal{A}((u,p),(v,q)) = a(u,v) + b(v,p) - b(u,q) + c(p,q) \qquad \forall (u,p),\,(v,q) \in V \times Q \tag{4.3.2}$$

and then requiring again that

$$\mathcal{A}((u,p),(v,q)) = \langle f, v\rangle_{V'\times V} - \langle g, q\rangle_{Q'\times Q} \qquad \forall (v,q) \in V \times Q. \tag{4.3.3}$$

□

We now want to look for conditions on a, b and c ensuring the existence and uniqueness of a solution to (4.3.1), together with the proper stability bounds.

Let us first consider a special case. We assume that $c(\cdot,\cdot)$ is coercive on Q, that is

$$\exists\, \gamma > 0 \text{ such that } c(q,q) \geq \gamma\, \|q\|_Q^2, \ \forall\, q \in Q \tag{4.3.4}$$

and that $a(\cdot,\cdot)$ is also coercive on V:

$$\exists\, \alpha > 0 \text{ such that } a(v,v) \geq \alpha\, \|v\|_V^2, \ \forall\, v \in V. \tag{4.3.5}$$

Then, we have the following proposition.

Proposition 4.3.1. *Together with Assumption $\mathcal{AB}$, assume that* (4.3.4) *and* (4.3.5) *hold. Then, for every $f \in V'$ and $g \in Q'$, problem* (4.3.1) *has a unique solution (u,p). Moreover, we have:*

$$\frac{\alpha}{2}\|u\|_V^2 + \frac{\gamma}{2}\|p\|^2 \leq \frac{1}{2\alpha}\|f\|_{V'}^2 + \frac{1}{2\gamma}\|g\|_{Q'}^2. \tag{4.3.6}$$

Proof. The proof is elementary (using, for instance, Lax-Milgram Lemma (4.1.6) on the bilinear form (4.3.2)). □

The estimate (4.3.6) is unsatisfactory. Actually, in many applications, we will deal with a bilinear form $c(\cdot,\cdot)$ defined by

$$c(p,q) = \lambda(p,q)_Q,\ \lambda \geq 0, \tag{4.3.7}$$

and we would like to get estimates that provide uniform bounds on the solution for λ small (say $0 \leq \lambda \leq 1$). Clearly, if $c(\cdot,\cdot)$ has the form (4.3.7), one has $\gamma = \lambda$ in (4.3.4) and the bound (4.3.6) explodes for vanishing λ. This fact has practical implications, as we shall see, on the numerical approximations of some problems, for instance when dealing with nearly incompressible materials. On the other hand, Proposition 4.3.1 makes no assumptions on $b(\cdot,\cdot)$ (except the usual (4.2.4)) and it is then quite natural for the choice $c \equiv 0$ to be forbidden.

It is then natural to start assuming, as we did in Sect. 3.6 of the previous chapter, that the corresponding *unperturbed* problem (corresponding to the case $c \equiv 0$) is well posed, and try to find sufficient conditions on c that ensure the well-posedness of the perturbed problem.

We shall start with the simplest case, generalising Proposition 3.3.1 in an obvious manner.

Proposition 4.3.2. *Together with Assumption $\mathcal{AB}$, assume that $A_{KK'}$ is an isomorphism from K to K' and that $\mathrm{Im}B = Q'$. Then, there exists an $\varepsilon_0 > 0$ such that, for every ε with $|\varepsilon| \leq \varepsilon_0$, condition $\|c\| \leq \varepsilon$ implies that problem* (4.3.1) *has a unique solution for every $f \in V'$ and for every $g \in Q'$.* □

As for Proposition 4.3.1, the proof is immediate, this time using the Kato Theorem 4.1.3.

The result of Proposition 4.3.2 is also unsatisfactory. For once, it does give us a result only for ε small enough. Besides, ε_0 will be very difficult to compute in practice. Without it, we basically never know, in every particular case, whether we are solving a well posed problem or not, which is clearly a quite unhappy situation.

We therefore have to look for better results. We could start, as in the previous subsection, by assuming that $\mathrm{Im}B = Q'$, and then try to adapt the results to the case in which $\mathrm{Im}B$ is closed but not equal to Q'. We remark, however, that, this time, the passage from the case when $\mathrm{Im}B = Q'$ (when $H = \{0\}$) and the case when $\mathrm{Im}B$ is simply closed is no longer so simple, as the bilinear form c could mix together the components of p in H and in $H^{\perp}$. Therefore, it is better to look directly at the case where we simply have $\mathrm{Im}B$ closed. On the other hand, we have already seen in the previous chapter that assuming symmetry of *both* a and c gives much better stability bounds. Hence, we decide to concentrate on that case. This is particularly reasonable since, in most applications, the symmetry assumptions are satisfied.

Therefore, to start with, we enlarge our Assumption $\mathcal{AB}$ to include the additional bilinear form c and the additional properties that we are going to use throughout this subsection.

Assumption $\mathcal{ABC}$: *Together with Assumption $\mathcal{AB}$, we assume that we are given a continuous bilinear form $c(\cdot,\cdot)$ on $Q \times Q$, and we denote by C its associated operator $Q \to Q'$. We assume, moreover, that $\mathrm{Im}B$ is closed, and that both $a(\cdot,\cdot)$ and $c(\cdot,\cdot)$ are symmetric and positive semi-definite:*

$$a(v,v) \geq 0, \ \forall\, v \in V \qquad c(q,q) \geq 0, \ \forall\, q \in Q. \tag{4.3.8}$$

We now introduce some additional notation, and a few related properties that hold when a and c are symmetric and positive semi-definite, and $\mathrm{Im}B$ is closed.

We define the semi-norms

$$|v|_a^2 := a(v,v) \qquad |q|_c^2 := c(q,q), \tag{4.3.9}$$

and we note that, thanks to the continuity of a and c,

$$|v|_a^2 \leq \|a\| \, \|v\|_V^2 \ \forall\, v \in V \quad \text{and} \quad |q|_c^2 \leq \|c\| \, \|q\|_Q^2 \ \forall\, q \in Q. \tag{4.3.10}$$

We also note that, from (4.2.29), we have

$$a(u, v) \leq |u|_a \, |v|_a \quad \text{and} \quad c(p, q) \leq |p|_c \, |q|_c, \tag{4.3.11}$$

and from (4.2.30),

$$\|Au\|^2 \leq \|a\| \, |u|_a^2 \quad \text{and} \quad \|Cp\|^2 \leq \|c\| \, |p|_c^2. \tag{4.3.12}$$

Setting again $K = \operatorname{Ker} B$ and $H = \operatorname{Ker} B^t$ as in (4.1.66), we can split each $v \in V$ and each $q \in Q$ as

$$v = v_0 + \overline{v} \qquad q = q_0 + \overline{q}, \tag{4.3.13}$$

with $v_0 \in K, \overline{v} \in K^{\perp}, q_0 \in H$, and $\overline{q} \in H^{\perp}$, and we note that

$$b(v, q) = b(\overline{v}, q) = b(\overline{v}, \overline{q}) = b(v, \overline{q}). \tag{4.3.14}$$

In a similar way, we can split each $f \in V'$ and each $g \in Q'$ as

$$f = f_0 + \overline{f} \qquad g = g_0 + \overline{g} \tag{4.3.15}$$

with $f_0 \in K', \overline{f} \in (K^{\perp})' \equiv K^0, g_0 \in H'$ and $\overline{g} \in (H^{\perp})' \equiv H^0$, and we note that

$$\langle f, v \rangle = \langle f_0, v_0 \rangle + \langle \overline{f}, \overline{v} \rangle \quad \langle g, q \rangle = \langle g_0, q_0 \rangle + \langle \overline{g}, \overline{q} \rangle \tag{4.3.16}$$

with obvious meaning of the duality pairings.

We therefore have the following result, in which the roles of a and c are perfectly interchangeable.

Theorem 4.3.1. *Together with Assumption $\mathcal{ABC}$, assume that $a(\cdot\,,\cdot)$ is coercive on K and $c(\cdot\,,\cdot)$ is coercive on H. Let therefore α_0, β, and γ_0 be positive constants such that*

$$\alpha_0 \|v_0\|_V^2 \leq a(v_0, v_0) \qquad \forall\, v_0 \in K, \tag{4.3.17}$$

$$\inf_{q \in H^{\perp}} \sup_{v \in V} \frac{b(v, q)}{\|q\|_Q \, \|v\|_V} = \inf_{v \in K^{\perp}} \sup_{q \in Q} \frac{b(v, q)}{\|q\|_Q \, \|v\|_V} = \beta > 0, \tag{4.3.18}$$

$$\gamma_0 \|q_0\|_Q^2 \leq c(q_0, q_0) \qquad \forall\, q_0 \in H. \tag{4.3.19}$$

Then, for every $f \in V'$ and $g \in Q'$, we have that the problem

$$\begin{cases} a(u, v) + b(v, p) = \langle f, v \rangle_{V' \times V}, \ \forall\, v \in V, \\ b(u, q) - c(p, q) = \langle g, q \rangle_{Q' \times Q}, \ \forall\, q \in Q \end{cases} \tag{4.3.20}$$

has a unique solution, that moreover satisfies

$$\|u\|_V + \|p\|_Q \le C\Big(\|f\|_{V'} + \|g\|_{Q'}\Big) \tag{4.3.21}$$

with C constant depending only on the stability constants α_0, β, γ_0 and on the continuity constants $\|a\|$ and $\|c\|$. More precisely, we have:

$$\|\overline{u}\|_V \le \frac{\|c\|\|\overline{f}\|}{\beta^2} + \frac{\sqrt{2\beta^2+\mu^2}\|c\|^{1/2}\,\|f_0\|}{\alpha_0^{1/2}\beta^2} + \frac{(\beta+\mu)\|\overline{g}\|}{\beta^2} + \frac{3\sqrt{\beta^2+\mu^2}\|c\|^{1/2}\|g_0\|}{\gamma_0^{1/2}\beta^2}, \tag{4.3.22}$$

$$\|u_0\|_V \le \frac{\|c\|\,\|a\|^{1/2}\,\|\overline{f}\|}{\alpha_0^{1/2}\beta^2} + \frac{2(\beta^2+\mu^2)\|f_0\|}{\alpha_0\beta^2} + \frac{(\beta+\mu)\|a\|^{1/2}\|\overline{g}\|}{\alpha_0^{1/2}\beta^2} + \frac{3\mu\sqrt{\beta^2+\mu^2}\|g_0\|}{\gamma_0^{1/2}\alpha_0^{1/2}\beta^2}, \tag{4.3.23}$$

$$\|\overline{p}\|_Q \le \frac{(\beta+\mu)\|\overline{f}\|}{\beta^2} + \frac{3\sqrt{\beta^2+\mu^2}\|a\|^{1/2}\,\|f_0\|}{\alpha_0^{1/2}\beta^2} + \frac{\|a\|\|\overline{g}\|}{\beta^2} + \frac{\sqrt{2\beta^2+\mu^2}\|a\|^{1/2}\,\|g_0\|}{\gamma_0^{1/2}\beta^2}, \tag{4.3.24}$$

$$\|p_0\|_Q \le +\frac{(\beta+\mu)\|c\|^{1/2}\|\overline{f}\|}{\gamma_0^{1/2}\beta^2} + \frac{3\mu\sqrt{\beta^2+\mu^2}\|f_0\|}{\alpha_0^{1/2}\gamma_0^{1/2}\beta^2} + \frac{\|a\|\,\|c\|^{1/2}\|\overline{g}\|}{\gamma_0^{1/2}\beta^2} + \frac{2(\beta^2+\mu^2)\|g_0\|}{\gamma_0\beta^2}, \tag{4.3.25}$$

where μ is defined by

$$\mu^2 := \|a\|\,\|c\|. \tag{4.3.26}$$

Proof. As the problem is symmetric, we just have to prove that the mapping $M : (u,p) \to (f,g)$ is bounding (that is, we have to prove that the bounds (4.3.22)–(4.3.25) hold true). Then, M will be injective and $M^t \equiv M$ will be surjective, and the theorem will be proved.

Then, we note that there is another (fundamental) symmetry in our assumptions when we exchange a with c, u with p and B with B^t. Hence, we can start proving our bounds for the case, say, $f = 0$. These bounds, due to the above symmetry, will imply similar ones for the case $g = 0$ (exchanging u with p, α_0 with γ_0, and so on). Then, by linearity, we will sum the estimates for $f = 0$ and those for $g = 0$, and obtain the final estimates for the general case.

Hence, we proceed by assuming $f = 0$. We first observe that, for $f = 0$, we have from the first equation

$$a(u, u_0) = -b(u_0, p) = 0 \tag{4.3.27}$$

since $u_0 \in \mathrm{Ker}B$. Hence, using (4.3.9), $u_0 = u - \overline{u}$, and (4.3.11),

$$|u_0|_a^2 = a(u_0, u_0) = -a(\overline{u}, u_0) \le |\overline{u}|_a \, |u_0|_a, \tag{4.3.28}$$

which, combined with the ellipticity condition (4.3.17) and then with (4.3.10), gives

$$\|u_0\|_V \le \frac{1}{\alpha_0^{1/2}} |u_0|_a \le \frac{1}{\alpha_0^{1/2}} |\overline{u}|_a . \le \frac{\|a\|^{1/2}}{\alpha_0^{1/2}} \|\overline{u}\|_V . \tag{4.3.29}$$

We also note that, in operator form, Eqs. (4.3.20), for $f = 0$, give

$$Au = -B^t p \tag{4.3.30}$$

and

$$Bu = Cp + g. \tag{4.3.31}$$

Moreover, taking in (4.3.20) $v = u$ in the first equation, $q = p$ in the second equation, and subtracting, we have

$$a(u, u) + c(p, p) = -\langle g, p\rangle, \tag{4.3.32}$$

implying through (4.3.12) that

$$\frac{\|Au\|_{V'}^2}{\|a\|} + \frac{\|Cp\|_{Q'}^2}{\|c\|} \le -\langle g, p\rangle. \tag{4.3.33}$$

At this point, it will be convenient to further distinguish the cases $g_0 = 0$ and $\overline{g} = 0$, to make the estimates separately, and then sum them. We start with the easier case $g_0 = 0$. Then, (4.3.33) becomes

$$\frac{\|Au\|_{V'}^2}{\|a\|} + \frac{\|Cp\|_{Q'}^2}{\|c\|} \le -\langle \overline{g}, \overline{p}\rangle. \tag{4.3.34}$$

On the other hand, $p - \overline{p} = p_0 \in H$ so that $B^t\overline{p} = B^t p$. Hence, using (4.3.30), then (4.3.34), and then (4.1.94), we have

$$\begin{aligned}\|B^t\overline{p}\|^2_{V'} &= \|B^t p\|^2_{V'} = \|Au\|^2_{V'} \\ &\le \|a\|\,\|\overline{g}\|_{Q'}\|\overline{p}\|_Q \le \|a\|\,\|\overline{g}\|_{Q'}\frac{1}{\beta}\|B^t\overline{p}\|_{V'}\end{aligned} \tag{4.3.35}$$

which, using again (4.1.94), gives

$$\|\overline{p}\|_Q \le \frac{1}{\beta}\|B^t\overline{p}\|_{V'} \le \frac{\|a\|}{\beta^2}\|\overline{g}\|_{Q'}. \tag{4.3.36}$$

At this point, we remark that, from the second equation of (4.3.20) tested on $q = p_0$, we have

$$c(p, p_0) = b(u, p_0) - \langle \overline{g}, p_0\rangle = 0 - 0 = 0, \tag{4.3.37}$$

since $B^t p_0 = 0$ and $\langle \overline{g}, p_0\rangle = 0$ as in (4.3.16). Proceeding exactly as in (4.3.27)–(4.3.29), we then have

$$|p_0|_c \le |\overline{p}|_c. \tag{4.3.38}$$

Using the ellipticity condition (4.3.19), then (4.3.38), (4.3.10) and the previous estimate (4.3.36) on $\overline{p}$, we have

$$\|p_0\|_Q \le \frac{1}{\gamma_0^{1/2}}|p_0|_c \le \frac{1}{\gamma_0^{1/2}}|\overline{p}|_c \le \frac{\|a\|\,\|c\|^{1/2}}{\gamma_0^{1/2}\beta^2}\|\overline{g}\|_{Q'}. \tag{4.3.39}$$

The estimates on $\overline{u}$ and u_0 can be obtained in a similar way: indeed we can use (4.3.31), then (4.3.34), and then again (4.3.36) to obtain

$$\|Bu - \overline{g}\|^2_{Q'} = \|Cp\|^2_{Q'} \le \|c\|\,\|\overline{g}\|_{Q'}\|\overline{p}\|_Q \le \frac{\|c\|\,\|a\|}{\beta^2}\,\|\overline{g}\|^2_{Q'},$$

giving

$$\|Bu\|_{Q'} \le \frac{(\|c\|\,\|a\|)^{1/2}}{\beta}\,\|\overline{g}\|_{Q'} + \|\overline{g}\|_{Q'} \le \frac{\mu+\beta}{\beta}\|\overline{g}\|_{Q'}, \tag{4.3.40}$$

where in the last step we used the definition of μ given in (4.3.26). Hence, using (4.1.93), $Bu = B\overline{u}$, and (4.3.40), we have

$$\|\overline{u}\|_V \le \frac{1}{\beta}\|B\overline{u}\|_{Q'} = \frac{1}{\beta}\|Bu\|_{Q'} \le \frac{\mu+\beta}{\beta^2}\|\overline{g}\|_{Q'}. \tag{4.3.41}$$

Finally, we can use (4.3.29) to obtain

$$\|u_0\|_V \leq \frac{\|a\|^{1/2}}{\alpha^{1/2}}\|\overline{u}\|_V \leq \frac{(\mu+\beta)\|a\|^{1/2}}{\alpha^{1/2}\beta^2}\|\overline{g}\|_{Q'}. \tag{4.3.42}$$

The estimates in the case $g_0 = 0$ are therefore completed.

We now consider the case $\overline{g} = 0$. Using the definition (4.3.9), then (4.3.13), and then the second equation of (4.3.20) with $q = p_0$, we have

$$\begin{aligned} |p_0|_c^2 &= c(p_0, p_0) = c(p, p_0) - c(\overline{p}, p_0) \\ &= b(u, p_0) - \langle g_0, p_0\rangle - c(\overline{p}, p_0) = -\langle g_0, p_0\rangle - c(\overline{p}, p_0) \\ &\leq \|g_0\|_{Q'}\|p_0\|_Q + |\overline{p}|_c\,|p_0|_c \end{aligned} \tag{4.3.43}$$

where the last equality holds since $p_0 \in Ker B^t$. We also note that, due to (4.3.19),

$$\|p_0\|_Q \leq \frac{1}{\gamma_0^{1/2}}|p_0|_c. \tag{4.3.44}$$

Joining (4.3.43) and (4.3.44), we then have

$$|p_0|_c^2 \leq \frac{\|g_0\|_{Q'}}{\gamma_0^{1/2}}|p_0|_c + |\overline{p}|_c|p_0|_c, \tag{4.3.45}$$

implying

$$|p_0|_c \leq \frac{\|g_0\|_{Q'}}{\gamma_0^{1/2}} + |\overline{p}|_c, \tag{4.3.46}$$

and using once more (4.3.44), and then (4.3.10),

$$\|p_0\|_Q \leq \frac{1}{\gamma_0^{1/2}}\left(\frac{\|g_0\|_{Q'}}{\gamma_0^{1/2}} + |\overline{p}|_c\right) \leq \frac{\|g_0\|_{Q'}}{\gamma_0} + \frac{\|c\|^{1/2}}{\gamma_0^{1/2}}\|\overline{p}\|_Q. \tag{4.3.47}$$

Proceeding as in (4.3.35), and then using (4.3.47), and finally (4.1.94), we now have

$$\begin{aligned} \|B^t\overline{p}\|_{V'}^2 &= \|B^t p\|_{V'}^2 = \|Au\|_{V'}^2 \leq \|a\|\,\|g_0\|_{Q'}\|p_0\|_Q \\ &\leq \frac{\|a\|\,\|g_0\|_{Q'}^2}{\gamma_0} + \frac{\|a\|\,\|g_0\|_{Q'}}{\gamma_0^{1/2}}\|c\|^{1/2}\|\overline{p}\|_Q \\ &\leq \frac{\|a\|\,\|g_0\|_{Q'}^2}{\gamma_0} + \frac{\|a\|\,\|g_0\|_{Q'}\|c\|^{1/2}}{\beta\gamma_0^{1/2}}\|B^t\overline{p}\|_{V'}. \end{aligned} \tag{4.3.48}$$

Using the classical inequality $xy \leq (x^2 + y^2)/2$ on the last term of (4.3.48), we obtain

$$\frac{1}{2}\|B^t\overline{p}\|^2_{V'} \le \frac{\|a\|\,\|g_0\|^2_{Q'}}{\gamma_0} + \frac{\|a\|^2\,\|g_0\|^2_{Q'}\|c\|}{2\beta^2\gamma_0} \le \frac{(2\beta^2+\mu^2)\|a\|\,\|g_0\|^2_{Q'}}{2\gamma_0\beta^2}, \tag{4.3.49}$$

implying easily

$$\|B^t\overline{p}\|_{V'} \le \frac{(2\beta^2+\mu^2)^{1/2}\|a\|^{1/2}\,\|g_0\|_{Q'}}{\gamma_0^{1/2}\beta}. \tag{4.3.50}$$

From (4.3.50), using once more (4.1.94), we obtain the estimate on $\overline{p}$

$$\|\overline{p}\|_Q \le \frac{(2\beta^2+\mu^2)^{1/2}\|a\|^{1/2}\,\|g_0\|_{Q'}}{\gamma_0^{1/2}\beta^2}, \tag{4.3.51}$$

which, using (4.3.47), also gives the bound on p_0:

$$\begin{aligned}\|p_0\|_Q &\le \frac{\|g_0\|_{Q'}}{\gamma_0} + \frac{\|c\|^{1/2}}{\gamma_0^{1/2}}\frac{(2\beta^2+\mu^2)^{1/2}\|a\|^{1/2}\,\|g_0\|_{Q'}}{\gamma_0^{1/2}\beta^2}\\ &= \frac{\beta^2+\mu(2\beta^2+\mu^2)^{1/2}}{\gamma_0\beta^2}\|g_0\|_{Q'} \le \frac{2(\beta^2+\mu^2)}{\gamma_0\beta^2}\|g_0\|_{Q'}.\end{aligned} \tag{4.3.52}$$

This, in turn, gives us a bound on $\|Cp\|_{Q'}$. Indeed, using (4.3.33) and remembering that in this case

$$\langle g, p\rangle = \langle g_0, p\rangle = \langle g_0, p_0\rangle \tag{4.3.53}$$

and then using (4.3.52), we easily have

$$\|Cp\|^2_{Q'} \le -\|c\|\,\langle g_0, p_0\rangle \le \frac{\|c\|\,2(\beta^2+\mu^2)}{\gamma_0\beta^2}\,\|g_0\|^2_{Q'}. \tag{4.3.54}$$

On the other hand, the second equation of (4.3.20) gives $Bu = Cp + g_0$, so that using (4.3.54),

$$\|Bu\|_{Q'} \le \left(\frac{\|c\|^{1/2}\sqrt{2(\beta^2+\mu^2)}}{\gamma_0^{1/2}\beta} + 1\right)\|g_0\|_{Q'} \le \frac{3\sqrt{\beta^2+\mu^2}\|c\|^{1/2}}{\gamma_0^{1/2}\beta}\|g_0\|_{Q'}, \tag{4.3.55}$$

where we used the fact that $\gamma_0 \le \|c\|$. We now note that $Bu = B\overline{u}$, so that, using (4.1.93) and (4.3.55), we have the estimate on $\overline{u}$

$$\|\overline{u}\|_V \le \frac{1}{\beta}\|Bu\|_{Q'} \le \frac{3\sqrt{\beta^2+\mu^2}\|c\|^{1/2}}{\gamma_0^{1/2}\beta^2}\|g_0\|_{Q'}. \tag{4.3.56}$$

The estimate on u_0 then follows from (4.3.29), that is,

$$\|u_0\|_V \le \frac{\|a\|^{1/2}}{\alpha_0^{1/2}}\|\overline{u}\|_V \le \frac{3\mu\sqrt{\beta^2+\mu^2}}{\gamma_0^{1/2}\alpha_0^{1/2}\beta^2}\|g_0\|_{Q'}. \tag{4.3.57}$$

As already discussed, the estimates for the cases $g = 0$ and $f = \overline{f}$ or $f = f_0$ are "symmetrical", and the proof is completed. □

Remark 4.3.3. Following the path of Theorem 3.6.1, we could have proved stability also for the case in which a or c are not symmetric (at least in the case $\mathrm{Im}B = Q'$). However, the dependence of the stability constants upon α_0 and β would have been *much worse.* □

A very particular (but important) case is met when c has the form, as in (4.3.7),

$$c(p,q) = \lambda(p,q)_Q,\ \lambda \ge 0 \tag{4.3.58}$$

where $(\cdot\,,\,\cdot)_Q$ is the scalar product in Q. We decided therefore to dedicate a theorem especially to it.

Theorem 4.3.2. *In the framework of Assumption $\mathcal{ABC}$, assume further that the inf-sup condition* (4.2.26) *and the ellipticity requirement (4.2.12) are satisfied, and that c is given by* (4.3.58) *with $\lambda > 0$. Then, for every $f \in V'$ and for every $g \in Q'$, problem* (4.3.20) *has a unique solution, and we have the estimate*

$$\|u\|_V \le \frac{\beta^2 + 4\lambda\,\|a\|}{\alpha_0\beta^2}\|f\|_{V'} + \frac{2\|a\|^{1/2}}{\alpha_0^{1/2}\beta}\|g\|_{Q'} \tag{4.3.59}$$

and

$$\|p\|_Q \le \frac{2\,\|a\|^{1/2}}{\alpha_0^{1/2}\beta}\|f\|_{V'} + \frac{4\,\|a\|}{\lambda\|a\| + 2\,\beta^2}\|g\|_{Q'}. \tag{4.3.60}$$

Proof. As we are already used to, we shall split the two cases $f = 0$ and $g = 0$, and then combine the estimates by linearity. Let us first consider the case $f = 0$, and assume that u, p and g satisfy

$$\begin{cases} a(u,v) + b(v,p) = 0, & \forall\, v \in V,\\ b(u,q) - \lambda(p,q)_Q = \langle g,q\rangle_{Q'\times Q}, & \forall\, q \in Q. \end{cases} \tag{4.3.61}$$

In operator form, (4.3.61) can be written as

$$\begin{cases} Au + B^t p = 0,\\ Bu - \lambda R_Q p = g, \end{cases} \tag{4.3.62}$$

where R_Q is the Ritz operator $Q \to Q'$ (see (4.1.37)).

Using (4.1.94) together with the first equation of (4.3.62), we obtain

$$\beta\|p\|_Q \leq \|B^t p\|_{Q'} = \|Au\|_{V'}. \tag{4.3.63}$$

On the other hand, we already noted (see (4.3.32)) that

$$a(u,u) + \lambda\|p\|_Q^2 = -\langle g, p\rangle_{Q'\times Q}. \tag{4.3.64}$$

Using (4.3.12), Eq. (4.3.64) and finally (3.4.17), we have

$$\|Au\|_{V'}^2 \leq \|a\|a(u,u) \leq \|a\|\|p\|_Q\,\|g\|_{Q'}, \tag{4.3.65}$$

which, combined with (4.3.63), yields

$$\|Au\|_{V'} \leq \frac{\|a\|}{\beta}\|g\|_{Q'}, \tag{4.3.66}$$

and using again (4.3.63),

$$\|p\|_Q \leq \frac{\|a\|}{\beta^2}\|g\|_{Q'}. \tag{4.3.67}$$

Using the lifting operator L_B defined in Theorem 4.1.5, we set

$$\tilde{u} := L_B(g + \lambda R_Q^{-1} p) \tag{4.3.68}$$

and we have from (3.4.43)

$$B\tilde{u} = g + \lambda R_Q^{-1} p. \tag{4.3.69}$$

Setting now

$$u_0 := u - \tilde{u}, \tag{4.3.70}$$

we have from (4.3.69) and the second equation of (4.3.62) that $u_0 \in K$. We then note that, testing the first equation of (4.3.61) with $v = u_0$, we have, as in (4.3.27):

$$a(u,u_0) = -b(u_0,p) = 0. \tag{4.3.71}$$

Moreover, using (4.3.70), (4.3.71) and (4.3.11), we have as in (4.3.28)

$$a(u_0,u_0) = -a(u_0,\tilde{u}) \leq |u_0|_a|\tilde{u}|_a, \tag{4.3.72}$$

which easily gives

$$|u_0|_a \leq |\tilde{u}|_a. \tag{4.3.73}$$

Hence, we can use (4.2.12) and (4.3.73) to obtain

$$\alpha_0\|u_0\|_V^2 \leq |u_0|_a^2 \leq |\tilde{u}|_a^2, \tag{4.3.74}$$

and finally from (4.3.74) and (4.3.10),

$$\|u_0\|_V \le \left(\frac{\|a\|}{\alpha_0}\right)^{1/2} \|\tilde{u}\|_V. \tag{4.3.75}$$

Finally, we can collect (4.3.70) and (4.3.75) and have an estimate for u:

$$\|u\|_V \le \|u_0\|_V + \|\tilde{u}\|_V \le \left(1 + \left(\frac{\|a\|}{\alpha_0}\right)^{1/2}\right) \|\tilde{u}\|_V. \tag{4.3.76}$$

We now consider the first equation of (4.3.61) with $v = u$, getting

$$a(u, u) + b(u, p) = 0. \tag{4.3.77}$$

Recalling that a is positive semi-definite (see (4.3.8)), we obtain

$$b(u, p) \le 0,$$

and substituting $p = \lambda^{-1} R_Q^{-1}(Bu - g)$:

$$\begin{aligned} 0 &\ge \langle Bu, \lambda^{-1} R_Q^{-1}(Bu - g)\rangle_{Q'\times Q} \\ &= \lambda^{-1}\Big(\|Bu\|_{Q'}^2 - \langle Bu, R_Q^{-1} g\rangle_{Q'\times Q}\Big), \end{aligned} \tag{4.3.78}$$

which easily implies

$$\|Bu\|_{Q'}^2 \le \langle Bu, R_Q^{-1} g\rangle_{Q'\times Q} \le \|Bu\|_{Q'} \, \|g\|_{Q'}, \tag{4.3.79}$$

giving

$$\|Bu\|_{Q'} \le \|g\|_{Q'}. \tag{4.3.80}$$

Using once more the *inf-sup* condition (4.1.93),

$$\|\tilde{u}\|_V \le \frac{1}{\beta} \|B\tilde{u}\|_{Q'} \le \frac{1}{\beta} \|Bu\|_{Q'} \le \frac{1}{\beta} \|g\|_{Q'}, \tag{4.3.81}$$

and inserting (4.3.81) in (4.3.76), then using $\alpha_0 \le \|a\|$, gives

$$\|u\|_V \le \left(1 + \left(\frac{\|a\|}{\alpha_0}\right)^{1/2}\right) \|\tilde{u}\|_V \le \frac{2\|a\|^{1/2}}{\beta \alpha_0^{1/2}} \|g\|_{Q'}. \tag{4.3.82}$$

We note at this point that we have another way to obtain an estimate for p, apart from (4.3.67); actually, from the second equation of (4.3.62), and (4.3.80):

$$\|p\|_Q \le \frac{1}{\lambda}\|Bu - g\|_{Q'} \le \frac{2}{\lambda}\|g\|_{Q'}. \tag{4.3.83}$$

With some manipulations, we see that (4.3.67) and (4.3.83) can be combined into

$$\|p\|_Q \le \frac{4\,\|a\|}{\|a\|\lambda + 2\,\beta^2}\|g\|_{Q'}. \tag{4.3.84}$$

We now consider the case in which $g = 0$ and assume that u, p and f satisfy

$$\begin{cases} a(u,v) + b(v,p) = \langle f, v\rangle_{V'\times V}, & \forall\, v \in V,\\ b(u,q) - \lambda(p,q)_Q = 0, & \forall\, q \in Q, \end{cases} \tag{4.3.85}$$

which in operator form reads:

$$\begin{cases} Au + B^t p = f\\ Bu - \lambda R_Q p = 0, \end{cases} \tag{4.3.86}$$

where again R_Q is the Ritz operator $Q \to Q'$ (see (4.1.37)). We use again the lifting operator L_B of Theorem (4.1.5), this time setting $\tilde{u} := L_B \lambda R_Q p$ so that

$$B\tilde{u} = Bu = \lambda R_Q p, \tag{4.3.87}$$

and, defining again u_0 as in (4.3.70), we still have $u_0 \in K$. Taking $v = \tilde{u}$ as test function in the first equation of (4.3.86), and substitute $p = R_Q^{-1}\lambda^{-1}Bu$:

$$a(u,\tilde{u}) + b(\tilde{u}, R_Q^{-1}\lambda^{-1}Bu) = \langle f, \tilde{u}\rangle. \tag{4.3.88}$$

As $B\tilde{u} = Bu$, we can rewrite (4.3.88) as follows

$$\lambda^{-1}\|Bu\|^2_{Q'} = \langle f, \tilde{u}\rangle - a(u,\tilde{u}) \le \|f\|_{V'}\,\|\tilde{u}\|_V - a(u,\tilde{u}). \tag{4.3.89}$$

We leave (4.3.89) for a while, and we estimate $-a(u,\tilde{u})$. Using the fact that $u = \tilde{u} + u_0$ and (4.3.11), we obtain

$$-\,a(u,\tilde{u}) = -a(\tilde{u} + u_0, \tilde{u}) \le -|\tilde{u}|_a^2 + |\tilde{u}|_a\,|u_0|_a. \tag{4.3.90}$$

On the other hand, testing the first equation with $v = u_0$, we get

$$a(u,u_0) = \langle f, u_0\rangle, \tag{4.3.91}$$

yielding

$$|u_0|_a^2 = a(u_0, u_0) = a(u, u_0) - a(\tilde{u}, u_0) \leq \|f\| \, \|u_0\|_V + |\tilde{u}|_a \, |u_0|_a. \tag{4.3.92}$$

On the other hand, (4.3.17) gives

$$\alpha_0 \|u_0\|_V^2 \leq |u_0|_a^2 \tag{4.3.93}$$

which, together with (4.3.92), yields

$$|u_0|_a \leq \frac{\|f\|_{V'}}{\alpha_0^{1/2}} + |\tilde{u}|_a. \tag{4.3.94}$$

Inserting this into (4.3.90), we have

$$-a(u, \tilde{u}) \leq -|\tilde{u}|_a^2 + |\tilde{u}|_a \left(\frac{1}{\alpha_0^{1/2}} \|f\|_{V'} + |\tilde{u}|_a\right) = |\tilde{u}|_a \frac{1}{\alpha_0^{1/2}} \|f\|_{V'}. \tag{4.3.95}$$

Inserting this into (4.3.89), then using (4.3.10), and finally using (4.1.93) gives

$$\begin{aligned}
\lambda^{-1} \|Bu\|_{Q'}^2 &\leq \|f\|_{V'} \, \|\tilde{u}\|_V + |\tilde{u}|_a \frac{1}{\alpha_0^{1/2}} \|f\|_{V'} \\
&\leq \left(1 + \frac{\|a\|^{1/2}}{\alpha_0^{1/2}}\right) \|f\|_{V'} \|\tilde{u}\|_V \leq \left(1 + \frac{\|a\|^{1/2}}{\alpha_0^{1/2}}\right) \|f\|_{V'} \frac{1}{\beta} \|Bu\|_{Q'} \\
&= \frac{\alpha_0^{1/2} + \|a\|^{1/2}}{\beta \alpha_0^{1/2}} \|f\|_{V'} \|Bu\|_{Q'}.
\end{aligned} \tag{4.3.96}$$

Using again (4.1.93) and then (4.3.96), we have therefore

$$\|\tilde{u}\|_V \leq \frac{1}{\beta} \|Bu\|_{Q'} \leq \frac{\lambda(\alpha_0^{1/2} + \|a\|^{1/2})}{\beta^2 \alpha_0^{1/2}} \|f\|_{V'}. \tag{4.3.97}$$

Using (4.3.93), (4.3.94), and (4.3.10) and then (4.3.97), we have then

$$\begin{aligned}
\|u_0\|_V &\leq \frac{1}{\alpha_0^{1/2}} |u_0|_a \leq \frac{\|f\|_{V'}}{\alpha_0} + \frac{|\tilde{u}|_a}{\alpha_0^{1/2}} \leq \frac{\|f\|_{V'}}{\alpha_0} + \left(\frac{\|a\|}{\alpha_0}\right)^{1/2} \|\tilde{u}\| \\
&\leq \left(\frac{1}{\alpha_0} + \frac{\lambda(\alpha_0^{1/2} + \|a\|^{1/2}) \|a\|^{1/2}}{\alpha_0 \beta^2}\right) \|f\|_{V'}.
\end{aligned} \tag{4.3.98}$$

From the second equation of (4.3.86) and (4.3.97), we also derive the estimate for p

$$\|p\|_Q = \|\lambda^{-1}Bu\|_{Q'} \leq \frac{\alpha_0^{1/2} + \|a\|^{1/2}}{\beta\alpha_0^{1/2}}\|f\|_{V'}. \tag{4.3.99}$$

We collect the results for $g = 0$, using the fact that $\alpha_0 \leq \|a\|$. From (4.3.97) and (4.3.98), we have the estimate on u

$$\begin{aligned}\|u\|_V &\leq \|\tilde{u}\|_V + \|u_0\|_V \\ &\leq \Big(\frac{\lambda\,(\|a\|^{1/2} + \alpha_0^{1/2})}{\alpha_0^{1/2}\beta^2} + \frac{1}{\alpha_0} + \frac{\lambda(\|a\| + (\|a\|\alpha_0)^{1/2})}{\alpha_0\beta^2}\Big)\|f\|_{V'} \\ &\leq \frac{\beta^2 + 4\lambda\|a\|}{\alpha_0\beta^2}\|f\|_{V'},\end{aligned} \tag{4.3.100}$$

while from (4.3.99) we have the estimate on p

$$\|p\|_Q \leq \frac{2\,\|a\|^{1/2}}{\alpha_0^{1/2}\beta}\|f\|_{V'}. \tag{4.3.101}$$

The final results can then be obtained collecting (4.3.82), (4.3.84), (4.3.100) and (4.3.101). □

Corollary 4.3.1. *In the framework of Assumption $\mathcal{ABC}$, assume that $\mathrm{Im}B$ is closed, that the ellipticity requirement (4.2.12) is satisfied and that c is given by* (4.3.58) *with $\lambda \geq 0$. Set $g = \overline{g} + g_0$ with $\overline{g} \in H^0$ and $g_0 \in H'$ (with $H := \ker B^t$, as usual), and set $p = \overline{p} + p_0$ with $\overline{p} \in H^\perp$ and $p_0 \in H$. Then, for every $f \in V'$ and for every $g \in Q'$, problem* (4.3.20) *has a unique solution, and we have the estimates*

$$\|u\|_V \leq \frac{\beta^2 + 4\lambda\,\|a\|}{\alpha_0\beta^2}\|f\|_{V'} + \frac{2\|a\|^{1/2}}{\alpha_0^{1/2}\beta}\|\overline{g}\|_{Q'}, \tag{4.3.102}$$

$$\|\overline{p}\|_Q \leq \frac{2\,\|a\|^{1/2}}{\alpha_0^{1/2}\beta}\|f\|_{V'} + \frac{4\,\|a\|}{\lambda\|a\| + 2\,\beta^2}\|\overline{g}\|_{Q'}, \tag{4.3.103}$$

$$\|p_0\|_Q \leq \frac{1}{\lambda}\|g_0\|_{Q'}. \tag{4.3.104}$$

Proof. It is immediate to check that, actually, the problem splits into two sub-problems: *find* $(u, \overline{p}) \in V \times H^\perp$ *such that*

$$\begin{cases} a(u, v) + b(v, p) = \langle f, v\rangle_{V'\times V}, & \forall\, v \in V, \\ b(u, q) - \lambda(\overline{p}, \overline{q})_Q = \langle g, \overline{q}\rangle_{(H^\perp)'\times H^\perp}, & \forall\, \overline{q} \in H^\perp, \end{cases} \tag{4.3.105}$$

and

$$\lambda p_0 = g_0. \tag{4.3.106}$$

For problem (4.3.105), we can apply the results of Theorem 4.3.2 using $H^{\perp}$ instead of Q (with the same norm). Problem (4.3.106) is trivial. □

In the case where c has the form (4.3.58), as in Theorem 4.3.2, it is also interesting to estimate the distance between the solution of the perturbed problem (4.3.20) and the solution of the limit problem, for $\lambda \to 0$.

We have in particular the following proposition.

Proposition 4.3.3. *Together with Assumption $\mathcal{AB}$, assume that $a(\cdot\,,\cdot)$ is symmetric, positive semi-definite and elliptic on K, and that* ImB *is closed. Let $f \in V'$, let $g \in \mathrm{Im}B$, and let (u^*, p^*) be the solution in $V \times H^{\perp}$ of the problem*

$$\begin{cases} a(u^*, v) + b(v, p^*) = \langle f, v\rangle_{V'\times V}, & \forall\, v \in V, \\ b(u^*, q) = \langle g, q\rangle_{Q'\times Q}, & \forall\, q \in Q. \end{cases} \tag{4.3.107}$$

Let moreover, for $\lambda > 0$, (u_λ, p_λ) be the solution in $V \times Q$ of

$$\begin{cases} a(u_\lambda, v) + b(v, p_\lambda) = \langle f, v\rangle_{V'\times V}, & \forall\, v \in V, \\ b(u_\lambda, q) - \lambda(p_\lambda, q)_Q = \langle g, q\rangle_{Q'\times Q}, & \forall\, q \in Q. \end{cases} \tag{4.3.108}$$

Then, we have

$$\|u^* - u_\lambda\|_V + \|p^* - p_\lambda\|_Q \le C\,\lambda, \tag{4.3.109}$$

where C is a constant depending only on α_0, $\|a\|$ and β.

Proof. Setting $\delta_u := u_\lambda - u^*$ and $\delta_p := p_\lambda - p^*$ and taking the difference of (4.3.108)–(4.3.107), we easily have

$$\begin{cases} a(\delta_u, v) + b(v, \delta_p) = 0, & \forall\, v \in V, \\ b(\delta_u, q) - \lambda(\delta_p, q) = \lambda\,(p^*, q)_Q, & \forall\, q \in Q. \end{cases} \tag{4.3.110}$$

Hence, we can apply estimates (4.3.59) and (4.3.60) with $g = \lambda R_Q p^*$. □

Remark 4.3.4. We point out that the validity of (4.3.109) *for* $\lambda \to 0$ could have been obtained directly from Theorem 4.3.2 and the Kato Theorem (4.1.3). □

We also point out the following result, that is particularly useful if one is not too keen on spotting the best dependence of the stability constants.

Proposition 4.3.4. *Together with Assumption $\mathcal{AB}$, assume that $a(\cdot\,,\cdot)$ is symmetric, positive semi-definite, and elliptic on K, and that* ImB *is closed. Then, for every $\chi > 0$, there exist a constant $\tilde{\alpha}$, depending on χ, $\|a\|$, α_0 and β (defined in* (4.3.18)*), such that*

$$\tilde{\alpha}\|v\|_V^2 \le a(v, v) + \chi\|Bv\|_{Q'}^2 \quad \forall\, v \in V. \tag{4.3.111}$$

Proof. It is easy to check that, for every $\varepsilon \in]0,1[$,

$$\begin{aligned}
a(v,v) + \chi\|Bv\|_{Q'}^2 &= |v_0|_a^2 + |\overline{v}|_a^2 + 2a(v_0,\overline{v}) + \chi\|Bv\|_{Q'}^2 \\
&\geq |v_0|_a^2 + |\overline{v}|_a^2 + \chi\beta\|\overline{v}\|_V^2 - 2|v_0|_a|\overline{v}|_a \\
&\geq |v_0|_a^2 + |\overline{v}|_a^2 + \chi\beta^2\|\overline{v}\|_V^2 - \varepsilon|v_0|_a^2 - \frac{1}{\varepsilon}|\overline{v}|_a^2 \\
&= (1-\varepsilon)|v_0|_a^2 + (1-\frac{1}{\varepsilon})|\overline{v}|_a^2 + \chi\beta^2\|\overline{v}\|_V^2 \\
&\geq (1-\varepsilon)|v_0|_a^2 + (\|a\| - \frac{\|a\|}{\varepsilon})|\overline{v}|_V^2 + \chi\beta^2\|\overline{v}\|_V^2 \\
&= (1-\varepsilon)|v_0|_a^2 + \frac{\chi\beta^2\varepsilon + \|a\|\varepsilon - \|a\|}{\varepsilon}\|\overline{v}\|_V^2 \\
&\geq \alpha_0(1-\varepsilon)\|v_0\|_V^2 + \frac{\chi\beta^2\varepsilon + \|a\|\varepsilon - \|a\|}{\varepsilon}\|\overline{v}\|_V^2,
\end{aligned}$$

and the result follows by taking $\varepsilon = \dfrac{\chi\beta^2 + 2\|a\|}{2\chi\beta^2 + 2\|a\|}$. □

Remark 4.3.5. It is clear that, conversely, the property (4.3.111) implies the ellipticity of a on the kernel K of B. □

Remark 4.3.6. Looking at the proof of Proposition 4.3.4, we can analyse the dependence of the constant $\tilde{\alpha}$ on $\chi\beta^2$, on $\|a\|$, and on α_0. Indeed, setting $k := \chi\beta^2$ and $m := \|a\|$, for $\varepsilon = \dfrac{k+2m}{2k+2m}$ we have

$$\frac{k\varepsilon + m\varepsilon - m}{\varepsilon} = \frac{(k/2) + m - m}{\varepsilon} = \frac{k/2}{\varepsilon} = \frac{k(k+m)}{k+2m} \tag{4.3.112}$$

while

$$\alpha_0(1-\varepsilon) = \frac{\alpha_0(2k+2m-k-2m)}{2k+2m} = \frac{\alpha_0 k}{2k+2m}. \tag{4.3.113}$$

On the other hand, since $\alpha_0 \leq m = \|a\|$, we have

$$\frac{k(k+m)}{k+2m} \geq \frac{k\alpha_0}{2k+2m} \tag{4.3.114}$$

which finally gives (looking at the last line of the proof of Proposition 4.3.4)

$$\tilde{\alpha} \geq \frac{\chi\beta^2\alpha_0}{2\chi\beta^2 + 2\|a\|}. \tag{4.3.115}$$

It is easy to see (taking the derivative) that the right-hand side of (4.3.115), as a function of χ, is monotonically increasing. Hence, we can say that, for every fixed $\chi_* > 0$, we have that for every $\chi \geq \chi_*$,

$$\tilde{\alpha} \geq \alpha_* := \frac{\chi_* \beta^2 \alpha_0}{2\chi_* \beta^2 + 2\|a\|}. \tag{4.3.116}$$

□

Remark 4.3.7. In the above theorem, there is no mention of any bilinear form c, and one may wonder why the theorem has been put in this subsection. However, the bilinear form $a(u, v) + \chi(Bu, Bv)_{Q'}$ is exactly what we get from problem (4.3.20) for $c(p, q) = \lambda(p, q)_Q$ (that is, in the case of the problem (4.3.108)). Indeed, in this case, the second equation of (4.3.108) can be written as: $Bu = \lambda R_Q p + g$ where R_Q is the Ritz operator in Q, as defined in Theorem 4.1.2. Solving for p and substituting in the first equation gives

$$a(u, v) + \frac{1}{\lambda}\langle R_Q^{-1} Bu, Bv\rangle_{Q\times Q'} = \langle f, v\rangle_{V'\times V} + \frac{1}{\lambda}\langle R_Q^{-1} g, Bv\rangle_{Q\times Q'}.$$

Then, we use the fact that $R_Q^{-1} \equiv R_{Q'}$, we set $\chi = 1/\lambda$, and we obtain that the problem (4.3.108) is **equivalent** to

$$\begin{cases} a(u, v) + \chi(Bu, Bv)_{Q'} = \langle f, v\rangle_{V'\times V} + \chi(g, Bv)_{Q'} & \forall\, v \in V, \\ p = \chi\, R_{Q'}(g - Bu), \end{cases} \tag{4.3.117}$$

where clearly the first equation can be solved by itself, and its solution u used to express the solution p of the second equation. □

We conclude the subsection on regular perturbations with the following general theorem, which is often useful in these kinds of problems.

Theorem 4.3.3 (The shadow solution). *Assume that $\mathcal{H}$ is a Hilbert space, and that $\mathbb{M}$ and $\mathbb{D}$ are linear continuous operators from $\mathcal{H}$ into its dual space. Assume that $\mathrm{Im}\mathbb{M}$ is closed and that there exists a $\lambda^* > 0$ such that, for every λ positive with $\lambda \leq \lambda^*$, we have*

$$\lambda \|\mathbf{x}\|^2_{\mathcal{H}} \leq C\, \langle \mathbb{M}\mathbf{x} + \lambda \mathbb{D}\mathbf{x}, \mathbf{x}\rangle_{\mathcal{H}'\times\mathcal{H}} \qquad \forall\, \mathbf{x} \in \mathcal{H}, \tag{4.3.118}$$

for some C independent of λ and $\mathbf{x}$. Let $\mathcal{F} \in \mathrm{Im}\mathbb{M}$ and consider, for every λ positive with $\lambda \leq \lambda^*$, the solution $\mathbf{x}_\lambda$ of the perturbed equation

$$\mathbb{M}\mathbf{x}_\lambda + \lambda \mathbb{D}\mathbf{x}_\lambda = \mathcal{F}. \tag{4.3.119}$$

Then, $\mathbf{x}_\lambda$ *has a unique limit* $\mathbf{x}_*$ *for* $\lambda \to 0+$ *and*

$$\|\mathbf{x}_\lambda - \mathbf{x}_*\|_{\mathcal{H}} \leq C\lambda \tag{4.3.120}$$

where C *is independent of* λ.

Proof. We give a hint of the proof. As $\mathcal{F} \in \mathrm{Im}\mathbb{M}$, we have $\mathcal{F} = \mathbb{M}\overline{\mathbf{x}}$ for some $\overline{\mathbf{x}} \in (\mathrm{Ker}\mathbb{M})^\perp$ with $\|\overline{\mathbf{x}}\|_{\mathcal{H}}$ bounded by $\|\mathcal{F}\|_{\mathcal{H}'}$. Then,

$$\langle \mathbb{M}(\overline{\mathbf{x}} - \mathbf{x}_\lambda), (\overline{\mathbf{x}} - \mathbf{x}_\lambda)\rangle + \lambda\langle \mathbb{D}(\overline{\mathbf{x}} - \mathbf{x}_\lambda), (\overline{\mathbf{x}} - \mathbf{x}_\lambda)\rangle = \lambda\,\langle \mathbb{D}\overline{\mathbf{x}}, (\overline{\mathbf{x}} - \mathbf{x}_\lambda)\rangle,$$

showing that

$$\|\mathbf{x}_\lambda - \overline{\mathbf{x}}\|^2_{\mathcal{H}} \leq C_0\,\langle \mathbb{D}\overline{\mathbf{x}}, (\overline{\mathbf{x}} - \mathbf{x}_\lambda)\rangle \leq C_1\|\mathcal{F}\|_{\mathcal{H}'}\|\mathbf{x}_\lambda - \overline{\mathbf{x}}\|_{\mathcal{H}}, \tag{4.3.121}$$

with C_0 and C_1 independent of λ. Hence, $\mathbf{x}_\lambda - \overline{\mathbf{x}}$ is bounded, and (up to the extraction of a subsequence) converges weakly in $\mathcal{H}$. We define then $\mathbf{x}_*$ as the weak limit (for $\lambda \to 0+$) of $\mathbf{x}_\lambda$. Then, we can go back to the first inequality in (4.3.121), and see that the convergence is strong. Now we remark that, for every λ, equation (4.3.119) gives that $\mathbb{D}\mathbf{x}_\lambda$ belongs to the image of $\mathbb{M}$. As the image is closed, its limit $\mathbb{D}\mathbf{x}_*$ is also in the image. Let $\mathbf{y}_* \in (\mathrm{Ker}\mathbb{M})^\perp$ be such that $\mathbb{M}\mathbf{y}_* = \mathbb{D}\mathbf{x}_*$. Set now $\mathbf{y}_\lambda := \mathbf{x}_* - \mathbf{x}_\lambda$, $\overline{\mathbf{y}} := \lambda\mathbf{y}_*$ and $\mathcal{G} := \mathbb{M}\overline{\mathbf{y}}$. We easily have that

$$\mathbb{M}\mathbf{y}_\lambda + \lambda\mathbb{D}\mathbf{y}_\lambda = \lambda\mathbb{D}\mathbf{x}_* = \mathbb{M}(\lambda\mathbf{y}_*) = \mathcal{G}. \tag{4.3.122}$$

Proceeding as in the previous part of the proof, we have then

$$\begin{aligned}\langle \mathbb{M}(\overline{\mathbf{y}} - \mathbf{y}_\lambda), (\overline{\mathbf{y}} - \mathbf{y}_\lambda)\rangle &+ \lambda\langle \mathbb{D}(\overline{\mathbf{y}} - \mathbf{y}_\lambda), (\overline{\mathbf{y}} - \mathbf{y}_\lambda)\rangle\\ &= \lambda\,\langle \mathbb{D}\overline{\mathbf{y}}, \overline{\mathbf{x}} - \mathbf{x}_\lambda\rangle = \lambda^2\,\langle \mathbb{D}\mathbf{y}_*, \overline{\mathbf{y}} - \mathbf{y}_\lambda\rangle,\end{aligned}$$

showing that

$$\|\mathbf{y}_\lambda - \overline{\mathbf{y}}\|^2_{\mathcal{H}} \leq C_0\lambda\,\langle \mathbb{D}(\mathbf{y}_*), (\overline{\mathbf{y}} - \mathbf{y}_\lambda)\rangle \leq \lambda C_2\|\overline{\mathbf{y}} - \mathbf{y}_\lambda\|_{\mathcal{H}}, \tag{4.3.123}$$

with C_2 independent of λ. Hence, $\|\mathbf{y}_\lambda - \overline{\mathbf{y}}\|_{\mathcal{H}} = O(\lambda)$. Recalling the definition of $\mathbf{y}_\lambda$ and $\overline{\mathbf{y}}$, we have then $\|\mathbf{x}_* - \mathbf{x}_\lambda - \lambda\mathbf{y}_*\|_{\mathcal{H}} = O(\lambda)$ and finally (4.3.120). □

Remark 4.3.8. We note that $\overline{\mathbf{x}}$ and $\mathbf{x}_*$ will both solve the limit equation $\mathbb{M}\mathbf{x} = \mathcal{F}$, and they have the same component in $(\mathrm{Ker}\mathbb{M})^\perp$. However, the perturbation $\lambda\mathbb{D}$, although vanishing in the limit, leaves a unique choice of the part of the solution that belongs to $(\mathrm{Ker}\mathbb{M})$: it is *the shadow* of the perturbation. □

Remark 4.3.9. The above theorem applies for instance to perturbed mixed formulations as (4.3.20) when a and c are positive definite, with $\mathcal{H} = V \times Q$. In this case, we can set $\mathbb{M}(u, p) = (Au + B^t p, -Bu)$ and $\mathbb{D}(u, p) = (0, Cp)$ and the theorem applies. Note that $\mathrm{Im}\mathbb{M}$ will be closed due to Remark (4.2.6). □

4.3.2 Singular Perturbations

An important variant of problem (4.3.20) will occur in applications (cf. Sect. 10.4). Assume that we are given a Hilbert space W continuously embedded in Q (that is $W \hookrightarrow Q$) and dense in Q. We recall that, as in (4.1.75), the continuous embedding means that $W \subseteq Q$ and, moreover,

$$\|w\|_Q \le C_{WQ}\|w\|_W \qquad \forall\, w \in W \tag{4.3.124}$$

(and without loss of generality we can assume here that $C_{WQ} = 1$). As discussed in Sect. 4.1.6, the density implies that $Q' \hookrightarrow W'$, that Q' is dense in W', the inequality

$$\|w\|_{W'} \le \|w\|_{Q'} \qquad \forall\, w \in Q', \tag{4.3.125}$$

and finally that

$$\langle g, q\rangle_{W'\times W} = \langle g, q\rangle_{Q'\times Q} \quad \text{whenever } g \in Q' \textbf{ and } q \in W. \tag{4.3.126}$$

Remark 4.3.10. Having assumed already that $C_{WQ} = 1$, and also in order to keep the formulae reasonably simple, throughout this subsection, we implicitly assume that the problem has been **adimensionalised**, so that all the quantities we deal with are pure numbers. □

We now consider for every $\lambda > 0$ a perturbation of the type $c(p,q) = \lambda\,(p,q)_W$, that is, we consider problems of the form: *find* (u_λ, p_λ) in $V \times W$ *such that*:

$$a(u_\lambda, v) + b(v, p_\lambda) = \langle f, v\rangle_{V'\times V}, \ \forall\, v \in V, \tag{4.3.127}$$

$$b(u_\lambda, q) - \lambda\,(p_\lambda, q)_W = \langle g_1, q\rangle_{Q'\times Q} + \langle g_2, q\rangle_{W'\times W}, \ \forall\, q \in W. \tag{4.3.128}$$

Depending on which space is identified to its dual space, we shall meet cases where $W \hookrightarrow Q \equiv Q' \hookrightarrow W'$ or where $Q' \hookrightarrow W' \equiv W \hookrightarrow Q$. In all cases, roughly speaking, *the solution of a problem in $V \times Q$ is approximated by the (smoother) solution of a problem in $V \times W$*. To put the problem in the right frame, we suppose first, for simplicity, that $a(\cdot,\cdot)$ is coercive on V and $b(\cdot,\cdot)$ continuous on $V \times Q$ (hence on $V \times W$) with $\mathrm{Im}B$ closed in Q'. We suppose in (4.3.128) that $g_1 \in \mathrm{Im}B$. Taking as usual (4.3.127) with $v = u_\lambda$ and subtracting (4.3.128) with $q = p_\lambda$, and then using the coercivity of a, we have immediately

$$\alpha\|u_\lambda\|^2 + \lambda\|p_\lambda\|_W^2 \le \|f\|_{V'}\|u_\lambda\|_V + \|g_1\|_{Q'}\|\overline{p}_\lambda\|_Q + \|g_2\|_{W'}\|p_\lambda\|_W, \tag{4.3.129}$$

where, as usual, $\overline{p}$ is the component of p in $H^\perp$, with $H = \mathrm{Ker}B^t$. On the other hand, with the usual arguments, one still has from (4.3.127) that $\beta\|\overline{p}_\lambda\|_Q \le \|a\|\ \|u_\lambda\| + \|f\|_{V'}$. By classical arguments, one then gets the estimate

$$\|u_\lambda\|_V^2 + \|\overline{p}_\lambda\|_Q^2 + \lambda\|p_\lambda\|_W^2 \le C\ (\|f\|_{V'}^2 + \|g_1\|_{Q'}^2 + \frac{1}{2\lambda}\|g_2\|_{W'}^2), \tag{4.3.130}$$

where (here and in the sequel of this subsection) we denote by C any constant that depends only on the bilinear forms a and b. If we have $g_2 = g_2(\lambda)$ with $\|g_2(\lambda)\|^2_{W'}/\lambda$ bounded independently of λ, then the solution will become unbounded in W for $\lambda \to 0$ but will remain bounded in $Q/\mathrm{Ker}B^t$, and we expect it to converge to the solution of problem (4.2.6).

Before discussing further this matter, we would like to relax the ellipticity condition on a, assuming ellipticity only in the kernel of B. This, however, will produce unnecessary technical difficulties, so that we will compromise on a slightly stronger condition: We have seen in Proposition 4.3.4 and in Remark 4.3.5 that, when $\mathrm{Im}B$ is closed and a is symmetric, the ellipticity in the kernel of a is equivalent to the property (4.3.111). Here, taking into account Remarks 4.3.7 and 4.3.6 as well, we are going to assume that *for every $\chi_* > 0$ there exists an $\alpha_* > 0$ such that:*

$$\forall \chi > \chi_* \ \exists \tilde{\alpha} > \alpha_* \text{ s. t. } \tilde{\alpha}\|u\|^2_V \leq a(v,v) + \chi\|Bv\|^2_{W'} \ \forall v \in V. \tag{4.3.131}$$

Note that, as W' is bigger than Q' (and has a smaller norm), condition (4.3.131) is *stronger* than the corresponding condition (4.3.111).

Finally, as we are interested in the case of λ small, we will not care about the possible behaviour for $\lambda \to +\infty$, and we can limit ourselves to the case $\lambda \leq \lambda_0$ (implying that χ is bigger that some fixed χ_*). In the next theorem, it will be convenient to take $\lambda_0 = 1/2$, just to have slightly nicer formulae.

Theorem 4.3.4. *Together with Assumption $\mathcal{AB}$, assume that* $\mathrm{Im}B$ *is closed in Q and that $a(\cdot,\cdot)$ is positive semi-definite and verifies* (4.3.131). *Assume moreover that W is a Hilbert space, continuously embedded in Q and dense in Q. Then, for every λ with $0 < \lambda \leq 1/2$, for every $f \in V'$, for every $g_1 \in ImB$, and for every $g_2 \in W'$, the problem* (4.3.127) *and* (4.3.128) *has a unique solution which, moreover, satisfies*

$$\|u_\lambda\|_V + \|\overline{p}_\lambda\|_Q + \lambda^{1/2}\|p_\lambda\|_W \leq C\,(\|f\|_{V'} + \|g_1\|_{Q'} + \frac{1}{\lambda^{1/2}}\|g_2\|_{W'}), \tag{4.3.132}$$

where $\overline{p}_\lambda$ is the component of p_λ in $H^\perp$.

Proof. Since we do not yet have the existence of the solution, we apply a *regularisation argument*. We first substitute a with a_ε given by

$$a_\varepsilon(u,v) = a(u,v) + \varepsilon(u,v)_V,$$

with $\varepsilon > 0$. Then, we prove a-priori bounds independent of ε and we have the solution in the limit for $\varepsilon \to 0+$. For brevity, we do not re-write problem (4.3.127) and (4.3.128) with a_ε in place of a, and we do not indicate the dependence of the solution of the regularised problem on ε. Taking the first equation (4.3.127) with $v = u_\lambda$, and subtracting the second equation (4.3.128) for $q = p_\lambda$, we get

$$\begin{aligned} \varepsilon\|u_\lambda\|^2_V + &a(u_\lambda,u_\lambda) + \lambda(p_\lambda,p_\lambda)_W \\ &= \langle f,u_\lambda\rangle + \langle g_1,p_\lambda\rangle_{Q'\times Q} + \langle g_2,p_\lambda\rangle_{W'\times W}. \end{aligned} \tag{4.3.133}$$

We note that we still have from the first equation that

$$\beta\|\overline{p}_\lambda\|_Q \le C(\|u_\lambda\|_V + \|f\|_{V'}), \tag{4.3.134}$$

and since we assumed $g_1 \in \operatorname{Im} B$, we have

$$\langle g_1, p_\lambda\rangle = \langle g_1, \overline{p}_\lambda\rangle \le C\|g_1\|_{Q'}(\|u_\lambda\|_V + \|f\|_{V'}). \tag{4.3.135}$$

On the other hand, we also have

$$\langle f, u_\lambda\rangle \le \|f\|_{V'}\,\|u_\lambda\|_V \tag{4.3.136}$$

and

$$\langle g_2, p_\lambda\rangle \le \frac{1}{\lambda^{1/2}}\|g_2\|_{W'}\lambda^{1/2}\|p_\lambda\|_W \le \frac{1}{2\lambda}\|g_2\|^2_{W'} + \frac{\lambda}{2}\|p_\lambda\|^2_W. \tag{4.3.137}$$

Inserting (4.3.135), (4.3.136), and (4.3.137) in (4.3.133) and dropping the term with the ε (which is positive), we then easily have

$$\begin{aligned} &a(u_\lambda, u_\lambda) + \lambda\|p_\lambda\|^2_W \\ &\le C\left(\|g_1\|_{Q'}(\|u_\lambda\|_V + \|f\|_{V'}) + \|f\|_{V'}\,\|u_\lambda\|_V + \frac{1}{\lambda}\|g_2\|^2_{W'}\right) \\ &\le C\left(\|u_\lambda\|_V(\|f\|_{V'} + \|g_1\|_{Q'}) + \|f\|^2_{V'} + \|g_1\|^2_{Q'} + \frac{1}{\lambda}\|g_2\|^2_{W'}\right). \end{aligned} \tag{4.3.138}$$

On the other hand, from the second equation we have that $\lambda\, R_W p_\lambda$ (where R_W is the Ritz operator in W, as in Theorem 4.1.2) is equal to $Bu_\lambda - g_1 - g_2$. Hence,

$$\lambda\|p_\lambda\|^2_W = \lambda\|R_W p_\lambda\|^2_{W'} = \frac{1}{\lambda}\|Bu_\lambda - g_1 - g_2\|^2_{W'}. \tag{4.3.139}$$

Hence, using $(a+b)^2 \le 2a^2 + 2b^2$, the assumption $\lambda \le 1/2$, (4.3.139) and (4.3.125), we have:

$$\begin{aligned} \|Bu_\lambda\|^2_{W'} &\le 2\|Bu_\lambda - g_1 - g_2\|^2_{W'} + 2\|g_1 + g_2\|^2_{W'} \\ &\le \frac{1}{\lambda}\|Bu_\lambda - g_1 - g_2\|^2_{W'} + 4\|g_1\|_{W'} + 4\|g_2\|_{W'} \\ &\le \lambda\|p_\lambda\|^2_W + 4\|g_1\|_{Q'} + 4\|g_2\|_{W'}, \end{aligned} \tag{4.3.140}$$

which, joined with (4.3.138), gives immediately

$$
\begin{aligned}
&a(u_\lambda, u_\lambda) + \|Bu_\lambda\|_{W'} + \lambda \|p_\lambda\|_W^2 \\
&\leq C\Big(\|u_\lambda\|_V(\|f\|_{V'} + \|g_1\|_{Q'}) + \|f\|_{V'}^2 + \|g_1\|_{Q'}^2 + \frac{1}{\lambda}\|g_2\|_{W'}^2\Big).
\end{aligned}
\tag{4.3.141}
$$

Finally, using (4.3.134) together with (4.3.131) and (4.3.141) gives

$$
\begin{aligned}
&\|u_\lambda\|_V^2 + \|\overline{p}_\lambda\|_Q^2 + \lambda \|p_\lambda\|_W^2 \\
&\leq C_1\Big(\|u_\lambda\|_V^2 + \|f\|_{V'}^2 + \lambda\|p_\lambda\|_W^2\Big) \\
&\leq C_2\Big(a(u_\lambda, u_\lambda) + \|Bu_\lambda\|_{W'} + \|f\|_{V'}^2 + \lambda\|p_\lambda\|_W^2\Big) \\
&\leq C_3\Big(\|u_\lambda\|_V(\|f\|_{V'} + \|g_1\|_{Q'}) + \|f\|_{V'}^2 + \|g_1\|_{Q'}^2 + \frac{1}{\lambda}\|g_2\|_{W'}^2\Big),
\end{aligned}
\tag{4.3.142}
$$

which easily yields the result (4.3.132) □

As we shall see, a particularly interesting case is met when both g_1 and g_2 are zero. In fact, it is remarkable that in this case we *do not* need the *inf-sup* condition (meaning that we do not need ImB to be closed). We have indeed the following proposition.

Theorem 4.3.5. *Together with Assumption $\mathcal{AB}$, assume that $a(\cdot,\cdot)$ is positive semi-definite and verifies* (4.3.131)*. Assume, moreover, that W is a Hilbert space, continuously embedded in Q and dense in Q. Then, for every λ with $0 < \lambda \leq 1/2$, and for every $f \in V'$, the problem: find (u_λ, p_λ) in $V \times W$ such that*

$$
a(u_\lambda, v) + b(v, p_\lambda) = \langle f, v\rangle_{V'\times V}, \ \forall\, v \in V, \tag{4.3.143}
$$

$$
b(u_\lambda, q) - \lambda\,(p_\lambda, q)_W = 0, \ \forall\, q \in W, \tag{4.3.144}
$$

has a unique solution, that moreover satisfies

$$
\tilde{\alpha}\|u_\lambda\|_V^2 + \lambda\|p_\lambda\|_W^2 \leq \frac{4\|f\|_{V'}^2}{\tilde{\alpha}}, \tag{4.3.145}
$$

where $\tilde{\alpha}$ is given in (4.3.131)*.*

Proof. Mimicking the proof of Theorem 4.3.4, we now have, using (4.3.131), then (4.3.139), and finally (4.3.133):

$$
\begin{aligned}
&\tilde{\alpha}\|u_\lambda\|_V^2 + \lambda\|p_\lambda\|_W^2 \\
&\qquad \leq a(u_\lambda, u_\lambda) + \frac{1}{\lambda}\|Bu_\lambda\|_{W'}^2 + \lambda\|p_\lambda\|_W^2 = a(u_\lambda, u_\lambda) + \frac{2}{\lambda}\|Bu_\lambda\|_{W'}^2 \\
&\qquad\qquad \leq 2\Big(a(u_\lambda, u_\lambda) + \frac{1}{\lambda}\|Bu_\lambda\|_{W'}^2\Big) \leq 2\langle f, u_\lambda\rangle_{V'\times V},
\end{aligned}
\tag{4.3.146}
$$

and the result follows immediately since

$$2\langle f, u_\lambda\rangle_{V'\times V} \leq \frac{2\|f\|_{V'}^2}{\tilde{\alpha}} + \frac{\tilde{\alpha}\|u_\lambda\|_V^2}{2}. \qquad \square$$

As we consider the augmented problem as a perturbation, we shall now try to get an estimate on $\|u - u_\lambda\|_V$ and $\|p - p_\lambda\|_Q$ as $\lambda \to 0+$.

Proposition 4.3.5. *With the same assumptions of Theorem 4.3.4, and assuming moreover that $g_2 = 0$, let $(u_\lambda, p_\lambda) \in V \times W$ be the solution of problem* (4.3.127)–(4.3.128) *and $(u, p) \in V \times Q$ be the solution of problem* (4.2.6). *We then have*

$$\|u - u_\lambda\|_V + \|\overline{p} - \overline{p}_\lambda\|_Q \leq C \inf_{p_w \in W}\Big[\|p - p_w\|_Q + \sqrt{\lambda}\,\|p_w\|_W\Big]. \qquad (4.3.147)$$

Proof. Subtracting (4.3.127)–(4.3.128) from (4.2.6) with $q \in W$, one easily has

$$\begin{cases} a(u - u_\lambda, v) + b(v, p - p_\lambda) = 0, \ \forall v \in V, \\ b(u - u_\lambda, q) = \lambda\,(p_\lambda, q)_W, \ \forall q \in W. \end{cases} \qquad (4.3.148)$$

The argument of Proposition 4.3.3 cannot be applied, for it would require (in the second equation of (4.3.148)) $q \in Q$. However, let p_w be any element of W. We rewrite (4.3.148) as

$$\begin{cases} a(u - u_\lambda, v) + b(v, p_w - p_\lambda) = b(v, p_w - p), \ \forall\, v \in V, \\ b(u - u_\lambda, q) + \lambda\,(p_w - p_\lambda, q)_W = -\lambda\,(p_w, q)_W, \ \forall\, q \in W. \end{cases} \qquad (4.3.149)$$

We can now apply Theorem 4.3.4 with $\langle g_2, q\rangle = \lambda(p_w, q)_W$, and use estimate (4.3.132) to get

$$\|u - u_\lambda\|_V^2 + \|\overline{p}_w - \overline{p}_\lambda\|_Q^2 \leq C\,(\|p_w - p\|_Q^2 + \lambda\,\|p_w\|_W^2). \qquad (4.3.150)$$

From the triangle inequality and the arbitrariness of p_w, one deduces (4.3.147). $\square$

Remark 4.3.11. The right-hand side of (4.3.147) will, in general, tend to zero with λ whenever p is more regular than just $p \in Q$. For instance, if $p \in W$, we can take $p_W = p$ and (4.3.147) will give

$$\|u - u_\lambda\|_V + \|\overline{p} - \overline{p}_\lambda\|_Q \leq C\,\sqrt{\lambda}. \qquad (4.3.151)$$

See also Remark 4.3.14 here below. $\square$

Remark 4.3.12. The above result is not optimal. For instance, it does not reduce to the estimate (4.3.109) of Proposition 4.3.3 when $W = Q$. Let us suppose however, for simplicity, that $\operatorname{Im} B = Q'$, and consider the space W^+ defined as

$$W^+ := R_W^{-1}(Q'), \tag{4.3.152}$$

where R_W is as usual the Ritz operator in W as defined in Theorem 4.1.2. Since Q' is a dense subspace of W', we easily have that W^+ is a dense subspace of W, and moreover,

$$W^+ \hookrightarrow W \hookrightarrow Q. \tag{4.3.153}$$

Furthermore, for every $p^+ \in W^+$, there exists, from the definition (4.3.152), a $g \in Q'$ such that

$$(p^+, q)_W = (R_W^{-1} g, q)_W = \langle g, q\rangle_{W' \times W} \tag{4.3.154}$$

and, using (4.1.76), we have, for every $q \in W$,

$$(p^+, q)_W = \langle g, q\rangle_{W' \times W} = \langle g, q\rangle_{Q' \times Q} \le \|g\|_{Q'} \|q\|_Q \quad \forall\, q \in W, \tag{4.3.155}$$

where we also used (4.3.126). We can think of W^+ as a subspace of W made of more regular functions. Taking now $p_w = p_{w_+} \in W^+$, we can now go back to (4.3.149), considering this time that the right-hand side of the second equation (that is $\lambda\, (p_w, q)_W$) corresponds to the choice $g_2 = 0$ and $\langle g_1, q\rangle = \lambda\, (p_{w_+}, q)_W$ when using Theorem 4.3.4. From (4.3.132), we now have

$$\|u - u_\lambda\|_V + \|p - p_\lambda\|_Q \le C \left(\inf_{p_{w_+} \in W^+} \|p - p_{w_+}\|_Q + \lambda\, \|p_{w_+}\|_{W^+} \right) \tag{4.3.156}$$

where we also took into account that we assumed $\operatorname{Im} B = Q'$ and hence $\overline{p}_\lambda = p_\lambda$. Now, (4.3.156) is optimal for $W^+ = Q$. □

Remark 4.3.13. The argument of the above remark can easily be extended to the case in which $\operatorname{Im} B$ is closed but does not coincide with Q'. We simply have to take $W^+ := R_W^{-1} H^0$ (where H^0 is the polar space of $H \equiv \operatorname{Ker} B^t$), so that $(p_{w+}, q)_W \le C\, \|\overline{q}\|_Q$. In general, such a W^+ will not be dense in W, but in many applications p (belonging to $H^\perp$) will still belong to its closure (that is, you can still approximate p with a sequence of elements in W^+). □

Remark 4.3.14. In the spirit of Remark 4.3.11, we observe that, here again, the right-hand side of (4.3.156) can be bounded in terms of λ whenever p is more regular. In particular for $p \in W^+$, we would have

$$\|u - u_\lambda\|_V + \|p - p_\lambda\|_Q \le C\, \lambda. \tag{4.3.157}$$

□

Remark 4.3.15. In both (4.3.147) and (4.3.156), an **intermediate regularity** between Q and W^+ can provide an **intermediate speed of convergence** for $\lambda \to 0$. More precisely, let us suppose that p belongs to $[W^+, Q]_{\theta,\infty}$ for $0 < \theta < 1$. The space $[W^+, Q]_{\theta,\infty}$ is an interpolation space between W^+ and Q. We refer the reader to [62] for more details on these spaces. Here, we just recall that

$$\|p\|_{[W^+,Q]_{\theta,\infty}} = \sup_{\lambda>0} \inf_{p_{w_+}\in W^+} (\lambda^{-\theta}\|p - p_{w_+}\|_Q + \lambda^{1-\theta}\;\|p_{w_+}\|). \qquad (4.3.158)$$

As a consequence, if $p \in [W^+, Q]_{\theta,\infty}$, then we have

$$\begin{aligned}\inf_{p_{w_+}\in W^+} &(\|p - p_{w_+}\| + \lambda\;\|p_{w_+}\|_{W^+})\\ &= \lambda^{\theta} \inf_{p_{w_+}\in W^+} (\lambda^{-\theta}\|p - p_{w_+}\|_Q + \lambda^{1-\theta}\;\|p_{w_+}\|)\\ &\leq \lambda^{\theta}\|p\|_{[W^+,Q]_{\theta,\infty}}\end{aligned}$$

(as in [62], Theorem 3.12). Hence, (4.3.156) can be written as

$$\|u - u_\lambda\|_V + \|p - p_\lambda\|_Q \leq C\;\lambda^{\theta}\|p\|_{[W^+,Q]_{\theta,\infty}}. \qquad (4.3.159)$$

Note that, in particular if $W^+ := H^1(\Omega)$ and $Q := L^2(\Omega)$, we have that

$$H^{\theta}(\Omega) \hookrightarrow [W^+, Q]_{\theta,\infty}.$$

Hence, if $p \in H^{\theta}(\Omega)$, we will also have $p \in [W^+, Q]_{\theta,\infty}$, and estimate (4.3.159) will hold true. Clearly, a similar argument could be applied to the estimate (4.3.147) for p having an intermediate regularity between Q and W. $\square$

Chapter 5
Approximation of Saddle Point Problems

This chapter concludes the abstract analysis of mixed formulations. After studying the finite dimensional case in Chap. 3 and the infinite-dimensional case in Chap. 4, we analyse here the problem of *approximating* the infinite dimensional case (that, in practice, will come from a PDE problem) by means of a finite dimensional one, treatable with a computer. We shall see that (apparently) reasonable approximations of a well-posed infinite dimensional problem could produce an ill-posed finite dimensional problem or, more generally, a problem whose solution is not an approximation of the original problem. Hence, the typical results of this chapter will be bounds for the difference between the *exact solution* (that is, the solution of the original, infinite dimensional problem) and the *approximate solution* (that is, the solution of the discretised, finite dimensional problem). The approximations considered in this chapter are evidently targeted to the Finite Element spaces introduced in Chap. 2. However, many results could be applied to other types of approximation such as spectral methods or Finite Volume methods.

In the first section, we shall present the basic assumptions that will be mostly used throughout this chapter and then discuss some basic relationships between differential operators and their realisations in finite dimensional spaces. In Sect. 5.2, we shall present, essentially, the main convergence and error estimates results, together with some simple examples related to the mixed formulation of a toy one-dimensional problem. In spite of the simplicity of this toy problem (and of its negligible practical interest), we recommend these examples (Sect. 5.2.4) for their enlightening capabilities. The third section will be devoted to a quick survey of the main tricks that can be used in order to prove the *inf-sup* condition for the discretised problem (once we know that it holds for the original infinite-dimensional problem). Section 5.5 will deal with extensions of the error estimates of Sect. 5.2, including perturbed problems, non conforming approximations, and dual error estimates. Section 5.6 will present some numerical issues related with the actual resolution of the discretised problems. Finally, the last section will anticipate, at an abstract level, some stabilising techniques that will be detailed, for each particular problem, in the following chapters.

D. Boffi et al., *Mixed Finite Element Methods and Applications*, Springer Series in Computational Mathematics 44, DOI 10.1007/978-3-642-36519-5_5,

5.1 Basic Results

5.1.1 The Basic Assumptions

We now turn to the approximation of problem (4.2.6). For this, we place ourselves in the framework of Assumption $\mathcal{AB}$ of Chap. 4, which we recall for the convenience of the reader.

Assumption $\mathcal{AB}$: *We are given two Hilbert spaces, V and Q, and two continuous bilinear forms: $a(\cdot,\cdot)$ on $V \times V$ and $b(\cdot,\cdot)$ on $V \times Q$. We denote by A and B, respectively, the linear continuous operators associated with them. We also set*

$$K := \mathrm{Ker}B \qquad \text{and} \qquad H := \mathrm{Ker}B^t. \tag{5.1.1}$$

□

Given $f \in V'$ and $g \in Q'$, we can consider our original problem of finding $u \in V$ and $p \in Q$ solution of

$$\begin{cases} a(u,v) + b(v,p) = \langle f, v\rangle_{V'\times V} \ \forall\, v \in V, \\ b(u,q) = \langle g, q\rangle_{Q'\times Q} \ \forall\, q \in Q. \end{cases} \tag{5.1.2}$$

Recalling Theorem 4.2.2, we see that the necessary and sufficient condition for the unique solvability of (5.1.2) is that the two following conditions are satisfied:

$$A_{KK'} \text{ is an isomorphism from } K \text{ to } K' \tag{5.1.3}$$

(where $A_{KK'}$ was defined in (4.2.17)) and

$$\mathrm{Im}B = Q'. \tag{5.1.4}$$

We also saw that (5.1.3) is equivalent to requiring that there exists an $\alpha_1 > 0$ such that

$$\begin{aligned} &\inf_{v_0\in K}\sup_{w_0\in K} \frac{a(v_0,w_0)}{\|v_0\|_V\,\|w_0\|_V} \geq \alpha_1 \\ &\inf_{w_0\in K}\sup_{v_0\in K} \frac{a(v_0,w_0)}{\|v_0\|_V\,\|w_0\|_V} \geq \alpha_1 \end{aligned} \tag{5.1.5}$$

and that (5.1.4) is equivalent to requiring that there exists a $\beta > 0$ such that

$$\inf_{q\in Q}\sup_{v\in V} \frac{b(v,q)}{\|v\|_V\,\|q\|_Q} \geq \beta. \tag{5.1.6}$$

As we have seen already several times, in many applications, condition (5.1.1) will be an immediate consequence of the (slightly) stronger *ellipticity in the kernel* condition

$$\exists\, \alpha_0 > 0 \text{ such that } a(v_0,v_0) \geq \alpha_0\|v_0\|^2_V \quad \forall\, v_0 \in K, \tag{5.1.7}$$

or of the even stronger *global ellipticity* condition:

$$\exists \alpha > 0 \text{ such that } a(v,v) \geq \alpha \|v\|_V^2 \quad \forall\, v \in V. \tag{5.1.8}$$

We also recall that in Theorem 4.2.3 we also provided *stability properties*, showing that the norm of the solution (u, p) of (5.1.2) can be bounded in terms of the norms of the data f and g, together with the values of the constants α and β and the norm $\|a\|$ of the bilinear form a.

Then let $V_h \subset V$ and $Q_h \subset Q$ be finite dimensional subspaces of V and Q respectively. The index h will eventually refer to a mesh from which these approximations are derived. We can obviously consider the restriction of the bilinear forms a and b to $V_h \times V_h$ and to $V_h \times Q_h$, respectively. Hence, we can consider the corresponding approximation of problem (5.1.2), looking for a couple (u_h, p_h) in $V_h \times Q_h$, solution of

$$\begin{cases} a(u_h, v_h) + b(v_h, p_h) = \langle f, v_h\rangle_{V'\times V} \;\forall\, v_h \in V_h, \\ b(u_h, q_h) = \langle g, q_h\rangle_{Q'\times Q} \;\forall\, q_h \in Q_h. \end{cases} \tag{5.1.9}$$

Remark 5.1.1. As we did in the previous chapter, to be precise, we should actually write $a(E_{V_h}u_h, E_{V_h}v_h)$ instead of $a(u_h, v_h)$, and $b(E_{V_h}v_h, E_{Q_h}q_h)$ instead of $b(v_h, q_h)$ (where obviously E_{V_h} and E_{Q_h} are the extension operators $V_h \to V$ and $Q_h \to Q$, respectively). However, we shall not do that, unless it is really helpful in order to clarify something. □

Following again the previous chapter, we can consider the restrictions $B_{V_h Q'_h}$, and $B^t_{Q_h V'_h}$ of the operators B and B^t respectively, which, for brevity we denote now by B_h and B^t_h. Recalling the notation (4.1.84), we then have

$$B_h\, v_h := \pi_{Q'_h} B E_{V_h}\, v_h \;\forall v_h \in V_h \quad B^t_h\, q_h := \pi_{V'_h} B^t E_{Q_h}\, q_h \;\forall q_h \in Q_h. \tag{5.1.10}$$

Similarly, A_h and A^t_h will be given by

$$A_h\, v_h := \pi_{V'_h} A E_{V_h}\, v_h \quad \text{and} \quad A^t_h\, v_h := \pi_{V'_h} A^t E_{V_h}\, v_h \quad \forall v_h \in V_h. \tag{5.1.11}$$

To avoid repeating the same assumptions in every statement, we condense additionally the above discrete framework in the following **Assumption** $\mathcal{AB}_h$.

Assumption $\mathcal{AB}_h$: *Together with* **Assumption** $\mathcal{AB}$, *we assume that we are given two finite dimensional spaces $V_h \subset V$ and $Q_h \subset Q$. Together with the kernels K and H, we consider then the discrete kernels*

$$K_h \equiv \operatorname{Ker} B_h := \{v_h \in V_h \text{ such that } b(v_h, q_h) = 0, \quad \forall\, q_h \in Q_h\}, \tag{5.1.12}$$

$$H_h \equiv \operatorname{Ker} B^t_h := \{q_h \in Q_h \text{ such that } b(v_h, q_h) = 0, \quad \forall\, v_h \in V_h\}. \tag{5.1.13}$$

Finally, for every element $g \in Q'$, we introduce the space

$$Z_h(g) := \{v_h \in V_h \text{ such that } b(v_h, q_h) = \langle g, q_h \rangle, \quad \forall\, q_h \in Q_h\}, \tag{5.1.14}$$

and for every element $f \in V'$, we introduce the space

$$Z_h^*(f) := \{q_h \in Q_h \text{ such that } b(v_h, q_h) = \langle f, v_h \rangle, \quad \forall\, v_h \in V_h\}. \tag{5.1.15}$$

We observe that in (5.1.14) we could write

$$B_h v_h = \pi_{Q_h'} g \quad \text{instead of} \quad b(v_h, q_h) = \langle g, q_h \rangle, \quad \forall\, q_h \in Q_h \tag{5.1.16}$$

while in (5.1.15) we could write

$$B_h^t q_h = \pi_{V_h'} f \quad \text{instead of} \quad b(v_h, q_h) = \langle f, v_h \rangle, \quad \forall\, v_h \in V_h. \tag{5.1.17}$$

The problems that we have to solve here concern both the existence and uniqueness of $\{u_h, q_h\}$ and the estimation of $\|u - u_h\|_V$ and $\|p - p_h\|_Q$. In view of the previous discussion, and considering that the two conditions (5.1.6) and (5.1.1) are *necessary*, it is then natural to assume that, *for every h*, there exists an $\alpha_1^h > 0$ such that:

$$\inf_{v_0^h \in K_h} \sup_{w_0^h \in K_h} \frac{a(v_0^h, w_0^h)}{\|v_0^h\|_V \, \|w_0^h\|_V} = \inf_{w_0^h \in K_h} \sup_{v_0^h \in K_h} \frac{a(v_0^h, w_0^h)}{\|v_0^h\|_V \, \|w_0^h\|_V} \geq \alpha_1^h, \tag{5.1.18}$$

and a $\beta_h > 0$ such that

$$\inf_{q_h \in Q_h} \sup_{v_h \in V_h} \frac{b(v_h, q_h)}{\|v_h\|_V \, \|q_h\|_Q} \geq \beta_h. \tag{5.1.19}$$

Remark 5.1.2. Here and in all this section, we will accept that β_h and α_1^h (as well as α_0^h or α_h here below) *depend on h*. Clearly, the desirable situation is that they are actually *independent from h*. However, the latter case can be immediately derived from the former. On the other hand, when one of the two constants (or both) depends on h, our error estimates might still provide a convergence result, although, in general, not an optimal one. □

As natural, in most applications, condition (5.1.18) could be replaced by the simpler *ellipticity in the kernel* condition

$$\exists\, \alpha_h^0 > 0 \text{ such that } a(v_0^h, v_0^h) \geq \alpha_h^0 \|v_0^h\|_V^2 \qquad \forall\, v_0^h \in K_h, \tag{5.1.20}$$

or by the even simpler *global ellipticity*

$$\exists\, \alpha_h > 0 \text{ such that } a(v_h, v_h) \geq \alpha_h \|v_h\|_V^2 \qquad \forall\, v_h \in V_h. \tag{5.1.21}$$

Remark 5.1.3. We also recall from the previous chapters (Lemma 4.2.2) that if the bilinear form a is *symmetric and positive semi-definite*, then the non-singularity on K_h given in (5.1.18) implies the ellipticity in K_h given in (5.1.20) with

$$\alpha_0^h = (\alpha_1^h)^2/M_a^h, \tag{5.1.22}$$

where

$$M_a^h := \sup_{v_h \in V_h} \sup_{w_h \in V_h} \frac{a(v_h, w_h)}{\|v_h\|_V \, \|w_h\|_V}. \tag{5.1.23}$$

□

5.1.2 The Discrete Operators

In (5.1.10) and (5.1.11), we defined the discrete operators A_h, B_h and their adjoints. It will be convenient to try to understand, from the very beginning, the relationships between the discrete operators and their continuous counterpart. Let us consider B_h (somehow, the most important) first. We recall that B_h maps V_h into Q'_h (while B maps V into Q'). Hence, to be picky, the comparison should not be made between B and B_h, but rather between BE_{V_h} and $E_{Q'_h}B_h$: for $v_h \in V_h$, we then consider the difference between

$$E_{Q'{}_h}\pi_{Q'_h}BE_{V_h}v_h \quad \text{and} \quad BE_{V_h}v_h. \tag{5.1.24}$$

It is clear that the operator $E_{Q'{}_h}\pi_{Q'_h}$ coincides with the identity operator only when applied to objects that, loosely speaking, are already in Q'_h. Hence, for every v_h^* in V_h, we have

$$B_h v_h^* = BE_{V_h}v_h^* (= Bv_h^*) \qquad \textbf{iff} \qquad Bv_h^* \in Q'_h \tag{5.1.25}$$

and in general,

$$\{B_h v_h = BE_{V_h}v_h \ \forall\, v_h \in V_h\} \quad \Leftrightarrow \quad B(V_h) \subseteq Q'_h. \tag{5.1.26}$$

We shall meet cases where this inclusion holds, but they are far from being the rule. When $B(V_h) \subseteq Q'_h$, the left-hand side of (5.1.26) could also be written, with a very minor abuse of language that we are going to use quite often, as $B_h v_h = Bv_h$. In this case (and only in this case), we might say that *B_h is the restriction of B to V_h*.

It is clear that similar considerations can be made for the operator B_h^t and for the operators A_h and A_h^t.

We also recall from Sect. 4.1.8 of the previous chapter that:

- The kernel K_h is not, in general, a subspace of K,
- The kernel H_h is not, in general, a subspace of H,
- As a consequence, property (5.1.1) will not imply (5.1.18),
- Similarly, property (5.1.6) will not imply (5.1.19).

In a certain number of applications, as we have seen, we have however that the bilinear form a is elliptic on the whole space V (that is, (5.1.8) holds). In these cases, it is clear that both (5.1.1) and (5.1.18) will be satisfied, independently of the nature of the kernels K and K_h.

Finally, we recall from Proposition 4.1.6 that

$$\mathrm{Ker}\,B_h \subseteq \mathrm{Ker}\,B \quad \textbf{iff} \quad \pi_{V_h'}(\mathrm{Im}\,B^t) \subseteq \mathrm{Im}\,B_h^t \tag{5.1.27}$$

and

$$\mathrm{Ker}\,B_h^t \subseteq \mathrm{Ker}\,B^t \quad \textbf{iff} \quad \pi_{Q_h'}(\mathrm{Im}\,B) \subseteq \mathrm{Im}\,B_h. \tag{5.1.28}$$

Remark 5.1.4. The lack of inclusion of the discrete kernels has different practical relevance in the two cases (K_h and H_h).

On one hand, the inclusion $K_h \subset K$ is a nice property, useful when you have it, but not *absolutely needed*.

On the other hand, the inclusion $H_h \subset H$ is extremely important. In the first place, as we typically assume that $H = \{0\}$, the lack of this last inclusion will imply that H_h will have dimension ≥ 1, so that problem (5.1.9) might fail to have a solution, and when it does, the solution will be determined up to an element of H_h. Secondly, even in cases (as we did in Sect. 4.2.4) when we do not have $H = \{0\}$, the elements of H should be considered as "physically natural": for instance, the pressure in some fluid mechanical problem is defined up to a constant, as it is often the case for the electric potential; in electromagnetic problems, the vector potential is often defined up to a gradient, and so on. As we have seen, the "solution" in these cases is to restrict the space Q, taking for instance, in its place, the space $Q_{/H}$ (or $H^\perp$, which is essentially the same): for instance, in the case of a pressure (when it is defined up to an additive constant), this will correspond to restrict Q to its subspace made of functions with zero mean value. Similarly, the electric potential is often assumed to vanish at a given point (chosen once and for all), the vector potential is assumed to be solenoidal, and so on. In these cases, you would like the solutions of the discretised problem to have the same gauge, which is, however, not true when the inclusion $H_h \subset H$ is not satisfied. Hence, the elements of H_h that do not belong to H should be regarded as *spurious numerical artefacts*, and in general, one does not like to have them around. □

The next results show how the inclusion of kernels is related to another interesting property, which will play an important role in the sequel.

Proposition 5.1.1. *In the Assumption $\mathcal{AB}_h$, suppose that there exists a linear operator $\Pi_h : V \to V_h$ such that*

$$b(v - \Pi_h v, q_h) = 0 \qquad \forall\, v \in V,\ \forall\, q_h \in Q_h. \tag{5.1.29}$$

Then, the properties in (5.1.28) *are verified.*

Proof. The proof is immediate. Indeed, (5.1.29) can also be written as

$$\pi_{Q'_h} Bv = \pi_{Q'_h} BE_{V_h} \Pi_h v \equiv B_h \Pi_h v \quad \forall v \in V, \tag{5.1.30}$$

and the right-hand side of (5.1.30) obviously belongs to $\mathrm{Im} B_h$. □

Remark 5.1.5. It can be easily seen that, conversely, the properties in (5.1.28) imply the existence of an operator $\Pi_h : V \to V_h$ that satisfies (5.1.29). Indeed, for every $v \in V$, we have from (5.1.28) that $\pi_{Q'_h} Bv \in \mathrm{Im} B_h$. As both V_h and Q_h are finite dimensional, we can therefore apply Corollary 3.1.1 and set $\Pi_h v := L_{B_h}(\pi_{Q'_h} Bv)$ which, using (3.1.40), will satisfy $B_h(\Pi_h v) = \pi_{Q'_h} Bv$, that is, (5.1.30). □

Exchanging B and B^t in the above discussion, we have the following proposition.

Proposition 5.1.2. *The following statements are equivalent:*

- *There exists a linear mapping $\Phi_h : Q \to Q_h$ such that*

$$b(v_h, q - \Phi_h q) = 0 \quad \forall q \in Q, \ \forall v_h \in V_h. \tag{5.1.31}$$

- $\pi_{V'_h} B^t q_h = B^t_h \Phi_h q_h \quad \forall q_h \in Q_h.$
- $\mathrm{Ker} B_h \subseteq \mathrm{Ker} B.$
- $\pi_{V'_h} \mathrm{Im} B^t \subseteq \mathrm{Im} B^t_h.$

□

*Remark 5.1.6 (B-**compatible operator**).* In the following, an operator satisfying (5.1.29) will be called a B**-compatible operator**. As we shall see later on, such operators will play an important role to obtain *inf-sup* conditions. □

Property (5.1.30) can be summarised by the fact that the following diagram commutes:

$$\begin{array}{ccc} V & \xrightarrow{\ B\ } & Q' \\ {\scriptstyle \Pi_h}\downarrow & & \downarrow{\scriptstyle \pi_{Q'_h}} \\ V_h & \xrightarrow[\ B_h\ ]{} & Q'_h \end{array} \tag{5.1.32}$$

and the corresponding property $\pi_{V'_h} B^t q_h = B^t_h \Phi_h q_h$ in Proposition 5.1.2 could also be summarised by a commuting diagram:

$$\begin{array}{ccc} Q & \xrightarrow{\ B^t\ } & V' \\ {\scriptstyle \Phi_h}\downarrow & & \downarrow{\scriptstyle \pi_{V'_h}} \\ Q_h & \xrightarrow[\ B^t_h\ ]{} & V'_h \end{array} \tag{5.1.33}$$

Remark 5.1.7. The case when B_h is the restriction of B to V_h (that is, when $B(V_h) \subseteq Q'_h$) is especially interesting. In this case, we can take $\Phi_h = \pi_{Q_h}$. Indeed, if $B(V_h) \subseteq Q'_h$, then as in (5.1.26), we have

$$BE_{V_h} \equiv E_{Q'_h}\pi_{Q'_h}BE_{V_h}. \tag{5.1.34}$$

Transposing both members of (5.1.34) and using the definition of B_h^t given in (5.1.10), we obtain

$$\pi_{V'_h}B^t \equiv \pi_{V'_h}B^t E_{Q_h}\pi_{Q_h} = B_h^t\pi_{Q_h}, \tag{5.1.35}$$

implying that the diagram (5.1.33) commutes taking $\Phi_h = \pi_{Q_h}$. □

Remark 5.1.8. The result of Proposition 5.1.1 is also directly linked to the *inf-sup* condition. It may be worthwhile to point out some facts. As $\mathrm{Im}B_h$ is finite dimensional, we always have, recalling that from (5.1.13) we have $H_h := \mathrm{Ker}B_h^t$,

$$\sup_{v_h\in V_h}\frac{b(v_h,q_h)}{\|v_h\|_V} \geq \beta_h\,\|q_h\|_{Q_h/H_h}. \tag{5.1.36}$$

However, as we stated in Remark 5.1.4, this may be useless if H_h is larger than H. The following result shows that the existence of the operator Π_h satisfying (5.1.29) implies a stronger version of (5.1.36). □

Corollary 5.1.1. *In the assumptions of Proposition 5.1.1, we have that property* (5.1.29) *is equivalent to*

$$\exists\,\beta_h > 0 \text{ such that } \sup_{v_h\in V_h}\frac{b(v_h,q_h)}{\|v_h\|_V} \geq \beta_h\,\|q_h\|_{Q/H}. \tag{5.1.37}$$

Proof. On the one hand, (5.1.29) implies the inclusion of kernels (5.1.27), which, joined to (5.1.36), gives immediately (5.1.37). Conversely, (5.1.37) implies that $\|q_h\|_{Q/H} = 0$ for any $q_h \in H_h$, which implies $H_h \subseteq H$. □

Remark 5.1.9. The reader should be aware of the differences between the (apparently similar) *inf-sup* conditions (5.1.19), (5.1.36) and (5.1.37). Indeed, (5.1.36) is always true as the finite-dimensional space $Im(B_h)$ is closed. Condition (5.1.19) means that B_h is surjective, that is, $\mathrm{Ker}B_h^t = \{0\}$. On the other hand, (5.1.37) implies the inclusion $H_h \subseteq H$. Thus, (5.1.19) coincides with (5.1.37) in the case of B being surjective. □

Remark 5.1.10. We already pointed out that since Q_h is finite dimensional, then $\mathrm{Im}B_h$ is closed. In particular, we can then apply Corollary 4.1.2 of the previous chapter, and obtain that there exists a lifting operator L_{B_h} that for any $g_h \in \mathrm{Im}B_h$ gives an $L_{B_h}(g_h)$ such that $B_h(L_{B_h}(g_h)) = g_h$ and

$$\|L_{B_h}g_h\|_{V_h} \leq \frac{1}{\beta_h}\|g_h\|_{Q'_h}. \tag{5.1.38}$$

From (5.1.28), we have that if $\mathrm{Ker}B_h^t \subseteq \mathrm{Ker}B^t$, then for any $u \in V$, we have $\pi_{Q_h'} Bu \in \mathrm{Im}B_h$ and therefore for every $u \in V$ we have

$$\|L_{B_h}(\pi_{Q_h'} Bu)\|_{V_h} \leq \frac{\|\pi_{Q_h'} Bu\|_{Q_h'}}{\beta_h} \leq \frac{\|Bu\|_{Q'}}{\beta_h} \leq \frac{\|b\|}{\beta_h}\|u\|_V. \tag{5.1.39}$$

□

Finally, from Proposition 5.1.1 and Remark 5.1.10, we obtain the following result that will be useful later on.

Proposition 5.1.3. *In the same assumptions as in Proposition 5.1.1, we have that, for any* $u \in V$,

$$\inf_{w_h \in Z_h(Bu)} \|u - w_h\|_V \leq \frac{2\|b\|}{\beta_h} \inf_{v_h \in V_h} \|u - v_h\|_V. \tag{5.1.40}$$

Proof. Let v_h be any element of V_h. We set

$$d_h := L_{B_h}(\pi_{Q_h'} B(u - v_h)). \tag{5.1.41}$$

We obviously have $B_h(d_h + v_h) = \pi_{Q_h'} Bu$ and therefore, from (5.1.16) and (5.1.14), we have $w_h := (d_h + v_h) \in Z_h(Bu)$. Moreover, from (5.1.39) we have that $\|d_h\|_{V_h} \leq (\|b\|/\beta_h)\|u - v_h\|_V$. Since $u - w_h = (u - v_h) - d_h$, we have from the triangle inequality

$$\|u - w_h\|_V \leq (1 + \frac{\|b\|}{\beta_h})\|u - v_h\|_V \tag{5.1.42}$$

and the result follows immediately since $\|b\| \geq \beta_h$. □

Remark 5.1.11. The above results show that bounding β_h from below, independently of h, will enable us to transform approximation estimates in $Z_h(g)$ into standard approximation estimates in V_h. This is also clearly related to a bound on the norm of the operator Π_h. We shall come back to this in Sect. 5.4. It must again be emphasised that the inclusion $H_h \subseteq H$ is essential for this result. Cases where this inclusion fail will, at best, require a special analysis when they are not totally doomed. □

5.2 Error Estimates for Finite Dimensional Approximations

5.2.1 Discrete Stability and Error Estimates

In the following sections, we are going to see some particular cases where conditions (5.1.18) and (5.1.19) are satisfied. Here, however, we first want to see how they can be used to prove error bounds on $\|u - u_h\|_V$ and $\|p - p_h\|_Q$. For this, we follow the strategy of [112]: we first consider general approximations u_I and p_I of u and p, respectively, in V_h and Q_h. You can think of them as suitable interpolations, or as the best approximations in the norms of V and Q, respectively, or just as general elements in V_h and Q_h. Indeed, the following argument will hold *for any* $u_I \in V_h$ *and for any* $p_I \in Q_h$. The idea is to first bound the distance of (u_h, p_h) from (u_I, p_I) in terms of the distance of (u_I, p_I) from (u, p), and part of the theorems of this chapter will express bounds of this type. The estimate of the distance of (u_h, p_h) from (u, p) will then follow by the triangle inequality, and another part of the theorems of this chapter will express the distance of (u_h, p_h) from (u, p) in terms of the distance of (u, p) from its best approximation (v_h, q_h) in $V_h \times Q_h$. It will then be convenient, from the very beginning, to define the **approximation errors**

$$E_u := \inf_{v_h \in V_h} \|u - v_h\|_V, \tag{5.2.1}$$

$$E_p := \inf_{q_h \in Q_h} \|p - q_h\|_Q. \tag{5.2.2}$$

Moreover, on several occasions, as intermediate steps, we will express the distance of (u_h, p_h) from (u, p) in terms of the distance of (u, p) from pairs (u_I, p_I) where one of the two (or both) are confined to belong to some affine manifold (typically, $u_I \in Z_h(Bu)$ and/or $p_I \in Z_h^*(B^t p)$, as defined in (5.1.14) and (5.1.15)). To this purpose, we introduce as well

$$E_u^Z := \inf_{v_h \in Z_h(Bu)} \|u - v_h\|_V, \tag{5.2.3}$$

$$E_p^Z := \inf_{q_h \in Z_h^*(B^t p)} \|p - q_h\|_Q. \tag{5.2.4}$$

We also explicitly point out that in this procedure we will not, in general, require us to assume the uniqueness of the solution (u, p) of the continuous problem, but only that of the discretised problem.

For this, we use the linearity of a and b to combine the continuous problem (5.1.2) with the discretised problem (5.1.9), adding and subtracting u_I and p_I. We obtain:

$$\begin{cases} a(u_h - u_I, v_h) + b(v_h, p_h - p_I) = a(u - u_I, v_h) + b(v_h, p - p_I) & \forall\, v_h \in V_h, \\ b(u_h - u_I, q_h) = b(u - u_I, q_h) \quad \forall\, q_h \in Q_h. \end{cases}$$

The following proposition is nothing more than an interpretation of the above formula. We state it explicitly since we are going to use it at several occasions in the sequel.

Proposition 5.2.1. *In the framework of Assumption* $\mathcal{AB}_h$, *let* (u, p) *and* (u_h, p_h) *be solutions of the continuous problem* (5.1.2) *and of the discretised problem* (5.1.9), *respectively. Then, for every* $(u_I, p_I) \in V_h \times Q_h$, *we have that* $(u_h - u_I, p_h - p_I)$ *is the solution, in* $V_h \times Q_h$, *of the variational problem*

$$\begin{cases} a(u_h - u_I, v_h) + b(v_h, p_h - p_I) = \langle \mathcal{F}, v_h \rangle_{V_h' \times V_h} & \forall\, v_h \in V_h, \\ b(u_h - u_I, q_h) = \langle \mathcal{G}, q_h \rangle_{Q_h' \times Q_h} & \forall\, q_h \in Q_h \end{cases} \tag{5.2.5}$$

where

$$\langle \mathcal{F}, v_h \rangle_{V_h' \times V_h} := a(u - u_I, v_h) + b(v_h, p - p_I) \quad \forall\, v_h \in V_h, \tag{5.2.6}$$

and

$$\langle \mathcal{G}, q_h \rangle_{Q_h' \times Q_h} := b(u - u_I, q_h) \quad \forall\, q_h \in Q_h. \tag{5.2.7}$$

□

The formulation (5.2.5), together with (5.2.6) and (5.2.7), will be the starting point of most of our estimates. Indeed, we can now go back to Theorem 4.2.3 and, applying it to the present finite dimensional case, we obtain the following result.

Theorem 5.2.1 (The basic estimate). *Under Assumption $\mathcal{AB}_h$, assume that V_h and Q_h verify* (5.1.18) *and* (5.1.19). *Let $f \in V'$ and $g \in Q'$. Assume that the continuous problem* (5.1.2) *has a solution (u, p) and let (u_h, p_h) be the unique solution of the discretised problem* (5.1.9). *Then, for every $u_I \in V_h$ and for every $p_I \in Q_h$, we have the estimates*

$$\|u_h - u_I\|_V \le \frac{1}{\alpha_1^h} \|\mathcal{F}\|_{V_h'} + \frac{2\|a\|}{\alpha_1^h \beta_h} \|\mathcal{G}\|_{Q_h'}, \tag{5.2.8}$$

$$\|p_h - p_I\|_Q \le \frac{2\|a\|}{\alpha_1^h \beta_h} \|\mathcal{F}\|_{V_h'} + \frac{2\|a\|^2}{\alpha_1^h \beta_h^2} \|\mathcal{G}\|_{Q_h'}, \tag{5.2.9}$$

where $\mathcal{F}$ and $\mathcal{G}$ are defined in (5.2.6) *and* (5.2.7). *If, moreover, $a(\cdot, \cdot)$ is symmetric and satisfies*

$$a(v_h, v_h) \ge 0 \qquad \forall\, v_h \in V_h, \tag{5.2.10}$$

then we have the improved estimates

$$\|u_h - u_I\|_V \le \frac{1}{\alpha_0^h} \|\mathcal{F}\|_{V_h'} + \frac{2\|a\|^{1/2}}{(\alpha_0^h)^{1/2} \beta_h} \|\mathcal{G}\|_{Q_h'}, \tag{5.2.11}$$

$$\|p_h - p_I\|_Q \le \frac{2\|a\|^{1/2}}{(\alpha_0^h)^{1/2} \beta_h} \|\mathcal{F}\|_{V_h'} + \frac{\|a\|}{\beta_h^2} \|\mathcal{G}\|_{Q_h'} \tag{5.2.12}$$

with α_0^h given by (5.1.22).

At this point, we just have to evaluate $\|\mathcal{F}\|_{V_h'}$ and $\|\mathcal{G}\|_{Q_h'}$:

$$\|\mathcal{F}\|_{V_h'} \le \|a\|\, \|u - u_I\|_V + \|b\|\, \|p - p_I\|_V, \tag{5.2.13}$$

$$\|\mathcal{G}\|_{Q_h'} \le \|b\|\, \|u - u_I\|_V, \tag{5.2.14}$$

and then apply the triangle inequality to obtain, from Theorem 5.2.1, the following error estimates.

Theorem 5.2.2 (The basic error estimate). *Under Assumption $\mathcal{AB}_h$, assume that V_h and Q_h verify* (5.1.18) *and* (5.1.19). *Let $f \in V'$ and $g \in Q'$. Assume that the continuous problem* (5.1.2) *has a solution (u, p) and let (u_h, p_h) be the unique solution of the discretised problem* (5.1.9). *Then, we have the estimate*

$$\|u_h - u\|_V \le \frac{4\|a\|\,\|b\|}{\alpha_1^h \beta_h} E_u + \frac{\|b\|}{\alpha_1^h} E_p, \tag{5.2.15}$$

$$\|p_h - p\|_Q \le \left(\frac{2\|a\|^2}{\alpha_1^h \beta_h} + \frac{2\|a\|\,\|b\|}{\beta_h^2}\right) E_u + \frac{3\|a\|\,\|b\|}{\alpha_1^h \beta_h} E_p. \tag{5.2.16}$$

If, moreover, $a(\cdot,\cdot)$ is symmetric and positive semi-definite (see (5.2.10)*), then we have the improved estimates*

$$\|u_h - u\|_V \le \left(\frac{2\|a\|}{\alpha_0^h} + \frac{2\|a\|^{1/2}\|b\|}{(\alpha_0^h)^{1/2}\beta_h}\right) E_u + \frac{\|b\|}{\alpha_0^h} E_p, \tag{5.2.17}$$

$$\|p_h - p\|_Q \le \left(\frac{2\|a\|^{3/2}}{(\alpha_0^h)^{1/2}\beta_h} + \frac{\|a\|\,\|b\|}{\beta_h^2}\right) E_u + \frac{3\|a\|^{1/2}\|b\|}{(\alpha_0^h)^{1/2}\beta_h} E_p \tag{5.2.18}$$

with α_0^h given by (5.1.22).

Remark 5.2.1. **Important:** in Theorems 5.2.1 and 5.2.2, we allowed, in principle, the constants β_h and α_1^h (or α_0^h) to depend on h. It is **obvious** (but still worth mentioning) that if there exist constants β_0 and α_0 such that $\beta_h \ge \beta_0$ and $\alpha_1^h \ge \alpha_0$ (or $\alpha_0^h \ge \alpha_0$) *for all* h, then the constants appearing in our estimates will be independent of h. In almost **all of this chapter** (with a few exceptions, including the next Theorem 5.2.5), we will keep allowing the stability constants to depend on h. We rely on the reader to understand what happens whenever one has a *uniform lower bound* for them. □

5.2.2 *Additional Error Estimates for the Basic Problem*

There is actually ample room for improving the result of Theorems 5.2.1 and 5.2.2. The principal source of non-optimality in their proof lies indeed in the rather poor job that we made in estimating $\|\mathcal{F}\|_{V_h'}$ and $\|\mathcal{G}\|_{Q_h'}$ in (5.2.13) and (5.2.14). Indeed, in the first place, we essentially estimated the norms in V' and Q', respectively. This is correct, but not optimal. Indeed, for instance, although the norms in V and in V_h are the same, from $V_h \subset V$, one easily deduces that for every $v' \in V'$:

$$\sup_{v \in V} \frac{\langle v', v\rangle}{\|v\|_V} \ge \sup_{v_h \in V_h} \frac{\langle v', v_h\rangle}{\|v_h\|_V}. \tag{5.2.19}$$

In the second place, we carried out our argument *for every* $(u_I, p_I) \in V_h \times Q_h$. This has surely the advantage of allowing the classical estimates in terms of E_u and E_p defined in (5.2.1) and (5.2.2). However, in particular cases, smarter choices of u_I and/or p_I could produce a better result.

In particular, the discrete *inf-sup* condition (5.1.19) gives that $\text{Ker}B_h^t = \{0\}$ and this, according to Remark 5.1.5, ensures the existence of an operator $\Pi_h : V \to V_h$ such that $b(u - \Pi_h u, q_h) = 0$ for every $q_h \in Q_h$. Hence, taking $u_I := \Pi_h u$, we will have

$$b(u - u_I, q_h) = 0 \qquad \forall\, q_h \in Q_h, \qquad (\text{that is, } B_h u_I = \pi_{Q_h'} Bu) \tag{5.2.20}$$

which, using the notation (5.1.14), can also be written as

$$u_I \in Z_h(Bu). \tag{5.2.21}$$

We now observe that for every $u_I \in Z_h(Bu)$ and for every p_I, the estimates (5.2.14) and (5.2.13) become

$$\|\mathcal{G}\|_{Q_h'} = 0, \tag{5.2.22}$$

$$\|\mathcal{F}\|_{V_h'} \le \|a\|\, \|u - u_I\|_V + \|b\|\, \|p - p_I\|_Q. \tag{5.2.23}$$

If, moreover, we can also choose a $p_I \in Z_h^*(B^t p)$, implying

$$b(v_h, p - p_I) = 0 \qquad \forall\, v_h \in V_h, \tag{5.2.24}$$

then the estimate (5.2.13) further simplifies to

$$\|\mathcal{F}\|_{V_h'} \le \|a\|\, \|u - u_I\|_V, \tag{5.2.25}$$

always for $u_I \in Z_h(Bu)$. Note that we will always be able to find such a p_I if $K_h \subseteq K$, as pointed out in Proposition 5.1.2. If, moreover, B_h is the restriction of B to V_h, then, according to Remark 5.1.7, we could even take $p_I := \pi_{Q_h} p$.

Therefore, again from Theorem 5.2.1, we have the following results.

Theorem 5.2.3 (Taking u_I in $Z_h(Bu)$). *Under Assumption $\mathcal{AB}_h$, assume that V_h and Q_h verify* (5.1.18) *and* (5.1.19). *Let $f \in V'$ and $g \in Q'$. Assume that the continuous problem* (5.1.2) *has a solution (u, p) and let (u_h, p_h) be the unique solution of the discretised problem* (5.1.9). *Then, we have the estimate*

$$\begin{aligned} \|u_h - u\|_V &\le \frac{1}{\alpha_1^h}\Big(2\|a\|\, E_u^Z + \|b\|\, E_p\Big) \\ &\le \frac{1}{\alpha_1^h}\left(\frac{4\,\|a\|\,\|b\|}{\beta_h}\, E_u + \|b\|\, E_p\right), \end{aligned} \tag{5.2.26}$$

$$
\begin{aligned}
\|p_h - p\|_Q &\le \frac{2\|a\|^2}{\alpha_1^h \beta_h}\left(E_u^Z + 2\frac{\|b\|}{\|a\|}\,E_p\right) \\
&\le \frac{2\|a\|^2}{\alpha_1^h \beta_h}\left(\frac{2\|b\|}{\beta_h}\,E_u + 2\frac{\|b\|}{\|a\|}\,E_p\right).
\end{aligned} \tag{5.2.27}
$$

Theorem 5.2.4 (Taking u_I in $Z_h(Bu)$ and p_I in $Z_h^*(B^t p)$). *Under Assumption $\mathcal{AB}_h$, assume that V_h and Q_h verify* (5.1.18) *and* (5.1.19). *Let $f \in V'$ and $g \in Q'$. Assume that the continuous problem* (5.1.2) *has a solution (u, p) and let (u_h, p_h) be the unique solution of the discretised problem* (5.1.9). *Assume moreover that $Z_h^*(B^t p)$ is not empty (so that there exists at least one $p_I \in Q_h$ that verifies* (5.2.24)*). Then,*

$$
\|u_h - u\|_V \le \frac{2\|a\|}{\alpha_1^h} E_u^Z \le \frac{4\|a\|\,\|b\|}{\alpha_1^h \beta_h} E_u, \tag{5.2.28}
$$

$$
\|p_h - p_I\|_Q \le \frac{2\|a\|^2}{\alpha_1^h \beta_h} E_u^Z \quad \forall p_I \in Z_h^*(B^t p) \tag{5.2.29}
$$

so that

$$
\|p_h - p\|_Q \le \frac{4\|a\|^2\|b\|}{\alpha_1^h \beta_h^2} E_u + E_p^Z. \tag{5.2.30}
$$

Remark 5.2.2. If $a(\cdot,\cdot)$ is symmetric and positive semi-definite, then, using (5.2.12) as in previous error estimates (as for instance in Theorem 5.2.2), we could slightly improve our error estimates, using a softer dependence on the constants. □

Following the above path, we could derive a number of possible other variants, choosing one of the many theorems of Chap. 3, and then using one of the many choices for u_I and p_I. We decided that this procedure is quite easy, and every reader could do it by her/himself, if necessary. It might however be convenient to state explicitly, for an easy use, the following theorem, which is just a particular case of the above results, but might be helpful if one wants something reasonably *simple*. In most cases, assumption (5.1.18) can be replaced by the (stronger) ellipticity in the kernel (5.1.20); in this case we have the following theorem.

Theorem 5.2.5 (Commonly used). *Let $(u, p) \in V \times Q$ and $(u_h, p_h) \in V_h \times Q_h$ be respectively solutions of problems,*

$$
\begin{cases}
a(u, v) + b(v, p) = \langle f, v\rangle, \ \forall\, v \in V, \\
b(u, q) = \langle g, q\rangle, \ \forall\, q \in Q,
\end{cases} \tag{5.2.31}
$$

and

$$
\begin{cases}
a(u_h, v_h) + b(v_h, p_h) = \langle f, v_h\rangle, \ \forall\, v_h \in V_h, \\
b(u_h, q_h) = \langle g, q_h\rangle, \ \forall\, q_h \in Q_h.
\end{cases} \tag{5.2.32}
$$

Assume that the inf-sup condition

$$\inf_{q_h \in Q_h} \sup_{v_h \in V_h} \frac{b(v_h, q_h)}{\|v_h\|_V \|q_h\|_Q} \geq \beta > 0 \tag{5.2.33}$$

is satisfied and let $a(\cdot,\cdot)$ be uniformly coercive on $K_h := \mathrm{Ker} B_h$, that is, there exists $\alpha_0 > 0$ such that

$$a(v_{0h}, v_{0h}) \geq \alpha_0 \|v_{0h}\|_V^2, \ \forall\, v_{0h} \in K_h. \tag{5.2.34}$$

Then, one has the following estimate, with a constant C depending on $\|a\|$, $\|b\|$, β, α_0 but independent of h:

$$\|u - u_h\|_V + \|p - p_h\|_Q \leq C\Big(\inf_{v_h \in V} \|u - v_h\|_V + \inf_{q_h \in Q_h} \|p - q_h\|_Q\Big). \tag{5.2.35}$$

Moreover, when we have the inclusion of kernels $K_h \subseteq K$, we have the better estimate

$$\|u - u_h\|_V \leq C \inf_{v_h \in V} \|u - v_h\|_V. \tag{5.2.36}$$

Remark 5.2.3. It is clear that, in all the previous theorems, if $H \equiv \mathrm{Ker} B^t$ is not zero, then the constant β_h must go to zero when h tends to zero. In these cases, when $g \in \mathrm{Im} B$, the theorems should be applied with $Q_{/H}$ instead of Q (while for $g \notin \mathrm{Im} B$ the solution (u, p) does not exist). □

5.2.3 Variants of Error Estimates

We have considered up to now the most basic form of mixed problems. Numerous variations are however possible. Some of them are too special to merit an abstract treatment and will be presented on specific examples in the subsequent chapters. We consider here some problems arising in a quite large number of practical situations.

The first pathology that we consider is the case where coerciveness on $\mathrm{Ker} B_h$ does not hold but can be replaced by a weaker condition.

Assume, in particular, that on V we have a (weaker) norm $\|\cdot\|_{V^*}$ such that

$$\exists\, \alpha_h^* > 0 \text{ such that } a(v, v) \geq \alpha_h^* \|v\|_{V^*}^2 \qquad \forall\, v \in K_h, \tag{5.2.37}$$

together with

$$\exists\, M_a^* \text{ such that } a(u, v) \leq M_a^* \|u\|_{V^*} \|v\|_{V^*} \qquad \forall\, u,\, v \in V. \tag{5.2.38}$$

We recall that by "a weaker norm" we mean that

$$\|v\|_{V^*} \leq \|v\|_V \qquad \forall\, v \in V. \tag{5.2.39}$$

Typically, to fix the ideas, the V^*-norm will be some kind of L^2 norm, opposed to an H^1-norm (or an $H(\mathrm{div})$-norm) in V, and $a(u, v)$ some kind of L^2 scalar product.

Remark 5.2.4. To be precise, (5.2.39) is not the usual definition of a *weaker norm*, which would allow the presence of a multiplicative constant giving, for instance, $\|v\|_{V^*} \leq C\,\|v\|_V$. However, in order to have a simpler notation, we forced the constant to be equal to one (or rather, we assumed that the constant was inserted in the expression of the V^*-norm). □

The situation described in (5.2.37) and (5.2.38) arises in several occasions. Let us see two different kinds of applications.

The first kind of application occurs when $\|\cdot\|_H$, on K, becomes equivalent to the original V-norm (so that (5.2.37) implies the usual coerciveness on K), but for the discretised problem, one does not have $K_h \subseteq K$. Condition (5.2.37) nevertheless ensures existence of the discrete solution by the equivalence of norms in a finite dimensional space (see below). Convergence properties are however likely to be altered: in particular, we might expect α_0^h to depend on h. In the mixed formulation of elasticity introduced in Chap. 1, we had $V := (H(\mathrm{div};\Omega))^2_s$ while the bilinear form $a(u,v) := \int_\Omega \underline{\underline{\sigma}} : \underline{\underline{\tau}}\,dx$ is coercive only on $(L^2(\Omega))^4_s =: V^*$. This is enough to have coerciveness on K but not in general on K_h unless one is clever and builds V_h and Q_h in order to have $K_h \subseteq K$. In general, the analysis of this problem is difficult as we shall see in Chap. 10.

A second kind of application occurs when one considers an ill-posed problem in the sense that the existence of (u,p) cannot be obtained directly in $V \times Q$ by the previous stability results, but only for instance through a regularity argument. Existence of a discrete solution however holds, and one would like to get error estimates. Such is the case in the $\psi - \omega$ mixed formulation of the biharmonic problem that we have seen in (1.3.65). For a more detailed analysis of this case, see Chap. 10.

Remark 5.2.5. In general, as we said, the V^*-norm and the V-norm will *not* be equivalent with a constant independent of h. Hence, introducing the quantity

$$S(h) := \sup_{v_h \in V_h} \frac{\|v_h\|_V}{\|v_h\|_{V^*}}, \tag{5.2.40}$$

we might expect that $S(h)$ tends to infinity when h goes to zero. Typically, in a finite element context, the value of $S(h)$ will be given by some suitable *inverse inequality*. □

We now note, and this is an important point, that we have been using, in deriving all the above results, only the stability in the finite dimensional spaces $V_h \times Q_h$, and the only appearance of functions not belonging to them has been through the right-hand sides $\mathcal{F}$ and $\mathcal{G}$. Hence, being in finite dimensional spaces, *we could use the norm* $\|\cdot\|_{V^*}$ *on* V_h, and nothing changes, apart from the definitions of $\|b\|$ and β_h that now should be replaced by

$$\|b\|_* = \sup_{q \in Q,\, v_h \in V_h} \frac{b(v_h,q)}{\|q\|_Q\,\|v_h\|_{V^*}} \tag{5.2.41}$$

and

$$\beta_h^* = \inf_{q_h \in Q_h} \sup_{v_h \in V_h} \frac{b(v_h, q_h)}{\|v_h\|_{V^*} \|q_h\|_Q}, \tag{5.2.42}$$

respectively. Note that, from (5.2.39), we immediately have $\beta_h^* \geq \beta_h$. Hence, we can take into account the *weaker condition*

$$\beta_h^* \equiv \inf_{q_h \in Q_h} \sup_{v_h \in V_h} \frac{b(v_h, q_h)}{\|v_h\|_{V^*} \|q_h\|_Q} > 0. \tag{5.2.43}$$

On the other hand, the shift from $\|b\|$ to $\|b\|_*$ will not be without a price, as in practice we shall almost always need some sort of inverse inequality and pay some power of h. Indeed, using (5.2.40), it is immediate to see that

$$\|b\|_* \leq S(h)\|b\| \tag{5.2.44}$$

and, in general, (5.2.44) cannot be improved.

In this framework, from Theorem 5.2.3 applied with the norms $\|\cdot\|_{V^*}$ and $\|\cdot\|_Q$, we have the following result.

Theorem 5.2.6 (Ellipticity in a weaker norm). *Under Assumption $\mathcal{AB}_h$, assume further that the inf-sup condition* (5.2.43) *is satisfied, and that the bilinear form a satisfies* (5.2.37) *and* (5.2.38)*. Let $f \in V'$ and $g \in Q'$. Assume that the continuous problem* (5.1.2) *has a solution (u, p), and let (u_h, p_h) be the solution of the discretised problem* (5.1.9)*. Then, for every $u_I \in Z_h(Bu)$ and for every $p_I \in Q_h$, we have the estimates*

$$\|u_h - u_I\|_{V^*} \leq \frac{1}{\alpha_h^*}(M_a^* \|u - u_I\|_{V^*} + \|b\|_* \|p - p_I\|_Q), \tag{5.2.45}$$

$$\|p_h - p_I\|_Q \leq \frac{2M_a^*}{\alpha_h^* \beta_h^*}(M_a^* \|u - u_I\|_{V^*} + \|b\|_* \|p - p_I\|_Q). \tag{5.2.46}$$

If, moreover, $K_h \subseteq K$, then we also have

$$\|u_h - u_I\|_{V^*} \leq \frac{M_a^*}{\alpha_h^*} \|u - u_I\|_{V^*} \tag{5.2.47}$$

and if, in addition, B_h is the restriction of B (see Remark 5.1.7), then

$$\|p_h - \pi_{Q_h} p\|_Q \leq \frac{2(M_a^*)^2}{\alpha_h^* \beta_h^*} \|u - u_I\|_{V^*}. \tag{5.2.48}$$

Remark 5.2.6. Adding and subtracting u in (5.2.45), and also using Proposition 5.1.3, we could derive from (5.2.45) an estimate of $\|u - u_h\|$ in terms of the infimum of $\|u - v_h\|$ and the infimum of $\|p - q_h\|$, as done, for instance, in

Theorem 5.2.3. Obviously, the same can be done for $\|p - p_h\|$. We leave these variants to the reader. □

Remark 5.2.7. The above "proof" of Theorem 5.2.6, made through the simple change of norm, might puzzle somebody. As an *exercise*, we can give a *direct proof* of (5.2.47) and (5.2.48) which is not based on the previous stability estimates for finite dimensional problems. Consider $u_I = \Pi_h u$ in $Z_h(Bu)$, and remember that

$$b(u - \Pi_h u, q_h) = 0 \qquad \forall\, q_h \in Q_h. \tag{5.2.49}$$

This implies in particular (using the second equations of the continuous and of the discretised problems) that the difference $u_h - \Pi_h u$ satisfies

$$b(u_h - \Pi_h u, q_h) = 0 \qquad \forall\, q_h \in Q_h, \tag{5.2.50}$$

and hence belongs to K_h. As we assumed that $K_h \subseteq K$, condition (5.2.49) then implies

$$b(u_h - \Pi_h u, q) = 0 \qquad \forall\, q \in Q. \tag{5.2.51}$$

Then: (1) we use (5.2.37), (2) we add and subtract u, (3) we use the first equations of the continuous and of the discrete problems, (4) we use (5.2.51), and finally, (5) we use (5.2.38):

$$\begin{aligned}
\alpha_h^* \|u_h - \Pi_h u\|_{V^*}^2 &\le a(u_h - \Pi_h u, u_h - \Pi_h u) \\
&= a(u_h - u, u_h - \Pi_h u) + a(u - \Pi_h u, u_h - \Pi_h u) \\
&= -b(u_h - \Pi_h u, p_h - p) + a(u - \Pi_h u, u_h - \Pi_h u) \\
&= 0 + a(u - \Pi_h u, u_h - \Pi_h u) \\
&\le M_a^* \|u - \Pi_h u\|_{V^*} \, \|u_h - \Pi_h u\|_{V^*},
\end{aligned} \tag{5.2.52}$$

and (5.2.47) follows simplifying $\|u_h - \Pi_h u\|_{V^*}$. Now, take a $v_h \in V_h$ different from 0 such that

$$b(v_h, p_h - \pi_{Q_h} p) \ge \beta_h \|v_h\|_V \, \|p_h - \pi_{Q_h} p\|_Q. \tag{5.2.53}$$

The existence of such a v_h is guaranteed from the *inf-sup* condition (5.1.19). Now, (1) use (5.2.53), (2) use Proposition 5.1.2 and Remark 5.1.7, (3) use the first equations of the continuous and of the discrete problems, (4) add and subtract $\Pi_h u$, (5) use (5.2.38) (twice) and (5.2.47), and finally, (6) compare (5.2.37) and (5.2.38) to get $\alpha_h^* \le M_a^*$:

$$\begin{aligned}
\beta_h \|v_h\|_V \, \|p_h - \pi_{Q_h} p\|_Q &\le b(v_h, p_h - \pi_{Q_h} p) \\
&= b(v_h, p_h - p) = a(u - u_h, v_h) \\
&= a(u - \Pi_h u, v_h) + a(\Pi_h u - u_h, v_h)
\end{aligned}$$

$$\leq (M_a^* + \frac{M_a^{*2}}{\alpha_h^*})\|u - \Pi_h u\|_{V^*} \|v_h\|_{V^*}$$

$$\leq \frac{2M_a^{*2}}{\alpha_h^*}\|u - \Pi_h u\|_{V^*} \|v_h\|_{V^*} \quad (5.2.54)$$

and (5.2.48) follows using (5.2.39) since we took $v_h \neq 0$. □

In Theorem 5.2.6, we used the constant β_h^*. We could, however, use the old β_h, as given by the *usual inf-sup* condition (5.1.19). Surprisingly enough, this often allows a better estimate, as shown in the following theorem.

Theorem 5.2.7 (First duality). *Under the same assumptions of Theorem 5.2.6, assume that, moreover, we have the following property: for every $\overline{q_h} \in Q_h$, the solution $(w_h, \psi_h) \in V_h \times Q_h$ of the problem*

$$\begin{cases} a(v_h, w_h) + b(v_h, \psi_h) = 0, & \forall\, v_h \in V_h, \\ b(w_h, q_h) = (\overline{q_h}, q_h)_Q, & \forall\, q_h \in Q_h \end{cases} \quad (5.2.55)$$

verifies

$$\|w_h\|_V \leq C\,\|\overline{q_h}\|_Q, \quad (5.2.56)$$

with C independent of h and of $\overline{q_h}$. Then, for every $u_I \in Z_h(Bu)$, we have

$$\|p_h - \pi_{Q_h} p\|_Q \leq C\Big(M_a^*\,\|u - u_I\|_{V^*} + \|b\|\,\|p - \pi_{Q_h} p\|_Q\Big). \quad (5.2.57)$$

Proof. Let $p_I := \pi_{Q_h} p$ and $u_I := \Pi_h u$ as in the previous theorem. Consider the auxiliary problem: *find* $(w_h, \psi_h) \in V_h \times Q_h$ *such that*

$$\begin{cases} a(v_h, w_h) + b(v_h, \psi_h) = 0, & \forall\, v_h \in V_h, \\ b(w_h, q_h) = (p_h - p_I, q_h)_Q, & \forall\, q_h \in Q_h. \end{cases} \quad (5.2.58)$$

Then, we have

$$\begin{aligned} \|p_h - p_I\|_Q^2 &= b(w_h, p_h - p_I) = b(w_h, p_h - p) + b(w_h, p - p_I) \\ &= a(u - u_h, w_h) + b(w_h, p - p_I) \\ &= a(u - u_I, w_h) + a(u_I - u_h, w_h) + b(w_h, p - p_I) \\ &= a(u - u_I, w_h) + b(u_h - u_I, \psi_h) + b(w_h, p - p_I) \\ &= a(u - u_I, w_h) + b(w_h, p - p_I) \\ &\leq M_a^* \|w_h\|_{V^*}\,\|u - u_I\|_{V^*} + \|b\|\,\|w_h\|_V\,\|p - p_I\|_Q \end{aligned} \quad (5.2.59)$$

and the result follows from (5.2.56). □

Remark 5.2.8. At first sight, the result (5.2.57) does not seem much better than the previous (5.2.46). However, looking more carefully, one notices that the $\|b\|_*$ appearing in (5.2.46) is actually replaced by $\|b\|$ in (5.2.46). In most applications, this means a factor $O(h^{-1})$ that is present in (5.2.46) and not in (5.2.57). □

Remark 5.2.9. Results of the type of Theorem 5.2.7 are a particular case of a more general class of estimates, called *dual estimates* that we will discuss in a while. □

Another variant that will be useful in the study of some hybrid methods is the following.

Let $|v|_V$ be a continuous *semi-norm* on V and let M denote its kernel (that is: M is the subspace of V made by those v that satisfy $|v|_V = 0$). We assume for simplicity that $M \subset V_h$. The semi-norm $|\cdot|_V$ is then a norm on the quotient space $V_{/M}$, as well as on $V_{h/M}$. We suppose that we have

$$\exists\, \alpha_M > 0 \text{ such that } a(v,v) \geq \alpha_M\, |v|_V^2, \quad \forall\, v \in V, \tag{5.2.60}$$

and

$$|a(u,v)| \leq \|a\|\; |u|_V\; |v|_V, \quad \forall\, u,\, v \in V. \tag{5.2.61}$$

Proposition 5.2.2. *Under Assumption $\mathcal{AB}_h$, assume further that the bilinear form a satisfies* (5.2.60) *and* (5.2.61)*. Let (u,p) and (u_h,p_h) be solutions of the continuous problem* (5.1.2) *and of the discretised problem* (5.1.9)*, respectively. Define*

$$\overline{Q}_h(p) := \{q_h \mid q_h \in Q_h,\; b(v_h, p-q_h) = 0\; \forall\, v_h \in M\}. \tag{5.2.62}$$

Then, we have the estimate

$$|u-u_h|_V \leq \left[1 + \frac{\|a\|}{\alpha}\right] \inf_{v_h \in Z_h(g)} |u-v_h|_V + \|b\| \inf_{q_h \in \overline{Q}_h(p)} \|p-q_h\|_Q. \tag{5.2.63}$$

Proof. The proof still follows from Theorem 5.2.1, working in $V_{h/M} \times Q_h$, and observing that, for every $q_h \in \overline{Q}_h(p)$, we have obviously

$$b(v_h, p-q_h) \leq \|b\|\; |v_h|_V\; \|p-q_h\|_Q \qquad \forall\, v_h \in V_h. \tag{5.2.64}$$

□

Remark 5.2.10. Note that we did not assume that $Z_h(g)$ and $\overline{Q}_h(p)$ are non empty. However, Eq. (5.2.63), if one of the two sets is empty, will give $|u-u_h|_V \leq +\infty$ (which is always true) since the infimum over the empty set is, by definition, $+\infty$. Hence, the result is "true" even when one of the two sets is empty (but in that case, it will be totally useless). Please forgive this little mathematical coquetry, which could have been used several times before. □

5.2.4 A Simple Example

We now want to present a *very simple* example which, however, could be very instructive if one reads it carefully. Similar considerations have been made for the corresponging eigenvalue problem in [77] (see, in particular, Sect. 5.4 of Part 1)

We consider the interval $I :=]-1, 1[$ and the problem of finding $p \in H_0^1(I)$ such that $p'' = g$, where g is a given function in, say, $L^2(I)$. Remember that the condition $p \in H_0^1(I)$ implies, among other things (such as the continuity of p), that $p(-1) = p(1) = 0$, so that our problem has clearly a unique solution. Particular attention will be devoted to the case $g = 1$ (whose solution is obviously $p(x) = (x^2 - 1)/2$).

The mixed formulation of our toy problem is easily reached by setting $u := p'$, and introducing the spaces $Q := L^2(I)$ and $V := H^1(I)$. In two dimensions, we would have $V := H(\text{div})$ which, however, in one dimension, coincides with H^1 (as the divergence coincides with the first derivative). We then set

$$a(u, v) := \int_{-1}^{1} u\, v\, dx, \quad \forall\, u, v \in V, \tag{5.2.65}$$

$$b(v, q) := \int_{-1}^{1} v' q\, dx, \quad \forall\, v \in V,\ \forall\, q \in Q, \tag{5.2.66}$$

and we easily recognise that (u, p) is the solution of

$$\begin{cases} a(u, v) + b(v, p) = 0, & \forall\, v \in V, \\ b(u, q) = \langle g, q\rangle_{Q'\times Q}, & \forall\, q \in Q. \end{cases} \tag{5.2.67}$$

We easily see that the operator B is now the first derivative (from $H^1(I)$ to $L^2(I)$) and that its kernel K is given by

$$K := \{\text{ constant functions}\}. \tag{5.2.68}$$

It is important to note that the bilinear form a (which is simply the L^2 inner product) *is not* elliptic in the whole $V = H^1(I)$. Indeed, no matter how small you take $\alpha > 0$, the inequality

$$a(v, v) \equiv \int_I v^2 dx \geq \alpha\Big(\int_I v^2 dx + \int_I (v')^2 dx\Big) \equiv \alpha \|v\|_V^2 \qquad \forall\, v \in V \tag{5.2.69}$$

is **false**. To be convinced of the falsity of (5.2.69), we recall the situation of Example 4.1.9 and consider, for $k \in \mathbb{N}$, the function $v_k(x) := \sin(\pi k x)$: we have

$$a(v_k, v_k) = 1, \qquad \|v_k\|_V^2 = 1 + k^2\pi^2/2, \tag{5.2.70}$$

and you *cannot* find an $\alpha > 0$, independent of k, such that

$$1 \geq \alpha(1 + k^2\pi^2/2) \quad \forall\, k \in \mathbb{N}. \tag{5.2.71}$$

However, inequality (5.2.69) is obviously true (with $\alpha = 1$) if restricted to $v \in K$ (see (5.2.68)), as constant functions have zero derivative. Hence, the ellipticity in the kernel (5.1.7) holds with $\alpha_0 = 1$.

On the other hand, the operator B is clearly surjective from V to Q. Indeed, for every $q \in Q$, we can find a $v_q \in V$, given by

$$v_q(x) := \int_0^x q(t)dt, \tag{5.2.72}$$

such that $B\, v_q \equiv v_q' = q$. It is also easy to see that

$$\|v_q\|_{L^2(I)} \leq \|v_q'\|_{L^2(I)} \equiv \|q\|_Q, \tag{5.2.73}$$

so that

$$\|v_q\|_V^2 \equiv \|v_q\|_{L^2(I)}^2 + \|v_q'\|_{L^2(I)}^2 \leq 2\|q\|_Q^2,$$

and therefore

$$\begin{aligned} \inf_{q\in Q}\sup_{v\in V} \frac{\int_I v'\, q}{\|v\|_V\, \|q\|_Q} &\geq \inf_{q\in Q} \frac{\int_I v_q'\, q}{\|v_q\|_V\, \|q\|_Q} \\ &= \inf_{q\in Q} \frac{\|q\|_Q^2}{\|v_q\|_V\, \|q\|_Q} = \inf_{q\in Q} \frac{\|q\|_Q}{\|v_q\|_V} \geq \inf_{q\in Q} \frac{\|q\|_Q}{\sqrt{2}\|q\|_Q} = \frac{1}{\sqrt{2}}, \end{aligned} \tag{5.2.74}$$

which is to say that the *inf-sup* condition (5.1.6) holds with a $\beta \geq 1/\sqrt{2}$. As we have already checked the ellipticity in the kernel, we can conclude that the mixed formulation of the continuous problem is well posed. So far, so good.

We can now tackle the problem of discretising (5.2.67). We therefore start by considering, for every positive integer N, a decomposition of the interval $I =]-1, 1[$ into N intervals of equal length $h = 2/N$. The spaces $\mathcal{L}_k^s$, with $s \in \{0, 1\}$, will then be the spaces of piecewise polynomials of local degree $\leq k$: globally continuous when $s = 1$, and discontinuous when $s = 0$, in agreement with the notation of Chap. 2. Our first choice is to take $V_h := \mathcal{L}_1^1$ and $Q_h := \mathcal{L}_0^0$. It is a simple and fortunate choice. Indeed, in the first place, we can note that the mapping $q \to v_q$, introduced in (5.2.72), applied to a function $q_h \in Q_h$ (hence, piecewise constant) produces a v_{q_h} that is continuous and piecewise linear, and hence belongs to V_h. Using exactly the same proof as before, we can now conclude that the *inf-sup* condition (5.1.19) on the bilinear form b still holds for this choice of subspaces, with a constant $\beta_h \geq 1/\sqrt{2}$ (hence, in particular, bounded from below independently of h). On the other hand, the discrete kernel K_h (see (5.1.12)) can easily be identified as being again made of global constants. Hence, we have $K_h \equiv K$. Moreover, we can use the fact that the bilinear form a coincides with the $L^2(I)$ inner product, so that (5.2.37) and (5.2.38) hold with $\alpha_h^* = M_a^* = 1$. Finally, we remark that we can construct the operator Π_h of Proposition 5.1.1 simply by taking $\Pi_h u$ as the usual nodal interpolant of u. We note indeed that if $\Pi_h u(x_k) = u(x_k)$ at each subdivision point x_k, then for every $q_h \in Q_h$ and for every interval $I_k = (x_k, x_{k+1})$ of our subdivision, we have from the fundamental theorem of Calculus,

$$\int_{I_k} (\Pi_h u - u)' \, q_h dx = q_{h|I_k} \int_{x_k}^{x_{k+1}} (\Pi_h u - u)' dx$$
$$= q_{h|I_k} \Big[(\Pi_h u - u)(x_{k+1}) - (\Pi_h u - u)(x_k) \Big] = 0. \quad (5.2.75)$$

We are happy, and we apply Theorem 5.2.6 with $u_I := \Pi_h u$.

From (5.2.47), (5.2.48) and usual interpolation estimates, we obtain

$$\|u_h - \Pi_h u\|_V \le \|u - \Pi_h u\|_V \le C\, h^2 |u|_{H^2(I)} \quad (5.2.76)$$

and

$$\|p_h - \pi_{Q_h} p\|_Q \le 2\sqrt{2} \|u - \Pi_h u\|_V \le C\, h^2\, |u|_{H^2(I)}, \quad (5.2.77)$$

which is, in fact, a *super-convergence result*, as Q_h is only made of piecewise constants. Indeed, using the triangle inequality and (5.2.76), we have

$$\|u_h - u\|_V \le 2\, \|u - \Pi_h u\|_V \le C\, h^2 |u|_{H^2(I)} \quad (5.2.78)$$

while from (5.2.77) we only have

$$\|p_h - p\|_Q \le \|p_h - p\|_Q + 2\sqrt{2}\|u - \Pi_h u\|_V \le C(h\, |p|_{H^1(I)} + h^2\, |u|_{H^2(I)}). \quad (5.2.79)$$

Everything works well, and the sun is shining for mixed formulations. We become greedy, and we would like to increase one of the two spaces, V_h or Q_h, in order to have an even better performance.

Let us start by increasing Q_h, and try $Q_h := \mathcal{L}_1^0$ (discontinuous piecewise polynomials of local degree ≤ 1). However, as soon as we look at this new choice, we immediately perceive the disaster. Indeed, the B_h operator goes from V_h, which has dimension equal to $N+1$, to Q_h', which has dimension $2N$. On the other hand, V_h contains the global constants, and B_h applied to one of them is 0. Hence, the dimension of the image of B_h can be at most N, and there is *no hope* that B_h could be surjective on a space of dimension $2N$. Hence, the *inf-sup* condition (5.1.19) will inevitably fail, and the discrete problem will have a singular matrix. To reduce Q_h to $\mathcal{L}_1^1$ (that is piecewise linear *continuous* functions) will not be enough either, as the dimension of $\mathcal{L}_1^1$ is $N+1$, and we are still down by one.

It seems therefore much more reasonable to increase instead the space V_h. For instance, we could take $V_h := \mathcal{L}_2^1$ (piecewise quadratic continuous functions). Indeed, increasing V_h for the same Q_h, we could only *improve* the *inf-sup* condition:

$$\inf_{q_h\in Q_h}\sup_{v_h\in\mathcal{L}_2^1}\frac{b(v_h,q_h)}{\|v_h\|_V\,\|q_h\|_Q}\geq\inf_{q_h\in Q_h}\sup_{v_h\in\mathcal{L}_1^1}\frac{b(v_h,q_h)}{\|v_h\|_V\,\|q\|_Q}\geq\frac{1}{\sqrt{2}}. \tag{5.2.80}$$

The situation, therefore, looks much better than before and even more so for the *monomaniacs of the inf-sup condition*, that consider it to be *the condition* ruling mixed formulations. As we do not belong to this group, we know that we still have to check the kernel K_h and the ellipticity in the kernel (5.1.20).

We start by observing that the new V_h can be thought as obtained by increasing $\mathcal{L}_1^1$ (which was the *previous* choice for V_h) with the addition of a quadratic bubble b_k in each element I_k ($k = 1, 2 \ldots, N$). Let us give a closer look at these bubbles that we are adding. We take, for simplicity, a model interval $I_h :=] - h/2, h/2[$. The "unit" quadratic bubble (with value 1 at the midpoint and vanishing at the endpoints) has equation $b(x) = 1 - (2x/h)^2$ and the mean value of its derivative $b'(x) = -8x/(h^2)$ over I_h vanishes (as it was to be expected as $b(x)$ vanishes at the endpoints of I_h). Hence, for every function $v_h^b \in \mathcal{L}_2^1$ vanishing at all the subdivision points x_k, and for every piecewise function $q_h \in \mathcal{L}_0^0$, we have

$$\int_I (v_h^b)'\, q_h\, dx = 0 \tag{5.2.81}$$

and the difference between $\mathcal{L}_2^1$ and the old $\mathcal{L}_1^1$ goes into the kernel K_h. This was, however, to be expected, as we started from a case in which B_h was already surjective, and we increased V_h.

We observe now that, as the kernel contains all quadratic bubbles, we cannot hope to have the ellipticity in the kernel (5.1.20) with a constant α_0^h which is independent of h. Indeed, we have, for instance,

$$\int_{I_h} b^2(x)dx = \frac{8h}{15} \qquad \int_{I_h}(b')^2(x)dx = \frac{16}{3h}, \tag{5.2.82}$$

so that when we sum over the N intervals (as $N = 2/h$),

$$a(b,b) = \int_{-1}^{1} b^2(x)dx = \frac{16}{15} \qquad |b|^2_{H^1(I_h)} = \int_{-1}^{1}(b')^2(x)dx = \frac{32}{3h^2}. \tag{5.2.83}$$

Hence, if we want $a(b,b) \geq \alpha_0^h \|b\|^2_{H^1(I)}$, that is,

$$\frac{16}{15} \geq \alpha_0^h\Big(\frac{16}{15} + \frac{32}{3h^2}\Big) \equiv 16\alpha_0^h\frac{h^2+10}{15h^2},$$

we cannot avoid taking

$$\alpha_0^h \leq \frac{h^2}{h^2+10} < \frac{h^2}{10}, \tag{5.2.84}$$

and α_0^h cannot be taken to be independent of h as h goes to zero. The situation looks dramatic: indeed, from the estimate (5.2.17) and usual interpolation estimates, we cannot have anything better than a $O(h^{-1})$ estimate in the V norm for u_h and, from (5.2.18), boundedness for the Q norm of p_h. The best we can do is to make use again of the fact that (5.2.37) and (5.2.38) still hold with $\alpha^* = M_a^* = 1$ and use the first part of Theorem 5.2.6, which does not require $K_h \subseteq K$. For this, however, we have to estimate $\|b\|_*$, given in (5.2.41). In our case, we have

$$\|b\|_* = \sup_{\substack{q\in Q\\ v_h\in V_h}} \frac{\int_I v_h' \, q \, dx}{\|q\|_{L^2(I)} \, \|v_h\|_{L^2(I)}}. \tag{5.2.85}$$

As $v_h' \in Q \equiv L^2(I)$ for every $v_h \in V_h$, this gives

$$\|b\|_* = \sup_{v_h\in V_h} \frac{\|v_h'\|_{L^2(I)}}{\|v_h\|_{L^2(I)}} = \frac{C}{h}, \tag{5.2.86}$$

(where, by the way, $C = 2\sqrt{15}$). This is bad news. Indeed, inserting (5.2.86) in estimates (5.2.45) and (5.2.46) (and noting that $\|p - p_I\|_Q$ cannot be better than $O(h)$), we cannot get anything better than boundedness for both $\|u_h - \Pi_h u\|_{V^*}$ and $\|p_h - \pi_{Q_h}\|_Q$.

We might still hope that our a priori estimates are not optimal. Indeed, if you do numerical experiments, the linear part of u_h and the whole p_h converge nicely.

See, in Fig. 5.1, the behaviour of p_h for $f = 1$. To make the picture clearer, we reconstructed a $\tilde{p}_h$ piecewise linear by taking the average of the true p_h at the subdivision points. The numerical convergence of $\tilde{p}_h$ is clear. However, the *worst news of all* is that (as it can be proved mathematically) $\tilde{p}_h$ (and p_h as well) is actually *converging to the wrong solution*! This can clearly be seen in Fig. 5.1, as we *know*, in the present case, that the exact solution $p = (x^2-1)/2$ has value -0.5 for $x = 0$, while our discrete solution seems, definitely, to converge to something slightly less than -0.08. A more careful analysis can show that, actually, we converge toward $p/6$, that is, in our case, toward $(x^2-1)/12$, so that its value at $\tilde{p}_h(0)$ converges to $-1/12 = -0.08\overline{3}$.

Remark 5.2.11. This super-simple case allows a detailed analysis, that works however in more general cases. Let us see it. We start by writing u_h as $u_L + u_B$, where u_L is the piecewise linear function that coincides with u_h at the endpoints of each subinterval, while the difference $u_B = u_h - u_L$ will be a piecewise quadratic polynomial that vanishes at the endpoints of each subinterval, and is therefore made of quadratic bubbles. In particular, from (5.2.81) we have

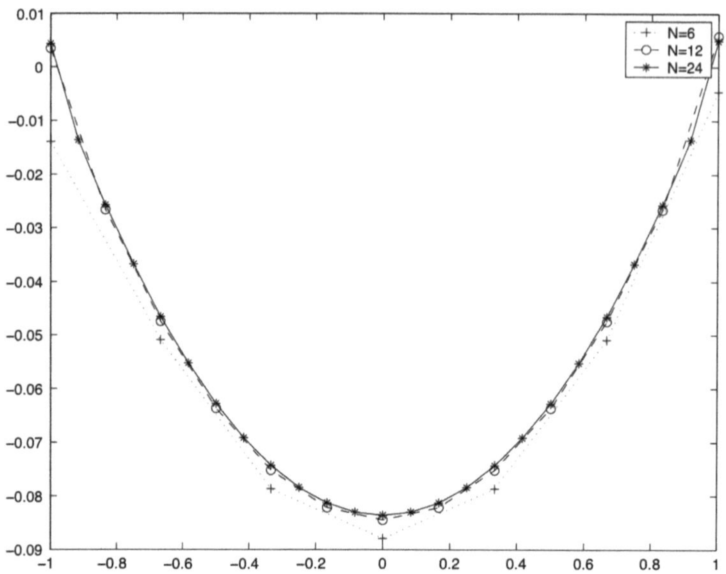

Fig. 5.1 Convergence of $\widetilde{p}_h$

$$(u'_B, q_h) = 0$$

for every piecewise constant q_h. Hence,

$$(u'_L, q_h) = (u'_h, q_h) = (1, q_h) \qquad \forall\, q_h \in \mathcal{L}^0_0$$

and from this we immediately have $u'_L = 1$. This, for symmetry reasons, immediately gives that $u_L = x$ in I. If you are picky and do not like symmetry reasons, check that, if (p_h, u_h) is a solution, then, setting $\widetilde{p_h}(x) = p_h(-x)$ and $\widetilde{u_h}(x) = -u_h(-x)$, you get another solution. As our discretised problem has a unique solution, this implies that p_h must be even and u_h odd. Having taken care of the picky ones, we can restart. For every subinterval $I_k =]x_k, x_{k+1}[$, we denote by b_k the unit quadratic bubble on I_k, that is

$$b_k(x) = 4(x - x_k)(x_{k+1} - x)/h^2.$$

Testing the first equation for $v_h = b_k$, we have

$$\int_{x_k}^{x_{k+1}} u_h\, b_k dx = 0 \int_{x_k}^{x_{k+1}} (u_B + x)\, b_k dx = 0. \tag{5.2.87}$$

After some computation, this gives us the value of u_h at the midpoint $x_{k+1/2}$ of I_k, that turns out to be

$$u_h(x_{k+1/2}) = -\frac{1}{8}\Big((u_h(x_k) + u_h(x_{k+1})\Big). \tag{5.2.88}$$

Note that this is not the value of u_B at the midpoint, which is obviously

$$\begin{aligned} u_B(x_{k+1/2}) &= u_h(x_{k+1/2}) - u_L(x_{k+1/2}) \\ &= -\frac{1}{8}\Big(u_h(x_k) + u_h(x_{k+1})\Big) - \frac{1}{2}\Big(u_h(x_k) + u_h(x_{k+1})\Big) \\ &= -\frac{5}{4}u_L(x_{k+1/2}). \end{aligned} \tag{5.2.89}$$

Hence, in I_k we have $u_B = B_k\, b_k$ with $B_k = -(5/4)u_L(x_{k+1/2})$. Remembering that the integral of b_k over I_k is $2h/3$, while the integral of u_L over I_k is $h\, u_L(x_{k+1/2})$, we conclude that

$$\int_{I_k} u_B dx = -\frac{5}{6}\int_{I_k} u_L dx \tag{5.2.90}$$

and finally (as $u_h = u_B + u_L$)

$$\int_{I_k} u_h dx = \frac{1}{6}\int_{I_k} u_L dx \tag{5.2.91}$$

and the magic coefficient $1/6$ shows up. Note that, from (5.2.87) to (5.2.91), we never used the fact that $f = 1$ or $u_L = x$. Hence, (5.2.91) applies to more general cases. Now take $\overline{p}_h := \pi_{Q_h} p/6$ (the piecewise constant projection of the *wrong limit*), and take ψ in $H^1_0(I)$ with $\psi'' = p_h - \overline{p}_h$. Then, take $w = \psi'$. Clearly, w will be piecewise linear. Taking $v_h = w$, we then have

$$\begin{aligned} \|p_h - \overline{p}_h\|^2_{L^2(I)} &= (p_h - \overline{p}_h, w') \\ &= (p_h - \frac{p}{6}, w') = (\frac{u}{6} - u_h, w) \\ &= \Big(\frac{u}{6} - u_h, w - \pi_{Q_h} w\Big) + \Big(\frac{u}{6} - u_h, \pi_{Q_h} w\Big). \end{aligned} \tag{5.2.92}$$

At this point, for simplicity, we use the fact that actually, in our case, $u_L = u = x$ and (5.2.91), so that the second term in the last line vanishes (but, otherwise, we could estimate it). We then have, from usual interpolation estimates,

$$\begin{aligned} \|p_h - \overline{p}_h\|^2_{L^2(I)} &= \Big(\frac{u}{6} - u_h, w - \pi_{Q_h} w\Big) \\ &\le \|\frac{u}{6} - u_h\|_{L^2(I)}\, C\, h\, \|w'\|_{L^2(I)} \\ &= \|\frac{u}{6} - u_h\|_{L^2(I)}\, C\, h\, \|p_h - \overline{p}_h\|_{L^2(I)}, \end{aligned} \tag{5.2.93}$$

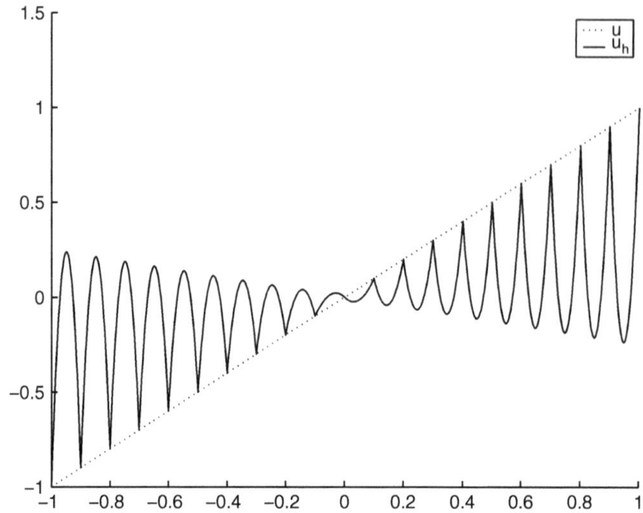

Fig. 5.2 u_h for h=1/10

and since we already know that u_h is bounded in L^2, we get a beautiful $O(h)$ convergence of p_h towards the wrong solution $p/6$.

With similar arguments, we can also prove that u_h converges weakly in L^2 towards $u/6$ and actually with order $O(h)$ in $H^{-1}(I)$. On the other hand, u_h in V (that is, in $H^1(I)$) *is* actually unbounded, as it can also be seen experimentally (see Fig. 5.2). □

Hence, our tentative to improve the results either by increasing Q_h or by increasing V_h are both *failures*. We observe, however, that, augmenting *at the same time* the local degree of V_h *and* the local degree of Q_h, we can restore optimality and improve the results of the initial choice (that was $V_h = \mathcal{L}_1^1$ and $Q_h = \mathcal{L}_0^0$). For instance, taking $V_h := \mathcal{L}_2^1$ and $Q_h := \mathcal{L}_1^0$, we have again that the *inf-sup* condition (5.1.19) is verified (still by the same proof, still with $\beta_h \geq 1/\sqrt{2}$) and that $K_h = K$ is given by the global constants. We can therefore apply Theorem 5.2.6 and we get

$$\|u_h - \Pi_h u\|_V \leq \|u - \Pi_h u\|_V \leq C\, h^3 |u|_{H^3(I)} \tag{5.2.94}$$

and

$$\|p_h - \pi_{Q_h} p\|_Q \leq 2\sqrt{2}\|u - \Pi_h u\|_V \leq C\, h^3\, |u|_{H^3(I)}, \tag{5.2.95}$$

which is again a *super-convergence result*, as Q_h is only made of piecewise linears. Indeed, using the triangle inequality and (5.2.76), we have

$$\|u_h - u\|_V \leq 2\,\|u - \Pi_h u\|_V \leq C\, h^3 |u|_{H^3(I)} \tag{5.2.96}$$

while, from (5.2.76), we only have

$$\|p_h - p\|_Q \le \|p_h - p\|_Q + 2\sqrt{2}\|u - \Pi_h u\|_V$$
$$\le C(h^2\,|p|_{H^1(I)} + h^3\,|u|_{H^3(I)}). \quad (5.2.97)$$

The sun is shining back on mixed formulations. However, in this case, since Q_h is made of discontinuous piecewise linears, we might well feel the *strong temptation* to take $Q_h := \mathcal{L}_1^1$, that is, piecewise linear *continuous functions*. Again, the *inf-sup* is trivially satisfied, in particular since $\mathcal{L}_1^1 \subset \mathcal{L}_1^0$. But, now, the discrete kernel K_h is no longer restricted to global constants, and in K_h one cannot expect any estimate better than

$$\alpha_0^h \simeq h^2, \quad (5.2.98)$$

as we had in (5.2.84). Again, this, inserted in (5.2.28), cannot give us anything better than

$$\|u_h - u_I\|_V \le C\,|u|_{H^2}. \quad (5.2.99)$$

Our only way out seems therefore to use (5.2.45) that, together with (5.2.86) and usual interpolation estimates, gives

$$\|u_h - \Pi_h u\|_{L^2} \le C\,h\,(|u|_{H^1} + |p|_{H^2}), \quad (5.2.100)$$

which is clearly suboptimal. Indeed, on a non-uniform grid, the slope of the error $\|u - u_h\|$ is not better than 1. See Fig. 5.3.

It is however puzzling to see (always in Fig. 5.3) that p_h converges to p as $O(h^2)$, which *cannot* be obtained from (5.2.46). Indeed, for this we have to apply Theorem 5.2.7. It is easy to see that its assumptions are verified, and that (5.2.57) gives the desired $O(h^2)$ rate. Note that, on a *uniform grid*, we would have an $O(h^2)$ convergence for $\|u - u_h\|_{L^2}$ as well (which decays to $O(h^{3/2})$) if we restrict further Q_h to be the subspace of $\mathcal{L}_1^1$ made of functions that vanish at the endpoints of I. These, however, are *super-convergence phenomena on uniform grids* that require a different type of analysis.

5.2.5 An Important Example: The Pressure in the Homogeneous Stokes Problem

Following the lines of Sect. 4.2.4 of the previous chapter, we will now briefly discuss the case in which $H_h \equiv \mathrm{Ker}B_h^t$ is *not* reduced to zero.

We start by considering the case in which H itself is not reduced to zero and H_h is a subspace of it. As we have partially seen in Remark 5.1.4, this is *a healthy case*. Indeed, in this case, the use of $Q_{/H} \equiv Q_{/\mathrm{Ker}B^t}$ instead of Q, and $Q_{h/H}$ instead of Q_h, will essentially fix the problem. We can also, if we want, apply Theorem 4.2.4 and deal with the standard case (the analogue of Theorem 5.2.1 of the present chapter). This might be a psychological help for some practitioner, used

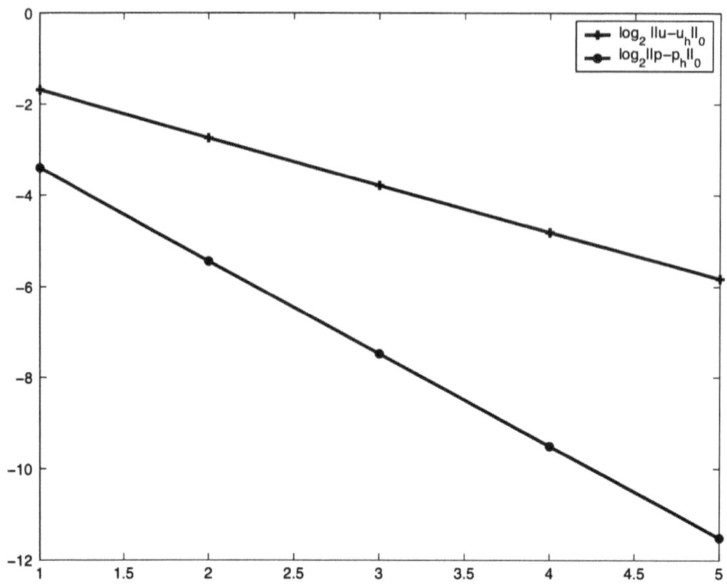

Fig. 5.3 log − log plot of $\|u - u_h\|_{L^2}$ and $\|p - p_h\|_{L^2}$ for $P_2 - P_1^{cont}$ element

for instance to work with pressures that do not have zero mean value, and escaping the lack of uniqueness just by fixing the pressure to be zero at a given node.

The above example, of pressures with zero mean value, is an important one, and deserves some further discussion dedicated to the readers that are less skilled in functional analysis. This subsection is dedicated to them. The other readers can just go on. Let us therefore abandon, for a while, the abstract framework, and stick to the Stokes problem with Dirichlet boundary conditions for the velocities all over the boundary of the domain: for instance, $\underline{u} := 0$. In this case, it is well known that the pressure is determined up to an additive constant. This corresponds to the fact that the operator B is the divergence (from $V := (H_0^1(\Omega))^d$, where $d = 2$ or 3 is the number of dimensions, to $Q := L^2(\Omega)$), and B^t is just the gradient. We all know that every constant function has its gradient equal to zero, and indeed $H \equiv \mathrm{Ker}B^t \equiv \mathbb{R}$ is the set of all constant functions.

You know that you cannot choose a g (as a right-hand side in the equation $\mathrm{div}\underline{u} = g$) which does not have zero mean value (Gauss doesn't want it, when $\underline{u} \cdot \underline{n} = 0$ on $\partial\Omega$), and this is what we mean, in this case, by asking that $g \in \mathrm{Im}B$: we ask that g have zero mean value.

To deal with the lack of uniqueness for the pressure, what we propose here is to work with $L^2(\Omega)_{/H} \equiv L^2(\Omega)_{/\mathbb{R}}$ as pressure space. Now, to be precise, we recall that $L^2(\Omega)_{/\mathbb{R}}$ is not a space of functions: its elements are *classes of functions*, and every class is made of functions that *differ from each other for an additive constant.* In a sense, the elements of $L^2(\Omega)_{/\mathbb{R}}$ could therefore be considered as "functions" that are determined up to an additive constant, just as our pressure unknown.

For instance, all the functions of the form $q(x) = x^2 + 3xy + c$, with c constant, will be one element of these classes, and hence *an element* of $L^2(\Omega)_{/\mathbb{R}}$. You surely remember seeing something very similar when you did indefinite integrals in first year Calculus:

$$\int x\,dx = \frac{x^2}{2} + c. \tag{5.2.101}$$

Here, it is just the same: $x^2/2 + c$ is not *a function*, but a bunch of functions that differ from each other for a constant.

In principle, when you choose Q_h to be a subspace of this mess, you should consider a space whose *elements* are sets of all possible functions that can be obtained by adding to a single piecewise polynomial function an arbitrary constant. For instance, consider $z_h(x, y)$ to be a specific piecewise constant (or a piecewise linear, depending on what you want); then, the set of all functions of the form $z_h(x, y) + c$ (where c is a *global* constant) will be *an element* of your Q_h.

Needless to say, this is not a nice space to work with on a computer. Hence, when you work with the computer program, you play smart. You fix a way to *select* one (and only one) *function* in each element of Q_h: for instance, you can specify "the one that has zero mean value on Ω". This works, because among all functions of the form, say, $z_h(x, y) + c$, there is always *one and only one* of them that has zero mean value on Ω. On the other hand, you could fix a point (for instance, the barycentre of a specified element, chosen once and for all), and specify "the one that vanishes at *this* point". This also works, because, among all functions of the form, say, $z_h(x, y) + c$, there is always *one and only one* of them that vanishes at *that* point you choose. Clearly, the second possibility is easier to implement. On the other hand, for reasons that are crystal clear to those who know functional analysis (but less clear to many other perfectly respectable researchers), the first choice is more convenient for the theoretical treatment. Basically, our religion forbids us to take the value of an L^2 function at a point. However, *in practice*, **there is no difference**. Hence, when you have to deal with a pressure that is determined up to an additive constant, please feel free to fix its value to be zero at a given point of your choice. However, when you read a theorem that speaks about pressures having zero mean value, don't be scared, and please do not say "Gasp! This does not apply to my case!": it is, essentially, *the same thing*. □

5.3 The Case of $\operatorname{Ker} B_h^t \neq \{0\}$

5.3.1 *The Case of* $\operatorname{Ker} B_h^t \subseteq \operatorname{Ker} B^t$

Coming back to our abstract framework, if H is not reduced to zero but $H_h \subseteq H$, we could either work with $\widetilde{Q} := Q_{/H}$, as we suggested before (several times), or use the following result.

Theorem 5.3.1. *Under the Assumption $\mathcal{AB}_h$, assume further that* ImB *is closed, that* (5.1.18) *is satisfied, and that $H_h \subset H$. Then, for every $f \in V'$ and for every $g \in \mathrm{Im}B$, the discrete problem* (5.1.9) *has a solution (u_h, p_h) in which u_h is uniquely determined, and p_h is determined up to an element of H_h. Moreover, setting*

$$\widetilde{\beta}_h := \inf_{q_h \in Q_h} \sup_{v_h \in V_h} \frac{b(v_h, q_h)}{\|v_h\|_V \, \|q_h\|_{Q/H}}, \tag{5.3.1}$$

we can say that: for every $u_I \in V_h$ and every $p_I \in Q_h$, we have the estimate

$$\|u_h - u_I\|_V \leq \frac{1}{\alpha_h^1}\|\mathcal{F}\|_{V_h'} + \frac{2\|a\|}{\alpha_1^h \widetilde{\beta}_h}\|\mathcal{G}\|_{Q_h'}, \tag{5.3.2}$$

$$\|p_h - p_I\|_{Q/H} \leq \frac{2\|a\|}{\alpha_h^1 \widetilde{\beta}_h}\|\mathcal{F}\|_{V_h'} + \frac{2\|a\|^2}{\alpha_h^1 \widetilde{\beta}_h{}^2}\|\mathcal{G}\|_{Q_h'} \tag{5.3.3}$$

where $\mathcal{F}$ and $\mathcal{G}$ are defined by (5.2.6) *and* (5.2.7), *respectively. If, moreover, $a(\cdot,\cdot)$ is symmetric, satisfies* (5.2.10) *and is positive semi-definite, then we have the improved estimates*

$$\|u_h - u_I\|_V \leq \frac{1}{\alpha_0^h}\|\mathcal{F}\|_{V_h'} + \frac{2\|a\|^{1/2}}{(\alpha_0^h)^{1/2}\widetilde{\beta}_h}\|\mathcal{G}\|_{Q_h'}, \tag{5.3.4}$$

$$\|p_h - p_I\|_{Q/H} \leq \frac{2\|a\|^{1/2}}{(\alpha_0^h)^{1/2}\widetilde{\beta}_h}\|\mathcal{F}\|_{V_h'} + \frac{\|a\|}{\widetilde{\beta}_h^2}\|\mathcal{G}\|_{Q_h'}. \tag{5.3.5}$$

where $\mathcal{F}$ and $\mathcal{G}$ are always defined by (5.2.6) *and* (5.2.7), *respectively.*

When applying Theorem 5.3.1, we can immediately make profit of the fact that in its assumptions we have $H_h \subseteq H$, which, thanks to Proposition 5.1.1, ensures the existence of an operator $\Pi_h : V \to V_h$ verifying (5.1.29). Hence, in the estimate of the terms $\|\mathcal{F}\|_{V_h'}$ and $\|\mathcal{G}\|_{Q_h'}$, we can immediately consider the choice $u_I = \Pi_h u$. We therefore have the following results, which can be seen as an extension of Theorems 5.2.3 and 5.2.4.

Theorem 5.3.2. *Under the same assumptions as in Theorem 5.3.1, we have the estimates*

$$\|u_h - u\|_V \leq \frac{1}{\alpha_h^1}\Big(\|a\| E_u^Z + \|b\|_h \inf_{q_h \in Q_h} \|p - q_h\|_{Q/H}\Big), \tag{5.3.6}$$

$$\|p_h - p\|_{Q/H} \leq \frac{2\|a\|}{\alpha_h^1 \widetilde{\beta}_h}\Big(\|a\| E_u^Z + 2\|b\|_h \inf_{q_h \in Q_h} \|p - q_h\|_{Q/H}\Big), \tag{5.3.7}$$

where $\|b\|_h$ is defined by

$$\|b\|_h := \sup_{\substack{q\in Q_{/H}\\ v_h\in V_h}} \frac{b(v_h,q)}{\|q\|_{Q_{/H}}\,\|v_h\|_V}. \tag{5.3.8}$$

Theorem 5.3.3. *Under the same assumptions as in Theorem 5.3.2, assume further that $Z_h^*(B^t p)$ is not empty. Then, we have the estimates*

$$\|u-u_h\|_V \le \frac{2\|a\|}{\alpha_h^1} E_u^Z, \tag{5.3.9}$$

$$\|p_h - p\|_{Q_{/H}} \le \frac{2\|a\|^2}{\alpha_h^1 \tilde{\beta}_h} E_u^Z + \inf_{p_I\in Z_h^*(B^t p)} \|p-p_I\|_{Q_{/H}} \tag{5.3.10}$$

Remark 5.3.1. We point out that even when $H \neq \{\mathbf{0}\}$ and $H_h \neq \{\mathbf{0}\}$, a requirement like $p_I \in Z_h^*(B^t p)$ makes sense, as it does not depend on the choices of p and p_I in their (respective) classes. Indeed, if p and p_I are such that $b(v_h, p-p_I) = 0$ for all $v_h \in V_h$, then for every $p^0 \in H$ and for every $p_h^0 \in H_h$ we also have $b(v_h, (p+p^0)-(p_I+p_h^0)) = 0$ for all $v_h \in V_h$. □

Remark 5.3.2. Needless to say, in the above estimates, we could have applied Proposition 5.1.3 and have gotten estimates in terms of the infimum of $\|u-v_h\|$ over the whole V_h. □

We could clearly keep going, repeating all the results we had for the case in which the *inf-sup* condition (5.1.19) is true, and just changing Q into $Q_{/H}$. We leave this to the reader.

5.3.2 *The Case of* $\mathrm{Ker}\,B_h^t \not\subseteq \mathrm{Ker}\,B^t$

We consider instead the case where H_h **is *not* a subspace of** H. For simplicity, we limit ourselves to the case where H is reduced to zero. We know that, otherwise, we can always change Q into $Q_{/H}$.

We shall see that as far as $g \in \mathrm{Im}B_h$, the situation is not too bad. Roughly speaking, we just have to filter the elements of H_h from our discrete solution. This however, in practice, can be done easily in some cases (when H_h is known and easy to deal with), and can be cumbersome in other cases. Similarly, the condition $g \in \mathrm{Im}B_h$ can be easy to check in some cases (surely, for instance, if $g = 0$), and much less easy in others.

Concerning error estimates, our advice, in general, is to change, whenever possible, the definition of Q_h using Q_{h/H_h} instead, and going back to the case in which the *inf-sup* condition (5.1.19) is satisfied.

We have for instance the following result, that could be seen as an immediate consequence of Theorem 5.2.1.

Theorem 5.3.4. *Under the Assumption $\mathcal{AB}_h$, assume further that* $\mathrm{Im}B = Q'$ *(hence the continuous inf-sup condition* (5.1.6) *is satisfied) and that* (5.1.18) *is satisfied. Then, for every* $f \in V'$ *and for every* $g \in \operatorname{Im} B_h$*, the discrete problem* (5.1.9) *has a solution* (u_h, p_h) *in which* u_h *is uniquely determined, and* p_h *is determined up to an element of* H_h*. Moreover, setting*

$$\widetilde{\beta_h} := \inf_{q_h \in Q_h} \sup_{v_h \in V_h} \frac{b(v_h, q_h)}{\|v_h\|_V \, \|q\|_{Q/H_h}}, \tag{5.3.11}$$

we have that: for every $u_I \in V_h$ *and every* $p_I \in Q_h$*, we have the estimate*

$$\|u_h - u_I\|_V \le \frac{1}{\alpha_h}\|\mathcal{F}\|_{V_h'} + \frac{2\|a\|}{\alpha_h \widetilde{\beta_h}}\|\mathcal{G}\|_{Q_h'}, \tag{5.3.12}$$

$$\|p_h - p_I\|_{Q/H_h} \le \frac{2\|a\|}{\alpha_h \widetilde{\beta_h}}\|\mathcal{F}\|_{V_h'} + \frac{2\|a\|^2}{\alpha_h \widetilde{\beta_h}^2}\|\mathcal{G}\|_{Q_h'}. \tag{5.3.13}$$

If, moreover, $a(\cdot,\cdot)$ *is symmetric and positive semi-definite, then we have the improved estimates*

$$\|u_h - u_I\|_V \le \frac{1}{\alpha_0^h}\|\mathcal{F}\|_{V_h'} + \frac{2\|a\|^{1/2}}{(\alpha_0^h)1/2\widetilde{\beta}_h}\|\mathcal{G}\|_{Q_h'}, \tag{5.3.14}$$

$$\|p_h - p_I\|_{Q/H_h} \le \frac{2\|a\|^{1/2}}{(\alpha_0^h)^{1/2}\widetilde{\beta}_h}\|\mathcal{F}\|_{V_h'} + \frac{\|a\|}{\widetilde{\beta}_h^2}\|\mathcal{G}\|_{Q_h'}. \tag{5.3.15}$$

Here also, we can specialise the choices of u_I and p_I to derive better special estimates for $\|\mathcal{F}\|_{V_h'}$ and $\|\mathcal{G}\|_{Q_h'}$. In particular, we have the following result.

Theorem 5.3.5. *Under the same assumptions as in Theorem 5.3.4, if* $Z_h(Bu)$ *is not empty, then*

$$\|u_h - u\|_V \le \frac{1}{\alpha_h}\Big(2\|a\|E_u^Z + \|b\|E_p\Big). \tag{5.3.16}$$

Remark 5.3.3. As we did before, we can use Proposition 5.1.3 to obtain

$$\|u_h - u\|_V \le \frac{1}{\alpha_h}\Big(\frac{4\|a\|\,\|b\|}{\widetilde{\beta}_h} E_u + \|b\|E_p\Big). \tag{5.3.17}$$

□

Remark 5.3.4. It is important to point out that, assuming that $Z_h(Bu)$ is not empty, we are already making an assumption on g as well. Indeed, requiring, for instance, the existence of an operator Π_h satisfying (5.1.29) would be equivalent (using Corollary 5.1.1) to requiring the *inf-sup* condition (5.1.37) (and, actually, (5.1.19), as we assumed that $H = \{\mathbf{0}\}$). □

5.3.3 The Case of β_h or $\tilde{\beta}_h$ going to zero

A possible source of (major) trouble can however arise when the constants β_h or $\tilde{\beta}_h$ tend to zero when h is tending to zero. In principle, we accepted that, and this is the reason why we keep using an index h for them. The idea is that, in certain applications, the bad behaviour of β_h could be partly compensated by the approximation error, leading possibly to a non-optimal error bound but still ensuring convergence. When this is not the case, however, the best strategy would be to give up and choose a different type of discretisation. In certain cases, however, the method in question is particularly appealing, and this might justify some "triple backwards somersault" in order to rescue it (or part of it). For instance, we have the following results.

Theorem 5.3.6. *Under the assumptions $\mathcal{AB}_h$, assume further, for simplicity, that a is elliptic on the whole V as in* (5.1.8) *and that* $\operatorname{Im} B = Q'$. *Suppose moreover that we have a couple of spaces $\hat{V} \subseteq V_h$ and $\hat{Q} \subseteq Q_h$ such that the pair $(\hat{V}, \hat{Q})$ satisfies the inf-sup condition*

$$\inf_{\hat{q}\in\hat{Q}} \sup_{\hat{v}\in\hat{V}} \frac{b(\hat{v},\hat{q})}{\|\hat{v}\,\|\hat{q}\|_Q} \geq \hat{\beta} > 0, \tag{5.3.18}$$

and that

$$b(\hat{v}, q_h) = b(\hat{v}, \pi_{\hat{Q}} q_h) \quad \forall \hat{v} \in \hat{V}\ \forall q_h \in Q_h. \tag{5.3.19}$$

Finally assume that **g** *is such that*

$$\langle \mathbf{g}, q_h \rangle = \langle \mathbf{g}, \pi_{\hat{Q}} q_h \rangle \quad \forall q_h \in Q_h. \tag{5.3.20}$$

Then, we have the estimate

$$\|u_h - u\|_V \leq \frac{4\|a\|\,\|b\|}{\alpha\hat{\beta}} \hat{E}_u + \frac{\|b\|}{\alpha} E_p, \tag{5.3.21}$$

where we set

$$\hat{E}_u := \inf_{\hat{v}\in\hat{V}} \|u - \hat{v}\| \textit{ and, for future use, } \hat{E}_p := \inf_{\hat{q}\in\hat{Q}} \|p - \hat{q}\|. \tag{5.3.22}$$

Proof. From (5.3.18) and (5.3.20), we easily deduce that

$$\hat{Z}_h(g) := \{\hat{v} \in \hat{V} \text{ s.t. } b(\hat{v},\hat{q}) = \langle g, \hat{q}\rangle\ \forall \hat{q} \in \hat{Q}\} \subseteq Z_h(g). \tag{5.3.23}$$

Therefore, we immediately have

$$\inf_{v_h\in Z_h(g)} \|u - v_h\| \leq \inf_{\hat{v}\in\hat{Z}_h(g)} \|u - \hat{v}\|_V \leq \frac{2\|b\|}{\hat{\beta}} \inf_{\hat{v}\in\hat{V}} \|u - \hat{v}\|_V, \tag{5.3.24}$$

where in the last step we applied Proposition 5.1.3. Hence, (5.3.21) follows, inserting (5.3.24) into (5.3.16). □

Always with the same assumptions as in Theorem 5.3.6, we can also prove some estimates on $p - p_h$. As, in a certain sense, we are playing the game that the pair (V_h, Q_h) is *unstable*, we have no hope to estimate $p - p_h$ flatly in Q. However, we can have estimates in *the only finite dimensional subspace of Q where we have stability*, and this is $\hat{Q}$.

Proposition 5.3.1. *Using the same assumptions as in Theorem 5.3.6, we have*

$$\|p - \pi_{\hat{Q}} p_h\|_Q \leq \frac{\|a\|}{\hat{\beta}}(\|u - u_h\|_V + \|u - \hat{u}\|_V) + \|p - \hat{p}\|_Q, \tag{5.3.25}$$

where $(\hat{u}, \hat{p}) \in \hat{V} \times \hat{Q}$ is the solution of the problem

$$a(\hat{u}, \hat{v}) + b(\hat{v}, \hat{p}) = (f, \hat{v}), \quad \forall \hat{v} \in \hat{V} \tag{5.3.26}$$

$$b(\hat{u}, \hat{q}) = (g, \hat{q}), \quad \forall \hat{v} \in \hat{V}. \tag{5.3.27}$$

Proof. From (5.3.18), we have that there exists a $\hat{v} \in \hat{V}$ with $\|\hat{v}\|_V = 1$ such that

$$\hat{\beta}\|\hat{p} - \pi_{\hat{V}} p_h\|_Q \leq b(\hat{v}, \hat{p} - \pi_{\hat{V}} p_h). \tag{5.3.28}$$

Using (5.3.26), (5.3.19), and the first equation of (5.1.9) in (5.3.28), then adding and subtracting u, and finally, using the continuity of the bilinear form a, we have

$$\begin{aligned}\hat{\beta}\|\hat{p} - \pi_{\hat{V}} p_h\|_Q &\leq b(\hat{v}, \hat{p} - \pi_{\hat{V}} p_h)\\ &= \langle f, \hat{v}\rangle - a(\hat{u}, \hat{v}) - b(\hat{v}, p_h) = a(u_h - \hat{u}, \hat{v})\\ &= a(u_h - u, \hat{v}) + a(u - \hat{u}, \hat{v})\\ &\leq \|a\|\,(\|u - u_h\|_V + \|u - \hat{u}\|_V)\|\hat{v}\|_V.\end{aligned} \tag{5.3.29}$$

Since $\|\hat{v}\|_V = 1$, we have

$$\|\hat{p} - \pi_{\hat{V}} p_h\|_Q \leq \frac{\|a\|}{\hat{\beta}}\,(\|u - u_h\|_V + \|u - \hat{u}\|_V), \tag{5.3.30}$$

and (5.3.25) easily follows. □

Remark 5.3.5. Using the ellipticity of a and the *inf-sup* condition (5.3.18), we can apply, for instance, Theorem 5.2.3 and have

$$\|u - \hat{u}\|_V \leq \frac{1}{\alpha}\Big(\frac{4\|a\|\,\|b\|}{\hat{\beta}}\,\hat{E}_u + \|b\|\,\hat{E}_p\Big), \tag{5.3.31}$$

$$\|p - \hat{p}\|_V \le \frac{2\|a\|}{\alpha\hat{\beta}}\Big(\frac{2\|a\|\,\|b\|}{\hat{\beta}}\hat{E}_u + 2\|b\|\,\hat{E}_p\Big), \tag{5.3.32}$$

which, inserted in (5.3.25) together with (5.3.21), provides a bound for $\|p - \pi_{\hat{Q}}\|$ in terms of approximation errors. □

Remark 5.3.6. We also have a more precise result on $\hat{u}_h$. Making $v_h = \hat{v}$ and $q_h = \hat{q}$ in the two equations of (5.1.9), and subtracting equations (5.3.26) and (5.3.27), we get

$$a(u_h - \hat{u}, \hat{v}) + b(\hat{v}, p_h - \hat{p}) = 0, \ \forall\, \hat{v} \in \hat{V}, \tag{5.3.33}$$

$$b(u_h - \hat{u}, \hat{q}) = 0, \ \forall\, \hat{q} \in \hat{Q}. \tag{5.3.34}$$

Taking $\hat{v} \in \mathrm{Ker}\hat{B}_h$ in (5.3.33), we have

$$\hat{u} = \pi_{\hat{Z}_h(g)} u_h, \tag{5.3.35}$$

meaning that $\hat{u}$ is the projection of u_h on $\hat{Z}_h(g)$. □

Finally, we must emphasise that this result will be useful only in certain special cases. Its application will generally rely on strong assumptions on the finite element meshes.

5.4 The *inf-sup* Condition: Criteria

5.4.1 Some Linguistic Considerations

In this section, we give some general strategies that can be used, in applications, to prove that the discrete *inf-sup* condition (5.1.19) is satisfied, possibly with a constant $\beta_h \ge \beta_0 > 0$, with β_0 independent of h.

It is worth mentioning, from the very beginning, that, in the literature on mixed formulations, there are various meanings for sentences of the type: *The inf-sup condition is satisfied* or *The inf-sup condition is not satisfied*. Indeed, we can distinguish, a priori, *two* different situations. The branching is obtained when we decide whether to look at the infimum of q_h over the whole Q_h, as in (5.1.19), or just over $Q_{h/H}$ (where, as usual, $H = \mathrm{Ker}B^t$) as in (5.3.11). Having made this choice (between (5.1.19) and (5.3.11)), it is then commonly accepted that the meaning of the sentence *The inf-sup condition is satisfied* is, when we choose (5.1.19), that β_h has a positive lower bound $\beta_h \ge \beta_0 > 0$, independent of h. If instead we chose (5.3.11), the meaning will be that $\tilde{\beta}_h$ has a positive lower bound $\tilde{\beta}_h \ge \tilde{\beta}_0 > 0$, independent of h.

This, we acknowledge, is not fully aligned with what is done in this book, where we often say *assume that the inf-sup condition* (5.1.19) *is satisfied*. To be consistent, we should say instead: *assume that* (5.1.19) *is satisfied*, recalling that the *inf-sup* condition should be considered to be satisfied only if the β_h appearing in (5.1.19) has a positive lower bound for $h \to 0$. We decided, however, as a didactic strategy, to instate (whenever possible and convenient) a sort of *nickname* for our formulae to easily refer to them. Indeed, we consider that a nickname could help the reader better than a number in recognising a formula and we failed to find for (5.1.19) a nickname better than *the inf-sup condition*.

5.4.2 General Considerations

Looking at the definition of the *inf-sup* condition (5.1.19), and also at the general considerations made in Chap. 3, one can see that, roughly speaking, the *inf-sup* condition requires the space V_h to be *big enough* with respect to Q_h. In other words, still roughly speaking, the bigger V_h is (or the smaller is Q_h), the more chances we have of satisfying the *inf-sup* condition. A more precise statement of this fact is expressed in the following proposition.

Proposition 5.4.1. *In the framework of Assumption $\mathcal{AB}_h$, assume that the inf-sup condition* (5.1.19) *is satisfied for a given choice of V_h and Q_h, and for a given value of the constant β_h. Then, for every subspace $\tilde{Q}_h \subseteq Q_h$, we have*

$$\inf_{\tilde{q}_h \in \tilde{Q}_h} \sup_{v_h \in V_h} \frac{b(v_h, \tilde{q}_h)}{\|v_h\|_V \, \|\tilde{q}_h\|_Q} \geq \beta_h, \tag{5.4.1}$$

and for every space $\tilde{V}_h$ with $V_h \subseteq \tilde{V}_h \subseteq V$, we have

$$\inf_{q_h \in Q_h} \sup_{\tilde{v}_h \in \tilde{V}_h} \frac{b(\tilde{v}_h, q_h)}{\|\tilde{v}_h\|_V \, \|q_h\|_Q} \geq \beta_h. \tag{5.4.2}$$

Proof. The proof is trivial. Indeed, if $\tilde{Q}_h \subseteq Q_h$, we have

$$\inf_{\tilde{q}_h \in \tilde{Q}_h} \Big(\sup_{v_h \in V_h} \frac{b(v_h, \tilde{q}_h)}{\|v_h\|_V \, \|\tilde{q}_h\|_Q} \Big) \geq \inf_{q_h \in Q_h} \Big(\sup_{v_h \in V_h} \frac{b(v_h, q_h)}{\|v_h\|_V \, \|q_h\|_Q} \Big) \geq \beta_h, \tag{5.4.3}$$

as the infimum on the smaller space is bigger (or equal) than the infimum on the bigger space. On the other hand, if $V_h \subseteq \tilde{V}_h$, we have

$$\inf_{q_h \in Q_h} \sup_{\tilde{v}_h \in \tilde{V}_h} \Big(\frac{b(\tilde{v}_h, q_h)}{\|\tilde{v}_h\|_V \, \|q_h\|_Q} \Big) \geq \inf_{q_h \in Q_h} \sup_{v_h \in V_h} \Big(\frac{b(v_h, q_h)}{\|v_h\|_V \, \|q_h\|_Q} \Big) \geq \beta_h, \tag{5.4.4}$$

as the supremum on a bigger space is bigger (or equal) than the supremum on a smaller space. □

The suggestion of Proposition 5.4.1 is clear: if your choice of spaces does not satisfy the *inf-sup* condition (or if you are uncertain and you want to play safe), then either take a bigger V_h or take a smaller Q_h, and the situation should, in principle, improve.

All this, however, is good and nice if you have a bilinear form $a(\cdot,\cdot)$ that is elliptic on the whole space V. Indeed, in this case, its ellipticity on the kernel K_h of B_h will be automatically guaranteed, no matter how big or how small K_h comes out to be. If, on the contrary, your bilinear form a is elliptic only on the kernel K of B (or, even worse, satisfies an *inf-sup* condition of type (5.1.1) on K), then the ellipticity on K_h (or the *inf-sup* for a on K_h as in (5.1.18)) will depend on the nature of K_h. The bigger the kernel is, the more difficult it will be to have (5.1.18) (or (5.1.20)) satisfied. Here start the *troubles*: indeed, for the same Q_h, a bigger V_h will, in general, have a bigger kernel K_h. For the same V_h, the smaller you take Q_h, the bigger becomes the kernel K_h.

Hence, the two conditions (5.1.18) and (5.1.19) *play one against the other*, and the two spaces V_h and Q_h have, somehow, to match perfectly to each other. We have seen all this at work in the example of Sect. 5.2.4.

These cases (in which the two conditions play against each other), in general, require to be dealt with on a case by case basis. In what follows we concentrate instead on the *inf-sup* condition for the bilinear form b, that is, in any case, a crucial ingredient in almost all mixed formulations.

5.4.3 *The* inf-sup *Condition and the B-Compatible Interpolation Operator Π_h*

In Proposition 5.1.1, we have seen that, if the continuous *inf-sup* condition (5.1.6) is satisfied (and hence $H = \mathrm{Ker}B^t = \{0\}$), and if there exists a B-compatible operator, that is an operator $\Pi_h : V \to V_h$ such that

$$b(v - \Pi_h v, q_h) = 0 \qquad \forall\, q_h \in Q_h, \tag{5.4.5}$$

then $H_h = \mathrm{Ker}B_h^t$ is also reduced to $\{0\}$, and hence the discrete *inf-sup* condition (5.1.19) is also satisfied. However, we have no indications on the value of β_h for each h, and β_h might as well tend to zero when h tends to zero.

In the next proposition, we connect the *numerical value* of β_h with the norm of the operator Π_h. Indeed, we have the following result.

Proposition 5.4.2. *Assume that, for each h, we are given an operator $\Pi_h : V \to V_h$ that satisfies* (5.4.5)*, and assume that there exists a constant $C_\Pi > 0$, independent of h, such that*

$$\|\Pi_h v\|_V \le C_\Pi \,\|v\|_V \qquad \forall\, v \in V. \tag{5.4.6}$$

Assume moreover that the continuous inf-sup condition (5.1.6) *is satisfied (for a certain value of* $\beta > 0$*). Then, we have*

$$\inf_{q_h \in Q_h} \sup_{v_h \in V_h} \frac{b(v_h, q_h)}{\|v_h\|_V \, \|q_h\|_Q} \geq \frac{\beta}{C_\Pi} > 0, \tag{5.4.7}$$

that is: the discrete inf-sup condition (5.1.19) *is satisfied with* $\beta_h = \beta / C_\Pi$.

Proof.

$$\sup_{v_h \in V_h} \frac{b(v_h, q_h)}{\|v_h\|_V} \geq \sup_{v \in V} \frac{b(\Pi_h v, q_h)}{\|\Pi_h v\|_V} = \sup_{v \in V} \frac{b(v, q_h)}{\|\Pi_h v\|_V}$$
$$\geq \sup_{v \in V} \frac{b(v, q_h)}{C_\Pi \, \|v\|_V} \geq \frac{\beta}{C_\Pi} \|q_h\|_Q. \tag{5.4.8}$$

□

Remark 5.4.1. Proposition 5.4.2 was first presented in [201], and is often called the *Fortin trick.* □

Remark 5.4.2. It is clear that, by the same proof, the result extends to the case where H is not $\{\mathbf{0}\}$ and, instead of the norm in Q, we use the norm in Q_H. □

Remark 5.4.3. Assume that, for every $w \in V$, the discrete problem (5.1.9) with $g = Bw$ and $f = Aw$ has a solution (w_h, p_h). If one *requires*, as we did in Theorem 3.5.1, that $\|w_h\|_V \leq C\|w\|_V$ for some constant C independent of h (which is to say that the mapping $w \to w_h$ is uniformly bounded), then one can set $w_h = \Pi_h w$. It is easy to see that (5.4.5) and (5.4.6) are both satisfied, and hence the *inf-sup* condition also holds with a constant independent of h. This shows that the existence of an operator Π_h which satisfies the assumptions of Proposition 5.4.2 (hence the validity of the *inf-sup* condition) is in a sense *necessary* if we want a reasonable behaviour of the discrete problem. However, the explicit construction of Π_h will be easy in some cases but very difficult in others. □

The following result generalises the above proposition, and, as we shall see later on, is often much easier to apply in concrete cases.

Proposition 5.4.3. *Assume that we are given a Banach space* $W \hookrightarrow V$*, with norm* $\|\cdot\|_W$ *and a linear subspace* $S_h \subseteq Q_h$ *with a semi-norm* $|\cdot|_S$. *Suppose that*

$$\sup_{w \in W} \frac{b(w, s_h)}{\|w\|_W} \geq \beta_W \, |s_h|_S \quad \forall\, s_h \in S_h, \tag{5.4.9}$$

and assume that there exists a family of uniformly continuous operators Π_h *from* W *into* V *satisfying*

$$\begin{cases} b(\Pi_h w - w, s_h) = 0, \ \forall\, w \in W, \quad \forall\, s_h \in S_h, \\ \|\Pi_h w\|_V \le C_W \|w\|_W \end{cases}, \tag{5.4.10}$$

with C_W independent of h. Then, we have

$$\sup_{v_h \in V_h} \frac{b(v_h, s_h)}{\|v_h\|_V} \ge \beta_0 \, |s_h|_S, \quad \forall\, s_h \in S_h, \tag{5.4.11}$$

with $\beta_0 = \beta_W / C_W$.

Proof. Indeed, we have

$$\sup_{v_h \in V_h} \frac{b(v_h, s_h)}{\|v_h\|_V} \ge \sup_{w \in W} \frac{b(\Pi_h w, s_h)}{\|\Pi_h w\|_V} = \sup_{w \in W} \frac{b(w, s_h)}{\|\Pi_h w\|_V}$$
$$\ge \sup_{w \in W} \frac{1}{C_W} \frac{b(w, s_h)}{\|w\|_W} \ge \frac{\beta_W}{C_W} |s_h|_S.$$

□

Remark 5.4.4. In most applications, we shall take $W = V$ and the semi-norm $|\cdot|_S := \|\cdot\|_{Q/H}$. In this case, the first condition of (5.4.10) indeed implies from Proposition 5.1.1 that $H_h \subseteq H$ and we can summarise Proposition 5.4.3 by saying that if the continuous *inf-sup* condition (5.1.6) holds, and if we have (5.4.10), then the discrete *inf-sup* condition holds. □

In some cases, it will be convenient to choose W to be a strict subspace of V. This will, for instance, be the case when V is not smooth enough to allow a simple construction of the operator Π_h. Obviously, we shall then have to check the *inf-sup* condition (5.4.9) on W, usually with $S = Q$ and $|\cdot|_S = \|\cdot\|_Q$ (or, exceptionally, with $|\cdot|_S = \|\cdot\|_{Q/H}$).

The more general statement of Proposition 5.4.3 will also be useful for some special cases where $\operatorname{Ker} B_h^t$ is larger than $\operatorname{Ker} B^t$ and where we would like to use $|\cdot|_S = \|\cdot\|_{Q/\operatorname{Ker} B_h^t}$. In those cases, (5.4.10) will hold only for an ad hoc choice of W and the main trouble will be to obtain (5.4.9) for this W.

Finally, there will still be other cases in which a special choice of $|\cdot|_S$ is needed. We shall meet, for example, cases where $H_h = H = \{\mathbf{0}\}$, where V is smooth enough to allow the construction of Π_h, but where the continuous *inf-sup* condition holds only if one takes a semi-norm $|\cdot|_S$ "small enough" (and in particular $\le \|\cdot\|_Q$).

5.4.4 Construction of Π_h

We can say that Proposition 5.4.2 and its generalisation Proposition 5.4.3 (that is, the idea of constructing suitable operators Π_h) are *the main instruments* to prove the *inf-sup* condition. On the other hand, the following result could be considered as *the main instrument* to construct the operators Π_h.

Proposition 5.4.4. *Let* $W \hookrightarrow V$ *be a subspace of* V *for which* (5.4.9) *holds. Let* $\Pi_1 \in \mathcal{L}(W, V_h)$ *and* $\Pi_2 \in \mathcal{L}(V, V_h)$ *be such that*

$$\begin{cases} \|\Pi_1 w\|_V \le c_1 \, \|w\|_W, & \forall w \in W, \\ b(\Pi_2 v - v,\ q_h) = 0, & \forall v \in V, \quad \forall\, q_h \in Q_h, \\ \|\Pi_2(I - \Pi_1)w\|_V \le c_2 \, \|w\|_W, & \forall w \in W. \end{cases} \tag{5.4.12}$$

Then, the operator $\Pi_h := \Pi_2(I - \Pi_1) + \Pi_1$ *satisfies* (5.4.10) *with* $c = c_1 + c_2$ *(and hence the inf-sup condition* (5.1.19) *holds with a* β_h *independent of* h*).*

Proof. It is easy to check that (5.4.10) holds. Indeed,

$$\begin{aligned} b(\Pi_h w, q_h) &= b(\Pi_2(w - \Pi_1 w), q_h) + b(\Pi_1 w, q_h) \\ &= b(w - \Pi_1 w, q_h) + b(\Pi_1 w, q_h) \\ &= b(w, q_h) \end{aligned} \tag{5.4.13}$$

and

$$\|\Pi_h w\|_V \le \|\Pi_2(w - \Pi_1 w)\|_V + \|\Pi_1 w\|_V \le (c_2 + c_1)\|w\|_W. \tag{5.4.14}$$

□

Remark 5.4.5. In the applications, Π_1 will be a kind of "best approximation" operator. To fix ideas, it will verify an estimate of type $\|\Pi_1 w - w\|_{V^*} \le ch^s \, \|w\|_W$ for some suitable norm $\|\cdot\|_{V^*}$. On the other hand, Π_2 will be a local adjustment (typically by bubble functions) in order to satisfy the first condition of (5.4.10). The third condition of (5.4.12) expresses the fact that we can allow the norm of Π_2 in $\mathcal{L}(V^*, V)$ to go to $+\infty$ (when h goes to zero), but not faster than h^{-s}. □

5.4.5 An Alternative Strategy: Switching Norms

An alternative way to prove the *inf-sup* condition can be summarised, in the present abstract context, as follows. Assume that we have two other Hilbert spaces V^* and Q^*, with $V \hookrightarrow V^*$ and $Q^* \hookrightarrow Q$, with dense embedding. For simplicity, we shall assume that the embedding constants are equal to 1, that is

$$\|v\|_{V^*} \le \|v\|_V \quad \forall\, v \in V, \text{ and } \quad \|q\|_Q \le \|q\|_{Q^*} \quad \forall\, q \in Q^*. \tag{5.4.15}$$

We also assume that there exists another continuous bilinear form $b^*(v,q)$, defined on $V^* \times Q^*$, which coincides with $b(v,q)$ whenever $v \in V$ and $q \in Q^*$:

$$b(v,q) = b^*(v,q) \qquad \forall\, v \in V,\ \forall\, q \in Q^*. \tag{5.4.16}$$

Remark 5.4.6. All this might look strange, but this situation is quite common: for instance, assume that $V := (H_0^1(\Omega))^2$, that $Q := L^2(\Omega)$, and that

$$b(v,q) := \int_\Omega \mathrm{div} v\, q\, dx.$$

Then, we could take $V^* := (L^2(\Omega))^2$ and $Q^* := H^1(\Omega)$, and the bilinear form b^* will simply be given by

$$b^*(v,q) := -\int_\Omega v \cdot \nabla q\, dx,$$

and clearly $b(v,q) = b^*(v,q)$ whenever $v \in V$ and $q \in Q^*$, by a simple integration by parts. Note that the density assumption implies that such a bilinear form b^*, if it exists, must be unique (essentially by the Hahn-Banach Theorem). □

We then set

$$\|b^*\| := \sup_{\substack{v \in V^* \\ q \in Q^*}} \frac{b^*(v,q)}{\|v\|_{V^*}\, \|q\|_{Q^*}}. \tag{5.4.17}$$

We finally assume that $Q_h \subset Q^*$, and that there exists a constant $\omega(h)$ such that we have both the approximation estimate in V_h

$$\|v - \pi_{V_h} v\|_{V^*} \leq \omega(h)\|v\|_V \qquad \forall\, v \in V \tag{5.4.18}$$

(where π_{V_h} is the projection operator, with the scalar product of V, over V_h) and the inverse inequalities in Q_h and V_h

$$\omega(h)\|q_h\|_{Q^*} \leq \|q_h\|_Q\ \forall\, q_h \in Q_h, \quad \omega(h)\|v_h\|_V \leq \|v_h\|_{V^*}\ \forall\, v_h \in V_h. \tag{5.4.19}$$

The main idea is then to prove, instead of the original *inf-sup* condition (5.1.19), a modified one, in the spaces V^* and Q^*:

$$\inf_{q_h \in Q_h} \sup_{v_h \in V_h} \frac{b^*(v_h,q_h)}{\|v_h\|_{V^*}\, \|q_h\|_{Q^*}} \geq \beta^* > 0. \tag{5.4.20}$$

We have indeed the following result.

Proposition 5.4.5. *Under the Assumption $\mathcal{AB}_h$, assume that the continuous inf-sup condition* (5.1.6) *holds. Assume moreover that all the assumptions of this subsection* *(namely* (5.4.15)–(5.4.20)*) are satisfied. Then, the discrete inf-sup condition* (5.1.19) *holds with* $\beta_h = \beta\beta^*/(1+\beta^*)$.

Proof.

$$\begin{aligned}
\beta\|q_h\|_Q \le \sup_{v\in V}\frac{b(v,q_h)}{\|v\|_V} &= \sup_{v\in V}\Big(\frac{b(\pi_{V_h}v,q_h)}{\|v\|_V}+\frac{b(v-\pi_{V_h}v,q_h)}{\|v\|_V}\Big)\\
&\le \sup_{v\in V}\frac{b(\pi_{V_h}v,q_h)}{\|v\|_V}+\sup_{v\in V}\frac{b(v-\pi_{V_h}v,q_h)}{\|v\|_V}\\
&\le \sup_{v\in V}\frac{b(\pi_{V_h}v,q_h)}{\|\pi_{V_h}v\|_V}+\sup_{v\in V}\frac{b^*(v-\pi_{V_h}v,q_h)}{\|v\|_V}\\
&\le \sup_{v_h\in V_h}\frac{b(v_h,q_h)}{\|v_h\|_V}+\sup_{v\in V}\frac{\|b^*\|\,\|v-\pi_{V_h}v\|_{V^*}\,\|q_h\|_{Q^*}}{\|v\|_V}\\
&\le \sup_{v_h\in V_h}\frac{b(v_h,q_h)}{\|v_h\|_V}+\omega(h)\|q_h\|_{Q^*}.
\end{aligned}$$

This shows, setting

$$S(q_h) := \sup_{v_h\in V_h}\frac{b(v_h,q_h)}{\|v_h\|_V}, \tag{5.4.21}$$

that

$$S(q_h) \ge \beta\|q_h\|_Q - \omega(h)\|q_h\|_{Q^*}. \tag{5.4.22}$$

On the other hand,

$$\begin{aligned}
S(q_h) = \sup_{v_h\in V_h}\frac{b(v_h,q_h)}{\|v_h\|_V} &\ge \sup_{v_h\in V_h}\frac{\omega(h)b(v_h,q_h)}{\|v_h\|_{V^*}}\\
&= \omega(h)\sup_{v_h\in V_h}\frac{b^*(v_h,q_h)}{\|v_h\|_{V^*}} \ge \omega(h)\beta^*\|q_h\|_{Q^*}.
\end{aligned}\tag{5.4.23}$$

Taking β^* times equation (5.4.22) plus equation (5.4.23), we get

$$(\beta^*+1)S(q_h) \ge \beta\beta^*\|q_h\|_Q \tag{5.4.24}$$

and the result follows. □

Remark 5.4.7. In (5.4.18) and (5.4.19) we used, for simplicity, the same quantity $\omega(h)$. Clearly, in each occurrence, we could have allowed, instead of $\omega(h)$, a quantity of the type $c_i\omega(h)$, where the c_i's are suitable constants independent of h. However, this would be the same as having exactly (5.4.18) and (5.4.19) after substituting $\|\cdot\|_{V^*}$ and $\|\cdot\|_{Q^*}$ with equivalent norms. □

Remark 5.4.8. Proposition 5.4.5 is a step-by-step remake, in abstract form, of a result due to Verfürth, and it often goes under the name of the *Verfürth trick* [375]. Its interest relies in the fact that, in some cases, the *inf-sup* condition with the "different norms" (5.4.20) is easier to prove than the original one (5.1.19). We shall see an application of it in the analysis of Hood-Taylor elements for the Stokes problem.

□

5.5 Extensions of Error Estimates

5.5.1 *Perturbed Problems*

We shall now study the approximation of the perturbed problems considered already several times, starting from Sect. 3.3, and seen last time in Sect. 4.2.2. There, we considered a general framework summarised in Assumption $\mathcal{ABC}$, that we repeat here for the convenience of the reader.

Assumption $\mathcal{ABC}$: *Together with Assumption $\mathcal{AB}$, we assume that we are given a continuous bilinear form $c(\cdot\,,\cdot)$ on $Q \times Q$, and we denote by C its associated operator $Q \to Q'$. We assume moreover that* ImB *is closed, and that both $a(\cdot\,,\cdot)$ and $c(\cdot\,,\cdot)$ are symmetric and positive semi-definite:*

$$a(v,v) \geq 0,\ \forall\, v \in V \qquad c(q,q) \geq 0,\ \forall\, q \in Q. \tag{5.5.1}$$

Together with assumption $\mathcal{ABC}$, we introduced $K :=$ Ker$B$ and $H :=$ KerB^t and defined a splitting of elements of V and Q of the form

$$v = v_0 + \overline{v} \qquad q = q_0 + \overline{q}, \tag{5.5.2}$$

with $v_0 \in K, \overline{v} \in K^{\perp}$, $q_0 \in H$ and $\overline{q} \in H^{\perp}$, together with a splitting of right-hand sides of the form

$$f = f_0 + \overline{f} \qquad g = g_0 + \overline{g}, \tag{5.5.3}$$

with $f_0 \in K', \overline{f} \in (K^{\perp})'$, $g_0 \in H'$ and $\overline{g} \in (H^{\perp})'$. We noted that

$$\langle f, v\rangle = \langle f_0, v_0\rangle + \langle \overline{f}, \overline{v}\rangle \quad \langle g, q\rangle = \langle g_0, q_0\rangle + \langle \overline{g}, \overline{q}\rangle,$$

with obvious meaning of the duality pairings. Then, we considered, for every $f \in V'$ and for every $g \in Q'$, the continuous problem: *find $u \in V$ and $p \in Q$ such that*

$$\begin{cases} a(u,v) + b(v,p) = \langle f, v\rangle_{V'\times V'}, & \forall\, v \in V, \\ b(u,q) - c(p,q) = \langle g, q\rangle_{Q'\times Q}, & \forall\, q \in Q. \end{cases} \tag{5.5.4}$$

We shall now consider the question of finite dimensional approximations and error estimation for this problem.

Following the path of the previous sections of the present chapter, we start by introducing an Assumption $\mathcal{ABC}_h$ to be used all over this section.

Assumption $\mathcal{ABC}_h$: *Together with Assumption $\mathcal{ABC}$, we assume that we are given two finite dimensional subspaces $V_h \subseteq V$ and $Q_h \subseteq Q$.*

The corresponding discretised problem will then be: *find $u_h \in V_h$ and $p_h \in Q_h$ such that*

$$\begin{cases} a(u_h, v_h) + b(v_h, p_h) = \langle f, v_h \rangle_{V' \times V'}, & \forall\, v_h \in V_h, \\ b(u_h, q_h) - c(p_h, q_h) = \langle g, q_h \rangle_{Q' \times Q}, & \forall\, q_h \in Q. \end{cases} \tag{5.5.5}$$

In agreement with the previous notation, each $v_h \in V_h$ and each $q_h \in Q_h$ might be split, when convenient, as

$$v = v_0^h + \overline{v}_h \qquad q_h = q_0^h + \overline{q}_h, \tag{5.5.6}$$

with $v_0^h \in K_h$, $\overline{v}_h \in K_h^{\perp}$, $q_0^h \in H_h$ and $\overline{q}_h \in H_h^{\perp}$. Similarly, the right-hand sides can be split as

$$f = f_0^h + \overline{f}_h \qquad g = g_0^h + \overline{g}_h \tag{5.5.7}$$

with $f_0^h \in K_h'$, $\overline{f}_h \in (K_h^{\perp})'$, $g_0^h \in H_h'$ and $\overline{g}_h \in (H_h^{\perp})'$. Note that the splitting (5.5.7) might, unfortunately, be different from the splitting (5.5.3) as, in general, $K_h \neq K$ and $H_h \neq H$. Note also that the spaces $K_h^{\perp}$ and $H_h^{\perp}$ should always be understood as subspaces of V_h and Q_h, respectively.

The following proposition is the counterpart of Proposition 5.2.1.

Proposition 5.5.1. *In the framework of Assumption $\mathcal{ABC}_h$, let (u, p) and (u_h, p_h) be solutions of the continuous problem* (5.5.4) *and of the discretised problem* (5.5.5), *respectively. Then, for every $(u_I, p_I) \in V_h \times Q_h$, we have that $(u_h - u_I, p_h - p_I)$ is the solution, in $V_h \times Q_h$, of the variational problem*

$$\begin{cases} a(u_h - u_I, v_h) + b(v_h, p_h - p_I) = \langle \mathcal{F}, v_h \rangle_{V_h' \times V_h}, & \forall\, v_h \in V_h, \\ b(u_h - u_I, q_h) - c(p_h - p_I, q_h) = \langle \mathcal{G}, q_h \rangle_{Q_h' \times Q_h}, & \forall\, q_h \in Q_h, \end{cases} \tag{5.5.8}$$

where

$$\langle \mathcal{F}, v_h \rangle_{V_h' \times V_h} := a(u - u_I, v_h) + b(v_h, p - p_I) \quad \forall\, v_h \in V_h, \tag{5.5.9}$$

and

$$\langle \mathcal{G}, q_h \rangle_{Q_h' \times Q_h} := b(u - u_I, q_h) - c(p - p_I, q_h) \quad \forall\, q_h \in Q_h. \tag{5.5.10}$$

A direct application of Theorem 4.3.1 produces a rather cumbersome result, but we shall simplify it later on in several particular cases.

Theorem 5.5.1. *Together with Assumption* $\mathcal{ABC}_h$*, assume that* $a(\cdot,\cdot)$ *is coercive on* K_h *and* $c(\cdot,\cdot)$ *is coercive on* H_h*. Let therefore* α_0^h*,* β_h*, and* γ_0^h *be such that*

$$\alpha_0^h\|v_0^h\|_V^2 \leq a(v_0^h, v_0^h) \qquad \forall\, v_0^h \in K_h, \tag{5.5.11}$$

$$\inf_{q_h\in H_h^{\perp}}\sup_{v_h\in V_h}\frac{b(v_h,q_h)}{\|q_h\|_Q\,\|v_h\|_V} = \inf_{v_h\in K_h^{\perp}}\sup_{q_h\in Q_h}\frac{b(v_h,q_h)}{\|q_h\|_Q\,\|v_h\|_V} = \beta_h > 0, \tag{5.5.12}$$

$$\gamma_0^h\|q_0^h\|_Q^2 \leq c(q_0^h, q_0^h) \qquad \forall\, q_0^h \in H_h. \tag{5.5.13}$$

Then, for every $f \in V'$ *and* $g \in Q'$*, we have that the discretised problem* (5.5.5) *has a unique solution. Moreover, if* (u, p) *is a solution of the continuous problem* (5.5.4)*, then for every* $u_I \in V_h$ *and for every* $p_I \in Q_h$*, we have the estimates*

$$\|\overline{u}_h - \overline{u}_I\|_V \leq \frac{\|c\|\|\overline{\mathcal{F}}_h\|}{\beta_h^2} + \frac{2\mu\|c\|^{1/2}\,\|\mathcal{F}_0^h\|}{(\alpha_0^h)^{1/2}\beta_h^2} + \frac{2\mu\|\overline{\mathcal{G}}_h\|}{\beta_h^2} + \frac{3\mu^2\,\|c\|^{1/2}\|\mathcal{G}_0^h\|}{(\gamma_0^h)^{1/2}\beta_h^2}, \tag{5.5.14}$$

$$\|(u_h - u_I)_0\|_V \leq \frac{\|c\|\,\|a\|^{1/2}\,\|\overline{\mathcal{F}}_h\|}{(\alpha_0^h)^{1/2}\beta_h^2} + \frac{2\mu^2\|\mathcal{F}_0^h\|}{\alpha_0^h\beta_h^2} + \frac{2\mu\|a\|^{1/2}\|\overline{\mathcal{G}}_h\|}{(\alpha_0^h)^{1/2}\beta_h^2} + \frac{3\mu^2\|\mathcal{G}_0^h\|}{(\gamma_0^h)^{1/2}(\alpha_0^h)^{1/2}\beta_h^2}, \tag{5.5.15}$$

$$\|\overline{p}_h - \overline{p}_I\|_Q \leq +\frac{2\mu\|\overline{\mathcal{F}}_h\|}{\beta_h^2} + \frac{3\mu^2\|\mathcal{F}_0^h\|}{(\gamma_0^h)^{1/2}(\alpha_0^h)^{1/2}\beta_h^2} + \frac{\|a\|\,\|\overline{\mathcal{G}}_h\|}{\beta_h^2} + \frac{2\mu\|a\|^{1/2}\,\|\mathcal{G}_0^h\|}{(\gamma_0^h)^{1/2}\beta_h^2}, \tag{5.5.16}$$

$$\|(p_h - p_I)_0\|_Q \leq +\frac{2\mu\|c\|^{1/2}\|\overline{\mathcal{F}}_h\|}{(\gamma_0^h)^{1/2}\beta_h^2} + \frac{3\mu^2\|a\|^{1/2}\|\mathcal{F}_0^h\|}{(\alpha_0^h)^{1/2}\beta_h^2} \quad \frac{\|a\|\,\|c\|^{1/2}\|\overline{\mathcal{G}}_h\|}{(\gamma_0^h)^{1/2}\beta_h^2} + \frac{2\mu^2\|\mathcal{G}_0^h\|}{\gamma_0^h\beta_h^2}, \tag{5.5.17}$$

where μ is defined by

$$\mu^2 := \|a\|\,\|c\| + \beta_h^2 \tag{5.5.18}$$

and where, referring to (5.5.6),

$$\langle \overline{\mathcal{F}}_h, v_h \rangle_{V_h' \times V_h} := a(u - u_I, \overline{v}_h) + b(\overline{v}_h, p - p_I) \quad \forall\, v_h \in V_h, \tag{5.5.19}$$

$$\langle \mathcal{F}_0^h, v_h \rangle_{V_h' \times V_h} := a(u - u_I, v_0^h) + b(v_0^h, p - p_I) \quad \forall\, v_h \in V_h, \tag{5.5.20}$$

$$\langle \overline{\mathcal{G}}_h, q_h \rangle_{Q_h' \times Q_h} := b(u - u_I, \overline{q}_h) - c(p - p_I, \overline{q}_h) \quad \forall\, q_h \in Q_h, \tag{5.5.21}$$

$$\langle \mathcal{G}_0^h, q_h \rangle_{Q_h' \times Q_h} := b(u - u_I, q_0^h) - c(p - p_I, q_0^h) \quad \forall\, q_h \in Q_h. \tag{5.5.22}$$

It is clear that all terms in (5.5.20) and (5.5.21) can be bounded, roughly, by

$$\begin{aligned} &\|\mathcal{F}_0^h\| + \|\mathcal{G}_0^h\| + \|\overline{\mathcal{F}}^h\| + \|\overline{\mathcal{G}}^h\| \\ &\qquad \le (2\|a\| + 4\|b\| + 2\|c\|)\,(\|u - u_I\|_V + \|p - p_I\|_Q). \end{aligned} \tag{5.5.23}$$

Hence, Theorem 5.5.1 immediately provides a rough estimate.

Proposition 5.5.2. *Under the same assumptions as in Theorem 5.5.1, we have:*

$$\begin{aligned} &\|u - u_h\|_V + \|p - p_h\|_Q \\ &\le K\Big(\|a\|,\ \|b\|,\ \|c\|,\ \frac{1}{\alpha_0^h},\ \frac{1}{\beta_h},\ \frac{1}{\gamma_0^h}\Big)\Big(E_u + E_p\Big) \end{aligned} \tag{5.5.24}$$

with K bounded on bounded subsets. □

Remark 5.5.1. Estimates of the type (5.5.23) or (5.5.24) are particularly ugly, as they require that all the quantities in play are adimensionalised (otherwise we are adding apples and oranges). They have, however, the merit to condense in a short sentence what would otherwise need a lengthy and complicated one. □

5.5.2 *Penalty Methods*

An improvement to the above situation can be obtained by requiring additional assumptions. For instance, we can assume that the bilinear form $c(\cdot,\cdot)$ is of the type (4.3.58) considered in Theorem 4.3.2, that is

$$c(p,q) = \lambda (p,q)_Q, \tag{5.5.25}$$

where λ is a positive real number (that we might possibly think as tending to zero). For this, for instance from Corollary 4.3.1, we have the following result.

Theorem 5.5.2. *Together with Assumption $\mathcal{ABC}_h$, assume that $a(\cdot,\cdot)$ is coercive on K_h and that $c(\cdot,\cdot)$ is given in* (5.5.25). *Let again α_0^h and β_h be defined as in* (5.5.11) *and* (5.5.12)*, respectively. Then, for every $f \in V'$ and $g \in Q'$, we have that the discretised problem* (5.5.5) *has a unique solution. Moreover, if (u,p) is a solution of the continuous problem* (5.5.4)*, then for every $u_I \in V_h$ and for every $p_I \in Q_h$ we have the estimates*

$$\|u_h - u_I\|_V \le \frac{{\beta_h}^2 + 4\lambda\,\|a\|}{\alpha_0^h {\beta_h}^2}\|\mathcal{F}^h\|_{V'} + \frac{2\|a\|^{1/2}}{{\alpha_0^h}^{1/2}\beta_h}\|\overline{\mathcal{G}}^h\|_{Q'}, \tag{5.5.26}$$

$$\|\overline{p_h} - \overline{p_I}\|_Q \le \frac{2\,\|a\|^{1/2}}{{\alpha_0^h}^{1/2}\beta_h}\|\mathcal{F}^h\|_{V'} + \frac{4\,\|a\|}{\lambda\|a\| + 2\,{\beta_h}^2}\|\overline{\mathcal{G}}^h\|_{Q'}, \tag{5.5.27}$$

$$\|(p_h - p_I)_0\|_Q \le \frac{1}{\lambda}\|\mathcal{G}_0^h\|_{Q'}. \tag{5.5.28}$$

At this point, we can specialise our results further, assuming for instance that $H_h = \mathrm{Ker}B_h^t = \{0\}$. In this case, $\mathcal{G}_0^h = 0$. Moreover, according to Proposition 5.1.1, we can take $u_I = \Pi_h u$ such that $b(u - u_I, q_h) = 0$ for all $q_h \in Q_h$.

In this case, we have

$$\|\mathcal{F}^h\|_{V'_h} \le \|a\|\,\|u - u_I\|_V + \|b\|\,\|p - p_I\|_Q, \tag{5.5.29}$$

$$\|\overline{\mathcal{G}}^h\|_{Q'_h} \le \lambda\|p - p_I\|_Q. \tag{5.5.30}$$

This would allow to make explicit the dependence of the estimates obtained in Proposition 5.5.2 on the various constants, without making the formulae too complicated.

Theorem 5.5.3. *Under the same assumptions as in Theorem 5.5.2, assume further that $H_h = H = \{0\}$ (implying in particular that the inf-sup condition* (5.1.19) *is satisfied). Then, for every u_I satisfying $b(u - u_I, q_h) = 0\ \forall\, q_h \in Q_h$, and for every $p_I \in Q_h$, we have:*

$$\begin{aligned}\|u_I - u_h\|_V \le{}& \frac{{\beta_h}^2 + 4\lambda\,\|a\|}{\alpha_0^h{\beta_h}^2}\|a\|\,\|u - u_I\|_V + \\ &+ \Big(\|b\|\frac{{\beta_h}^2 + 4\lambda\,\|a\|}{\alpha_0^h{\beta_h}^2} + \frac{2\lambda\,\|a\|^{1/2}}{{\alpha_0^h}^{1/2}\beta_h}\Big)\|p - p_I\|_Q\end{aligned} \tag{5.5.31}$$

and

$$\begin{aligned}\|p_I - p_h\|_Q \le{}& \frac{2\,\|a\|^{3/2}}{{\alpha_0^h}^{1/2}\beta_h}\,\|u - u_I\|_V \\ &+ \Big(\frac{2\,\|a\|^{1/2}\|b\|}{{\alpha_0^h}^{1/2}\beta_h} + \frac{4\lambda\,\|a\|}{\lambda\,\|a\| + 2\,\beta^2}\Big)\|p - p_I\|_Q.\end{aligned} \tag{5.5.32}$$

Remark 5.5.2. The results of Theorem 5.5.3 imply, as usual, error estimates for $\|u - u_h\|$ and $\|p - p_h\|$. In particular, setting

$$\mu^2 := \lambda\|a\| + \|b\|^2, \tag{5.5.33}$$

they can be expressed in a more compact way

$$\|u - u_h\|_V \leq \frac{6\mu^2}{\alpha_0^h \beta_h^2}(\|a\|\, E_u + \|b\|\, E_p) \tag{5.5.34}$$

$$\|p - p_h\|_Q \leq \frac{6\|a\|^{1/2}}{(\alpha_0^h)^{1/2}\beta_h}(\|a\|\, E_u + \|b\|\, E_p). \tag{5.5.35}$$

□

An alternative possibility, without using condition (5.3.1), is to proceed as in Proposition 4.3.1. In this direction, we have the following result.

Proposition 5.5.3. *Together with Assumption $\mathcal{ABC}_h$, assume that the bilinear form a is coercive on the whole V (that is, it verifies* (5.1.8)*) and that the bilinear form c is of the form* (5.5.25)*. Let, for every $\lambda > 0$, (u, p) and (u_h, p_h) be the solutions of the continuous problem* (5.5.4) *and of the discretised problem* (5.5.4)*, respectively. Then, we have the estimate*

$$\alpha\|u - u_h\|_V^2 + \lambda\|p - p_h\|_Q^2 \leq \frac{3\|a\|\mu^2}{\alpha\lambda}E_u^2 + \frac{3\mu^2}{\alpha}E_p^2, \tag{5.5.36}$$

where μ^2 is still given by (5.5.33)*.*

Proof. Using the estimate (4.3.6) in Proposition 5.5.1, then the expressions of $\mathcal{F}$ and $\mathcal{G}$ in (5.5.9) and (5.5.10), and then some algebra, we have

$$\begin{aligned}
\alpha\|u_h - u_I\|_V^2 + \lambda\|p_h - p_I\|_Q^2 &\leq \frac{1}{\alpha}\|\mathcal{F}\|_{V_h'}^2 + \frac{1}{\lambda}\|\mathcal{G}\|_{Q_h'}^2 \\
&\leq \frac{1}{\alpha}\Big(\|a\|\|u - u_I\|_V + \|b\|\,\|p - p_I\|_Q\Big)^2 + \frac{1}{\lambda}\Big(\|b\|\|u - u_I\|_V + \lambda\|p - p_I\|_Q\Big)^2 \\
&\leq 2\Big(\frac{\|a\|^2}{\alpha} + \frac{\|b\|^2}{\lambda}\Big)\|u - u_I\|_V^2 + 2\Big(\frac{\|b\|^2}{\alpha} + \lambda\Big)\|p - p_I\|_V^2,
\end{aligned} \tag{5.5.37}$$

and the result follows easily. □

Remark 5.5.3. Usually, estimates of the type of Proposition 5.5.3 are interesting when the term $\lambda(p, q)_Q$ is used as *penalty* in order to stabilise a choice $V_h \times Q_h$ for which the *inf-sup* condition does not hold. Then, λ is chosen as a suitable power of h in such a way that the right-hand-side of (5.5.36) still converges to zero (since, roughly speaking, $\|u - u_I\|_V^2$ tends to zero faster than λ). We shall see such a situation in more details in Sect. 8.13. □

Remark 5.5.4. The result of Proposition 5.5.3, as well as the ones of the following subsection, assume that the problem has been **adimensionalised**. This has been done in order to have more concise formulae. □

5.5.3 Singular Perturbations

We now go back to the situation that was considered at the end of Chap. 4, that we (very) briefly recall for the convenience of the reader. We assumed that we were given a Hilbert space W continuously embedded in Q (that is $W \hookrightarrow Q$) and dense in Q. For simplicity, we also assumed the embedding constant to be equal to 1, so that

$$\|w\|_Q \leq \|w\|_W \quad \forall\, w \in W, \qquad \text{and} \qquad \|w\|_{W'} \leq \|w\|_{Q'} \quad \forall\, w \in Q'. \tag{5.5.38}$$

We then considered, for every $\lambda > 0$, a perturbation of the type

$$c(p,q) = \lambda\,(p,q)_W, \tag{5.5.39}$$

that is, we considered problems of the form: *find* (u_λ, p_λ) in $V \times W$ *such that*

$$a(u_\lambda, v) + b(v, p_\lambda) = \langle f, v\rangle_{V'\times V}, \quad \forall\, v \in V, \tag{5.5.40}$$

$$b(u_\lambda, q) - \lambda\,(p_\lambda, q)_W = \langle g, q\rangle_{Q'\times Q}, \quad \forall\, q \in W. \tag{5.5.41}$$

Note that, in the previous chapter, we allowed the presence of an additional term in the right-hand side of (5.5.41), of the form $\langle g_2, q\rangle_{W'\times W}$. Here, for simplicity, we only consider the case when $g_2 = 0$.

Assuming that $Q_h \subseteq W$, we can consider the discretised problem: *find* (u_h, p_h) in $V_h \times Q_h$ *such that*

$$\begin{cases} a(u_h, v_h) + b(v_h, p_h) = \langle f, v_h\rangle_{V_h'\times V_h}, & \forall\, v_h \in V_h, \\ b(u_h, q_h) - \lambda\,(p_h, q_h)_W = \langle g, q_h\rangle_{Q_h'\times Q_h}, & \forall\, q_h \in W_h. \end{cases} \tag{5.5.42}$$

As in Proposition 5.5.1, we now have, with obvious notation,

$$\begin{cases} a(u_h - u_I, v_h) + b(v_h, p_h - p_I) = \langle \mathcal{F}, v_h\rangle_{V_h'\times V_h}, & \forall\, v_h \in V_h, \\ b(u_h - u_I, q_h) - c(p_h - p_I, q_h) = \langle \mathcal{G}, q_h\rangle_{Q_h'\times Q_h}, & \forall\, q_h \in Q_h \end{cases} \tag{5.5.43}$$

where

$$\langle \mathcal{F}, v_h \rangle_{V_h' \times V_h} := a(u - u_I, v_h) + b(v_h, p - p_I), \ \forall\, v_h \in V_h \tag{5.5.44}$$

and

$$\langle \mathcal{G}, q_h \rangle_{Q_h' \times Q_h} := b(u - u_I, q_h) - \lambda(p - p_I, q_h)_W, \ \forall\, q_h \in Q_h. \tag{5.5.45}$$

In the previous chapter, we derived error estimates with the additional assumption that for every $\chi > 0$ there exists a positive $\tilde{\alpha}$ such that

$$\tilde{\alpha}\|u\|_V^2 \le a(v, v) + \chi \|Bv\|_{W'}^2, \quad \forall\, v \in V. \tag{5.5.46}$$

Using the stability result of Theorem 4.3.4, we now have the following error estimate.

Theorem 5.5.4. *Together with Assumption $\mathcal{AB}_h$, assume that W is a Hilbert space, continuously embedded in Q and dense in Q. Assume moreover that the inf-sup condition* (5.1.19) *holds and that $a(\cdot\,,\cdot)$ is positive semi-definite and verifies* (5.5.46)*. Assume finally that $Q_h \subset W$. For every λ with $0 < \lambda \le 1/2$, let (u, p) and (u_h, p_h) be the solutions of* (5.5.40)–(5.5.41) *and* (5.5.42) *respectively. Then, for every $(u_I, p_I) \in V_h \times Q_h$, we have*

$$\|u_h - u_I\|_V + \|p_h - p_I\|_Q + \lambda^{1/2}\|p_h - p_I\|_W \le C\,(\|\mathcal{F}\|_{V_h'} + \|\mathcal{G}\|_{Q_h'} + \lambda^{1/2}\|\tilde{\mathcal{G}}\|_{W_h'}), \tag{5.5.47}$$

where C depends on the constant $\tilde{\alpha}$ in (5.5.46)*, on the constant β_h in* (5.1.19)*, on $\|a\|$, and on $\|b\|$, and where*

$$\langle \mathcal{F}, v_h \rangle_{V_h' \times V_h} := a(u - u_I, v_h) + b(v_h, p - p_I), \quad \forall\, v_h \in V_h, \tag{5.5.48}$$

$$\langle \mathcal{G}, q_h \rangle_{Q_h' \times Q_h} := b(u - u_I, q_h), \quad \forall\, q_h \in Q_h, \tag{5.5.49}$$

$$\langle \tilde{\mathcal{G}}, q_h \rangle_{Q_h' \times Q_h} := (p - p_I, q_h)_W, \quad \forall\, q_h \in Q_h. \tag{5.5.50}$$

Remark 5.5.5. Following the typical strategy of this chapter (as we did for instance with Theorem 5.2.2 or in Proposition 5.5.1), the use of Theorem 5.5.4, in the applications, is to provide the necessary error estimates after using interpolation estimates to find bounds for $\mathcal{F}$, $\mathcal{G}$ and $\tilde{\mathcal{G}}$, and then using the triangle inequality. Here, however, we have an additional difficulty, concerning the estimate of $\lambda^{1/2}\|\tilde{\mathcal{G}}\|_{W_h'}$. Indeed, using (5.5.50), we have

$$\|\tilde{\mathcal{G}}\|_{W_h'}\| := \sup_{q_h \in Q_h} \frac{(p - p_I, q_h)_W}{\|q_h\|_W} \le \|p - p_I\|_W. \tag{5.5.51}$$

Hence, we get

$$
\begin{aligned}
&\|u - u_h\|_V^2 + \|p - p_h\|_Q^2 + \lambda \|p - p_h\|_W^2 \\
&\quad \leq C\Big(\inf_{v_h \in V_h} \|u - v_h\|_V^2 + \inf_{q_h \in Q_h} \{\|p - q_h\|_Q^2 + \lambda \|p - q_h\|_W^2\}\Big).
\end{aligned}
\tag{5.5.52}
$$

□

Remark 5.5.6. As in Theorem 4.3.4, we could avoid assuming the surjectivity of B, assuming only that $\operatorname{Im} B$ is closed, and working in Q_{H_h} instead of Q. In this case, we would need $g \in \operatorname{Im} B$ in order to have a solution that is uniformly bounded in λ. Moreover, the constant C would now depend on the constant $\tilde{\beta}_h$ in (5.3.11) instead of β_h in (5.1.19). □

In some applications, including the important case of Reissner-Mindlin plates, it will be however much more interesting (and powerful) to use Theorem 4.3.5 which, in particular, does not use the *inf-sup* condition. Let us see how this can be done.

Theorem 5.5.5. *Together with Assumption $\mathcal{AB}_h$, assume that W is a Hilbert space, continuously embedded in Q and dense in Q. Assume moreover that $a(\cdot,\cdot)$ is positive semi-definite and verifies* (5.5.46)*. For every λ with $0 < \lambda \leq 1/2$, let (u, p) and (u_h, p_h) be the solutions of* (5.5.40)*–*(5.5.41) *and* (5.5.42) *respectively. Then, for every pair $(u_I, p_I) \in V_h \times Q_h$ which satisfies*

$$
b(u - u_I, q_h) - \lambda (p - p_I, q_h)_W = 0, \tag{5.5.53}
$$

we have

$$
\begin{aligned}
\tilde{\alpha} \|u_I - u_h\|_V^2 + \lambda \|p_I - p_h\|_W^2 &\leq \frac{4\|\mathcal{F}\|_{V'}^2}{\tilde{\alpha}} \\
&\leq \frac{4}{\tilde{\alpha}} \Big(\|a\| \, \|u - u_I\|_V + \|b\| \, \|p - p_I\|_Q \Big)^2
\end{aligned}
\tag{5.5.54}
$$

where $\tilde{\alpha}$ is given in (5.5.46)*.*

The proof is obvious, using (5.5.43)–(5.5.45) in Theorem 4.3.5.

5.5.4 Nonconforming Methods

We shall now rapidly consider the effect on error estimates of changing problem (5.1.9) into a perturbed problem of the form

$$
\begin{cases}
a_h(u_h, v_h) + b_h(v_h, p_h) = \langle f, v_h \rangle_h, & \forall\, v_h \in V_h, \\
b_h(u_h, q_h) = \langle g, g_h \rangle_h, & \forall\, q_h \in Q_h,
\end{cases}
\tag{5.5.55}
$$

where $a_h(\cdot,\cdot)$ and $b_h(\cdot,\cdot)$ are, in a sense to be made precise, approximations of $a(\cdot,\cdot)$ and $b(\cdot,\cdot)$, and where $\langle\cdot,\cdot\rangle_h$ denotes an approximation of the duality brackets $\langle\cdot,\cdot\rangle_{V'\times V}$ or $\langle\cdot,\cdot\rangle_{Q'\times Q}$.

Formulations of the type (5.5.55) arise when *nonconforming* approximations are introduced. In this case we no longer have $V_h \subset V$ and $Q_h \subset Q$, so that the problem must be embedded in larger spaces. We shall give an alternative treatment of nonconforming methods using domain decomposition methods in Chap. 7. However, their importance is worth their presence in our abstract discussion.

We suppose that there exist spaces X and Y such that V_h and V are closed subspaces of X and, similarly, Q_h and Q are closed subspaces of Y. We suppose that $a_h(\cdot,\cdot)$ and $b_h(\cdot,\cdot)$ satisfy

$$a_h(u_h, v_h) \le C_a^{nc}\ \|u_h\|_X\ \|v_h\|_X, \quad \forall\, u_h\, v_h\ \in X, \tag{5.5.56}$$

$$b_h(v_h, q_h) \le C_b^{nc}\ \|v_h\|_X\ \|q_h\|_Y, \quad \forall\ v_h\ \in X\ \forall\, q_h\ \in Y. \tag{5.5.57}$$

We denote by B_h^{nc} the operator $V_h \to Q_h'$ associated with the bilinear form b_h (that actually, with our notation, should be denoted by $\pi_{Y'\to Q_h'} B_{X\to Y'} E_{V_h\to X}$). Moreover, we set

$$K_h^{nc} := \{v_h \mid v_h \in V_h,\ b_h(v_h, q_h) = 0\ \forall\, q_h \in Q_h\}, \tag{5.5.58}$$

and we suppose that $a_h(\cdot,\cdot)$ is coercive on K_h^{nc}, that is, we suppose that there exists a positive constant α_0^{nc} such that

$$a_h(v_{0h}, v_{0h}) \ge \alpha_0^{nc}\ \|v_{0h}\|_X^2,\ \forall\, v_{0h} \in K_h^{nc}. \tag{5.5.59}$$

We suppose, moreover, that there exists a constant $\beta^{nc} > 0$, independent of h, such that

$$\sup_{v_h\in V_h} \frac{b_h(v_h, q_h)}{\|v_h\|_X} \ge \beta^{nc}\ \|q_h\|_Y, \tag{5.5.60}$$

implying that B_h is surjective from V_h to Q_h' (we shall consider the general case later). Finally, we define

$$\|f\|_h = \sup_{v_h\in V_h} \frac{\langle f, v_h\rangle_h}{\|v_h\|_X}, \quad \|g\|_h = \sup_{q_h\in Q_h} \frac{\langle g, q_h\rangle_h}{\|q_h\|_Y}. \tag{5.5.61}$$

We obviously have the following result.

Proposition 5.5.4. *Under hypotheses* (5.5.56) *through* (5.5.61)*, problem* (5.5.55) *has a unique solution and there exists a constant C independent of h such that*

$$\|u_h\|_X + \|p_h\|_Y \le C(C_a^{nc}, C_b^{nc}, \frac{1}{\alpha_0^{nc}}, \frac{1}{\beta^{nc}})\ (\|f\|_h + \|g\|_h), \tag{5.5.62}$$

where C is bounded on bounded subsets. □

We now want, as in Theorem 5.2.1, to use this stability result to obtain an error estimate. For this, we first mimic Proposition 5.2.1.

Proposition 5.5.5. *Under hypotheses* (5.5.56)–(5.5.60)*, let (u, p) and (u_h, p_h) be solutions of the continuous problem* (5.1.2) *and of the discretised problem* (5.5.55)*, respectively. Then, for every $(u_I, p_I) \in V_h \times Q_h$, we have that $(u_h - u_I, p_h - p_I)$ is the solution, in $V_h \times Q_h$, of the variational problem*

$$\begin{aligned} a_h(u_h - u_I, v_h) + b_h(v_h, p_h - p_I) &\\ = [a_h(u - u_I, v_h) + b_h(v_h, p - p_I)] &\\ - [a_h(u, v_h) + b_h(v_h, p) - \langle f, v_h\rangle] &\\ + [\langle f, v_h\rangle_h - \langle f, v_h\rangle], \quad \forall\, v_h \in V_h, & \end{aligned} \tag{5.5.63}$$

$$\begin{aligned} b_h(u_h - u_I, q_h) &\\ = b_h(u - u_I, q_h) &\\ - [b(u, q_h) - \langle g, q_h\rangle] &\\ + [\langle g, q_h\rangle_h - \langle g, q_h\rangle], \quad \forall\, q_h \in Q_h. & \end{aligned} \tag{5.5.64}$$

Remark 5.5.7. Proposition 5.5.5 covers the case of *nonconforming approximations*. If instead we wanted to deal with numerical integration, we would have to change the setting as, in general, numerical integration will not be allowed for generic elements of V and Q: you will not have enough regularity to allow point-wise values. Formulae (5.5.63) and (5.5.64) will, however, be true (after all, they are just obtained by adding and subtracting) whenever u and p will be smooth enough to allow all the terms to be meaningful. □

Using the same arguments as in Theorem 5.2.1, we then have the following proposition.

Proposition 5.5.6. *Assume that the hypotheses of Proposition 5.5.4 hold, let (u, p) be the solution of problem* (5.1.2) *and let (u_h, p_h) be the solution of problem* (5.5.55)*. Then, we have*

$$\|u - u_h\|_X + \|p - p_h\|_Y \leq C\left(\inf_{v_h \in V_h} \|u - v_h\|_X + \inf_{q_h \in Q_h} \|p - q_h\|_Y + \sum_{i=1}^{4} M_{ih}\right) \tag{5.5.65}$$

where $C = C(C_a^{nc}, C_b^{nc}, \frac{1}{\alpha_0^{nc}}, \frac{1}{\beta^{nc}})$ is bounded on bounded subsets, and where the "consistency terms" $M_{1h}, \dots, M_{4h}$ are defined as

$$M_{1h} := \sup_{v_h \in V_h} \frac{|a_h(u, u_h) + b_h(v_h, p) - \langle f, v_h \rangle|}{\|v_h\|_X}, \tag{5.5.66}$$

$$M_{2h} := \sup_{v_h \in V_h} \frac{|\langle f, v_h \rangle - \langle f, v_h \rangle_h|}{\|v_h\|_X}, \tag{5.5.67}$$

$$M_{3h} := \sup_{q_h \in Q_h} \frac{|b_h(u, q_h) - \langle g, q_h \rangle|}{\|n q_h\|_Y}, \tag{5.5.68}$$

$$M_{4h} := \sup_{q_h \in Q_h} \frac{|\langle g, q_h \rangle - \langle g, q_h \rangle_h|}{\|q_h\|_Y}. \tag{5.5.69}$$

Remark 5.5.8. Using Proposition 5.5.6 in practice means to find proper bounds for the terms M_{1h}, M_{2h}, M_{3h}, M_{4h}. In some problems, it will be natural to use a nonconforming approximation of V but a conforming one on Q. For instance in the Stokes problem (Chap. 8), we have $Q = L^2(\Omega)$ and it is rather hard to think of a non conforming approximation to this space. If we then suppose that $b_h(u, q_h) = b(u, q_h)$, then we have $M_{3h} = 0$. On the other hand, the terms M_{2h} and M_{4h} normally come from the use of numerical quadrature formulae for the right-hand sides, and they can be handled by standard techniques [147, 148]. Finally, the term M_{1h} will be treated with the usual techniques of non conforming methods. □

Remark 5.5.9. An important case is the use of conforming approximations where $a(\cdot,\cdot)$ and $b(\cdot,\cdot)$ are computed by numerical quadrature. In this case, we have (if u is smooth enough to give sense to $a_h(u, v_h)$)

$$a(u, v_h) + b(v_h, p) - \langle f, v_h \rangle = 0 \tag{5.5.70}$$

and we can transform M_{1h} to

$$\hat{M}_{1h} = \sup_{v_h \in V_h} \frac{|a(u, v_h) - a_h(u, v_h)|}{\|v_h\|_V} + \sup_{v_h \in V_h} \frac{|b(v_h, p) - b_h(v_h, p)|}{\|v_h\|_V} \tag{5.5.71}$$

and M_{3h} to

$$\hat{M}_{3h} = \sup_{q_h \in Q_h} \frac{|b(u, q_h) - b_h(u, q_h)|}{\|q_h\|_Q}. \tag{5.5.72}$$

□

Nonconforming methods can also be used for perturbed problems, and even for singularly perturbed problems. The basic ideas remain unchanged, and the way to derive error estimates from stability results (applied to $\|u_h - u_I\|$ and $\|p_h - p_I\|$) plus continuity and approximation results (applied to $\|u - u_I\|$ and $\|p - p_I\|$) still works. Here, we do not want to cover all the possible variants. We just present an example of "nonconforming" approximations of singularly perturbed problems, that extends to nonconforming methods the result of Theorem 5.5.5.

For this, consider again the singularly perturbed problem (5.5.40)–(5.5.41), and consider the following nonconforming approximation

$$\begin{cases} a(u_h, v_h) + \tilde{b}_h(v_h, p_h) = \langle f, v_h \rangle_{V_h' \times V_h}, & \forall\, v_h \in V_h, \\ \tilde{b}_h(u_h, q_h) - \lambda\,(p_h, q_h)_W = \langle g, q_h \rangle_{Q_h' \times Q_h}, & \forall\, q_h \in W_h, \end{cases} \tag{5.5.73}$$

where $\tilde{b}_h$ is a continuous bilinear form from $V \times Q$ in $\mathbb{R}$. We then have the following proposition.

Proposition 5.5.7. *Together with Assumption $\mathcal{AB}_h$, assume $\tilde{b}_h$ is a continuous bilinear form from $V \times Q$ in $\mathbb{R}$. Assume further that W is a Hilbert space, continuously embedded in Q and dense in Q, and that $a(\cdot\,,\cdot)$ is positive semi-definite and verifies the following analogue of* (5.5.46)*: for every $\chi > 0$, there exists a positive $\tilde{\alpha}$ such that*

$$\tilde{\alpha}\|u\|_V^2 \le a(v, v) + \chi\|\tilde{B}_h v\|_{W'} \quad \forall\, v \in V. \tag{5.5.74}$$

For every λ with $0 < \lambda \le 1/2$, let (u, p) and (u_h, p_h) be the solutions of (5.5.40)–(5.5.41) *and* (5.5.73) *respectively. Then, for every pair $(u_I, p_I) \in V_h \times Q_h$ which satisfies*

$$\tilde{b}_h(u_I, q_h) - \lambda(p_I, q_h)_W = b(u, q_h) - \lambda(p, q_h)_W \quad \forall\, q_h \in Q_h, \tag{5.5.75}$$

we have

$$\tilde{\alpha}\|u_I - u_h\|_V^2 + \lambda\|p_I - p_h\|_W^2 \le \frac{4}{\tilde{\alpha}}\Big(\|a\|\,\|u - u_I\|_V + \sup_{v_h \in V_h} \frac{b(v_h, p) - \tilde{b}_h(v_h, p_I)}{\|v_h\|_V}\Big)^2, \tag{5.5.76}$$

where $\tilde{\alpha}$ is given in (5.5.74).

Proof. We first observe that the difference $(u_h - u_I, p_h - p_I)$ verifies

$$\begin{cases} a(u_h - u_I, v_h) + \tilde{b}_h(v_h, p_h - p_I) = \langle \tilde{\mathcal{F}}, v_h \rangle_{V_h' \times V_h}, & \forall\, v_h \in V_h, \\ \tilde{b}_h(u_h - u_I, q_h) - \lambda(p_h - p_I, q_h)_W = 0, & \forall\, q_h \in Q_h, \end{cases} \tag{5.5.77}$$

where this time

$$\langle \tilde{\mathcal{F}}, v_h \rangle_{V_h' \times V_h} := a(u - u_I, v_h) + b(v_h, p) - \tilde{b}_h(v_h, p_I), \quad \forall\, v_h \in V_h. \tag{5.5.78}$$

The result then follows immediately by applying Theorem 4.3.5 to problem (5.5.77). □

Remark 5.5.10. It is not clear whether the discrete problem (5.5.42) should be considered as a *non conforming approximation* of (4.3.127)–(4.3.128). In a sense, we shouldn't, since we have $V_h \subset V$ and $Q_h \subset W$. Possibly, we should say that (5.5.40)–(5.5.41) *introduces a consistency term in the error* which, however, is not very satisfactory either. We decided to follow the most common usage, though unhappily. □

To end this section, we briefly discuss how we can check an inf-sup condition (5.5.60) for the bilinear form $b_h(\cdot,\cdot)$. We first give a criterion that will be useful for some applications, in particular in Sect. 8.12.3 in the context of numerical quadrature.

Proposition 5.5.8. *Let us suppose that we have two bilinear forms $b_1(\cdot,\cdot)$ and $b_2(\cdot,\cdot)$ on $V_h \times Q_h$, and suppose that there exists a continuous operator $\Pi_h : V_h \to V_h$ such that*

$$\|\Pi_h v_h\|_X \le c_0 \, \|v_h\|_X, \tag{5.5.79}$$

and such that

$$b_1(\Pi_h v_h, q_h) = b_2(v_h, q_h), \quad \forall q_h \in Q_h. \tag{5.5.80}$$

Then, if $b_1(\cdot,\cdot)$ satisfies the inf-sup condition (5.5.60)*, then $b_2(\cdot,\cdot)$ also does.*

The proof is the same as for Proposition 5.4.2.

Remark 5.5.11. In practice, for an approximation using numerical quadrature, we have $b_1(\cdot,\cdot) = b(\cdot,\cdot)$ and $b_2(\cdot,\cdot) = b_h(\cdot,\cdot)$, so that (5.5.80) means that the numerical quadrature is not exact for the computation of $b(v_h, q_h)$ but rather integrates $b(\Pi_h v_h, q_h)$ with $\Pi_h v_h$ near enough to v_h. □

It is also useful to consider the following result.

Proposition 5.5.9. *Under the Assumption $\mathcal{AB}_h$, suppose that $B_h \equiv \pi_{Q'_h} B E_{V_h}$ satisfies the inf-sup condition* (5.1.19) *with a constant $\beta_h \ge \beta_0 > 0$. Let b_h be a bilinear form on $V_h \times Q_h$, and assume that there exists a constant $C(h)$, with $C(h) \to 0$ when $h \to 0$, such that*

$$|b(v_h, q_h) - b_h(v_h, q_h)| \le C(h) \, \|v_h\|_V \, \|q_h\|_{Q_h}. \tag{5.5.81}$$

Then, for h small enough, $b_h(\cdot,\cdot)$ also satisfies the inf-sup condition (5.1.19) *with a constant $\beta_h \ge \beta_0/2$.*

Proof. Indeed, one may write $b(v_h, q_h) = b_h(v_h, q_h) + (b(v_h, q_h) - b_h(v_h, q_h))$ and thus

$$\sup_{v_h \in V_h} \frac{b(v_h, q_h)}{\|v_h\|} \le \sup_{v_h \in V_h} \frac{b_h(v_h, q_h)}{\|v_h\|} + \sup_{v_h \in V_h} \frac{|b(v_h, q_h) - b_h(v_h, q_h)|}{\|v_h\|}. \tag{5.5.82}$$

Using (5.5.81) and the inf-sup condition (5.1.19) for $b(\cdot,\cdot)$, we get

$$\sup_{v_h \in V_h} \frac{b_h(v_h, q_h)}{\|v_h\|} \geq (\beta_0 - C(h))\|q_h\|_{Q_h} \tag{5.5.83}$$

which is the desired result. □

5.5.5 *Dual Error Estimates*

We now present, to end this section on error estimation, an extension of the Aubin-Nitsche duality technique [37, 313] to the analysis of problem (5.1.2). We consider an abstract setting that will be general enough to include most cases where we will like to use such techniques, for instance in Chap. 7 for Dirichlet's problem (to get H^{-1}-estimates) or in Chap. 8 for the Stokes problem (to get $L^2(\Omega)$-estimates). We refer to [192] where similar, and in some cases more general, results are presented.

Let us then consider two spaces V_- and Q_- (the minus index intuitively meaning a "less regular" space) with the *dense* inclusions

$$V \hookrightarrow V_- \text{ and } Q \hookrightarrow Q_-. \tag{5.5.84}$$

We would like to estimate $\|u - u_h\|_{V_-}$ and $\|p - p_h\|_{Q_-}$. Let us denote

$$V'_+ = (V_-)',\ Q'_+ = (Q_-)'. \tag{5.5.85}$$

In a sense, the above notation, with the plus index, suggests that we have "more regular" spaces. Indeed, we have from (5.5.84) and (4.1.80)

$$V'_+ \hookrightarrow V',\ Q'_+ \hookrightarrow Q'. \tag{5.5.86}$$

It is then reasonable to make the following hypothesis.

Hypothesis H1: For any $f_+ \in V'_+,\ g_+ \in Q'_+$, the solution (w, s) of the problem

$$\begin{cases} a(v, w) + b(v, s) = \langle f_+, v\rangle,\ \forall\, v \in V, \\ b(w, q) = \langle g_+, q\rangle,\ \forall\, q \in Q, \end{cases} \tag{5.5.87}$$

belongs to $V_{++} \times Q_{++}$, where $V_{++} \hookrightarrow V$ and $Q_{++} \hookrightarrow Q$, and there exists a constant $\hat{C}$ (independent of f_+ and g_+), such that

$$\|w\|_{V_{++}} + \|s\|_{Q_{++}} \leq \hat{C}\ (\|f\|_{V'_+} + \|g\|_{Q'_+}). \tag{5.5.88}$$

□

Remark 5.5.12. The above Hypothesis H1 evidently means, in practice, that we have a regularity property and that $f \in V'_+$, $g \in Q'_+$ yield a more regular solution. □

Remark 5.5.13. Note that, in the first equation of (5.5.87), we actually used $a(v,w)$ and not $a(w,v)$. This means that, in a sense, (5.5.87) is the *adjoint problem* of our original problem (5.1.2), as it is common when doing duality estimates. □

We then have the following result, which we consider under the (more commonly used) assumptions of Theorem 5.2.5. Clearly, we could also extend many other previous theorems of this chapter.

Theorem 5.5.6. *Under the hypotheses of Theorem 5.2.5, assume that Hypothesis H1 holds. Then, there exists a constant C_1 (independent of h), such that if (u,p) and (u_h,p_h) are the solutions of problem (5.1.2) and of problem (5.2.31), respectively, then we have*

$$\|u-u_h\|_{V_-}+\|p-p_h\|_{Q_-} \le C_1\,(\|u-u_h\|_V+\|p-p_h\|_Q)(m(h)+n(h)), \quad (5.5.89)$$

where $m(h)$ and $n(h)$ are defined as

$$m(h) := \sup_{w\in V_{++}} \inf_{w_h\in V_h} \frac{\|w-w_h\|_V}{\|w\|_{V_{++}}}, \quad (5.5.90)$$

and

$$n(h) := \sup_{s\in Q_{++}} \inf_{s_h\in Q_h} \frac{\|s-s_h\|_Q}{\|s|_{Q_{++}}}. \quad (5.5.91)$$

Proof. Let us choose $f_+ \in V'_+$ and $g_+ \in Q'_+$ with $\|f_+\|_{V'_+} = 1$ and $\|g_+\|_{Q'_+} = 1$ such that one has

$$\begin{cases} \langle f_+, u-u_h\rangle_{V'_+\times V_-} = \|u-u_h\|_{V_-}, \\ \langle g_+, p-p_h\rangle_{Q'_+\times Q_-} = \|p-p_h\|_{Q_-}, \end{cases} \quad (5.5.92)$$

and let $(w,s) \in V_{++}\times Q_{++}$ be the corresponding solution of (5.5.87), therefore bounded by (5.5.88). We may thus write

$$\|w\|_{V_{++}} + \|s\|_{V_{++}} \le 2\,\hat{C}. \quad (5.5.93)$$

Making $v = u-u_h$ and $q = p-p_h$ in (5.5.87), we thus have from (5.5.92):

$$\|u-u_h\|_{V_-}+\|p-p_h\|_{Q_-} = a(u-u_h,w)+b(u-u_h,s)+b(w,p-p_h). \quad (5.5.94)$$

However, we know that one also has, subtracting (5.2.31) and (5.2.32),

$$\begin{cases} a(u-u_h, w_h) + b(w_h, p-p_h) = 0, \ \forall\, w_h \in V_h, \\ b(u-u_h, s_h) = 0, \ \forall\, s_h \in Q_h. \end{cases} \tag{5.5.95}$$

We may thus write in (5.5.94),

$$\|u-u_h\|_{V_-} + \|p-p_h\|_{Q_-} \\ = a(u-u_h, w-w_h) + b(u-u_h, s-s_h) + b(w-w_h, p-p_h) \tag{5.5.96}$$

for all $w_h \in V_h$ and for all $s_h \in Q_h$. We now note that (5.5.90) and (5.5.91) imply

$$\inf_{w_h\in V_h} \|w-w_h\|_V \leq m(h)\ \|w\|_{V_{++}}, \tag{5.5.97}$$

and

$$\inf_{q_h\in Q_h} \|s-q_h\|_Q \leq n(h)\ \|s\|_{Q_{++}}, \tag{5.5.98}$$

respectively, so that

$$\inf_{w_h\in V_h} \|w-w_h\|_V + \inf_{q_h\in Q_h} \|s-q_h\|_Q \leq 2\,\hat{C}\,(m(h)+n(h)), \tag{5.5.99}$$

and (5.5.89) follows easily from (5.5.96). □

Remark 5.5.14. We shall also use in Chap. 7 a super-convergence result that can be extended to the abstract setting of Theorem 5.5.6. □

The following result generalises the result of Theorem 5.2.7.

Proposition 5.5.10. *Under the same assumptions as in Theorem 5.5.6, assume further that $K_h \subseteq K$. Assume moreover that Q_h can be identified with a subspace of $R_{Q'}(Q'_+)$ where $R_{Q'}$ is the Ritz operator $Q' \to Q$. Then, we have*

$$\|\Phi_h p - p_h\|_Q \leq \|u-u_h\|_V\ \hat{C}\left(\|a\| \sup_{w\in V_{++}} \frac{\|w-\Pi_h w\|_V}{\|w\|_{V_{++}}} + \|b\| n(h)\right), \tag{5.5.100}$$

where $n(h)$ is still given by (5.5.91), *and Π_h and Φ_h satisfy* (5.1.29) *and* (5.1.31), *respectively.*

Proof. As a first step, let us point out that the assumption that Q_h can be identified with a subspace of $R_{Q'}(Q'_+)$ means, in other (less technical) terms, that for every $\overline{q}_h \in Q_h$, the solution $(z, \phi) \in V \times Q$ of the problem

$$\begin{cases} a(v,z) + b(v,\phi) = 0, \ \forall\, v \in V, \\ b(z,q) = (\overline{q}_h, q)_Q, \ \forall\, q \in Q \end{cases} \tag{5.5.101}$$

actually belongs to $V_{++} \times Q_{++}$ and, moreover,

$$(\|z\|_{V_{++}} + \|\phi\|_{V_{++}}) \leq \hat{C}\ \|\overline{q}_h\|_Q. \tag{5.5.102}$$

It is clear that this is strictly related to (5.2.55), which was used in the proof of Theorem 5.2.7. Now, consider problem (5.5.101) with

$$\overline{q}_h := \Phi_h p - p_h. \tag{5.5.103}$$

From (5.5.101) with (5.5.103), then (5.1.29), and then (5.1.31), we have

$$\|\Phi_h p - p_h\|_Q^2 = b(z, \Phi_h p - p_h) = b(\Pi_h z, \Phi_h p - p_h) = b(\Pi_h z, p - p_h). \tag{5.5.104}$$

Using (5.1.31) in (5.5.104), we then have

$$\begin{aligned} \|\Phi_h p - p_h\|^2 &= b(\Pi_h z, p - p_h) \\ &= a(u_h - u, \Pi_h z) \\ &= a(u_h - u, \Pi_h z - z) + a(u_h - u, z). \end{aligned} \tag{5.5.105}$$

Taking $v = u_h - u$ in (5.5.101), this becomes, for all $q_h \in Q_h$,

$$\begin{aligned} \|\Phi_h p - p_h\|^2 &= a(u_h - u, \Pi_h z - z) - b(u_h - u, \phi) \\ &= a(u_h - u, \Pi_h z - z) - b(u_h - u, \phi - q_h). \end{aligned} \tag{5.5.106}$$

Finally, from (5.5.106), we have, always for all $q_h \in Q_h$,

$$\|\Phi_h p - p_h\|^2 \leq \|u_h - u\|_V (\|a\| \|z - \Pi_h z\|_V + \|b\| \|\phi - q_h\|_Q), \tag{5.5.107}$$

and the result follows using (5.5.91) and (5.5.102) with (5.5.103). □

This result uses the strong assumption $K_h \subset K$ and its use is rather technical. Anyhow, the above analysis shows when it can be expected to hold, besides the example of Chap. 7.

5.6 Numerical Properties of the Discrete Problem

This section will present a few general facts related to numerical computations with the previously described problem. As we are still in a rather abstract setting, we will not be able to obtain directly usable results. However, some basic facts are common to a large number of methods and presenting them in a unified frame may help understanding the relations existing between apparently different methods.

5.6.1 The Matrix Form of the Discrete Problem

We shall first consider problem (5.1.9) and develop a matrix form suited to numerical computation. We shall set, for the finite dimensional spaces V_h and Q_h,

$$\begin{cases} N := \dim V_h, \\ M := \dim Q_h, \end{cases} \tag{5.6.1}$$

and we shall use a basis of theses spaces, namely $\{v_{ih} |\ 1 \leq i \leq N\}$ for V_h and $\{q_{ih} |\ 1 \leq i \leq M\}$ for Q_h. We can now define,

$$\mathbb{A}_{ij} := a(v_{jh}, v_{ih}), \tag{5.6.2}$$

$$\mathbb{B}_{ij} := b(v_{jh}, q_{ih}), \tag{5.6.3}$$

$$f_i := \langle f, v_{ih} \rangle, \tag{5.6.4}$$

$$g_i := \langle g, q_{ih} \rangle. \tag{5.6.5}$$

We set $\mathbb{A}_{N\times N} := (\mathbb{A}_{ij})$, $\mathbb{B}_{M\times N} := (\mathbb{B}_{ij})$, $\mathbf{f}_N := \{f_i\}$, $\mathbf{g}_M := \{g_i\}$ and we denote by $\mathbf{u} := \{u_i\}$ and $\mathbf{p} := \{p_i\}$ the vectors of $\mathbb{R}^N$ and $\mathbb{R}^M$ (respectively) formed by the coefficients of u_h and p_h in the expressions

$$u_h = \sum_{i=1}^{N} u_i \ v_{ih}, \tag{5.6.6}$$

$$p_h = \sum_{i=1}^{M} p_i \ q_{ih}. \tag{5.6.7}$$

Problem (5.1.9) can now be written in matrix form as

$$\begin{cases} \mathbb{A}\mathbf{u} + \mathbb{B}^T\mathbf{p} = \mathbf{f}, \\ \mathbb{B}\mathbf{u} = \mathbf{g}, \end{cases} \tag{5.6.8}$$

or

$$\begin{pmatrix} \mathbb{A} & \mathbb{B}^T \\ \mathbb{B} & 0 \end{pmatrix} \begin{pmatrix} \mathbf{u} \\ \mathbf{p} \end{pmatrix} = \begin{pmatrix} \mathbf{f} \\ \mathbf{g} \end{pmatrix}. \tag{5.6.9}$$

In practice, the bases $\{v_{ih}\}$ and $\{q_{ih}\}$ will be built using a finite element technique. This will impose additional structure on problem (5.6.9). We can however see that, for a symmetric bilinear form $a(\cdot\,,\cdot)$, we have to solve a symmetric but in general indefinite linear system. The fact that we have positive and negative eigenvalues is, of course, directly related to the fact that we discretise a saddle point problem. As an alternative, we can change the sign of the second equation of (5.6.9), getting

$$\begin{pmatrix} \mathbb{A} & \mathbb{B}^T \\ -\mathbb{B} & 0 \end{pmatrix} \begin{pmatrix} \mathbf{u} \\ \mathbf{p} \end{pmatrix} = \begin{pmatrix} \mathbf{f} \\ -\mathbf{g} \end{pmatrix}, \tag{5.6.10}$$

having now a matrix which is not symmetric but that is positive semi-definite.

Problem (5.6.8) is clearly an example of what we had in Chap. 3 (see problem (3.0.1)), and all the results obtained there apply here.

We observe that *if the matrix* $\mathbb{A}$ *is invertible, one can eliminate the variable* $\mathbf{u}$ *from this linear system.* Indeed, one gets from the first equation of (5.6.8),

$$\mathbf{u} = \mathbb{A}^{-1}\mathbf{f} - \mathbb{A}^{-1}\mathbb{B}^T\mathbf{p} \tag{5.6.11}$$

and thus inserting it in the second equation of (5.6.8),

$$\mathbb{B}\mathbf{u} = \mathbb{B}\mathbb{A}^{-1}\mathbf{f} - \mathbb{B}\mathbb{A}^{-1}\mathbb{B}^T\mathbf{p} = \mathbf{g}. \tag{5.6.12}$$

that is:

$$\mathbb{B}\mathbb{A}^{-1}\mathbb{B}^T\mathbf{p} = \mathbb{B}\mathbb{A}^{-1}\mathbf{f} - \mathbf{g}. \tag{5.6.13}$$

If $\mathbb{B}\mathbb{A}^{-1}\mathbb{B}^T$ is non-singular, problem (5.6.13) can be solved for $\mathbf{p}$. Once $\mathbf{p}$ has been computed, one can go back to (5.6.11) to get $\mathbf{u}$.

This is a discrete form of the *dual problem* of Sect. 1.3. Let us consider the matrix $\mathbb{B}\mathbb{A}^{-1}\mathbb{B}^T$. If the matrix $\mathbb{A}$ is positive definite, then $\mathbb{B}\mathbb{A}^{-1}\mathbb{B}^T$ is also positive semi-definite. Indeed, one has

$$\mathbf{p}^T\mathbb{B}\mathbb{A}^{-1}\mathbb{B}^T\mathbf{p} = (\mathbb{B}^T\mathbf{p})^T\mathbb{A}^{-1}(\mathbb{B}^T\mathbf{p}) \geq \alpha\ \|\mathbb{B}^T\ \mathbf{p}\|^2, \tag{5.6.14}$$

and we see that $\mathbb{B}\mathbb{A}^{-1}\mathbb{B}^T$ is positive definite if and only if $\mathrm{Ker}\mathbb{B}^T = \{0\}$ (that is if the *inf-sup* condition (5.1.19) is satisfied). Problem (5.6.13) seems therefore easier to solve than problem (5.6.9), as numerical methods for positive definite systems are more efficient and more stable.

Unfortunately, this simplification of the problem cannot, in general, be done in practice. The trouble comes from $\mathbb{A}^{-1}$ which is likely to be a full matrix even if $\mathbb{A}$ is sparse. The system (5.6.13) is then too large to be stored and handled. *We shall, however, meet some cases where such a reduction of the problem can be done*, thus providing an efficient solution method.

An interesting special case concerning the matrix $\mathbb{B}\,\mathbb{A}^{-1}\mathbb{B}^T$ arises when one can identify the matrix $\mathbb{A}$ defined by (5.6.2) with the matrix associated with the scalar product on V_h, while the norm in Q_h is equivalent to the Euclidean norm. In this case (see [207]), if $\mu_1 \geq \mu_2 \geq \ldots . \geq \mu_k$ are the singular values of the matrix $\mathbb{S}_Y\,\mathbb{B}\,\mathbb{S}_X$ discussed in Proposition 3.4.5, we have

$$\mathrm{Cond}(\mathbb{B}\mathbb{A}^{-1}\mathbb{B}^T) = \frac{\mu_{\max}}{\mu_{\min}}. \tag{5.6.15}$$

5.6.2 *And if the* inf-sup *Condition Does Not Hold?*

One of the most frustrating things in the analysis of mixed finite element methods is often the apparent discrepancy between experience and theory. To quote [202], "Computations were done, (with success!), using theoretically dubious elements or, at best, using elements on which theory remained silent". This is especially the case for the Stokes problem of Chap. 8 where velocity results are generally quite good, even with elements not satisfying the *inf-sup* condition, while reasonable pressure results can often be recovered after a filtering post-treatment of the raw results. The singular value decomposition introduced in Chap. 3 allows us to get a better understanding of those disconcerting behaviours.

Let us go back to Proposition 3.4.5 and to the singular value decomposition that brings the matrix $\tilde{\mathbb{B}} := \mathbb{S}_Y\, \mathbb{B}\, \mathbb{S}_X$ into the pseudo-diagonal form:

$$\tilde{\mathbb{B}} = \begin{pmatrix} \mu_1 & \cdot & \cdot & \cdot & 0 \cdot 0 \\ \cdot & \mu_2 & \cdot & \cdot & 0 \cdot 0 \\ \cdot & \cdot & \cdot & \cdot & \cdot \cdot \cdot \\ \cdot & \cdot & \cdot & \mu_k & 0 \cdot 0 \\ \cdot & \cdot & 0 & \cdot & 0 \cdot 0 \\ \cdot & \cdot & \cdot & \cdot & \cdot \cdot \cdot \\ \cdot & \cdot & 0 & \cdot & 0 \cdot 0 \end{pmatrix}, \tag{5.6.16}$$

where we suppose again that the singular values μ_i are written in decreasing order. To simplify the matter even further, let us assume that on V_h and on Q_h the norms $\|\cdot\|_V$ and $\|\cdot\|_Q$, respectively, are associated to the identity matrix, so that we can consider that $\tilde{\mathbb{B}} \equiv \mathbb{B}$.

The solution of our problem will depend directly on the behaviour of those singular values in a way which we shall now try to describe. Let us first note that in (5.6.16), columns of zeros (i.e. $j > k$) correspond to the kernel of $\mathbb{B}$ while rows of zeros correspond to the kernel of $\mathbb{B}^T$. Rows of zeros imply that it is possible to solve $\mathbb{B}\mathbf{u} = \mathbf{g}$ only if $\mathbf{g}$ takes the form

$$\mathbf{g} = \begin{pmatrix} g_1 \\ g_2 \\ \vdots \\ g_k \\ 0 \\ \vdots \\ 0 \end{pmatrix}, \tag{5.6.17}$$

that is, if $\mathbf{g}$ has no component in $\mathrm{Ker}\mathbb{B}^T$. We have already discussed the importance of the dimension of $\mathrm{Ker}\mathbb{B}^T$. If this dimension happens to be larger than the dimension of the kernel of the corresponding infinite dimensional operator, we have spurious zero energy modes in $\mathbf{p}$ together with artificial (non-physical) constraints on $\mathbf{g}$.

Another important point is the dimension of Ker$\mathbb{B}$, that is, the number of zero columns. Think, for a while, of a problem where $g = 0$. Then, the exact solution u will belong to the kernel $K = \mathrm{Ker}B$ of the differential operator B, and the discrete solution u_h will belong to $K_h = \mathrm{Ker}B_h$ (here represented by the kernel of the matrix $\mathbb{B}$). In order to get a u_h that is a good approximation of the infinite dimensional u, the dimension of K_h should then grow when the number of degrees of freedom increases. Whenever this growth is not occurring properly, we shall have a *locking phenomenon*, which may be:

- *total* when

$$\mathbb{B}\mathbf{u} = \mathbf{0} \text{ implies } \mathbf{u} = \mathbf{0}, \tag{5.6.18}$$

- *partial* when $\mathbf{u}$ is constrained into too small a subspace. This will happen whenever the space Q_h is chosen too large, thus over-constraining the solution.

To complete our picture, we shall now divide the singular values of $\mathbb{B}$ into three sets, writing

$$\mathbb{B} = \begin{pmatrix} \Sigma_1 & 0 & 0 \\ 0 & \Sigma_2 & 0 \\ 0 & 0 & 0 \end{pmatrix}, \tag{5.6.19}$$

where Σ_1 contains the "stable part of $\mathbb{B}$" (i.e., $\mu_i > \beta_0 \geq 0$), and Σ_2 contains the singular values vanishing when h gets small. As we noted above, the columns (u_3) associated to zero singular values correspond to Ker$\mathbb{B}$ while the rows correspond to Ker$\mathbb{B}^T$. We can now write system (5.6.9) as

$$\begin{pmatrix} \mathbb{A}_{11} & \mathbb{A}_{12} & \mathbb{A}_{13} & \Sigma_1 & 0 & 0 \\ \mathbb{A}_{21} & \mathbb{A}_{22} & \mathbb{A}_{23} & 0 & \Sigma_2 & 0 \\ \mathbb{A}_{31} & \mathbb{A}_{32} & \mathbb{A}_{33} & 0 & 0 & 0 \\ \Sigma_1 & 0 & 0 & 0 & 0 & 0 \\ 0 & \Sigma_2 & 0 & 0 & 0 & 0 \\ 0 & 0 & 0 & 0 & 0 & 0 \end{pmatrix} \begin{pmatrix} u_1 \\ u_2 \\ u_3 \\ p_1 \\ p_2 \\ p_3 \end{pmatrix} = \begin{pmatrix} f_1 \\ f_2 \\ f_3 \\ g_1 \\ g_2 \\ g_3 \end{pmatrix}. \tag{5.6.20}$$

If we want to solve (5.6.20), we must first have $g_3 = 0$, leaving p_3, the component of $\mathbf{p}$ in Ker$\mathbb{B}^T$, indeterminate. As we have already discussed, this condition may imply artificial constraints if Ker$\mathbb{B}^T$ is too large: they could then eventually be satisfied by suitably modifying the data, and the question would then be whether this can be done without losing precision. Supposing that this point can be settled, we can now proceed in (5.6.20) to solve for u_1, u_2 and u_3,

$$\begin{cases} u_1 = \Sigma_1^{-1} g_1, \\ u_2 = \Sigma_2^{-1} g_2, \\ u_3 = \mathbb{A}_{33}^{-1} f_3 - \mathbb{A}_{31}^{-1} u_1 - \mathbb{A}_{32}^{-1} u_2. \end{cases} \tag{5.6.21}$$

The potential trouble obviously lies in u_2, which depends on the inverse of the unstable part Σ_2. Again, if g_2 is null (or sufficiently small), u_2 will be null (or negligible) while u_1 and u_3 will behave correctly. This can happen either because we can set g_2 to zero without losing precision, or because "normal data" contain only a small g_2 component corresponding, for example, to "high frequency components" which are small for regular functions. In such a case, one can expect reasonable results even if the *inf-sup* condition is not satisfied and $\mathbb{B}$ contains an unstable part Σ_2.

Finally, u_1, u_2 and u_3 being known, we get from (5.6.20)

$$\begin{cases} p_1 = -\Sigma_1^{-1}(\mathbb{A}_{11}^{-1}u_1 + \mathbb{A}_{12}^{-1}u^2 + \mathbb{A}_{13}^{-1}u_3 - f_1), \\ p_2 = -\Sigma_2^{-1}(\mathbb{A}_{21}^{-1}u_1 + \mathbb{A}_{22}^{-1}u_2 + \mathbb{A}_{23}^{-1}u_3 - f_2). \end{cases} \tag{5.6.22}$$

Here, p_2 depends on the inverse of the unstable part Σ_2, (and in fact, looking at (5.6.21), can even grow as Σ_2^{-2}, if g_2 is not zero). Even for $g_2 = 0$, p_2 cannot be expected to be correct, but p_1 will then remain stable. If this stable part of $\mathbf{p}$ is rich enough to approximate the exact infinite-dimensional solution, a post processing procedure filtering out p_2 will produce good results. This is indeed what happens in many situations. One may however think that relying on such borderline conditions is likely to lead to unreliable results at times, and, more generally, to a method that is not *robust*.

The complete analysis of an approximation should therefore identify how well a "normal problem" can be approximated by the "good part" u_1, u_3, p_1 of the numerical solution. This would imply the knowledge of the singular decomposition, which is a rather strong requirement.

5.6.3 Solution Methods

As it was stated in Sect. 5.6.1, the matrix associated with a mixed formulation as in (5.6.9) is indefinite. This is a problem both for direct methods (which may require pivoting) and for many iterative methods. We shall describe, in Chap. 7, Sect. 7.2, how the 'Hybridisation technique', using inter-element multipliers and 'static condensation', can be used to recover a positive definite problem. However, this is not applicable, for example, to the important case of incompressible problems described in Chap. 8. Nevertheless, the situation is far from desperate and there exist methods which can solve efficiently those cases.

5.6.3.1 Solution by Penalty Methods

Penalty methods have been quite popular for the numerical solution of some saddle point problems, especially the Stokes problem described in Chap. 8. The basic idea behind these methods is general, and we believe it is worth presenting it in an abstract setting.

The idea is very simple and is quite classical in the theory of mathematical programming. Let us consider problem (5.6.8) with $\mathbb{A}$ symmetric and positive definite. This system is equivalent to

$$\inf_{\mathbb{B}\mathbf{v}=\mathbf{g}} \left\{ \frac{1}{2} \mathbf{v}^T \mathbb{A} \mathbf{v} - \mathbf{f} \right\}. \tag{5.6.23}$$

Let then $\mathbb{S}$ be any positive definite matrix in $\mathbb{R}^N$. We can replace (5.6.23) by

$$\inf_{\mathbf{v}} \left\{ \frac{1}{2} \mathbf{v}^T \mathbb{A} \mathbf{v} + \frac{1}{2\varepsilon} (\mathbb{B}\mathbf{v} - \mathbf{g})^T \mathbb{S}^{-1} (\mathbb{B}\mathbf{v} - \mathbf{g}) - \mathbf{v}^T \mathbf{f} \right\}, \tag{5.6.24}$$

whose minimiser is the solution of

$$\mathbb{A}\mathbf{u}_\varepsilon + \frac{1}{\varepsilon} \mathbb{B}^T \mathbb{S}^{-1} \mathbb{B} \mathbf{u}_\varepsilon = \mathbf{f} + \frac{1}{\varepsilon} \mathbb{B}^T \mathbb{S}^{-1} \mathbf{g}. \tag{5.6.25}$$

If the matrix $\mathbb{S}$ is "easy to invert", (in particular if $\mathbb{S}^{-1}$ is a sparse matrix, by preference block diagonal), (5.6.24) provides a way to reduce our problem to a more standard quadratic unconstrained problem. This is a widely used technique and it is indeed quite efficient.

Remark 5.6.1. It is clear that (5.6.24) can be used also when $\mathbb{A}$ is not symmetric and positive definite, although in this case, the non-singularity of the matrix $\mathbb{A} + (1/\varepsilon)\mathbb{B}^T\mathbb{S}^{-1}\mathbb{B}$ will not be automatically true. □

Remark 5.6.2. Setting $\mathbf{p}_\varepsilon = (1/\varepsilon)\mathbb{S}^{-1}(\mathbb{B}\mathbf{u} - \mathbf{g})$, we have that Problem (5.6.25) can be written in the form

$$\begin{cases} \mathbb{A}\mathbf{u}_\varepsilon + \mathbb{B}^T \mathbf{p}_\varepsilon = \mathbf{f}, \\ \mathbb{B}\mathbf{u}_\varepsilon - \varepsilon \mathbb{S} \mathbf{p}_\varepsilon = \mathbf{g}, \end{cases} \tag{5.6.26}$$

as already pointed out on several occasions (see Remarks 3.6.4, 4.3.7, or 5.5.3). □

Remark 5.6.3. One of the main drawbacks of penalty methods is the fact that the penalty term $(1/\varepsilon)\, \mathbb{B}^T\mathbb{S}^{-1}\mathbb{B}$ *has a strong negative impact on the condition number* of the linear system (5.6.25). For instance, *using a penalty method is almost impossible if an iterative method is used for the solution of the linear system* (5.6.25), since iterative methods are in general quite sensitive to the condition number of the matrix at hand. For this reason, penalty methods (which have been *the standard* for two-dimensional problems, where direct solvers are used) are much less suitable in the three-dimensional case where, so far, iterative methods are the rule. □

Remark 5.6.4 (Reduced penalty). In Proposition 4.3.3, we saw that the solution of a penalised problem converges to the solution of the original (non penalised) problem. The result can of course be applied to a discretised problem. The reader should notice that discretising a penalised problem is not, in general, equivalent to penalising a discrete problem. In this last case, a choice of spaces $V_h \subset V$ and $Q_h \subset Q$ is explicitly done and the penalty method is to be considered as a solution procedure. Discretising the penalised problem is in general equivalent to choosing $Q_h = B(V_h)$ which is in general a poor choice. *Reduced penalty methods* have been introduced to circumvent these difficulties and their equivalence with mixed method will be discussed in Chap. 8 in the context of Stokes' problem and in Chap. 10 for moderately thick plates *à la* Mindlin, in a slightly more general setting. Let us see here the general flavour of these methods. Let us go back to the continuous problem (4.3.108) (with $\lambda = \varepsilon$), and assume for simplicity that $g = 0$ and Q is identified with its dual space Q'. Then, the problem will be equivalent to finding $u \in V$ such that

$$a(u, v) + \frac{1}{\varepsilon}(Bu, Bv)_Q = \langle f, v\rangle \quad \forall\, v \in V. \tag{5.6.27}$$

To discretise (5.6.27), one normally takes $V_h \subset V$ and looks for $u_h \in V_h$ such that

$$a(u_h, v_h) + \frac{1}{\varepsilon}(Bu_h, Bv_h)_Q = \langle f, v_h\rangle \quad \forall\, v_h \in V_h. \tag{5.6.28}$$

As we said before, this is equivalent to having discretised the mixed formulation

$$\begin{cases} a(u, v) + (Bv, p)_Q = \langle f, v\rangle & \forall v \in V, \\ (Bu, q)_Q - \varepsilon(p, q)_Q = 0 & \forall q \in Q \end{cases} \tag{5.6.29}$$

with $Q_h := B(V_h)$ (that, we repeat, is often a poor choice). Reduced penalty, instead, would introduce a *reduction operator* P_h, linear from $B(V_h)$ to Q, and consider, instead of (5.6.28), the problem

$$a(u_h, v_h) + \frac{1}{\varepsilon}(P_h\, Bu_h, P_h\, Bv_h)_Q = \langle f, v_h\rangle \quad \forall\, v_h \in V_h. \tag{5.6.30}$$

It many cases, P_h will be a *projection operator* (meaning that $P_h^t = P_h$ and $P_h^2 = P_h$) from Q into itself. We could then define $Q_h := P_h(Q)$, define p_h as $p_h := (1/\varepsilon) P_h\, B\, u_h$, and discover that we are actually discretising problem (5.6.29) by means of $V_h \subset V$ and $Q_h \subset Q$. □

5.6.3.2 Iterative Solution Methods

In recent years, efficient iterative methods for problems of the form (5.6.8) have been developed. We would like to give here a quick overview and some references which may guide the reader.

Classically, the first iterative algorithm for (5.6.8), in the case where the matrix $\mathbb{A}$ is invertible, was Uzawa's algorithm. When $\mathbb{A}$ is invertible, we have already seen that $\mathbf{u}$ can be eliminated and we get, in matrix form, problem (5.6.13). The matrix

$\mathbb{B}\mathbb{A}^{-1}\mathbb{B}^T$ that appears in this problem is positive semi-definite and is well suited for a solution by a descent method such as the gradient method or the conjugate gradient method. Moreover, in many important cases, *the condition number* of $\mathbb{B}\mathbb{A}^{-1}\mathbb{B}^T$ will not grow as the discretisation mesh is reduced so that convergence properties will be independent of the mesh, which is a very desirable feature. We refer to [205] for more details and convergence proofs of the algorithms described below that are nothing but a gradient method applied to (5.6.13) or a variant of (5.6.13). Multigrid versions of the method can also be found in [374].

The basic algorithm can be written as:

- Let $\mathbf{p}^0$ be chosen arbitrarily,
- For $n \geq 0$, until convergence:
  - Find $\mathbf{u}^{n+1}$ solution of

$$\mathbb{A}\mathbf{u}^{n+1} = -\mathbb{B}^T\mathbf{p}^n + \mathbf{f} \tag{5.6.31}$$

  - Compute $\mathbf{p}^{n+1}$ using, with ρ properly chosen,

$$\mathbf{p}^{n+1} = \mathbf{p}^n + \rho[\mathbb{B}\mathbf{u}^{n+1} - \mathbf{g}]. \tag{5.6.32}$$

This algorithm can be improved by changing it to a conjugate gradient method. It can also be married with the penalty method described above. In this case, it can be considered as a way to eliminate the penalty error and to obtain the true solution of the underlying limit problem. This extension of Uzawa's algorithm is called the *augmented Lagrangian algorithm* and it was introduced by Hestenes [246] and Powell [325]. Its properties are discussed in detail in [205]. The algorithm essentially consists in changing (5.6.31) into

$$\mathbb{A}\mathbf{u}^{n+1} + \frac{1}{\varepsilon}\mathbb{B}^T\mathbb{S}^{-1}\mathbb{B}\mathbf{u}^{n+1} = -\mathbb{B}^T\mathbf{p}^n + \mathbf{f} + \frac{1}{\varepsilon}\mathbb{B}^T\mathbb{S}^{-1}\mathbf{g}. \tag{5.6.33}$$

Remark 5.6.5. When using (5.6.33), we can then take $\rho = 1/\varepsilon$ in (5.6.32), which is very close to the optimal value for ε small, and rewrite the method as

$$\begin{cases} \mathbb{A}\mathbf{u}^{n+1} + \mathbb{B}^T\mathbf{p}^{n+1} = \mathbf{f}, \\ \mathbb{B}\mathbf{u}^{n+1} - \varepsilon\,\mathbb{S}(\mathbf{p}^{n+1} - \mathbf{p}^n) = \mathbf{g}. \end{cases} \tag{5.6.34}$$

Taking $\mathbf{p}^n = 0$, this is the standard penalty method already seen in (5.6.26), and, according to Proposition 4.3.3, for ε small, $\mathbf{p}^{n+1}$ is already a good approximation of $\mathbf{p}$. In general, two or three iterations will be sufficient to completely eliminate the error due to the penalty term. □

Remark 5.6.6. Although the Augmented Lagrangian algorithm is a powerful tool for the numerical solution of Stokes' problem (Chap. 8), it suffers from the same problem that plagued the penalty method, namely the ill conditioning of (5.6.33). Moreover, the cost associated to the solution at each iteration of this equation to obtain $\mathbf{u}^{n+1}$ is important. □

The idea of more efficient methods was already introduced in [205], where an iterative method acting simultaneously on both variables $\mathbf{u}^n$ and $\mathbf{p}^n$ was presented. However, the algorithms proposed were doomed by the absence of an automatic method for the choice of parameters. Fortunately, some progresses in this direction have been made in the meantime. In [186], the reader can find an analysis of the Minimum Residual algorithm applied to indefinite problems of the form (5.6.9). Another approach using the Generalised Conjugate Residual has been applied to large three-dimensional incompressible elasticity problems in [185] where other references can also be found. Hence, we are now allowed to say that solving the discrete problems arising from mixed methods is not any more a drawback to their use.

5.7 Concluding Remarks

In this chapter, we tried to present the basic facts that will serve throughout the book to the analysis of various applications. Many cases have not been treated. However, we feel that our presentation should enable the readers to easily master the different extensions that can be found in the literature and even to build by themselves the variants that would be necessary to cover new problems. Some important problems have not been treated in our presentation. This is the case, in particular, of eigenvalue problems for which mixed and hybrid methods can provide an alternate approach. These will be treated separately in Chap. 6.

Chapter 6
Complements: Stabilisation Methods, Eigenvalue Problems

In this chapter, we shall consider two special topics related to the approximation of saddle-point problems. The first one is about stabilised methods, which are more and more widely used in many applications where it is difficult to build approximations satisfying both the ellipticity in the kernel and the *inf-sup* properties. The second section will be devoted to an abstract presentation of eigenvalue problems for mixed problems, where an emphasis will be put on both necessary and sufficient conditions.

6.1 Augmented Formulations

6.1.1 An Abstract Framework for Stabilised Methods

Stabilisation techniques have become quite popular and new methods have been introduced along many avenues. Taking into account the enormous variety of possible applications, stabilisation techniques would require a book of their own. On the other hand, we might conceive stabilisation techniques as an arsenal of tricks to manipulate the problem and transform it into one for which the general stability theories (as the ones described in this book), can be applied. Hence, we just give the flavour of some of these tricks, and refer to the specialised literature for applications on the various particular problems. We will start with some general considerations regarding *augmented formulations* (that are the basis of the so-called "stabilisations à la Hughes-Franca"). Then, following [123], we shall describe a general framework for the study of stability issues, in which one tries to reduce the stabilising modifications at the strictly necessary minimum (whence the name "minimal stabilisations").

We have seen previously, in Sect. 1.5, that some augmented formulations cannot be written as Euler's equations of a Lagrangian but rather through an antisymmetric bilinear form. To include these formulations, among others, in our framework, we

D. Boffi et al., *Mixed Finite Element Methods and Applications*, Springer Series in Computational Mathematics 44, DOI 10.1007/978-3-642-36519-5_6,

start by introducing an abstract framework, that contains the mixed methods studied in the previous chapters as a special case.

Let therefore $\mathcal{W}$ be a Hilbert space, let $\mathcal{A}$ be in $\mathcal{L}(\mathcal{W}, \mathcal{W}')$ (the space of linear continuous operators from $\mathcal{W}$ to $\mathcal{W}'$ as defined in Sect. 4.1.4), and let F be in $\mathcal{W}'$. We consider the problem: *find* $X \in \mathcal{W}$ *such that,*

$$\mathcal{A}X = F, \tag{6.1.1}$$

which in variational formulation can be written as

$$\langle \mathcal{A}X, Y\rangle_{\mathcal{W}'\times\mathcal{W}} = \langle F, Y\rangle_{\mathcal{W}'\times\mathcal{W}} \qquad \forall\, Y \in \mathcal{W}. \tag{6.1.2}$$

From now on, we shall always assume that the bilinear form associated to $\mathcal{A}$ is positive semi-definite, that is

$$\langle \mathcal{A}Y, Y\rangle \geq 0 \qquad \forall\, Y \in \mathcal{W}. \tag{6.1.3}$$

Remark 6.1.1. As we are mostly interested in **mixed problems**, it is worth showing that this abstract formalism contains the usual theory for these problems. Indeed, let $\mathcal{W} := V \times Q$, with $X := (u, p)$, and $Y := (v, q)$, and define

$$\begin{cases} \langle \mathcal{A}X, Y\rangle := a(u, v) + b(v, p) - b(u, q), \\ \langle F, Y\rangle := \langle f, v\rangle_{V'\times V} - \langle g, q\rangle_{Q'\times Q}. \end{cases} \tag{6.1.4}$$

In this context, it is clear that (6.1.2) is just another way of writing

$$\begin{cases} a(u, v) + b(v, p) = (f, v)_V & \forall\, v \in V, \\ b(u, q) = (g, q) & \forall\, q \in Q. \end{cases} \tag{6.1.5}$$

It must however be noted that we are implicitly using the non symmetric form

$$\begin{pmatrix} A & B^t \\ -B & 0 \end{pmatrix} \begin{pmatrix} u \\ p \end{pmatrix} = \begin{pmatrix} f \\ -g \end{pmatrix} \tag{6.1.6}$$

rather than the symmetric one

$$\begin{pmatrix} A & B^t \\ B & 0 \end{pmatrix} \begin{pmatrix} u \\ p \end{pmatrix} = \begin{pmatrix} f \\ g \end{pmatrix}. \tag{6.1.7}$$

As a consequence of this choice, assuming

$$a(v, v) \geq 0 \qquad \forall\, v \in V, \tag{6.1.8}$$

we clearly have that (6.1.3) holds. □

6.1.2 Stabilising Terms

We shall consider here a very wide class of stabilisations, the so-called *augmented formulations*. Philosophically, we could think of them as based on the following observation. Suppose that we are given a general problem of the type: *find* $X \in \mathcal{W}$ *such that*

$$\langle \mathcal{A}X, Y\rangle_{\mathcal{W}'\times\mathcal{W}} = \langle F, Y\rangle_{\mathcal{W}'\times\mathcal{W}} \qquad \forall\, Y \in \mathcal{W}, \tag{6.1.9}$$

and assume that $\mathcal{A}$ is an isomorphism from $\mathcal{W}$ to $\mathcal{W}'$ (so that our problem has a unique solution). In general, as we observed already in Remark 1.5.1, given a subspace $\mathcal{W}_h \subset \mathcal{W}$, we cannot be sure that the discretised problem: *find* $X_h \in \mathcal{W}_h$ *such that*

$$\langle \mathcal{A}X_h, Y\rangle_{\mathcal{W}'\times\mathcal{W}} = \langle F, Y\rangle_{\mathcal{W}'\times\mathcal{W}} \qquad \forall\, Y \in \mathcal{W}_h \tag{6.1.10}$$

has a unique solution as well. On the other hand, still following Remark 1.5.1, if we assume that we have an ellipticity condition of the form

$$\exists\,\alpha > 0 \;\text{ such that }\; \langle \mathcal{A}Y, Y\rangle_{\mathcal{W}'\times\mathcal{W}} \geq \alpha\|Y\|^2_{\mathcal{W}} \qquad \forall\, Y \in \mathcal{W}, \tag{6.1.11}$$

(that clearly implies stability, with constant $1/\alpha$, and unique solvability of (6.1.9)) then *for every subspace* $\mathcal{W}_h \subset \mathcal{W}$, we will immediately have

$$\exists\,\alpha > 0 \;\text{ such that }\; \langle \mathcal{A}Y_h, Y_h\rangle_{\mathcal{W}'\times\mathcal{W}} \geq \alpha\|Y_h\|^2_{\mathcal{W}} \qquad \forall\, Y_h \in \mathcal{W}_h, \tag{6.1.12}$$

and we have unique solvability of (6.1.10) (with the same stability constant $1/\alpha$) without any need to be smart. Hence, although simple-minded, the general idea is: given the problem (6.1.9), try to present its solution as the solution of *another problem* for which an ellipticity condition of the type (6.1.11) holds true. In these precise terms, this is very easy. Indeed, the solution X of (6.1.9) will also be a solution of the problem: *find* $X \in \mathcal{W}$ *such that*:

$$(\mathcal{A}X, \mathcal{A}Y)_{\mathcal{W}'} = (F, \mathcal{A}Y)_{\mathcal{W}'} \qquad \forall Y \in \mathcal{W}. \tag{6.1.13}$$

Note that if $\mathcal{A}$ is an isomorphism between $\mathcal{W}$ and its dual $\mathcal{W}'$, then for every $Y \in \mathcal{W}$, we obviously have $Y = \mathcal{A}^{-1}(\mathcal{A}Y)$ and problem (6.1.13) will satisfy the ellipticity condition

$$(\mathcal{A}Y, \mathcal{A}Y)_{\mathcal{W}'} = \|\mathcal{A}Y\|^2_{\mathcal{W}'} \geq \frac{\|Y\|^2_{\mathcal{W}}}{\|\mathcal{A}^{-1}\|^2} \qquad \forall\, Y \in \mathcal{W}. \tag{6.1.14}$$

At this level of generality, it is difficult to explain why, in several applications, we are not happy with this "solution", and we still want to look for some different trick.

Just to make an example, if $\mathcal{A}$ is a differential operator (say, the Laplace operator), then problem (6.1.14) will correspond to a differential operator in which the order is doubled (in our example: the biharmonic operator) and for which discretisation would produce a matrix that is more ill-conditioned than the original one discretising $\mathcal{A}$. In this subsection, we will see some of these possible alternative techniques. We develop our discussion in the general setting of [46] but we shall mostly restrict our examples to the specific case of mixed methods.

We come back to the operator $\mathcal{A}$. At a general level, the operator $\mathcal{A}$ has a *symmetric part* $\mathcal{A}_s$, defined as

$$\mathcal{A}_s = (\mathcal{A} + \mathcal{A}^t)/2 \tag{6.1.15}$$

and an *antisymmetric part* $\mathcal{A}_a$, defined as

$$\mathcal{A}_a = (\mathcal{A} - \mathcal{A}^t)/2. \tag{6.1.16}$$

It is immediate to see that, for every $Y \in \mathcal{W}$, we have

$$\langle \mathcal{A}_s Y, Y\rangle = \langle \mathcal{A} Y, Y\rangle \quad \text{and} \quad \langle \mathcal{A}_a Y, Y\rangle = 0. \tag{6.1.17}$$

We point out that, keeping the assumption (6.1.3), we now have that $\mathcal{A}_s$ is symmetric and non-negative. Hence, we can use Lemma 4.2.1 and then (6.1.17) to obtain

$$\|\mathcal{A}_s Y\|^2_{\mathcal{W}'} \le \langle \mathcal{A}_s Y, Y\rangle \, \|\mathcal{A}_s\| = \langle \mathcal{A} Y, Y\rangle \, \|\mathcal{A}_s\| \qquad \forall\, Y \in \mathcal{W}. \tag{6.1.18}$$

We then define, for $t \in \mathbb{R}$,

$$\mathcal{A}_t = \mathcal{A}_a + t\mathcal{A}_s \tag{6.1.19}$$

and we consider for $\mu > 0$ the following augmented problem: *find* $X \in \mathcal{W}$ *such that*

$${}_{\mathcal{W}'}\langle \mathcal{A}X - F, Y\rangle_{\mathcal{W}} + \mu(\mathcal{A}X - F, \mathcal{A}_t Y)_{\mathcal{W}'} = 0 \quad \forall\, Y \in \mathcal{W}. \tag{6.1.20}$$

Remark 6.1.2. We call the attention of the reader on the difference between $\mathcal{A}^t$ (the transposed operator of $\mathcal{A}$) and $\mathcal{A}_t$, defined by (6.1.19). We apologise for the similarity of these two symbols that have, however, a totally different meaning. □

It is clear that every solution X of the original problem (6.1.9) will also be a solution of the augmented problem (6.1.20). A possible advantage of the formulation (6.1.20) over (6.1.13) is that we can hope to be allowed to take μ small enough, so that the condition number of the resulting matrix will not be much worse than the condition number of the matrix coming from the discretisation of $\mathcal{A}$.

Example 6.1.1. As we already stated, we shall restrict our examples here to the case of mixed methods, that we write again in the form (6.1.6):

$$\mathcal{A} = \begin{pmatrix} A & B^t \\ -B & 0 \end{pmatrix}. \tag{6.1.21}$$

We shall assume here that the bilinear form $a(\cdot,\cdot)$ defining A is symmetric and non-negative, as in (5.5.1) (or in (4.2.28)). As already pointed out, the non-negativity of $a(\cdot,\cdot)$ will imply, in particular, that (6.1.3) is satisfied. The symmetry of $a(\cdot,\cdot)$, on the other hand, will imply that the symmetric part of the operator $\mathcal{A}$ is given by

$$\mathcal{A}_s = \begin{pmatrix} A & 0 \\ 0 & 0 \end{pmatrix} \tag{6.1.22}$$

and the antisymmetric part is given by

$$\mathcal{A}_a = \begin{pmatrix} 0 & B^t \\ -B & 0 \end{pmatrix}. \tag{6.1.23}$$

From the symmetry and non-negativity of a, using (4.2.30) we have

$$\|Au\|^2_{V'} \le \|a\| a(u,u) = \|A\| \langle Au, u\rangle_{V'\times V} \tag{6.1.24}$$

that represents (6.1.18) in our particular case. It is not difficult to check that the stabilising term $(\mathcal{A}X, \mathcal{A}_t Y)_{\mathcal{W}'}$, for $X = (u, p)$ and $Y = (v, q)$ now becomes

$$\begin{aligned}(\mathcal{A}X, \mathcal{A}_t Y)_{\mathcal{W}'} &= (Au + B^t p, tAv + B^t q)_{V'} + (Bu, Bv)_{Q'} \\ &= t(Au + B^t p, Av)_{V'} + (Bu, Bv)_{Q'} + (Au + B^t p, B^t q)_{V'}. \end{aligned} \tag{6.1.25}$$

□

Example 6.1.2. The treatment of advection dominated equations is surely outside the scope of this book. However, it might be interesting to see how the general setting above can deal with a problem of the type: *find* $u \in H^1_0(\Omega)$ *such that*

$$-\varepsilon\Delta u + \mathbf{c}\cdot \underline{\mathrm{grad}}\, u = f \qquad \text{in } \Omega \tag{6.1.26}$$

where ε is a given positive and "small" number, $\mathbf{c}$ is a given smooth vector field (that, for simplicity, we assume to be *divergence-free*), and f is a given forcing term, say, in $L^2(\Omega)$. In this case, the stabilising term would be

$$(\mathcal{A}u, \mathcal{A}_t v)_{\mathcal{W}'} = (-\varepsilon\Delta u + \mathbf{c}\cdot\underline{\mathrm{grad}}\, u, -t\varepsilon\Delta v + \mathbf{c}\cdot\underline{\mathrm{grad}}\, v)_{H^{-1}(\Omega)}. \tag{6.1.27}$$

□

Remark 6.1.3. The structure of an augmented problem can be described as follows. First, we observe that the equation $\mathcal{A}X - F = 0$ takes place in the dual space $\mathcal{W}'$. Indeed, in its variational formulation (6.1.2), the equation is tested on a generic element $Y \in \mathcal{W}$, giving $\langle \mathcal{A}X - F, Y\rangle = 0$ for all Y. In the augmented problem, we keep this term, and we sum to it (with a suitable multiplier μ) a term containing the same equation, but this time tested on a term of the type $\mathcal{A}_t Y$ (always for Y generic in $\mathcal{W}$). Since the term $\mathcal{A}_t Y$ is itself in $\mathcal{W}'$ (as the difference $\mathcal{A}X - F$), this new term cannot be written as a duality between $\mathcal{W}'$ and $\mathcal{W}$, and we must take the scalar product of the two terms in $\mathcal{W}'$. The idea of *adding a term made by the scalar product of the equation times a suitable operator acting on the test function* Y is, somehow, the essence of the original idea of Hughes and Brooks, that has been extended and exploited in a more general setting by Hughes, Franca, and various co-authors, and became popular under the name of *stabilisation à la Hughes-Franca*. However, as we shall see, to take the inner product in $\mathcal{W}'$ is, in general, not so easy and the stabilising terms that are found in the literature (starting from the earliest ones by Hughes and his group) do not have exactly this form. Indeed, a big variety of different stabilising terms have been introduced, studied, and used in the literature of the last two or three decades (see, for instance, [10], [193], [214], [250] and [339]), all (or almost all) based on L^2 inner products (possibly multiplied by some suitable power of the mesh-size h) rather than on the $\mathcal{W}'$ inner product. However, as pointed out in [46], we could think at most of these variants as being different attempts to mimic, in one way or another, the effect of $\mu(\mathcal{A}X - F, \mathcal{A}_t Y)_{\mathcal{W}'}$. □

6.1.3 *Stability Conditions for Augmented Formulations*

Now, we want to study the behaviour of augmented problems of the type of (6.1.20). To start with, we look for sufficient conditions on t and μ ensuring that the augmented problem (6.1.20) has a unique solution.

Theorem 6.1.1. *Let $\mathcal{W}$ be a Hilbert space, and $\mathcal{A} \in \mathcal{L}(\mathcal{W}, \mathcal{W}')$ be an isomorphism which verifies* (6.1.3). *If $t \in \mathbb{R}$ and $\mu > 0$ verify*

$$\mu(1-t)^2 < 4\|\mathcal{A}_s\|^{-1}, \tag{6.1.28}$$

then there exists $\alpha_{stab} > 0$ such that

$$\langle \mathcal{A}Y, Y\rangle_{\mathcal{W}'\times\mathcal{W}} + \mu(\mathcal{A}Y, \mathcal{A}_t Y)_{\mathcal{W}'} \geq \alpha_{stab}\|\mathcal{A}Y\|^2_{\mathcal{W}'} \qquad \forall\, Y \in \mathcal{W}, \tag{6.1.29}$$

where $\mathcal{A}_t$ is defined in (6.1.19).

Proof. We apply (6.1.18) and (6.1.19), and then Cauchy-Schwarz to obtain

$$
\begin{aligned}
&\langle \mathcal{A}Y, Y\rangle_{\mathcal{W}'\times\mathcal{W}} + \mu(\mathcal{A}Y, \mathcal{A}_t Y)_{\mathcal{W}'} \\
&\geq \frac{1}{\|\mathcal{A}_s\|}\|\mathcal{A}_s Y\|^2_{\mathcal{W}'} + \mu\Big(\|\mathcal{A}_a Y\|^2_{\mathcal{W}'} + t\,\|\mathcal{A}_s Y\|^2_{\mathcal{W}'} + (1+t)\,(\mathcal{A}_a Y, \mathcal{A}_s Y)_{\mathcal{W}'}\Big) \\
&\geq \Big(\frac{1}{\|\mathcal{A}_s\|} + \mu t\Big)\|\mathcal{A}_s Y\|^2_{\mathcal{W}'} + \mu\|\mathcal{A}_a Y\|^2_{\mathcal{W}'} - \mu|1+t|\,\|\mathcal{A}_a Y\|_{\mathcal{W}'}\,\|\mathcal{A}_s Y\|_{\mathcal{W}'}.
\end{aligned}
\tag{6.1.30}
$$

The last line of (6.1.30) is a quadratic form in $\|\mathcal{A}_s Y\|$ and $\|\mathcal{A}_a Y\|$. Hence, the desired (6.1.29) will be satisfied for some $\alpha_{stab} > 0$ if

$$4\mu\Big(\frac{1}{\|\mathcal{A}_s\|} + \mu t\Big) > (\mu(1+t))^2. \tag{6.1.31}$$

This can be written as

$$\frac{4}{\|\mathcal{A}_s\|} > \mu(1+t)^2 - 4\mu t = \mu(t-1)^2, \tag{6.1.32}$$

and the result follows. □

Remark 6.1.4. We note that condition (6.1.28) implies in particular that the coefficient of $\|\mathcal{A}_s Y\|$ in the last line of (6.1.30) is positive. This is clear if we note that (6.1.31) is actually *equivalent* to (6.1.28). □

Remark 6.1.5. It is immediate to see that, for $t = 1$, we have that (6.1.28) is satisfied for every value of $\mu > 0$. This is not so unreasonable since, for $t = 1$, we have $\mathcal{A}_t = \mathcal{A}$. One could then argue that $t = 1$ is the best choice and that other values for t have no interest. However, as we shall see, in several applications, including mixed formulations and advection dominated elliptic equations, both the choices $t = 0$ and $t = -1$ have been abundantly used. □

Essentially with the same proof, one has the following result, which is slightly more general.

Theorem 6.1.2. *Under the same assumptions as in Theorem 6.1.1, let $\mathcal{M}$ be a continuous, bilinear form on $\mathcal{W}' \times \mathcal{W}'$ and let M and μ_0 be positive constants such that*

$$\mathcal{M}(X', Y') \leq M\|X'\|_{\mathcal{W}'}\,\|Y'\|_{\mathcal{W}'} \qquad \forall\, X',\, Y' \in \mathcal{W}' \tag{6.1.33}$$

and

$$\mu_0\|Y'\|^2_{\mathcal{W}'} \leq \mathcal{M}(Y', Y') \qquad \forall\, Y' \in \mathcal{W}'. \tag{6.1.34}$$

If $t \in \mathbb{R}$ and $\mu > 0$ verify

$$\mu\Big(M^2(1+t)^2 - 4\mu_0^2 t\Big) < \frac{4\mu_0}{\|\mathcal{A}_s\|}, \tag{6.1.35}$$

then there exists $\alpha_{stab} > 0$ such that

$$\langle \mathcal{A}Y, Y\rangle_{\mathcal{W}'\times\mathcal{W}} + \mu\,\mathcal{M}(\mathcal{A}Y, \mathcal{A}_t Y) \geq \alpha_{stab}\|\mathcal{A}Y\|^2_{\mathcal{W}'} \qquad \forall\, Y \in \mathcal{W}, \tag{6.1.36}$$

where $\mathcal{A}_t$ is always defined in (6.1.19).

Proof. We use (6.1.18), (6.1.19), (6.1.34), and then (6.1.33) to obtain

$$\begin{aligned}
&\langle \mathcal{A}Y, Y\rangle_{\mathcal{W}'\times\mathcal{W}} + \mu\mathcal{M}(\mathcal{A}Y, \mathcal{A}_t Y)\\
&\geq \frac{1}{\|\mathcal{A}_s\|}\|\mathcal{A}_s Y\|^2_{\mathcal{W}'} + \mu\Big(\mu_0\|\mathcal{A}_a Y\|^2_{\mathcal{W}'} + t\mu_0\,\|\mathcal{A}_s Y\|^2_{\mathcal{W}'} + (1+t)\,\mathcal{M}(\mathcal{A}_a Y, \mathcal{A}_s Y)_{\mathcal{W}'}\Big)\\
&\geq \Big(\frac{1}{\|\mathcal{A}_s\|} + \mu\mu_0 t\Big)\|\mathcal{A}_s Y\|^2_{\mathcal{W}'} + \mu\mu_0\|\mathcal{A}_a Y\|^2_{\mathcal{W}'} - \mu|1+t|\,M\,\|\mathcal{A}_a Y\|_{\mathcal{W}'}\,\|\mathcal{A}_s Y\|_{\mathcal{W}'}.
\end{aligned} \tag{6.1.37}$$

The last line of (6.1.30) is a quadratic form in $\|\mathcal{A}_s Y\|$ and $\|\mathcal{A}_a Y\|$. Hence, the desired (6.1.36) will be satisfied for some $\alpha_{stab} > 0$ if

$$4\mu\mu_0\Big(\frac{1}{\|\mathcal{A}_s\|} + \mu\mu_0 t\Big) > \mu^2 M^2(1+t)^2, \tag{6.1.38}$$

and the result follows. □

Remark 6.1.6. We note that condition (6.1.35) implies in particular that the coefficient of $\|\mathcal{A}_s Y\|$ in the last line of (6.1.37) is positive. This is again clear if we note that (6.1.38) is actually *equivalent* to (6.1.35). □

Remark 6.1.7. Looking at the proof of Theorems 6.1.1 and 6.1.2, we see that we could also write more specialised estimates, of the type

$$\alpha\Big(\frac{\|\mathcal{A}_s Y\|^2_{\mathcal{W}'}}{\|\mathcal{A}_s\|} + \mu\|\mathcal{A}_a Y\|^2_{\mathcal{W}'}\Big) \leq \langle \mathcal{A}Y, Y\rangle_{\mathcal{W}'\times\mathcal{W}} + \mu\mathcal{M}(\mathcal{A}Y, \mathcal{A}_t Y). \tag{6.1.39}$$

This would prove relevant in cases like advection dominated problems (6.1.26), where $\|\mathcal{A}_s\| \simeq \varepsilon$ and $\|\mathcal{A}_s v\|^2_{\mathcal{W}'} \simeq \|\varepsilon\Delta v\|^2_{\mathcal{W}'} \simeq \|\varepsilon v\|^2_{H^1}$. □

Remark 6.1.8. Theorem 6.1.2 reproduces Theorem 6.1.1 when we use $\mathcal{M}$ to define the scalar product in $\mathcal{W}'$ (so that $M = \mu_0 = 1$). □

Remark 6.1.9. It is also obvious that the exact solution X of (6.1.2) will also satisfy the *augmented formulation* of the problem: *find $X \in \mathcal{W}$ such that*

$$\langle \mathcal{A}X - F, Y\rangle_{\mathcal{W}'\times\mathcal{W}} + \mu\,\mathcal{M}(\mathcal{A}X - F, \mathcal{A}_t Y) = 0 \qquad \forall Y \in \mathcal{W}. \qquad (6.1.40)$$

□

Remark 6.1.10. In the same assumptions as in Theorem 6.1.2, and essentially with the same proof, one could show that there exists an ω_0, depending only on $\|\mathcal{A}_s\|$, t, M, and μ_0, and an $\alpha_0 > 0$ such that, *for every* ω with $0 < \omega \le \omega_0$ and for every $Y \in \mathcal{W}$, one has

$$\langle \mathcal{A}Y, Y\rangle_{\mathcal{W}'\times\mathcal{W}} + \omega^2\,\mathcal{M}(\mathcal{A}Y, \mathcal{A}_t Y) \ge \alpha_0\Big(\|\mathcal{A}_s Y\|^2_{\mathcal{W}'} + \omega^2\|\mathcal{A}_a Y\|^2_{\mathcal{W}'}\Big). \quad (6.1.41)$$

The interest of this variant, as we shall see, is that we would be allowed to take an $\omega = \omega(h)$ that goes to zero with h. □

We now consider our main example, that is, mixed formulations (6.1.5) inserted in the present framework through (6.1.4). We have seen that the general stabilising term takes the form (6.1.25). We point out that, in particular,

$$(\mathcal{A}Y, \mathcal{A}_t Y)_{\mathcal{W}'} = t\|Av\|^2_{V'} + (1+t)(Av, B^t q)_{V'} + \|B^t q\|^2_{V'} + \|Bv\|^2_{Q'}, \quad (6.1.42)$$

while

$$\|\mathcal{A}_s Y\|_{\mathcal{W}'} = \|Av\|_{V'} \quad \text{and} \quad \|\mathcal{A}_a Y\|^2_{\mathcal{W}'} = \|B^t q\|^2_{V'} + \|Bv\|^2_{Q'}. \qquad (6.1.43)$$

It is clear that the general philosophy, requiring that the stabilising term vanishes when X is the exact solution, would still be respected by taking a more general term, instead of $(\mathcal{A}X - F, \mathcal{A}_t Y)$. Hence, in some sense, we could specialise the result of Theorem 6.1.2 and adapt it to the case (here most interesting) of mixed methods. For instance, for general positive constants μ_1 and μ_2, we could consider a stabilising term of the form

$$(Au + B^t p - f, tAv + \mu_1 B^t q)_{V'} + (Bu - g, \mu_2 Bv)_{Q'}. \qquad (6.1.44)$$

It is clear that if (u, p) is a solution of (6.1.5), then it is also a solution of

$$\begin{aligned} &{}_{V'}\langle Au + B^t p - f, v\rangle_V - {}_{Q'}\langle Bu - g, q\rangle_{Q'} \\ &\quad + \mu\Big((Au + B^t p - f, tAv + \mu_1 B^t q)_{V'} + (Bu - g, \mu_2 Bv)_{Q'}\Big) = 0 \end{aligned} \quad (6.1.45)$$

for all $v \in V$ and for all $q \in Q$.

Concerning the stability (and hence, in particular, the uniqueness of the solution of (6.1.45)), we note that

$$\begin{aligned} &(Av + B^t q, tAv + \mu_1 B^t q)_{V'} + (Bv, \mu_2 Bv)_{Q'} \\ &\quad = t\|Av\|^2_{V'} + (t+\mu_1)(Av, B^t q)_{V'} + \mu_1\|B^t q\|^2_{V'} + \mu_2\|Bv\|^2_{Q'}. \end{aligned} \quad (6.1.46)$$

Hence, mimicking the proof of Theorem 6.1.1, we easily have that

$$
\begin{aligned}
&{}_{V'}\langle Av+B^t q, v\rangle_V - {}_{Q'}\langle Bv, q\rangle_Q + \mu(Av+B^t q, tAv+\mu_1 B^t q)_{V'} + \mu(Bv, \mu_2 Bv)_{Q'} \\
&\geq \frac{1}{\|\mathcal{A}_s\|}\|Av\|^2_{V'} + \mu\Big(t\|Av\|^2_{V'} + (t+\mu_1)(Av, B^t q)_{V'} + \mu_1\|B^t q\|^2_{V'} + \mu_2\|Bv\|^2_{Q'}\Big) \\
&\geq \Big(\frac{1}{\|\mathcal{A}_s\|} + \mu t\Big)\|Av\|^2_{V'} + \mu\Big(\mu_1\|B^t q\|^2_{V'} + \mu_2\|Bv\|^2_{Q'} - |t+\mu_1|\,\|Av\|_{V'}\|\,\|B^t q\|_{V'}\Big) \\
&\qquad\qquad \geq C\Big(\|Av\|^2_{V'} + \mu_1\|B^t q\|^2_{V'} + \mu_2\|Bv\|^2_{Q'}\Big)
\end{aligned}
\tag{6.1.47}
$$

whenever μ is small enough, and precisely,

$$
\mu(t-\mu_1)^2 < \frac{4\mu_1}{\|\mathcal{A}_s\|}. \tag{6.1.48}
$$

We can make this result more explicit in the following theorem.

Theorem 6.1.3. *Let V and Q be Hilbert spaces, and a and b bilinear forms on $V \times V$ and $V \times Q$, respectively, as in Assumption $\mathcal{AB}$ of Chap. 4 (Sect. 4.2.1). Assume that a is symmetric and positive semi-definite as in* (6.1.8)*, and assume that the continuous problem* (6.1.5) *is well posed (that is, a is elliptic on the kernel of B, and b satisfies the inf-sup condition). Let $t \in \mathbb{R}$ and let μ, μ_1, and μ_2 be positive real numbers. If* (6.1.48) *is satisfied, then there exists an $\alpha_M > 0$ such that, for every $(v,q) \in V \times Q$, we have*

$$
\begin{aligned}
&\alpha_M\Big(\|Av\|^2_{V'} + \mu_1\|B^t q\|^2_{V'} + \mu_2\|Bv\|^2_{Q'}\Big) \\
&\quad \leq {}_{V'}\langle Av + B^t q, v\rangle_V - {}_{Q'}\langle Bv, q\rangle_Q \\
&\quad + \mu(Av + B^t q, tAv + \mu_1 B^t q)_{V'} + \mu(Bv, \mu_2 Bv)_{Q'}.
\end{aligned}
\tag{6.1.49}
$$

6.1.4 Discretisations of Augmented Formulations

The augmented formulations (6.1.40) or (6.1.45) can then be transported into the discretised problem.

Starting from the more general case of (6.1.40), we consider therefore the discrete stabilised problem: *find $X_h \in \mathcal{W}_h$ such that*

$$
\begin{aligned}
&\langle \mathcal{A}X_h, Y_h\rangle_{\mathcal{W}'\times\mathcal{W}} + \mu\mathcal{M}(\mathcal{A}X_h, \mathcal{A}_t Y_h) \\
&\qquad = \langle F, Y_h\rangle_{\mathcal{W}'\times\mathcal{W}} + \mu\mathcal{M}(F, \mathcal{A}_t Y_h) \qquad \forall\, Y_h \in \mathcal{W}_h.
\end{aligned}
\tag{6.1.50}
$$

It is clear that, whenever (6.1.35) holds true, the ellipticity property (6.1.36) will be inherited by the discrete problem, that will therefore be stable. Hence, we immediately have the following result.

Theorem 6.1.4. *Let $\mathcal{W}$ be a Hilbert space, and $\mathcal{A} \in \mathcal{L}(\mathcal{W}, \mathcal{W}')$ be an isomorphism which verifies* (6.1.3)*. Let moreover $\mathcal{M}$ be a continuous, bilinear form on $\mathcal{W}' \times \mathcal{W}'$ and let M and μ_0 be positive constants such that* (6.1.33) *and* (6.1.34) *are satisfied. Finally, let $t \in \mathbb{R}$ and $\mu > 0$ verify* (6.1.35)*, and let $\mathcal{A}_t$ be defined as in* (6.1.19)*. Then, for every $F \in \mathcal{W}'$ and for every finite dimensional subspace $\mathcal{W}_h$, denoting by X and X_h the solutions of the continuous problem* (6.1.40) *and of the stabilised-discretised one* (6.1.50)*, respectively, we have*

$$\|X - X_h\|_{\mathcal{W}} \leq C \inf_{Y_h \in \mathcal{W}_h} \|X - Y_h\|_{\mathcal{W}}, \tag{6.1.51}$$

where C is a constant depending on $\|\mathcal{A}^{-1}\|$, $\|\mathcal{A}\|$, μ, M, t and on the constant α_{stab} appearing in (6.1.36)*, bounded on bounded subsets, but independent of the choice of $\mathcal{W}_h$.*

Proof (Hint). As usual, for every $X_I \in \mathcal{W}_h$, we apply the stability estimate (6.1.36) to the difference $\delta X := X_h - X_I$. Then, in the right-hand side, we substitute X *in lieu* of X_h, using the fact that they are the solutions of (6.1.2) and (6.1.50), respectively. Finally, we use the continuity of $\mathcal{A}$, of $\mathcal{A}^{-1}$ and $\mathcal{M}$ to have an estimate of $\|X_h - X_I\|_{\mathcal{W}}$ in terms of $\|X - X_I\|_{\mathcal{W}}$. Then, we add and subtract X and use the triangle inequality. Finally, since X_I is generic in $\mathcal{W}_h$, we replace $\|X - X_I\|$ with the infimum of $\|X - Y_h\|$ for Y_h varying in $\mathcal{W}_h$. □

Remark 6.1.11. In the simplified case where $\mathcal{M}$ is the scalar product in $\mathcal{W}'$, the above problem (6.1.50) could *formally* be obtained by writing

$$\langle \mathcal{A}X_h - F, Y_h + \mu\mathcal{A}_t Y_h \rangle = 0 \quad \forall\, Y_h \in \mathcal{W}_h \tag{6.1.52}$$

and we could call this a "Petrov-Galerkin" method as the test functions are not in the same space as the solution. However, unless $\mathcal{W}$ can be identified to $\mathcal{W}'$, (6.1.52) has no sense. One must make a certain number of additional manipulations in order to reach a viable formulation. □

Shifting now to the particular case of mixed formulations, and considering (6.1.45), we assume that V_h and Q_h are finite dimensional subspaces of V and Q, respectively. It might be convenient to recall some definitions from the previous chapters. We do it quickly:

$$B_h := \pi_{Q_h'} B E_V \quad B_h^t := \pi_{V_h'} B E_Q \quad A_h := \pi_{V_h'} A E_V \tag{6.1.53}$$

$$\begin{aligned} K &:= \operatorname{Ker} B = \{v \in V \text{ s.t. } Bv = 0\}, \\ K_h &:= \operatorname{Ker} B_h = \{v_h \in V_h \text{ s.t. } B_h v_h = 0\}. \end{aligned} \tag{6.1.54}$$

We now consider the discretised problem: *find* $(u_h, p_h) \in V_h \times Q_h$ *such that*

$$\begin{aligned}
&{}_{V'}\langle Au_h + B^t p_h - f, v_h\rangle_V + {}_{Q'}\langle Bu_h - g, q_h\rangle_{Q'} \\
&+ \mu(Au_h + B^t p_f - f, tAv_h + \mu_1 B^t q_h)_{V'} \\
&+ \mu(Bu_h - g, \mu_2 Bv_h)_{Q'} = 0 \qquad \forall\, v_h \in V_h\, \forall q_h \in Q_h.
\end{aligned} \tag{6.1.55}$$

It is clear that the stability result of Theorem 6.1.3 can now be used to get the following error estimate.

Theorem 6.1.5. *In the same assumptions as in Theorem 6.1.3, assume further that the continuous problem* (6.1.5) *is stable (that is, a is elliptic in the kernel, and b satisfies the inf-sup condition). Let* $V_h \subset V$ *and* $Q_h \subset Q$ *be finite dimensional subspaces, and for* $f \in V'$ *and* $g \in Q'$, *let* (u, p) *and* (u_h, p_h) *be the solutions of the continuous problem* (6.1.45) *and of* (6.1.55), *respectively. Then, we have*

$$\|u - u_h\|_V^2 + \|p - p_h\|_Q^2 \le C\Big(\inf_{v_h \in V_h} \|u - v_h\|_V + \inf_{v_h \in V_h} \|u - v_h\|_V\Big) \tag{6.1.56}$$

where C *is a constant depending on* $\|A^{-1}\|$, $\|A\|$, μ, M, t *and on the constant* α_M *appearing in* (6.1.49), *bounded on bounded subsets, but independent of the choices of* V_h *and* Q_h.

Proof. The proof follows exactly the same lines as the proof of Theorem 6.1.4, and the classical form of all the "stability+consistency" error bound. □

We shall now make explicit problem (6.1.45) in a few special cases. It is not difficult to see that (6.1.45) corresponds to have a linear "augmented operator" of the type

$$\begin{aligned}
\mathbb{M}_{stab} &= \begin{pmatrix} A & B^t \\ -B & 0 \end{pmatrix} \\
&+ \mu\Big(t\begin{pmatrix} A^t A & A^t B^t \\ 0 & 0 \end{pmatrix} + \mu_1 \begin{pmatrix} 0 & 0 \\ BA & BB^t \end{pmatrix} + \mu_2 \begin{pmatrix} B^t B & 0 \\ 0 & 0 \end{pmatrix}\Big).
\end{aligned} \tag{6.1.57}$$

Let us see, for $\mu = 1$, three typical values of t, namely $t = 1$, $t = 0$ and $t = -\mu_1$.

(i) **Case t = 1**

The augmented system is:

$$\begin{cases}
\langle Au_h + B^t p_h - f, v_h\rangle_{V'\times V} + (Au_h + B^t p_h - f, Av_h)_{V'} \\
\qquad + \mu_2(Bu_h - g, Bv_h)_{Q'} = 0 \quad \forall\, v_h \in V_h, \\
\langle -Bu_h + g, q\rangle_{Q'\times Q} \\
\qquad + \mu_1(Au_h + B^t p_h - f, B^t q_h)_{V'} = 0 \quad \forall\, q_h \in Q_h.
\end{cases} \tag{6.1.58}$$

(ii) **Case t = 0**
The augmented system is:

$$\begin{cases} \langle Au_h + B^t p_h - f, v_h\rangle_{V'\times V} \\ \qquad + \mu_2(Bu_h - g, Bv_h)_{Q'} = 0 \quad \forall\, v_h \in V_h, \\ \langle -Bu_h + g, q\rangle_{Q'\times Q} \\ \qquad + \mu_1(Au_h + B^t p_h - f, B^t q_h)_{V'} = 0 \quad \forall\, q_h \in Q_h. \end{cases} \tag{6.1.59}$$

(iii) **Case t** $= -\mu_1$
The augmented system is:

$$\begin{cases} \langle Au_h + B^t p_h - f, v_h\rangle_{V'\times V} - \mu_1(Au_h + B^t p_h - f, Av_h)_{V'} \\ \qquad + \mu_2(Bu_h - g, Bv_h)_{Q'} = 0 \quad \forall\, v_h \in V_h, \\ \langle -Bu_h + g, q\rangle_{Q'\times Q} \\ \qquad + \mu_1(Au_h + B^t p_h - f, B^t q_h)_{V'} = 0 \quad \forall\, q_h \in Q_h. \end{cases} \tag{6.1.60}$$

Remark 6.1.12. From the point of view of "economy", the case $t = 0$ implies the smallest number of extra terms and would be our favourite. On the other hand, we have seen that, in several cases, the choice $t = 1$ guarantees stability *for every value of the stabilisation parameter* μ *in* (6.1.50), and this is also a nice feature. Finally, for the choice $t = -\mu_1$, we can see that the final expression of $\mathbb{M}_{stab}$ in (6.1.57) is

$$\mathbb{M}_{stab} = \begin{pmatrix} A - \mu_1 A^t A + \mu_2 B^t B & B^t - \mu_1 A^t B^t \\ -B + \mu_1 BA & \mu_1 BB^t \end{pmatrix} \tag{6.1.61}$$

which, changing the sign of the second equation, becomes *symmetric* since obviously we have $(B^t - \mu_1 A^t B^t)^t = B - \mu_1 BA$. In conclusion, all the three choices present some interesting aspects. □

Remark 6.1.13. It is not too difficult to spot the role of each of the extra terms in (6.1.58) and (6.1.57). Indeed, we can easily see that if A is coercive on the kernel of B (a property that, in general, will not be inherited by the discretised problem), then, according to Proposition 4.3.4,

$$\langle Au, u\rangle_{V'\times V} + \mu_2(Bu, Bu)_{Q'\times Q'} \geq \tilde{\alpha}\|u\|_V^2 \quad \forall\, u \in V, \tag{6.1.62}$$

for a suitable constant $\tilde{\alpha}$, a property that will be inherited by the discretised problem. It is then clear that the extra term on the first equation (that is, the term containing μ_2) will allow to bypass problems related to the coercivity of the bilinear form a. On the other hand, the extra term in the second equation will help in controlling p as the (continuous) *inf-sup* condition implies

$$\mu_1(B^t q, B^t q)_{V'\times V'} \equiv \mu_1\|B^t q\|_{V'}^2 \geq \mu_1\beta^2\|q\|_Q^2. \tag{6.1.63}$$

The choice between the three possibilities above will obviously depend on the type of discretisation that we want to use, as well as on many other possible considerations. We will see some of them in the following chapters. □

Remark 6.1.14. We point out that, in most applications, things are usually not *totally bad*: in general, we will either have a lack of coercivity on the kernel but a good *inf-sup* condition or the reverse. Very roughly speaking, the lack of the *inf-sup* condition will occur when V_h is not big enough, compared with Q_h (so that, for instance, the image of B_h will not fill Q'_h). On the other hand, if V_h, compared with Q_h, is too big, then the kernel K_h of B_h will contain elements that are not in the kernel K of the operator B, and the ellipticity in the kernel, for the discrete problem, would fail. In these cases, we can limit ourselves to a *lighter* stabilisation. Two typical cases are

1. To retrieve coercivity of a: in (6.1.61), take $t = \mu_1 = 0$

$$\begin{cases} \langle Au_h + B^t p_h - f, v_h\rangle_{V'\times V} + \mu_2(Bu_h - g, Bv_h)_{Q'} = 0 \quad \forall\, v_h \in V_h, \\ \langle -Bu_h + g, q_h\rangle_{Q'\times Q} = 0 \quad \forall\, q_h \in Q_h. \end{cases} \tag{6.1.64}$$

2. To retrieve the *inf-sup* condition for b: in (6.1.61) take $t = \mu_2 = 0$

$$\begin{cases} \langle Au_h + B^t p_h - f, v_h\rangle_{V'\times V} = 0 \quad \forall\, v_h \in V_h, \\ \langle -Bu_h + g, q_h\rangle_{Q'\times Q} + \mu_1(Au_h + B^t p_h - f, B^t q_h)_{V'} = 0 \;\forall\, q_h \in Q_h. \end{cases} \tag{6.1.65}$$

It is easily seen, following the path of Theorem 6.1.1, that the above problems are stable under suitable conditions. The simplest case would be that A is defined by a bilinear form which is coercive on V such as in the Stokes problem. This will be developed in Chap. 8. In that case, a stabilisation such as in (6.1.65) would be sufficient. The first case (6.1.64) is nothing but the discretised version of (1.5.10). We will come back to this line of thought in the next section. □

Remark 6.1.15 (Caveat emptor). We recall that we have used the **exact** norms in V' and Q'. In many cases, (e.g. when this would imply the use of the H^{-1} norm) this may well be impossible (or very difficult) to implement numerically, and we shall have to introduce an approximation of our stabilised problem. This will typically be done by applying, in the discretised problem, the differential operators *element-wise*, and then substituting the H^{-1} scalar product with h^2 times the L^2 scalar product. □

6.1.5 *Stabilising with the "Element-Wise Equations"*

To give an idea of the techniques mentioned in the above remark, we consider the following variant of Theorem 6.1.4. As we shall see, the variant follows the spirit of Remark 6.1.10, and is closely connected with the family of methods of the next subsection. For this, however, we have to introduce some new objects. We assume that we have a space $\mathcal{W}^+$ (made of smoother functions) and a Hilbert space $\mathcal{H}$ (that we identify with its dual space $\mathcal{H}'$) such that

$$\mathcal{W}^+ \subset \mathcal{W} \subseteq \mathcal{H} \equiv \mathcal{H}' \subseteq \mathcal{W}' \tag{6.1.66}$$

and

$$\mathcal{A}_s(\mathcal{W}^+) \subseteq \mathcal{H}, \quad \mathcal{A}_a(\mathcal{W}^+) \subseteq \mathcal{H}, \tag{6.1.67}$$

and for all h

$$\mathcal{A}_a(\mathcal{W}_h) \subseteq \mathcal{H}. \tag{6.1.68}$$

We also assume that we have, for all h, a linear operator

$$\mathcal{S}_h : \mathcal{H} + \mathcal{A}_s(\mathcal{W}_h) + \mathcal{A}_a(\mathcal{W}_h) \to \mathcal{H} \tag{6.1.69}$$

such that

$$\|\mathcal{S}_h Y\|_{\mathcal{H}} = \|Y\|_{\mathcal{H}} \qquad \forall Y \in \mathcal{H}, \tag{6.1.70}$$

and we note that, together with (6.1.68), this gives

$$\|\mathcal{S}_h(\mathcal{A}_a Y_h)\|_{\mathcal{H}} \geq \|\mathcal{A}_a Y_h\|_{\mathcal{H}} \qquad \forall Y_h \in \mathcal{W}_h. \tag{6.1.71}$$

We assume further that there exists a monotonically increasing function $\omega : \mathbb{R}_+ \to \mathbb{R}_+$ such that

$$\omega(h)\|\mathcal{S}_h(\mathcal{A}_r Y_h)\|_{\mathcal{H}} \leq \|\mathcal{A}_r Y_h\|_{\mathcal{W}'} \quad \text{where } r = s \text{ or } a, \ \forall\, Y_h \in \mathcal{W}_h. \tag{6.1.72}$$

Remark 6.1.16. In the applications that we have in mind, the space $\mathcal{H}$ will be either L^2 or a Cartesian product of several copies of L^2, and the operator $\mathcal{S}_h$ will be the one that allows to take the element-by-element derivatives of functions that are smooth (typically, polynomials) inside each element but might be discontinuous from one element to the next (or are continuous but not C^1, when you take second derivatives). In mathematical words, $\mathcal{S}_h(\chi)$ would take the restriction $\chi_{|T}$ to each individual *open* triangle T, and then consider the L^2 function that in each triangle T is equal to $\chi_{|T}$. In this way, possible Dirac masses concentrated on the inter-element boundaries would be dropped. Having this in mind, it should be clear that the assumption in (6.1.68) is a *very strong* one, and in all the applications that we considered, it requires either that the antisymmetric part of $\mathcal{A}$ is an operator of *lower order* (as it happens for advection dominated flows) or that the elements of $\mathcal{W}_h$ have, in a certain sense, *more continuity* than strictly necessary (as when using continuous pressures in the Stokes problem). □

Assuming further that

$$F \in \mathcal{H} \tag{6.1.73}$$

where F is the right-hand side of (6.1.1), we can consider the discretised problem: *find* $X_h \in \mathcal{W}_h$ *such that*

$$\langle \mathcal{A}X_h - F, Y_h\rangle_{\mathcal{W}'\times\mathcal{W}} + \mu\omega^2(h)(\mathcal{S}_h(\mathcal{A}X_h - F), \mathcal{S}_h(\mathcal{A}_t Y_h))_{\mathcal{H}} = 0 \tag{6.1.74}$$

for all $Y_h \in \mathcal{W}_h$. Proceeding as in Theorems 6.1.1 and 6.1.2, it is not difficult to see that if

$$(1-t)^2\mu\,\omega^2(h) < \frac{4}{\|\mathcal{A}_s\|}, \tag{6.1.75}$$

then there exists $\alpha_0 > 0$ such that

$$\begin{aligned}\langle \mathcal{A}Y_h, Y_h\rangle_{\mathcal{W}'\times\mathcal{W}} &+ \mu\,\omega^2(h)\,(\mathcal{S}_h(\mathcal{A}Y_h), \mathcal{S}_h(\mathcal{A}_t Y_h))_{\mathcal{H}} \\ &\geq \alpha_0\Big(\|\mathcal{A}_s Y_h\|^2_{\mathcal{W}'} + \mu\,\omega^2(h)\|\mathcal{S}_h(\mathcal{A}_a Y_h)\|^2_{\mathcal{H}}\Big) \quad \forall Y_h \in \mathcal{W}_h.\end{aligned} \tag{6.1.76}$$

We can now apply the above estimate to have a bound on the error.

Theorem 6.1.6. *Let* $\mathcal{W}$ *be a Hilbert space and* $\mathcal{A} \in \mathcal{L}(\mathcal{W}, \mathcal{W}')$ *be an isomorphism which verifies* (6.1.3). *Assume that all the additional assumptions* (6.1.66)–(6.1.73) *are satisfied, and assume further that the solution* X *of problem* (6.1.2) *belongs to* $\mathcal{W}^+$. *For* $t \in \mathbb{R}$ *and for* $\mu > 0$, *let* X_h *be the solution of* (6.1.74). *If* (6.1.75) *is satisfied, then there exists a constant* C, *depending only on* α_0, t, *and* μ, *such that*

$$\begin{aligned}\|\mathcal{A}_s(X - X_h)\|_{\mathcal{W}'} &+ \omega(h)\|\mathcal{S}_h(\mathcal{A}_a(X - X_h))\|_{\mathcal{H}} \\ &\leq C \inf_{Y_h\in\mathcal{W}_h}\Big((\|X - Y_h\|_{\mathcal{W}} + \omega^{-1}(h)\|X - Y_h\|_{\mathcal{H}} \\ &+ \omega(h)\|\mathcal{S}_h(\mathcal{A}_a(X - Y_h))\|_{\mathcal{H}}\Big).\end{aligned} \tag{6.1.77}$$

Proof. We first observe that the Galerkin orthogonality equation

$$\langle \mathcal{A}(X - X_h), Y_h\rangle_{\mathcal{W}'\times\mathcal{W}} + \mu\omega^2(h)(\mathcal{S}_h(\mathcal{A}(X - X_h)), \mathcal{S}_h(\mathcal{A}_t Y_h))_{\mathcal{H}} = 0 \tag{6.1.78}$$

holds for all $Y_h \in \mathcal{W}_h$. Then let X_I be a generic element of $\mathcal{W}_h$, and set as before $\delta X := X_h - X_I$ and $\delta_I := X - X_I$. We apply the estimate (6.1.76) to δX and then we add and subtract X and use (6.1.78) to obtain

$$
\begin{aligned}
&\alpha_0\Big(\|\mathcal{A}_s\delta X\|^2_{\mathcal{W}'} + \mu\,\omega^2(h)\|\mathcal{S}_h(\mathcal{A}_a\delta X)\|^2_{\mathcal{H}}\Big)\\
&\quad\le \langle \mathcal{A}\,\delta X, \delta X\rangle_{\mathcal{W}'\times\mathcal{W}} + \mu\,\omega^2(h)\,(\mathcal{S}_h(\mathcal{A}\delta X), \mathcal{S}_h(\mathcal{A}_t\delta X))_{\mathcal{H}} \qquad (6.1.79)\\
&\quad= \langle \mathcal{A}\,\delta_I, \delta X\rangle_{\mathcal{W}'\times\mathcal{W}} + \mu\,\omega^2(h)\,(\mathcal{S}_h(\mathcal{A}\delta_I), \mathcal{S}_h(\mathcal{A}_t\delta X))_{\mathcal{H}}.
\end{aligned}
$$

The first term in the last line of (6.1.79), using (6.1.19), (6.1.68), and then (6.1.71), can be estimated by

$$
\begin{aligned}
&\langle \mathcal{A}\,\delta_I, \delta X\rangle_{\mathcal{W}'\times\mathcal{W}} \le \|\delta_I\|_{\mathcal{W}}\cdot\|\mathcal{A}_s\delta X\|_{\mathcal{W}'} + \|\delta_I\|_{\mathcal{H}}\cdot\|\mathcal{A}_a\delta X\|_{\mathcal{H}},\\
&\le \|\delta_I\|_{\mathcal{W}}\cdot\|\mathcal{A}_s\delta X\|_{\mathcal{W}'} + \omega^{-1}(h)\|\delta_I\|_{\mathcal{H}}\cdot\omega(h)\|\mathcal{S}_h(\mathcal{A}_a\delta X)\|_{\mathcal{H}}, \qquad (6.1.80)\\
&\le (\|\delta_I\|_{\mathcal{W}} + \omega^{-1}(h)\|\delta_I\|_{\mathcal{H}})\cdot(\|\mathcal{A}_s\delta X\|_{\mathcal{W}'} + \omega(h)\|\mathcal{S}_h(\mathcal{A}_a\delta X)\|_{\mathcal{H}})
\end{aligned}
$$

while the second term, using (6.1.19) and then (6.1.72), is easily estimated by

$$
\begin{aligned}
&\omega^2(h)\,(\mathcal{S}_h(\mathcal{A}\delta_I), \mathcal{S}_h(\mathcal{A}_t\delta X))_{\mathcal{H}}\\
&\quad\le \omega(h)\,\|\mathcal{S}_h(\mathcal{A}\delta_I)\|_{\mathcal{H}}\cdot\omega(h)\|\mathcal{S}_h(\mathcal{A}_t\delta X)\|_{\mathcal{H}} \qquad (6.1.81)\\
&\quad\le \omega(h)\,\|\mathcal{S}_h(\mathcal{A}\delta_I)\|_{\mathcal{H}}\cdot\Big(|t|\,\|\mathcal{A}_s\delta X\|_{\mathcal{W}'} + \omega(h)\|\mathcal{S}_h(\mathcal{A}_a\delta X)\|_{\mathcal{H}}\Big).
\end{aligned}
$$

The result (6.1.77) now follows easily by a repeated use of the arithmetic-geometric mean inequality and finally, the use of the triangle inequality to estimate $X - X_h$ in terms of δX and δ_I. Note that the last term in the right-hand side of (6.1.77) appears only in this final step (using the triangle inequality).

Remark 6.1.17. In most applications, the constant $\omega(h)$ corresponds to some *inverse inequality* applied to piecewise polynomial functions. The same constant (in terms of powers of h) will often appear if we compare the best approximation of a smooth function X taken in the norm of $\mathcal{H}$ rather than in the (stronger) norm of $\mathcal{W}$. As a result, the first two terms appearing in the right-hand side of (6.1.77) will, in general, be of *the same order*, and the third will be either of the same order or smaller. □

Remark 6.1.18. As we can see, the strong assumption (6.1.68) has been used only to estimate the term $\langle \mathcal{A}_a\delta_I, \delta X\rangle$ in (6.1.80). In a certain number of applications, one could take advantage of some particular feature of the problem at hand, and survive without it. To do so when dealing with the abstract problem would be, however, very complicated. Hence, we defer the analysis of the different applications of the above theory to the following chapters, mostly to Chap. 8 concerning the Stokes problem, and we just consider here below some example of the possible stabilisations of Laplace operator in mixed form. □

Example 6.1.3 (Stabilisation of the mixed Poisson problem). In Sect. 1.5.1 of Chap. 1, we have considered many augmented methods for the mixed formulation of the Dirichlet problem. Most of these methods can be written in the framework that we have just developed. For simplicity, we refer to the simplest formulations (1.5.2) and (1.5.9), that we briefly recall for the convenience of the reader:

$$(\underline{u},\underline{v})+(\operatorname{div}\underline{v},p)-(\operatorname{div}\underline{u}+f,q)=0,\quad \forall\,\underline{v}\in H(\operatorname{div};\Omega)\;\forall\,q\in L^2(\Omega) \tag{6.1.82}$$

and

$$(\underline{u}-\underline{\operatorname{grad}}\,p,\underline{v})+(\underline{u},\underline{\operatorname{grad}}\,q)-(f,q)=0\quad \forall\,\underline{v}\in L^2(\Omega)\;\forall\,q\in H^1_0(\Omega). \tag{6.1.83}$$

Other examples will be seen in the following chapters. We have therefore, for the formulation (6.1.82), $\mathcal{W}=H(\operatorname{div};\Omega)\times L^2(\Omega)$ and, for the formulation (6.1.83), $\mathcal{W}=(L^2(\Omega))^d\times H^1_0(\Omega)$. In all cases, we take $\mathcal{H}:=(L^2(\Omega))^d\times L^2(\Omega)$, and we use the symbol $(\cdot\,,\cdot)$ to denote the inner product in $L^2(\Omega)$ or in $(L^2(\Omega))^d$. We also assume, for simplicity, that the solution $(\underline{u},p)$ belongs to $(H^1(\Omega))^d\times H^2(\Omega)\cap H^1_0(\Omega)$. Finally, following the common usage, we denote by $\underline{\operatorname{grad}}_h\,q$ the element-wise gradient $\mathcal{S}_h(\underline{\operatorname{grad}}\,q)$ and by $\operatorname{div}_h\underline{v}$ the element-wise divergence $\mathcal{S}_h(\operatorname{div}\underline{v})$. In the first case (that is when using the formulation (6.1.82)), we have

$$\begin{aligned}(\underline{u},\underline{v})&+(\operatorname{div}\underline{v},p)-(\operatorname{div}\underline{u}+f,q)\\&+\mu h^2\Big(t((\underline{u},\underline{v})+(\operatorname{div}\underline{v},p))\\&-\mu_1(\underline{u}-\underline{\operatorname{grad}}_h\,p,\underline{\operatorname{grad}}_h\,q)+\mu_2(\operatorname{div}\underline{u}+f,\operatorname{div}\underline{v})\Big)=0,\end{aligned} \tag{6.1.84}$$

while in the second case (that is when using the formulation (6.1.83)), we have instead

$$\begin{aligned}(\underline{u}&-\underline{\operatorname{grad}}\,p,\underline{v})+(\underline{u},\underline{\operatorname{grad}}\,q)-(f,q)\\&+\mu h^2\Big(t(\underline{u}-\underline{\operatorname{grad}}\,p,\underline{v})\\&-\mu_1(\underline{u}-\underline{\operatorname{grad}}\,p,\underline{\operatorname{grad}}\,q)+\mu_2(\operatorname{div}_h\underline{u}+f,\operatorname{div}_h\underline{v})\Big)=0.\end{aligned} \tag{6.1.85}$$

Let us see some particular cases related to this last example. In all cases, we will take, for simplicity, $\mu=1$.

(i) **Case t = 1**. In this case, the augmented formulation is

$$\begin{aligned}&(1+h^2)(\underline{u}-\underline{\operatorname{grad}}\,p,\underline{v})+\mu_2h^2(\operatorname{div}_h\underline{u}+f,\operatorname{div}_h\underline{v})=0\quad\forall\,\underline{v}\in(L^2(\Omega))^2,\\&(1-\mu_1h^2)(\underline{u},\underline{\operatorname{grad}}\,q)+\mu_1h^2(\underline{\operatorname{grad}}\,p,\underline{\operatorname{grad}}\,q)-(f,q)=0\quad\forall\,q\in H^1_0(\Omega).\end{aligned} \tag{6.1.86}$$

Note that stability holds for every choice of $\mu_2\geq 0$.

(ii) **Case** $\underline{t = 0}$. In this case, the augmented formulation is

$$
\begin{aligned}
&(\underline{u} - \underline{\mathrm{grad}}\, p, \underline{v}) + \mu_2 h^2 (\mathrm{div}_h\, \underline{u} + f, \mathrm{div}_h\, \underline{v}) = 0 \quad \forall\, \underline{v} \in (L^2(\Omega))^2,\\
&(1 - \mu_1 h^2)(\underline{u}, \underline{\mathrm{grad}}\, q) + \mu_1 h^2 (\underline{\mathrm{grad}}\, p, \underline{\mathrm{grad}}\, q) - (f, q) = 0 \quad \forall\, q \in H_0^1(\Omega).
\end{aligned}
\tag{6.1.87}
$$

This formulation, in particular with $\mu_1 = 0$, is particularly appealing for discretisations in which the *inf-sup* condition holds already but the ellipticity in the kernel is lacking.

(iii) **Case** $\mathbf{t} = -\boldsymbol{\mu}_1$. In this case, the augmented formulation is

$$
\begin{aligned}
&(1 - \mu_1 h^2)(\underline{u} - \underline{\mathrm{grad}}\, p, \underline{v}) + \mu_2 h^2 (\mathrm{div}_h\, \underline{u} + f, \mathrm{div}_h\, \underline{v}) = 0 \ \forall\, \underline{v} \in (L^2(\Omega))^2,\\
&(1 - \mu_1 h^2)(\underline{u}, \underline{\mathrm{grad}}\, q) + \mu_1 h^2 (\underline{\mathrm{grad}}\, p, \underline{\mathrm{grad}}\, q) - (f, q) = 0 \quad \forall\, q \in H_0^1(\Omega).
\end{aligned}
\tag{6.1.88}
$$

Note that, changing the sign of the second equation, we reach a *symmetric* problem, as already pointed out in Remark 6.1.12. □

6.2 Other Stabilisations

In this subsection, we still want to deal with methods for transforming the problem in a stable one, but not necessarily reaching a formulation where ellipticity holds. In particular, here, we want to analyse methods to fix discretisations that have already some sort of stability, in a spirit similar to the one of Remark 6.1.14.

6.2.1 General Stability Conditions

We go back to our original abstract formulation (6.1.1) which we re-write for the convenience of the reader. We consider the problem: *find* $X \in \mathcal{W}$ *such that*

$$
\mathcal{A}X = F, \tag{6.2.1}
$$

together with its variational formulation

$$
\langle \mathcal{A}X, Y \rangle_{\mathcal{W}' \times \mathcal{W}} = \langle F, Y \rangle_{\mathcal{W}' \times \mathcal{W}} \qquad \forall\, Y \in \mathcal{W}. \tag{6.2.2}
$$

We also recall that we assumed the non-negativity condition (6.1.3) that we also repeat here

$$
\langle \mathcal{A}Y, Y \rangle_{\mathcal{W}' \times \mathcal{W}} \geq 0, \qquad \forall Y \in \mathcal{W}. \tag{6.2.3}
$$

The following result is an exercise of functional analysis, but, for the convenience of the readers, we sketch a proof.

Proposition 6.2.1. *If* (6.2.3) *holds, then the two following conditions are equivalent:*

$$(i) \quad \mathcal{A} \text{ is an isomorphism from } \mathcal{W} \text{ onto } \mathcal{W}' \tag{6.2.4}$$

$$(ii) \quad \exists\, \Phi \in \mathcal{L}(\mathcal{W},\mathcal{W}) \text{ and a constant } \alpha_\Phi > 0 \text{ such that}$$

$$\langle \mathcal{A}Y, \Phi(Y)\rangle_{\mathcal{W}'\times\mathcal{W}} \geq \alpha_\Phi \|Y\|^2_{\mathcal{W}} \qquad \forall\, Y \in \mathcal{W}. \tag{6.2.5}$$

Proof. Let $J = R_{\mathcal{W}'}$ be the Ritz operator from $\mathcal{W}'$ to $\mathcal{W}$ as defined in Theorem 4.1.2. The implication $(i) \implies (ii)$ follows by taking $\Phi = J\mathcal{A}$. To prove the converse implication, we denote by Id the identity operator in $\mathcal{W}$, and we remark that, if (6.1.3) holds, then for every positive real number s, we have, for all $Y \in \mathcal{W}$,

$$\langle (s\Phi + Id)^t \mathcal{A}Y, Y\rangle_{\mathcal{W}'\times\mathcal{W}} = \langle \mathcal{A}Y, (s\Phi + Id)Y\rangle_{\mathcal{W}'\times\mathcal{W}} \geq s\,\alpha_\Phi \|Y\|^2_{\mathcal{W}}.$$

This easily implies that $(s\Phi + Id)^t\mathcal{A}$ is an isomorphism from $\mathcal{W}$ onto $\mathcal{W}'$. On the other hand, we know that $s\Phi + Id$ is an isomorphism for s small enough (see for instance Theorem 4.1.3), so that $(s\Phi + Id)^t$ will also be an isomorphism, as well as its inverse $(s\Phi + Id)^{-t}$. Hence, $\mathcal{A} = [(s\Phi + Id)^{-t}][(s\Phi + Id)^t\mathcal{A}]$ (as product of two isomorphisms) is also an isomorphism, and (i) holds. □

Remark 6.2.1. If we further assume $\mathcal{A} = \mathcal{A}^t$ (that is, if we assume the bilinear form $\langle \mathcal{A}Y, Y\rangle$ to be *symmetric*), then, using Lemma 4.2.2, we see that in (6.2.5) we could always use $\Phi = Id$, and the equivalence would still hold. □

Remark 6.2.2. If (6.1.3) is not satisfied, we always have $(i) \implies (ii)$ but the converse is false. This can be seen by considering in $L^2(]0,+\infty[)$ the mapping:

$$\begin{cases} (\mathcal{A}u)(x) = u(x-1) \text{ for } x > 1 \\ (\mathcal{A}u)(x) = 0 \text{ for } 0 < x \leq 1 \end{cases}$$

(corresponding to shifting the graph of u to the right by 1, and inserting 0 in the interval $(0,1)$). Clearly, (ii) is satisfied by taking $\Phi u := \mathcal{A}u$, but (i) is not, as $\mathcal{A}$ is injective but not surjective. For an operator that does not satisfy (6.1.3), we would need two conditions instead of (6.2.5), that is: $\exists\, \Phi_1, \Phi_2 \in \mathcal{L}(\mathcal{W},\mathcal{W})$ such that, for all $Y \in \mathcal{W}$,

$$\begin{cases} \langle \mathcal{A}Y, \Phi_1(Y)\rangle_{\mathcal{W}'\times\mathcal{W}} \geq \alpha_1 \|Y\|^2_{\mathcal{W}}, \\ \langle \Phi_2(Y), \mathcal{A}^tY\rangle_{\mathcal{W}\times\mathcal{W}'} \geq \alpha_2 \|Y\|^2_{\mathcal{W}}, \end{cases} \tag{6.2.6}$$

implying that $\mathcal{A}$ is both injective and surjective. □

Remark 6.2.3. It must be noted that the **stability constant** of Problem (6.1.2), that is, the smallest constant C such that

$$\|X\| \leq C\|\mathcal{A}X\| \qquad \forall\, X \in \mathcal{W}, \tag{6.2.7}$$

is not $1/\alpha_\Phi$ (see (6.2.5)) but rather

$$C = \|\Phi\|/\alpha_\Phi. \tag{6.2.8}$$

□

Remark 6.2.4. We now consider again the case of (6.1.4), in which the abstract problem (6.2.2) is just a different way of writing the mixed problem (6.1.5). For this case, we want to get an explicit construction of some Φ that satisfies (6.2.5) starting from the usual stability conditions developed previously in Chap. 4. In other words, we are going to see the equivalence of (6.2.5) with the *ellipticity in the kernel* and *inf-sup* conditions. We thus consider, for any given $X^* = (u^*, p^*)$ in $V \times Q$, two auxiliary problems, which have a unique solution if the mixed problem (6.1.5) is well posed:

– *Find* (u_1, p_1)*, solution of*

$$\begin{cases} a(v, u_1) + b(v, p_1) = (u^*, v)_V & \forall\, v \in V, \\ b(u_1, q) = 0 & \forall\, q \in Q, \end{cases} \tag{6.2.9}$$

– *Find* (u_2, p_2)*, solution of*

$$\begin{cases} a(v, u_2) + b(v, p_2) = 0 & \forall\, v \in V, \\ b(u_2, q) = (p^*, q)_Q & \forall\, q \in Q. \end{cases} \tag{6.2.10}$$

In other words, we take $(u_1, p_1) = \mathcal{A}^{-1}(R_V u^*, 0)$ and $(u_2, p_2) = \mathcal{A}^{-1}(0, R_Q p^*)$, where R_V and R_Q are the Ritz operators from V to V' and from Q to Q', respectively (see (4.1.37)). We now set $\Phi((u^*, p^*)) := (u_1 + u_2, -p_1 - p_2)$ and we have:

$$\begin{aligned} A(X^*, \Phi(X^*)) &= a(u^*, u_1 + u_2) + b(u_1 + u_2, p^*) + b(u^*, p_1 + p_2) \\ &= \|u^*\|_V^2 + \|p^*\|_Q^2 = \|X\|_{\mathcal{W}}^2. \end{aligned} \tag{6.2.11}$$

□

Remark 6.2.5. Problems (6.2.9) and (6.2.10) could, by linearity, be combined into one. We preferred to make more explicit the separate control of $\|u^*\|_V$ and $\|p^*\|_Q$. One should also note that (see Remark 6.2.3) the stability constant in (6.2.7) (which, using (6.2.8), is now equal to $\|\Phi\|$, since by (6.2.11) we have $\alpha_\Phi = 1$) depends through (6.2.9) and (6.2.10) on the usual constants defining, for example, the coercivity in the kernel and the *inf-sup* condition. No *free lunch.* □

6.2.2 Stability of Discretised Formulations

Let us now turn to the discretisation of problem (6.1.2). For a given sequence of subspaces $\mathcal{W}_h$ of $\mathcal{W}$ (usually of finite dimension), we consider, for each h, the discrete problem: *find* $X_h \in \mathcal{W}_h$ *such that*

$$\langle \mathcal{A}X_h, Y_h\rangle = \langle F, Y_h\rangle \qquad \forall\, Y_h \in \mathcal{W}_h. \tag{6.2.12}$$

In general, for an arbitrary choice of the sequence $\{\mathcal{W}_h\}$, (6.2.12) will not be stable, that is, we cannot ensure that there exists a sequence of linear operators $\Phi_h \in \mathcal{L}(\mathcal{W}_h, \mathcal{W}_h)$, uniformly bounded in h, such that for some $\alpha_\Phi > 0$ independent of h:

$$\langle \mathcal{A}Y_h, \Phi_h(Y_h)\rangle \geq \alpha_\Phi \|Y_h\|^2_{\mathcal{W}} \qquad \forall\, Y_h \in \mathcal{W}_h. \tag{6.2.13}$$

We suppose that we have, for each h, a stabilising term R with the structure

$$R(X_h, Y_h) := L(X_h, Y_h) + \langle N, Y_h\rangle \tag{6.2.14}$$

where N, which is possibly null, will depend on F, and where $L(X_h, Y_h)$ is a continuous bilinear form on $\mathcal{W}_h$ with a continuity constant c_L,

$$|L(X_h, Y_h)| \leq c_L \|X_h\|_{\mathcal{W}} \|Y_h\|_{\mathcal{W}}. \tag{6.2.15}$$

In practice, we shall build $R(X_h, Y_h)$ in such a way that it can be used as a *stabilising term* in a sense that will be defined in hypothesis **H.0** below. All the stabilisations of the previous section (see, for instance, (6.1.20) or (6.1.40)) had indeed the above structure. Here, however, we shall often use just the bilinear part $L(X_h, Y_h)$.

We shall now consider an abstract error estimate based on the following hypothesis.

H.0 We have:

(i) A continuous problem

$$\langle \mathcal{A}X, Y\rangle = \langle F, Y\rangle \qquad \forall\, Y \in \mathcal{W}, \tag{6.2.16}$$

which we assume to have a unique solution,

(ii) A sequence of stabilised discrete problems

$$\langle \mathcal{A}X_h, Y_h\rangle + rR(X_h, Y_h) = \langle F, Y_h\rangle \qquad \forall\, Y_h \in \mathcal{W}_h, \tag{6.2.17}$$

where $R(X_h, Y_h)$ is of the form (6.2.14) and $r > 0$ is a scalar,

(iii) Two constants $\tilde{c}_\Phi$ and $\tilde{\alpha}_\Phi$, and an operator $\tilde{\Phi}_h \in \mathcal{L}(\mathcal{W}_h, \mathcal{W}_h)$ such that

$$\|\tilde{\Phi}_h(Y_h)\| \le \tilde{c}_\Phi \|Y_h\| \quad \forall\, Y_h \in \mathcal{W}_h \tag{6.2.18}$$

and

$$\langle \mathcal{A}Y_h, \tilde{\Phi}_h(Y_h)\rangle_{\mathcal{W}'\times\mathcal{W}} + rL(Y_h, \tilde{\Phi}_h(Y_h)) \ge \tilde{\alpha}_\Phi \|Y_h\|^2_{\mathcal{W}}. \tag{6.2.19}$$

□

Under the assumption **H.0**, we have the following error bound.

Proposition 6.2.2. *Assume that* **H.0** *holds, and let* X *and* X_h *be the solutions of* (6.2.16) *and* (6.2.17) *respectively. For every* $X_I \in \mathcal{W}_h$*, let us set*

$$\mathcal{R}(X_I) := \sup_{Y_h \in \mathcal{W}_h} \frac{R(X_I, Y_h)}{\|Y_h\|}. \tag{6.2.20}$$

We then have

$$\frac{\tilde{\alpha}_\Phi}{\tilde{c}_\Phi}\|X_I - X_h\| \le \|\mathcal{A}\|\, \|X - X_I\| + r\mathcal{R}(X_I), \tag{6.2.21}$$

and consequently

$$\|X - X_h\| \le \frac{\tilde{c}_\Phi \|\mathcal{A}\| + \tilde{\alpha}_\Phi}{\tilde{\alpha}_\Phi}\|X - X_I\| + \frac{\tilde{c}_\Phi r \mathcal{R}(X_I)}{\tilde{\alpha}_\Phi}. \tag{6.2.22}$$

Proof. Set $\delta X := X_I - X_h$ and $\tilde{Y}_h := \tilde{\Phi}_h(\delta X)$. From (6.2.18), we immediately have

$$\|\tilde{Y}_h\| \le \tilde{c}_\Phi \,\|\delta X\|. \tag{6.2.23}$$

On the other hand, using (6.2.19), adding and subtracting X and using (6.2.14), then using (6.2.16) and (6.2.17), and finally (6.2.20), we obtain:

$$\begin{aligned}
\tilde{\alpha}_\Phi \,\|\delta X\|^2 &\le \langle \mathcal{A}\delta X, \tilde{Y}_h\rangle + r\, L(\delta X, \tilde{Y}_h) \\
&= \langle \mathcal{A}(X_I - X), \tilde{Y}_h\rangle + \langle \mathcal{A}X, \tilde{Y}_h\rangle - \langle \mathcal{A}X_h, \tilde{Y}_h\rangle - rR(X_h, \tilde{Y}_h) + rR(X_I, \tilde{Y}_h) \\
&= \langle \mathcal{A}(X_I - X), \tilde{Y}_h\rangle + rR(X_I, \tilde{Y}_h) \\
&\le \|\tilde{Y}_h\| \left(\|\mathcal{A}\|\, \|X_I - X\| + r\mathcal{R}(X_I)\right)
\end{aligned}$$

and (6.2.21) follows immediately using (6.2.23). Finally, (6.2.22) follows from (6.2.21) using the triangle inequality. □

On many occasions, as we have seen in the previous section, the perturbation term R can be chosen in such a way that the strong consistency property (usually called *Galerkin orthogonality*) still holds. In these cases, the solution X of (6.2.16) would verify

$$R(X, Y_h) = 0 \qquad \forall\, Y_h \in \mathcal{W}_h, \tag{6.2.24}$$

implying that for the discrete stabilised problem we have

$$\langle \mathcal{A}X, Y_h\rangle + rR(X, Y_h) = \langle F, Y_h\rangle \qquad \forall\, Y_h \in \mathcal{W}_h. \tag{6.2.25}$$

In this case, we have, essentially by the same proof as in Proposition 6.2.2, the following corollary.

Corollary 6.2.1. *Assume that* **H.0** *holds, and let X and X_h be the solutions of* (6.2.16) *and* (6.2.17) *respectively. Assume moreover that X satisfies the strong consistency condition* (6.2.24). *Then, we have*

$$\|X - X_h\| \leq \frac{\tilde{c}_\Phi(\|\mathcal{A}\| + rc_L) + \tilde{\alpha}_\Phi}{\tilde{\alpha}_\Phi} \inf_{Y_h \in \mathcal{W}_h} \|X - Y_h\| \tag{6.2.26}$$

where c_L is defined in (6.2.15).

Remark 6.2.6. It is clear that the above results, and in particular Corollary 6.2.1, could be applied to the methods of the previous section. □

The results of Proposition 6.2.2 and of Corollary 6.2.1 are of a *general* nature and, in order to obtain sharper results, we shall have to specialise somehow the construction of $R(X_h, Y_h)$ and its properties. This will be done in the following subsection.

6.3 Minimal Stabilisations

In several applications, we will have that there exists a subspace $\overline{\mathcal{W}}_h \subseteq \mathcal{W}$ and a positive constant $\overline{\alpha}$ such that

$$\langle \mathcal{A}Z, Z\rangle \geq \overline{\alpha}\|\pi_{\overline{\mathcal{W}}_h} Z\|^2 \qquad \forall Z \in \overline{\mathcal{W}}_h, \tag{6.3.1}$$

or, more generally,

$${}_{\mathcal{W}'}\langle \mathcal{A}Z, \overline{\Phi}_h(Z)\rangle_{\mathcal{W}} \geq \overline{\alpha}\|\pi_{\overline{\mathcal{W}}_h} Z\|_{\mathcal{W}}\, \|\overline{\Phi}_h(Z)\| \qquad \forall Z \in \overline{\mathcal{W}}_h, \tag{6.3.2}$$

for some linear mapping $\overline{\Phi}_h$ from $\mathcal{W}_h$ to $\mathcal{W}_h$. In these cases, we can consider that the part of the solution that belongs to $\overline{W}_h$ will somehow be "*under control*" and will not need to be stabilised.

In these cases, the stabilising term R in (6.2.14) could be chosen of the form

$$R(X_h, Y_h) = (G_h X_h - N, G_h Y_h)_{\mathcal{H}} \tag{6.3.3}$$

for some suitable Hilbert space $\mathcal{H}$, a suitable N in $\mathcal{H}$ (either equal to 0 or depending on f), and a suitable linear operator G_h from $\mathcal{W}_h$ to $\mathcal{H}$. Conditions for $(G_h X_h, G_h Y_h)_{\mathcal{H}}$ to be stabilising will be given in the subsections below. Often, roughly speaking, one would take $N = 0$ and use a G_h with

$$\operatorname{Ker} G_h = \overline{\mathcal{W}}_h, \tag{6.3.4}$$

so that G_h will act only on the part of $\mathcal{W}_h$ that is *not* in $\overline{\mathcal{W}}_h$. In other cases, as we already did at the end of the previous section, we have to deal with several equations, and G_h will act differently on each of them. Moreover, in several cases, $\overline{\mathcal{W}}_h$ will contain all the *low frequencies* of $\mathcal{W}_h$, so that a smooth solution $X \in \mathcal{W}$ could be approximated fairly well by elements $\overline{X}_h \in \overline{W}_h$. Then, for every $X_I \in \mathcal{W}_h$ and for every $\overline{X}_h \in \overline{\mathcal{W}}_h$, we will have that the term $\mathcal{R}(X_I)$ in (6.2.21) can be estimated by

$$\begin{aligned}\frac{R(X_I, Y_h)}{\|Y_h\|} &= \frac{R(X_I - \overline{X}_h, Y_h)}{\|Y_h\|}\\ &\leq c_L \|X_I - \overline{X}_h\| \leq c_L(\|X_I - X\| + \|X - \overline{X}_h\|),\end{aligned}$$

so that, from (6.2.21), we have in this case

$$\frac{\tilde{\alpha}_\Phi}{\tilde{c}_\Phi}\|X_I - X_h\| \leq (\|\mathcal{A}\| + rc_L)\,\|X - X_I\| + r\,c_L\|X - \overline{X}_h\| \tag{6.3.5}$$

and the error estimate will depend on the approximation properties of both $\mathcal{W}_h$ and $\overline{\mathcal{W}}_h$, on the value of c_L and on the choice of r. We shall now provide a precise and sharper analysis of some of these situations.

We still suppose that the discrete problem defined by (6.2.12) is not stable. We may however suppose that a partial stability holds for some semi-norm $[Y_h]_h$ on $\mathcal{W}_h$.

Remark 6.3.1. In general, the "biggest semi-norm" one could consider is clearly

$$[X]_h := \sup_{Y_h \in \mathcal{W}_h} \frac{{}_{\mathcal{W}'}\langle \mathcal{A}X, Y_h\rangle_{\mathcal{W}}}{\|Y_h\|_{\mathcal{W}}}. \tag{6.3.6}$$

However, in many applications, simpler (and more explicit) norms can be preferred. □

The following assumption expresses in a precise way the fact that a certain semi-norm $[X_h]_h$ is "under control":

H.1 *For every h, there exists*

(i) A semi-norm $[\cdot]_h$ *on* $\mathcal{W}$,
(ii) An operator $\Phi_h \in \mathcal{L}(\mathcal{W}_h, \mathcal{W}_h)$,
(iii) A constant c_Φ *such that*

$$\|\Phi_h(Y_h)\|_{\mathcal{W}} \le c_\Phi \|Y_h\|_{\mathcal{W}} \quad \forall\, Y_h \in \mathcal{W}_h, \tag{6.3.7}$$

(iv) A constant $\alpha_\Phi > 0$ *such that*

$$\langle A Y_h, \Phi_h(Y_h)\rangle \ge \alpha_\Phi [Y_h]_h^2 \quad \forall\, Y_h \in \mathcal{W}_h. \tag{6.3.8}$$

□

Assumption **H.1** might seem cumbersome or difficult to realise in practice. This is not the case. Indeed, before proceeding, we point out that Assumption **H.1** is indeed verified in a number of applications. In particular, we consider the following (rather typical) situations.

Minimal stabilisation of mixed formulations. Assume that $\langle \mathcal{A}X, Y\rangle$ is defined as in (6.1.4), and recall the definitions (6.1.53) and (6.1.54).

For every fixed $X_h \equiv (u_h, p_h) \in \mathcal{W}_h$, we consider, in the spirit of Remark 6.3.1,

$$S(X_h) := \sup_{(v_h,q_h)\in V_h\times Q_h} \frac{{}_{\mathcal{W}'}\langle \mathcal{A}(u_h, p_h), (v_h, q_h)\rangle_{\mathcal{W}}}{\|(v_h, q_h)\|_{\mathcal{W}}}. \tag{6.3.9}$$

Assuming that (6.1.8) holds, we will always have

$$\begin{aligned} S(X_h) &\ge \frac{{}_{\mathcal{W}'}\langle \mathcal{A}(u_h, p_p), (u_h, p_h)\rangle_{\mathcal{W}}}{\|(u_h, p_h)\|_{\mathcal{W}}} \\ &\ge \frac{a(u_h, u_h)}{\|(u_h, p_h)\|_{\mathcal{W}}} =: \frac{|u_h|_a^2}{\|(u_h, p_h)\|_{\mathcal{W}}}. \end{aligned} \tag{6.3.10}$$

Similarly, we have (always without any assumptions on V_h or Q_h)

$$S(X_h) \ge \sup_{(v_h,0)\in K_h\times\{0\}} \frac{a(u_h, v_h)}{\|v_h\|_V} = \|\pi_{K_h'} A\, u_h\|_{V'}, \tag{6.3.11}$$

$$S(X_h) \ge \sup_{(0,q_h)\in \{0\}\times Q_h} \frac{b(u_h, q_h)}{\|q_h\|_Q} = \|B_h\, u_h\|_{Q'}, \tag{6.3.12}$$

$$S(X_h) \ge \sup_{(v_h,0)\in K_h^{\perp}\times\{0\}} \frac{a(u_h, v_h) + b(v_h, p_h)}{\|v_h\|_V} = \|\pi_{K_h^0} A\, v_h + B_h^t\, q_h\|_{V'}, \tag{6.3.13}$$

but, in general, we would not be able to get estimates of the type

$$S(X_h) \geq C\|A_h u_h\|_V \quad \text{or} \quad S(X_h) \geq \|B_h^t\|_{Q'} \tag{6.3.14}$$

separately. In most particular cases, however, one might have some good property and exploit it. We have seen in the previous chapters that sufficient conditions that ensure stability and error estimates are the *ellipticity in the kernel* K_h *(elker)* of the bilinear form $a(\cdot\,,\cdot)$ and the *discrete inf-sup condition for the bilinear form* $b(\cdot\,,\cdot)$. We also pointed out that the two conditions play, in a certain sense, one against the other: taking a bigger V_h helps in ensuring the *inf-sup condition* but increases the kernel and makes *elker* more at risk and the other way round. It is therefore not unreasonable to assume that we already took care of *one of the two conditions* (just by increasing or decreasing one of the two spaces), and ask the help of some stabilising trick in order to take care of the other. More precisely, we assume, to start with, that we have a continuous problem that is well posed.

A.1 *We suppose that* $\langle \mathcal{A}X, Y\rangle$ *is defined as in* (6.1.4) *and that*

(*i*) *The bilinear form* $a(\cdot,\cdot)$ *is* K*-elliptic, that is,*

$$\exists\, \alpha_0 > 0 \; s.t. \quad a(v,v) \geq \alpha\|v\|_V^2 \qquad \forall\, v \in K = \mathrm{Ker}B; \tag{6.3.15}$$

(*ii*) *The bilinear form* $b(v,q)$ *satisfies the inf-sup condition in* $V \times Q$ *where*

$$\beta := \inf_{v\in V}\sup_{q\in Q} \frac{b(v,q)}{\|v\|_V\,\|q\|_Q}. \tag{6.3.16}$$

□

In what follows, we will discuss the cases in which one of the two conditions is not satisfied for the discretised problem.

Example 6.3.1 (Minimal stabilisation of the inf-sup condition). For simplicity, we further assume that the ellipticity condition holds on the whole V, that is,

$$\exists\, \alpha > 0 \; s.t. \quad a(v,v) \geq \alpha\|v\|_V^2 \qquad \forall\, v \in V. \tag{6.3.17}$$

We know that the full ellipticity in V (6.3.17) implies automatically the full ellipticity in V_h. On the other hand, this is not true, in general, for the *inf-sup* condition. Hence, as we consider methods in need to be stabilised, we suppose that the discrete *inf-sup* condition *does not hold* with a constant independent of h. In order to see that, however, Assumption **H.1** is satisfied; we consider the following semi-norm:

$$[Y_h]_h^2 = [(v_h, q_h)]^2 := \|v_h\|_V^2 + [\![q_h]\!]_h^2 \tag{6.3.18}$$

where

$$[\![q_h]\!]_h := \sup_{v_h \in V_h} \frac{b(v_h, q_h)}{\|v_h\|_V} \equiv \|B_h^t q_h\|_{V'} \qquad \forall\, q_h \in Q_h. \tag{6.3.19}$$

The following proposition states that, in this framework, Assumption **H.1** holds.

Proposition 6.3.1. *Let $\mathcal{A}$ be of the form* (6.1.4) *and assume that* **A.1** *holds, together with the full ellipticity* (6.3.17). *Then,* **H.1** *also holds. In particular,* (6.3.7) *and* (6.3.8) *hold with*

$$\alpha_\Phi = \frac{\alpha}{2}\, min \left(1, \frac{1}{\|a\|^2}\right), \tag{6.3.20}$$

$$c_\Phi = 1 + \frac{\alpha}{\|a\|^2} \tag{6.3.21}$$

and with the semi-norm $[\cdot]_h$ *defined in* (6.3.18) *and* (6.3.19).

Proof. For a given $Y_h := (v_h, q_h)$, let $v_h^* \in V_h$ be such that

$$\frac{b(v_h^*, q_h)}{\|v_h^*\|_V} = \sup_{w_h \in V_h} \frac{b(w_h, q_h)}{\|w_h\|_V} =: [\![q_h]\!]_h \tag{6.3.22}$$

scaled in such a way that

$$\|v_h^*\|_V = [\![q_h]\!]_h. \tag{6.3.23}$$

We now choose

$$\Phi_h(Y_h) = (v_h + \delta v_h^*, q_h), \tag{6.3.24}$$

with δ a positive real number to be specified later on. We have from (6.1.4) and (6.3.24):

$$\begin{aligned}
\langle \mathcal{A} Y_h, \Phi_h(Y_h)\rangle &= a(v_h, v_h) + \delta a(v_h, v_h^*) \\
&\quad + b(v_h, q_h) + \delta b(v_h^*, q_h) - b(v_h, q_h) \\
&\geq \alpha \|v_h\|_V^2 - \delta \|a\| \|v_h\|_V \|v_h^*\|_V + \delta [\![q_h]\!]_h \|v_h^*\|_V \\
&= \alpha \|v_h\|_V^2 - \delta \|a\| \|v_h\|_V [\![q_h]\!]_h + \delta [\![q_h]\!]_h^2,
\end{aligned} \tag{6.3.25}$$

having used (6.3.17), (6.3.22), and, in the last step, (6.3.23). It is now clear that, choosing $\delta = \alpha / \|a\|^2$, (6.3.25) implies

$$\langle A Y_h, \Phi_h(Y_h)\rangle \geq \frac{\alpha}{2} \|v_h\|_V^2 + \frac{\delta}{2} [\![q_h]\!]_h^2 \tag{6.3.26}$$

having used $2ab \le a^2 + b^2$. Hence, we have (6.3.8) with the constant α_Φ given by (6.3.20). On the other hand, (6.3.23) and the choice of δ imply (6.3.7) and (6.3.21) since

$$\|v_h - \delta v_h^*\| \le \|v_h\| + \delta \|v_h^*\| = \|v_h\| + \delta [\![q_h]\!]_h. \qquad \square$$

Remark 6.3.2. Looking at the above proof, we can see that, actually, we proved, instead of (6.3.7), the stronger inequality

$$\|\Phi_h(Y_h)\| \le c_\Phi [Y_h]_h \quad \forall\, Y_h \in \mathcal{W}_h. \tag{6.3.27}$$

□

Example 6.3.2 (Minimal stabilisations of the ellipticity condition). Another possible case in which **H.1** is satisfied is the following one, in which we suppose, this time, that the discrete *inf-sup* condition *does hold* with a constant independent of h, but the ellipticity in the kernel does not. For instance, we might have that a is elliptic on the kernel K of B, but the kernel K_h of B_h is not a subset of K, and ellipticity does not hold for all $v_h \in K_h$.

In particular, we assume that **A1** holds, that the discrete *inf-sup* condition

$$\exists \beta^* > 0 \text{ such that } := \inf_{v \in V} \sup_{q \in Q} \frac{b(v,q)}{\|v\|_V \, \|q\|_Q} \ge \beta_* \tag{6.3.28}$$

holds with β^* independent of h, and that, moreover, as in (5.2.37) and (5.2.38), there exists a Hilbert space V^* with $V \hookrightarrow V^*$ such that

$$\exists\, \alpha^* > 0 \text{ such that } a(v,v) \ge \alpha^* \|v\|_{V^*}^2 \qquad \forall\, v \in V, \tag{6.3.29}$$

together with

$$\exists\, M_a^* \text{ such that } a(u,v) \le M_a^* \|u\|_{V^*} \|v\|_{V^*} \qquad \forall\, u,\, v \in V. \tag{6.3.30}$$

Then, we consider the following semi-norm:

$$[Y_h]_h^2 = [(v_h, q_h)]^2 := \|v_h\|_{V^*}^2 + \|q_h\|_Q^2. \tag{6.3.31}$$

The following proposition states that in this framework, Assumption **H.1** holds.

Proposition 6.3.2. *Let $\mathcal{A}$ be of the form* (6.1.4) *and assume that* **A.1** *holds, together with assumptions* (6.3.28)–(6.3.30)*. Then,* **H.1** *also holds. In particular,* (6.3.7) *and* (6.3.8) *hold with*

$$\alpha_\Phi = \frac{\alpha}{2} \min\left(1, \frac{\beta_*^2}{(M_a^*)^2}\right), \tag{6.3.32}$$

$$c_\Phi = 1 + \frac{\alpha^* \beta_*}{(M_a^*)^2} \tag{6.3.33}$$

and with the semi-norm $[\cdot]_h$ *defined in* (6.3.31).

Proof. For a given $Y_h := (v_h, q_h)$, we use (6.3.28) to choose $v_h^* \in V_h$ such that

$$\frac{b(v_h^*, q_h)}{\|v_h^*\|_V} = \sup_{w_h \in V_h} \frac{b(w_h, q_h)}{\|w_h\|_V} \geq \beta_* \|q_h\|_Q^2, \tag{6.3.34}$$

scaled in such a way that

$$\|v_h^*\|_V = \|q_h\|_Q. \tag{6.3.35}$$

We now choose

$$\Phi_h(Y_h) = (v_h + \delta v_h^*, q_h), \tag{6.3.36}$$

with δ a positive real number to be specified later on. We have from (6.1.4) and (6.3.36):

$$\begin{aligned}
\langle \mathcal{A}Y_h, \Phi_h(Y_h)\rangle &= a(v_h, v_h) + \delta a(v_h, v_h^*) \\
&\quad + b(v_h, q_h) + \delta b(v_h^*, q_h) - b(v_h, q_h) \\
&\geq \alpha^* \|v_h\|_{V^*}^2 - \delta M_a^* \|v_h\|_{V^*} \|v_h^*\|_{V^*} + \delta\beta_* \|q_h\|_Q \|v_h^*\|_V \\
&= \alpha^* \|v_h\|_V^2 - \delta M_a^* \|v_h\|_V \|q_h\|_Q + \delta\beta_* \|q_h\|_Q^2,
\end{aligned} \tag{6.3.37}$$

having used (6.3.15), (6.3.34), and, in the last step, (6.3.35). It is now clear that, choosing $\delta = \alpha^* \beta_* / (M_a^*)^2$, (6.3.25) implies

$$\langle \mathcal{A}Y_h, \Phi_h(Y_h)\rangle \geq \frac{\alpha^*}{2} \|v_h\|_V^2 + \frac{\delta\beta_*}{2} [\![q_h]\!]_h^2, \tag{6.3.38}$$

having used $2ab \leq a^2 + b^2$. Hence, we have (6.3.8) with the constant α_Φ given by (6.3.32). On the other hand, (6.3.35) and the choice of δ imply (6.3.7) with (6.3.33), since

$$\|v_h - \delta v_h^*\| \leq \|v_h\| + \delta \|v_h^*\| = \|v_h\| + \delta \|q_h\|. \qquad \square$$

Remark 6.3.3. Looking at the above proof, we can see that, together with (6.3.7), we could also prove the inequality

$$[\Phi_h(Y_h)]_h \le c_\Phi [Y_h]_h \quad \forall\, Y_h \in \mathcal{W}_h, \tag{6.3.39}$$

which in this case is stronger than (6.3.7). □

Remark 6.3.4. It is important to note that $\Phi_h(v_h, q_h)$, defined by (6.3.24) or (6.3.36), leaves the **second component** of (v_h, q_h) **unchanged**. This property can be useful in several circumstances. □

General results on minimal stabilisations. Having seen that Assumption **H.1** is indeed a reasonable one, we are now going to see how to use it in order to stabilise the problem. Roughly speaking, as we have already mentioned, we are going to add a bilinear form $L(X_h, Y_h)$ on $\mathcal{W}_h \times \mathcal{W}_h$, assuming that it could take care of "the remaining part of the $\mathcal{W}$ norm", that is "the part of the $\mathcal{W}$ norm which is not controlled by the semi-norm $[\cdot]_h$".

For technical reasons, we are going to make this assumption *in two steps*: we shall first assume in **H.2** that $L(Y_h, Y_h)$ controls a suitable intermediate term $\|G_h Y_h\|^2_{\mathcal{H}}$ (to be discussed later on), and then we shall assume, in **H.3**, that this intermediate term, together with the semi-norm $[\cdot]_h$, can control the whole $\mathcal{W}$ norm. Let us see this in a more precise way.

H.2 *There exist a Hilbert space $\mathcal{H}$, a bilinear form $L \in \mathcal{L}(\mathcal{H}, \mathcal{H})$, three positive constants c_G, c_L and α_G, and, for every h, an operator $G_h \in \mathcal{L}(\mathcal{W}_h, \mathcal{H})$, with$\|G_h\|_{\mathcal{L}(\mathcal{W}_h,\mathcal{H})} \le c_G$, such that*

$$L(G_h Z_h, G_h Y_h) \le c_L \|G_h Z_h\|_{\mathcal{H}} \|G_h Y_h\|_{\mathcal{H}} \qquad \forall\, Y_h,\ Z_h \in \mathcal{W}_h, \tag{6.3.40}$$

$$L(G_h Y_h, G_h Y_h) \ge \alpha_G \|G_h Y_h\|^2_{\mathcal{H}} \qquad \forall\, Y_h \in \mathcal{W}_h. \tag{6.3.41}$$

□

Remark 6.3.5. It is clear that hypothesis **H2** is tailored for using a stabilising term R of the form (6.3.3). □

We now consider, for some positive real number r, the stabilised operator $\tilde{\mathcal{A}}$ defined as

$$\langle \tilde{\mathcal{A}} X_h, Y_h \rangle := \langle \mathcal{A} X_h, Y_h \rangle + rL(G_h X_h, G_h Y_h) \qquad \forall\, X_h, Y_h \in \mathcal{W}_h \tag{6.3.42}$$

and the corresponding regularised problem

$$\langle \tilde{\mathcal{A}} X_h, Y_h \rangle = rL(N, G_h Y_h) + \langle F, Y_h \rangle \qquad \forall\, Y_h \in \mathcal{W}_h. \tag{6.3.43}$$

We have the following result.

Lemma 6.3.1. *Assume that* **H.1** *and* **H.2** *hold, and assume moreover that, for the mapping Φ_h considered in* **H.1** *and the map G_h considered in* **H.2**, *we have*

$$\|G_h(\Phi_h(Y_h))\|_{\mathcal{H}} \le c_{G\Phi} \|G_h(Y_h)\|_{\mathcal{H}} \quad \forall Y_h \in \mathcal{W}_h \tag{6.3.44}$$

for some constant $C_{G\Phi}$ *independent of h. Then, there exist a linear mapping* $\Phi_h^* \in \mathcal{L}(\mathcal{W}_h, \mathcal{W}_h)$ *and two constants* α_Φ^* *and* c_Φ^**, depending only on* α_Φ*,* c_Φ α_R*, and* c_R *such that, for every h, for every r and for every* $Y_h \in \mathcal{W}_h$*, we have*

$$\|\Phi_h^*(Y_h)\|_{\mathcal{W}} \leq c_\Phi^* \|Y_h\|_{\mathcal{W}}, \tag{6.3.45}$$

$$\|G_h(\Phi_h^*(Y_h))\|_{\mathcal{H}} \leq c_\Phi^* \|G_h(Y_h)\|_{\mathcal{H}}, \tag{6.3.46}$$

and

$$\langle \tilde{A} Y_h, \Phi_h^*(Y_h)\rangle \geq \alpha_\Phi^* \Big([Y_h]_h^2 + r\|G_h(Y_h)\|_{\mathcal{H}}^2\Big), \tag{6.3.47}$$

where $\tilde{A}$ *is given in* (6.3.42). □

Proof. We set

$$\Phi_h^*(Y_h) := Y_h + \delta \Phi_h(Y_h) \tag{6.3.48}$$

with δ to be chosen later on. Then, using first (6.3.42) and (6.3.48), then (6.3.8), (6.3.41), and using (6.3.44) to bound the last term, we have:

$$\begin{aligned} &\langle \tilde{A} Y_h, \Phi_h^*(Y_h)\rangle \\ &\geq \delta \alpha_\Phi [Y_h]_h^2 + r\alpha_G \|G_h(Y_h)\|_{\mathcal{H}}^2 - r c_L \|G_h(Y_h)\|_{\mathcal{H}} \delta c_{G\Phi} \|G_h(Y_h)\|_{\mathcal{H}}, \end{aligned} \tag{6.3.49}$$

and the result follows easily for $r\delta$ smaller than $4\alpha_G \alpha_\Phi/(c_L\, c_{G\Phi})^2$. □

Remark 6.3.6. It is easy to check that (6.3.44) holds easily whenever

$$L(X_h, \Phi_h(Y_h)) = L(X_h, Y_h), \tag{6.3.50}$$

implying that $L(X_h, Y_h)$ depends only on the part of Y_h which is left unchanged by Φ_h. □

We finally need a further assumption that connects the right-hand side of (6.3.47) with the norm in $\mathcal{W}_h$.

H.3 *With the notation of assumptions* **H.1** *and* **H.2**, *we further assume that there exist two positive constants* γ_2 *and* γ_3 *such that*

$$[Y_h]_h^2 + \gamma_2 \|G_h Y_h\|_{\mathcal{H}}^2 \geq \gamma_3 \|Y_h\|_{\mathcal{W}}^2 \qquad \forall\, Y_h \in \mathcal{W}_h. \tag{6.3.51}$$

□

It is clear that, if Assumption **H.3** is also verified, then (6.3.47) will give a stability result of type (6.2.13), where the explicit value of the constant α_Φ can be easily deduced from the values of the other constants. On the other hand, the estimate (6.3.47) will also be used in the sequel in cases when some constant

(r mostly, and sometimes γ_2) might depend on h, so that it is convenient to leave it in its present form.

We now consider the problem of error estimates. As we introduced sufficient conditions to ensure stability, the question will be to check consistency, and in particular the effect on consistency of the extra stabilising terms. As we said, in many applications, the constants r, γ_2, and γ_3 appearing in (6.3.51) might be allowed to depend on h. Hence, it is important that we keep track of them in our abstract estimates. Mimicking (6.2.20), we now set

$$\mathcal{R}(X_I) := \sup_{Y_h \in \mathcal{W}_h} \frac{L(G_h X_I - N, G_h Y_h)}{\|G_h Y_h\|_{\mathcal{H}}}. \tag{6.3.52}$$

Theorem 6.3.1. *Let X and X_h be the solutions of* (6.2.16) *and (6.3.43) respectively. Assume that* **H.1**, **H.2**, *and* **H.3** *hold. Then, for every $X_I \in \mathcal{W}_h$, we have*

$$[X_I - X_h]_h^2 + r\|G_h(X_I - X_h)\|_{\mathcal{H}}^2 \\ \leq C\left(\frac{r+\gamma_2}{r\gamma_3}\|\mathcal{A}\|^2\|X_I - X\|^2 + r(\mathcal{R}(X_I))^2\right), \tag{6.3.53}$$

where the constant C depends on α_G, α_Φ, c_L and c_Φ, but does not depend on the other parameters.

Proof. We set $\delta X := X_I - X_h$ and $Y_h := \Phi_h^*(\delta X)$, with Φ_h^* given in Lemma 6.3.1. Using Lemma 6.3.1, then the continuous equation (6.2.16) and the stabilised discrete one (6.3.43), and then (6.3.40), (6.3.45), and (6.3.46), we get

$$\begin{aligned} \alpha_\Phi[\delta X]_h^2 + r\alpha_G\|G_h\delta X\|^2 &\leq \langle \mathcal{A}(\delta X), \Phi_h^*(\delta X)\rangle + rL(G_h(\delta X), G_h(\Phi_h^*(\delta X))) \\ &= \langle \mathcal{A}(X_I - X), \Phi_h^*(\delta X)\rangle + rL(G_h X_I - N, G_h(\Phi_h^*(\delta X))) \\ &\leq c_\Phi^*\|\mathcal{A}\|\,\|X_I - X\|\,\|\delta X\| + r\mathcal{R}(X_I)\|G_h\delta X\|. \end{aligned} \tag{6.3.54}$$

We now use **H.3** to bound $\|\delta X\|$:

$$\|\delta X\| \leq \left(\frac{[\delta X]_h^2 + \gamma_2\|G_h\delta X\|^2}{\gamma_3}\right)^{1/2} \leq \frac{[\delta X]_h + \gamma_2^{1/2}\|G_h\delta X\|}{\gamma_3^{1/2}}. \tag{6.3.55}$$

At this point, we need, just for a while, a lighter notation. We denote one of the two terms on the right-hand side of (6.3.53) by $D_1 := \|\mathcal{A}\|\,\|X_I - X\|$ and the other by $D_2 := \|G_h(X_I)\|_{\mathcal{H}}$. We also denote the second term in the left-hand side by $g := \|G_h(\delta X)\|_{\mathcal{H}}$. With this notation, the inequality that we have to prove becomes

$$[\delta X]_h^2 + rg^2 \le C\Big(\frac{r+\gamma_2}{r\gamma_3} D_1^2 + rD_2^2\Big) \tag{6.3.56}$$

and what we have got, inserting (6.3.55) into (6.3.54) and using the new notation, can be written as

$$[\delta X]_h^2 + rg^2 \le C\Big(\frac{D_1}{\gamma_3{}^{1/2}}[\delta X]_h + D_1\Big(\frac{\gamma_2}{r\gamma_3}\Big)^{1/2} r^{1/2}g + r^{1/2}D_2 r^{1/2}g\Big). \tag{6.3.57}$$

Then, we apply the inequality $ab \le \frac{c}{2}a^2 + \frac{1}{2c}b^2$ with suitable choices of c, move three terms to the left and multiply the resulting equation by a suitable constant to get (6.3.53). □

As an immediate consequence of Theorem 6.3.1, we have the following error estimate.

Theorem 6.3.2. *Let X and X_h be the solutions of* (6.2.16) *and (6.3.43) respectively. Assume that* **H.1**, **H.2** *and* **H.3** *hold, and assume that the operator G_h could be extended to a space $\mathcal{W}(h) \subseteq \mathcal{W}$ containing both $\mathcal{W}_h$ and X. Then, there exists a constant $C = C(\alpha_G, \alpha_\Phi, c_L, c_\Phi)$ such that*

$$\begin{aligned}[X - X_h]_h^2 &+ r\|G_h(X - X_h)\|_{\mathcal{H}}^2 \\ &\le C \inf_{X_I \in \mathcal{W}_h} \Big(\frac{r+\gamma_2}{r\gamma_3}\|\mathcal{A}\|^2\|X - X_I\|^2 + r(\mathcal{R}(X_I))^2\Big).\end{aligned} \tag{6.3.58}$$

Moreover, we have the following important corollary.

Corollary 6.3.1. *Keep the same assumptions as in Theorem 6.3.2, and assume moreover that for every h we have a space $\mathcal{W}(h)$ containing both $\mathcal{W}_h$ and the exact solution X of* (6.2.16) *such that G_h could be extended to an operator in $\mathcal{L}(\mathcal{W}(h), \mathcal{H})$, with norm c_G uniformly bounded in h, and such that* (6.3.40) *and* (6.3.41) *still hold for Y_h and Z_h in $\mathcal{W}(h)$. Finally, assume that, for the exact solution X of* (6.2.16)*, we have*

$$L(G_hX - N, G_hY_h) = 0 \quad \forall Y_h \in \mathcal{W}_h, \tag{6.3.59}$$

so that, with the notation of (6.3.52)*, we have $\mathcal{R}(X) = 0$. Then, there exists a constant $C = C(\alpha_G, \alpha_\Phi, c_L, c_\Phi)$ such that*

$$\begin{aligned}[X - X_h]_h^2 &+ r\|G_h(X - X_h)\|_{\mathcal{H}}^2 \\ &\le C \inf_{X_I \in \mathcal{W}_h} \Big(\frac{r+\gamma_2}{r\gamma_3}\|\mathcal{A}\|^2\|X - X_I\|^2 + r\|G_h(X - X_I)\|_{\mathcal{H}}^2\Big).\end{aligned} \tag{6.3.60}$$

Remark 6.3.7. It is not difficult to see that the approach described here, and in particular the result of Corollary 6.3.1, has much in common with the ones described and analysed in the previous section. Actually, many stabilising techniques available on the market can be set equally well in the framework of the previous section as in that of the present one. Still, in certain cases, one of the two would be easier to use, and in other cases, only one of these two approachcs will be usable. □

As we mentioned already, in different applications, we might allow some of the constants (and mainly r and γ_2) to depend on h. Hereafter, we shall rapidly see some typical cases, taking as simplest example the one-dimensional version of mixed formulations for the Poisson problem, already seen in (6.1.82) and (6.1.83). Other examples will be seen in the following chapters, for different applications.

The first, and main difference, is whether the constant γ_2 can be assumed to tend to zero (and fast enough) when h tends to zero. We first consider the case in which it is more convenient to take a γ_2 that does not depend on h. In a certain number of cases, Theorem 6.3.2 or Corollary 6.3.1 can be applied with all the constants (r, γ_2, and γ_3) independent of h.

Example 6.3.3 (Both r and γ_2 are independent of h). It is clear that Corollary 6.3.1 is the natural candidate to be applied in these cases. Most augmented formulations and their variants can be analysed in this way. Just to see an example, consider the one-dimensional Poisson problem (6.1.82), and assume that we take $V_h := \mathcal{L}_1^2$ and $Q_h := \mathcal{L}_0^0$. We have already seen in the previous Chapter (in Sect. 5.2.4) that this choice leads to a total disaster, due to the failure of the *elker* condition. However, adding a term

$$R(X_h, Y_h) \equiv R((u_h, p_h), (v_h, q_h)) = (u_h' + f, v_h') \tag{6.3.61}$$

will restore the full ellipticity and give a good solution. On the other hand, the bound (6.3.12) tells us, in this case, that the projection of $Bu_h \equiv u_h'$ onto Q_h is already under control, and a further analysis would show that indeed the term in (6.3.61) could be multiplied by h^2 and still provide a sufficient stabilisation (see [125]). □

Example 6.3.4 (Taking γ_2 fixed and r depending on h). In this case, we are allowed to use directly Theorem 6.3.1. The bound (6.3.53) will provide (for r "small") an estimate of the type

$$[\delta X]_h^2 + r\|G_h(\delta X)\|_{\mathcal{H}}^2 \le C\left(\frac{1}{r}\|X_I - X\|^2 + r\|G_h(X_I)\|_{\mathcal{H}}^2\right), \tag{6.3.62}$$

which, when (6.3.4) holds, can become

$$\begin{aligned}[\delta X]_h^2 + r\|G_h\delta X\|^2 &\le C\left(\frac{1}{r}\|X_I - X\|^2 + r\|\overline{X}_h - X\|^2\right)\\ &\le C\left(\frac{1}{r}h^{s_1} + rh^{s_2}\right),\end{aligned} \tag{6.3.63}$$

by usual interpolation estimates with, in general, $s_1 \geq s_2 \geq 0$. Then, by taking $r = h^s$, we get

$$[\delta X]_h^2 + h^s \|G_h \delta X\|^2 \leq C \left(h^{s_1 - s} + h^{s_2 + s}\right), \tag{6.3.64}$$

with the optimal choice given by $s = (s_1 - s_2)/2$. We further develop such a case in the following example. □

Example 6.3.5 (Penalty methods for the inf-sup condition). Coming back to the case of Proposition 6.3.1, a large class of stabilisations to cure methods where the *inf-sup* condition fails can be built taking a subspace $\tilde{Q}_h$ of Q_h, denoting by $\tilde{P}$ the projection operator on $\tilde{Q}_h$, and setting

$$\begin{cases} G_h((v_h, q_h)) := \tilde{P}(q_h), \\ R((u_h, p_h), (v_h, q_h)) := (\tilde{P} p_h, \tilde{P} q_h)_Q. \end{cases} \tag{6.3.65}$$

The subspace $\tilde{Q}_h$ will be chosen so that **H.3** holds. This means that, using the notation (6.3.19), we should have

$$[\![q_h]\!]_h^2 + \gamma_2 \|P_{\tilde{Q}_h} q_h\|^2 \geq \gamma_3 \|q_h\|_Q^2, \tag{6.3.66}$$

for some positive constants γ_2 and γ_3. The stabilised problem then becomes

$$\begin{cases} a(u_h, v_h) + b(v_h, p_h) = (f, v_h) & \forall\, v_h \in V_h, \\ b(u_h, q_h) - r(\tilde{P} p_h, \tilde{P} q_h) = (g, q_h) & \forall\, q_h \in Q_h. \end{cases} \tag{6.3.67}$$

In this case, the estimate (6.3.53) would yield

$$\begin{aligned} \|u_I - u_h\|_V^2 &+ [\![p_I - p_h]\!]_h^2 + r\|\tilde{P}(p_I - p_h)\|_Q^2 \\ &\leq C \frac{r + \gamma_2}{r} (\|A\|^2 \|u_I - u\|_V^2 + \|p_I - p\|_Q^2) \\ &\qquad + r\|\tilde{P}(p_I - p)\|_Q^2 + r\|\tilde{P}(p)\|_Q^2. \end{aligned} \tag{6.3.68}$$

We now consider three cases:

(i) **Stable penalty**. The *inf-sup* condition is satisfied. In this case, we obviously have $[\![q_h]\!]_h \simeq \|q_h\|_Q$. We can then take $\gamma_2 = 0$ and $\tilde{Q}_h = Q_h$, which means that $\tilde{P} = I$. We can take r as small as we want and there is an $O(r)$ term in the right-hand side. This means that the penalty is used as a computational trick to obtain an otherwise good solution.

(ii) **Brute force penalty**. We have no (usable) *inf-sup* condition (roughly speaking: $[\![q_h]\!]_h = 0$ for any q_h). We take again $\tilde{Q}_h = Q_h$ and $\tilde{P} = I$ but we now need $\gamma_2 = O(1)$. The error estimate becomes,

$$\|u_I - u_h\|_V^2 + r\|p_I - p_h\|_Q^2$$
$$\leq C\frac{1}{r}(\|A\|^2\|u_I - u\|_V^2 + \|p_I - p\|_Q^2) + r\|p\|_Q^2. \quad (6.3.69)$$

In that case, we only get a bound on $\|u_I - u_h\|$. Suppose, to fix ideas, that we would expect an $O(h^k)$ from the usual error estimates. It is clear that the best that we can do is to take $r = O(h^k)$ and get a final bound in $O(h^{k/2})$ instead of $O(h^k)$.

(iii) **Clever-penalty**. Let us suppose that there is a subspace $\overline{Q}_h \subset Q_h$ such that $V_h \times \overline{Q}_h$ satisfies the *inf-sup* condition and let $\overline{P}$ be the projection onto $\overline{Q}_h$. We set $P = (I - \overline{P})$ and we now need $\gamma_2 = O(1)$. The bound would now be

$$\|u_I - u_h\|_V^2 + \|\bar{p}_I - \overline{p}_h\|^2 + r\|(I - \overline{P})(p_I - p_h)\|_Q^2$$
$$\leq C\frac{1}{r}(\|A\|^2\|u_I - u\|_V^2 + \|p_I - p\|_Q^2) + r\|(I - \overline{P})p\|_Q^2. \quad (6.3.70)$$

Two possibilities arise, depending on the approximation properties in $\overline{Q}_h$. If all the terms $\|u_I - u\|_V$, $\|p_I - p\|_Q$, and $\|(I - \overline{P})p\|_Q$ in the right-hand side have a similar order in h, then any positive r will provide an estimate of the best possible order. Such stabilising methods have been considered in [349]. On the other hand, suppose that the last term is of lower order than the other ones. One could use a small value of r to get a better accuracy but to the expense of loosing on the first terms. If, for instance, we expect

$$(\|u_I - u\|_V^2 + \|p_I - p\|_Q^2) = O(h^4) \quad \text{and} \quad \|(I - \overline{P})(p)\|_Q^2 = O(h^2),$$

then the choice $r = O(h)$ will yield an estimate of $O(h^{3/2})$ on $\|u - u_h\|_V$ and of $O(h)$ on $\|p - p_h\|_Q$. Such a procedure was introduced by Lovadina and Auricchio [284] for the Stokes problem. □

Example 6.3.6 (Using a γ_2 that depends on h). We now consider the cases in which it is possible, and convenient, to use a γ_2 that tends to zero with h. At first sight, one might think that this *never* (or almost never) occurs. However, this is not true. Assume, for instance, that

$$[q_h]_h \geq \|\overline{q}_h\|_{L^2}, \quad (6.3.71)$$

where $\overline{q}_h$ is the L^2 projection of q_h onto the space of piecewise constant functions. It is elementary, by the Poincaré inequality, to see that, for a q_h piecewise in H^1, we have

$$\|q_h\|_{L^2}^2 \leq \|\overline{q}_h\|_{L^2}^2 + C\,h^2\|\,\underline{\mathrm{grad}}_h\, q_h\|_{L^2}^2, \quad (6.3.72)$$

where, as before, $\underline{\text{grad}}_h$ is the piecewise gradient and C is a constant that depends only on the minimum angle of the decomposition (and where, for simplicity, we assumed a quasi-uniform decomposition). Hence, in this case, we would have

$$[q_h]_h^2 + h^2 \| \underline{\text{grad}}_h \, q_h \|_{L^2}^2 \geq \gamma_3 \| q_h \|_{L^2}^2 \tag{6.3.73}$$

with a constant γ_3 independent of h: a formula of the type of (6.3.51) with $\gamma_2 = h^2$. Looking at (6.3.53), it seems natural, when γ_2 is small, to take r as small as γ_2 (that is, going to zero with the same order). It is what we consider in the next example. □

Example 6.3.7 (Both r and γ_2 depend on h). We consider the case in which both r and γ_2 depend on h, and go to zero. For simplicity, we assume that we take, brutally, $r \geq \gamma_2$, but everything will work just taking, say, $r \geq \kappa\gamma_2$ with a constant κ independent of h. Then, we have

$$[\delta X]_h^2 + r \| G_h(\delta X) \|_{\mathcal{H}}^2 \geq [\delta X]_h^2 + \gamma_2 \| G_h(\delta X) \|_{\mathcal{H}}^2 \geq \gamma_3 \| \delta X \|^2 \tag{6.3.74}$$

so that applying (6.3.53) to the left-hand side of (6.3.74) gives

$$\gamma_3 \, \| \delta X \|^2 \leq C \left(\frac{2}{\gamma_3} \| \mathcal{A} \|^2 \| X_I - X \|^2 + r \| G_h(X_I) \|_{\mathcal{H}}^2 \right). \tag{6.3.75}$$

For instance, dealing with the one-dimensional version of (6.1.82) and starting from $V_h := \mathcal{L}_1^1$ and $Q_h := \mathcal{L}_2^1$, it is easy to see that $[(v_h, q_h)]_h^2 \geq \| v_h \|_1^2 + \| \overline{q}_h \|_0^2$, where again $\overline{q}_h$ is the projection of q_h on piecewise constants. In view of (6.3.72), we can then take

$$R((u, p), (v, q)) = (p', q') \tag{6.3.76}$$

and $r = \gamma_2 = h^2$. This will give linear convergence for both u and p. □

6.3.1 Another Form of Minimal Stabilisation

We now develop a more sophisticated variant of the previous case (where γ_2 depends on h) that is based, instead of (6.3.71), on a (possible) estimate of the type

$$[q_h]_h \geq \| q_h \|_{L^2} - C \, h^2 \| \underline{\text{grad}} \, q_h \|_{L^2}^2. \tag{6.3.77}$$

Estimates of this type are met in situations like the one analysed in Sect. 5.4.5 (see, in particular Eq. (5.4.22)) and related to the technique known as *Verfürth's trick* [375] that we discussed in the previous chapter. We still suppose that Assumption **A.1** holds and we complete it by the following assumption.

A.2 There exists a Hilbert space H with $V \hookrightarrow H \equiv H' \hookrightarrow V'$ such that

$$B^t(Q_h) \subset H \tag{6.3.78}$$

(where $B^t : Q \longrightarrow V'$ is, as usual, the linear operator associated with the bilinear form $b(v,q)$), and there exist a monotone function $\omega : \mathbb{R}^+ \to \mathbb{R}^+$ and a positive constant σ, independent of h, such that

$$\omega(h)\|v_h\|_V \le \|v_h\|_H \qquad \forall\, v_h \in V_h, \tag{6.3.79}$$

$$\omega(h)\|B^t q_h\|_H \le \|q_h\|_Q \quad \forall q_h \in Q_h \tag{6.3.80}$$

and

$$\|v - \pi_{V_h} v\|_H \le \sigma\omega(h)\|v\|_V \quad \forall\, v \in V. \tag{6.3.81}$$

□

We note that, setting

$$[\![q_h]\!]_h := \sup_{v_h\in V_h} \frac{b(v_h,q_h)}{\|v_h\|_V} = \sup_{v_h\in V_h} \frac{(v_h, B^t q_h)_H}{\|v_h\|_V}, \tag{6.3.82}$$

from (6.3.79) we have

$$[\![q_h]\!]_h = \sup_{v_h\in V_h} \frac{(v_h, B^t q_h)_H}{\|v_h\|_H} \frac{\|v_h\|_H}{\|v_h\|_V} \ge \omega(h)\, \|\pi_{V_h'} B^t q_h\|_H \equiv \omega(h)\|B_h^t q_h\|_H . \tag{6.3.83}$$

In agreement with the general procedure of this section, we can now take $\mathcal{H} = H$ with

$$G_h((v_h,q_h)) = B^t q_h - B_h^t q_h = (I - \pi_{V_h'})B^t q_h \tag{6.3.84}$$

and define:

$$R((u_h,p_h),(v_h,q_h)) = \left(B^t p_h - B_h^t p_h, B^t q_h - B_h^t q_h\right)_H . \tag{6.3.85}$$

It is clear that both (6.3.40) and (6.3.41) will hold with constants independent of h, so that **H.2** holds. We are left with **H.3** which will be proved in the next two propositions using essentially the so-called Verfürth's trick [375] that we already discussed in Sect. 5.4.5.

Lemma 6.3.2. *Assume that* **A.1** *and* **A.2** *hold. Then,*

$$[\![q_h]\!]_h := \sup_{v_h\in V_h} \frac{b(v_h,q_h)}{\|v_h\|_V} \ge \beta\|q\|_Q - \sigma\omega(h)\|B^t q_h\|_H \quad \forall\, q_h \in Q_h, \tag{6.3.86}$$

where β is the inf-sup constant appearing in (6.3.16), *$\omega(h)$ is given in* (6.3.79)–(6.3.81), *and σ is given in* (6.3.81).

Proof. The proof is essentially the same as the one that was used to prove (5.4.22) in the last chapter. Let us see it briefly. We start from the *inf-sup* condition (6.3.16), we add and subtract the projection $\pi_{V_h} v$ of v over V_h and then use (6.3.82) and (6.3.81):

$$
\begin{aligned}
\beta\|q_h\|_Q &\le \sup_{v\in V}\frac{b(v,q_h)}{\|v\|_V} = \sup_{v\in V}\left(\frac{b(\pi_{V_h}v,q_h)}{\|v\|_V} + \frac{b(v-\pi_{V_h},q_h)}{\|v\|_V}\right)\\
&\le \sup_{v\in V}\frac{b(\pi_{V_h}v,q_h)}{\|\pi_{V_h}v\|_V} + \sup_{v\in V}\frac{(v-\pi_{V_h}v,B^tq_h)_H}{\|v\|_V}\\
&\le \sup_{v_h\in V_h}\frac{b(v_h,q_h)}{\|v_h\|_V} + \sup_{v\in V}\frac{\|v-\pi_{V_h}v\|_H\|B^t(q_h)\|_H}{\|v\|_V}\\
&\le [\![q_h]\!]_h + \sigma\omega(h)\|B^tq_h\|_H \quad \forall\, q_h\in Q_h.
\end{aligned}
\tag{6.3.87}
$$

□

We can now easily get the following result.

Lemma 6.3.3. *Under Assumptions* **A.1** *and* **A.2**, *there exists a constant $\tilde\beta$, independent of h, such that*

$$
[\![q_h]\!]_h^2 + \omega^2(h)\|B^tq_h - \pi_{V_h}B^tq_h\|_H^2 \ge \tilde\beta\|q_h\|_Q^2 \quad \forall\, q_h\in Q_h. \tag{6.3.88}
$$

Proof. Indeed, by the triangle inequality, we have, for every $q_h \in Q_h$:

$$
\|B^tq_h - \pi_{V_h'}B^tq_h\|_H + \|\pi_{V_h'}B^tq_h\|_H \ge \|B^tq_h\|_H. \tag{6.3.89}
$$

On the other hand, summing (6.3.86) plus σ times (6.3.83), we have

$$
(1+\sigma)[\![q_h]\!]_h \ge \beta\|q_h\|_Q + \sigma\omega(h)\Big(\|\pi_{V_h'}B^tq_h\|_H - \|B^tq_h\|_H\Big) \tag{6.3.90}
$$

so that

$$
(1+\sigma)[\![q_h]\!]_h + \sigma\omega(h)\|B^tq_h - \pi_{V_h'}B^tq_h\|_H \ge \beta\|q_h\|_Q \tag{6.3.91}
$$

and the result follows easily. □

Remark 6.3.8. Actually, by Pythagora's theorem, we obviously have, for every $q_h \in Q_h$:

$$
\|B^tq_h\|_H^2 = \|B^tq_h - \pi_{V_h'}B^tq_h\|_H^2 + \|\pi_{V_h'}B^tq_h\|_H^2. \tag{6.3.92}
$$

However, as we have seen, the triangle inequality (6.3.89) is enough for our proof. □

Lemma 6.3.3 implies that **H.3** holds, with the above choices for $[\cdot]_h$ and G_h, with a constant γ_3 independent of h, and with $\gamma_2 = \omega^2(h)$.

Remark 6.3.9. In certain cases, it would be more convenient to introduce another finite element space $\tilde{V}_h \subseteq H$, and use a stabilising term like

$$R((u_h, p_h), (v_h, q_h)) = \left(B^t p_h - \pi_{\tilde{V}_h'} B^t p_h, B^t q_h - \pi_{\tilde{V}_h'} B^t q_h\right)_H \qquad (6.3.93)$$

instead of (6.3.85). Then, Lemma 6.3.3 will still hold (and consequently **H.3** will also hold), provided that we have some additional result that guarantees, in the particular case under study, that

$$\|B^t q_h\|_H \geq C(\|B^t q_h - \pi_{\tilde{V}_h'} B^t q_h\|_H + \|\pi_{V_h'} B^t q_h\|_H) \quad \forall q_h \in Q_h \qquad (6.3.94)$$

for some constant C independent of h. □

In view of the above remark, it might be convenient to treat the two cases (that is: using (6.3.85) or (6.3.93) when (6.3.94) also holds) together. For this, we introduce the following assumption.

A.3 *With the notation of Assumption* **A.2**, *we consider a space* $\tilde{V}_h \subseteq H$ *and we assume that there exists a positive constant* $\tilde{\kappa}$, *independent of* h, *such that*

$$\|\pi_{V_h} B^t q_h\|_H + \|B^t q_h - \pi_{\tilde{V}_h} B^t q_h\|_H \geq \tilde{\kappa}\, \|B^t q_h\|_H \quad \forall\, q_h \in Q_h. \qquad (6.3.95)$$

□

Assumption **A.3** obviously holds, for instance, if $\tilde{V}_h = \{0\}$, or more generally whenever $\tilde{V}_h \subseteq V_h$. The case of a $\tilde{V}_h$ larger than V_h, instead, will work only in some special case, and will require an *ad hoc* (and sometimes delicate) proof. We can collect the result of Lemma 6.3.3 and the above discussion in the following theorem.

Theorem 6.3.3. *Assume that Assumptions* **A.1**, **A.2**, *and* **A.3** *hold. Assume moreover that the full ellipticity condition* (6.3.17) *holds. Assume that we are given subspaces* $V_h \subset V$ *and* $Q_h \subset Q$, *and we take* $\mathcal{W}_h := V_h \times Q_h$ *with* (6.3.18) *and* (6.3.19). *Set*

$$G_h((v_h, q_h)) := B^t q_h - \pi_{\tilde{V}_h} B^t q_h. \qquad (6.3.96)$$

Then, **H.3** *holds with a constant* γ_3 *independent of* h, *and with* $\gamma_2 = \omega^2(h)$.

We are therefore in a situation very similar to that of Example 6.3.6. A very reasonable choice would then be to use an r that also behaves as $\omega(h)^2$ as in Example 6.3.7. Then using Theorem 6.3.1 as in (6.3.75), we have the following theorem.

Theorem 6.3.4. *Assume that* **A.1**, **A.2**, *and* **A.3** *hold, and let* (u, p) *be the solution of Problem* (6.1.5). *Assume that, in (6.2.17), R is defined through (6.3.85), and that r is a positive number $\geq \omega(h)^2$. Then, Problem (6.2.17) has a unique solution*

(u_h, p_h) and there exists a constant C, independent of h and r, such that, for every $(u_I, p_I) \in V_h \times Q_h$, we have:

$$
\begin{aligned}
&\|u_I - u_h\|_V^2 + \|p_I - p_h\|_Q^2 \\
&\qquad \leq C \left(\|u - u_I\|_V^2 + \|p - p_I\|_Q^2 + r\|G_h(0, p_I)\|^2 \right).
\end{aligned} \tag{6.3.97}
$$

Proof. The proof is an immediate consequence of (6.3.53) as in (6.3.74) and (6.3.75). □

Remark 6.3.10. The convenience in using $r > \omega^2(h)$ (including the case of an r fixed that does not depend on h) can occur when the term $\|G_h(0, p_I)\|_{\mathcal{H}}$ in (6.3.97) is already small. Indeed, with a choice of the type (6.3.93), we would have

$$
\begin{aligned}
\|G_h(0, p_I)\|_{\mathcal{H}} &= \|B^t p_I - \pi_{\tilde{V}_h} B^t p_I\|_{\mathcal{H}} \\
&\leq \|B^t(p_I - p) - \pi_{\tilde{V}_h} B^t(p_I - p)\|_{\mathcal{H}} + \|B^t p - \pi_{\tilde{V}_h} B^t p\| \\
&\leq C\|B^t(p - p_I)\|_{\mathcal{H}} + \|(I - \pi_{Vth})B^t p\|_{\mathcal{H}}
\end{aligned} \tag{6.3.98}
$$

that could be small whenever $B^t p$ is smooth and $\tilde{V}_h$ has good approximation properties. □

Remark 6.3.11. It is clear that the condition $r \geq \omega^2(h)$ could be replaced with $r \geq \kappa\omega^2(h)$ for some $\kappa > 0$ independent of h. This, indeed, will make the notation in the proof heavier, but the final result will end up in a different value of the constant C in (6.3.97). □

Remark 6.3.12. As usual, we can then take u_I and p_I as the best approximations of u and p, respectively (in the respective norms), and then deduce an estimate for $\|u - u_h\|_V + \|p - p_I\|_Q$ by the triangle inequality. □

We now consider the case of an r *smaller* that $\omega^2(h)$.

Theorem 6.3.5. *In the same assumptions as in Theorem 6.3.4, taking $r \leq \omega^2(h)$, we have:*

$$
\begin{aligned}
&\|u_I - u_h\|_V^2 + [\![p_I - p_h]\!]^2 + r\|p_I - p_h\|_Q^2 \\
&\qquad \leq C\left(\frac{2\omega^2(h)}{r}(\|u - u_I\|_V^2 + \|p - p_I\|_Q^2) + r\|G_h(0, p_I)\|^2\right).
\end{aligned} \tag{6.3.99}
$$

Proof. The proof is again an easy consequence of (6.3.53) in Theorem 6.3.1.

Remark 6.3.13. In applications, the choice of the form of $r(h)$ will be done in order to get the best possible estimate. In particular, the choice $r = \kappa\omega(h)^2$ will be the best choice when $\tilde{V}_h = 0$ and first order approximations are employed.

This situation will be met for instance in the Brezzi-Pitkäranta stabilisation for the Stokes problem considered in Chap. 8, that however could be treated by the simpler estimates of Example 6.3.7. □

6.4 Enhanced Strain Methods

The so-called "enhanced strain" methods have become popular as a stabilising device. We shall try to give a feeling of how they work and the way they can be applied to mixed methods. We shall however first consider a more classical setting. Let us thus consider two Hilbert spaces V and Q. To simplify the notation, we assume, from the very beginning, that Q is identified with its dual space, that is $Q \equiv Q'$. We then assume that we have a continuous operator B from V to $Q = Q'$ and a continuous isomorphism $\mathfrak{C}$ from Q onto Q. We want to solve a variational problem of the form,

$$\inf_{v \in V} \frac{1}{2}(\mathfrak{C} Bv, Bv)_Q - \langle f, v\rangle_{V' \times V}. \tag{6.4.1}$$

As usual, we consider an analogous problem in subspaces V_h and Q_h where we make the assumption that $B(V_h) \subseteq Q_h$. We want to solve

$$\inf_{v_h \in V_h} \frac{1}{2}(\mathfrak{C} Bv_h, Bv_h)_Q - \langle f, v_h\rangle_{V' \times V}. \tag{6.4.2}$$

In some cases, for instance in an almost incompressible elasticity problem, the numerical solution may behave badly: a locking phenomenon can occur when problem (6.4.2) is too stiff. To build an enhanced method, we introduce a new space $E_h \subset Q$ and we change the problem into

$$\inf_{v_h \in V, \eta \in E_h} \frac{1}{2}(\mathfrak{C}(Bv_h + \eta), (Bv_h + \eta))_Q - \langle f, v_h\rangle_{V' \times V}, \tag{6.4.3}$$

where $\eta \in E_h$ is some "enhancement" of Bv_h. In terms of mathematical programming, this would be called a "slack variable". The optimality conditions of (6.4.3) are

$$\begin{aligned} &(\mathfrak{C}(Bu_h + \eta), Bv_h)_Q - \langle f, v_h\rangle_{V' \times V} = 0, \quad \forall\, v_h \in V_h, \\ &(\mathfrak{C}(Bu_h + \eta), \delta)_Q = 0, \quad \forall\, \delta \in E_h. \end{aligned} \tag{6.4.4}$$

Assuming, for simplicity, that $\mathfrak{C}(E_h) \subseteq E_h$ (as it is almost always the case in practice), and denoting by P_E the projection on E_h, the last equation of (6.4.4) can be read:

$$\mathfrak{C}\eta = -P_E \mathfrak{C} Bu_h \tag{6.4.5}$$

and taking this expression into the first equation, we obtain

$$(\mathfrak{C}(I - P_E)Bu_h, Bv_h)_Q - \langle f, v_h\rangle_{V'\times V} = 0, \quad \forall\, v_h \in V_h, \tag{6.4.6}$$

which is clearly a weaker formulation of the original problem. This idea has been used to obtain stable formulations for a variety of problems such as nearly incompressible elasticity or simulation of very thin structures.

We shall now see briefly, following [284], how this idea can be extended to the case of a mixed formulation (that we repeat once more for the sake of convenience)

$$\begin{cases} a(u,v) + b(v,p) = (f,v)_V & \forall\, v \in V,\\ b(u,q) = (g,q)_Q & \forall\, q \in Q, \end{cases} \tag{6.4.7}$$

assuming again that Q is identified with its dual space.
We shall not try to cover all cases: to avoid unnecessary technicalities, we shall concentrate on the *inf-sup* condition. We shall then suppose that the bilinear form $a(u,v)$ can be decomposed as

$$a(u,v) = a_D(u,v) + \chi(Bu, Bv), \tag{6.4.8}$$

where $a_D(u,v)$ is coercive on the kernel of B so that, according to Proposition 4.3.4, $a(u,v)$ is coercive on the whole space V if $\chi > 0$. We write explicitly:

$$\begin{cases} a_D(u_h,v_h) + \chi(Bu_h, Bv_h) + (Bv_h, p_h) = (f, v_h) & \forall\, v_h \in V_h,\\ (Bu_h, q_h) = (g, q_h) & \forall\, q_h \in Q_h. \end{cases} \tag{6.4.9}$$

Following the idea of enhanced methods, we introduce a subspace E_h of Q and we change the problem into

$$\begin{cases} a_D(u_h,v_h) + \chi(Bu_h + \eta, Bv_h) + (Bv_h, p_h) = (f, v_h) & \forall\, v_h \in V_h,\\ \chi(Bu_h + \eta, \delta) + (\delta, p_h) = 0 & \forall\, \delta \in E_h,\\ (Bu_h + \eta, q_h) = (g, q_h) & \forall\, q_h \in Q_h. \end{cases} \tag{6.4.10}$$

The second equation of (6.4.10) can be read as

$$\eta = -P_E Bu_h - (1/\chi) P_E p_h. \tag{6.4.11}$$

Bringing (6.4.11) into the first and the last equation of (6.4.10), we get:

$$\begin{cases} a_D(u_h,v_h) + \chi((I - P_E)Bu_h, Bv_h) + b(v_h, (I - P_E)p_h) = (f, v_h) & \forall\, v_h \in V_h,\\ b(u_h, (I - P_E)q_h) - (1/\chi)(P_E p_h, P_E q_h) = (g, q_h) & \forall\, q_h \in Q_h, \end{cases} \tag{6.4.12}$$

where in the second equation we used the fact that $((I - P_E)Bu_h, q_h)$ is equal to $b(u_h, (I - P_E)q_h)$.

We can now consider two interesting special cases. In the first one, we choose E_h so that $P_E B v_h = 0$ for any v_h. Equations (6.4.12) now simplify to

$$\begin{cases} a_D(u_h, v_h) + \chi(Bu_h, Bv_h) + b(v_h, p_h) = (f, v_h) \ \forall\, v_h \in V_h, \\ b(u_h, q_h) - (1/\chi)(P_E p_h, P_E q_h) = (g, q_h) \qquad \forall\, q_h \in Q_h. \end{cases} \tag{6.4.13}$$

This is clearly like using a penalty method to stabilise p_h, as we have seen already in Remark 4.3.7. We have already seen these methods in Example 6.3.5, and we shall discuss their applications at various occasions in the following chapters, and in particular in Chap. 8.

It is also interesting to give a look at the case where we choose as E_h a subspace of Q_h. Let us denote by $\overline{Q}_h$ the orthogonal complement of E_h and by $\overline{P} \equiv I - P_E$ the projection onto $\overline{Q}_h$. We can now write (6.4.12) as

$$\begin{cases} a_D(u_h, v_h) + \chi(\overline{P} Bu_h, \overline{P} Bv_h) + b(v_h, \overline{P} p_h) = (f, v_h) \ \forall\, v_h \in V_h, \\ b(u_h, \overline{P} q_h) - (1/\chi)(P_E p_h, P_E q_h) = (g, q_h) \qquad \forall\, q_h \in Q_h. \end{cases} \tag{6.4.14}$$

Writing the second equation for $q_h = \overline{q}_h \in \overline{Q}_h$ and then for $q_h = \delta \in E_h$, we have

$$b(u_h, \overline{q}_h) = (g, \overline{q}_h) \quad \forall\, \overline{q}_h \in \overline{Q}_h \tag{6.4.15}$$

plus

$$c P_E p_h = P_E g. \tag{6.4.16}$$

Thus, we have simply written a discrete problem in $V_h \times \overline{Q}_h$:

$$\begin{cases} a_D(u_h, v_h) + \chi(\overline{B} u_h, \overline{B} v_h) + b(v_h, \overline{p}_h) = (f, v_h) \ \forall\, v_h \in V_h, \\ b(u_h, \overline{q}_h) = (g, \overline{q}_h) \quad \forall\, \overline{q}_h \in \overline{Q}_h, \end{cases} \tag{6.4.17}$$

with $\overline{B} = \overline{P}(B)$, and then corrected $p_h = \overline{p}_h + (1/\chi) P_E g$.

Is that all? By no means! *There are more things in Heaven and Earth, Horatio, than are dreamt of in your philosophy.*
And among all those things, "stabilisation methods" hold a non negligible place.

6.5 Eigenvalue Problems

We shall consider in this section a general setting for the approximation of eigenvalue problems associated with the mixed problems introduced in Sect. 5.1. To make the presentation clearer, we recall some basic assumptions. We thus have two Hilbert spaces, V and Q. Moreover, $a(v,v)$ and $b(v,q)$ are continuous bilinear forms on $V \times V$ and $V \times Q$,

$$\begin{aligned} &\exists\, M_a > 0 \quad \forall u, v \in V \quad a(u,v) \le M_a ||u||_V ||v||_V \\ &\exists\, M_b > 0 \quad \forall v \in V,\ \forall q \in Q \quad b(v,q) \le M_b ||v||_V ||q||_Q. \end{aligned} \tag{6.5.1}$$

To simplify the presentation, we also assume that

$$a(\cdot,\cdot) \text{ is symmetric and positive semi-definite.} \tag{6.5.2}$$

Setting $||v||_a := (a(v,v))^{1/2}$ (which in general will only be a semi-norm on V), this immediately gives

$$\forall u, v \in V \quad a(u,v) \le ||u||_a ||v||_a. \tag{6.5.3}$$

These properties will be assumed to hold throughout all the section. For any given pair (f, g) in $V' \times Q'$, the standard mixed problem is then to find (u, p) in $V \times Q$ such that

$$\begin{cases} a(u,v) + b(v,p) = \langle f, v\rangle & \forall v \in V \\ b(u,q) = \langle g, q\rangle & \forall q \in Q. \end{cases} \tag{6.5.4}$$

We now know that in order to have existence, uniqueness and continuous dependence from the data for problem (6.5.4), it is necessary and sufficient that the bilinear forms $a(\cdot,\cdot)$ and $b(\cdot,\cdot)$ satisfy conditions (5.1.6) and (5.1.1). We thus suppose to have on $b(\cdot,\cdot)$ the *inf-sup condition*.

There exists $\beta > 0$ such that

$$\inf_{q\in Q} \sup_{v\in V} \frac{b(v,q)}{||v||_V ||q||_Q} \ge \beta. \tag{6.5.5}$$

We shall, for simplicity, assume the *ellipticity on the kernel* (5.1.7) instead of (5.1.1).

There exists $\alpha > 0$ such that

$$a(v,v) \ge \alpha ||v||_V^2 \quad \forall v \in \mathrm{Ker}B \tag{6.5.6}$$

where the kernel $\mathrm{Ker}B$ is defined as:

$$\mathrm{Ker}B = \{v \in V \text{ such that } b(v,q) = 0 \quad \forall q \in Q\}. \tag{6.5.7}$$

In Chap. 1 (see Sect. 1.3.4), we have seen many examples of mixed formulations of boundary value problems related to various applications in fluid mechanics and in continuous mechanics and we have shown that there are eigenvalue problems associated with most of them. We shall be interested, here, in the approximation of these eigenvalue problems. We thus consider the discrete analogue of (6.5.4).

We assume that we are given two families of finite dimensional subspaces V_h and Q_h of V and Q, respectively, and we consider the discretised problem: *find* (u_h, p_h) *in* $V_h \times Q_h$ *such that*

$$\begin{cases} a(u_h, v_h) + b(v_h, p_h) = \langle f, v_h \rangle \ \forall v_h \in V_h \\ b(u_h, q_h) = \langle g, q_h \rangle \qquad\qquad\qquad \forall q_h \in Q_h. \end{cases} \tag{6.5.8}$$

We have seen in Chap. 5 that discrete analogues of (6.5.5) and (6.5.6) are sufficient to ensure solvability of the discrete problem together with optimal error bounds. More precisely, the spaces V_h and Q_h should satisfy two conditions:

- The *discrete ellipticity on the kernel*: there exists $\alpha > 0$, independent of h, such that

$$a(v_h, v_h) \geq \alpha ||v_h||_V^2 \quad \forall v_h \in \mathrm{Ker} B_h, \tag{6.5.9}$$

  where the discrete kernel $\mathrm{Ker} B_h$ is defined as

$$\mathrm{Ker} B_h = \{v_h \in V_h \text{ such that } b(v_h, v_h) = 0 \ \forall v_h \in V_h\},$$

- The *discrete inf-sup* condition: there exists $\beta > 0$, independent of h, such that

$$\inf_{q_h \in Q_h} \sup_{v_h \in V_h} \frac{b(v_h, q_h)}{||v_h||_V ||q_h||_Q} \geq \beta. \tag{6.5.10}$$

  Then, we have unique solvability of (6.5.8) and the following error estimate

$$||u - u_h||_V + ||p - p_h||_V \leq C \left(\inf_{v \in V_h} ||u - v||_V + \inf_{q \in Q_h} ||p - q||_Q \right). \tag{6.5.11}$$

We now turn to the eigenvalue problems. As we have seen, the eigenvalue problem which is naturally associated with the corresponding boundary value problem in strong form does not correspond to taking $(\lambda u, \lambda p)$ as right-hand side of (6.5.4). Instead, according to the different cases, the *natural* eigenvalue problem is obtained by taking $(\lambda u, 0)$ or $(0, -\lambda p)$ as right-hand side of (6.5.4). One expects, as for instance in [299], that (6.5.9) and (6.5.10), together with suitable compactness properties, are sufficient to ensure good convergence of the eigenvalues. However, when the problem is set in mixed variational form, compactness is more delicate to deal with. It was shown in [82] that, for the particular case of (1.3.85) for the mixed Poisson problem, even if the operator mapping g into u is clearly compact, assumptions (6.5.9) and (6.5.10) are not sufficient to avoid, for instance, the presence of spurious eigenvalues in the discrete spectrum. Here, we address a more general problem, in abstract form, and we look for sufficient (and, possibly, necessary) conditions in order to have good approximation properties for the eigenvalue problems having either $(\lambda\, u, 0)$ or $(0, -\lambda\, p)$ at the right-hand side. As we shall see, in each of the two cases, (6.5.9) and (6.5.10) might be neither necessary nor sufficient for that.

Our approach will be more similar to the one of [188] than to the one of [112] or [40]. Important references for the study of eigenvalue problems in mixed form are [43, 299, 316]. With respect to *sufficient* conditions, our development introduces minor differences. For instance, our bilinear form $a(\cdot,\cdot)$ is not supposed to be positive definite. Moreover, previous related papers deal mostly with cases in which the two components of the solution of the direct problem are both convergent, while we accept discretisations that can produce singular global matrices. On the other hand, having assumed symmetry of $a(\cdot,\cdot)$, we do not have to consider adjoint problems as in [188]. However, in practical cases, the actual gain is negligible. The major interest of the present setting consists in showing that our sufficient conditions are, mostly, also *necessary*, thus providing a severe test for assessing whether a given discretisation is suitable for computing eigenvalues or not. This justifies, in our opinion, the apparently excessive generality of our abstract approach. Indeed, as we shall see, convergence of discrete eigenvalues does not even imply, for mixed formulations, the non-singularity of the corresponding global matrices.

Finally, we point out that we do not look here for a priori estimates for eigenvalues and eigenvectors, but only deal with convergence. This is somehow in agreement with the fact that necessary conditions are a major issue here. However, in most cases, a priori error estimates can be readily deduced, checking the last step in the proofs of sufficient conditions and/or applying the general instruments of, say, [43, 109, 299] (see also [76] for a review).

6.5.1 Some Classical Results

Before considering the case of eigenvalue problems in mixed form, we need to recall some classical facts. Let H be a Hilbert space and $T : H \to H$ be a self-adjoint compact operator. To simplify the presentation, we assume that T is non-negative.

We are interested in the eigenvalues $\lambda \in \mathbb{R}$ defined by

$$\lambda Tu = u, \quad \text{with } u \in H \setminus \{0\}. \tag{6.5.12}$$

In the above assumptions, it is well-known that there exists a sequence $\{\lambda_i\}$ and an associated orthonormal basis $\{u_i\}$ such that

$$\begin{aligned} \lambda_i T u_i &= u_i, \\ 0 \le \lambda_1 \le \lambda_2 &\le \cdots \le \lambda_i \le \cdots, \\ \lim_{i\to\infty} \lambda_i &= +\infty. \end{aligned} \tag{6.5.13}$$

We also set, for $i \in \mathbb{N}$, $E_i = \text{span}(u_i)$.

The following mapping will be useful. Let $m : \mathbb{N} \to \mathbb{N}$ be the application which to every N associates the dimension of the space generated by the eigenspaces of the first N distinct eigenvalues; that is

$$m(1) = \dim\{\oplus_i E_i : \lambda_i = \lambda_1\},$$
$$m(N+1) = m(N) + \dim\left\{\oplus_i E_i : \lambda_i = \lambda_{m(N)+1}\right\}. \tag{6.5.14}$$

Clearly, $\lambda_{m(1)}, \ldots, \lambda_{m(N)}$ ($N \in \mathbb{N}$) will now be the first N *distinct* eigenvalues of (6.5.12).

Assume that we are given, for every $h > 0$, a self-adjoint non-negative operator $T_h : H \to H$ with finite range. We denote by $\lambda_i^h \in \mathbb{R}$ the eigenvalues of the problem

$$\lambda\, T_h u = u, \quad \text{with } u \in H \setminus \{0\}. \tag{6.5.15}$$

Let H_h be the finite-dimensional range of T_h and $\dim H_h =: N(h)$; then, T_h admits $N(h)$ real eigenvalues denoted λ_i^h such that

$$0 \le \lambda_1^h \le \cdots \le \lambda_i^h \le \ldots \lambda_{N(h)}^h. \tag{6.5.16}$$

The associated discrete eigenfunctions u_i^h, $i = 1, \ldots, N(h)$, give rise to an orthonormal basis of H_h with respect to the scalar product of H. Let $E_i^h :=$ span(u_i^h).

We assume that

$$\lim_{h \to 0} ||T - T_h||_{\mathcal{L}(H)} = 0. \tag{6.5.17}$$

It is a classical result in spectrum perturbation theory that (6.5.17) implies the following convergence property for eigenvalues and eigenvectors:

$$\begin{aligned} &\forall \epsilon > 0,\ \forall N \in \mathbb{N} \quad \exists h_0 > 0 \text{ such that } \forall h \le h_0 \\ &\max_{i=1,\ldots,m(N)} |\lambda_i - \lambda_i^h| \le \epsilon, \\ &\hat{\delta}(\oplus_{i=1}^{m(N)} E_i, \oplus_{i=1}^{m(N)} E_i^h) \le \epsilon, \end{aligned} \tag{6.5.18}$$

where $\hat{\delta}(E, F)$, for E and F linear subspaces of H, represents the gap between E and F and is defined by

$$\begin{aligned} \hat{\delta}(E, F) &= \max[\delta(E, F), \delta(F, E)], \\ \delta(E, F) &= \sup_{u \in E,\ ||u||_H = 1} \ \inf_{v \in F} ||u - v||_H. \end{aligned} \tag{6.5.19}$$

Vice versa, it is not difficult to prove that (6.5.18) is a sufficient condition for (6.5.17).

6.5.2 Eigenvalue Problems in Mixed Form

Let us go back to the abstract framework introduced above. In particular, assume, for the moment, that (6.5.5) and (6.5.6) are satisfied and that (6.5.8) has a solution for every (f, g) in $V' \times Q'$. Problems (6.5.4) and (6.5.8) then define, in a natural way, two operators $S(f, g) = (u, p)$ (solution of (6.5.4)) and $S_h(f, g) = (u_h, p_h)$ (solution of (6.5.8)). To make things precise, we introduce, for every $h > 0$, the dual norms:

$$||f||_{V_h'} = \sup_{v_h \in V_h} \frac{\langle f, v_h \rangle}{||v_h||_V} \qquad ||g||_{Q_h'} = \sup_{q_h \in Q_h} \frac{\langle g, q_h \rangle}{||q_h||_Q}. \tag{6.5.20}$$

From Theorem 3.4.4 of Chap. 3, we know that (6.5.10) and (6.5.9) imply that the discrete operator S_h is bounded from $V_h' \times Q_h'$ to $V \times Q$, uniformly in h, and we have the bounds (3.4.103) and (3.4.104) (with $\mathbf{x} = u_h$ and $\mathbf{y} = p_h$). Moreover, Lemma 3.5.2 tells us that the converse holds true.

Lemma 6.5.1. *If there exists a constant $C > 0$ such that, for every $h > 0$ and for every quadruplet $(u_h, p_h, f, g) \in V_h \times Q_h \times V' \times Q'$ satisfying* (6.5.8)*, one has*

$$||S_h(f, g)||_{V \times Q} \leq C(||f||_{V_h'} + ||g||_{Q_h'}), \tag{6.5.21}$$

then (6.5.10) *and* (6.5.9) *are verified with $\beta = 1/C$ and $\alpha = 1/(C^2 M_a)$. Then,* (6.5.8) *has a solution for all $f \in V_h'$ and $g \in Q_H'$.*

Proof. This is a mere rewriting of Lemma 3.5.2. □

We now consider the eigenvalue problem. For the sake of simplicity, let us assume for the moment that there exist two Hilbert spaces H_V and H_Q such that we can identify

$$\begin{aligned} H_V &\equiv H_V', \\ H_Q &\equiv H_Q' \end{aligned} \tag{6.5.22}$$

and such that

$$\begin{aligned} V \subseteq H_V \subseteq V' \\ Q \subseteq H_Q \subseteq Q' \end{aligned} \tag{6.5.23}$$

hold with dense and continuous embedding, in a compatible way.

The restrictions of S and S_h to $H_V \times H_Q$ now define two operators from $H_V \times H_Q$ into itself.

As a consequence of (6.5.11) and Lemma 6.5.1, it is immediate to prove the following proposition.

Proposition 6.5.1. *Assume that* (6.5.10) *and* (6.5.9) *hold. Then, S_h converges uniformly to S in $\mathcal{L}(H_V \times H_Q)$ if and only if S (from $H_V \times H_Q$ into itself) is compact.* □

This proposition concludes the convergence analysis for the eigenvalue problems associated to (6.5.4) and (6.5.8). However, in the applications, one usually finds eigenvalue problems associated to (6.5.4) and (6.5.8) when one of the two components of the datum is zero. Let us set these eigenvalue problems in their appropriate abstract framework introducing the following operators:

$$\begin{aligned} &C_V : V' \to V' \times Q' \qquad && C_Q : Q' \to V' \times Q' \\ &C_V(f) = (f, 0) && C_Q(g) = (0, g) \end{aligned} \tag{6.5.24}$$

and their adjoints

$$\begin{aligned} &C_V^* : V \times Q \to V \qquad && C_Q^* : V \times Q \to Q \\ &C_V^*(v, q) = v && C_Q^*(v, q) = q. \end{aligned} \tag{6.5.25}$$

We shall say that (6.5.4) is a *problem of the type* $(f,\ 0)$ if the right-hand side in (6.5.4) satisfies $g = 0$. Similarly, we shall say that (6.5.4) is a *problem of the type* $(0,\ g)$ if the right-hand side in (6.5.4) satisfies $f = 0$. Correspondingly, we shall study the approximation of the eigenvalues of the following operators:

$$\begin{aligned} T_V &= C_V^* \circ S \circ C_V : V' \to V, \quad \text{for problems of the type } (f,\ 0), \\ T_Q &= C_Q^* \circ S \circ C_Q : Q' \to Q, \text{ for problems of the type } (0,\ g). \end{aligned} \tag{6.5.26}$$

Whenever the associated discrete problems are solvable, we can introduce the discrete counterparts of T_V and T_Q as:

$$\begin{aligned} T_V^h &= C_V^* \circ S_h \circ C_V : V' \to V, \quad \text{for problems of the type } (f,\ 0), \\ T_Q^h &= C_Q^* \circ S_h \circ C_Q : Q' \to Q, \text{ for problems of the type } (0,\ g). \end{aligned} \tag{6.5.27}$$

6.5.3 Special Results for Problems of Type $(f, 0)$ and $(0, g)$

In the remaining part of this section, we recall the results obtained in Sect. 3.5.3 on the solvability and boundedness of the discrete operators with either the discrete *inf-sup* condition or the discrete ellipticity on the kernel for the special type of data associated with our eigenvalue problems.

Problems of the type $(f, 0)$: From Proposition 3.5.2, we have the following result.

Proposition 6.5.2. *If the discrete ellipticity on the kernel* (6.5.9) *holds and* $g = 0$, *then problem* (6.5.8) *has at least one solution* (u_h, p_h). *Moreover,* u_h *is uniquely determined by* f *and*

$$||u_h||_V \leq \frac{1}{\alpha} ||f||_{V_h'}, \tag{6.5.28}$$

where α *is the constant appearing in* (6.5.9). □

We also have the reciprocal from Proposition 3.5.3.

Proposition 6.5.3. *Assume that there exists a constant $C > 0$ such that for every $h > 0$ and for every quadruplet $(u_h, p_h, f, 0) \in V_h \times Q_h \times V' \times Q'$ satisfying* (6.5.8), *one has*

$$||u_h||_V \leq C||f||_{V_h'}, \tag{6.5.29}$$

then the operator T_V^h is defined in all V' and the discrete ellipticity on the kernel (6.5.9) *holds with $\alpha = 1/(C^2 M_a)$, M_a being the continuity constant of $a(\cdot,\cdot)$ (see* (6.5.1)*).* □

Problems of the form $(0, g)$: In the same way, we have from Proposition 3.5.5 the following result.

Proposition 6.5.4. *Assume that the following* weak discrete *inf-sup* condition *holds: for every $h > 0$, there exists a constant $\beta_h > 0$ such that*

$$\inf_{q_h \in Q_h} \sup_{v_h \in V_h} \frac{b(v_h, q_h)}{||v_h||_V ||q_h||_Q} \geq \beta_h. \tag{6.5.30}$$

Then, for every $g \in V'$ and $f = 0$, problem (6.5.8) *has at least one solution (u_h, p_h) and p_h is uniquely determined by g.* □

Proposition 6.5.5. *Assume that there exists a constant $C > 0$ such that for every $h > 0$ and for every quadruplet $(u_h, p_h, 0, g) \in V_h \times Q_h \times V' \times Q'$ satisfying* (6.5.8), *one has*

$$||p_h||_V \leq C||g||_{V_h'}. \tag{6.5.31}$$

Then, the operator T_Q^h is defined in all Q' and the weak discrete inf-sup condition (6.5.30) *holds. In general,* (6.5.31) *does not imply the discrete inf-sup condition* (6.5.10)*.* □

Proof. As in Proposition 6.5.5, the assumption (6.5.31) implies that, with obvious notation, B_h^t is injective, therefore B_h will be surjective and this implies (6.5.30).

However, (6.5.10) cannot be deduced in general: consider the case when $a(\cdot,\cdot) \equiv 0$, $V_h = Q_h$ and $b(\cdot,\cdot)$ is h times the scalar product in V_h. □

Proposition 6.5.6. *Assume that there exists a constant $C > 0$ such that for every $h > 0$ and for every quadruplet $(u_h, p_h, 0, g) \in V_h \times Q_h \times V' \times Q'$ satisfying* (6.5.8), *one has*

$$||u_h||_V + ||p_h||_Q \leq C||g||_{Q_h'}, \tag{6.5.32}$$

then both T_Q^h and $C_V^* \circ S_h \circ C_Q$ are defined on Q' and (6.5.10) *holds with $\beta = 1/C$.* □

Proof. This results directly from Proposition 6.5.6. □

Moreover, we have the following proposition.

Proposition 6.5.7. *If there exists $C > 0$ such that*

$$||C_Q^* \circ S_h \circ C_V||_{\mathcal{L}(V_h', V_h)} \leq C \tag{6.5.33}$$

for every $h > 0$, then (6.5.10) *holds with $\beta = 1/C$.* □

Proof. The proof can be done as we did for Lemma 3.5.2. □

We therefore see from Propositions 6.5.3 and 6.5.7, that for problems of the type $(f, 0)$, the estimate (6.5.29) on u_h implies (6.5.9) and the estimate (6.5.33) on p_h implies (6.5.10). Analogue properties do not entirely hold for problems of the type $(0, g)$.

6.5.4 Eigenvalue Problems of the Type $(f, 0)$

In this section, together with (6.5.1) and (6.5.2), we assume that ellipticity on the kernel (6.5.6) and the *inf-sup* condition (6.5.5) are verified. We also assume that we are given a Hilbert space H_V (that we shall identify with its dual space H_V') such that

$$V \subseteq H_V \subseteq V' \tag{6.5.34}$$

with continuous and dense embeddings. We consider the eigenvalue problem: *find (λ, u) in $\mathbb{R} \times V$, with $u \neq 0$ such that there exists $p \in V$ verifying*

$$\begin{aligned} &a(u, v) + b(v, p) = \lambda(u, v)_{H_V} \quad \forall v \in V, \\ &b(u, q) = 0 \qquad\qquad\qquad\qquad\quad\; \forall q \in Q. \end{aligned} \tag{6.5.35}$$

In the formalism of Sect. 6.5.2, this can be written as

$$\lambda\, T_V u = u. \tag{6.5.36}$$

We assume that the operator T_V is compact from H_V to V.

Suppose now that we are given two finite dimensional subspaces V_h and Q_h of V and Q, respectively. Then, the approximation of (6.5.35) is: *find (λ_h, u_h) in $\mathbb{R} \times V_h$, with $u_h \neq 0$ such that there exists $p_h \in Q_h$ verifying*

$$\begin{aligned} &a(u_h, v_h) + b(v_h, p_h) = \lambda_h(u_h, v_h)_{H_V} \quad \forall v_h \in V_h, \\ &b(u_h, q_h) = 0 \qquad\qquad\qquad\qquad\qquad\quad\; \forall q_h \in Q_h, \end{aligned} \tag{6.5.37}$$

which can be written as

$$\lambda_h\, T_V^h u_h = u_h. \tag{6.5.38}$$

We are now looking for necessary and sufficient conditions that ensure the uniform convergence of T_V^h to T_V in $\mathcal{L}(H_V, V)$ which, as we have seen, implies the convergence of eigenvalues and eigenvectors (see (6.5.18)).

To start with, we look for sufficient conditions and for this, we introduce some notation. Let V_0^H and Q_0^H be the subspaces of V and Q, respectively, containing all the solutions $u \in V$ and $p \in V$, respectively, of problem (6.5.4) when $g = 0$; that is, with the formalism of the Sect. 6.5.2,

$$\begin{aligned} V_0^H &= C_V^* \circ S \circ C_V(H_V) = T_V(H_V) \\ Q_0^H &= C_Q^* \circ S \circ C_V(H_V). \end{aligned} \tag{6.5.39}$$

Notice that the following inclusion holds true:

$$V_0^H \subseteq \mathrm{Ker} B.$$

The spaces V_0^H and Q_0^H will be endowed with the *natural* norm: that is, for instance,

$$\begin{aligned} ||v||_{V_0^H} &:= \inf\{||\eta||_{H_V},\ T_V\eta = v\}; \\ ||q||_{Q_0^H} &:= \inf\{||\eta||_{H_V},\ C_Q^* \circ S \circ C_V\eta = q\}. \end{aligned} \tag{6.5.40}$$

Definition 6.5.1. We say that the **weak approximability** of Q_0^H is verified if there exists $\omega_1(h)$, tending to zero as h tends to zero, such that for every $p \in Q_0^H$,

$$\sup_{v_h \in \mathrm{Ker} B_h} \frac{b(v_h, p)}{||v_h||_V} \leq \omega_1(h)||p||_{Q_0^H}. \tag{6.5.41}$$

Notice that, in spite of its appearance, (6.5.41) is indeed an approximability property. Actually, as $v_h \in \mathrm{Ker} B_h$, we have $b(v_h, p) = b(v_h, p - p_I)$ for every $p_I \in Q_h$, which has, usually, to be used to verify (6.5.41).

Definition 6.5.2. We say that the **strong approximability** of V_0^H is verified if there exists $\omega_2(h)$, tending to zero as h tends to zero, such that for every $u \in V_0^H$, there exists $u_I \in \mathrm{Ker} B_h$ such that

$$||u - u_I||_V \leq \omega_2(h)||u||_{V_0^H}. \tag{6.5.42}$$

Theorem 6.5.1. *Let us assume that the discrete ellipticity on the kernel* (6.5.9) *is verified. Assume moreover the* weak approximability *of* Q_0^H *and the* strong approximability *of* V_0^H. *Then, the sequence* T_V^h *converges uniformly to* T_V *in* $\mathcal{L}(H_V, V)$, *that is, there exists* $\omega_3(h)$, *tending to zero as h tends to zero, such that*

$$||T_V f - T_V^h f||_V \leq \omega_3(h)||f||_{H_V}, \quad \textit{for all } f \in H_V. \tag{6.5.43}$$

Proof. Let $f \in H_V$ and let $(u, p) \in V_0^H \times Q_0^H$ be solution of (6.5.4): $(u, p) = S(f, 0)$. As we assumed (6.5.9), Proposition 6.5.2 ensures that T_V^h is well defined on V'. Recall that $u := T_V(f)$. Let $u_h := T_V^h(f)$ and let p_I be such that (u_h, p_I) is a solution of (6.5.8) (such a p_I might not be unique). In order to prove the uniform convergence of T_V^h to T_V, we have to estimate the difference $||u - u_h||_V$. We do it by bounding the term $||u_I - u_h||_V$, where u_I is given by (6.5.42), and then by using the triangular inequality. We have

$$\begin{aligned}
\alpha||u_I - u_h||_V^2 &\le a(u_I - u_h, u_I - u_h)\\
&= a(u_I - u, u_I - u_h) + a(u - u_h, u_I - u_h)\\
&\le M_a||u_I - u||_V||u_I - u_h||_V - b(u_I - u_h, p - p_h)\\
&\le \left(M_a||u_I - u||_V + \sup_{v_h \in \mathrm{Ker} B_h} \frac{b(v_h, p - p_h)}{||v_h||_V}\right)||u_I - u_h||_V \qquad (6.5.44)\\
&= \left(M_a||u_I - u||_V + \sup_{v_h \in \mathrm{Ker} B_h} \frac{b(v_h, p)}{||v_h||_V}\right)||u_I - u_h||_V .
\end{aligned}$$

The result then follows immediately from the strong approximability of V_0^H and the *weak approximability* of Q_0^H. In particular, we can take $\omega_3(h) = (1 + M_a/\alpha)\omega_2(h) + \omega_1(h)/\alpha$. □

In the following theorem, we shall see that the assumptions of Theorem 6.5.1 are also, in a sense, necessary for the uniform convergence of T_V^h to T_V in $\mathcal{L}(H_V, V)$.

Theorem 6.5.2. *Assume that the sequence T_V^h is bounded in $\mathcal{L}(V', V)$, and converges uniformly to T_V in $\mathcal{L}(H_V, V)$ (see* (6.5.43)*). Then, the ellipticity in the kernel property* (6.5.9) *holds true. Moreover, both the* strong approximability *of V_0^H and the* weak approximability *of Q_0^H are satisfied.*

Proof. Condition (6.5.9) can be obtained applying Proposition 6.5.3. Let u be an element of V_0^H. Then, by definition of V_0^H, there is $f \in H_V$ such that $u = T_V f$. Define $u_I := T_V^h f$. Uniform convergence implies the *strong approximability* of V_0^H.

In a similar way, let p be an element of Q_0^H. Then, by definition of Q_0^H, $p = C_Q^* \circ S \circ C_V f$ for some $f \in H_V$. There might be more than one such f. We choose $\overline{f}$ such that $||\overline{f}||_{H_V} \le \frac{3}{2} \inf f\{||f||_{H_V} : C_Q^* \circ S \circ C_V f = p\} = \frac{3}{2}||p||_{Q_0^H}$. Let $u := T_V \overline{f}$. Correspondingly, let $u_h := T_V^h \overline{f}$ and let p_h be such that (u_h, p_h) is a solution of (6.5.8) with the same right-hand side (such a p_h might not be unique). Then, we obtain

$$\sup_{v_h \in \mathrm{Ker} B_h} \frac{b(v_h, p)}{||v_h||_V} = \sup_{v_h \in \mathrm{Ker} B_h} \frac{b(v_h, p - p_h)}{||v_h||_V} = \sup_{v_h \in \mathrm{Ker} B_h} \frac{a(u - u_h, v_h)}{||v_h||_V}$$

$$\leq M_a||u-u_h||_V \leq M_a\omega_3(h)||f||_{H_V} \leq \frac{3}{2}M_a\omega_3(h)||p||_{Q_0^H},$$

which gives (6.5.41) with $\omega_1(h) = \frac{3}{2}M_a\omega_3(h)$, that is, the *weak approximability* of Q_0^H. □

Remark 6.5.1. We shall present examples of eigenvalue problems of type $(f, 0)$ for the Stokes problem in Sect. 8.11. □

6.5.5 Eigenvalue Problems of the Form (0, *g*)

In this section, together with (6.5.1) and (6.5.2), we assume that, for every given $g \in Q'$ and $f = 0$, problem (6.5.4) has a unique solution (u, p) and that there exists a constant C (independent of g) such that

$$||u||_V + ||p||_Q \leq C||g||_{Q'}. \tag{6.5.45}$$

It is easy to see that this implies the *inf-sup* condition (6.5.5) but not the ellipticity on the kernel (6.5.6).

Remark 6.5.2. An example of this situation can be found in Sect. 10.1.1 for the $\psi - \omega$ formulation of the biharmonic problem. □

In the following, we assume that we are given a Hilbert space H_Q (that we shall identify with its dual space H'_Q) such that

$$Q \subseteq H_Q \subseteq Q' \tag{6.5.46}$$

with continuous and dense embeddings. For simplicity, we assume that for every $q \in Q$, we have $||q||_{H_Q} \leq ||q||_Q$ (with constant equal to 1).

We consider the eigenvalue problem: *find* (λ, p) *in* $\mathbb{R} \times V$, *with* $p \neq 0$ *such that there exists* $u \in V$ *verifying*

$$\begin{cases} a(u,v) + b(v,p) = 0 & \forall v \in V \\ b(u,q) = -\lambda(p,q)_{H_Q} & \forall q \in Q \end{cases} \tag{6.5.47}$$

which in the formalism of Sect. 6.5.2 can be written as

$$\lambda\, T_Q p = -p. \tag{6.5.48}$$

As we shall see, problems of the type (0, g) are more closely related to the abstract theory of [188] than problems of the previous type $(f, 0)$.

From now on, we assume that the operator T_Q is compact from H_Q into Q.

We introduce two finite dimensional subspaces V_h and Q_h of V and Q, respectively. Then, the approximation of (6.5.47) reads: *find* (λ_h, p_h) *in* $\mathbb{R} \times Q_h$, *with* $p_h \neq 0$ *such that there exists* $u_{\in} V_h$ *verifying*

$$\begin{aligned} a(u_h, v_h) + b(v_h, p_h) &= 0 \quad \forall v_h \in V_h \\ b(u_h, q_h) &= -\lambda_h (p_h, q_h)_{H_Q} \quad \forall q_h \in Q_h, \end{aligned} \tag{6.5.49}$$

that is,

$$\lambda_h T_Q^h p_h = -p_h. \tag{6.5.50}$$

We are now looking for necessary and sufficient conditions that ensure the uniform convergence of T_Q^h to T_Q in $\mathcal{L}(H_Q, Q)$, which implies the convergence of eigenvalues and eigenvectors (see (6.5.18)).

To start with, we look for sufficient conditions.

We introduce some notation. Let V_H^0 and Q_H^0 be the subspaces of V and Q respectively, containing all the solutions $u \in V$ and $p \in Q$, respectively, of problem (6.5.4) when $f = 0$; that is, with the formalism of Sect. 6.5.2,

$$\begin{aligned} V_H^0 &= C_V^* \circ S \circ C_Q(H_Q) \\ Q_H^0 &= C_Q^* \circ S \circ C_Q(H_Q) = T_Q(H_Q). \end{aligned} \tag{6.5.51}$$

It will also be useful to define the space $V_{Q'}^0$ as the image of $C_V^* \circ S \circ C_Q$ (from Q' to V).

As before, the spaces V_H^0, Q_H^0 and $V_{Q'}^0$ will be endowed with their natural norms (see for instance (6.5.40)).

Definition 6.5.3. We say that the *weak approximability of* Q_H^0 *with respect to* $a(\cdot,\cdot)$ is verified if there exists $\omega_4(h)$, tending to zero as h goes to zero, such that for every $p \in Q_0^H$ and for every $v_h \in \operatorname{Ker} B_h$,

$$b(v_h, p) \leq \omega_4(h) ||p||_{Q_H^0} ||v_h||_a. \tag{6.5.52}$$

Notice that (6.5.52) is indeed an approximation property, as we already pointed out for its counterpart (6.5.41).

Definition 6.5.4. We say that the *strong approximability* of Q_H^0 is verified if there exists $\omega_5(h)$, tending to zero as h goes to zero, such that for every $p \in Q_H^0$ there exists $p^I \in Q_h$ such that

$$||p - p^I||_V \leq \omega_5(h) ||p||_{Q_H^0}. \tag{6.5.53}$$

Notice that (6.5.52) and (6.5.53) are (much) weaker forms of assumption H.7 of [188].

Definition 6.5.5. Following Sect. 5.4.3, we say that an operator Π_h from V (or from a subspace of it) into V_h is a B-compatible operator with respect to the bilinear form $b(\cdot,\cdot)$ and the subspace $Q_h \subset Q$ if it verifies, for all v in its domain,

$$b(v - \Pi_h v, q_h) = 0 \quad \forall q_h \in Q_h, \tag{6.5.54}$$

and there exists a constant C_Π, independent of h, such that:

$$||\Pi_h||_{\mathcal{L}(V^0_{Q'},V)} \le C_\Pi. \tag{6.5.55}$$

We now introduce a stronger form of (6.5.55).

Definition 6.5.6. *B-Id-compatible operator*
We shall say that the operator Π_h is B-Id-compatible if it satisfies (6.5.54), (6.5.55) and if moreover it converges to the Identity operator in norm, that is, if there exists $\omega_6(h)$, tending to zero as h tends to zero, such that for every $v \in V^0_H$, we have

$$||v - \Pi_h v||_a \le \omega_6(h)||v||_{V^0_H}. \tag{6.5.56}$$

Remark 6.5.3. In Sect. 5.4.3, we have seen that (6.5.54) and (6.5.55) imply the discrete *inf-sup* condition (6.5.10). Notice that (6.5.56) is strongly related to assumption H.5 of [188]. As we shall see, the condition that $\omega_6(h)$ goes to 0 with h is actually necessary for the convergence of eigenvalues. In [188], it is only assumed to be bounded, that is, essentially (6.5.55). Indeed, their interest was in a priori bounds (and not on necessity) and, moreover, they were dealing with direct problems (and not with eigenvalues). In particular, (6.5.56) is not necessary to obtain point-wise convergence of T^h_Q to T_Q where the discrete ellipticity on the kernel (6.5.9) and the discrete *inf-sup* condition (6.5.10) are sufficient. Notice that, from Remark 5.4.3, the *inf-sup* (6.5.5) and its discrete counterpart (6.5.10) imply (6.5.55), but not (6.5.56). □

We can now prove the following result.

Theorem 6.5.3. *Let us assume that there exists a B-Id-compatible operator Π_h : $V^0_{Q'} \to V_h$, that is satisfying* (6.5.54)–(6.5.56)*. Assume moreover that the* strong approximability *of Q^0_H is verified (see* (6.5.53)*) as well as the* weak approximability of Q^0_H with respect to *a (see* (6.5.52)*). Then, the sequence T^h_Q converges to T_Q uniformly from H_Q into Q, that is, there exists $\omega_7(h)$, tending to zero as h goes to zero, such that*

$$||T_Q g - T^h_Q g||_V \le \omega_7(h)||g||_{H_Q}, \quad \textit{for all } g \in H_Q. \tag{6.5.57}$$

Proof. As we recalled above (6.5.5) and (6.5.54)–(6.5.55) imply the discrete *inf-sup* condition (6.5.10). Thanks to Proposition 6.5.4, T_Q^h is then well defined.

Let $g \in H_Q$ and let $(u, p) \in V_H^0 \times Q_H^0$ be the solution of (6.5.4) with $f = 0$. Recall that $p = T_Q g$. Let $p_h := T_Q^h g$ and let u_h be such that (u_h, p_h) is a solution of (6.5.8) (such a u_h might be not unique). In order to prove the uniform convergence of T_Q^h to T_Q, we have to find a priori estimates for the error $||p - p_h||_Q$. Let $\tilde{g} \in Q'$ be such that $\langle \tilde{g}, p - p_h \rangle = ||p - p_h||_Q$ and $||\tilde{g}||_{Q'} = 1$. Take $vt := C_V^* \circ S \circ C_Q \tilde{g}$, hence $||vt||_{V^0_{Q'}} \le ||\tilde{g}||_{Q'} = 1$ (see (6.5.40)). Then, we have

$$\begin{aligned} ||p - p_h||_Q &= \langle \tilde{g}, p - p_h \rangle = b(vt, p - p_h) \\ &= b(vt - \Pi_h vt, p - p_h) + b(\Pi_h vt, p - p_h) \\ &= b(vt - \Pi_h vt, p - pI) - a(u - u_h, \Pi_h vt). \end{aligned} \tag{6.5.58}$$

Let us estimate separately the two terms in the right-hand side:

$$\begin{aligned} b(vt - \Pi_h vt, p - pI) &\le M_b ||vt - \Pi_h vt||_V ||p - pI||_Q \\ &\le M_b \left(||vt||_V + ||\Pi_h vt||_V\right) ||p - pI||_Q \\ a(u - u_h, \Pi_h vt) &\le ||\Pi_h vt||_a ||u - u_h||_a. \end{aligned} \tag{6.5.59}$$

Using (6.5.55), we obtain the following estimate for $\Pi_h vt$

$$||\Pi_h vt||_V \le C_\Pi ||vt||_{V^0_{Q'}} \le C_\Pi. \tag{6.5.60}$$

Putting together (6.5.58)–(6.5.60) and using (6.5.53), we obtain

$$\begin{aligned} ||p - p_h||_Q &\le M_b(1 + C_\Pi)||p - pI||_Q + C_\Pi ||u - u_h||_a \\ &\le M_b(1 + C_\Pi)\omega_5(h)||p||_{Q_H^0} + C_\Pi ||u - u_h||_a. \end{aligned} \tag{6.5.61}$$

To conclude the proof, there remains to estimate $||u - u_h||_a$. Thanks to the triangular inequality and to (6.5.56), we bound only $||\Pi_h u - u_h||_a$ using also (6.5.52) and (6.5.54). Notice that $\Pi_h u - u_h$ belongs to $\mathrm{Ker} B_h$

$$\begin{aligned} ||\Pi_h u - u_h||_a^2 &= a(\Pi_h u - u, \Pi_h u - u_h) + a(u - u_h, \Pi_h u - u_h) \\ &\le ||u - \Pi_h u||_a ||\Pi_h u - u_h||_a - b(\Pi_h u - u_h, p - p_h) \\ &= ||u - \Pi_h u||_a ||\Pi_h u - u_h||_a - b(\Pi_h u - u_h, p) \\ &\le ||\Pi_h u - u_h||_a \left(||u - \Pi_h u||_a + \omega_4(h)||p||_{Q_H^0}\right), \end{aligned} \tag{6.5.62}$$

which, due to (6.5.56), gives

$$||u - u_h||_a \leq 2||u - \Pi_h u||_a + \omega_4(h)||p||_{Q_H^0} \leq 2\omega_6(h)||u||_{V_H^0} + \omega_4(h)||p||_{Q_H^0} \tag{6.5.63}$$

and (6.5.57) holds with $\omega_7(h) = M_b(1 + C_\Pi)\omega_5(h) + 2C_\Pi\omega_6(h) + C_\Pi\omega_4(h)$. □

Remark 6.5.4. In Theorem 6.5.3, we have proved the uniform convergence of T_Q^h to T_Q in $\mathcal{L}(H_Q, Q)$. However, in Sect. 6.5.2, we have seen that the convergence of the spectrum is equivalent to the uniform convergence of T_Q^h to T_Q in $\mathcal{L}(H_Q)$. Indeed, the latter holds under the weaker assumption that there exists a B-compatible operator satisfying only (6.5.56) as we shall see in the following theorem. □

Theorem 6.5.4. *Let us assume that there exists a B-compatible operator (see* (6.5.54)*)* $\Pi_h : V_{Q'}^0 \to V_h$ *satisfying* (6.5.56)*. Assume moreover that both the* strong approximability *of* Q_H^0 *(see* (6.5.53)*) and the* weak approximability of Q_H^0 with respect to $a(\cdot,\cdot)$ *(see* (6.5.52)*) are verified. Then, the sequence* T_Q^h *converges uniformly to* T_Q *in* H_Q.

Proof. We observe that (6.5.5) and (6.5.54) imply the weak discrete *inf-sup* condition (6.5.30). Thanks to Proposition 6.5.4, T_Q^h is then well defined.

Let $g \in H_Q$ and let $(u, p) \in V_H^0 \times Q_H^0$ be the solution of (6.5.4) with $f = 0$. Recall that $p = T_Q g$. Let $p_h := T_Q^h g$ and let u_h be such that (u_h, p_h) is a solution of (6.5.8) with right-hand side $(0, g)$ (such a u_h might be not unique). We estimate $||p - p_h||_{H_Q}$. Using a duality argument, let $(ut, pt) \in V \times V$ be defined by $(ut, pt) := S(0, p - p_h)$. Due to the definition (6.5.51), ut belongs to V_H^0 with the following estimate $||ut||_{V_H^0} \leq ||p - p_h||_{H_Q}$ (see (6.5.40))

$$\begin{aligned}
||p - p_h||_{H_Q}^2 &= (p - p_h, p - p_h) = b(ut, p - p_h)\\
&= b(ut - \Pi_h ut, p) + b(\Pi_h ut, p - p_h)\\
&= -a(u, ut - \Pi_h ut) - a(u - u_h, \Pi_h ut)\\
&\leq ||u||_a||ut - \Pi_h ut||_a + ||u - u_h||_a||\Pi_h ut||_a\\
&\leq ||u||_a\omega_6(h)||ut||_{V_H^0} + 2||u - u_h||_a||ut||_{V_H^0}\\
&\leq (\omega_6(h)||u||_a + 2||u - u_h||_a)\,||p - p_h||_{H_Q},
\end{aligned}$$

having assumed $\omega_6(h) \leq 1$. Hence,

$$||p - p_h||_{H_Q} \leq \omega_6(h)||u||_a + 2||u - u_h||_a.$$

The rest of the proof follows the same lines as the one of Theorem 6.5.3, using (6.5.52) and (6.5.56) (see (6.5.62) and (6.5.63)). □

The remaining part of this section is devoted to see what one can deduce from the uniform convergence of T_Q^h to T_Q.

Theorem 6.5.5. *Assume that the sequence T_Q^h is bounded in $\mathcal{L}(Q', Q)$. Then, there exists a B-compatible operator (see* (6.5.54)*) $\Pi_h : V_{Q'}^0 \to V_h$ such that*

$$||u - \Pi_h u||_a \leq C||u||_{V_{Q'}^0}. \tag{6.5.64}$$

Proof. Let u belong to $V_{Q'}^0$. Then, by definition, $u = C_V^* \circ S \circ C_Q g$ for some $g \in Q'$. There is only one g in this condition, and therefore, by definition, $||u||_{V_{Q'}^0} = ||g||_{Q'}$ (see (6.5.40)). Let $p \in Q$ be such that $(u, p) = S(0, g)$. Let $p_h := T_Q^h g$; notice that, by assumption, $||p_h||_Q \leq C||g||_{Q'}$. By Propositions 6.5.5 and 6.5.4, there exists at least one u_h such that $(u_h, p_h) \in V_h \times Q_h$ is a corresponding discrete solution of (6.5.8). If such a u_h is unique, we define $\Pi_h u := u_h$. Otherwise, we still define $\Pi_h u$ as the u_h having minimum norm in V. By construction, we have (6.5.54) and

$$||\Pi_h u||_a^2 = \langle g, p_h\rangle \leq ||g||_{Q'}||T_Q^h g||_Q \leq C||g||_{Q'}^2 = C||u||_{V_{Q'}^0}^2. \tag{6.5.65}$$

Let us bound $||u - \Pi_h u||_a$:

$$\begin{aligned} ||u - \Pi_h u||_a^2 &= a(u - \Pi_h u, u - \Pi_h u) \\ &= a(u, u - \Pi_h u) - a(\Pi_h u, u - \Pi_h u) \\ &= -b(u - \Pi_h u, p) - a(u - \Pi_h u, \Pi_h u). \end{aligned} \tag{6.5.66}$$

The first term in the right-hand side can be handled as follows:

$$\begin{aligned} b(u - \Pi_h u, p) &= b(u - \Pi_h u, p - p_h) \\ &= \langle g, p - p_h\rangle - b(\Pi_h u, p - p_h) \\ &= \langle g, p - p_h\rangle + a(u - \Pi_h u, \Pi_h u). \end{aligned} \tag{6.5.67}$$

Inserting (6.5.67) in (6.5.66), we obtain

$$\begin{aligned} ||u - \Pi_h u||_a^2 &= -\langle g, p - p_h\rangle - 2a(u - \Pi_h u, \Pi_h u) \\ &\leq ||g||_{Q'}||p - p_h||_Q + 2||u - \Pi_h u||_a||\Pi_h u||_a \\ &\leq ||g||_{Q'}\left(||p||_Q + ||p_h||_Q\right) + 2||u - \Pi_h u||_a||\Pi_h u||_a, \end{aligned} \tag{6.5.68}$$

and then the boundedness of T_Q^h and (6.5.65) imply (6.5.64). □

Theorem 6.5.6. *Assume that the sequence T_Q^h converges to T_Q uniformly in $\mathcal{L}(H_Q, Q)$. Then, for all $p \in Q_H^0$, there exists $p^I \in Q_h$ such that* (6.5.53) *holds true.*

Proof. Let p belong to Q_H^0, then $p = T_Q g$ for a suitable g in H_Q. Let $p_h := T_Q^h g$ be the corresponding discrete solution, then we define $p^I := p_h$ and the inequality (6.5.53) is an easy consequence of the uniform convergence of $T_Q^h g$ to $T_Q g$ in Q. □

Theorem 6.5.7. *Let us assume that the sequence* T_Q^h *is bounded in* $\mathcal{L}(Q', Q)$ *and converges uniformly to* T_Q *in* $\mathcal{L}(H_Q, Q)$*. In addition, we assume that the following bound holds for the solutions of* (6.5.8) *with* $f = 0$

$$||u_h||_V \leq C||g||_{Q'}. \tag{6.5.69}$$

Then, there exists a B-compatible operator $\Pi_h : V_{Q'}^0 \to V_h$ *satisfying* (6.5.55) *and* (6.5.56)*. Moreover, we have the discrete inf-sup condition* (6.5.10) *and the weak approximability of* Q_H^0 *with respect to* $a(\cdot,\cdot)$ *(see* (6.5.52)*) holds.*

Proof. From Proposition 6.5.5, we have that $C_V^* \circ S \circ C_Q$ is also well defined and (6.5.10) holds. Let us check (6.5.55). For $u \in V_{Q'}^0$, there exists $g \in Q'$ and $p \in Q$ such that $(u, p) = S(0, g)$. We set $\Pi_h u := C_V^* \circ S_h \circ C_Q g$. As we have seen, (6.5.54) holds trivially, and now (6.5.55) also holds in virtue of (6.5.69), with $C_\Pi := C$.

Now, let us check (6.5.56). Let u belong to V_H^0; by definition $u = C_V^* \circ S \circ C_Q g$ for some $g \in H_Q$. As in the proof of Theorem 6.5.5, g is unique, and $||u||_{V_H^0} = ||g||_{H_Q}$. Let $p := T_Q g$; clearly, $p \in Q_H^0$. Let $p_h := T_Q^h g$. By construction, $(\Pi_h u, p_h)$ solves (6.5.8) with the right-hand side $(0, g)$. Moreover, by the same computations as above, we arrive at (see the first line in (6.5.68))

$$||u - \Pi_h u||_a^2 = -\langle g, p - p_h\rangle - 2a(u - \Pi_h u, \Pi_h u).$$

From this, we have

$$\begin{aligned} ||u - \Pi_h u||_a^2 &= -\langle g, p - p_h\rangle - 2b(\Pi_h u, p - p_h) \\ &\leq \left(||g||_{Q'} + 2M_b||\Pi_h u||_V\right)||p - p_h||_Q \\ &\leq (1 + 2M_b C)||g||_{Q'}\omega_7(h)||g||_{H_Q} \\ &\leq (1 + 2M_b C)\omega_7(h)||g||_{H_Q}^2 \\ &= (1 + 2M_b C)\omega_7(h)||u||_{V_H^0}^2, \end{aligned} \tag{6.5.70}$$

where we used (6.5.69) and the uniform convergence of T_Q^h to T_Q in $\mathcal{L}(H_Q, V)$ (see (6.5.57)). The bound (6.5.70) gives (6.5.56) with:

$$\omega_6(h) = ((1 + 2M_b C)\omega_7(h))^{1/2}.$$

Now, let us check (6.5.52). If $p \in Q_H^0$, then $p = T_Q g$ for a suitable g in H_Q. Let u be such that $(u, p) = S(0, g)$ and set $p_h := T_Q^h g$ and $u_h := \Pi_h u$. Then we get, for every $v_h \in \mathrm{Ker} B_h$,

$$\begin{aligned} b(v_h, p) &= b(v_h, p - p_h) \\ &= a(\Pi_h u - u, v_h) \leq M_a ||\Pi_h u - u||_a ||v_h||_a \end{aligned} \tag{6.5.71}$$

and (6.5.56) (already proved) ends the proof since, by definition:

$$||u||_{V_H^0} = ||g||_{H_Q} = ||p||_{Q_H^0}. \qquad \square$$

Examples for the mixed formulation of second order linear elliptic problems will be presented in Chap. 7, Sect. 7.1.3. We also refer to Sect. 10.1.2 for an example with the $\psi - \omega$ approximation of the biharmonic problems.

Chapter 7
Mixed Methods for Elliptic Problems

This chapter will present a first set of applications of the theory developed in the previous chapters. It will provide us with the occasion of introducing many ideas which often have a more general scope than the simple considered case. We shall indeed consider the most simple cases of non-standard methods for Dirichlet's problem, including hybrid methods. We then concentrate on numerical issues for the solution of the discrete problems arising from the previous constructions. In the following section, we sketch miscellaneous results on error estimates in different norms. Section 7.6 is dedicated to an example of application to semiconductor devices simulation. Section 7.7 discusses the sensitivity of low order mixed formulations to mesh deformation. We shall then consider in Sect. 7.8 the relations between mixed methods and the Finite Volume Method. A related idea, using a nonconforming element, will then be discussed in Sect. 7.9 and shown not to be convergent. Finally, Sect. 7.10 presents some applications of augmented formulations introduced in Sect. 1.5.

Stabilised methods will also be developed. All this should provide the reader with a first overview of non-standard methods and open the way to more complex problems.

7.1 Non-standard Methods for Dirichlet's Problem

7.1.1 Description of the Problem

This section presents a unified framework for the analysis of non-standard methods for problems involving an elliptic, Laplacian-like, equation in $\mathbb{R}^n$. Although we shall mainly consider the case $n=2$, most results can be extended to the case $n=3$ using the construction developed in Chap. 2. We shall use the notation employed in the simulation of Darcy's law in reservoir simulation. The primal variable p will represent a pressure and $\underline{u} := \underline{\mathrm{grad}}\, p$ will be a velocity. We thus consider a problem of the following type:

D. Boffi et al., *Mixed Finite Element Methods and Applications*, Springer Series in Computational Mathematics 44, DOI 10.1007/978-3-642-36519-5_7,

$$\begin{cases} -\operatorname{div} A(x)\underline{\operatorname{grad}} p = f \text{ in } \Omega, \\ p| = g_1 \text{ on } \Gamma_D \\ A(x)\underline{\operatorname{grad}} p \cdot \underline{n} = g_2 \text{ on } \Gamma_N, \end{cases} \tag{7.1.1}$$

where Ω is a bounded domain in $\mathbb{R}^n$ and $\Gamma = \Gamma_D \cup \Gamma_N = \partial\Omega$. We assume $A(x)$ to be an $n \times n$ positive definite matrix and that its smallest eigenvalue is bounded away from zero, uniformly with respect to x, that is,

$$\langle A(x)\underline{v}, \underline{v}\rangle \geq \alpha |\,\underline{v}|^2_{\mathbb{R}^n}, \ \forall\, \underline{v} \in \mathbb{R}^n, \tag{7.1.2}$$

with α independent of x. We have already introduced this problem in Chap. 1 with $A(x) = I$. Restricting ourselves temporarily to the case $g_1 = 0$, the standard variational formulation is the following minimisation problem (when $A(x)$ is symmetric),

$$\inf_{q \in H^1_{0,\Gamma_D}(\Omega)} \frac{1}{2}\int_\Omega A\underline{\operatorname{grad}} q.\underline{\operatorname{grad}} q \, dx - \int_\Omega f\, q \, dx - \int_{\Gamma_N} g_2\, q \, d\sigma, \tag{7.1.3}$$

where (cf. Chap. 2)

$$H^1_{0,\Gamma_D}(\Omega) := \{q \in H^1(\Omega) |\, q|_{\Gamma_D} = 0\}. \tag{7.1.4}$$

It is classical that there exists a unique solution to this problem. We shall call problem (7.1.3) the *Primal Formulation.*

Using duality methods, we also transformed, in Chap. 1, this problem to get a *Mixed Formulation*, namely for $f \in L^2(\Omega)$ and $g_2 = 0$,

$$\inf_{\underline{v} \in H_{0,\Gamma_N}(\operatorname{div};\Omega)} \sup_{q \in L^2(\Omega)} \frac{1}{2}\int_\Omega A^{-1}\underline{v}\cdot\underline{v}\, dx + \int_\Omega (\operatorname{div} \underline{v} + f)\, q\, dx + \int_{\Gamma_D} g_1\, \underline{v}\cdot\underline{n}\, ds, \tag{7.1.5}$$

where one has (cf. Sect. 2.1.1)

$$H_{0,\Gamma_N}(\operatorname{div};\Omega) := \{\underline{v} \,|\, \underline{v} \in H(\operatorname{div};\Omega),\ \underline{v}\cdot\underline{n}|_{\Gamma_N} = 0\}, \tag{7.1.6}$$

the sense of $\underline{v}\cdot\underline{n}|_{\Gamma_N} = 0$ being defined as in Sect. 2.1.1. Later, we shall come back to the non-homogeneous case $\underline{v}\cdot\underline{n} = g_2 \neq 0$. Problem (7.1.5) is equivalent to the *Dual Formulation*

$$\inf_{\substack{\underline{v} \in H_{0,\Gamma_N}(\operatorname{div};\Omega) \\ \operatorname{div}\underline{v}+f=0}} \int_\Omega A^{-1}\underline{v}\cdot\underline{v}\, dx + \int_{\Gamma_D} g_1\, \underline{v}\cdot\underline{n}\, ds. \tag{7.1.7}$$

Problem (7.1.7) is a constrained problem (in the sense of mathematical programming). The mixed formulation uses the Lagrange multiplier q to deal with the linear constraint $\operatorname{div}\underline{v} + f = 0$.

It must be remarked that problem (7.1.7) is not, strictly speaking, the dual of problem (7.1.3). That dual problem should be written with $\underline{v} \in L^2(\Omega)^n$ and $f \in H^{-1}(\Omega)$. Here, we use a modified form using a stronger space for $\underline{v}$ while the regularity of q has been weakened. It must also be said that the approximation of this problem is not the main interest. The reason for such a detailed study is that it provides a simple framework that will later be generalised to other important problems.

This section will thus be entirely devoted to the study of problems (7.1.3), (7.1.5) and (7.1.7). We shall first consider approximations of problem (7.1.5), that is *mixed finite element methods*. To work out such an approximation, we shall have to use the finite element spaces approximating $H(\operatorname{div};\Omega)$ built in Chap. 2.

To approximate the *dual problem*, we shall need to build vector functions $\underline{v}$ satisfying the condition

$$\operatorname{div}\underline{v} + f = 0. \tag{7.1.8}$$

This condition is the analogue of the *equilibrium condition* in elasticity theory and approximations satisfying it will be called *equilibrium methods*. Finally, *domain decomposition methods* will lead us *to hybrid finite element* methods. Hybrid methods will be called primal or dual, depending on the formulation being used. This distinction corresponds to assumed stress or assumed displacement hybrid methods in elasticity theory.

Our analysis will rely directly on the properties of $H^1(\Omega)$ and $H(\operatorname{div};\Omega)$ and of their approximations considered in Chap. 2.

7.1.2 Mixed Finite Element Methods for Dirichlet's Problem

We are now able to consider in details the approximation of the mixed formulation

$$\inf_{\underline{v}}\sup_{q} \frac{1}{2}\int A^{-1}\,\underline{v}\cdot\underline{v}\,dx + \int_\Omega (\operatorname{div}\underline{v} + f)q\,dx + \int_{\Gamma_D} g_1\,\underline{v}\cdot\underline{n}\,ds, \tag{7.1.9}$$

with $\underline{v} \in H_{0,\Gamma_N}(\operatorname{div};\Omega)$ and $q \in L^2(\Omega)$. We can now see, by the results of Sect. 7.1.1, that the last term of (7.1.9) makes sense if $g_1 \in H^{[1/2](\Gamma_D)}$ and that the boundary integral must be read as a formal way of writing the duality between $H^{\frac{1}{2}}$ and $H^{-\frac{1}{2}}$. Problem (7.1.9) is a saddle point problem. In the notations of Chaps. 3–5,

$$a(\underline{u},\underline{v}) = \int_\Omega A^{-1}\underline{u}\cdot\underline{v}\,dx \tag{7.1.10}$$

and

$$b(\underline{v}, q) = \int_{\Omega} q \text{ div } \underline{v}\, dx. \tag{7.1.11}$$

The optimality conditions for (7.1.9) can be written as

$$\begin{cases} a(\underline{u}, \underline{v}) + b(\underline{v}, p) = \langle g_1, \underline{v} \cdot \underline{n} \rangle \ \forall\, \underline{v} \in H_{0,\Gamma_N}(\text{div}; \Omega), \\ \qquad b(\underline{u}, q) = -\displaystyle\int_{\Omega} f\, q\, dx \ \forall\, q \in L^2(\Omega). \end{cases} \tag{7.1.12}$$

We work with the spaces $V := H_{0,\Gamma_N}(\text{div}, \Omega)$, $Q := L^2(\Omega)$. It is natural here to identify Q and its dual space Q'. The operator B is then the divergence operator from V into Q. This operator is surjective. Indeed, if $f \in L^2(\Omega) = Q$ is given, we can solve the problem,

$$\begin{cases} -\Delta\phi &= f \text{ in } \Omega, \\ \phi|_{\Gamma_D} &= 0, \\ \frac{\partial\phi}{\partial n}|_{\Gamma_N} &= 0, \end{cases} \tag{7.1.13}$$

to find $\phi \in H^1(\Omega)$. Taking $\underline{u} = \underline{\text{grad}}\, \phi$, we have found $\underline{u} \in H_{0,\Gamma_N}(\text{div}; \Omega)$ with $\text{div } \underline{u} + f = 0$.

Remark 7.1.1. Note also that such a $\underline{u}$ will belong, for instance, to the space $(L^s(\Omega))^n$ for some $s > 2$. Setting

$$W := \{\underline{v} \,|\, \underline{v} \in (L^s(\Omega))^n, \text{ div } \underline{v} \in L^2\} \cap H_{0,\Gamma_N}(\text{div}; \Omega), \tag{7.1.14}$$

we have $\|\underline{u}\|_W \le c\, \|f\|_Q$. Hence, B has a continuous lifting from Q into W.

Moreover, we have coerciveness of $a(\cdot, \cdot)$ on $\text{Ker}B$, although not on V. Using assumption (7.1.2), we have, in fact, whenever $\text{div } \underline{v} = 0$,

$$a(\underline{v}_0, \underline{v}_0) \ge \alpha\, |\underline{v}_0|^2_{(L^2(\Omega))^n} = \alpha\, \|\underline{v}_0\|^2_{H(\text{div};\Omega)}. \tag{7.1.15}$$

The theory of Chap. 4 then applies in a straightforward way and we obtain *existence and uniqueness of a solution* $\{\underline{u}, p\}$ *to this problem.* □

Remark 7.1.2. Uniqueness of the Lagrange multiplier p is a consequence of the surjectivity of B which implies $\text{Ker}B^t = \{0\}$. □

The above results also enable us to consider a non-homogeneous problem, that is, the case $g_2 \ne 0$ in (7.1.1). To do so, we consider any $\underline{\tilde{u}}$ such that

$$A^{-1}\underline{\tilde{u}} \cdot \underline{n} = g_2 \text{ on } \Gamma_N. \tag{7.1.16}$$

This is possible and can be done explicitly by considering a classical solution to problem (7.1.1) with $f = 0$ and $g_1 = 0$ and then taking $\underline{\tilde{u}} = \underline{\text{grad}}\, p$. We then look for $\underline{u} = \underline{\tilde{u}} + \underline{u}_0$ with $\underline{u}_0 \in H_{0,\Gamma_N}(\text{div};\Omega)$. This leads us to the problem

$$\begin{cases} a(\underline{u}_0, \underline{v}_0) + b(\underline{v}_0, p) = \langle g_1, \underline{v}_0 \cdot \underline{n}\rangle - a(\underline{\tilde{u}}, \underline{v}_0), \ \forall\, \underline{v} \in H_{0,\Gamma_N}(\text{div};\Omega), \\ \qquad b(\underline{u}_0, q) = -\displaystyle\int_\Omega f\, q\, dx - b(\underline{\tilde{u}}, q), \ \forall\, q \in L^2(\Omega). \end{cases} \tag{7.1.17}$$

This means that considering $g_2 \neq 0$ can be reduced to changing the right-hand side of (7.1.2).

We are therefore ready to consider the approximation of the mixed formulation.

In Chap. 2, we built function spaces for the purpose of approximating $H(\text{div};\Omega)$. We can now use to discretise problem (7.1.12) or (7.1.17) anyone of the spaces $\underline{M}_k(\Omega, \mathcal{T}_h)$ introduced in Sect. 2.5.2. The approximation of $Q = L^2(\Omega)$ is then implicitly done: Q_h must be $\mathcal{L}^0(D_k, \mathcal{T}_h)$, where, for example,

$$\begin{aligned} D_k &= \mathcal{L}^0_k \ \text{ for } \mathcal{RT}_k \text{ and } \mathcal{BDFM}_k, \\ D_k &= \mathcal{L}^0_{k-1} \ \text{ for } \mathcal{BDM}_k. \end{aligned} \tag{7.1.18}$$

To fix ideas, we shall use, following [331], $\mathcal{RT}_k(\Omega, \mathcal{T}_h)$ and we define

$$V_h := \{\underline{v}_h \,|\, \underline{v}_h \in \mathcal{RT}_k(\Omega, \mathcal{T}_h),\ \underline{v}_h \cdot \underline{n}|_{\Gamma_N} = 0\}. \tag{7.1.19}$$

Such a definition is possible if the partition into elements is made in such a way that there is no element across the interface between Γ_D and Γ_N on Γ. Having chosen V_h as in (7.1.19), we must take

$$Q_h := \mathcal{L}^0_k(\Omega) = \{q_h \,|\, q_h|_K \in P_k(K)\}. \tag{7.1.20}$$

We could replace this choice with any of the elements listed in Sect. 2.5. In order to apply results of Chap. 5 without unnecessary technicalities, we shall restrict ourselves to the case of affine elements.

We may now introduce the discrete problem: *find* $(\underline{u}_h, p_h) \in V_h \times Q_h$ *such that*

$$\begin{cases} \displaystyle\int_\Omega A^{-1}\underline{u}_h \cdot \underline{v}_h\, dx + \int_\Omega p_h \,\text{div}\, \underline{v}_h\, dx = \langle \underline{\tilde{g}}, \underline{v}_h\rangle \ \forall\, \underline{v}_h \in V_h, \\ \displaystyle\int_\Omega q_h \,\text{div}\, \underline{u}_h\, dx + \langle \underline{\tilde{f}}, q_h\rangle = 0 \ \forall\, q_h \in Q_h, \end{cases} \tag{7.1.21}$$

where $\tilde{f}$ and $\tilde{g}$ eventually include non-homogeneous boundary conditions as in problem (7.1.17), that is,

$$\langle \underline{\tilde{g}}, \underline{v}\rangle = \langle g_1, \underline{v} \cdot \underline{n}\rangle - a(\underline{\tilde{u}}, \underline{v}), \tag{7.1.22}$$

$$\langle \tilde{f}, q\rangle = \int_\Omega f q\, dx - b(\underline{\tilde{u}}, q). \tag{7.1.23}$$

To apply the results of Chap. 5, we must check that the bilinear form $a(\cdot,\cdot)$ is coercive on $\mathrm{Ker}B_h$ and that we have the *inf-sup* condition. These properties will be an easy consequence of the commutative diagram (2.5.27). In particular, we already know from (2.5.28) that

$$\mathrm{div}\ V_h = Q_h. \tag{7.1.24}$$

This shows that B_h is nothing but the restriction to V_h of the divergence operator and that it is surjective so that $\mathrm{Ker}B_h^t = \{0\}$. Moreover, we have

$$B_h = B|_{V_h} = \mathrm{div}\,|_{V_h}, \tag{7.1.25}$$

and this obviously implies that we are in the special and interesting case where

$$\mathrm{Ker}B_h \subset \mathrm{Ker}B. \tag{7.1.26}$$

We can then rewrite (2.5.27) in the abstract form

$$\begin{array}{ccc} W & \xrightarrow{\ B\ } & Q \equiv Q' \\ {\scriptstyle \Pi_h}\big\downarrow & & {\scriptstyle P_h}\big\downarrow \\ V_h & \xrightarrow{\ B_h\ } & Q_h \equiv Q'_h \end{array} \tag{7.1.27}$$

with P_h the L^2-projection from Q onto Q_h. From Remark 7.1.1, we know that B has a continuous lifting from Q to W. Since the operators Π_h are uniformly bounded from W to V_h, we have

$$\begin{cases} \displaystyle\int_\Omega (\mathrm{div}\ \underline{v} - \mathrm{div}\ \Pi_h\underline{v})q_h\,dx = 0,\ \forall\, q_h\ \in Q_h, \\ \|\Pi_h\underline{v}\|_V \le c\ \|\underline{v}\|_W. \end{cases} \tag{7.1.28}$$

The first part of (7.1.28) is a direct consequence of the commuting property of diagram (7.1.27). Using (7.1.28) and Proposition 5.4.3, we obtain that the discrete *inf-sup* condition is satisfied with a constant independent of h.

On the other hand, (7.1.26) implies that the coercivity of $a(\cdot,\cdot)$ on $\mathrm{Ker}B_h$ is trivial and follows directly from (7.1.15).

We can now apply our abstract results to obtain the following proposition.

Proposition 7.1.1. *Problem* (7.1.21) *has a unique solution. Moreover, if* $(\underline{u}, p)$ *is the solution of problem* (7.1.17)*, we have the estimates*

$$\|\underline{u} - \underline{u}_h\|_V \le c \inf_{\underline{v}_h \in V} \|\underline{v} - \underline{v}_h\|_{V_h}, \tag{7.1.29}$$

$$\|p - p_h\|_Q \le c\Big(\inf_{q_h \in Q_h} \|p - q_h\|_Q + \inf_{\underline{v} \in V_h} \|\underline{u} - \underline{v}_h\|_V\Big). \tag{7.1.30}$$

Proof. We just apply Theorem 5.2.5, and in particular (5.2.36) that deals with the case where $\mathrm{Ker} B_h \subseteq \mathrm{Ker} B$. □

Remark 7.1.3. It would be possible to use the results of Chap. 5 in a more precise way to make explicit the constants in (7.1.29) and (7.1.30). □

This direct use of Chap. 5 is optimal when the spaces $\mathcal{RT}$ or $\mathcal{BDFM}$ are used but not with $\mathcal{BDM}$. This comes from the fact that, for $\mathcal{RT}_k(\Omega, \mathcal{T})$, we have an estimate on $\inf_{\underline{v}_h \in V_h} \|\underline{v} - \underline{v}_h\|_0$ and $\inf_{\underline{v}_h \in V_h} \|\operatorname{div} \underline{v} - \operatorname{div} \underline{v}_h\|_0$ of the same order $O(h^{k+1})$, whereas the latter is only $O(h^k)$ for $\mathcal{BDM}_k(\Omega; \mathcal{T}_h)$ (Proposition 2.5.4). We must, however, not despair. Denoting

$$V^* := (L^2(\Omega))^n, \tag{7.1.31}$$

we have, as in Sect. 5.2.3,

$$\begin{cases} a(\underline{u}, \underline{v}) \le \|\underline{u}\|_{V^*} \|\underline{v}\|_{V^*}, \\ a(\underline{v}, \underline{v}) \ge \alpha \|\underline{v}\|^2_{V^*}, \end{cases} \tag{7.1.32}$$

and indeed $\|\underline{v}\|_{V^*} = \|\underline{v}\|_V$ for any $\underline{v} \in \mathrm{Ker} B$. We can thus apply estimate (5.2.47) of Theorem 5.2.6 which yields,

$$\|\underline{u} - \underline{u}_h\|_{V^*} \le c \inf_{\underline{v}_h \in Z_h(g)} \|\underline{u} - \underline{v}_h\|_{V^*} \le c \|\underline{u} - \Pi_h \underline{u}\|_{V^*}, \tag{7.1.33}$$

which is now optimal. Estimate (7.1.29) is now optimal for any of the spaces considered in Chap. 2.

We can now join the above results with the approximation results (2.5.29) and (2.5.30).

Proposition 7.1.2. *Let $\underline{M}_k(\Omega, \mathcal{T}_h)$ be any of the spaces defined from* (2.5.22). *Let $\mathcal{L}^0(D_k, \mathcal{T}_h)$ be the corresponding space given by* (2.5.23). *Let $(\underline{u}, p)$ be the solution of problem* (7.1.17). *Let $(\underline{u}_h, p_h)$ be the solution in $V_h \times Q_h = \underline{M}_k(\Omega, \mathcal{T}_h) \times \mathcal{L}^0(D_k, \mathcal{T}_h)$ of problem* (7.1.21). *Then, we have the estimates*

$$\|\underline{u} - \underline{u}_h\|_0 \le c h^s \|\underline{u}\|_{s,\Omega}, \tag{7.1.34}$$

for $s \le k + 1$. Moreover, we also have

$$\|p - p_h\|_0 \le c h^s (\|p\|_s + \|\underline{u}\|_s), \tag{7.1.35}$$

for $s \le k+1$ for the spaces $\mathcal{RT}$ and $\mathcal{BDFM}$, and $s \le k$ for the spaces $\mathcal{BDM}$. □

Remark 7.1.4. The case of non-affine elements is somewhat more tricky. In that case, the contravariant transformation $\mathcal{G}$ of (2.1.69) no longer has a constant Jacobian and we no longer have $B_h = \operatorname{div}|_{V_h}$ because

$$\text{div}\ \underline{v} = \text{div}(\mathcal{G}\hat{\underline{v}}) = \mathcal{F}\Big(\frac{\text{div}\ \hat{\underline{v}}}{J}\Big), \tag{7.1.36}$$

where $\mathcal{F}$ is the standard change of variables (2.1.59). As the Jacobian is not constant in the general case, div $\hat{\underline{v}} \notin Q_h$. It can however be checked that $\text{Ker}B_h \hookrightarrow \text{Ker}B$. We refer to [366] for a study of this case. The reader should be warned that using non-affine elements can be very dangerous for what the approximation properties of finite elements spaces are concerned (see Sect. 2.2.4 and 2.2.5). □

Remark 7.1.5. In the affine case (where div $V_h = Q_h$), a direct subtraction of the second equations in problems (7.1.17) and (7.1.21) yields

$$\int_\Omega (\text{div}\ \underline{u} - \text{div}\ \underline{u}_h) q_h\, dx = 0, \ \forall\, q_h \in Q_h. \tag{7.1.37}$$

This means that div $\underline{u}_h$ is the $L^2(\Omega)$-projection of div $\underline{u}$ onto Q_h. An estimate of $\|\,\text{div}\ \underline{u} - \text{div}\ \underline{u}_h\|$ then directly follows. □

We shall come back to this mixed method in Sect. 7.2.2. We shall then consider sharper estimates and introduce Lagrange multipliers to deal with continuity of $\underline{u}_h \cdot \underline{n}$ at interfaces. This will allow us in particular to build an efficient solution method and to obtain from the results a better approximation of $\underline{u}$. This method of Lagrange multipliers is in fact quite general and will lead to a more standard interpretation of otherwise non-standard methods. In particular, $\mathcal{BDM}$ spaces will recover in the scalar variable the same order of convergence as the other methods.

7.1.3 Eigenvalue Problem for the Mixed Formulation

We shall now consider the results of Sect. 1.2.1 in the context of the formulation of the Dirichlet problem described in the previous subsection. This eigenvalue problem has already been introduced in (1.3.85). We recall it here for the convenience of the reader. We thus consider a mixed formulation for the eigenvalue problem

$$-\Delta p = \lambda p, \quad p \in H_0^1(\Omega) \tag{7.1.38}$$

and we want to solve

$$\begin{cases} \displaystyle\int_\Omega \underline{u}\cdot\underline{v}\, dx + \int_\Omega p\ \text{div}\ \underline{v}\, dx = 0, \ \forall\, \underline{v} \in H(\text{div};\Omega), \\ \displaystyle\int_\Omega \text{div}\ \underline{u}\, q\, dx = -\lambda \int_\Omega p\, q\, dx, \ \forall\, q \in L^2(\Omega). \end{cases} \tag{7.1.39}$$

We now have $V := H(\text{div};\Omega)$ and $Q := L^2(\Omega)$. As usual, we identify $L^2(\Omega)$ with its dual space so that in the notation of Sect. 6.5.2, we have $Q = H_Q = Q' = L^2(\Omega)$. Referring to definitions (6.5.39), it is easy to see (using e.g. [233]) that if Ω is, for instance, a convex polygon, then V_H^0, the space where eigenvectors of the

Laplace operator are to be found, is contained in $H^2(\Omega) \cap H^1_0(\Omega)$ that $V^0_H \equiv V^0_{Q'}$ and that $Q^0_H = \underline{\text{grad}}(V^0_H) \subseteq (H^1(\Omega))^2$.

Let V_h and Q_h be finite dimensional subspaces of V and Q respectively. We can now define the discrete eigenvalue problem,

$$\begin{cases} \displaystyle\int_\Omega \underline{u}_h \cdot \underline{v}_h \, dx + \int_\Omega p_h \text{ div } \underline{v}_h \, dx = 0, \ \forall \, \underline{v}_h \in V_h, \\ \displaystyle\int_\Omega \text{div } \underline{u}_h \, q_h \, dx = -\lambda_h \int_\Omega p_h \, q_h \, dx, \forall \, q_h \in Q_h. \end{cases} \tag{7.1.40}$$

We consider, first, classical approximations of $H(\text{div};\Omega)$; for instance, we can choose as V_h the spaces of the elements $\mathcal{RT}_k$, $\mathcal{BDM}_k$ and $\mathcal{BDFM}_k$ introduced in Chap. 2.

Correspondingly, Q_h will be the space $D_k = \text{div } V_h$ defined in (7.1.18).

As stated in the previous section (see (7.1.28)), a B-compatible operator satisfying (6.5.55) can be constructed for all these choices of finite element spaces.

Moreover, we have, by Proposition 2.5.4, that (6.5.56) holds true for the approximations that we consider. Since $Q_h = \text{div } V_h$, then $\text{Ker}B_h \subseteq \text{Ker}B$, hence the discrete ellipticity on the kernel (6.5.9) and (6.5.52) trivially hold. To apply the theory of Sect. 6.5.5, there remains only to verify the strong approximability of Q^0_H, that is

$$||q - q^I||_0 \leq \omega_5(h)||q||_1 \quad \text{for all } q \in H^1(\Omega), \tag{7.1.41}$$

which also holds thanks to standard approximation properties of piecewise polynomial spaces. We may thus state that the eigenvalue problem (7.1.40) is a good approximation of (7.1.39) when proper finite element spaces are employed.

However, for various reasons, see for instance [52, 378], one might want to approximate V_h by continuous functions, for instance the space $\mathcal{L}^1_1$ introduced in Chap. 2, using therefore finite element spaces that are not especially fit for mixed formulations.

In constructing these new spaces, one might believe that the standard discrete ellipticity and discrete *inf-sup* conditions would be sufficient in order to approximate correctly eigenvalues and eigenvectors, once Q_h satisfies the *strong approximability in* Q^0_H, assumption of (6.5.53). However, while conditions (6.5.52), (6.5.54) and (6.5.55), defining a B-compatible operator, can be deduced from the discrete ellipticity on the kernel and the discrete *inf-sup* condition, the bound (6.5.56) does not, as it is shown by the following choice of the so-called $P_1-\text{div}(P_1)$ element on a criss–cross mesh. For this example, we shall see that discrete ellipticity in the kernel and discrete *inf-sup* condition are satisfied, while the eigenvalues are not correctly approximated.

Example 7.1.1. We introduce an approximation which will be considered later, in Example 8.10.2 of Chap. 8. As we shall then see, this is a case where the discrete *inf-sup* condition is not satisfied for the couple of spaces $(V, Q) = (H^1_0(\Omega), L^2(\Omega))$ which we have to consider for the Stokes problem. However, we are in a different situation, with $V := H(\text{div};\Omega)$ and an *inf-sup* condition that depends on the norms of the spaces at hand.

Let us then assume that Ω is a square, which is divided into $2N \times 2N$ sub-squares, each of them partitioned into four triangles K by its diagonals, thus defining a triangulation $\mathcal{T}_h$. Then, we set

$$\begin{aligned} V_h &:= \mathcal{L}_1^1, \\ Q_h &:= \text{div}(V_h) \subset \{q_h \mid q_h|_K \in P_0(K)\ \forall K \in \mathcal{T}_h\}. \end{aligned} \tag{7.1.42}$$

In [82], it was proved that the pair (V_h, Q_h) defined in (7.1.42) satisfies both the discrete ellipticity and the discrete *inf-sup* conditions but that the sequence T_Q^h does not converge uniformly to T_Q in $L^2(\Omega)$. This fact produces in the numerical computations spurious eigenvalues which converge to points not belonging to the resolvent set of T_Q. It was also proved in [82] that on the same regular mesh as above, the $\underline{Q}_1 - P_0$ approximation of Sect. 8.10.2 also satisfies a discrete *inf-sup* condition on $(H(\text{div};\Omega), L^2(\Omega))$ and leads to the same kind of spurious eigenvalues.The point is that the proof of the *inf-sup* condition does not yield (6.5.56). □

Hence, (6.5.56), which we have seen to be necessary, has to be checked independently of the discrete ellipticity and the discrete *inf-sup* conditions. On the other hand, as we have seen in Sect. 1.2.1, discrete ellipticity on the kernel is not necessary, and we can obtain convergence of eigenvalues with finite element spaces that fail to satisfy it, as we see in the following example.

Example 7.1.2. Let us consider, on a quasi-uniform triangulation, an approximation of our eigenvalue problem, taking $V_h := \mathcal{BDM}_3$ and $Q_h^1 := \mathcal{L}_1^0$ instead of $\mathcal{L}_2^0$, which would correspond to $Q_h = \text{div}\, V_h$ as in (7.1.18) and would provide, as we noted above, a correct approximation. Now, however, having chosen a smaller Q_h, we obtain a bigger $\text{Ker}B_h$ (not any more contained in $\text{Ker}B$). This will not jeopardise property (6.5.56) (the Π_h operator working for the pair $(\mathcal{BDM}_3, \mathcal{L}_2^0)$ will also work for the pair $(\mathcal{BDM}_3, \mathcal{L}_1^0)$) but (6.5.52) is now at risk. However, by the inverse inequality (see (2.2.60)),

$$b(u_h, q) = b(u_h, q - q^I) \le c\|u_h\|_1 \|q - q^I\|_0 \le ch^{-1}\|u_h\|_a\, h^2 \|q\|_2, \tag{7.1.43}$$

for $u_h \in \text{Ker}B_h$ and $q^I = L^2$-projection of q onto $\mathcal{L}_1^0$. Notice that this argument will work for any pair $(\mathcal{BDM}_k, \mathcal{L}_r^0)$ provided $k \ge (r+2) > 2$ (in particular, $(\mathcal{BDM}_2, \mathcal{L}_0^0)$ will not work). □

7.1.4 Primal Hybrid Methods

We now consider for the first time a non-standard method (cf. [332]) based on domain decomposition. We place ourselves in the framework of Example 1.3.4.

To avoid complicating our presentation, we shall restrict ourselves to problem (7.1.1) in which $\Gamma_D = \Gamma$, that is, with Dirichlet boundary conditions on the whole of Γ. This restriction is in no way essential and does not diminish the generality of our results. We thus want to find $p \in H^1_0(\Omega)$ solution of the minimisation problem,

$$\inf_{q \in H^1_0(\Omega)} \frac{1}{2}\int_\Omega A\underline{\mathrm{grad}}q \cdot \underline{\mathrm{grad}}q\,dx - \int_\Omega fq\,dx, \tag{7.1.44}$$

or equivalently of the variational problem

$$\int_\Omega A\underline{\mathrm{grad}}p \cdot \underline{\mathrm{grad}}q\,dx = \int_\Omega fq\,dx,\ \forall\, q \in H^1_0(\Omega). \tag{7.1.45}$$

Introducing now a partition of Ω into elements, it is natural (this is in fact one of the basic ideas of the finite element method) to define p on each element and to impose continuity conditions at the interfaces. The standard assembly process is based on this idea. We now follow a slightly different route. We use $X(\Omega) = \prod_r H^1(K_r)$ as defined by (2.1.43) with the product norm (2.1.44). $H^1_0(\Omega)$ is then a closed subspace of $X(\Omega)$ and the fact of belonging to $H^1_0(\Omega)$ can be considered as a linear constraint on p. From this, we can transform (7.1.44) into a saddle point problem:

$$\inf_{q \in X(\Omega)}\ \sup_{\underline{v} \in H(\mathrm{div};\Omega)} \sum_K \Big\{\frac{1}{2}\int_K A\underline{\mathrm{grad}}q \cdot \underline{\mathrm{grad}}q\,dx - \int_{\partial K} q\,\underline{v}\cdot\underline{n}\,ds - \int_K f\,q\,dx\Big\}, \tag{7.1.46}$$

where we formally write $\int_{\partial K} q\,\underline{v} \cdot \underline{n}\,ds$ for the duality between $H^{\frac{1}{2}}(\partial K)$ and $H^{-\frac{1}{2}}(\partial K)$. The optimality conditions of problem (7.1.46) are indeed:

$$\sum_K \Big\{\int_K A\underline{\mathrm{grad}}p \cdot \underline{\mathrm{grad}}q\,dx - \int_{\partial K} q\,\underline{u}\cdot\underline{n}\,ds - \int_K f\,q\,dx\Big\} = 0\ \forall q \in X(\Omega),$$

$$\sum_K \Big\{\int_{\partial K} p\,\underline{v}\cdot\underline{n}\,ds\Big\} = 0\ \ \forall\, \underline{v} \in H(\mathrm{div};\Omega). \tag{7.1.47}$$

From Proposition 2.1.1, we then have $p \in H^1_0(\Omega)$ so that p satisfies (7.1.45). Let us now set our problem in the framework of the general theory of Chap. 4. Taking $V := X(\Omega)$ and $Q := H(\mathrm{div};\Omega)$, we then define

$$a(p,q) := \sum_K \Big\{\int_K A\underline{\mathrm{grad}}p \cdot \underline{\mathrm{grad}}q\,dx\Big\}\quad \forall\, p,q \in V \tag{7.1.48}$$

and

$$b(q,\underline{v}) := \sum_K \Big\{-\int_{\partial K} q\,\underline{v}\cdot\underline{n}\,d\sigma\Big\}\quad \forall\, q \in V,\ \forall\, \underline{v} \in Q, \tag{7.1.49}$$

always using the formal integral notation for the duality between $H^{-\frac{1}{2}}(\partial K)$ and $H^{\frac{1}{2}(\partial K)}$. The bilinear form $b(q, \underline{v})$ defines an operator B from V into Q'. We have, from Propositions 2.1.1 and 2.1.2,

$$\mathrm{Ker} B = H_0^1(\Omega) \tag{7.1.50}$$

and

$$\begin{aligned}\mathrm{Ker}\, B^t &= \{\underline{v} \mid \underline{v} \in H(\mathrm{div}; \Omega), \underline{v} \cdot \underline{n}|_{\partial K} = 0, \ \forall\, K \in \mathcal{T}_h\} \\ &= \prod_K H_{0,\partial K}(\mathrm{div}; K).\end{aligned} \tag{7.1.51}$$

Loosely speaking, the operator B associates to p its jumps on inter-element interfaces. We could also have defined it from V *onto* the space

$$\mathcal{M}^{\frac{1}{2}} = \prod_K H^{\frac{1}{2}}(\partial K). \tag{7.1.52}$$

We thus want to check the closedness of $\mathrm{Im} B$ by obtaining an inequality of the form

$$\sup_{q \in V} \frac{b(q, \underline{v})}{\|q\|_V} \geq k\, \|\underline{v}\|_{Q/\mathrm{Ker} B^t}. \tag{7.1.53}$$

In the present case, it is obvious that one has

$$\sup_{q \in V} \frac{b(q, \underline{v})}{\|q\|_V} = \frac{1}{2}\Big\{\sum_K (\|\underline{v} \cdot \underline{n}\|_{-1/2,\partial K})^2\Big\}^{1/2} \tag{7.1.54}$$

and to obtain (7.1.53), it is sufficient to show that one has (on each element)

$$\inf_{\underline{v}_0 \in H_{0,\partial K}(\mathrm{div}; K)} \|\underline{v} + \underline{v}_0\|_{H(\mathrm{div}; K)} \leq \|\underline{v} \cdot \underline{n}\|_{-1/2,\partial K}. \tag{7.1.55}$$

However, (7.1.55) is readily obtained by solving a Neumann problem

$$\int_K \underline{\mathrm{grad}}\, \phi \cdot \underline{\mathrm{grad}} q \, dx + \int_K \phi q \, dx = \int_{\partial K} \underline{v} \cdot \underline{n}\, q \, ds. \tag{7.1.56}$$

Setting $\hat{\underline{v}} = \underline{\mathrm{grad}}\, \phi$, we have $\hat{\underline{v}} \cdot \underline{n} = \underline{v} \cdot \underline{n}$ and $\mathrm{div}\, \hat{\underline{v}} = \phi \in L^2(K)$. Moreover, we have

$$\|\hat{\underline{v}}\|_{H(\mathrm{div}; K)} = \|\phi\|_{H^1} \leq \|\underline{v} \cdot \underline{n}\|_{-1/2,\partial K} \tag{7.1.57}$$

and (7.1.55) follows.

Proposition 7.1.3. *Let $f \in L^2(\Omega)$ be given. There exists a solution $(p, \underline{u})$ to problem* (7.1.47)*. The first component is unique and the second one is defined up to an element of* $\mathrm{Ker} B^t$ *as defined by* (7.1.51)*.*

Proof. Assumption (7.1.2) made on A implies that $a(\cdot,\cdot)$ is coercive on $\mathrm{Ker}B = H_0^1(\Omega)$. The result follows by the closedness of $\mathrm{Im}B$ and Theorem 4.2.2. □

Remark 7.1.6. The first component p is of course the unique solution of problem (7.1.1). The second component *can be chosen* so that div $\underline{u} + f = 0$. Indeed, taking $q = 1$ on K and 0 elsewhere, we have from (7.1.37) for any solution $\underline{u}_0$,

$$\int_{\partial K} \underline{v}_0 \cdot \underline{n}\, d\sigma = \int_K \operatorname{div} \underline{v}_0\, dx = -\int_k f\, dx. \tag{7.1.58}$$

It is then possible to solve on K the Neumann problem,

$$\begin{cases} -\Delta\phi = f + \operatorname{div} \underline{u}_0, \\ \left.\dfrac{\partial\phi}{\partial n}\right|_{\partial K} = 0. \end{cases} \tag{7.1.59}$$

The solution exists and is defined up to an additive constant, as the right-hand side is compatible. Then, $\underline{v}_0 = \underline{\mathrm{grad}}\,\phi \in \mathrm{Ker}B^t$ and $\underline{u}_1 = \underline{u}_0 + \underline{v}_0$ satisfies div $\underline{u}_1 + f = 0$. □

Remark 7.1.7. It is moreover possible to choose $\underline{u} = A\underline{\mathrm{grad}}\,p$. Indeed, there comes from the first equation of (7.1.47) that $\underline{u}\cdot\underline{n}|_{\partial K} = A\,\underline{\mathrm{grad}}\,p\cdot\underline{n}|_{\partial K}$ on any $K \in \mathcal{T}_h$. □

We are now able to consider a discretisation of problem (7.1.47). We shall use, as an example,

$$V_h := \mathcal{L}^0_{k+1}(\Omega) \subset V = X(\Omega) \tag{7.1.60}$$

$$Q_h := \{\underline{v}_h |\, \underline{v}_h \in H(\mathrm{div};\Omega),\ \underline{v}_h\cdot\underline{n} \in R_k(\partial K),\ \forall\, K \in \mathcal{T}_h\}. \tag{7.1.61}$$

Note that only the traces of vectors in Q_h are polynomials. Our space Q_h is in fact infinite dimensional. This is no problem in practice as only the (finite dimensional) traces are used in computing. We then solve the discrete problem

$$\begin{cases} \displaystyle\sum_K \left\{ \int_K A\underline{\mathrm{grad}}\,p_h \cdot \underline{\mathrm{grad}}\,q_h\, dx - \int_{\partial K} q_h\, \underline{u}_h\cdot\underline{n}\, ds - \int_K f\, q_h\, dx \right\} = 0 \\ \qquad\qquad\qquad\qquad\qquad\qquad\qquad\qquad\qquad \forall\, q_h \in V_h, \\ \displaystyle\sum_K \left\{ \int_{\partial K} p_h\, \underline{v}_h\cdot\underline{n}\, ds \right\} = 0\ \forall\, \underline{v}_h \in Q_h. \end{cases} \tag{7.1.62}$$

The first step in the analysis of such a discretisation is to examine the properties of the operator B_h associated with the bilinear form $b(q_h, \underline{v}_h)$.

The first point that comes out is that we do not have, as in the previous example, $\mathrm{Ker}B_h \subset \mathrm{Ker}B$; that is, functions in $\mathrm{Ker}B_h$ do not belong to $H_0^1(\Omega)$. It is, however, easy to see that their moments up to order k are continuous across inter-element boundaries. This in turn implies, as the traces are polynomials of degree $k+1$, that we have continuity of the functions in $\mathrm{Ker}B_h$ at the $k+1$ Gauss-Legendre points, associated to a quadrature formula of degree $k+2$, on every interface. Eliminating the Lagrange multiplier $\underline{v}_h$ thus yields a nonconforming approximation of problem (7.1.47), namely: *find* $u_h \in \mathrm{Ker}B_h$ *solution of*

$$\sum_K \Big\{ \int_K A \underline{\mathrm{grad}}\, p_h \cdot \underline{\mathrm{grad}}_h \, dx \Big\} = \int_\Omega f\, q_h \, dx \;\forall\, q_h \in \mathrm{Ker}B_h. \tag{7.1.63}$$

We have already considered such approximations in Chap. 2 and their analysis is fairly well established [142, 147, 148, 165, 211, 358, 360].

One can therefore say that primal hybrid methods are another way of introducing nonconforming methods. The new point is to introduce an approximation of $\underline{u} = A\underline{\mathrm{grad}}\, p$ which is more regular than the approximation deduced directly from p_h. Moreover, this approximation can be built in order to satisfy the equilibrium conditions. Finally, the convergence analysis through the saddle point approach is simpler than the standard one and permits to introduce correctly the "patch test" arising in the analysis of nonconforming methods.

Before coming to this point, we first have to show existence and uniqueness of a solution. With respect to the existence and uniqueness of the solution p_h of (7.1.63), we fortunately have no problem. It is obvious that

$$|q_h|_{V_h} = \sqrt{a(q_h, q_h)} \tag{7.1.64}$$

defines on V_h a continuous semi-norm. The kernel of this semi-norm is

$$M := \{q_h \,|\, q_h \in L^2(\Omega),\; q_h|_K \in P_0(K)\} = \mathcal{L}_0^0, \tag{7.1.65}$$

and we have $M \cap \mathrm{Ker}B_h = 0$ so that $a(p_h, q_h)$ is coercive on $\mathrm{Ker}B_h$:

$$a(q_{0h}, q_{0h}) \geq \alpha_h \, |q_{0h}|^2_{V_h} \;\forall\, q_{0h} \in \mathrm{Ker}B_h. \tag{7.1.66}$$

We do not know, however, how α_h depends on h. This would require a discrete Poincaré inequality and is quite technical to prove.

We shall rather obtain an error bound in the semi-norm $|q_h|_{V_h}$ using Proposition 5.2.2. In order to do so, we must build an interpolate $\Pi_h \underline{u}$ of $\underline{u}$ such that

$$b(q_h, \underline{u} - \Pi_h \underline{u}) = 0 \;\forall\, q_h \in M. \tag{7.1.67}$$

However, this is immediate by taking $\Pi_h \underline{u}$ as defined by (2.5.26), provided $\underline{u}$ is at least in W, defined by (7.1.14). We thus obtain the error bound

$$|p - p_h|_{V_h} \leq c \Big(\inf_{q_h \in \mathrm{Ker}B_h} |p - q_h|_{V_h} + \|\underline{u} - \Pi_h \underline{u}\|_Q \Big). \tag{7.1.68}$$

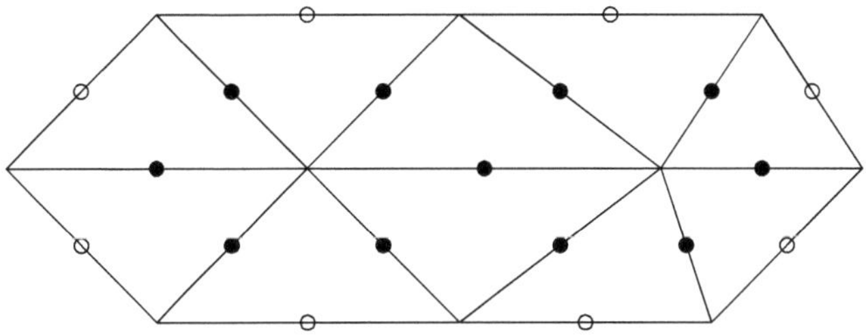

Fig. 7.1 ker B_h

Such an estimate is typical of nonconforming methods. The first term is readily estimated by standard methods, that is, by the use of some interpolation operator. The second one has already been considered. It must be remarked that as $\underline{u}$ is defined only up to an element of $\mathrm{Ker}B^t$, this norm depends in fact only on the values of $\underline{u} \cdot \underline{n}$ and $\underline{u}_h \cdot \underline{n}$ on the boundary of the elements. If $\underline{u}$ is regular, we get the same order of accuracy as in the first term. We therefore recognise here a form of the classical patch test: moments up to order k must be continuous to get the optimal convergence rate [142]. This corresponds to the choice of multipliers belonging to $P_k(e_i)$ on interfaces and thus to the choice (7.1.61) for Q_h. *Consistency terms that appear in the analysis of nonconforming methods are nothing but the contribution of the dual variable to error estimates.* Choosing a poorer approximation would destroy convergence properties. The main difficulty in the present situation will be to study the convergence of $\underline{u}_h$. To do so, we now have to check the *inf-sup* condition. We shall try to do it by the criterion of Proposition 5.4.3, that is, by building a proper interpolation operator for $p \in V$: *given* $p \in V = X(\Omega)$, *one must find* $\tilde{p}_h \in V_h$ *such that*

$$b(p - \tilde{p}_h, \underline{v}_h) = 0, \ \forall \, \underline{v}_h \in Q_h, \tag{7.1.69}$$

and depending continuously on p. This would prove $\mathrm{Ker}B_h^t \subset \mathrm{Ker}B^t$ and the *inf-sup* condition. We must then distinguish between two cases depending on whether k is even or odd. To make things simpler, we shall restrict our presentation to $k = 1$ or 2 (which are, by far, the most important in practice).

Example 7.1.3 (Hybrid method $k = 1$). This is the simplest case of a primal hybrid method (or nonconforming method). Functions of V_h are piecewise linear and $\mathrm{Ker}B_h$ contains those of them which are continuous at mid-side points on interfaces (Fig. 7.1).

The space $Q_h/\mathrm{Ker}B_h^t$ can be assimilated here to $\mathcal{RT}_0(\Omega)$. Now taking $p \in X(\Omega)$, one readily builds $\tilde{p}_h$ by taking on each K

$$\int_{e_i} \tilde{p}_h \, ds = \int_{e_i} p \, ds, \ i = 1, 2, 3. \tag{7.1.70}$$

Fig. 7.2 Gauss-Legendre points and the nonconforming elliptic bubble

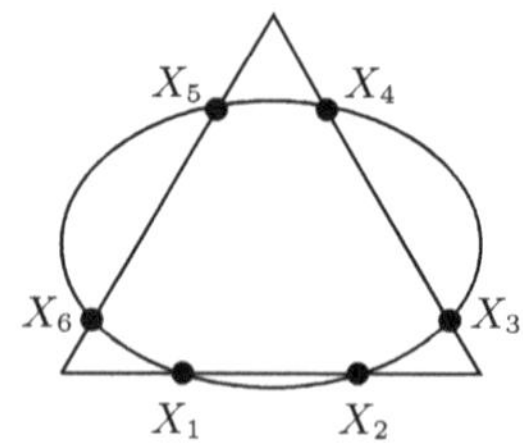

It is then obvious that (7.1.69) holds. Moreover, checking continuity is straightforward so that we have the error bound:

$$\|\underline{u} - \underline{u}_h\|_{Q/\mathrm{Ker}B^t} \leq c\,(\|\underline{u} - \underline{v}_h\|_{Q/\mathrm{Ker}B^t} + |p - p_h|_{V_h}),\ \forall\, \underline{v}_h \in Q_h. \qquad (7.1.71)$$

In practice, this means that one can extract from such a nonconforming formulation an approximation of $\underline{\mathrm{grad}}\,p$ that is better than the direct one $\underline{\mathrm{grad}}\,p_h$. We shall see later (Remark 7.4.2) how this approximation can be easily deduced from the standard one by a simple post-processing trick [291]. □

Example 7.1.4 (Hybrid Method, $k = 2$). This hybrid formulation yields the next simpler case of a nonconforming method. Its use was long rejected because of a problem in the choice of the degrees of freedom. Although the functions of $\mathrm{Ker}B_h$ are continuous at two Gauss–Legendre points on each side, these points cannot be used as degrees of freedom for their values are linked by a linear relation. Indeed, let a_i $(1 \leq i \leq 6)$ be the six values of a second degree polynomial on the six Gauss-Legendre points of the sides of a triangle (Fig. 7.2) that is $a_i = p_2(x_i)$.

One then has

$$(a_6 - a_5) + (a_4 - a_3) + (a_2 - a_1) = \int_{\partial K} \frac{\partial p_2}{\partial t} ds = 0 \qquad (7.1.72)$$

[209]. In Example 2.2.5, we called *nonconforming bubble* the second-degree function vanishing at the six Gauss–Legendre points ($a_i = 0$) and taking value 1 at the barycentre of K. There also follows from (7.1.72) that one cannot define $\tilde{p}_h|_K$ by the six moments

$$\int_{e_i} \tilde{p}_h\, \phi_i\, ds,\ \phi_i \in P_1(e_i) \qquad (7.1.73)$$

and this precludes checking (7.1.69) by the simple method of the previous example. Considering the problem a little more thoroughly, one then sees that $\mathrm{Ker}B_h^t \not\subset \mathrm{Ker}B^t$ and that (7.1.71) cannot hold.

Indeed, $\mathrm{Ker}B_h^t$ contains one vector $\underline{v}_h^0$ that does not lie in $\mathrm{Ker}B^t$. It is sketched in Fig. 7.3, where the symbols + and – represent equal absolute values of the normal component of $\underline{v}_h$.

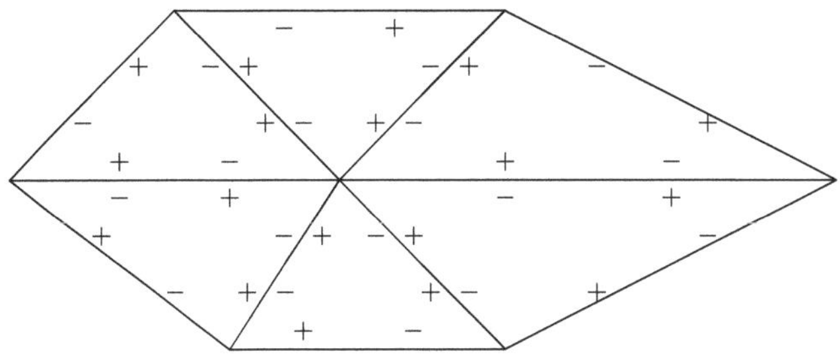

Fig. 7.3 The vector $\underline{q}_h^0$

This is the first occurrence of a pathological situation where the *inf-sup* condition does not hold. In principle, this should imply some compatibility condition on the data. However, in the present case, the second equation of (7.1.47) is always solved with a zero right-hand side and $Z_h(g) = Z_h(0) = \mathrm{Ker}B_h$ is always non-empty.

It must be noted that, contrarily to other cases of spurious modes that we shall meet, for instance in Chap. 8, the existence of $\underline{u}_h^0$ does not depend on the mesh. Moreover, we know that its existence does not compromise the error estimates on p_h. One may therefore wonder if some convergence of $\underline{u}_h$ could not be obtained, modulo $\underline{v}_h^0$, that is, using an *inf-sup* condition of the form

$$\sup_{q_h \in V_h} \frac{b(q_h, \underline{v}_h)}{\|q_h\|_V} \geq k_0 \, \|\underline{v}_h\|_{Q_h/\mathrm{Ker}B_h^t}. \tag{7.1.74}$$

From Proposition 5.4.3, this will hold if, given $p \in V$, $b(p, \underline{v}_h^0) = 0$, one can build $\tilde{p}_h \in V_h$ such that (7.1.69) holds.

This can, indeed, be done through a construction that is not local and for which we do not know how to prove that the operation $\Pi_h : p \to p_h$ is uniformly continuous (with respect to h). We shall however be able to prove a partial result: $\underline{u}_h$ will converge in a quotient space Q_h/M_h with $\mathrm{Ker}B_h^t \subset M_h$. In order to see this, we first define on every element K

$$\underline{v}_K^0 = \underline{v}_h^0|_K, \tag{7.1.75}$$

and we denote by Q_K^0 the one-dimensional space generated by $\underline{v}_K^0$. We then define $M_h := \sum_K Q_h^0$ and

$$Q_h^* := Q_h + M_h. \tag{7.1.76}$$

It must be noted that $Q_h^* \not\subset H(\mathrm{div}; \Omega)$ so that we must now consider a nonconforming framework replacing Q by $Q^* = \prod_K H(\mathrm{div}; K)$ as in Sect. 5.5.4.

Let us first remark that the proof given for $\mathrm{Im}B$ to be closed is directly extended to the operator $B^* : V \to Q^*$ now associated with the bilinear form $b(\cdot,\cdot)$ as this proof did not rely on any continuity property. It is also easy to check that one now has

$$\text{Ker} B_h^{*t} = M_h + \text{Ker} B^t, \tag{7.1.77}$$

where B_h^* is evidently defined by the extension of $b(\cdot, \cdot)$ to $V_h \times Q_h^*$.

Now considering the problem

$$\begin{cases} a(p_h^*, q_h) + b(q_h, \underline{u}_h^*) = \langle f, q_h \rangle, \ \forall\, q_h \in V_h, \\ \qquad\qquad\qquad b(p_h^*, \underline{v}_h^*) = 0, \ \forall\, \underline{v}_h^* \in Q_h^*, \end{cases} \tag{7.1.78}$$

it is easy to see that $p_h^* = p_h$. This comes from

$$b(p_h, \underline{m}_h) = 0, \ \forall\, \underline{m}_h \in M_h, \tag{7.1.79}$$

which is a direct consequence of (7.1.78). We thus have increased the indeterminacy of $\underline{v}_h$ without changing p_h. To prove convergence, we shall use Proposition 5.3.1 which is directly suitable. Let us thus define in the notations of this proposition,

$$\begin{aligned} \tilde{Q}_h &:= M_h, \\ \hat{Q}_h &:= Q_h^* \backslash M_h, \\ \hat{V}_h &:= V_h. \end{aligned} \tag{7.1.80}$$

From (7.1.79), we have $b(q_h, \tilde{\underline{v}}_h) = 0, \ \forall\, q_h \in V_h, \ \forall\, \tilde{\underline{v}}_h \in \tilde{Q}_h$, and there remains to prove that $b(\cdot, \cdot)$ satisfies an *inf-sup* condition on $V_h \times \hat{Q}_h$.

To do so, by Proposition 5.4.3, one must build in a continuous way $p_h = \Pi_h p$ such that

$$b(\tilde{p}_h - p, \hat{\underline{v}}_h) = 0 \ \forall\, \hat{\underline{v}}_h \in \hat{Q}_h. \tag{7.1.81}$$

Working in $\hat{Q}_h$ (from which components $\underline{v}_K^0$ have been removed) now enables us to do it in a local way, that is, element by element. It is indeed sufficient, as in the previous example ($k = 1$), to interpolate $\underline{u}$ using its moments. This does not determine $\tilde{p}_h$ in a unique way and a minimum norm solution has to be chosen to get the desired uniform continuity property. We thus have the *inf-sup* condition (k being independent of h),

$$\sup_{q_h \in V_h} \frac{b(q_h, \underline{v}_h^*)}{\|q_h\|_{V_h}} \geq k_0 \, \|\underline{v}_h^*\|_{Q_h^*/M_h} \ \forall\, \underline{v}_h^* \in Q_h^*. \tag{7.1.82}$$

From the results of Sect. 5.3.3, we obtain that if the exact solution p satisfies

$$b(p, \underline{m}_h) = 0 \ \forall\, \underline{m}_h \in M_h = \tilde{Q}_h, \tag{7.1.83}$$

we have the estimate

$$\|\underline{u} - \underline{u}_h^*\|_{Q_h^*/M_h} \leq \left(\inf_{\underline{v}_h \in Q_h} ||\underline{u} - \underline{v}_h||_Q + |p - p_h|_{V_h} \right) + \inf_{\underline{v}_h^* \in Q_h^*} ||\underline{u} - \underline{v}_h^*||_{Q^*}. \tag{7.1.84}$$

□

Remark 7.1.8. It must be noted that condition (7.1.83) is not so stringent as it may appear. Indeed, given $p \in V$ and replacing Bp by $B\hat{p}_h$, with $\hat{p}_h$ the interpolate of p in V_h, introduces a perturbation of the problem which now has $\hat{p}_h$ as a solution. This means that, by a slight modification of the data, it is possible to switch from a non-compatible problem to a compatible one without really changing the solution. □

Remark 7.1.9. Knowing that $\underline{v}_h^* \in Q_h^*$ implies that $\underline{v}_h \cdot \underline{n}$ is continuous at mid-points of the interfaces. It can be checked, using the results of [209], that the converging part of $\underline{u}_h$ is sometimes in fact equal to $\underline{\text{grad}}\, p_h$ which satisfies the same continuity properties for some right-hand sides. However, the procedure sketched above can be extended to higher approximations, the case $k = 4$ for instance, where this equality will no longer hold. □

Remark 7.1.10. In the case $k = 2$, it is possible to build the solution $\underline{u}_h$ of (7.1.70) starting from $\underline{\text{grad}}\, \tilde{p}_h$ with $\tilde{p}_h \in \tilde{Q}_h$. The trick is to use a spanning tree of the elements: starting from the root, one can then adjust $\alpha_K \underline{v}_K^0$ on each element so that $\underline{v}_h \cdot \underline{n}$ is continuous on the interfaces with previously visited elements. The properties of $\underline{\text{grad}}\, p_h$ shown in [209] enable us to do so in a unique way as $\alpha_K \underline{v}_K^0$ can be chosen arbitrarily on the root of the spanning tree. This is obviously not a local construction. Its continuity depends on the diameter of the spanning tree and thus on h and this leads us to believe that our result is probably optimal. This is not the case for the construction for $k = 1$ described earlier, which is local. □

7.1.5 Primal Macro-hybrid Methods and Domain Decompositions

We consider again problem (7.1.45), but this time we suppose that we have a two level-decomposition of Ω: a coarse-level decomposition into *macrolelements* (or *subdomains*) $\Omega_1, \dots, \Omega_S$ of a rather arbitrary shape, and then a further decomposition of each macroelelement into more traditional finite elements (let's say "triangles" to fix the ideas).

We now start by considering, this time, $X(\Omega) = \prod_s H^1(\Omega_s)$ (again as defined by (2.1.43) with the product norm (2.1.44)). Mimicking (7.1.47) we can now consider the problem

$$\sum_s \left\{ \int_{\Omega_s} A\underline{\text{grad}}\, p \cdot \underline{\text{grad}}\, q \, dx - \int_{\partial\Omega_s} q \, \underline{u} \cdot \underline{n} \, d\ell - \int_{\Omega_s} f \, q \, dx \right\} = 0 \quad \forall q \in X(\Omega),$$

$$\sum_s \left\{ \int_{\partial\Omega_s} p \, \underline{v} \cdot \underline{n} \, d\ell \right\} = 0 \quad \forall \underline{v} \in H(\text{div}; \Omega). \qquad (7.1.85)$$

(again in the unknowns p and $\underline{u} \cdot \underline{n}$) where we keep writing $\int_{\partial\Omega_s} q\,\underline{v} \cdot \underline{n}\,d\ell$ for the duality between $H^{\frac{1}{2}}(\partial\Omega_s)$ and $H^{-\frac{1}{2}}(\partial\Omega_s)$. This time, however, the final discretization (with the passage to the finite dimensional case) is not made by choosing a polynomial space in each subdomain Ω_s, but rather discretizing each $H^1(\Omega_s)$ with a space V_h^s made of piecewise polynomials on the finer-level subgrid, while $\underline{u}\cdot\underline{n}$ (and $\underline{v}\cdot\underline{n}$) are instead discretized by suitable piecewise polynomials spaces $M_h(e)$ on each edge e (or face, in three dimensions). A convenient choice of this last discretization (on the interfaces) gives rise to the popular "mortar method" of Bernardi-Maday-Patera [69] in the framework of *domain decomposition methods*:

$$\sum_s \left\{ \int_{\Omega_s} A\underline{\text{grad}}\,p_h \cdot \underline{\text{grad}}\,q_h\,dx - \int_{\partial\Omega_s} q_h\,\underline{u_h} \cdot \underline{n}\,d\ell \right.$$
$$\left. - \int_{\Omega_s} f\,q_h \quad dx \right\} = 0 \quad \forall q_h \in \prod_s V_h^s, \tag{7.1.86}$$
$$\sum_s \left\{ \int_{\partial\Omega_s} p_h\,\underline{v_h} \cdot \underline{n}\,d\ell \right\} = 0 \;\; \forall\,\underline{v_h} \in \prod_e M_h(e).$$

(with the obvious convention that an element $\underline{v_h} \cdot \underline{n} \in \prod_e M_h(e)$ changes sign when considered as being defined on a certain $\partial\Omega_s$ or on the boundary of the abutting macro-element)

Remark 7.1.11. It is clear that, in the mortar-like approach as well, the interelement solution $\underline{u_h} \cdot \underline{n}$ will be an approximation of the flux $\underline{u} \cdot \underline{n} = A\underline{\text{grad}}\,p \cdot \underline{n}$ at the interelement boundaries. Substituting (brutally) $A\underline{\text{grad}}\,p_h \cdot \underline{n}$ in place of $\underline{u_h} \cdot \underline{n}$ in the first equation of (7.1.86), and considering for simplicity that the two decompositions (coarser and finer) coincide, we find then

$$\sum_K \left\{ \int_K A\underline{\text{grad}}\,p_h \cdot \underline{\text{grad}}\,q_h\,dx \right.$$
$$\left. - \int_{\partial K} q_h\,A\underline{\text{grad}}\,p_h \cdot \underline{n}\,d\ell - \int_{\Omega_s} f\,q_h\,dx \right\} = 0 \quad \forall q_h \in \prod V_h, \tag{7.1.87}$$

that is exactly the *non stabilised* version of the most elementary Discontinuous Galerkin approach [13] (often called Incomplete Interior Penalty Galerkin, or IIPG [361]).

7.1.6 Dual Hybrid Methods

We now turn to another use of domain decomposition, this time to solve the dual formulation (7.1.7) [333, 365]. In this formulation, the main difficulty is to work in the affine subspace of $H(\text{div};\Omega)$,

$$W_f := \{\underline{v}_f \,|\, \underline{v}_f \in H(\text{div}\ \Omega),\ \text{div}\ \underline{v}_f + f = 0\}. \tag{7.1.88}$$

If Neumann conditions were imposed on $\Gamma_N \subset \Gamma$, it would also be necessary to ask for $\underline{v}_f$ to satisfy

$$\underline{v}_f \cdot \underline{n}|_{\Gamma_N} = g_2. \tag{7.1.89}$$

The idea of the dual hybrid formulation will again be to relax continuity, this time for the normal trace $\underline{v}\cdot\underline{n}$ at interfaces between elements. Condition (7.1.89) will also be treated weakly. We thus transform problem (7.1.7) into

$$\inf_{\underline{v}_f \in V_f}\ \sup_{q_{g_1} \in Q_{g_1}} \frac{1}{2}\int_\Omega A^{-1}\underline{v}_f\cdot\underline{v}_f\,dx + \sum_K \int_{\partial K} \underline{v}_f\cdot\underline{n}\,q_{g_1}\,ds - \int_{\Gamma_N} g_2\,q_{g_1}\,ds, \tag{7.1.90}$$

where, denoting $Y(\Omega) = \prod\limits_K H(\text{div};K)$, one sets

$$V_f := \{\underline{v} \,|\, \underline{v} \in Y(\Omega),\ \text{div}\ \underline{v}|_K + f = 0,\ \forall\, K\}, \tag{7.1.91}$$

$$Q_{g_1} := \{q \,|\, q \in H^1(\Omega),\ q|_{\Gamma_D} = g_1\}. \tag{7.1.92}$$

Taking an arbitrary element $\hat{\underline{v}}_f$ of V_f and an arbitrary element $\hat{q}_{g_1}$ of Q_{g_1}, one may write (7.1.90) as

$$\inf_{\underline{v}_0 \in V_0}\ \sup_{q_0 \in Q_0} \frac{1}{2}\int_\Omega A^{-1}(\underline{v}_0+\hat{\underline{v}}_f)\cdot(\underline{v}_0+\hat{\underline{v}}_f)\,dx$$
$$+\sum_K \int_{\partial K} (\underline{v}_0+\hat{\underline{v}}_f)\cdot\underline{n}\,(q_0+\hat{q}_{g_1})ds - \int_{\Gamma_N} g_2(q_0+\hat{q}_{g_1})ds \tag{7.1.93}$$

where V_0 and Q_0 are defined by (7.1.91) and (7.1.92) with $f = 0$ and $g_1 = 0$. Denoting as in the previous section

$$b(\underline{v},q) = \sum_K \int_{\partial K} \underline{v}\cdot\underline{n}\,q\,ds, \tag{7.1.94}$$

problem (7.1.93) is equivalent to finding $(\underline{u}_0, p_0) \in V_0 \times Q_0$ solution of

$$\int_\Omega A^{-1}\underline{u}_0\cdot\underline{v}_0\,dx + b(\underline{v}_0, p_0) = -\int_\Omega A^{-1}\hat{\underline{v}}_f\cdot\underline{v}_0 - b(\underline{v}_0,\hat{q}_{g_1})\ \forall\,\underline{v}_0 \in V_0, \tag{7.1.95}$$

$$b(\underline{u}_0, q_0) = -b(\hat{\underline{v}}_f, q_0) + \int_N g_2\,q_0\,ds\ \forall\, q_0 \in Q_0. \tag{7.1.96}$$

This is now in standard form and we shall try to apply the general theory. First note that we have

$$\mathrm{Ker} B^t = \prod_K H_0^1(K) \tag{7.1.97}$$

and

$$\mathrm{Ker} B = \{\underline{v} \mid \underline{v} \in H_{0,\Gamma_N}(\mathrm{div};\Omega),\ \mathrm{div}\ \underline{v} = 0\}. \tag{7.1.98}$$

It is then clear that

$$a(\underline{u}, \underline{v}) = \int_\Omega A^{-1}\underline{u}\cdot\underline{v}\,dx \tag{7.1.99}$$

is coercive on V_0, and in order to apply our general existence result, one must show an *inf-sup* condition, that is, for all $v \in Q \equiv Q_0$,

$$\sup_{\underline{v}_0 \in V_0} \frac{b(\underline{v}_0, q_0)}{\|\underline{v}_0\|_{H(\mathrm{div}\ \Omega)}} \geq k_0 \|q_0\|_{Q/\mathrm{Ker}B^t}. \tag{7.1.100}$$

To obtain this, q being given, we select $q_0 \in Q$ such that

$$\begin{cases} -\Delta q_0 = 0 \text{ on each element } K \in \mathcal{T}_h, \\ q_0|_{\partial K} = q|_{\partial K}. \end{cases} \tag{7.1.101}$$

Now, we take $\underline{v}_0 = \underline{\mathrm{grad}} q_0$ and we have

$$\int_\Omega |\underline{\mathrm{grad}} q_0|^2\,dx = \sum_K \int_K |\underline{\mathrm{grad}} q_0|^2\,dx = \sum_K \int_{\partial K} q_0\,\underline{v}_0\cdot\underline{n}\,ds = b(\underline{u}_0, q_0). \tag{7.1.102}$$

Moreover, div $\underline{v}_0 = 0$ and, using Poincaré's inequality, we may write

$$\|\underline{u}_0\|_{(H\ \mathrm{div};\Omega)} = \|\underline{u}_0\|_0 = ||\underline{\mathrm{grad}} q_0||_0 \geq \frac{1}{C(\Omega)}\|q_0\|_{1,\Omega}, \tag{7.1.103}$$

provided the domain is bounded and Dirichlet conditions are imposed on a part of $\partial\Omega$ (that is $\Gamma_D \neq \emptyset$). From (7.1.101) and (7.1.103), we then have

$$\|q\|_{Q/\mathrm{Ker}B^t} \leq \|q_0\|_{1,\Omega} \leq \frac{b(\underline{u}_0, v)}{\|\underline{u}_0\|_{H(\mathrm{div};\Omega)}} \leq \sup_{\underline{v}_0} \frac{b(\underline{v}_0, q)}{\|\underline{v}_0\|_{H(\mathrm{div};\Omega)}}, \tag{7.1.104}$$

which is the desired result.

We now know that problem (7.1.95) and (7.1.96) has a unique solution (up to an element of KerB^t for q_0). Our concern is now to introduce a discretisation and, to do so, we shall again use the spaces defined in Chap. 2. We define, using the notation of Propositions 2.5.1 and 2.5.2,

$$V_h := \prod_K \underline{M}_k(K), \tag{7.1.105}$$

where $\underline{M}_k(K)$ is one of the approximations of $H(\text{div}; K)$ introduced in Sect. 2.3. We suppose $f|_K \in D_k = \text{div}(\underline{M}_k(K))$ so that it is possible to find $\hat{\underline{v}}_{hf}$ satisfying $\text{div}\ \hat{\underline{q}}_{hf} + f = 0$ in each K. In general, f can be approximated on D_k without loss of precision.

Using (2.2.6), we also set

$$\begin{aligned} Q_h &:= \{q_h \,|\, q_h \in H^1(\Omega), q_h|_{\partial K} \in T_{k+1}(\partial K),\ \forall\, K \in \mathcal{T}_h\}, \\ Q_{0h} &:= \{q_h \,|\, q_h \in Q_h, q_h|_D = 0\}. \end{aligned} \tag{7.1.106}$$

We now have again the unusual situation where the approximation Q_{0h} is infinite dimensional. However, only the traces on K are relevant to computation and we do not really have to worry about this. Moreover, the choice

$$V_{0h} := \{\underline{v}_h \,|\, \underline{v}_h \in V_h, \text{div}\ \underline{v}_h|_K = 0,\ \forall\, K \in \mathcal{T}_h\} \tag{7.1.107}$$

ensures that we have no problem with coerciveness of $a(\cdot,\cdot)$. This comes from the inclusion $V_{0h} \subset V_0$. The only crucial point with respect to convergence is thus to get a discrete *inf-sup* condition. We must now show that for any $q_{0h} \in Q_{0h}$, we have with k independent of h and q_{0h}

$$\sup_{\underline{v}_0 \in V_{0h}} \frac{b(\underline{v}_0, q_{0h})}{\|\underline{v}_0\|} \geq k\|q_{0h}\|_{Q/\text{Ker}B_h^t}. \tag{7.1.108}$$

The correct situation is of course obtained for $\text{Ker}B_h^t \subset \text{Ker}B^t$. To prove (7.1.108), we shall try, as usual, to use Proposition 5.4.3, that is to build a B-compatible interpolation. To do so, $\underline{v}_0 \in V_0$ being given, we should be able to build $\underline{v}_{0h} = \Pi_h \underline{v}_0$ such that

$$\begin{cases} b(\underline{v}_0 - \underline{v}_{0h}, q_{0h}) = 0 \ \ \forall\, q_{0h} \in Q_{0h} \\ \|\underline{v}_{0h}\|_0 \leq C\|\underline{v}_0\|_0, \end{cases} \tag{7.1.109}$$

with a constant C independent of h. To get this result, we shall build, for any $\underline{v} \in V$, $\underline{v}_h = \Pi_h \underline{v}$ in such a way that $\Pi_h \underline{v} \in V_{0h}$ if $\underline{v} \in V_0$, satisfying

$$b(\underline{v} - \underline{v}_h, q_h) = 0 \ \ \forall\, q_h \in Q_h. \tag{7.1.110}$$

From the definition of $b(\cdot,\cdot)$, this will, a fortiori, hold whenever one has

$$\int_{\partial K} (\underline{v} - \Pi_h \underline{v}) \cdot \underline{n}\, q_h\, ds = 0 \quad \forall\, q_h \in V_h\ \forall\, K \in \mathcal{T}_h. \tag{7.1.111}$$

Condition (7.1.110) is, however, nothing but a small linear system:

$$\int_{\partial K} \underline{v}_h \cdot \underline{n}\, q_h\, ds = \int_{\partial K} \underline{v} \cdot \underline{n}\, q_h\, ds \quad \forall\, q_h \in T_{k+1}(\partial K). \tag{7.1.112}$$

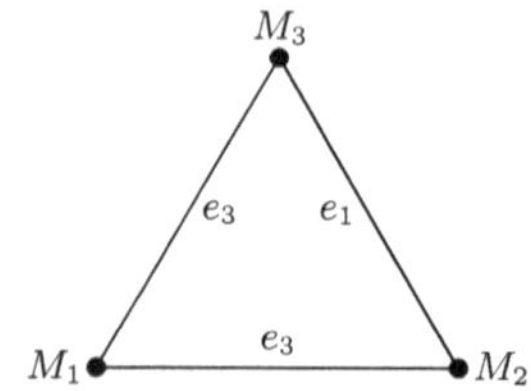

Fig. 7.4 The case k=0

We have again the same problem as in the previous section: solving (7.1.112) for $\underline{v}_h$ depends on the degree of polynomials at hand. As we shall see, the cure is, however, much simpler here. To fix ideas, we shall therefore consider two simple examples.

Example 7.1.5 ($k = 0$ triangular elements). This is the simplest case, and it is easily seen that system (7.1.112) can always be solved. Indeed, the degrees of freedom of $M_0(K)$ are the constant values $q_i = (\underline{v}_h \cdot \underline{n})_i$ on each side e_i of length ℓ_i of K when $D_0 = P_0(K)$. System (7.1.112) takes the form (cf. Fig. 7.4)

$$\begin{cases} \dfrac{1}{2}[(v_2)\ell_2 + (v_3)\ell_3] = \displaystyle\int_{\partial K} (\underline{v} \cdot \underline{n})\lambda_1 \, ds, \\ \dfrac{1}{2}[(v_3)\ell_3 + (v_1)\ell_1] = \displaystyle\int_{\partial K} (\underline{v} \cdot \underline{n})\lambda_2 \, ds, \\ \dfrac{1}{2}[(v_2)\ell_2 + (v_1)\ell_1] = \displaystyle\int_{\partial K} (\underline{v} \cdot \underline{n})\lambda_3 \, ds, \end{cases} \tag{7.1.113}$$

which can always be solved. Moreover, (7.1.113) implies, by summing the three equations,

$$v_1\ell_1 + v_2\ell_2 + v_3\ell_3 = \int_{\partial K} \underline{v}_h \cdot \underline{n} \, ds = \int_{\partial K} \underline{v} \cdot \underline{n} \, ds \tag{7.1.114}$$

so that $\underline{v} \in V_0$ implies $\underline{v}_h \in V_{0h}$. □

Remark 7.1.12. Let us recall that any divergence-free function of $\mathcal{RT}_k(K)$ is the curl of a stream function $\psi_h \in P_{k+1}$. If we want to check (7.1.112) only for divergence-free functions, which is sufficient to get the *inf-sup* condition, we can write, with $\partial\psi/\partial\tau$ denoting the tangential derivative of ψ on ∂K,

$$\int_{\partial K} \frac{\partial \psi_h}{\partial \tau} q_h \, ds = \int_{\partial K} \frac{\partial \psi}{\partial \tau} q_h \, ds, \tag{7.1.115}$$

where ψ is the stream function associated to $\underline{v}$.

System (7.1.115) is then always singular as, for q_h = constant on ∂K, both sides vanish. For $k = 0$ (and all even k), the system can always be solved. For the case $k = 1$ of our next example, an extra linear dependence will appear among the equations. We refer to Lemma 10.2.2, where a similar situation will be encountered in the analysis of hybrid methods for fourth-order problems. □

Example 7.1.6 (Case $k = 1$). We now use the space $M_1(K) = \mathcal{RT}_1(K)$ with $D_1 = P_1(K)$ (but this is not the only possible choice). The degrees of freedom of $\underline{v}_h$ are now given following Chap. 2 by two values (or moments) of the linear normal trace $\underline{v}_h \cdot \underline{n}$ on each side of K, plus two internal nodes which will be used to obtain the divergence-free condition on $\underline{v}_h$.

When trying to solve (7.1.112), we are again facing the same pathology that we had already met when studying primal hybrid methods: there exists a second-degree polynomial ϕ_K, which we already called the "nonconforming bubble" such that

$$\int_{\partial K} \underline{v}_h \cdot \underline{n}\, \phi_K \, ds = 0, \ \forall \, \underline{v}_h \in V_h. \tag{7.1.116}$$

This implies that system (7.1.112) is not of maximal rank and cannot, in general, be solved for a general $\underline{v}$. Our *local construction* thus fails. It can, however, be checked that $\mathrm{Ker} B_h^t \subseteq \mathrm{Ker} B^t$ because no function of Q_h can be built from nonconforming bubbles, satisfying Dirichlet conditions on a part of $\partial\Omega$. However, we know of no way to prove an *inf-sup* condition (if one holds).

The standard cure in such a situation is to use a richer space for V_0: we shall add to $M_1(K)$ one element of the next member of the family, that is, $\mathcal{RT}_2(K)$. Let us then define $\psi_{3K} \in P_3(K)$ such that,

$$\frac{\partial \psi_{3K}}{\partial \tau} = \phi_K. \tag{7.1.117}$$

We can now take $\psi_h \in P_2(K) \oplus$ span (ψ_{3K}) and the system (7.1.115) becomes of maximal rank and always has a solution. It is not unique and we may select the solution of minimal norm. □

Remark 7.1.13. Let $b_{3,K}$ be the cubic bubble on K. Taking $\underline{\tilde{v}}_h = \mathrm{curl}\, b_{3K}$ and working in the space $\mathcal{RT}_1(K) \oplus$ span $\underline{\tilde{v}}_h$, we could have found a solution of (7.1.112) and made it divergence-free using the internal nodes of $\mathcal{RT}_1(K)$. □

The above examples are quite representative of situations generally encountered in hybrid methods: construction of approximations differ for odd or even degrees. Whenever a difficulty arises, enrichment of V_h can be used to cure the trouble. From a *computational point of view*, this enrichment is not troublesome, since degrees of freedom of $\underline{v}_h$ are *internal* to the element ($Y(\Omega)$ satisfies no continuity on interfaces). The standard practice is then to use "static condensation" and to reduce the problem to degrees of freedom in V_h. Dual hybrid methods can then be seen as a variant of standard conforming methods in which the shape of approximations inside K is not specified. We still have a technical point to set. In order to apply Proposition 5.4.3, we must show that Π_h is continuous (uniformly in h) from V into V_{0h}. However, this reduces to continuity in $L^2(K)$ as div $\underline{v} = 0$ implies div $\underline{v}_{0h} = 0$. This is easily obtained by a scaling argument or, equivalently, by transforming the problem to a reference element. We must, however, do this through the Piola transformation (2.1.69) to make $\underline{\hat{v}}_0$ divergence-free. Continuity of Π_h is then easily deduced from (2.1.75) and (2.1.76) with a standard condition on the shape of elements.

Remark 7.1.14. For an application of dual hybrid methods to the problem of the torsion of an elastic bar the reader may refer to [320] and [113] for the corresponding mathematical analysis. □

7.2 Numerical Solutions

7.2.1 *Preliminaries*

In this section, we present some additional results on the application of mixed finite element methods to linear elliptic problems. In particular, we shall discuss some aspects of the numerical techniques that can be used for solving the linear system of equations that one obtains after discretisation. The procedure suggested here is essentially due (to our knowledge) to Fraeijs de Veubeke and, as we shall see, involves the introduction of suitable inter-element Lagrange multipliers λ. Such a trick has the remarkable effect of reducing the total number of unknowns and leads to solving a linear system for a matrix which is symmetric and positive definite instead of the original indefinite one. A rough analysis of the computational effort that this procedure requires for the various elements is presented in Sect. 7.3. Moreover, as we shall see in Sect. 7.4, the new unknown λs that are obtained by such a procedure allow the construction of a new approximation p_h^* of p, depending on λ and p_h, which is usually much closer to p. For the sake of simplicity, we shall present the arguments on the homogeneous isotropic model case.

$$\begin{cases} -\Delta p = f \text{ in } \Omega, \\ \quad p = g \text{ on } \Gamma_D, \\ \dfrac{\partial p}{\partial n} = 0 \text{ on } \Gamma_N, \end{cases} \tag{7.2.1}$$

although the range of generality is much wider. Similarly, we shall discuss in detail the simplest case of the approximation by means of the $\mathcal{RT}_0 \equiv \mathcal{BDFM}_1$ element and give statements and references for the proofs of the other cases.

7.2.2 *Inter-element Multipliers*

As we have seen in Sect. 7.1, the mixed formulation of (7.2.1) is

$$\begin{cases} (\underline{u}, \underline{v}) + (p, \operatorname{div} \underline{v}) = \langle g, \underline{v} \cdot \underline{n} \rangle \ \forall\, \underline{v} \in H_{0,\Gamma_N}(\operatorname{div}; \Omega), \\ (q, \operatorname{div} \underline{u}) = (f, q) \ \forall\, q \in L^2(\Omega), \end{cases} \tag{7.2.2}$$

where $p,\ f,\ g$ are the same as in (7.2.1) and $\underline{u} = \underline{\text{grad}}\, p$. Assume, for the sake of simplicity, that Ω is a polygon and let $\mathcal{T}_h$ be a triangulation of Ω. We recall, following Sect. 2.5.2, that the use of the $\mathcal{RT}_0$ element for the approximation of (7.2.2) proceeds through the following steps. We had

$$\mathcal{RT}_0(K) = \{(a + bx_1, c + bx_2),\ a,\ b,\ c\ \in \mathbb{R}\} \subseteq (P_1(K))^2, \tag{7.2.3}$$

$$\mathcal{M}^0 = \{\underline{v} \mid \underline{v}\ \in L^2(\Omega),\ \underline{v}\big|_K \in \mathcal{RT}_0(K),\ \forall\ K\ \in \mathcal{T}_h\}, \tag{7.2.4}$$

$$\mathcal{M} = \mathcal{M}^0 \cap H(\text{div}; \Omega) = \{\underline{v}\ \in \mathcal{RT}_0(\Omega, \mathcal{T}_h)\}. \tag{7.2.5}$$

Thus, the elements of $\mathcal{M}$ are the elements of $\mathcal{M}^0$ such that $\underline{v} \cdot \underline{n}$ is continuous across the inter-element boundaries. The discretised version of (7.2.2) is now

$$\begin{cases} (\underline{u}_h, \underline{v}_h) + (p_h, \text{div}\ \underline{v}_h) = \langle g, \underline{v}_h \cdot \underline{n}\rangle\ \ \forall\ \underline{v}_h\ \in \mathcal{M}, \\ (q_h, \text{div}\ \underline{u}_h) = -(f, q_h)\ \ \forall\, q_h\ \in \mathcal{L}^0_0, \end{cases} \tag{7.2.6}$$

where, clearly, $\underline{u}_h$ is sought in $\mathcal{M}$ and p_h in $\mathcal{L}^0_0$. We remind the reader that $\mathcal{L}^0_0$ is the space of piecewise constant functions. The linear system of equations associated with (7.2.6) has the form (see (5.6.9))

$$\begin{pmatrix} A & B^t \\ B & 0 \end{pmatrix} \begin{pmatrix} U \\ P \end{pmatrix} = \begin{pmatrix} G \\ F \end{pmatrix} \tag{7.2.7}$$

and its matrix is indefinite. This is definitely a considerable source of trouble. Therefore, following essentially the ideas of [210], we introduce the space

$$\Lambda := \mathcal{L}^0_0(\mathcal{E}_h) \tag{7.2.8}$$

of functions μ_h which are constant on each edge of the decomposition $\mathcal{T}_h$. For any function $\chi\ \in L^2(\Gamma_D)$, we consider

$$\Lambda_{\chi,D} := \{\mu_h \mid \mu_h\ \in \Lambda,\ \int_e (\mu_h - \chi)\, ds = 0\ \forall\, e\ \in \mathcal{E}_h \cap \Gamma_D\}. \tag{7.2.9}$$

It will finally be convenient to set, for $\underline{v}_h\ \in \mathcal{M}^0$ and $\mu_h\ \in \Lambda$,

$$c(\mu_h, \underline{v}_h) := \sum_K \int_{\partial K} \mu_h\, \underline{v}_h \cdot \underline{n}\, ds. \tag{7.2.10}$$

The following lemma is a direct consequence of the definition (7.2.5).

Lemma 7.2.1. *Assume that* $\underline{v}_h\ \in \mathcal{M}^0$. *Then,*

$$(c(\mu_h, \underline{v}_h) = 0\ \forall\, \mu_h\ \in \Lambda_{0,D}) \Leftrightarrow \underline{v}_h\ \in \mathcal{M}. \tag{7.2.11}$$

Now let $(\underline{u}_h, p_h)$ be the solution of (7.2.6) and consider the linear mapping

$$\phi : \underline{v}_h \to (\underline{u}_h, \underline{v}_h) + (p_h, \text{div } \underline{v}_h)_h - \langle g, \underline{v}_h \cdot \underline{n} \rangle, \tag{7.2.12}$$

where $(\chi, \psi)_h = \sum_K \int_K \chi \psi \, dx$. It is clear that $\phi(\underline{v}_h) = 0$ for all $\underline{v}_h \in \mathcal{M}$. Therefore, (7.2.11) implies that there exists a $\lambda_{0h} \in \Lambda_{0,D}$ such that, by Proposition 4.1.5,

$$\phi(\underline{v}_h) = c(\lambda_{0h}, \underline{v}_h) \ \forall \, \underline{v}_h \in \mathcal{M}^0. \tag{7.2.13}$$

Let us show that such a λ_{0h} is unique. This will be an immediate consequence of the following lemma.

Lemma 7.2.2. *If* $\mu_h \in \Lambda_{0,D}$ *and*

$$c(\mu_h, \underline{v}_h) = 0 \ \forall \, \underline{v}_h \in \mathcal{M}^0, \tag{7.2.14}$$

then $\mu_h \equiv 0$.

Proof. Let e^* be an edge in $\mathcal{E}_h$ and $K^* \in \mathcal{T}_h$ be a triangle such that $e^* \subset \partial K^*$. Let $\underline{v}_h^* \in \mathcal{M}^0$ be such that

$$\underline{v}_h^*|_K = 0 \ \forall \, K \neq K^* \tag{7.2.15}$$

and defined on K^* by

$$\begin{cases} \underline{v^*}_h \cdot \underline{n} = 0 \text{ on the edges } e \neq e^*, \\ \underline{v^*}_h \cdot \underline{n} = 1 \text{ on } e^*. \end{cases} \tag{7.2.16}$$

Then, $c(\mu_h, \underline{v}_h^*) = \int_{e^*} \mu_h \, ds$ and (7.2.14) implies that $\mu_h = 0$ on e^*. Since e^* was any edge in $\mathcal{E}_h$, this concludes the proof. □

Let us now define λ_h by means of

$$\lambda_h \in \Lambda_{g,D}, \ \lambda_h \equiv \lambda_{0h} \text{ on } \mathcal{E}_h \backslash \Gamma_D. \tag{7.2.17}$$

Then, (7.2.12) and (7.2.13) imply that

$$(\underline{u}_h, \underline{v}_h) + (p_h, \text{div } \underline{v}_h)_h = c(\lambda_h, \underline{v}_h) \ \forall \, \underline{v}_h \in \mathcal{M}^0. \tag{7.2.18}$$

We can summarise the results obtained so far in the next theorem.

Theorem 7.2.1. *Let* $(\underline{u}_h, p_h)$ *be the solution of* (7.2.6) *and let* λ_h *be defined through* (7.2.13) *and* (7.2.17)*. Then, the triplet* $(\underline{u}_h, p_h, \lambda_h)$ *is the unique solution of the following problem: find* $(\underline{u}_h, p_h, \lambda_h)$ *in* $\mathcal{M}^0 \times \mathcal{L}_0^0 \times \Lambda_{g,D}$ *such that*

$$\begin{cases} (\underline{u}_h, \underline{v}_h) + (p_h, \operatorname{div} \underline{v}_h)_h + c(\lambda_h, \underline{v}_h) = 0 \ \ \forall \underline{v}_h \in \mathcal{M}^0, \\ (q_h, \operatorname{div} \underline{u}_h)_h = -(f, q_h) \ \ \forall q_h \in \mathcal{L}_0^0, \\ c(\mu_h, \underline{u}_h) = 0 \ \ \forall \mu_h \in \Lambda_{0,D}. \end{cases} \tag{7.2.19}$$

Proof. The matrix associated with (7.2.18) has now the form

$$\begin{pmatrix} \bar{A} & \bar{B}^t & \bar{C}^t \\ \bar{B} & O & O \\ \bar{C} & O & O \end{pmatrix} \begin{pmatrix} \bar{U} \\ \bar{P} \\ \bar{\Lambda} \end{pmatrix} = \begin{pmatrix} \bar{G} \\ \bar{F} \\ O \end{pmatrix} \tag{7.2.20}$$

and we still do not see any improvement on (7.2.7). However, consider that now, the space $\mathcal{M}^0$ is completely discontinuous from one element to another. This was not the case with $\mathcal{M}$, which required the continuity of $\underline{v}_h \cdot \underline{n}$. As a consequence, we can choose in $\mathcal{M}^0$ a basis, made of vectors $\underline{v}_h$ which are different from zero only on one triangle (as was the vector $\underline{v}_h^*$ in (7.2.15) and (7.2.16)). Then, matrix $\bar{A}$ becomes *block diagonal*, each block being a 3×3 matrix corresponding to a single element, and we can eliminate the unknown $\bar{U}$ *at the element level* by solving

$$\bar{U} = \bar{A}^{-1}(\bar{G} - \bar{B}^t \bar{P} - \bar{C}^t \bar{\Lambda}). \tag{7.2.21}$$

We are left with the system

$$\begin{pmatrix} -\bar{B}\bar{A}^{-1}\bar{B}^t & -\bar{B}\bar{A}^{-1}\bar{C}^t \\ -\bar{C}\bar{A}^{-1}\bar{B}^t & -\bar{C}\bar{A}^{-1}\bar{C}^t \end{pmatrix} \begin{pmatrix} \bar{P} \\ \bar{\Lambda} \end{pmatrix} = \begin{pmatrix} -\bar{B}\bar{A}^{-1}\bar{G} + \bar{F} \\ 0 \end{pmatrix}. \tag{7.2.22}$$

Now recall that $\mathcal{L}_0^0$ is made of piecewise constants. This means that the matrix $\bar{B}\bar{A}^{-1}\bar{B}^t$ is *diagonal* (in a more general case, it will be block diagonal, each block corresponding again to a single element). This means that we can eliminate the unknown $\bar{P}$ at the element level by solving

$$\bar{P} = (\bar{B}\bar{A}^{-1}\bar{B}^t)^{-1}[-\bar{B}\bar{A}^{-1}\bar{C}^t\bar{\Lambda} + \bar{B}\bar{A}^{-1}\bar{G} - \bar{F}]. \tag{7.2.23}$$

We are finally left with a system of the form

$$H\bar{\Lambda} = R \tag{7.2.24}$$

with

$$H = \bar{C}\bar{A}^{-1}\bar{B}^t(\bar{B}\bar{A}^{-1}\bar{B}^t)^{-1}\bar{B}\bar{A}^{-1}\bar{C}^t - \bar{C}\bar{A}^{-1}\bar{C}^t \tag{7.2.25}$$

and

$$R = \bar{C}\bar{A}^{-1}\bar{B}^t(\bar{B}\bar{A}^{-1}\bar{B}^t)^{-1}[\bar{B}\bar{A}^{-1}\bar{G} - \bar{F}]. \tag{7.2.26}$$

It is clear that H is symmetric and positive definite. It is easy to see that the procedure for getting from (7.2.20) to (7.2.24) is exactly the most common procedure for eliminating internal degrees of freedom, better known as *static condensation*.

Clearly, all that we have described so far applies to the various $\mathcal{RT}_k$, $\mathcal{BDM}_k$ and $\mathcal{BDFM}_k$ elements described in Chap. 2 for the mixed approximation of elliptic problems, as well as to their corresponding elements for quadrilaterals. More generally (and more philosophically), this procedure can be applied to systems of the form (7.2.7) whenever the matrix A corresponds to a bilinear continuous form on a space V which does not have continuity requirements at the vertices. See [23, 118–120] for the corresponding proofs in cases more general than the present one. Other examples in which the procedure applies will be presented in Chap. 10. □

7.3 A Brief Analysis of the Computational Effort

As we have seen, the introduction of the inter-element multiplier λ_h is in general the most effective way of solving a discrete version of (7.2.2). This implies that a comparison among different kinds of discretisations, as far as the computational effort is concerned, must be done by the light of the "λ-procedure". In this respect, two basic steps must be taken into account. The *first step* is the work which has to be done *at the element level*: basically, the hard part of this work is the inversion of the matrix $\bar{A}$ (see (7.2.21)), and, if p_h has many d.o.f. per element, also the inversion of $\bar{B}\bar{A}^{-1}\bar{B}^t$ (see (7.2.23)). This, in our example, was trivial since $\bar{A}$, on each element, was a 3×3 matrix and $\bar{B}\bar{A}^{-1}\bar{B}^t$ a 1×1 matrix (that is, a scalar). In more general cases, those numbers can be bigger. Therefore, it is always a good feature, for a space $\mathcal{M}$ approximating $H(\text{div}; K)$, to have a basis in which the two components are independently assumed. Let us make this clearer with a simple example. If K is a triangle and $\hat{K}$ the reference element, a reasonable choice for a basis in $\mathcal{RT}_0(\hat{K})$ is

$$\underline{u}^1 = (1, 0);\ \underline{u}^2 = (0, 1);\ \underline{u}^3 = (x, y). \tag{7.3.1}$$

Now, since $\underline{u}^3$ has two components which are both different from zero, the corresponding local matrix

$$A_{ij}^K = \int_K \underline{u}^i \cdot \underline{u}^j \, dx\, dy \tag{7.3.2}$$

will have the structure

$$\begin{pmatrix} \otimes & 0 & \otimes \\ 0 & \otimes & \otimes \\ \otimes & \otimes & \otimes \end{pmatrix} \tag{7.3.3}$$

where $\otimes$ means, a priori, a non-zero element. On the other hand, if K is a rectangle, one will choose as local basis for $\mathcal{RT}_{[0]}(K)$:

$$\underline{u}^1 = (1,0);\ \underline{u}^2 = (x,0);\ \underline{u}^3 = (0,1);\ \underline{u}^4 = (0,y). \tag{7.3.4}$$

Now, each element of the basis (7.3.4) has one component identically zero, and the corresponding matrix A^K

$$A^K = \begin{pmatrix} \otimes & \otimes & 0 & 0 \\ \otimes & \otimes & 0 & 0 \\ 0 & 0 & \otimes & \otimes \\ 0 & 0 & \otimes & \otimes \end{pmatrix} \tag{7.3.5}$$

is block diagonal. An elementary inspection on the spaces $\mathcal{RT}_k$, $\mathcal{BDM}_k$ and $\mathcal{BDFM}_k$ gives the following outcome:

$$\begin{cases} \text{for K} = & \textit{triangle}\text{, then } \mathcal{BDM}_k (\equiv (P_k)^2) \text{ gives rise to a block diagonal} \\ & \text{elementary matrix } A^k \text{ while } \mathcal{RT}_{[k]} \text{ and } \mathcal{BDFM}_{[k]} \text{ do not;} \end{cases} \tag{7.3.6}$$

$$\begin{cases} \text{for K} = & \textit{rectangle}\text{, then } \mathcal{RT}_{[k]} \text{ and } \mathcal{BDFM}_{[k]} \text{ give rise to block} \\ & \text{diagonal elementary matrices while } \mathcal{BDM}_{[k]} \text{ does not.} \end{cases} \tag{7.3.7}$$

It must also be noted that the total dimension of A^K also comes into play. For instance, for K = rectangle and $k = 2$, then $\mathcal{RT}_{[k]}$ produces a matrix A^K (24×24) which is block diagonal and each of the two blocks is a 12×12 matrix, while $\mathcal{BDM}_{[k]}$ produces a matrix A^K which is 14×14 (actually made from a 12×12 block diagonal matrix with two 6×6 blocks, plus two full rows and columns). On the other hand, $\mathcal{BDFM}_{[k+1]}$ gives two 9×9 blocks. Obviously, on a uniform mesh with constant coefficients, one just performs *one* inversion *once*, so that the total cost is negligible. However, in a general case, the inversion of A^K on each K might be expensive.

As far as the matrix $\bar{B}\bar{A}^{-1}\bar{B}^t$ is concerned, usually, one gets, on each element, a full matrix so that the total dimension of it (that is, the number of degrees of freedom for p_h in each K) is the only way of comparison. Let us now consider the second step which is the solution of the final system (7.2.23) in the unknown λ_h. It is easy to observe that the total number of degrees of freedom for λ_h equals the total number of degrees of freedom for $\underline{u}_h$ which lie on the edges in $\mathcal{E}_h$. In this respect, $\mathcal{BDM}_k$ produces the same number of λ_h unknowns as $\mathcal{RT}_k$ while $\mathcal{BDFM}_k$ produces the same number of λ_h unknowns as $\mathcal{RT}_{k-1}$. The same is true for both triangles and rectangles.

We have used, for the comparison, $\mathcal{BDFM}_{k+1}$ rather than $\mathcal{BDFM}_k$ because, as we shall see in the next section, the order of convergence of $\mathcal{BDFM}_{k+1}$ is essentially the same as $\mathcal{BDM}_k$ or $\mathcal{RT}_k$.

It must also be pointed out that the splitting of the vector space into two (or three) independent components is a crucial starting point for the use of ADI solvers. See for instance [118, 119, 173, 174, 176].

7.4 Error Analysis for the Multiplier

Let us consider again, for the sake of simplicity, the approximation of (7.2.2) by means of the discretisation (7.2.6). We consider the case of homogeneous Dirichlet conditions, that is, we assume from now on that $\Gamma_N = \emptyset$ and $g_1 = 0$ in the notation of Sect. 7.1. This, if Ω is for instance a convex polygon, will ensure at least H^2-regularity. We have seen in Sect. 7.1.2 error estimates (7.1.29) and (7.1.30) on $\|\underline{u} - \underline{u}_h\|_{H(\mathrm{div};\Omega)}$ and $\|p - p_h\|_0$. From the interpolation estimates of Chap. 2, this yields

$$\|p - p_h\|_0 + \|\underline{u} - \underline{u}_h\|_{H(\mathrm{div};\Omega)} \leq ch\ (\|p\|_2 + \|f\|_1). \tag{7.4.1}$$

Now, if we are going to solve (7.2.6) through the introduction of the inter-element multiplier λ_h, we compute the λ_h unknown first, from (7.2.24), and then p_h and $\underline{u}_h$ out of it (this is done element by element). However we still have computed λ_h which physically must be an approximation of p and we seek some further use of it. In order to do that, we first need an estimate which is somehow better than (7.4.1) and was proved first by Douglas and Roberts [177]. If $\bar{p}_h$ is the L^2-projection of p onto $\mathcal{L}^0_0$, then

$$\|\bar{p}_h - p_h\|_0 \leq ch^2\ (\|p\|_2 + \|f\|_1). \tag{7.4.2}$$

Estimates of this kind can be obtained from the abstract duality results of Chap. 5. However, we found it more convenient to sketch a direct proof. To do so, let $\phi \in H^2(\Omega) \cap H^1_0(\Omega)$ be the solution of $\Delta\phi = \bar{p}_h - p_h$. Clearly, we have

$$\|\phi\|_2 \leq c\ \|\bar{p}_h - p_h\|_0. \tag{7.4.3}$$

Now set $\underline{z} = \underline{\mathrm{grad}}\,\phi$ and let $\Pi_h\underline{z}$ be the interpolate of $\underline{z}$ in $\mathcal{RT}_0$. Recall that $(\mathrm{div}\ (\Pi_h\underline{z} - \underline{z}), q_h) = 0, \quad \forall q_h \in \mathcal{L}^0_0$ (see Sect. 2.5.2), so that, in particular, $\mathrm{div}\ \Pi_h\underline{z} = \bar{p}_h - p_h$. Then, we have

$$\begin{aligned}
\|\bar{p}_h - p_h\|_0^2 &= (\mathrm{div}\ \Pi_h\underline{z}, \bar{p}_h - p_h) = (\mathrm{div}\ \Pi_h\underline{z}, p - p_h) \\
&= (\underline{u}_h - \underline{u}, \Pi_h\underline{z}) \\
&= (\underline{u}_h - \underline{u}, \Pi_h\underline{z} - \underline{z}) + (\underline{u}_h - \underline{u}, \underline{z}) \\
&= (\underline{u}_h - \underline{u}, \Pi_h\underline{z} - \underline{z}) + (\underline{u}_h - \underline{u}, \underline{\mathrm{grad}}\,\phi) \\
&= (\underline{u}_h - \underline{u}, \Pi_h\underline{z} - \underline{z}) + (\mathrm{div}\ (\underline{u}_h - \underline{u}), \phi).
\end{aligned} \tag{7.4.4}$$

Remember that $(\mathrm{div}\ (\underline{u}_h - \underline{u}), q_h) = 0, \ \forall q_h \in \mathcal{L}^0_0$. Hence, if $\bar{\phi}_h = L^2$-projection of ϕ onto $\mathcal{L}^0_0$, then (7.4.4) yields

$$\|\bar{p}_h - p_h\|_0^2 = (\underline{u}_h - \underline{u}, \Pi_h\underline{z} - \underline{z}) + (\mathrm{div}(\underline{u}_h - \underline{u}),\ \phi - \bar{\phi}_h). \tag{7.4.5}$$

Since

$$\|\underline{z} - \Pi_h \underline{z}\|_0 + \|\phi - \bar{\phi}_h\|_0 \leq ch\ \|\phi\|_2, \tag{7.4.6}$$

(7.4.2) follows from (7.4.1), (7.4.3), (7.4.5) and (7.4.6).

We are now ready to get some extra information from λ_h. First, if $(\underline{u}, p)$ is the solution of (7.2.2) and $\underline{v}_h \in \mathcal{M}^0$, then by Green's formula on each K, we have

$$(\underline{u}, \underline{v}_h) + (\text{div } \underline{v}_h, p)_h = \sum_K \int_{\partial K} p\ \underline{v}_h \cdot \underline{n}\ ds = c(p, \underline{v}_h). \tag{7.4.7}$$

From (7.4.7) and the first equation of (7.2.18), one gets

$$(\underline{u} - \underline{u}_h, \underline{v}_h) + (\text{div } \underline{v}_h, \bar{p}_h - p_h)_h = c(p - \lambda_h, \underline{v}_h)\ \forall\, \underline{v}_h \in \mathcal{M}^0, \tag{7.4.8}$$

where we were allowed to use $\bar{p}_h$ instead of p since div $\underline{v}_h$ is constant in each element. Let us now define p_h^* and $\tilde{p}_h$ to be the interpolate in $\mathcal{L}_1^{NC}$ (Sect. 2.2.3) of λ_h and p, respectively, by means of

$$\int_e (p_h^* - \lambda_h)\, ds = \int_e (\tilde{p}_h - p)\, ds = 0\ \forall\, e \in \mathcal{E}_h. \tag{7.4.9}$$

Equation (7.4.8) now implies

$$\sum_K \int_{\partial K} (\tilde{p}_h - p_h^*)\ \underline{v}_h \cdot \underline{n}\ ds = (\underline{u} - \underline{u}_h, \underline{v}_h) + (\text{div } \underline{v}_h, \bar{p}_h - p_h)_h \qquad \forall\, \underline{v}_h \in \mathcal{M}^0. \tag{7.4.10}$$

On the other hand, we have, by Green's formula,

$$\int_{\partial K} (\tilde{p}_h - p_h^*)\ \underline{v}_h \cdot \underline{n}\ ds = \int_K \underline{\text{grad}}(\tilde{p}_h - p_h^*) \cdot \underline{v}_h\ dx + \int_K (\tilde{p}_h - p_h^*)\ \text{div } \underline{v}_h\ dx. \tag{7.4.11}$$

A simple scaling argument shows that, for any $\tilde{q}_h \in \mathcal{L}_1^{NC}$ and for any K in $\mathcal{T}_h$,

$$\|\tilde{q}_h\|_{0,K} \leq c \sup_{\underline{v}_h \in \mathcal{RT}_0(K)} \frac{\int_K \underline{\text{grad}}\tilde{q}_h \cdot \underline{v}_h\ dx + \int_K \tilde{q}_h\ \text{div } \underline{v}_h\ dx}{h_K^{-1} \|\underline{v}_h\|_{0,K} + \|\ \text{div } \underline{v}_h\|_{0,K}} \tag{7.4.12}$$

so that from (7.4.12), (7.4.11) and (7.4.10), we have

$$\|\tilde{p}_h - p_h^*\|_{0,K} \leq c\left[h_K\ \|\underline{u} - \underline{u}_h\|_{0,K} + \|\bar{p}_h - p_h\|_{0,K}\right],\ \forall\, K \in \mathcal{T}_h, \tag{7.4.13}$$

which together with (7.4.1) and (7.4.2) gives

$$\|\tilde{p}_h - p_h^*\|_0 \leq ch^2 \,(\|p\|_2 + \|f\|_1). \tag{7.4.14}$$

Since $\|p - \tilde{p}_h\|_0 \leq ch^2 \,\|p\|_2$, we get, by the triangle inequality, that

$$\|p - p_h^*\|_0 \leq ch^2 \,(\|p\|_2 + \|f\|_1). \tag{7.4.15}$$

We can now summarise the above results in a theorem.

Theorem 7.4.1. *Let* $(\underline{u}_h, p_h, \lambda_h)$ *be the solution of* (7.2.19), *let* p *be the solution of* (7.2.1) *and let* p_h^* *be the* $\mathcal{L}_1^{NC}$*-interpolant of* λ_h *defined by* (7.4.9). *Then,*

$$\|p - p_h^*\|_0 \leq ch^2 \,(\|p\|_2 + \|f\|_1) \tag{7.4.16}$$

with c *independent of* h *and* u.

Remark 7.4.1. The proof that we have given of Theorem 7.4.1 is somehow "unconventional". The traditional proof (see for instance [23]) will, as an intermediate step, estimate first the distance of λ_h from the $L^2(\mathcal{E}_h)$ projection $\bar{\lambda}$ of p onto Λ, defined by

$$\int_e (p - \bar{\lambda})\, ds = 0 \;\; \forall\, e \,\in \mathcal{E}_h. \tag{7.4.17}$$

In particular, in our case, one would get

$$\|\bar{\lambda} - \lambda_h\|_{h,-1/2} \leq ch^2 \,(\|p\|_2 + \|f\|_1), \tag{7.4.18}$$

where

$$\|\mu_h\|_{h,-1/2} := \Big(\sum_e |e| \; \|\mu_h\|_{o,e}^2\Big)^{\frac{1}{2}}. \tag{7.4.19}$$

Then, (7.4.15) would follow from (7.4.18) by extending λ_h in the interior of each K (in our case, such an extension is p_h^*). □

Results of type (7.4.18) hold in much more general cases. For instance, one has

$$\|\bar{\lambda} - \lambda_h\|_{h,-1/2} \leq ch^{k+2} \tag{7.4.20}$$

for $\mathcal{RT}_k$ or $\mathcal{RT}_{[k]}$ or $\mathcal{BDFM}_{k+1}$ or $\mathcal{BDFM}_{[k+1]}$, whereas for $\mathcal{BDM}_k$ or $\mathcal{BDM}_{[k]}$, one has

$$\|\bar{\lambda} - \lambda_h\|_{h,-1/2} \leq ch^{k+2} \;(k \geq 2), \tag{7.4.21}$$

$$\|\bar{\lambda} - \lambda_h\|_{h,-1/2} \leq ch^2 \quad (k = 1). \tag{7.4.22}$$

In (7.4.21) and (7.4.22), λ_h is still the inter-element multiplier, now in $\Lambda = \mathcal{L}_k^0(\mathcal{E}_h)$, whereas $\bar{\lambda}$ is the $L^2(\mathcal{E}_h)$-projection of p onto Λ. For the proofs, we refer to [23, 120] and [119]. One has now to extend λ_h in the interior of each element in order to derive from (7.4.20) to (7.4.22) estimates of type (7.4.15). This can be done in several ways. We shall indicate here one possible choice.

If K is a triangle and k is *even*, we can define $p_h^* \in P_{k+1}(K)$ simply by setting

$$\int_{e_i} (p_h^* - \lambda_h)\, p_k = 0\ \forall\, p_k \in P_k(e_i),\ i = 1, 2, 3, \tag{7.4.23}$$

$$\int_K (p_h^* - p_h) p_{k-2}\, dx = 0 \quad \forall\, p_{k-2} \in P_{k-2}(K), \quad (k \geq 2). \tag{7.4.24}$$

It is easy to check that (7.4.23) and (7.4.24) determine $p_h^* \in P_{k+1}(K)$ in a unique way. In order to show this, check first that the number of conditions in (7.4.23) and (7.4.24) matches correctly the dimension of P_{k+1}:

$$3(k+1) + \frac{(k-1)k}{2} = \frac{(k+2)(k+3)}{2}. \tag{7.4.25}$$

Then, it is enough to show that if $\lambda_h = 0$ and $p_h = 0$, formulae (7.4.23) and (7.4.24) yield $p_h^* = 0$. First note that (7.4.23) (for $\lambda_h = 0$) implies that p_h^*, on each e_i, coincides with $\ell_{k+1}(e_i)$, the Legendre polynomial of degree $k+1$, up to a scaling factor. The continuity of p_h^* at the corners and the fact that for $k+1$ odd, ℓ_{k+1} is antisymmetric will then give $p_h^*|_{\partial K} = 0$. Hence, for $k \geq 2$, this means $p_h^* = b_3 p_{k-2}$ for some $p_{k-2} \in P_{k-2}(K)$ where b_3 is the cubic bubble on K. Condition (7.4.24) will now easily give $p_h^* \equiv 0$.

Let us now go to the case where K is a triangle and k is odd. Here, the construction (7.4.23), (7.4.24) does not work any more. We shall indicate another choice that works. Other possible choices can be found in [23, 120]. Let us define, for k odd ≥ 1, ϕ_{k+2} as the polynomial $\in P_{k+2}$ such that $\phi_{k+2} = 0$ at the vertices of K and:

$$\frac{\partial \phi_{k+2}}{\partial t}|_{e_i} = \ell_{k+1}(e_i), \ \ i = 1, 2, 3, \tag{7.4.26}$$

$$\int_K \phi_{k+2}\, p_{k-1} = 0 \qquad\qquad \forall\, p_{k-1} \in P_{k-1}(K). \tag{7.4.27}$$

Note that in (7.4.26), $\partial/\partial t$ is the anticlockwise tangential derivative and $\ell_{k+1}(e_i)$ is the Legendre polynomial of degree $k+1$ taking the value 1 at the endpoints. We also define $\psi_{k+1} \in P_{k+1}(K)$ by

$$\psi_{k+1} = \frac{\partial \phi_{k+2}}{\partial t} \text{ on } \partial K, \tag{7.4.28}$$

$$\int_K \psi_{k+1}\; p_{k-2}dx = 0 \qquad \forall\; p_{k-2}\; \in P_{k-2}(K),\;\; k \geq 3. \tag{7.4.29}$$

Note that from (7.4.28) to (7.4.29), $\psi_{k+1}|_{e_i} = \ell_{k+1}(e_i)$ $(1 = 1, 2, 3)$. Now we can set

$$S_{k+1} = P_{k+1} \oplus \{\phi_{k+2}\}. \tag{7.4.30}$$

Our extension p_h^* will be defined as the unique (we have to prove that!) element of S_{k+1} such that

$$\int_{e_i} (p_h^* - \lambda_h)\; p_k ds = 0\; \forall\; p_k\; \in P_k(e_i)\; (i = 1, 2, 3), \tag{7.4.31}$$

$$\int_K (p_h^* - p_h)\; p_{k-2}dx = 0\; \forall\; p_{k-2}\; \in P_{k-2}(K),\; k \geq 3, \tag{7.4.32}$$

$$\int_K (p_h^* - p_h)\; \Delta\psi_{k+1}dx = 0. \tag{7.4.33}$$

Note that the dimensional count (7.4.25), being independent of the parity of k, still holds since both sides are increased by one. Assume therefore that $\lambda_h = p_h = 0$ and let us see that (7.4.31)–(7.4.33) imply $p_h^* = 0$. For this, first note that, for every $p_{k+1} \in P_{k+1}(K)$, we have

$$\int_{\partial K} p_{k+1} \frac{\partial \psi_{k+1}}{\partial t} ds = -\int_{\partial K} \frac{\partial p_{k+1}}{\partial t} \psi_{k+1} ds = 0. \tag{7.4.34}$$

On the contrary,

$$\int_{\partial K} \phi_{k+2} \frac{\partial \psi_{k+1}}{\partial t} ds = -\int_{\partial K} (\psi_{k+1})^2 ds \neq 0. \tag{7.4.35}$$

Hence, (7.4.31), with $\lambda_h = 0$, will first give $p_h^* \in P_{k+1}(K)$ (by taking $p_k|_{e_i} = \partial\psi_{k+1}/\partial t|_{e_i}$ and summing over (i)); then again, (7.4.31) will imply that

$$p_h^* = \alpha\; \psi_{k+1} + b_{k+1} = \alpha\; \psi_{k+1} + b_3\; q_{k-2} \tag{7.4.36}$$

(where b_{k+1} is a bubble of degree $k + 1$ and b_3 is the cubic bubble) for some $\alpha \in \mathbb{R}$ and some $q_{k-2} \in P_{k-2}(K)$. Now using (7.4.36), (7.4.29) and (7.4.32) with $p_h = 0$, we easily get $q_{k-2} = 0$. Finally, (7.4.33) gives $\alpha = 0$.

A different approach for reconstructing an approximation $p_h^* \in P_{k+1}(K)$ of p which converges to p faster than p_h can be found for instance in [354]. Basically, one solves, in every K, a Neumann problem with $\underline{u}_h \cdot \underline{n}$ as boundary data by using p_h^* in order to fix the mean value in each element.

Remark 7.4.2. Let us go back to the simplest case of the lowest-order element $\mathcal{RT}_0 \equiv \mathcal{BDFM}_1$. It has been proved by Marini [291] that if we consider the space $\mathcal{L}_1^{NC}$ and define $p_h^* \in \mathcal{L}_1^{NC}$ to be the solution of

$$\sum_K \int_K \underline{\mathrm{grad}}\, p_h^* \cdot \underline{\mathrm{grad}}\, q_h \, dx = \int_\Omega f q_h \, dx, \quad \forall\, q_h \in \mathcal{L}_1^{NC}, \tag{7.4.37}$$

then, for f piecewise constant, one can compute, a posteriori, the solution $(\underline{u}_h, p_h)$ of (7.2.6) through the formulae

$$\underline{u}_h(x)|_K = \underline{\mathrm{grad}}\, p_h^* + (\underline{x} - \underline{x}_K)\, \frac{f}{2}\Big|_K, \qquad \forall\, K \in \mathcal{T}_h, \tag{7.4.38}$$

$$p_h|_K = \frac{1}{|\mathrm{area}(K)|} \int_K p_h^* \, dx + O(h^2), \ \forall\, K \in \mathcal{T}_h, \tag{7.4.39}$$

where $\underline{x}_K$ is the barycentre of K. Formulae (7.4.38) and (7.4.39) (in particular (7.4.38)) are especially interesting because the *principle* of (7.4.37), and therefore, its implementation and use is much simpler than the principle of (7.2.6). On the other hand, experimental results show that, in some applications, the accuracy of (7.2.6), as far as $\underline{u}_h$ is concerned, is much superior to the accuracy of the traditional methods (see e.g. [294]) and that the correction (7.4.38) away from $\underline{x}_K$ (say, at ∂K) has a relevant improving effect on the accuracy. □

7.5 Error Estimates in Other Norms

We have seen in Sect. 7.1.2 that all the families of mixed finite element methods for the Laplace operator (and hence for more general elliptic problems) satisfy well the *inf-sup* condition and therefore provide optimal error estimates in the "natural norms", which are here the $H(\mathrm{div};\Omega)$ norm for $\underline{u} - \underline{u}_h$ and the $L^2(\Omega)$-norm for $p - p_h$. We have also seen in this chapter that if one introduces Lagrange multipliers λ_h in order to solve (7.2.6) (and in general one does want to do so), then it is possible to obtain some additional information which allows to construct a new approximation p_h^* of p that provides some extra accuracy for p in the $H^1(\Omega)$-norm. In this section, we will present some other error estimates for $\underline{u} - \underline{u}_h$, $p - p_h$ and $p - p_h^*$ in other norms which might be interesting for applications. In particular, we shall deal with L^∞-norms and $H^{-s}(\Omega)$-norms. The interest of using L^∞-norms (especially for $\underline{u} - \underline{u}_h$) is quite obvious in the applications: a large stress field in a very small region can have a small L^2-norm but will be very dangerous for safety reasons. The interest of having dual estimates, like the estimates in $H^{-s}(\Omega)$ $(s > 0)$, can only be understood as a prerequisite to the use of a "smoothing post-processor" (see, e.g., [110, 111]). We shall not present here such smoothing post-processors; however, we can describe their features: if you have a continuous solution (say p) and an approximate solution (say p_h) such that $p - p_h$ is small (say $O(h^{s+k})$) in some dual norm $\| \cdot \|_{-s}$, then you can operate some "local" and relatively simple averages on p_h in order to produce a new approximation p_h^s such that $\|p - u_h^s\|_0 = O(h^{s+k})$. We refer to [110, 111] for more precise information. Let us now list the error estimates which have been proved so far in the L^∞-norms. We shall only report the ones obtained by Gastaldi and Nochetto [218, 219]. Previous results were obtained by Johnson and Thomée [265], Douglas and Roberts [177, 178] and Scholz [343, 345, 346].

Let us see, for instance, the spaces $\mathcal{RT}_k$ or $\mathcal{RT}_{[k]}$. Then, one has

$$\|\underline{u} - \underline{u}_h\|_{L^\infty} + \|p - p_h\|_{L^\infty} \le ch^{k+1}. \tag{7.5.1}$$

Note that, for $k = 0$, (7.5.1) holds only if f is smooth enough inside each element. Moreover, for $k \ge 0$, the assumption $p \in W^{k+2,\infty}(\Omega)$ is obviously required. We also have a super-convergence result for $p_h - P_h p$ (here $P_h p := L^2(\Omega)$-projection of p onto $\mathcal{L}_k^0$):

$$\|P_h p - p_h\|_{L^\infty} \le ch^{k+2} \,|\log h|, \tag{7.5.2}$$

where again p is assumed in $W^{k+2,\infty}(\Omega)$ and some extra regularity for f is needed for $k = 0$. Finally, for the case of rectangular elements, one gets, for $k \ge 0$,

$$|\underline{u}(S) - \underline{u}_h(S)| \le ch^{k+2} \,|\log h|^2 \,(\|f\|_{k,\infty,\Omega} + \delta_{k,0} \,\|f\|_{H^1}) \tag{7.5.3}$$

at the Gauss-Legendre points S of each element. It is also possible to study the error $p - p_h^*$. One has, for $k \ge 0$,

$$\|p - p_h^*\|_{L^\infty} \le ch^{k+2} \,|\log h|^2 \,(\|f\|_{k,\infty,\Omega} + \delta_{k,0}\|f\|_{H^1}), \tag{7.5.4}$$

and, for $p \in W^{k+2,\infty}$ (and for smoother f if $k = 0$),

$$\|p - p_h^*\|_{L^\infty} \le ch^{k+2} \,|\log h|. \tag{7.5.5}$$

Similar results hold for $\mathcal{BDM}$ and $\mathcal{BDFM}$ spaces and for their analogues in three dimensions. As far as the dual norms are concerned, we have for $\mathcal{RT}_k$ or $\mathcal{RT}_{[k]}$ of $\mathcal{BDFM}_{k+1}$ or $\mathcal{BDFM}_{[k+1]}$ elements, in two and three variables,

$$\|p - p_h\|_{-s} + \|\underline{u} - \underline{u}_h\|_{-s} + \|\operatorname{div}(\underline{u} - \underline{u}_h)\|_{-s} \le ch^{k+s+1}, \; 0 \le s \le k+1, \tag{7.5.6}$$

whereas for $\mathcal{BDM}_k$ or $\mathcal{BDM}_{[k]}$ elements, in two or three variables, one has

$$\|p - p_h\|_{-s} + \|\operatorname{div}(\underline{u} - \underline{u}_h)\|_{-s} \le ch^{k+s}, \; 0 \le s \le k, \tag{7.5.7}$$

and

$$\|\underline{u} - \underline{u}_h\|_{-s} \le ch^{k+s+1}, \; 0 \le s \le k-1. \tag{7.5.8}$$

For more precise estimates involving explicitly the regularity of the solution, we refer to [118–120, 178]. Interior estimates can be found for instance in [175].

7.6 Application to an Equation Arising from Semiconductor Theory

We now consider a special case of application of mixed finite element methods that is interesting in the simulation of semiconductor devices. Let us assume that we have to solve an equation of the type

$$\operatorname{div}(\varepsilon \, \underline{\operatorname{grad}}\, p + p \, \underline{\operatorname{grad}}\, \psi) = f \text{ in } \Omega \tag{7.6.1}$$

and assume, for the sake of simplicity, that we have Dirichlet boundary conditions

$$p = g \text{ on } \partial\Omega. \tag{7.6.2}$$

Note, however, that, in practice, we will always have a Neumann boundary condition $(\varepsilon \, \underline{\operatorname{grad}}\, u + p \, \underline{\operatorname{grad}}\, \psi) \cdot \underline{n} = 0$ on a part of $\partial\Omega$. In (7.6.1), we may assume ψ to be known, and in the computations, we shall also assume that ψ is piecewise linear; this is realistic since, in practice, ψ will be the discretised solution of another equation (coupled with (7.6.1)). Assume moreover that ε is constant and small. In order to present the mixed exponential fitting approximation of (7.6.1) and (7.6.2), [128–130], we first introduce the Slotboom variable

$$\rho := e^{\psi/\varepsilon} p \tag{7.6.3}$$

with its boundary value

$$\chi := e^{\psi/\varepsilon} g. \tag{7.6.4}$$

In order to simplify the notation, we shall often write

$$\phi := \psi/\varepsilon. \tag{7.6.5}$$

Using unknown ρ, problem (7.6.1) and (7.6.2) becomes

$$\begin{cases} \varepsilon \operatorname{div}(e^{-\phi} \, \underline{\operatorname{grad}}\, \rho) = f \text{ in } \Omega, \\ \rho = \chi \text{ on } \partial\Omega. \end{cases} \tag{7.6.6}$$

Note that the quantity

$$p = \varepsilon e^{-\phi} \, \underline{\operatorname{grad}}\, \rho = \varepsilon \, \underline{\operatorname{grad}}\, p + p \, \underline{\operatorname{grad}}\, \psi \tag{7.6.7}$$

(which has here the physical meaning of the electric current $\underline{J}$ through the device) is the most relevant unknown of the problem.

We now apply a mixed method to the solution of (7.6.6). By choosing the lowest-order Raviart–Thomas method, formulation (7.2.19) becomes: *find* $(\underline{u}_h, \rho_h, \lambda_h) \in \mathcal{M}^0 \times \mathcal{L}_0^0 \times \Lambda_\chi$ *such that*

$$\begin{cases} (\varepsilon^{-1} e^{\phi} \underline{u}_h, \underline{v}_h) + (\rho_h, \operatorname{div} \underline{v}_h)_h = c(\lambda_h, \underline{v}_h) \; \forall \, \underline{v}_h \in \mathcal{M}^0, \\ (\sigma_h, \operatorname{div} \underline{u}_h)_h = (f, \sigma_h) \; \forall \, \sigma_h \in \mathcal{L}_0^0, \\ c(\mu_h, \underline{u}_h) = 0 \; \forall \, \mu_h \in \Lambda_0, \end{cases} \tag{7.6.8}$$

where $(\cdot,\cdot)_h$ and $c(\mu, q_h)$ are defined as in (7.2.19). By static condensation, (7.6.8) can be reduced as in (7.2.24) to the form

$$H\lambda = R, \tag{7.6.9}$$

with H symmetric and positive definite. We also point out that H will be an M-matrix (see for instance [372]) provided that the triangulation is of weakly acute type. However, the scheme (7.6.8) (and the unknown ρ) are not suitable for actual computations. Indeed, one can see from (7.6.3) to (7.6.4) that ρ can become very large or very small in different parts of the domain Ω when ε is very small. Hence, we go back to the variable p. Since, as we have seen, λ_h in (7.6.8) will be an approximation of ρ at the inter-element boundaries, we can use the inverse transformation of (7.6.3) in the form

$$p_h := e^{-\bar{\phi}_h} \lambda_h \tag{7.6.10}$$

where $\bar{\phi}_h \in \mathcal{L}_0^0(\mathcal{E}_h)$ is defined as

$$\int_{e_i} e^{\bar{\phi}_h} ds = \int_{e_i} e^{\phi_h} ds \; \forall \, e_i \in \mathcal{E}_h. \tag{7.6.11}$$

Problem (7.6.8) now becomes: find $(\underline{u}_h, \rho_h, p_h) \in \mathcal{M}^0 \times \mathcal{L}_0^0 \times \Lambda_g$ such that

$$\begin{cases} (\varepsilon^{-1} e^{\bar{\phi}_h} \underline{u}_h, \underline{v}_h) + (\rho_h, \operatorname{div} \underline{v}_h)_h = c(e^{\bar{\phi}_h} p_h, \underline{q}_h) \; \forall \, \underline{v}_h \in \mathcal{M}^0, \\ (\sigma_h, \operatorname{div} \underline{u}_h)_h = (f, \sigma_h) \; \forall \, \sigma_h \in \mathcal{L}_0^0, \\ c(\mu_h, \underline{u}_h) = 0 \; \forall \, \mu \in \Lambda_0. \end{cases} \tag{7.6.12}$$

The static condensation procedure applied to (7.6.12) now produces a system in the sole unknown p_h of the form

$$\tilde{H} p_h = \tilde{R}, \tag{7.6.13}$$

where the unknown, the coefficients, and the right-hand side have a reasonable size. Moreover, it is easy to check that the passage from H to $\tilde{H}$ involves only the multiplication of each row by a factor of the type $e^{-\bar{\phi}_h}$, which does not alter the M-character of the matrix. Hence, if the decomposition is of weakly acute type, $\tilde{H}$ will be an M-matrix.

The most relevant feature of this approach is, however, that the approximation $\underline{u}_h$ of the current obtained by (7.6.12) will now have continuous normal components at the inter-element boundaries. We therefore have a strong conservation of the current.

Remark 7.6.1. Problem (7.6.6) could also be discretised by dual hybrid methods. However, in this case, the conservation of the current will hold only in a weak sense [130]. □

Remark 7.6.2. It is easy to check that the one-dimensional version of this approach reproduces the celebrated Sharfetter-Gummel method, also known as the exponential fitting method. The use and the analysis of non-standard formulations (involving the harmonic average of the coefficients) in one dimension can be found in [42]. □

Remark 7.6.3. It can be checked that, for very small ε, the scheme (7.6.13) produces an up-wind discretisation of (7.6.1). See [129] for this kind of analysis. □

Remark 7.6.4. If (7.6.1) contains a zero-order term

$$\operatorname{div}(\varepsilon \underline{\operatorname{grad}}\, u + p \underline{\operatorname{grad}}\, \psi) + cu = f, \tag{7.6.14}$$

then, in general, the matrix H in (7.6.9) will not be an M-matrix any longer, and the same will be true for the matrix $\tilde{H}$ in (7.6.13). To circumvent this difficult, one can change the choice of the space $\mathcal{M}^0$. We refer to [292] for a general theory of *nonconforming mixed methods* and to [293] for applications to semiconductor devices. □

7.7 Using Anisotropic Meshes

In some mesh adaptation procedures [239, 252, 301], anisotropic meshes are generated in regions where the solution varies slowly in some direction and rapidly in the orthogonal one. Using such a procedure with mixed formulations can lead to bad results for some choices of approximations of the space $H(\operatorname{div}, \Omega)$. We shall illustrate this by the same simple model problem that we used in the previous section. We suppose that we want to solve on the interval $(0, 1)$ the problem

$$p'' = 2,\ p(0) = p(1) = 0. \tag{7.7.1}$$

The exact solution is evidently

$$p(x) = x^2 - x,\ p' = 2x - 1.$$

Instead of solving (7.7.1), we rather try to find $p(x, y)$ on $\Omega = (0, 1) \times (0, b)$, solution of

$$\Delta p = 2, \quad p(0, y) = p(1, y) = 0,\ p_y(x, 0) = p_y(x, b) = 0. \tag{7.7.2}$$

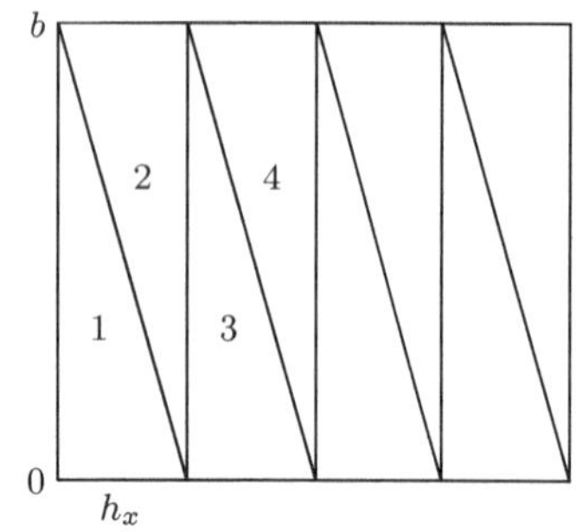

Fig. 7.5 Elongated mesh

Of course, using a bi-dimensional formulation to solve (7.7.1) is silly. What we want to model is the situation where the solution of a bi-dimensional computation happens to be more or less constant in some direction. A clever anisotropic mesh adaptation procedure [239,301] would then create elements with a strong elongation in this direction. In our model case, let us suppose that this resulted in a mesh of regular triangles such as in Fig. 7.5.

If we use a *standard formulation* and piecewise linear elements to solve (7.7.2), it is easy to see that we get a correct solution: $p_h(x, y)$ is independent of y and converges if we refine the mesh in the direction x. Let us now consider a mixed formulation of the problem,

$$\begin{cases} (\underline{u}, \underline{v}) + (p, \text{div } \underline{v}) = 0, & \forall\, \underline{v} \in H_{0N}(\text{div}, \Omega) \\ (q, \text{div } \underline{u}) = (f, q) \;\; \forall\, q \in L^2(\Omega), \end{cases} \tag{7.7.3}$$

Γ_N being the lower and upper boundaries of our rectangle and $f = 2$. We discretise this problem using a $\mathcal{RT}_0$ approximation. We thus look for $(\underline{u}_h, p_h)$, solution of the discrete problem defined by

$$\begin{cases} (\underline{u}_h, \underline{v}_h) + (p_h, \text{div } \underline{v}_h) = 0, & \forall\, \underline{v}_h \in \mathcal{M}, \\ (q_h, \text{div } \underline{u}_h) = (f, q_h) \;\; \forall\, q_h \in \mathcal{L}_0^0, \end{cases} \tag{7.7.4}$$

where we define

$$\mathcal{M} := \{\underline{v} \,|\, \underline{v} \in H(div, \Omega),\; \underline{q} \cdot \underline{n} = 0 \text{ on } \Gamma_N\}. \tag{7.7.5}$$

The exact solution for $\underline{u}$ is

$$\underline{u} = \begin{pmatrix} 2x - 1 \\ 0 \end{pmatrix}. \tag{7.7.6}$$

Taking, on every element K of $\mathcal{T}_h$, the interpolate $\Pi_{\mathcal{RT}}(\underline{u}) \in \mathcal{RT}_0(K)$ of $\underline{u}$, it is easily seen that we have

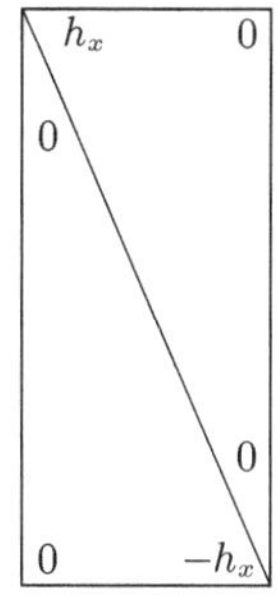

Fig. 7.6 $(\Pi_{\mathcal{RT}}(\underline{u})_1 - \underline{u}_1)$

1. In element 1

$$\Pi_{\mathcal{RT}}(\underline{u}) = \begin{pmatrix} x-1 \\ y \end{pmatrix}, \tag{7.7.7}$$

2. In element 2

$$\Pi_{\mathcal{RT}}(\underline{u}) = \begin{pmatrix} x+h_x-1 \\ y-b \end{pmatrix}, \tag{7.7.8}$$

3. In element 3

$$\Pi_{\mathcal{RT}}(\underline{u}) = \begin{pmatrix} x+h_x-1 \\ y \end{pmatrix}, \tag{7.7.9}$$

4. In element 4

$$\Pi_{\mathcal{RT}}(\underline{u}) = \begin{pmatrix} x+2h-1 \\ y-h_y \end{pmatrix}, \tag{7.7.10}$$

and so on. We now claim that the solution $\underline{u}_h$ of our mixed formulation is nothing but $\Pi_{\mathcal{RT}}(\underline{u})$. Indeed, as $\operatorname{div} \Pi_{\mathcal{RT}}(\underline{u}) = \operatorname{div} \underline{u}_h = 2$, $\underline{u}_h$ can only differ from $\Pi_{\mathcal{RT}}(\underline{u})$ by curl ϕ_h where ϕ_h is a piecewise linear function. Given the boundary conditions ($\underline{u} \cdot \underline{n} = 0$ on Γ_N), this function must take constant values for $y = 0$ and $y = b$ and is thus of the form $\alpha + \beta y$. This means that curl $\phi = \{\beta, 0\}$, which implies that $(\Pi_{\mathcal{RT}}(\underline{u}) - \underline{u}_h)_2 = 0$, but the first components could differ by a constant. However, one checks that

$$\int_{\Omega} (\Pi_{\mathcal{RT}}(\underline{u}) - \underline{u}_h)_1 \, dx = 0 \tag{7.7.11}$$

so that the first components also coincide. Indeed, we may take $\underline{v}_h = \{1, 0\}$ in the first equation of (7.7.4) and conclude that $\int_{\Omega} (\underline{u}_h)_1 \, dx = 0$. On the other hand, one can check directly that $\int_{\Omega} \Pi_{\mathcal{RT}}(\underline{u})_1 \, dx = 0$. This can also be seen if we consider, on the rectangle formed by two adjacent elements, the difference $(\Pi_{\mathcal{RT}}(\underline{u})_1 - \underline{u}_1)$. We represent the nodal values of this difference in Fig. 7.6. The pattern is the same for all patches and the integral is obviously null.

We also know that the average value $\bar{p}$ of p is equal to $-1/6$ as is easily checked from the exact solution. This is also easily obtained by taking $\underline{v} = \underline{u}$ in the first equation of (7.7.3) to obtain

$$2\int_{\Omega} p dx = -\int_{\Omega} |\underline{u}|^2 \, dx. \tag{7.7.12}$$

On the other hand, taking $\underline{v}_h = \underline{u}_h = \Pi_{\mathcal{RT}}(\underline{u})$ in the first equation of (7.7.4), we get

$$2\int_{\Omega} p_h dx = -\int_{\Omega} |\underline{u}_h|^2 \, dx. \tag{7.7.13}$$

This integral can be computed explicitly but it is more interesting to start from

$$\int_{\Omega} |\underline{u}_h|^2 dx = -\int_{\Omega} |\underline{u}|^2 dx + \int_{\Omega} |\underline{u} - \underline{u}_h|^2 dx + 2\int_{\Omega} \underline{u} \cdot \underline{u}_h dx. \tag{7.7.14}$$

Taking $\underline{v} = \underline{u}_h$ in the first equation of (7.7.3), one sees that

$$\int_{\Omega} \underline{u} \cdot \underline{u}_h dx = \int_{\Omega} |\underline{u}|^2 dx, \tag{7.7.15}$$

so that (7.7.14) becomes

$$\int_{\Omega} |\underline{u}_h|^2 \, dx = \int_{\Omega} |\underline{u}|^2 \, dx + \int_{\Omega} |\underline{u} - \underline{u}_h|^2 \, dx. \tag{7.7.16}$$

We deduce from (7.7.12) to (7.7.13) that

$$2\int_{\Omega} p_h \, dx = 2\int_{\Omega} p \, dx - \int_{\Omega} |\underline{u} - \underline{u}_h|^2 \, dx. \tag{7.7.17}$$

Referring to Fig. 7.6, it is easy to compute the last term

$$\int_{\Omega} |\underline{u} - \underline{u}_h|^2 dx = \frac{b\, h_x^2}{6} + \frac{b^3}{6} \tag{7.7.18}$$

and we get

$$\bar{p}_h = -\frac{1}{6} - \frac{h_x^2}{12} - \frac{b^2}{12}. \tag{7.7.19}$$

Thus, even if h_x becomes small, we do not get a correct value for $\bar{p}_h$. However, we could also have discretised the problem with many layers of thickness h_y as in Fig. 7.7.

It is intuitively obvious, and can be checked from the equations, that the solution of our problem on this mesh can be obtained by piling the solution described above,

Fig. 7.7 Isotropic mesh

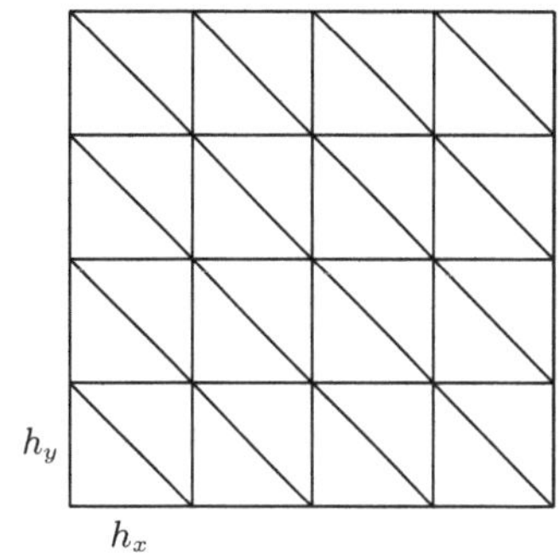

taking $b = h_y$. One would then obtain

$$\bar{p}_h = -\frac{1}{6} - \frac{h_x^2}{12} - \frac{h_y^2}{12} \tag{7.7.20}$$

and we thus recover convergence if the mesh is refined in an isotropic way (as indeed the general theory predicts!).

Remark 7.7.1. If we had used a $\mathcal{BDM}_1$ approximation, that is, a full linear approximation for both components of p_h, we would have

$$\Pi_{\mathcal{BDM}}(\underline{u}) = \underline{u}$$

and we would get a correct solution, even for elongated elements.Thus, even if a first glance to the error estimates might lead one to believe that there is no advantage to using the richer space $\mathcal{BDM}_1$ instead of $\mathcal{RT}_0$, there are indeed situations where such an advantage definitely exists. □

Remark 7.7.2. It is also remarkable that the finite Volume method that we discuss in Sect. 7.8 would yield, for the elongated mesh, a correct solution for p_h at the barycentre of the rectangles formed by two triangles sharing an oblique diagonal in $\mathcal{T}_h$. □

7.8 Relations with Finite Volume Methods

We shall rapidly present in this section some relations between the mixed finite element methods considered in this chapter and some finite volume methods. The kind of approximation employed in mixed methods of low degree is indeed very close to what is used in finite volume formulations: constant values of the unknown on elements and fluxes on interfaces. Moreover, both these methods impose the same conservation of the fluxes on every element. The difference lies in the procedure to compute the fluxes, which is local in finite volume methods and global in mixed methods. We shall see that if we introduce an approximation of the bilinear form $a(\underline{u}, \underline{v})$ to make the associated matrix diagonal, the computation of the fluxes reduces to a finite volume procedure. This then enables the elimination of the variable $\underline{u}$, leading to a final system of linear algebraic equations $MP = F$ where each unknown P_i represents the constant value of the approximated solution p_h in a single element. Similar ideas can be found in [11, 383] or [266].

7.8.1 The One and Two-Dimensional Cases

We consider, from now on, the discrete problem (7.1.21), using the lowest degree Raviart-Thomas element. We have already seen, in Sect. 7.2.2, how introducing a Lagrange multiplier to ensure the inter-element continuity of the fluxes unknowns $\underline{u}_h$ can lead to the elimination of these unknowns, leaving a positive definite system for the multipliers, a system which can be seen as a nonconforming standard finite element method. We now consider another way of reducing the system by eliminating the fluxes. Contrarily to the method of Sect. 7.2.2, this will necessitate a modification of the problem. We shall first illustrate this by a simple one-dimensional example. We use the model problem

$$\begin{aligned} \int_0^1 u\, v \,\mathrm{d}\mathbf{x} + \int_0^1 p v' \,\mathrm{d}\mathbf{x} &= 0 \\ \int_0^1 u' q \,\mathrm{d}\mathbf{x} + \int_0^1 f q \,\mathrm{d}\mathbf{x} &= 0 \end{aligned} \tag{7.8.1}$$

which we discretise using piecewise linear elements for u and piecewise constants for p on a partition of $(0, 1)$ into n subintervals, separated by nodes $x_i = ih$, $0 \leq i \leq n$.

Denoting by P and U the vectors of degrees of freedom of p_h and u_h, the first equation gives rise to a matrix equation

$$AU + B^t P = 0 \tag{7.8.2}$$

with A being a mass matrix of the form

$$\frac{h}{6}\begin{pmatrix} 2 & 1 & & & & \\ 1 & 4 & 1 & & & \\ & 1 & 4 & 1 & & \\ & & \cdots & & & \\ & & & & 1\ 4\ 1 & \\ & & & & & 1\ 2 \end{pmatrix}. \tag{7.8.3}$$

This is a global problem which makes a *local* elimination of U impossible. A standard trick to bypass this is to use *mass lumping*. In this procedure, matrix

A is made diagonal by summing all the terms of each line to the diagonal. This corresponds to using the trapezoidal rule to integrate (approximately) $\int_0^1 u\, v\, d\mathbf{x}$. We then obtain explicitly at node i, denoting $P_{i+1/2}$, $P_{i-1/2}$, the values of p_h in the intervals at the right and the left of x_i and by P_0, P_n the given values at $x = 0$ and $x = 1$,

$$\begin{cases} U_i = \dfrac{P_{i+1/2} - P_{i-1/2}}{h}, \text{ for an internal node,} \\ U_1 = \dfrac{2}{h}(P_{1/2} - P_0), \\ U_n = \dfrac{2}{h}(P_n - Pn - 1/2). \end{cases} \tag{7.8.4}$$

The form of the last two equations comes from the fact that $h/2$ instead of h is employed to compute the difference formula. One can then eliminate the unknown U_i by taking the values of u_h obtained from (7.8.4) into the second equation of (7.8.1). One easily sees that the final result is a standard discretisation of $-p'' = f$ by the usual three-point formula for the second derivative.

7.8.2 The Two-Dimensional Case

We now try to follow a similar idea for two-dimensional problems. This means that we must find a way to make the matrix A in (7.2.20) diagonal without impairing the precision of the method. Historically, the first successful attempt to do so was made by Baranger-Maitre-Oudin in [49]. Inspired by a previous result of Haugazeau-Lacoste [241] concerning $H(curl, \Omega)$ spaces, they decided to look, in every element K, for a suitable bilinear form $a_{K,h}(\underline{u}_h, \underline{v}_h)$ of the type

$$a_{K,h}(\underline{u}_h, \underline{v}_h) = \sum_{k=1}^{3} \omega_i\, (\underline{u}_h(M_i) \cdot \underline{n}_k^i)\, (\underline{v}_h(M_i) \cdot \underline{n}_k^i). \tag{7.8.5}$$

In (7.8.5), M_i represents the midpoint of the i-th edge, and $\underline{n}_K^i$ is the unit outward normal to that edge ($i = 1, 2, 3$). The weights ω_i must be chosen in order to have

$$a_{K,h}(\underline{u}_h, \underline{v}_h) = a_K(\underline{u}_h, \underline{v}_h) \qquad \forall\, \underline{u}_h\, \underline{v}_h \text{ constant on } K. \tag{7.8.6}$$

After some manipulations, one discovers that a bilinear form satisfying (7.8.6) indeed exists, and that the weights ω_i can be computed in a simple geometrical way. Referring to Fig. 7.8, let C be the circumcentre of K (that is, the centre of the unique circle that passes through the vertices of K), and for each $i = 1, 2, 3$, let H_i denote the distance of C from the straight line ℓ_i containing the i-th edge e_i (this is in fact the distance from C to the midpoint of the edge). The straight line ℓ_i clearly splits the whole plane into two half-planes. If C belongs to the same half-plane

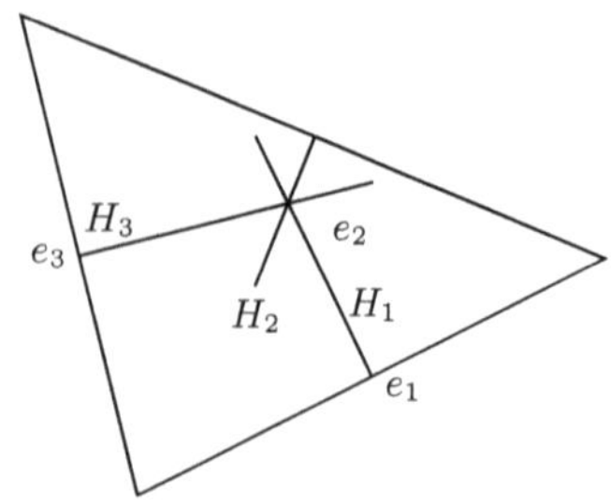

Fig. 7.8 Circumcenter and weights

containing K, then we set $\omega_i = H_i$. Otherwise, we set $\omega_i = -H_i$. It is easy to check that if for instance all the angles of K are acute, then C falls inside K, and the choice $\omega_i = H_i$ will be made for all i's. In this case, all the weights come out to be positive. If, however, the edge e_i is opposite to an obtuse angle, then ω_i turns out to be $-H_i$, and it will be negative. Up to a certain extent, this could be tolerated (see [49] for further details). When e_i is opposite to a right angle, then H_i is zero, and so is ω_i.

Coming back to the $\mathcal{RT}_0$ space, we have seen in Chap. 2 that a basis for it could be obtained in the following way. For each edge e_k in $\mathcal{T}_h$, we choose a unit vector $\underline{n}^k$ normal to e_k. We do it for $k = 1, \cdots, NE$ where NE is the number of edges in $\mathcal{T}_h$. Then, for each k, we define the vector $\underline{u}^k$ as the unique vector in $\mathcal{RT}_0$ that satisfies

$$\underline{u}^k \cdot \underline{n}^k = 1 \qquad \text{and} \qquad \underline{u}^k \cdot \underline{n}^r = 0 \ \forall\, r \neq k. \tag{7.8.7}$$

It is immediate to see that, with respect to this basis, the matrix

$$A_{r,k} := \sum_{K \in \mathcal{T}_h} a_{K,h}(\underline{u}^k, \underline{v}^r) \tag{7.8.8}$$

is diagonal. The idea is then the following one: change the original bilinear form $a(\cdot\,,\cdot)$ into

$$a_h(\underline{u}_h, \underline{v}_h) := \sum_{K \in \mathcal{T}_h} a_{K,h}(\underline{u}_h, \underline{v}_h), \tag{7.8.9}$$

then change the original mixed formulation (7.2.6) into: *find $\underline{u}_h \in \mathcal{RT}_0$ and $p_h \in Q_h$ such that*

$$a_h(\underline{u}_h, \underline{v}_h) = b(p_h, \underline{v}_h) \qquad \forall\, \underline{v}_h \in \mathcal{RT}_0, \tag{7.8.10}$$

$$b(p_h, q_h) = -(f\, q_h) \qquad \forall\, q_h \in Q_h. \tag{7.8.11}$$

In the sequel, we shall refer to this procedure as the BMO formulation. We remark that in the discrete system associated to this formulation, A is a diagonal matrix. Then, eliminate $U = -A^{-1}P$ to reach the form

$$B(A^{-1})B^t P = -F, \tag{7.8.12}$$

Fig. 7.9 Shaded L_k

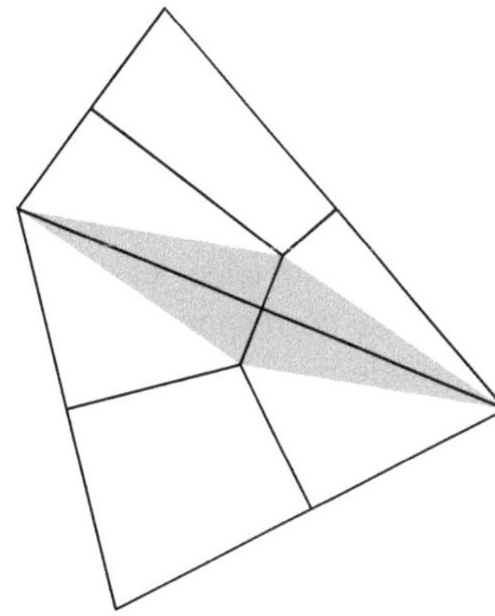

where now A^{-1} is explicitly known. Using the fact that the bilinear form $a_h(\cdot,\cdot) = a(\cdot,\cdot)$ for piecewise constant vectors, it is proven in [49] that the consistency error originated by the change of the bilinear form $a(\cdot,\cdot)$ can be kept under control, in the sense that the consistency error introduced by the numerical integration procedure is of the same order as the approximation error in $\mathcal{RT}_0$. Hence, we still have optimal a priori error estimates. The resulting scheme coincides with a classical finite volume scheme for diffusion operators (see e.g. [187]), where the flux on each edge e, common to the triangles T_1 and T_2, is defined by dividing the jump $p_h^1 - p_h^2$ by the distance $C_1 - C_2$ between the circumcentres of T_1 and T_2, respectively.

It is also possible to see that construction as a different mixed formulation that allows a simpler analysis. It is worth looking at it since, as we shall see in the next section, the interpretation of [49] does not hold in three dimensions.

Assume, for simplicity, that all the angles of all the triangles are acute. This is not strictly necessary (in the sense that the condition can be weakened) but makes the exposition much simpler. In this case, all the circumcentres will be internal to their respective triangles. Split every triangle K in three sub-triangles using the circumcentres. Every internal edge e_k is adjacent to two such sub-triangles: take the union of the two, and call it L_k as in Fig. 7.9.

For the boundary edges, we shall have just one sub-triangle, that we still call L_k. The union of all the L_k $(k = 1, \dots, NE)$ obtained in that way is still equal to Ω. Consider now the new vector space

$$V_L := \{\underline{v}_h |\ \underline{v}_{h|L_k} = c\underline{n}^k \text{ with } c \in \mathbb{R}\ \forall\, k = 1, \dots NE\} \tag{7.8.13}$$

where, as before, $\underline{n}^k$ is the chosen unit vector normal to e_k. For vectors $\underline{v}_h \in V_L$ and scalars $q_h \in Q_h$, the bilinear form $b(q_h, \underline{v}_h)$ still makes sense, provided we write it as

$$b(q_h, \underline{v}_h) = \sum_K \int_{\partial K} q_h\, \underline{v}_h \cdot \underline{n}_K. \tag{7.8.14}$$

One must not confuse $\underline{n}^k$ (normal to the edge e_k) and $\underline{n}_K$ (outward unit normal to ∂K). In what follows, it will be sometimes convenient to rearrange terms in the sum appearing in (7.8.14), making the sum over the edges rather than over the triangles. In order to do so, we first introduce the jumps of a piecewise constant function $q_h \in Q_h$ on an edge e_k in the following way. Let T^1 and T^2 be the two triangles having e_k as an edge, let q_h^1 and q_h^2 be the corresponding values of q_h, and let $\underline{n}_{T^1}$ and $\underline{n}_{T^2}$ be the corresponding unit outward normals. If e_k is a boundary edge belonging to a single triangle K, we set $q_h^1 = q_h|K$, $q_h^2 = 0$, $\underline{n}_{T^1} = \underline{n}_K$ and $n_{T^2} = -\underline{n}^K$. The jump of q_h over e_k is the *vector*

$$[\![q]\!]_k := q_h^1\,\underline{n}_{T^1} + q_h^2\,\underline{n}_{T^2}. \tag{7.8.15}$$

It is now easy to see that, whenever convenient, the bilinear form b can be written (for $\underline{v}_h \in V_L$ and $q_h \in Q_h$) as

$$b(q_h, \underline{v}_h) := \sum_{k=1}^{NE} \int_{e_k} [\![q]\!] \cdot \underline{v}_h. \tag{7.8.16}$$

It will be useful to introduce, for every piecewise constant function q_h, its "gradient" $\mathbf{g}(q_h)$ defined as the unique element in V_L such that

$$2\int_{L_k} \mathbf{g}(q_h)\,\mathrm{d}\mathbf{x} = \int_{e_k} [\![q]\!]_k \qquad \forall\, k = 1, \ldots, NE. \tag{7.8.17}$$

Introducing, for every e_k, the quantity h_k defined as

$$h_k := 2\frac{\mathrm{meas}(L_k)}{\mathrm{meas}(e_k)} \tag{7.8.18}$$

(that is, for internal edges, the distance of the two circumcentres), it is easy to see that (7.8.17) can be written as

$$\mathbf{g}(q_h)_{|L_k} = [\![q]\!]_k / h_k. \tag{7.8.19}$$

In order to reproduce the BMO formulation, we also need to introduce for $\underline{u}_{Lh}$ and $\underline{v}_{Lh}$ in V_L

$$a_L(\underline{u}_{Lh}, \underline{v}_{Lh}) = 2a(\underline{u}_{Lh}, \underline{v}_{Lh}). \tag{7.8.20}$$

The reason for this factor 2 is that, in a sense, $\underline{u}_L$ contains only one component of $\underline{\mathrm{grad}}\,p$. If we define $P_L\underline{u}_h \in \mathcal{RT}_0$ to V_L by taking on every edge e_k

$$P_L|_{L_k} = \underline{u}_h \cdot \underline{n}_k,$$

we have

$$a_L(P_L\underline{u}_h, P_L\underline{v}_h) = a_h(\underline{u}_h, \underline{v}_h), \tag{7.8.21}$$

where $a_h(\underline{u}_h, \underline{v}_h)$ is defined by (7.8.9). We can now consider the mixed formulation: *find* $\underline{u}_{Lh} \in V_L$ *and* $p_h \in Q_h$ *such that*

$$a_L(\underline{u}_{Lh}, \underline{v}_{Lh}) = b(p_h, \underline{v}_{Lh}), \qquad \forall\, \underline{v}_{Lh} \in V_L, \tag{7.8.22}$$

$$b(q_h, \underline{u}_{Lh}) = -(f, q_h), \qquad \forall\, q_h \in Q_h, \tag{7.8.23}$$

which, indeed, is just a different (and apparently a little more cumbersome) way of writing the BMO formulation (7.8.10) and (7.8.11). The analysis, however, can come out simpler. We simply give a quick outline of it.

We consider two different approximations of the exact flux $\underline{u}$. The first one, that we call $\underline{u}_I$, is defined as the unique element in V_L that satisfies

$$\int_{e_k} (\underline{u} - \underline{u}_I) \cdot \underline{n}^k = 0 \quad \forall\, k = 1, \ldots, NE. \tag{7.8.24}$$

Using the divergence theorem and (7.8.23), we immediately get

$$b(q_h, \underline{u}_{Lh} - \underline{u}_I) = 0 \qquad \forall\, q_h \in Q_h, \tag{7.8.25}$$

which is a useful property. The second approximation for $\underline{u}$, that we call $\underline{u}_2$, will be obtained by considering first $p_C \in Q_h$ as the unique piecewise constant that verifies

$$p_C(C_K) = p(C_K) \quad \text{for } C_K\text{=circumcentre of } K \quad \forall\, K \in \mathcal{T}_h. \tag{7.8.26}$$

We then set

$$\underline{u}_2 := -\mathbf{g}(p_C), \tag{7.8.27}$$

where we used the operator $q_h \mapsto \mathbf{g}(q_h)$ as defined in (7.8.17) or (7.8.19). Using (7.8.16) and (7.8.27), it is an elementary matter to verify that

$$a_L(\underline{u}_2, \underline{v}_{Lh}) = b(p_C, \underline{v}_{Lh}) \qquad \forall\, \underline{v}_{Lh} \in V_L. \tag{7.8.28}$$

The error estimate now goes easily: set $\underline{w} := \underline{u}_{Lh} - \underline{u}_I$ and use (7.8.25) to see that $b(q_h, \underline{w}) = 0$ for all $q_h \in Q_h$. This implies, using (7.8.22) and (7.8.28), that

$$a_L(\underline{u}_{Lh} - \underline{u}_2, \underline{w}) = b(p_h, \underline{w}) - b(p_C, \underline{w}) = 0 - 0 = 0. \tag{7.8.29}$$

Adding and subtracting $\underline{u}_2$ and using the above property, you get

$$\|\underline{w}\|^2 = a_L(\underline{w}, \underline{w}) = a_L(\underline{u}_{Lh} - \underline{u}_2, \underline{w}) + a_L(\underline{u}_2 - \underline{u}_I, \underline{w}) = a_L(\underline{u}_2 - \underline{u}_I, \underline{w}), \tag{7.8.30}$$

which easily implies $\|\underline{w}\| \le \|\underline{u}_2 - \underline{u}_I\|$. The proof ends by remarking that the line joining two circumcentres C_1 and C_2 of two triangles T^1 and T^2 having an edge e_k in common is perpendicular to e_k. This implies, using the definition (7.8.26) of p_I

and the definition of **g** (7.8.19), that $\underline{u}_2$ equals the value of the normal component of $\underline{u} \equiv -\,\underline{\text{grad}}\, p$ on a point of the segment joining C_1 and C_2. On the other hand, $\underline{u}_I$ is the value of $\underline{u} \cdot \underline{n}^k$ on a point of the edge e_k, and the difference of the two is easily bounded.

7.8.3 The Three-Dimensional Case

We now consider the three-dimensional problem. The definition of the local spaces $\mathcal{RT}(K)$ remains formally unchanged, as well as the definitions of the bilinear forms $a(\cdot,\cdot)$ and $b(\cdot,\cdot)$. The first change with respect to the two-dimensional case is that a formula of the type BMO (7.8.5), that we used to diagonalise the approximated bilinear form $a_h(\cdot,\cdot)$ (see (7.8.9)), does not exist, unless K is a very special tetrahedron, as for instance a regular tetrahedron (see always [49]). The three-dimensional analogue of the "quadrature formula" (7.8.5) was given in [241] and reads

$$a_{K,h}(\underline{u}_h, \underline{v}_h) = \sum_{i=1}^{6} \beta^i_K (\underline{u}_h(M_i) \cdot \underline{t}^i)(\underline{v}_h(M_i) \cdot \underline{t}^i), \tag{7.8.31}$$

where, now, M_i represents the midpoint of the edge e_i and $\underline{t}^i$ the tangent direction to e_i (the sign, for the moment, is immaterial). The theorem in [241] states that it is possible to find the coefficients β^i_K in such a way that, as in (7.8.6), $a_{K,h}(\underline{u}_h, \underline{v}_h)$ coincides with the exact integral $a_K(\underline{u}_h, \underline{v}_h)$ whenever both $\underline{u}_h$ and $\underline{v}_h$ are constant on K. Unfortunately, for the moment, this has no use for our purpose.

On the other hand, we have seen that, in two dimensions, the BMO trick could be written in a different way, using the space V_L (7.8.13) and the related mixed formulation (7.8.22) and (7.8.23). Even though the BMO trick itself does not work in three dimensions, its equivalent formulation using V_L generalises rather easily to the three dimensional case.

Assume again, for simplicity, that for every tetrahedron K, the centre C_K of its circumsphere (that is, the unique sphere that passes through the four vertices of K) lies inside K. As in the two-dimensional case, this assumption can be relaxed, but to the expenses of the simplicity of the presentation. We also point out that this condition is stricter than assuming that the projection of each vertex on the opposite face falls inside the face.

Using the assumption $C_K \in K$, we can now split every tetrahedron in four parts and attach to each face f_k a region L_k as we did for triangles. This allows us to define the space V_L, formally as in (7.8.13), and to proceed with the corresponding mixed formulation (7.8.22) and (7.8.23).

The jump of a piecewise constant q can still be defined as in (7.8.15), and the alternative way of writing b given in (7.8.16) still holds.

It is not difficult to check that the analysis sketched in the previous section works practically with no changes. It can be seen that this gives back a classical finite volume scheme for diffusion operators (see e.g. [187]).

7.9 Nonconforming Methods: A Trap to Avoid...

We shall rapidly present, in this section, a formulation which appears at first sight as very clever and promising but which turns out to be catastrophic. The idea is to introduce a nonconforming approximation of $H(\text{div};\Omega)$. Precisely, we use the space Λ defined in (7.2.8) and the bilinear form $c(\mu_h,\underline{v}_h)$ defined in (7.2.10), that is,

$$c(\mu_h,\underline{v}_h) = \sum_K \int_{\partial K} \mu_h \, \underline{v}_h \cdot \underline{n} \, ds. \tag{7.9.1}$$

We write

$$H_{nc} := \{\underline{v}_h \mid \underline{v}_h \in (\mathcal{L}_0^1)^2, c(\mu_h,\underline{v}_h) = 0, \quad \forall\, \mu_h \in \Lambda\}. \tag{7.9.2}$$

The vectors of H_{nc} are therefore piecewise linear and their normal component is continuous at mid-edge points of the triangulation. We complete our approximation by a piecewise constant approximation for p_h, that is,

$$p_h \in \mathcal{L}_0^0. \tag{7.9.3}$$

In order to obtain a discrete problem using this nonconforming approximation, we must introduce

$$b_h(p_h,\underline{v}_h) = \sum_K \int_K p_h \, \operatorname{div} \underline{v}_h \, d\mathbf{x}, \tag{7.9.4}$$

so that our discrete problem becomes, taking $g = 0$ to simplify,

$$a(\underline{u}_h,\underline{v}_h) + b_h(p_h,\underline{v}_h) = 0, \tag{7.9.5}$$

$$b_h(q_h,\underline{u}_h) = -\langle f, q_h\rangle. \tag{7.9.6}$$

Now, the nice thing is that there exists a numerical quadrature formula which is exact for second degree polynomials and uses only mid-edge points. Denoting by m_{iK} the three such point of K and (somewhat inaccurately) by $\underline{n}$ and $\underline{t}$ the normal and tangential vectors at these points, we can thus write for $\underline{u}_h$ and $\underline{v}_h$ in H_{nc}

$$a(\underline{u}_h,\underline{v}_h) = \sum_K \frac{area(K)}{3} \sum_i (\underline{u}_h(m_{iK}) \cdot \underline{n} \, \underline{v}_h(m_{iK}) \cdot \underline{n} + \underline{u}_h(m_{iK}) \cdot \underline{t} \, \underline{v}_h(m_{iK}) \cdot \underline{t}). \tag{7.9.7}$$

In the present case, it is clearly natural to employ as degrees of freedom of H_{nc} the normal and tangential components. The first are shared by adjacent triangles while the latter are internal to each element. Given those degrees of freedom, we are in a situation quite similar to the case of the finite volume method of the previous section: the matrix associated to $a(\underline{u}_h, \underline{v}_h)$ is diagonal. We now consider the first equation of formulation (7.9.5),

$$a(\underline{u}_h, \underline{v}_h) + b_h(p_h, \underline{v}_h) = 0. \tag{7.9.8}$$

It is easy to see that, using (7.9.7), this yields a local formula for the normal components,

$$\frac{area(K)}{3}\underline{u}_h(m_{iK}) \cdot \underline{n} = meas(e_k)[\![p_h]\!]. \tag{7.9.9}$$

Moreover, (7.9.7) implies

$$\underline{u}_h(m_{iK}) \cdot \underline{t} = 0. \tag{7.9.10}$$

What we have is a method very similar to the one which we have studied in the previous section. The difference is that the construction is based on barycentres instead of circumcentres. Another important difference is that we do not introduce an approximation for the bilinear form $a(\underline{u}_h, \underline{v}_h)$, which is computed exactly, but for the bilinear form $b(p_h, \underline{v}_h)$, which has been approximated by $b_h(p_h, \underline{v}_h)$ defined in (7.9.4). This looks nice! The trouble arises when we try to get an error estimate. To do so, we have to introduce $\underline{v}_h \in H_{nc}$ in (7.2.2) and subtract the result from (7.9.5). However, this cannot be done directly as $b(p, \underline{v}_h)$ is not defined. We rather start from the strong form,

$$\int_\Omega \underline{u} \cdot \underline{v}_h \, d\mathbf{x} = \int_\Omega \underline{\text{grad}}\, p \cdot \underline{v}_h \, d\mathbf{x} \tag{7.9.11}$$

and integrate the right-hand side by parts to obtain

$$\int_\Omega \underline{u} \cdot \underline{v}_h \, d\mathbf{x} = -\sum_K \int_K p \ \text{div}\, \underline{v}_h \, d\mathbf{x} + \sum_K \int_{\partial K} p \ \underline{v}_h \cdot \underline{n} \, ds, \tag{7.9.12}$$

which yields

$$a(\underline{u}, \underline{v}_h) + b_h(p, \underline{v}_h) - \sum_K \int_{\partial K} p \ \underline{v}_h \cdot \underline{n} \, ds. \tag{7.9.13}$$

We can now proceed as usual for the error estimate. We obviously get

$$a(\underline{u} - \underline{u}_h, \underline{v}_h) + b_h(p - p_h, \underline{v}_h) - \sum_K \int_{\partial K} p \ \underline{v}_h \cdot \underline{n} \, ds = 0, \tag{7.9.14}$$

$$b_h(q_h, \underline{u} - \underline{u}_h) = 0. \tag{7.9.15}$$

With respect to the standard case, we shall have to bound the extra consistency term

$$\sum_K \int_{\partial K} p \ \underline{v}_h \cdot \underline{n} ds.$$

This is classical for nonconforming methods. Moreover, we know that

$$\sum_K \int_{\partial K} p\, \underline{v}_h \cdot \underline{n} = \sum_K \int_{\partial K} (p - \mu_h)\, \underline{v}_h \cdot \underline{n} \quad \forall \mu_h \in \mathcal{L}_0^0 \tag{7.9.16}$$

and that

$$\int_{\partial K} (p - \mu_h)\, \underline{v}_h \cdot \underline{n} \leq \inf_{\mu_h} \|p - \mu_h\|_{1,K} \|\underline{v}_h\|_{H(\mathrm{div},K)}. \tag{7.9.17}$$

Unfortunately, $\mu_h \in \mathcal{L}_0^0$ cannot be a good approximation in $H^1(K)$. The best that can be done is the bound

$$\inf_{\mu_h} \|p - \mu_h\|_{1,K} \leq C \tag{7.9.18}$$

which does not imply any convergence. Numerical evidence shows that this is true and not a lack in the technique of proof.

7.10 Augmented Formulations (Galerkin Least Squares Methods)

We have seen in the previous sections the importance of both conditions of the general theory of Chap. 5, namely, the coerciveness on the kernel and the *inf-sup* condition. We shall now apply the ideas of Sect. 1.5 to bypass one or both of these conditions. The methods presented should be seen as models to more complex situations and they do not have practical importance by themselves. We shall consider a similar idea in Sect. 8.13 when studying Stokes problems.

Let us first consider the simplest modification, enabling us to obtain coerciveness on the whole space and not only on the kernel. We shall use the augmented formulation (1.5) for which we write the optimality conditions:

$$\begin{cases} \displaystyle\int_\Omega \underline{u} \cdot \underline{v}\, dx + \int_\Omega p \operatorname{div} \underline{v}\, dx + \beta \int_\Omega (\operatorname{div} \underline{u} + f) \operatorname{div} \underline{v}\, dx = 0 \ \forall \underline{v} \in H(\mathrm{div};\Omega), \\ \displaystyle\int_\Omega \operatorname{div} \underline{u} q\, dx + \int_\Omega f q\, dx = 0 \quad \forall q \in L^2(\Omega). \end{cases} \tag{7.10.1}$$

The bilinear form $a(\underline{u}, \underline{v})$ is now defined by

$$a(\underline{u}, \underline{v}) = \int_\Omega \underline{u} \cdot \underline{v}\, dx + \beta \int_\Omega \operatorname{div} \underline{u} \operatorname{div} \underline{v}\, dx \tag{7.10.2}$$

and we obviously have, for any $\beta > 0$ with $\gamma = \min(1, \beta)$,

$$a(\underline{u}, \underline{u}) \geq \gamma ||\underline{u}||^2_{H(\mathrm{div};\Omega)}. \tag{7.10.3}$$

Hence, for the discretisation of (7.10.1), we only have to worry about the *inf-sup* condition.

Similarly, in more general two-dimensional cases, if we were interested in employing a continuous approximation for $\underline{u}$, we might choose, for instance, the MINI element or any other element well suited for the Stokes problem introduced in Chap. 8.

If we now want to avoid as well the problem of the *inf-sup* condition, we can use formulation (1.5.14) for which the optimality condition can be written as

$$\int_{\Omega} \underline{u}\cdot\underline{v}\,dx + \int_{\Omega} p \,\mathrm{div}\,\underline{v}\,dx + \beta\int_{\Omega}(\mathrm{div}\,\underline{u} + f)\,\mathrm{div}\,\underline{v}\,dx \\ -\alpha\int_{\Omega}(\underline{u} - \underline{\mathrm{grad}}\,p)\cdot\underline{v} = 0,\ \forall\,\underline{v}\ \in H(\mathrm{div};\Omega), \tag{7.10.4}$$

$$\alpha\int_{\Omega}(\underline{\mathrm{grad}}\,p - \underline{u})\cdot\underline{\mathrm{grad}}\,q\,dx - \int_{\Omega} q\,\mathrm{div}\,\underline{u}\,dx \\ -\int_{\Omega} fq\,dx = 0,\ \forall\,q\ \in H^1_0(\Omega). \tag{7.10.5}$$

It is easy to check that we now have a problem of the form (4.3.1)

$$\begin{cases} a(\underline{u},\underline{v}) + b(p,\underline{v}) = \langle F,\underline{v}\rangle & \forall\,\underline{v}\ \in H(\mathrm{div};\Omega) \\ b(q,\underline{u}) - c(p,q) = \langle G,q\rangle & \forall\,q\ \in H^1_0(\Omega), \end{cases} \tag{7.10.6}$$

where

$$a(\underline{u},\underline{v}) = (1-\alpha)\int_{\Omega}\underline{u}\cdot\underline{v}\,dx + \beta\int_{\Omega}\mathrm{div}\,\underline{u}\,\mathrm{div}\,\underline{v}\,dx, \tag{7.10.7}$$

$$b(p,\underline{v}) = (1-\alpha)\int_{\Omega} p\,\mathrm{div}\,\underline{v}\,dx, \tag{7.10.8}$$

$$c(p,q) = \alpha\int_{\Omega}\underline{\mathrm{grad}}\,p\cdot\underline{\mathrm{grad}}\,q,dx. \tag{7.10.9}$$

If we choose $0 < \alpha < 1$ and $\beta > 0$, conditions (4.3.4) and (4.3.5) are satisfied and optimal error estimates follow.

Remark 7.10.1. It is obviously also possible to use the variational formulation (7.6.11) instead of (7.6.6) and to obtain convergence for every $\alpha > 0$ and $\beta > 0$. □

We have rapidly presented here the basic idea of using augmented formulations. We refer for more details to [125].

7.11 A Posteriori Error Estimates

During the recent years, one of the important developments in applications of the finite element methods has been mesh adaptation. Presenting this technique would need a book in itself and indeed some were written. We refer to [376] for a general presentation. Mesh adaptation grossly means the capacity of doing the work where it must be done, to obtain a desired level of accuracy. This capacity has radically changed the range of feasible computations. Many mesh adaptation methods have been proposed and most of them permit to locally refine the mesh where the error is large. But some methods are more powerful and permit to build new meshes which are optimal under some criterion. This can be done through remeshing, as in [221], where adapted meshes are built to fit a metric which is obtained from a post-processing of the finite element solution. Another approach [96, 97, 235] applies some transformations to a given mesh in order to minimise some measure of the error.

Adapting a mesh must rely on some criterion to decide where elements have to be added, deleted and eventually oriented and stretched. This is then imbedded in an iterative process: given a mesh and a solution, some a posteriori error estimation leads to a new mesh, a new solution and so on, until the desired accuracy, or the "best possible" accuracy for a given number of degrees of freedom, is obtained.

For standard finite element approximations, this is now a well mastered procedure. Many types of error estimators have been proposed. We just mention the most popular.

- Residual methods were initiated by the work of [45]. They estimate the error using a discrete norm of the residual obtained when the discrete solution is fed to the continuous equation.
- Hierarchical methods [48] estimate the error by obtaining an approximate higher-order solution.
- Another type of estimators is based on the recovery of derivatives through some re-interpolation procedure [308]. In particular, obtaining an estimate of the Hessian permits to build a metric and this is widely employed for solution by piece-wise linear

All this has yielded hundreds of publications and any attempt to review this would be, by far, out of the scope of this book. We just want to point out that these ideas have also been applied to mixed methods.

A review of some methods can be found in [380]. Other results can be found in [5, 137, 274, 285] or [220]. For elasticity problems, one can refer to[140] and [282]. Hierarchical methods have been applied to mixed methods in [1]. Discontinuous Galerkin methods have been considered in [56].

This is, by far, not exhaustive and should only be considered as a starting point for interested readers.

Chapter 8
Incompressible Materials and Flow Problems

Although the approximation of incompressible flows by finite element methods has grown quite independently of the main stream of mixed and hybrid methods, it was soon recognised that a precise analysis requires the framework of mixed methods. In many cases, one may directly apply the techniques and results of Chaps. 4 and 5. In particular, the elements used are often standard elements or simple variants of standard elements. The specificity of the Stokes problem has however led to the development of special techniques; we shall present some of them that seem particularly interesting. Throughout this study, the main point will be to make a clever choice of elements leading to the satisfaction of the *inf-sup* condition which is here the important one as coercivity considerations are almost always straightforward.

This chapter, after a quick description of the problem, will present some simple examples of elements and techniques of proof which can be used as an introduction to the subject. This will be followed by a more detailed presentation. It will not be possible to analyse in detail all the elements for which results are known; we shall try to group them by families which can be treated by similar methods. These families will be arbitrary and will overlap in many cases.

Besides this presentation of elements, we shall also consider solution techniques by penalty methods and we will develop the related problem of almost incompressible elastic materials. We shall consider the equivalence of penalty methods and mixed methods and some questions arising from it. Stabilisation techniques will also be considered.

Finally, a section will be devoted to numerical considerations and to the choice of elements.

D. Boffi et al., *Mixed Finite Element Methods and Applications*, Springer Series in Computational Mathematics 44, DOI 10.1007/978-3-642-36519-5_8,

8.1 Introduction

We have already considered, in Example 1.3.1, the Stokes problem or creeping flow problem for an incompressible fluid. We had written it as a system of variational equations: *find* $\underline{u} \in V$ *and* $p \in Q$ *such that*

$$\begin{cases} 2\mu \displaystyle\int_\Omega \underline{\underline{\varepsilon}}(\underline{u}) : \underline{\underline{\varepsilon}}(\underline{v})\, dx - \int_\Omega p \operatorname{div} \underline{v}\, dx = \int_\Omega \underline{f} \cdot \underline{v}\, dx \quad \forall\, \underline{v} \in V, \\ \displaystyle\int_\Omega q \operatorname{div} \underline{u}\, dx = \int_\Omega g\, q\, dx, \quad \forall\, q \in Q, \end{cases} \tag{8.1.1}$$

where $V := (H_0^1(\Omega))^n$ and Q is the subspace of $L^2(\Omega)$ consisting of functions with zero mean value on Ω. In this formulation, $\underline{u}$ is the velocity of the fluid and p is its pressure. A similar problem arises for the displacement of an incompressible elastic material.

Remark 8.1.1. Although incompressibility corresponds to the case $g = 0$, we shall see in Remark 8.2.2 that non zero boundary conditions correspond implicitly to introducing $g \neq 0$. □

Remark 8.1.2 (Almost incompressible materials). For a linear elastic material, following Example 1.2.2, we have to solve the variational equation

$$2\mu \int_\Omega \underline{\underline{\varepsilon}}(\underline{u}) : \underline{\underline{\varepsilon}}(\underline{v})\, dx + \lambda \int_\Omega \operatorname{div} \underline{u}\, \operatorname{div} \underline{v}\, dx = \int_\Omega \underline{f} \cdot \underline{v}\, dx, \quad \forall\, v \in V. \tag{8.1.2}$$

The case where λ is large, or equivalently, when the Poisson ratio $\nu = \lambda/2(\lambda + \mu)$ approaches $1/2$, can be considered as an approximation of (8.1.1) by a penalty method as in Sect. 5.6.3. The limiting case is exactly (8.1.1) up to the fact that $\underline{u}$ is a displacement instead of a velocity. Problems where λ is large are quite common and correspond to almost incompressible materials. Results of Sect. 5.5.2 can be applied and give conditions under which error estimates can be found that do not depend on λ. Problem (8.1.2) will be considered in detail in Sect. 8.12. □

It is also worth recalling that, defining

$$A\underline{u} := \begin{cases} \dfrac{\partial^2 u_1}{\partial x_1^2} + \dfrac{1}{2}\dfrac{\partial}{\partial x_2}\left(\dfrac{\partial u_1}{\partial x_2} + \dfrac{\partial u_2}{\partial x_1}\right) \\ \dfrac{\partial^2 u_2}{\partial x_2^2} + \dfrac{1}{2}\dfrac{\partial}{\partial x_1}\left(\dfrac{\partial u_1}{\partial x_2} + \dfrac{\partial u_2}{\partial x_1}\right), \end{cases} \tag{8.1.3}$$

that is, $A\underline{u} = \underline{\operatorname{div}}\underline{\underline{\varepsilon}}(\underline{u})$, problems (8.1.1) and (8.1.2) are then respectively equivalent to

$$
\begin{cases}
-2\mu A\underline{u} + \underline{\text{grad}}\, p = \underline{f}, \\
\text{div}\, \underline{u} = g, \\
\underline{u}|_\Gamma = 0
\end{cases}
\tag{8.1.4}
$$

and

$$
-2\mu A\underline{u} - \lambda\, \underline{\text{grad}}\, \text{div}\, \underline{u} = \underline{f}. \tag{8.1.5}
$$

Remark 8.1.3. The above problems are sometimes written in a simplified way. Indeed, we have, for incompressible materials,

$$
\mu A\underline{u} = \mu\Delta\underline{u} + \mu\ \underline{\text{grad}}\, \text{div}\, \underline{u} = \mu\Delta\underline{u}. \tag{8.1.6}
$$

However, this simplification of the operator is valid only if Dirichlet conditions are considered everywhere. Otherwise, the natural Neumann conditions are different and those associated with the simplified operator are unphysical. □

Remark 8.1.4. The problems described above are, of course, physically unrealistic, as they involve body forces and homogeneous Dirichlet boundary conditions. The aim of doing so is to avoid purely technical difficulties and this implies no loss of generality. The results obtained will be valid, unless otherwise stated, for all acceptable boundary conditions. □

To approximate the Stokes problem, two approaches follow quite naturally from the preceding considerations. *The first* one is to use system (8.1.1) and to discretise $\underline{u}$ and p by standard (or less standard) finite element spaces. *The second* one is to use formulation (8.1.2) with λ large as a penalty approximation to system (8.1.1).

It rapidly became clear that both these approaches could yield strange results. In particular, the first one often led to non convergence of the pressure (see Sect. 8.3.1) and the second one to a *locking mechanism*, the numerical solution being uniformly zero, or unnaturally small for λ large. For velocity-pressure approximations, empirical cures were found by Hughes and Allik [255], Hood and Taylor [249] and others. At about the same time, some elements using discontinuous pressure fields were shown to work properly [165,200] from the mathematical point of view.

For the penalty method, the cure was found in selective or reduced integration procedures. This consisted in evaluating terms like $\int_\Omega \text{div}\, \underline{u}\ \text{div}\, \underline{v}\, dx$ by quadrature formulae of low order. This *sometimes* led to good results.

It was finally stated [287], even if the result was implicit in earlier works [59], that the analysis underlying the two approaches is the same. Penalty methods are often equivalent to some mixed methods. In such cases, the penalty method works if and only if the associated mixed method works [60]. This will be developed in Sect. 8.12.

First, we must give a more precise framework to our problem.

8.2 The Stokes Problem as a Mixed Problem

8.2.1 Mixed Formulation

We shall describe in this section how the Stokes problem (8.1.1) can be analysed in the general framework of Chaps. 4 and 5. Defining $V := (H_0^1(\Omega))^n$, $Q := L^2(\Omega)$ and

$$a(\underline{u}, \underline{v}) := 2\mu \int_\Omega \underline{\underline{\varepsilon}}(\underline{u}) : \underline{\underline{\varepsilon}}(\underline{v})\, dx \tag{8.2.1}$$

$$b(\underline{v}, q) := -\int_\Omega q \operatorname{div} \underline{v}\, dx, \tag{8.2.2}$$

problem (8.1.1) can clearly be written in the form: *find $\underline{u} \in V$ and $p \in Q$ such that*

$$\begin{cases} a(\underline{u}, \underline{v}) + b(\underline{v}, p) = (\underline{f}, \underline{v}) & \forall\, \underline{v} \in V, \\ b(\underline{u}, q) = (g, q) & \forall\, q \in Q, \end{cases} \tag{8.2.3}$$

which is a saddle point problem in the sense of Chap. 4. Indeed, we have already seen that p is the Lagrange multiplier associated with the incompressibility constraint.

Remark 8.2.1. It is apparent, from the definition (8.2.2) of $b(\cdot,\cdot)$ and from the boundary conditions of the functions in V, that p, if exists, is defined up to a constant. Therefore, we change the definition of the space Q into

$$Q := L_0^2(\Omega) = L^2(\Omega)/\mathbb{R}, \tag{8.2.4}$$

where two elements $q_1, q_2 \in L^2(\Omega)$ are identified if their difference is constant. It is not difficult to show that Q is isomorphic to the subspace of $L^2(\Omega)$ consisting of functions with zero mean value on Ω. □

With this choice, our problem reads: *find $\underline{u} \in V$ and $p \in Q$ such that*

$$\begin{cases} a(\underline{u}, \underline{v}) + b(\underline{v}, p) = (\underline{f}, \underline{v}) & \forall\, \underline{v} \in V, \\ b(\underline{u}, q) = (g, q) & \forall\, q \in Q. \end{cases} \tag{8.2.5}$$

Let us check the hypotheses of Theorem 4.2.2 to ensure that our problem is well-posed. Using the notation of Chap. 4, we can write

$$B = -\operatorname{div} : (H_0^1(\Omega))^n \to L^2(\Omega)/\mathbb{R} \tag{8.2.6}$$

and

$$B^t = \underline{\operatorname{grad}} : L^2(\Omega)/\mathbb{R} \to (H^{-1}(\Omega))^n. \tag{8.2.7}$$

It can be shown (see, e.g., [362]) that

$$\mathrm{Im}B = Q \cong \left\{ q \mid q \in L^2(\Omega), \int_\Omega q\,dx = 0 \right\}, \tag{8.2.8}$$

hence the operator B has a continuous lifting and the continuous *inf-sup* condition (4.2.26) holds. We also notice that, with our definition of the space Q, the kernel $\mathrm{Ker}B^t$ reduces to zero.

The bilinear form $a(\cdot,\cdot)$ is coercive on V: there exists α such that

$$a(\underline{v},\underline{v}) \geq \alpha \|\underline{v}\|_V^2. \tag{8.2.9}$$

This is the well known Korn inequality (see [183, 362]), whence (4.2.12) also will follow (i.e., A is invertible on $\mathrm{Ker}B$).

We state the well-posedness of problem (8.2.5) in the following theorem. The proof follows from Theorem 4.2.1.

Theorem 8.2.1. *Let $\underline{f}$ be given in $(H^{-1}(\Omega))^n$ and g in $Q = L_0^2(\Omega)$. Then, there exists a unique $(\underline{u}, p) \in V \times Q$, solution to problem* (8.2.5)*, which satisfies*

$$||\underline{u}||_V + ||p||_Q \leq C(||\underline{f}||_{H^{-1}} + \|g\|_Q). \tag{8.2.10}$$

Now, choosing an approximation $V_h \subset V$ and $Q_h \subset Q$ yields the discrete problem

$$\begin{cases} 2\mu \displaystyle\int_\Omega \underline{\underline{\varepsilon}}(\underline{u}_h) : \underline{\underline{\varepsilon}}(\underline{v}_h)\,dx - \int_\Omega p_h \;\mathrm{div}\,\underline{v}_h\,dx = \int_\Omega \underline{f} \cdot \underline{v}_h\,dx \quad \forall\, \underline{v}_h \in V, \\ \displaystyle\int_\Omega q_h \;\mathrm{div}\,\underline{u}_h\,dx = (g, q_h) \quad \forall\, q_h \in Q_h. \end{cases} \tag{8.2.11}$$

The bilinear form $a(\cdot,\cdot)$ is coercive on V; hence, according to the theory developed in Chaps. 3 and 4, there is no problem for the *existence* of a solution $\{\underline{u}_h, p_h\}$ to problem (8.2.11), at least with $g = 0$. Indeed, we have a finite-dimensional problem where $\mathrm{Im}B$ is closed and the right-hand side of the second equation of (8.2.11) is zero. It should be noted, however, that we might have trouble with the *uniqueness* of p_h and that there might be compatibility conditions on g for some approximations.

We thus try to obtain estimates of the errors $||\underline{u} - \underline{u}_h||_V$ and $||p - p_h||_Q$.

First, we observe that, even for $g = 0$, the discrete solution $\underline{u}_h$ *needs not be divergence-free*. Indeed, the bilinear form $b(\cdot,\cdot)$ defines a discrete divergence operator

$$B_h = -\mathrm{div}_h : V_h \to Q_h \tag{8.2.12}$$

(it is convenient here to identify $Q = L^2(\Omega)/\mathbb{R}$ and $Q_h \subset Q$ with their dual spaces). In fact, we have

$$(\mathrm{div}_h \underline{u}_h, q_h)_Q = \int_\Omega q_h \,\mathrm{div}\, \underline{u}_h \, dx \tag{8.2.13}$$

and thus $\mathrm{div}_h \underline{u}_h$ turns out to be the L^2-projection of $\mathrm{div}\, \underline{u}$ onto Q_h.

The discrete divergence operator coincides with the standard divergence operator if $\mathrm{div}\, V_h \subset Q_h$. Referring to Chap. 5, we see that obtaining error estimates requires a careful study of the properties of the operator $B_h = -\mathrm{div}_h$ and of its transpose that we denote by $\underline{\mathrm{grad}}_h$.

The first issue is to characterise the kernel $\mathrm{Ker} B_h^t = \mathrm{Ker}(\underline{\mathrm{grad}}_h)$. It might happen that $\mathrm{Ker} B_h^t$ contains non-trivial functions. In these cases, $\mathrm{Im} B_h = \mathrm{Im}(\mathrm{div}_h)$ will be *strictly smaller* than $Q_h = P_{Q_h}(\mathrm{Im} B)$; this may lead to pathologies. In particular, if we consider a modified problem, like the one that usually originates when dealing with Dirichlet boundary conditions, the strict inclusion $\mathrm{Im} B_h \subset Q_h$ may even imply trouble with the existence of the solution. This situation is made clearer with the following example.

Remark 8.2.2. Let us consider problem (8.1.4) with *non-homogeneous boundary conditions*, that is, let $\underline{r}$ be such that

$$\underline{u}|_\Gamma = \underline{r}, \qquad \int_\Gamma \underline{r} \cdot \underline{n} \, ds = 0. \tag{8.2.14}$$

It is classical to reduce this case to a problem with homogeneous boundary conditions by first introducing a function $\underline{\tilde{u}} \in (H^1(\Omega))^2$ such that $\underline{\tilde{u}}|_\Gamma = \underline{r}$. Setting $\underline{u} = \underline{u}_0 + \underline{\tilde{u}}$ with $\underline{u}_0 \in (H_0^1(\Omega))^2$, we have to solve

$$\begin{cases} -2\mu A \underline{u}_0 + \underline{\mathrm{grad}}\, p = \underline{f} + 2\mu A \underline{\tilde{u}} = \underline{\tilde{f}}, \\ \mathrm{div}\, \underline{u}_0 = -\ \mathrm{div}\, \underline{\tilde{u}} = g, \quad \underline{u}_0|_\Gamma = 0, \end{cases} \tag{8.2.15}$$

with A defined in (8.1.3). We thus find a problem with a constraint $B\underline{u}_0 = g$ where $g \neq 0$. We have seen in Chap. 5 that the associated discrete problem may fail to have a solution because $g_h = P_{Q_h} g$ does not necessarily belong to $\mathrm{Im} B_h$, whenever $\mathrm{Ker} B_h^t \not\subset \mathrm{Ker} B^t$. Discretisations where $\mathrm{Ker}(\underline{\mathrm{grad}}_h)$ is non-trivial *can therefore lead to ill-posed problems* in particular for some non-homogeneous boundary conditions. Examples of such conditions can be found in [340, 341]. In general, any method that relies on extra compatibility conditions is a source of trouble when applied to more complicated (non-linear, time-dependent, etc.) problems. □

For a first attempt to error estimates, we shall use Theorem 5.2.2. Since the bilinear form $a(\cdot,\cdot)$ is coercive on V, we only have to worry about the *inf-sup* condition. The following proposition will be the starting point for the analysis of any finite element approximation of (8.2.5).

Proposition 8.2.1. *Let* $(\underline{u}, p) \in V \times Q$ *be the solution of* (8.2.5) *and suppose the following inf-sup condition holds true*

$$\inf_{q_h \in Q_h} \sup_{\underline{v}_h \in V_h} \frac{\int_\Omega q_h \operatorname{div} \underline{v}_h \, dx}{||q_h||_Q ||\underline{v}_h||_V} \geq k_h. \tag{8.2.16}$$

Then, there exists a unique $(\underline{u}_h, p_h) \in V_h \times Q_h$, *solution to* (8.2.11), *and the following estimate holds*

$$\|u_h - u\|_V \leq \left(\frac{2\|a\|}{\alpha} + \frac{2\|a\|^{1/2}\|b\|}{(\alpha)^{1/2} k_h} \right) E_u + \frac{\|b\|}{\alpha} E_p, \tag{8.2.17}$$

$$\|p_h - p\|_Q \leq \left(\frac{2\|a\|^{3/2}}{(\alpha)^{1/2} k_h} + \frac{\|a\| \, \|b\|}{k_h^2} \right) E_u + \frac{3\|a\|^{1/2}\|b\|}{(\alpha)^{1/2} k_h} E_p \tag{8.2.18}$$

with α *given by* (8.2.9). □

Remark 8.2.3. Actually, as it has been already observed, the existence of the discrete solution $(\underline{u}_h, p_h)$ (when the right-hand side in the second equation of (8.2.5) is zero) is not a consequence of the *inf-sup* condition (8.2.16). However, we should not forget about the possible situation presented in Remark 8.2.2. □

Of course, we shall be looking for cases where

$$k_h \geq k_0 > 0. \tag{8.2.19}$$

In this case, it may be useful to summarise the estimates (8.2.17) and (8.2.18) in the following result.

Proposition 8.2.2. *With the same hypotheses as in Proposition 8.2.1, let us suppose that* (8.2.19) *holds. Then, there exists* C, *independent of* h, *such that*

$$\|u_h - u\|_V + \|p_h - p\|_Q \leq C(E_u + E_p). \tag{8.2.20}$$

□

Remark 8.2.4. We shall also meet cases in which the constant k_h is not bounded below by k_0. We shall then try to know precisely how it depends on h and to see whether a lower-order convergence can be achieved. When $\mathrm{Ker}(\underline{\mathrm{grad}}_h)$ is non-trivial, we are interested in a weaker form of (8.2.16)

$$\sup_{\underline{v}_h \in V_h} \frac{\int_\Omega q_h \operatorname{div} \underline{v}_h \, dx}{||\underline{v}_h||_V} \geq k_h \inf_{q \in \mathrm{Ker}(\underline{\mathrm{grad}}_h)} ||q_h - q||_{L^2(\Omega)}, \tag{8.2.21}$$

and in the dependence of k_h in terms of h. From (8.2.17) and (8.2.18), one sees that the effect will be stronger on the error $\|p - p_h\|_Q$. □

8.3 Some Examples of Failure and Empirical Cures

This section will present some classical troubles associated to the approximation of incompressible materials. We shall thus recall the difficulties associated with some 'obvious' approximations. We shall consider some examples of possible choices for the spaces V_h and Q_h, namely the $\underline{P}_1 - P_1$ element, a case of continuous pressure, and the $\underline{P}_1 - P_0$ element, a case of discontinuous pressure. These elements do not satisfy the *inf-sup* condition (8.2.16) and are not applicable in practice. We shall introduce some cures which will be developed and eventually justified later in this chapter.

8.3.1 Continuous Pressure: The $\underline{P}_1 - P_1$ Element

As we stated in the introduction, the development of finite element methods for incompressible problems was done, at least in the beginning, independently of the theory of mixed methods. The natural idea when attempting to solve a problem involving incompressibility would be to employ the same approximation for both velocity and pressure, in the simplest case a P_1 continuous interpolation

$$V_h := (\mathcal{L}_1^1)^n \cap V, \quad Q_h := \mathcal{L}_1^1 \cap Q. \tag{8.3.1}$$

In the two-dimensional case, it is easy to check that if the number of triangles is large enough, then there exist non-trivial functions satisfying the *discrete* divergence-free condition. Thus, no locking will occur and a solution can be computed. Indeed, this method would not provide an optimal approximation of the pressures by virtue of the unbalanced approximation properties of the discrete spaces (while Q_h achieves second order in L^2, V_h gives only first order in H^1). On the other hand, users of such methods (you can think of using also, for instance, $(\underline{P}_2 - P_2)$, $(\underline{Q}_1 - Q_1)$, etc.) soon became aware that their results were strongly mesh dependent. In particular, the computed pressures exhibited a very strange instability. This comes from the fact that for some meshes, the kernel of the discrete gradient operator, $\mathrm{Ker}(\underline{\mathrm{grad}}_h)$, is not the subspace of constant functions, as one would expect from the continuous problem, but is a larger subspace. This means that the solution obtained is determined only up to a given number of *spurious pressure modes* [340, 341] and that, at best, some filtering will have to be done before accurate results are available. We shall come back later on to this phenomenon also named chequerboarding in Sect. 8.10. We have already stated in Remark 8.2.2 that such spurious modes can impose non physical conditions on the data. To better understand the nature of spurious pressure modes, the reader may check the results of Fig. 8.1 in which different symbols denote points where functions in $\mathrm{Ker}(\underline{\mathrm{grad}}_h)$ must have equal values for a $(\underline{P}_1 - P_1)$ approximation.

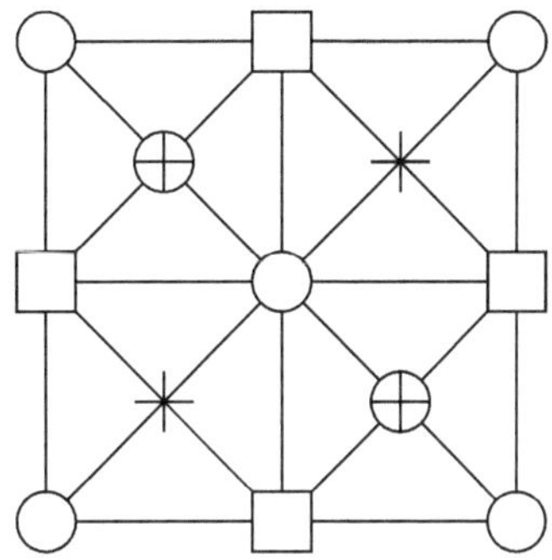

Fig. 8.1 Spurious modes for the $\underline{P}_1 - P_1$ case

In this case, we have *three* spurious pressure modes. This also shows that there exists on this mesh one non-trivial discrete divergence-free function whereas a direct count would predict locking.

Remark 8.3.1 (Possible cures). A cure for this problem was found empirically [249]: good results can be obtained using a $\underline{P}_2$ approximation for velocity but a P_1 approximation for pressure. It is also clear that this does not impair the order of accuracy. This will be analysed in Sect. 8.8.

Another possibility to obtain stable elements is to add some internal degrees of freedom. The simplest case is the MINI element of [25] which we present in Sect. 8.4.2 and in a more general form in Sect. 8.5.5. □

8.3.2 Discontinuous Pressure: The $\underline{P}_1 - P_0$ Approximation

A second natural approach would be to try imposing directly the divergence-free condition. The simplest element one can imagine for the approximation of an incompressible flow would use a standard $\underline{P}_1$ approximation for the velocities and a piecewise constant approximation for the pressures. With the notation of Chap. 2, this would read, again in the two-dimensional case,

$$V_h := (\mathcal{L}_1^1)^2 \cap V, \quad Q_h := \mathcal{L}_0^0 \cap Q. \tag{8.3.2}$$

As the divergence of a $\underline{P}_1$ velocity field is piecewise constant, this would lead to a truly divergence-free approximation. Moreover, this would give a well-balanced $O(h)$ approximation in estimates (8.2.17) and (8.2.18).

However, it is easy to see that such an element will not work for a general mesh. Indeed, consider a triangulation of a (simply connected) domain Ω and let us denote by

- t the number of triangles,
- v_I the number of internal vertices,
- v_B the number of boundary vertices.

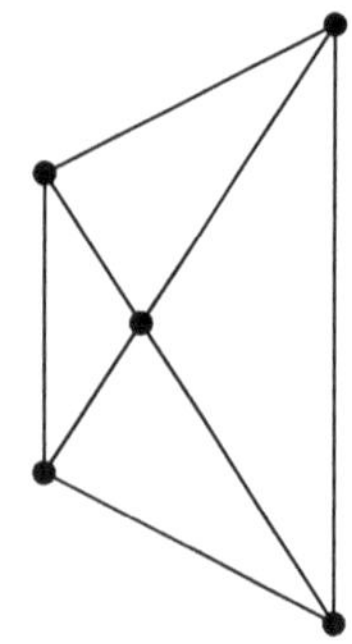

Fig. 8.2 The cross-grid element

We shall thus have $2v_I$ degrees of freedom (d.o.f.) for the space V_h (since the velocities vanish on the boundary) and $(t-1)$ d.o.f. for Q_h (because of the zero mean value of the pressures) leading to $(t-1)$ independent divergence-free constraints. By Euler's relations, we have

$$t = 2v_I + v_B - 2 \tag{8.3.3}$$

and thus

$$(t-1) \geq 2(v_I - 1). \tag{8.3.4}$$

A function $\underline{u}_h \in V_h$ is thus over-constrained and a *locking phenomenon* is likely to occur: in general, the only divergence-free discrete function is $\underline{u}_h \equiv 0$.

When the mesh is built under certain restrictions, it is however possible that some linear constraints become dependent: this will be the case for the cross-grid macro-element (Fig. 8.2) which will be analysed in Example 8.10.3.

As we shall see in Sect. 8.8.1, in general, obtaining truly divergence-free elements requires high degree approximations and some conditions on the mesh. We shall give, in the next section, the simplest example of a stable discontinuous pressure element, the $\underline{P}_2 - P_0$ element.

8.4 Building a B-Compatible Operator: The Simplest Stable Elements

We shall first recall here some of the results of Sect. 5.4.4 as applied to our incompressible problems. Then, we present a complete analysis of the MINI and $\underline{P}_2 - P_0$ elements and of the nonconforming $\underline{P}_1 - P_0$ elements. We shall obtain in Sect. 8.5.5 a more general proof for the MINI element.

It is not recommended to use the element $\underline{P}_2 - P_0$ because of its "unbalanced" approximation properties ($O(h^2)$ for V_h in the V-norm and only $O(h)$ for Q_h in the norm of Q), so that estimate (8.2.20) turns out to be suboptimal. However, the analysis of this element contains basic issues for getting familiar with the approximation of the Stokes problem. Moreover, the stability properties of this element will often be used as an intermediate step for the analysis of other, more efficient, elements. It must also be said that this element is not directly generalisable to the three-dimensional case.

8.4.1 Building a B-Compatible Operator

An efficient way (sometimes known as Fortin's trick) of proving the *inf-sup* condition (8.2.16) consists in building a B-compatible interpolation operator Π_h like in Sect. 5.4 (see [201]). We write down, here, how the hypotheses of Proposition 5.4.3 read in this particular situation.

Proposition 8.4.1. *If there exists a linear operator* $\Pi_h : V \to V_h$ *such that*

$$\int_\Omega \operatorname{div}(\underline{u} - \Pi_h \underline{u}) q_h \, dx = 0 \quad \forall\, \underline{v} \in V,\ q_h \in Q_h, \tag{8.4.1}$$

$$||\Pi_h \underline{u}||_V \leq c ||\underline{u}||_V, \tag{8.4.2}$$

then the inf-sup condition (8.2.16) *holds true.* □

Remark 8.4.1. As it is shown in Chap. 5, condition (8.4.1) is equivalent to $\mathrm{Ker}(\underline{\mathrm{grad}}_h) \subset \mathrm{Ker}(\underline{\mathrm{grad}})$. An element with this property will present no spurious pressure modes. □

In several cases, the operator Π_h can be constructed in two steps as in Proposition 5.4.4. This was the case, for instance, in the proof of Proposition 8.4.3. In general, it will be enough to build two operators $\Pi_1, \Pi_2 \in \mathcal{L}(V, V_h)$ such that

$$||\Pi_1 \underline{v}||_V \leq c_1 ||\underline{v}||_V \qquad \forall\, \underline{v} \in V, \tag{8.4.3}$$

$$||\Pi_2 (I - \Pi_1) \underline{v}||_V \leq c_2 ||\underline{v}||_V \qquad \forall\, \underline{v} \in V, \tag{8.4.4}$$

$$\int_\Omega \operatorname{div}(\underline{v} - \Pi_2 \underline{v}) q_h = 0 \qquad \forall\, \underline{v} \in V,\ \forall\, q_h \in Q_h, \tag{8.4.5}$$

where the constants c_1 and c_2 are independent of h. Then, the operator Π_h satisfying (8.4.1) and (8.4.2) will be found as

$$\Pi_h \underline{u} = \Pi_1 \underline{u} + \Pi_2 (\underline{u} - \Pi_1 \underline{u}). \tag{8.4.6}$$

In many cases, Π_1 will be the interpolation operator of [154] (cf. Chap. 2) defined in $H^1(\Omega)$.

On the contrary, the choice of Π_2 will vary from one case to the other, according to the choice of V_h and Q_h. However, the common feature of the various choices for

Π_2 will be the following one: the operator Π_2 is constructed on each element K in order to satisfy (8.4.5). In many cases, it will be such that

$$||\Pi_2\underline{v}||_{1,K} \leq c(h_K^{-1}||\underline{v}||_{0,K} + |\underline{v}|_{1,K}). \tag{8.4.7}$$

We can summarise this result in the following proposition.

Proposition 8.4.2. *Let V_h be such that a "Clément's operator": $\Pi_1 : V \to V_h$ exists and satisfies* (8.4.24). *If there exists an operator $\Pi_2 : V \to V_h$ such that* (8.4.5) *and* (8.4.7) *hold, then the operator Π_h defined by* (8.4.6) *satisfies* (8.4.1) *and* (8.4.2) *and therefore the discrete inf-sup condition* (8.2.16) *holds.* □

We now consider some simple examples where this construction can be used.

8.4.2 A Stable Case: The MINI Element

We now show how we can enrich the space V_h of Example 8.3.1 so that, in the end, the new choice will yield a stable and convergent approximation to the Stokes problem (8.2.3). We set, as in (2.2.28),

$$B_3 := \{b(\underline{x}) \mid b(\underline{x})|_T \in P_3(T) \cap H_0^1(T), \ \forall\, T \in \mathcal{T}_h\}. \tag{8.4.8}$$

Hence, each $b(\underline{x})$ of B_3, on each triangle T, has the form $\alpha(T)\ \lambda_1(\underline{x})\ \lambda_2(\underline{x})\ \lambda_3(\underline{x})$ with $\alpha(T)$ constant in T. Following [25], we set

$$V_h := \{\mathcal{L}_1^1(\mathcal{T}_h) \oplus B_3\}^2 \cap V \quad Q_h := \mathcal{L}_1^1(\mathcal{T}_h) \cap Q \tag{8.4.9}$$

and we want to show that (8.4.9) leads to a stable and convergent approximation of the Stokes problem. For this, we are going to apply Proposition 8.4.2. We therefore have to construct an operator Π_h such that

$$\int_\Omega \operatorname{div}(\underline{v} - \Pi_h\underline{v})\, q_h\, dx = 0 \ \forall\, q_h \in Q_h \ \forall\, \underline{v} \in V, \tag{8.4.10}$$

$$\|\Pi_h\underline{v}\|_V \leq c\, \|\underline{v}\|_V \ \forall\, \underline{v} \in V. \tag{8.4.11}$$

Following Proposition 8.4.2, we first take for Π_1 the operator r_h of Proposition 2.2.1 and Corollary 2.2.1. We set

$$\Pi_1\underline{v}|_K = r_h\underline{v}|_K \tag{8.4.12}$$

which, from (2.2.20), yields

$$|\underline{v} - \Pi_1\underline{v}|_{m,K} \leq c\Big(\sum_{\bar{K}'\cap\bar{K}\neq\emptyset} h_{K'}^{1-m}\,|\underline{v}|_{1,K'}\Big). \tag{8.4.13}$$

In particular, (8.4.13) implies the first condition of (5.4.12)

$$\|\Pi_1 \underline{v}\|_V \le c\,\|\underline{v}\|_V. \tag{8.4.14}$$

We now define the operator $\Pi_2 : V \to (B_3)^2$ by means of

$$\int_\Omega \operatorname{div}(\Pi_2 \underline{v} - \underline{v}) q_h\,dx = \int_\Omega (\underline{v} - \Pi_2 \underline{v}) \cdot \underline{\operatorname{grad}}\, q_h\,dx = 0 \;\forall\, q_h \in Q_h. \tag{8.4.15}$$

Since $\underline{\operatorname{grad}}\, q_h$ is piecewise constant, (8.4.15) is easily satisfied by choosing, in each K, bubbles with the same mean value as $\underline{v}$. It is easy to check that (under a minimum angle condition)

$$\|\Pi_2 \underline{v}\|_{r,K} \le c h_K^{-r}\,\|\underline{v}\|_{0,K} \;\forall\, \underline{v} \in V,\; r = 0, 1. \tag{8.4.16}$$

Indeed, for $r = 1$, this is the inverse inequality of Sect. 2.2.7.

From (8.4.15), it is then immediate to check that the second condition of (5.4.12) is fulfilled and, from (8.4.16) and (8.4.13), we easily have the third condition.

We can thus apply Proposition 8.4.2 and the *inf-sup* condition holds. Now, we apply Proposition 8.2.2 (or the more complete result of Proposition 8.2.1) and we obtain

$$\|\underline{u} - \underline{u}_h\|_V + \|p - p_h\|_Q \le c h\,(\|\underline{u}\|_{2,\Omega} + \|p\|_1), \tag{8.4.17}$$

that is, an optimal error estimate for $\underline{u}$.

8.4.3 *Another Stable Approximation: The Bi-dimensional $\underline{P}_2 - P_0$ Element*

Let us now move, in the two-dimensional case, to the *stable* $\underline{P}_2 - P_0$ element. Precisely, we use continuous piecewise quadratic vectors for the approximation of the velocities and piecewise constants for the pressures.

The discrete divergence-free condition can then be written as

$$\int_K \operatorname{div} \underline{u}_h\,dx = \int_{\partial K} \underline{u}_h \cdot \underline{n}\,ds = 0, \quad \forall\, K \in \mathcal{T}_h, \tag{8.4.18}$$

that is, as a conservation of mass on every element. This is intuitively an approximation of $\operatorname{div} \underline{u}_h = 0$, directly related to the physical meaning of this condition. It is clear from error estimates (8.2.17), (8.2.18) and standard approximation results (cf. Chap. 2) that such an approximation will lead to the loss of one order of accuracy due to the poor approximation of the pressures. However, an augmented Lagrangian technique can be used in order to recover a part of the accuracy loss (see Remark 8.4.4).

Fig. 8.3 The $\underline{P}_2$ element

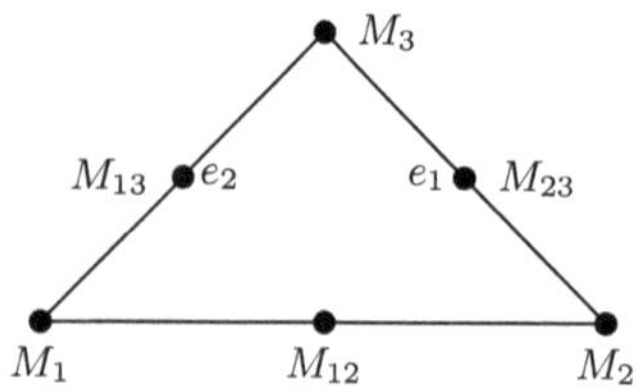

Proposition 8.4.3. *The choice*

$$V_h := (\mathcal{L}_2^1)^2 \cap V, \quad Q_h := \mathcal{L}_0^0 \cap Q \tag{8.4.19}$$

fulfils the inf-sup condition (8.2.16).

Proof. Before giving the rigorous proof of Proposition 8.4.3, we are going to sketch the main argument.

If we try to check the *inf-sup* condition by building an operator Π_h satisfying (5.4.10), then, given $\underline{u}$, we have to build $\underline{u}_h = \Pi_h \underline{u}$ such that

$$\int_\Omega \operatorname{div}(\underline{u} - \underline{u}_h) q_h \, dx = 0 \quad \forall q_h \in Q_h. \tag{8.4.20}$$

Since q_h is constant on every element $K \in \mathcal{T}_h$, this is equivalent to

$$\int_K \operatorname{div}(\underline{u} - \underline{u}_h) \, dx = \int_{\partial K} (\underline{u} - \underline{u}_h) \cdot \underline{n} \, ds = 0. \tag{8.4.21}$$

This last condition would be satisfied if $\underline{u}_h$ could be built in the following way. Let us denote by M_i and e_i, $i = 1, 2, 3$, the vertices and the sides of the triangular element K (Fig. 8.3); the mid-side nodes are denoted by M_{ij}.

We then define

$$\underline{u}_h(M_i) = \underline{u}(M_i), \quad i = 1, 2, 3 \tag{8.4.22}$$

$$\int_{e_i} \underline{u}_h \, ds = \int_{e_i} \underline{u} \, ds. \tag{8.4.23}$$

Condition (8.4.23) can be fulfilled by a correct choice of $\underline{u}_h(M_{ij})$. Moreover, this construction can be done at element level as the choice of $\underline{u}_h(M_{ij})$ is compatible on adjacent elements (that is, with this definition, $\underline{u}_h$ turns out to be continuous).

Although this is the basic idea, some technicalities must be introduced before a real construction is obtained. Indeed, for $\underline{u} \in (H_0^1(\Omega))^2$, condition (8.4.22) does not make sense.

Let us then give a rigorous proof of Proposition 8.4.3. We shall rely again on Proposition 8.4.2. Denoting by $\Pi_1 : V \to V_h$ the Clément interpolant [154] described in Proposition 2.2.1, we then have

$$\sum_K h_K^{2r-2}|\underline{v} - \Pi_1 \underline{v}|_{r,K}^2 \leq c||\underline{v}||_{1,\Omega}^2, \quad r = 0, 1. \tag{8.4.24}$$

Setting $r = 1$ and using the triangular inequality $||\Pi_1 \underline{v}|| \leq ||\underline{v} - \Pi_1 \underline{v}|| + ||\underline{v}||$ gives

$$||\Pi_1 \underline{v}||_V \leq c_1 ||\underline{v}||_V \quad \forall\, \underline{v} \in V. \tag{8.4.25}$$

We now modify Π_1 in a suitable way. Let us define $\Pi_2 : V \to V_h$ in the following way:

$$\Pi_2 \underline{v}|_K(M) = 0 \quad \forall\, M \text{ vertex of } K, \tag{8.4.26}$$

$$\int_e \Pi_2 \underline{u}\, ds = \int_e \underline{u}\, ds \quad \forall\, e \text{ edge of } K. \tag{8.4.27}$$

By construction, Π_2 satisfies

$$\int_\Omega \operatorname{div}(\underline{v} - \Pi_2 \underline{v}) q_h\, dx = 0 \quad \forall\, \underline{v}_h \in V_h,\ q_h \in Q_h \tag{8.4.28}$$

and a scaling argument (see Sect. 2.2.7) gives

$$|\Pi_2 \underline{v}|_{1,K} = |\widehat{\Pi_2 \underline{v}}|_{1,\hat{K}} < c(K, \theta_0)||\hat{\underline{v}}||_{1,\hat{K}} \leq c(K, \theta_0)(h_K^{-1}|\underline{v}|_{0,K} + |\underline{v}|_{1,K}). \tag{8.4.29}$$

We can now define, as in Proposition 5.4.4,

$$\Pi_h \underline{u} = \Pi_1 \underline{u} + \Pi_2(\underline{u} - \Pi_1 \underline{u}) \tag{8.4.30}$$

and observe that (8.4.29) and (8.4.24) imply

$$||\Pi_2(I - \Pi_1)\underline{u}||_V \leq c_2 ||\underline{u}||_V \quad \forall\, \underline{v} \in V, \tag{8.4.31}$$

since

$$||\Pi_2(I - \Pi_1)\underline{v}||_{1,\Omega}^2 = \sum_K ||\Pi_2(I - \Pi_1)\underline{v}||_{1,K}^2$$

$$\leq c \sum_K \{h_K^{-2}||(I - \Pi_1)\underline{v}||_{0,K}^2 + |(I - \Pi_1)\underline{v}|_{1,K}^2\} \leq c||\underline{v}||_{1,\Omega}^2. \tag{8.4.32}$$

Hence, Proposition 5.4.4 applies and the proof is concluded. □

The above proof can easily be extended to more general cases. It applies to the $(Q_2)^2 - P_0$ quadrilateral element, provided that the usual regularity assumptions on quadrilateral meshes are made.

Remark 8.4.2 (The 2D SMALL element). The proof will hold for elements in which only the normal component of velocity is used as a d.o.f. at the mid-side nodes

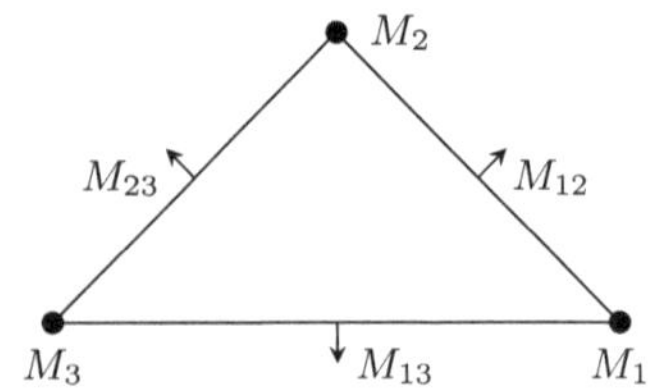

Fig. 8.4 The 2d SMALL element

[70, 198, 202]. Indeed, if only the normal component of $\underline{u}_h$ is used as a degree of freedom, the $(P_2)^2 - P_0$ element becomes the element of Fig. 8.4 in which, on each side, the normal component of $\underline{u}_h$ is quadratic, whereas the tangential component is only linear.

In this case, we can define $\Pi_2\underline{v}$ by setting

$$\int_e (\Pi_2\underline{v} \cdot \underline{n})\, ds = \int_e \underline{v} \cdot \underline{n}\, ds. \tag{8.4.33}$$

The same remark is valid for the $(Q_2)^2 - P_0$ quadrilateral element. □

Remark 8.4.3. The philosophical idea behind the $(P_2)^2 - P_0$ element is that we need one degree of freedom per each interface (actually, the normal component of the velocity) in order to control the jump of the pressures. This is basically the meaning of Green's formula (8.4.21). For three-dimensional elements, however, we would need a **mid-face** node instead of a mid-side node in order to control the normal flux from one element to the other.

In particular, we point out that adding *internal degrees of freedom* to the velocity space *cannot stabilise* elements with *piecewise constant pressures* which do not satisfy the *inf-sup* condition. □

Remark 8.4.4. To reduce the loss of accuracy due to the unbalanced approximation properties of the spaces V_h and Q_h, we can employ the augmented Lagrangian technique of Sect. 5.6.3. The discrete scheme reads: *find* $(\underline{u}_h, p_h) \in V_h \times Q_h$ *such that*

$$\begin{aligned} \int_\Omega \underline{\underline{\varepsilon}}(\underline{u}_h) : \underline{\underline{\varepsilon}}(\underline{v}_h)\, dx + h^{-1/2} \int_\Omega \operatorname{div} \underline{u}_h \operatorname{div} \underline{v}_h\, dx \\ - \int_\Omega p_h \operatorname{div} \underline{v}_h\, dx = \int_\Omega \underline{f} \cdot \underline{v}_h\, dx, \quad \forall\, \underline{v}_h \in V_h, \end{aligned} \tag{8.4.34}$$

$$\int_\Omega q_h \operatorname{div} \underline{u}_h\, dx = 0, \quad \forall\, q_h \in Q_h.$$

Following [95], we have the following error estimate

$$||\underline{u} - \underline{u}_h||_V + ||p - p_h||_Q \le c h^{3/2} \inf_{\underline{v} \in V_h,\, q \in Q_h} \left(||\underline{u} - \underline{v}||_V + ||p - q||_Q \right). \tag{8.4.35}$$

□

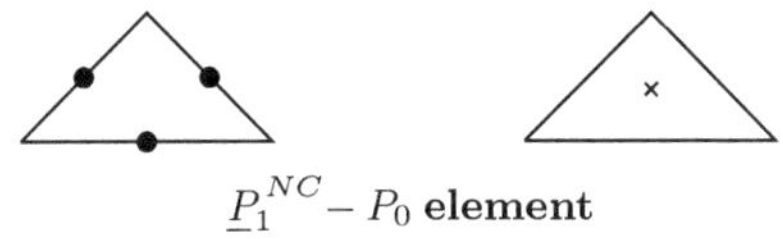

Fig. 8.5 $\underline{P}_1^{NC} - P_0$ element

8.4.4 The Nonconforming $\underline{P}_1 - P_0$ Approximation

Finally, to conclude this section about simple examples, we consider the classical (almost) stable nonconforming triangular element introduced in [165], in which mid-side nodes are used as degrees of freedom for the velocities. This generates a piecewise linear nonconforming approximation; pressures are taken constant on each element as illustrated in Fig. 8.5. It is also possible to build a three-dimensional version of this element, using mid-face nodes as degrees of freedom. We thus choose again $Q_h := \mathcal{L}_0^0 \cap Q$ and

$$V_h := \{v_h \mid v_h \in \mathcal{L}^{1,NC}(P_1, \mathcal{T}_h)^2 \text{ vanishing at the boundary midpoints.}\} \quad (8.4.36)$$

We remark that this method is attractive for several reasons. In particular, the restriction to an element K of the solution $\underline{u}_h \in V_h$ is exactly divergence-free, since $\operatorname{div} V_h \subset Q_h$.

As we have a nonconforming element, we must define discrete bilinear forms,

$$a_h(\underline{u}_h, \underline{v}_h) := \sum_K \int_K \underline{\underline{\operatorname{grad}}}\, \underline{u}_h : \underline{\underline{\operatorname{grad}}}\, \underline{v}_h \, dx, \quad (8.4.37)$$

$$b_h(\underline{v}_h, q_h) := \sum_K \int_K \operatorname{div} \underline{v}_h q_h \, dx \quad (8.4.38)$$

and consider the problem

$$a_h(\underline{u}_h, \underline{v}_h) + b_h(\underline{v}_h, p_h) = (\underline{f}, \underline{v}_h) \ \forall\, \underline{v}_h \in V_h, \quad (8.4.39)$$

$$b_h(\underline{u}_h, q_h) = 0 \qquad \forall\, q_h \in \mathcal{L}_0^0. \quad (8.4.40)$$

Remark 8.4.5 (Problem with coercivity). It must also be recalled that coercivity is a problem for the $\underline{P}_1^{NC} - P_0$ element. The trouble is that the bilinear form (8.2.1) is not coercive on the nonconforming space V_h and we do not have the discrete version of Korn's inequality. This issue has been deeply investigated and clearly illustrated in [16]. It is important to note that (8.4.37) is not the same as in (8.2.1). As we stated earlier, the modified problem is valid only for Dirichlet boundary conditions. Even for the Stokes problem, the *inf-sup* condition is not always the only relevant one. □

Nevertheless, let us see how we can show the *inf-sup* condition. We may now construct $\Pi_h : V \to V_h$ by

$$\int_{\partial K} (\Pi_h \underline{v} - \underline{v}) \cdot \underline{\phi}\, ds = 0 \;\forall\, \underline{\phi} \in R_0(\partial K) \tag{8.4.41}$$

and, again, it is easy to see that

$$|\Pi_h \underline{v}|_{1,h} \leq c\, \|\underline{v}\|_{1,h} \tag{8.4.42}$$

(where as usual $|v_h|^2_{1,h} = \sum_K |v_h|^2_{1,K}$) and

$$b_h(\underline{v} - \Pi_h \underline{v}, q_h) = 0 \;\forall\, q_h \in \mathcal{L}^0_0, \tag{8.4.43}$$

which implies, by Proposition 5.4.2,

$$\inf_{q \in Q_h/\mathbb{R}} \sup_{\underline{v} \in V_h} \frac{b_h(\underline{v}, q)}{|\underline{v}|_{1,h} \|q\|_{0/\mathbb{R}}} \geq c > 0. \tag{8.4.44}$$

On the other hand, we also have

$$a_h(\underline{v}_h, \underline{v}_h) \geq \alpha\, \|\underline{v}_h\|^2_{1,h} \;\forall\, v_h \in V_h. \tag{8.4.45}$$

We may now apply Proposition 5.5.6 and get

$$|\underline{u} - \underline{u}_h|_{1,h} + \|p - p_h\|_{0,\mathbb{R}} \leq ch + E_h(\underline{u}, p), \tag{8.4.46}$$

where

$$\begin{aligned} E_h(\underline{u}, p) &= \sup_{\underline{v}_h \in V_h} |\underline{v}_h|^{-1}_{1,h} \{a_h(\underline{u}, \underline{v}_h) + b_h(\underline{v}_h, p) - (\underline{f}, \underline{v}_h)\} \\ &= \sup_{\underline{v}_h \in V_h} |\underline{v}_h|^{-1}_{1,h} \sum_K \int_{\partial K} [(\underline{\operatorname{grad}}\, \underline{u}) \cdot \underline{n}] \cdot \underline{v}_h\, ds \\ &\leq ch\, \|\underline{u}\|_{2,\Omega}, \end{aligned} \tag{8.4.47}$$

so that, in the end, we have the optimal estimate

$$|\underline{u} - \underline{u}_h|_{1,h} + \|p - p_h\|_{0/\mathbb{R}} \leq ch\, \|\underline{u}\|_{2,\Omega}. \tag{8.4.48}$$

The present element has been generalised to second order in [209]. In this case, there is no problem with coercivity.

Remark 8.4.6. The generalisation of nonconforming finite elements to quadrilaterals is not straightforward. In particular, approximation properties of the involved spaces are not obvious. More details can be found in [330]. □

8.5 Other Techniques for Checking the *inf-sup* Condition

Having presented a few simple examples, we now consider, in a more systematic way, standard techniques for the proof of the *inf-sup* stability condition (8.2.16) that can be applied to a large class of elements. For ease of presentation, in this section, we develop the theory and postpone the examples to Sects. 8.6 and 8.7, for two- and three-dimensional schemes, respectively. However, after the description of each technique, we list some schemes to which that technique can be applied.

8.5.1 *Projection onto Constants*

Following [116], we now consider a modified *inf-sup* condition

$$\inf_{q_h \in Q_h} \sup_{\underline{v}_h \in V_h} \frac{\int_\Omega q_h \operatorname{div} \underline{v}_h \, dx}{||\underline{v}_h||_V ||q_h - \bar{q}_h||_Q} \geq k_0 > 0, \tag{8.5.1}$$

where $\bar{q}_h$ is the L^2-projection of q_h onto $\mathcal{L}_0^0$ (that is, piecewise constant functions).

Proposition 8.5.1. *Let us suppose that the modified inf-sup condition* (8.5.1) *holds with k_0 independent of h. Assume moreover that V_h is such that, for any $q_h \in \mathcal{L}_0^0 \cap Q$,*

$$\sup_{\underline{v}_h \in V_h} \frac{\int_\Omega q_h \operatorname{div} \underline{v}_h \, dx}{||\underline{v}_h||_V} \geq \gamma_0 ||q_h||_Q, \tag{8.5.2}$$

with γ_0 independent of h. Then, the inf-sup condition (8.2.16) *holds true.*

Proof. For any $q_h \in Q_h$, one has

$$\begin{aligned}
\sup_{\underline{v}_h \in V_h} \frac{b(\underline{v}_h, q_h)}{||\underline{v}_h||_V} &= \sup_{\underline{v}_h \in V_h} \left\{ \frac{b(\underline{v}_h, q_h - \bar{q}_h)}{||\underline{v}_h||_V} + \frac{b(\underline{v}_h, \bar{q}_h)}{||\underline{v}_h||_V} \right\} \\
&\geq \sup_{\underline{v}_h \in V_h} \frac{b(\underline{v}_h, \bar{q}_h)}{||\underline{v}_h||_V} - \sup_{\underline{v}_h \in V_h} \frac{b(\underline{v}_h, q_h - \bar{q}_h)}{||\underline{v}_h||_V} \\
&\geq \gamma_0 ||\bar{q}_h||_Q - ||q_h - \bar{q}_h||_0,
\end{aligned} \tag{8.5.3}$$

which implies

$$\sup_{\underline{v}_h \in V_h} \frac{b(\underline{v}_h, q_h)}{||\underline{v}_h||_V} \geq \frac{k_0 \gamma_0}{1+k_0} ||\bar{q}_h||_Q. \tag{8.5.4}$$

Putting together (8.5.1) and (8.5.4) proves the proposition. □

Remark 8.5.1. In the case of *continuous pressures* schemes, hypothesis (8.5.2) can be replaced with the following approximation assumption: for any $\underline{v} \in V$ there exists $\underline{v}^I \in V_h$ such that

$$||\underline{v} - \underline{v}^I||_{L^2(\Omega)} \leq c_1 h ||\underline{v}||_V, \quad ||\underline{v}^I||_V \leq c_2 ||\underline{v}||_V. \tag{8.5.5}$$

The details of the proof can be found in [116] when the mesh is quasi-uniform. The quasi-uniformity assumption is actually not needed, as it can be shown with an argument similar to the one which will be presented in the next subsection (see, in particular, Remark 8.5.2). □

Example 8.5.1. The technique presented in this section will be used, for instance, for the stability proof of the generalised two-dimensional Hood–Taylor element (see Sect. 8.8.2 and Theorem 8.8.1). □

8.5.2 *Verfürth's Trick*

Verfürth's trick [375], already presented in Sect. 5.4.5, applies to *continuous pressure* approximations and is essentially based on two steps. The first step is quite general and can be summarised in the following Lemma.

Lemma 8.5.1. *Let Ω be a bounded domain in $\mathbb{R}^n$ with Lipschitz continuous boundary. Let $V_h \subset (H^1_0(\Omega))^n =: V$ and $Q_h \subset H^1(\Omega) \cap Q$ be closed subspaces. Assume that there exists a linear operator Π^0_h from V into V_h and a constant c (independent of h) such that*

$$||\underline{v}_h - \Pi^0_h \underline{v}||_r \leq c \sum_{K \in \mathcal{T}_h} \left(h_K^{2-2r} ||\underline{v}||^2_{1,K} \right)^{1/2} \quad \forall\, \underline{v} \in V,\ r = 0, 1. \tag{8.5.6}$$

Then, there exist two positive constants c_1 and c_2 such that, for every $q_h \in Q_h$,

$$\sup_{\underline{v} \in V_h} \frac{\int_\Omega q_h \operatorname{div} \underline{v}_h \, dx}{||\underline{v}_h||_V} \geq c_1 ||q_h||_Q - c_2 \left(\sum_{K \in \mathcal{T}_h} h_K^2 ||\underline{\operatorname{grad}}\, q_h||^2_{0,K} \right)^{1/2}. \tag{8.5.7}$$

Proof. Given $q_h \in Q_h$, let $\underline{\bar{v}} \in V$ be such that

$$\frac{\int_\Omega q_h \operatorname{div} \underline{\bar{v}}\, dx}{||\underline{\bar{v}}||_V ||q_h||_Q} \geq \beta > 0, \tag{8.5.8}$$

where β is the continuous *inf-sup* constant. Then,

$$\begin{aligned}
\sup_{\underline{v}_h \in V_h} \frac{\int_\Omega q_h \operatorname{div} \underline{v}_h\, dx}{||\underline{v}_h||_V} &\geq \frac{\int_\Omega q_h \operatorname{div} \Pi_h^0 \underline{\bar{v}}\, dx}{||\Pi_h^0 \underline{\bar{v}}||_V} \geq \frac{1}{2c} \frac{\int_\Omega q_h \operatorname{div} \Pi_h^0 \underline{\bar{v}}\, dx}{||\underline{\bar{v}}||_V} \\
&= \frac{1}{2c} \frac{\int_\Omega q_h \operatorname{div} \underline{\bar{v}}\, dx}{||\underline{\bar{v}}||_V} + \frac{1}{2c} \frac{\int_\Omega q_h \operatorname{div}(\Pi_h^0 \underline{\bar{v}} - \underline{\bar{v}})\, dx}{||\underline{\bar{v}}||_V} \\
&\geq \beta/4c||q_h||_Q + \frac{1}{2c} \frac{\int_\Omega \underline{\operatorname{grad}}\, q_h \cdot (\Pi_h^0 \underline{\bar{v}} - \underline{\bar{v}})\, dx}{||\underline{\bar{v}}||_V} \\
&\geq \beta/4c||q_h||_Q - \left(\frac{1}{2} \sum_{K \in \mathcal{T}_h} h_K^2 || \underline{\operatorname{grad}}\, q_h ||_{0,K}^2 \right)^{1/2} .
\end{aligned} \tag{8.5.9}$$

□

Remark 8.5.2. Indeed, via a scaling argument, it can be shown that the last term in the right-hand side of equation (8.5.7) is equivalent to $||q_h - \bar{q}_h||_0$, where $\bar{q}_h$ denotes, as in the previous subsection, the L^2-projection onto the piecewise constants. □

We are now in the position of stating the main result of this subsection. Note that Verfürth's trick consists in proving a kind of *inf-sup* condition where the zero norm of q_h is substituted by $h|q_h|_1$.

Proposition 8.5.2. *Suppose that the hypotheses of Lemma 8.5.1 hold true. Assume, moreover, that there exists a constant c_3 such that, for every $q_h \in Q_h$,*

$$\sup_{\underline{v}_h \in V_h} \frac{\int_\Omega q_h \operatorname{div} \underline{v}_h}{||\underline{v}_h||_V} \geq c_3 \left(\sum_{K \in \mathcal{T}_h} h_K^2 |q_h|_{1,K}^2 \right)^{1/2} . \tag{8.5.10}$$

Then, the standard inf-sup condition (8.2.16) *holds true.*

Proof. Let us multiply (8.5.7) by c_3 and (8.5.10) by c_2 and sum up the two equations. We have

$$(c_3 + c_2) \sup_{\underline{v}_h \in V_h} \frac{\int_\Omega q_h \operatorname{div} \underline{v}_h\, dx}{||\underline{v}_h||_V} \geq c_1 c_3 ||q_h||_Q, \tag{8.5.11}$$

that is, the *inf-sup* condition (8.2.16). □

Example 8.5.2. The Verfürth trick has been designed for the stability analysis of the Hood–Taylor method. It will be used for this purpose in Sect. 8.8.2 (see Theorem 8.8.1). □

8.5.3 Space and Domain Decomposition Techniques

Sometimes, the spaces V_h and Q_h decompose into the sum (direct or not) of subspaces for which it might be easier to prove an *inf-sup* condition. This is the case, for instance, when a *domain decomposition* technique is employed. Some of the results we are going to present can be viewed as a particular case of the macro-element technique which will be introduced in Sect. 8.5.4.

The next result has been presented and proved in [223].

Proposition 8.5.3. *Suppose Ω can be decomposed as the union of disjoint subdomains with Lipschitz continuous boundaries*

$$\Omega := \bigcup_{r=1}^{R} \Omega_r. \tag{8.5.12}$$

We make use of the following notation:

$$\begin{aligned} V_{0,r} &:= \{\underline{v} \mid \underline{v} \in V_h, \underline{v} = 0 \textit{ in } \Omega \setminus \Omega_r\}, \\ Q_{0,r} &:= \{q \mid q \in Q_h, \int_{\Omega_r} q\, dx = 0\}, \\ K &:= \{q \mid q \in Q, q|_{\Omega_r} \textit{ is constant}, \ r = 1, \dots, R\}. \end{aligned} \tag{8.5.13}$$

Suppose, moreover, that the spaces $V_{0,r}$ and $Q_{0,r}$ satisfy the following inf-sup condition

$$\inf_{q_h \in Q_{0,r}} \sup_{\underline{v}_h \in V_{0,r}} \frac{\int_{\Omega_r} q_h \operatorname{div} \underline{v}_h \, dx}{||q_h||_Q ||\underline{v}_h||_V} \geq k_r > 0, \tag{8.5.14}$$

with k_r independent of h ($r = 1, \dots, R$) and that the following inf-sup condition between V_h and K holds true

$$\inf_{q_h \in K} \sup_{\underline{v}_h \in V_h} \frac{\int_{\Omega} q_h \operatorname{div} \underline{v}_h \, dx}{||q_h||_Q ||\underline{v}_h||_V} \geq k_K > 0, \tag{8.5.15}$$

with k_K independent of h. Then, the spaces V_h and Q_h satisfy the inf-sup condition (8.2.16). □

Sometimes, it is not possible (or it is not the best choice) to partition Ω into *disjoint* sub-domains. Let us describe the case of two overlapping sub-domains. The following proposition can be checked by a direct computation.

Proposition 8.5.4. *Let Ω be the union of two sub-domains Ω_1 and Ω_2 with Lipschitz continuous boundaries. With the notation of the previous proposition, suppose that the spaces $V_{0,r}$ and $Q_{0,r}$ satisfy the inf-sup conditions*

$$\inf_{q_h \in Q_{0,r}} \sup_{\underline{v}_h \in V_{0,r}} \frac{\int_{\Omega_r} q_h \operatorname{div} \underline{v}_h \, dx}{||q_h||_Q ||\underline{v}_h||_V} \geq k_r > 0, \tag{8.5.16}$$

for $r = 1, 2$. Then, the spaces V_h and Q_h satisfy the condition

$$\inf_{q_h \in Q_h} \sup_{\underline{v}_h \in V_h} \frac{\int_{\Omega} q_h \operatorname{div} \underline{v}_h \, dx}{||q_h - \overline{q}_h||_Q ||\underline{v}_h||_V} \geq \frac{1}{\sqrt{2}} \min(k_1, k_2), \tag{8.5.17}$$

where, as in Sect. 8.5.1, we have denoted by $\overline{q}_h$ the L^2 projection of q_h onto the space $\mathcal{L}_0^0$. □

Another useful technique for proving the *inf-sup* condition can be found in [328]. This result is quite general; in particular, the decomposition of the spaces V_h and Q_h does not rely on a decomposition of the domain Ω. In [328], the following proposition is stated for a two-subspaces decomposition, but it obviously extends to more general situations.

Proposition 8.5.5. *Let Q_1 and Q_2 be subspaces of Q_h such that*

$$Q_h := Q_1 + Q_2. \tag{8.5.18}$$

If V_1, V_2 are subspaces of V_h and α_1, α_2 positive constants such that

$$\inf_{q_h \in Q_i} \sup_{\underline{v}_h \in V_i} \frac{\int_{\Omega} q_h \operatorname{div} \underline{v}_h \, dx}{||q_h||_Q ||\underline{v}_h||_V} \geq \alpha_i, \quad i = 1, 2 \tag{8.5.19}$$

and β_1, β_2 are non-negative constants such that

$$\begin{aligned} \left| \int_{\Omega} q_1 \operatorname{div} \underline{v}_2 \, dx \right| &\leq \beta_1 ||q_1||_Q ||\underline{v}_2||_V \quad \forall q_1 \in Q_1, \ \forall \, \underline{v}_2 \in V_2, \\ \left| \int_{\Omega} q_2 \operatorname{div} \underline{v}_1 \, dx \right| &\leq \beta_2 ||q_2||_Q ||\underline{v}_1||_V \quad \forall q_2 \in Q_2, \ \forall \, \underline{v}_1 \in V_1, \end{aligned} \tag{8.5.20}$$

with

$$\beta_1 \beta_2 < \alpha_1 \alpha_2, \tag{8.5.21}$$

then the inf-sup condition (8.2.16) *holds true with* k_0 *depending only on* α_i, β_i, $i = 1, 2$. □

Remark 8.5.3. Condition (8.5.21) is trivially true, for instance, when $\beta_1\beta_2 = 0$ and $\alpha_1\alpha_2 > 0$. □

Example 8.5.3. Most of the techniques presented in this section can be seen as a particular case of the macro-element technique (see Sect. 8.5.4). Proposition 8.5.4 will be used in Theorem 8.8.1 for the stability proof of the Hood–Taylor scheme. □

8.5.4 Macro-element Technique

In this section we present a technique introduced by Stenberg (see [351–355]) which, under suitable hypotheses, reduces the matter of checking the *inf-sup* condition (8.2.16) to an algebraic problem. We also refer to [98] for related results in a somewhat different setting.

The present technique is based on a decomposition of the triangulation $\mathcal{T}_h$ into disjoint macro-elements, where we refer to a *macro-element* as an open polygon (resp., polyhedron in $\mathbb{R}^3$) which is the union of adjacent elements.

Let us introduce some notation.

A macro-element$M$ is said to be *equivalent* to a reference macro-element $\hat{M}$ if there exists a mapping $F_M : \hat{M} \to M$ such that

1. F_M is continuous and invertible;
2. $F_M(\hat{M}) = M$;
3. If $\hat{M} = \cup \hat{K}_j$, where K_j, $j = 1, \dots m$, are the elements defining $\hat{M}$, then $K_j = F_M(\hat{K}_j)$, $j = 1, \dots m$, are the elements of M;
4. $F_M|_{\hat{K}_j} = F_{K_j} \circ F^{-1}_{\hat{K}_j}$, $j = 1, \dots m$, where F_K denotes the affine mapping from the reference element to a generic element K.

We denote by $\mathcal{E}_{\hat{M}}$ the equivalence class of $\hat{M}$. We now introduce the discrete spaces associated with V_h and Q_h on the generic macro-element M (N is the dimension of Ω):

$$
\begin{aligned}
V_{0,M} &:= \left\{\underline{v} \mid \underline{v} \in (H^1_0(M))^N, \underline{v} = \underline{w}|_M \text{ with } \underline{w} \in V_h\right\}, \\
Q_{0,M} &:= \left\{p \mid p \in L^2(\Omega), \int_M p\,dx = 0,\ p = q|_M \text{ with } q \in Q_h\right\}.
\end{aligned}
\tag{8.5.22}
$$

We finally introduce a space which corresponds to the kernel of B_h^t on the macro-element M:

$$
K_M := \left\{p \mid p \in Q_{0,M}, \int_M p \ \operatorname{div} \underline{v}\,dx = 0,\ \forall\, \underline{v} \in V_{0,m}\right\}. \tag{8.5.23}
$$

The *macro-elements condition* reads

$$K_M = \{0\}, \tag{8.5.24}$$

that is, the analogous (at a macro-element level) of the necessary condition for the discrete Stokes problem to be well-posed that the kernel of B_h^t reduces to the zero function.

Proposition 8.5.6. *Suppose that each triangulation $\mathcal{T}_h$ can be decomposed into disjoint macro-elements belonging to a* fixed *number (independent of h) of equivalence classes $\mathcal{E}_{\hat{M}_i}$, $i = 1, \dots n$. Suppose, moreover, that the pair $V_h - \mathcal{L}_0^0/\mathbb{R}$ is a stable Stokes element, that is,*

$$\inf_{q_h \in \mathcal{L}_0^0/\mathbb{R}} \sup_{\underline{v}_h \in V_h} \frac{\int_\Omega q_h \operatorname{div} \underline{v}_h \, dx}{||q_h||_Q ||\underline{v}_h||_V} \geq \beta > 0, \tag{8.5.25}$$

with β independent of h. Then, the macro-element condition (8.5.24) *(for every $M \in \mathcal{E}_{\hat{M}_i}$, $i = 1, \dots n$) implies the inf-sup condition* (8.2.16)*.*

Proof. We do not give the technical details of the proof, for which we refer to [351]. The basic arguments of the proof are sketched in Remark 8.5.4. □

Remark 8.5.4. The macro-element condition (8.5.24) is strictly related to the *patch test* used by engineers (cf., e.g., [388]). However, the count of the degrees of freedom is clearly insufficient by itself. Hence, let us point out how the hypotheses of Proposition 8.5.6 are important.

Hypothesis (8.5.24) (the macro-element condition) implies, via a compactness argument, that a discrete *inf-sup* condition holds true between the spaces $V_{0,M}$ and $Q_{0,M}$. The *finite* number of equivalent macro-elements classes is sufficient to conclude that the corresponding *inf-sup* constants are uniformly bounded below by a positive number.

Then, we are basically in the situation of the domain decomposition technique of Sect. 8.5.3. We now use hypothesis (8.5.25) to control the constant functions on each macro-element and to conclude the proof. □

Remark 8.5.5. Hypothesis (8.5.25) is satisfied in the two-dimensional case whenever V_h contains piecewise quadratic functions (see Sect. 8.3). In the three-dimensional case, things are not so easy: to control the constants, we need extra degrees of freedom on the faces, as observed in Remark 8.4.3. For this reason, let us state the following proposition which can be proved with the technique of Sect. 8.5.1 (see Remark 8.5.1) and which applies to the case of *continuous pressures* approximations. □

Proposition 8.5.7. *Let us make the same assumptions as in Proposition 8.5.6 with* (8.5.25) *replaced by the condition of Remark 8.5.1 (see* (8.5.5)*). Then, provided $Q_h \subset C^0(\Omega)$, the inf-sup condition* (8.2.16) *holds true.* □

Remark 8.5.6. The hypothesis that the macro-element partition of $\mathcal{T}_h$ is *disjoint* can be weakened, in the spirit of Proposition 8.5.4, by requiring that each element K of $\mathcal{T}_h$ belongs at most to a finite number N of macro-elements with N independent of h. □

Example 8.5.4. The macro-element technique can be used in order to prove the stability of several schemes. Among those, we recall the $\underline{Q}_2 - P_1$ element (see Sect. 8.6.3) and the three-dimensional generalised Hood-Taylor scheme (see Theorem 8.8.2). □

8.5.5 Making Use of the Internal Degrees of Freedom

This subsection presents a general framework providing a general tool for the analysis of finite element approximations to problems of incompressible materials.

The basic idea has been used several times on particular cases, starting from [165] for discontinuous pressures and from [24] and [25] for continuous pressures. We are going to present it in its final general form given by Brezzi and Pitkiäranta [131]. It consists essentially in stabilising an element by adding suitable bubble functions to the velocity field.

In order to do that, following the notation of Remark 2.2.4, we first associate to every finite element discretisation $Q_h \subset Q$ the space

$$B(b_K \,\underline{\mathrm{grad}}\, Q_h) := \left\{\underline{\beta} \mid \underline{\beta} \in V, \underline{\beta}|_K = b_K \,\underline{\mathrm{grad}}(q_h|_K) \text{ for some } q_h \in Q_h\right\}, \tag{8.5.26}$$

where b_K is a bubble function defined in K. In particular, we can take $b_K = b_{3,K}$ as the standard cubic bubble if K is a triangle, or a bi-quadratic bubble if K is a square or other obvious generalisations in 3D. In other words, the restriction of a $\underline{\beta} \in B(b_K \,\underline{\mathrm{grad}}\, Q_h)$ to an element K is the product of the bubble functions b_K times the gradient of a function of $Q_h|_K$.

Remark 8.5.7. In (8.5.26), we take the gradient on K so that $B(b_K \,\underline{\mathrm{grad}}\, Q_h)$ is well defined even if Q_h is a space of discontinuous pressures. □

Remark 8.5.8. Notice that the space $B(b_k \,\underline{\mathrm{grad}}\, Q_h)$ is not defined through a basic space $\hat{B}$ on the reference element. This could be easily done in the case of *affine* elements, for all the reasonable choices of Q_h. However, this is clearly *unnecessary*: if we know how to compute q_h on K, we also know how to compute $\underline{\mathrm{grad}}\, q_h$ and there is no need for a reference element. □

We can now prove our basic results, concerning the two cases of continuous or discontinuous pressures.

Proposition 8.5.8 (Stability of continuous pressure elements). *Assume that there exists an operator $\Pi_1 \in \mathcal{L}(V, V_h)$ satisfying the property of the Clément interpolant* (8.4.24)*. If $Q_h \subset C^0(\Omega)$ and V_h contains the space $B(b_k \,\underline{\mathrm{grad}}\, Q_h)$, then* *the pair* (V_h, Q_h) *is a stable element, in the sense that it satisfies the inf-sup condition* (8.2.16).

Proof. We shall build a B-compatible operator, like in Proposition 8.4.2. We only need to construct the operator Π_2. We define $\Pi_2 : V \to B(b_k \,\underline{\mathrm{grad}}\, Q_h)$, on each element, by requiring

$$\begin{aligned} &\Pi_2\underline{v}|_K \in B(b_K \,\underline{\mathrm{grad}}\, Q_h)|_K, \\ &\int_K (\Pi_2\underline{v} - \underline{v}) \cdot \underline{\mathrm{grad}}\, q_h \, dx = 0, \quad \forall\, q_h \in Q_h. \end{aligned} \tag{8.5.27}$$

Problem (8.5.27) has obviously a unique solution and Π_2 satisfies (8.4.5). Finally, (8.4.7) follows by a scaling argument. Hence, Proposition 8.4.2 gives the desired result. □

Corollary 8.5.1. *Assume that* $Q_h \subset Q$ *is a space of continuous piecewise smooth functions. If* V_h *contains* $(\mathcal{L}_1^1)^2 \oplus B(b_K \,\underline{\mathrm{grad}}\, Q_h)$*, then the pair* (V_h, Q_h) *satisfies the inf-sup condition* (8.2.16).

Proof. Since V_h contains piecewise linear functions, there exists a Clément interpolant Π_1 satisfying (8.4.24). Hence, we can apply Proposition (8.5.8). □

We now consider the case of discontinuous pressure elements.

Proposition 8.5.9 (Stability of discontinuous pressure elements). *Assume that there exists an operator* $\tilde{\Pi}_1 \in \mathcal{L}(V, V_h)$ *satisfying*

$$\begin{aligned} &||\tilde{\Pi}_1\underline{v}||_V \le c||\underline{v}||_V \quad \forall\, \underline{v} \in V, \\ &\int_K \mathrm{div}(\underline{v} - \tilde{\Pi}_1\underline{v})\, dx = 0 \quad \forall\, \underline{v} \in V \;\forall\, K \in \mathcal{T}_h. \end{aligned} \tag{8.5.28}$$

If V_h *contains* $B(b_K \,\underline{\mathrm{grad}}\, Q_h)$*, then the pair* (V_h, Q_h) *is a stable element, in the sense that it satisfies the inf-sup condition* (8.2.16).

Proof. We are going to use Proposition 8.5.8. We take $\tilde{\Pi}_1$ as operator Π_1. We are not defining Π_2 on the whole V, but only in the subspace

$$V^0 := \left\{ \underline{v} \mid \underline{v} \in V, \int_K \mathrm{div}\, \underline{v}\, dx = 0 \quad \forall\, K \in \mathcal{T}_h \right\}. \tag{8.5.29}$$

This will be enough, since we need to apply Π_2 to the difference $\underline{v} - \tilde{\Pi}_1\underline{v}$ which is in V^0 by (8.5.28).

For every $\underline{v} \in V^0$, we define $\Pi_2\underline{v} \in B(b_K \,\underline{\mathrm{grad}}\, Q_h)$ by requiring that, in each element K,

$$\begin{aligned} &\Pi_2\underline{v}|_K \in B(b_K \,\underline{\text{grad}}\, Q_h)|_K, \\ &\int_K \text{div}(\Pi_2\underline{v} - \underline{v})q_h \, dx = 0 \quad \forall\, q_h \in Q_h|_K. \end{aligned} \tag{8.5.30}$$

Note that (8.5.30) is uniquely solvable, if $\underline{v} \in V^0$, since the divergence of a bubble function has always zero mean value (hence, the number of non-trivial equations is equal to $\dim(Q_h|_K) - 1$, which is equal to the number of unknowns; the non-singularity then follows easily). It is obvious that Π_2, as given by (8.5.30), will satisfy (8.4.5) for all $\underline{v} \in V^0$. We have to check that

$$\|\Pi_2\underline{v}\|_1 \leq c\|\underline{v}\|_V, \tag{8.5.31}$$

which actually follows by a scaling argument making use of the following bound

$$|\widehat{\Pi_2\underline{v}}|_{0,\hat{K}} \leq c(\theta_0)|\hat{\underline{v}}|_{1,\hat{K}}. \tag{8.5.32}$$

□

Corollary 8.5.2 (Two-dimensional case). *Assume that $Q_h \subset Q$ is a space of piecewise smooth functions. If V_h contains $(\mathcal{L}_2^1)^2 \oplus B(b_K \,\underline{\text{grad}}\, Q_h)$, then the pair (V_h, Q_h) satisfies the inf-sup condition* (8.2.16).

Proof. The stability of the $(P_2)^2 - P_0$ element (see Sect. 8.3) implies the existence of $\tilde{\Pi}_1$ as in Proposition 8.5.9. □

Propositions 8.5.8 and 8.5.9 are worth a few comments. They show that almost any element can be stabilised by using bubble functions. For continuous pressure elements, this procedure is mainly useful in the case of affine elements. For discontinuous pressure elements, it is possible to stabilise elements which are already stable for a piecewise constant pressure field. Examples of such a procedure can be found in [199]. Stability with respect to piecewise constant pressure implies that at least one degree of freedom on each side or face of the element is linked to the normal component of velocity (see [202] and Remark 8.4.3).

Example 8.5.5. Internal degrees of freedom can be used in the stability analysis of several methods. For instance, we use it for the analysis of the MINI element (see Sects. 8.6.1 and 8.7.1) in the case of continuous pressures and of the Crouzeix-Raviart element (see Proposition 8.6.2 and Sect. 8.7.2) in the case of discontinuous pressures. □

8.6 Two-Dimensional Stable Elements

In this section, we shall make use of the techniques presented in Sect. 8.5 to prove the stability for some of the most popular two-dimensional Stokes elements. The degrees of freedom corresponding to some of those are collected in Fig. 8.6.

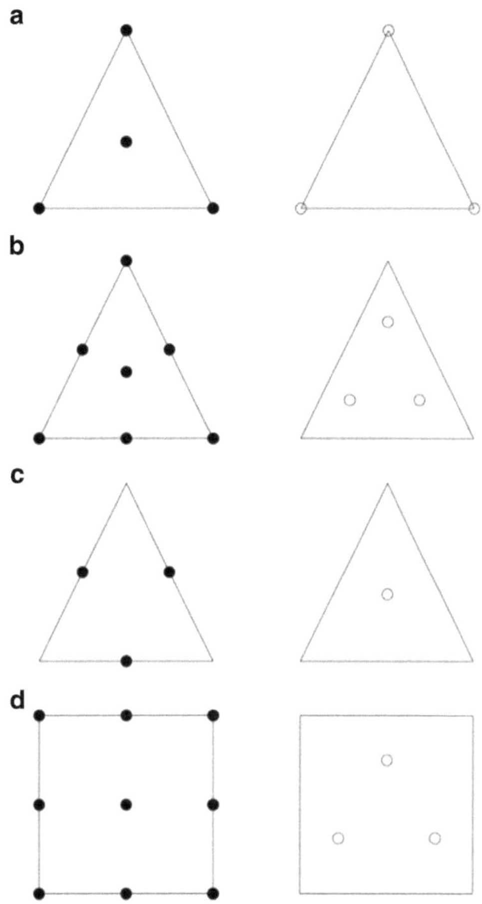

Fig. 8.6 Some stable two-dimensional Stokes elements: (**a**) the MINI element, (**b**) the Crouzeix–Raviart element, (**c**) the $\underline{P}_1^{NC} - P_0$ element, (**d**) the $\underline{Q}_2 - P_1$ element

We start with triangular elements and then we present schemes based on quadrilaterals.

The Hood–Taylor element (two- and three-dimensional) and its generalisation will be presented in Sect. 8.8. Figure 8.6 presents the most simple cases of the elements that we shall discuss.

8.6.1 Continuous Pressure Elements

We have already presented in Sect. 8.4.2 the MINI element. This element, which is probably the simplest one for the approximation of the Stokes equation, has been introduced in [25]. Using the results of Sect. 8.5.5, in particular Corollary 8.5.1, we easily deduce the following result.

Proposition 8.6.1. *The following pair of spaces are stable for any $k \geq 1$*

$$V_h := (\mathcal{L}_1^k \oplus B(b_{3,K}\,\underline{\text{grad}}\,Q_h))^2 \cap V, \qquad Q_h := \mathcal{L}_1^k \cap Q, \tag{8.6.1}$$

$$V_h := (\mathcal{L}_1^{k+1} \oplus B(b_{3,K}\,\underline{\text{grad}}\,Q_h))^2 \cap V, \qquad Q_h := \mathcal{L}_1^k \cap Q. \tag{8.6.2}$$

□

For $k = 1$, (8.6.1) is the MINI element while (8.6.2) defines a variant of the Hood-Taylor element, which we shall consider in Sect. 8.8, enriched by bubbles. This produces an element with a slightly better k_h in the *inf-sup* condition.

8.6.2 Discontinuous Pressure Elements

We have already considered in Sect. 8.4.3 the element $\underline{P}_2 - P_0$. Using a P_0 pressure ensures an element-wise conservation of mass which is an advantage in some situations. We now rely on Proposition 8.5.9 and more precisely on Corollary 8.5.2.

Example 8.6.1 (The Crouzeix-Raviart element). This element, presented in [165], is an enrichment to the $\underline{P}_2 - P_0$ scheme which provides well-balanced approximation properties. Given a mesh of triangles, the approximating spaces are

$$V_h := (\mathcal{L}_2^1 \oplus B_3)^2 \cap V, \qquad Q_h := \mathcal{L}_1^0 \cap Q. \tag{8.6.3}$$

The proof of the stability for this element is a direct consequence of Proposition 8.5.8. □

The Crouzeix-Raviart element is the simplest one of a general family. Indeed, the construction of the Crouzeix-Raviart element relies on the fact that, as we have seen in Proposition 8.5.9 and Corollary 8.5.2, adding enough bubbles (that is, internal degrees of freedom) to an element which is stable for pressures in $\mathcal{L}_0^0$ can make it stable for pressures in $\mathcal{L}_k^0$. We can thus state the following proposition.

Proposition 8.6.2. *For $k \geq 2$, let*

$$V_h := (\mathcal{L}_k^1 \oplus B_{k+1})^2 \cap V, \quad Q_h = \mathcal{L}_{k-1}^0 \cap Q. \tag{8.6.4}$$

Then, the couple $V_h \times Q_h$ is stable. □

Remark 8.6.1 (A nonconforming version). It can easily be checked [209] that one obtains a stable element of second order accuracy by replacing the standard bubble by the nonconforming bubble of (2.2.39), that is, taking

$$V_h := (\mathcal{L}_2^1 \oplus B_{NC})^2 \cap V, \quad Q_h = \mathcal{L}_1^0 \cap Q. \tag{8.6.5}$$

Besides some nice continuity properties of the stress at mid-side nodes, the fact that second order polynomials are employed simplifies things for the numerical integration when building the discrete problem. □

Remark 8.6.2. Instead of enriching V_h, stability can be obtained by taking a smaller Q_h, in general to the expense of accuracy. It can easily be checked that if we take, for $k \geq 2$, $V_h = (\mathcal{L}_k^1)^2$ and $Q_h = \mathcal{L}_{k-2}^0$, we have enough internal degrees of freedom in V_h to ensure stability. This is not true in the three-dimensional case where one would need $Q_h = \mathcal{L}_{k-3}^0$ and, evidently, $k \geq 3$. This choice of elements would have a severe impact on accuracy. □

8.6.3 Quadrilateral Elements, $\underline{Q}_k - P_{k-1}$ Elements

We now discuss the stability and convergence of a family of quadrilateral elements. The lowest order of this family, the $\underline{Q}_2 - P_1$ element, is one of the most popular Stokes elements. These elements are *discontinuous pressure* elements and they originate from attempts to use the reduced integration technique which will be analysed in Sect. 8.12. Let us first consider what would appear to be a natural construction. Given $k \geq 1$, the discrete spaces are defined as follows:

$$V_h := (\mathcal{L}_{[k]}^1)^2 \cap V, \qquad Q_h := \mathcal{L}_{[k-1]}^0 \cap Q. \tag{8.6.6}$$

For $k = 1$, this yields the unstable $\underline{Q}_1 - P_0$ which will be considered in detail in Sect. 8.10. For $k = 2$, we have the $\underline{Q}_2 - Q_1$ element which appears quite naturally in the use of reduced integration penalty methods (see [60]). This element is not stable and suffers from the same problems as the $\underline{Q}_1 - P_0$ element.

Let us now consider, instead of (8.6.6),

$$V_h := (\mathcal{L}_{[k]}^1)^2 \cap V, \qquad Q_h := \mathcal{L}_{k-1}^0 \cap Q. \tag{8.6.7}$$

Using P_k pressure instead of Q_k is not a natural choice, although it is the good one. If the mesh is built of *rectangles*, the stability proof is an immediate consequence of Proposition 8.5.9, since (8.5.28) is satisfied for V_h (indeed, the $\underline{Q}_2 - P_0$ is a stable Stokes element, see Remark 8.4.3). In the case of a general *quadrilateral* mesh, things are not so easy; even the definition of the space Q_h is not so obvious and there have been different opinions, during the years, about two possible natural definitions. Following [90], we discuss in detail the case $k = 2$. We shall see that in this case there are important issues related to the approximation properties of finite elements on non-affine meshes.

8.6.3.1 The $\underline{Q}_2 - P_1$ Element

This element was apparently discovered around a blackboard at the Banff Conference on Finite Elements in Flow Problems (1979). Two different proofs of stability can be found in [223] and [351] for the rectangular case. This element is a relatively late comer in the field; the reason for this is that, as we stated earlier, using a P_1 pressure on a quadrilateral is not a standard procedure. It appeared as a cure for the instability of the $\underline{Q}_2 - Q_1$ element which appears quite naturally in the use of reduced integration penalty methods (see [60]). Another cure can be obtained by adding internal nodes (see [199]).

On a general quadrilateral mesh, the space Q_h can be defined in two different ways: either Q_h consists of (discontinuous) piecewise linear functions, or it is built by considering three linear shape functions on the reference unit square and mapping them to the general elements like it is usually done for continuous finite elements (see (2.1.59)). We point out that since the mapping F_K from the reference element $\hat{K}$ to the general element K in this case is bilinear but not affine, the two constructions are not equivalent. We shall refer to the first possibility as the **unmapped** pressure approach and to the second one as the **mapped** pressure approach.

In order to analyse the stability of either scheme, we use the macro-element technique presented in Sect. 8.5.4 with macro-elements consisting of one single element.

The **unmapped** pressure approach yields the original proof presented in [351]. Let M be a macro-element and $q_h = a_0 + a_x x + a_y y \in Q_{0,M}$ an arbitrary function in K_M. If $b(x, y)$ denotes the bi-quadratic bubble function on K, then $\underline{v}_h = (a_x b(x, y), 0)$ is an element of $V_{0,M}$ and

$$0 = \int_M q_h \operatorname{div} \underline{v}_h \, dx\, dy = -\int_M \underline{\operatorname{grad}}\, q_h \cdot \underline{v}_h \, dx\, dy = -a_x \int_M b(x, y)\, dx\, dy$$

implies $a_x = 0$. In a similar way, we get $a_y = 0$ and, since the average of q_h on M vanishes, we have the macro-element condition $q_h = 0$.

We now consider the **mapped** pressure approach, following the proof presented in [90]. There, it is recalled that the macro-element condition (8.5.24) can be related to an algebraic problem in which we are led to prove that a two-by-two matrix is non-singular. Actually, it turns out that the determinant of such a matrix is a multiple of the Jacobian determinant of the function mapping the reference square $\hat{K}$ onto M, evaluated at the barycentre of $\hat{K}$. Since this number must be non-zero for any element of a well-defined mesh, we can deduce that the macro-element condition is also satisfied in this case, and we can then conclude that the stability holds thanks to Proposition 8.5.6.

So far, we have shown that both the **unmapped** and the **mapped** pressure approach give rise to a stable $Q_2 - P_1$ scheme. However, as a consequence of the results proved in [20], we have that the mapped pressure approach *cannot achieve optimal approximation order*. Namely, the unmapped pressure space provides a second order convergence in L^2, while the mapped one achieves only $O(h)$ in the same norm. In [90], several numerical experiments have been reported, showing that on general quadrilateral meshes (with constant distortion), the unmapped pressure approach provides a second order convergence (for both velocity in H^1 and pressure in L^2), while the mapped approach is only sub-optimally first order convergent.

It is interesting to remark that, in this case also, the convergence of the velocities is suboptimal, according to the error estimate (8.2.17).

8.7 Three-Dimensional Stable Elements

Many elements presented in Sect. 8.6 have a three-dimensional extension. Some of them are schematically plotted in Fig. 8.7. However, there are important differences between the two and three-dimensional cases. One is that *bubbles* are at least of fourth degree in the three-dimensional case. Another difference is that the $\underline{P}_2 - P_0$ element is not stable: to control piecewise constant pressure, we need some degrees of freedom on the faces. It would indeed be possible to prove that the $\underline{P}_3 - P_0$ is stable but this yields a highly unbalanced approximation.

8.7.1 *Continuous Pressure 3-D Elements*

The most important continuous pressure element is the Hood-Taylor element, and its generalisations, which will be presented in the next section.

The families associated with the MINI element introduced in (8.6.1) and (8.6.2) can be generalised to the three-dimensional with an appropriate choice of bubbles. Consider a regular sequence of decompositions of Ω into tetrahedra.

Proposition 8.7.1. *The following pair of spaces are stable for any* $k \geq 1$

$$V_h := (\mathcal{L}_1^k \oplus B(b_{4,K}\,\underline{\text{grad}}\,Q_h))^3 \cap V, \qquad Q_h := \mathcal{L}_1^k \cap Q, \tag{8.7.1}$$

$$V_h := (\mathcal{L}_1^{k+1} \oplus B(b_{4,K}\,\underline{\text{grad}}\,Q_h))^3 \cap V, \qquad Q_h := \mathcal{L}_1^k \cap Q. \tag{8.7.2}$$

The proof follows easily, like in the 2D case, from Corollary 8.5.1. □

The first member of the first family is the MINI element and the first member of the second family is a version of the Hood-Taylor element where the velocities are enriched by quartic bubbles. Paradoxically, this may increase the precision on pressure through a better *inf-sup* constant.

8.7.2 *Discontinuous Pressure 3-D Elements*

As we stated earlier, the situation for discontinuous pressure elements is less favourable than in the two-dimensional case. We first consider an example.

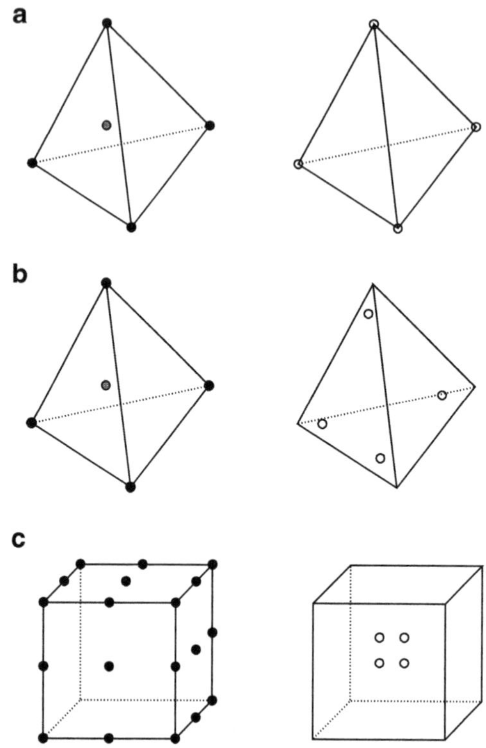

Fig. 8.7 Some stable three-dimensional Stokes elements: (**a**) the MINI element, (**b**) the Crouzeix–Raviart element, (**c**) the $\underline{Q}_2 - P_1$ element

Example 8.7.1. The SMALL element is the smallest three-dimensional one if one wants to use piecewise constant pressure. It is the analogue of the two-dimensional construction of Remark 8.4.2, but we now have to work on the faces and not on the edges. Let us thus consider, on a tetrahedral element, the cubic bubble $\underline{b}_{3,F}$ associated to the face $F = (i, j, k)$ and defined, using barycentric coordinates, by

$$b_{3,F} := \lambda_i \, \lambda_j \, \lambda_k. \tag{8.7.3}$$

We define our new space by adding on each face of a standard piecewise linear element such a cubic bubble. We shall denote this extra space on an element K by

$$BF_3 := \left\{ v_h \mid v_h \in P_3(K), v_h = \sum_F \alpha_F b_{3,F} \right\}. \tag{8.7.4}$$

The final space is thus

$$V_h := (\mathcal{L}_1^1 + BF_3)^3 \cap V, \qquad Q_h := \mathcal{L}_0^0 \cap Q.$$

This provides the control of the flux on the face and one can easily check that we have stability for piecewise constant pressure.

Remark 8.7.1. In fact, the bubble on the face is needed only for the normal component of the velocity. This implies some complexity for the implementation but greatly reduces the global number of degrees of freedom. We could thus use, instead of (8.7.4), denoting $\underline{n}_F$ the normal to the face,

$$BF_3^n := \left\{ v_h \mid v_h \in P_3(K), v_h = \sum_F \alpha_F b_{3,F} \underline{n}_F \right\} \tag{8.7.5}$$

and $V_h := ((\mathcal{L}_1^1)^3 + BF_3^n) \cap V$. □

We must retain that discontinuous pressure elements in 3D require third degree polynomials. □

Example 8.7.2 (3D analogues of the Crouzeix–Raviart element). In order to generalise the Crouzeix-Raviart element, we must first get a stable element for constant pressure. The previously defined SMALL element provides only first order accuracy. We therefore start from a quadratic approximation and enrich it by face bubbles to control the fluxes on the faces and by internal bubbles to control linear pressure. This yields

$$V_h := (\mathcal{L}_2^1 + BF_3 + B_4)^3 \cap V, \qquad Q_h := \mathcal{L}_1^0 \cap Q.$$

The stability is an easy consequence of Proposition 8.5.9.

The face bubbles do not increase the order of accuracy and we could employ the normal bubbles of (8.7.5) However, if one wants to stick to more standard elements, the natural thing would be to start from the stable $\underline{P}_3 - P_0$ element. To get a balanced precision, we define

$$V_h := (\mathcal{L}_3^1 + B_5)^3 \cap V, \qquad Q_h := \mathcal{L}_2^0 \cap Q. \tag{8.7.6}$$

We would then have third order accuracy but to the price of a quite large number of degrees of freedom. This can obviously be extended to higher degrees. □

Example 8.7.3 (Nonconforming elements). A possibility to reduce the order of polynomials needed to obtain stable elements is to use nonconforming elements. The triangular $P_1^{NC} - P_0$ easily generalises to tetrahedra in 3D. Also in this case, since div $V_h \subset Q_h$, the restriction of the discrete solution to every element is truly divergence-free. The problems of coercivity are still there.

However, it is also possible to obtain second order without this problem. We have already seen in Remark 8.6.1 that one could get a variant of the Crouzeix-Raviart element using nonconforming bubbles. The construction can be extended to the three-dimensional case. Indeed, one can replace the face-bubble of (8.7.3) by its nonconforming version of (2.2.39), that is,

$$b_{NC,F} = (\lambda_i^2 + \lambda_j^2 + \lambda_k^2) - 1. \tag{8.7.7}$$

The internal bubble is also replaced by $(\lambda_i^2 + \lambda_j^2 + \lambda_k^2 + \lambda_l^2) - 1$ and we now manipulate only second degree polynomials, which is definitely an advantage. However, things are a little more complicated: we have too many degrees of freedom and one must remove the vertices. We refer to [203] for details. □

Example 8.7.4 (Quadrilateral $\underline{Q}_k - P_{k-1}$ elements). Given a mesh of hexahedra, we define

$$V_h := (\mathcal{L}_k^1)^3 \cap V, \qquad Q_h := \mathcal{L}_{k-1}^0 \cap Q, \tag{8.7.8}$$

for $k \geq 2$. We refer to the two-dimensional case (see Sect. 8.6.3) for the definition of the pressure space. In particular, we recall that Q_h on each element consists of true polynomials and is not defined via the reference element. With the correct definition of the pressure space, the proof of stability for this element is a simple generalisation of the corresponding two-dimensional version. □

8.8 $\underline{P}_k - P_{k-1}$ Schemes and Generalised Hood–Taylor Elements

The main result of this section (see Theorems 8.8.1 and 8.8.2) consists in showing that a family of popular Stokes elements satisfies the *inf-sup* condition (8.2.16). The first element of this family has been introduced in [249] and for this reason, the members of the whole family are usually referred to as *generalised* Hood–Taylor elements.

This section is organised in two subsections. In the first one, we discuss discontinuous pressure approximations for the $\underline{P}_k - P_{k-1}$ element in the two-dimensional triangular case; it turns out that this choice is not stable in the lower order cases and requires suitable conditions on the mesh sequences for the stability of the higher order elements.

The last subsection deals with the generalised Hood–Taylor elements, which provide a continuous pressure approximation in the plane (triangles and quadrilaterals) and in the three-dimensional space (tetrahedra and hexahedra).

8.8.1 *Discontinuous Pressure $\underline{P}_k - P_{k-1}$ Elements*

In this subsection, we shall recall the statement of a basic result by Scott and Vogelius (see [347]) which, roughly speaking, says: *under suitable assumptions on the decomposition $\mathcal{T}_h$ (in triangles), the pair $V_h := (\mathcal{L}_k^1)^2 \cap V$, $Q_h := \mathcal{L}_{k-1}^0 \cap Q$ satisfies the inf-sup condition for $k \geq 4$.*

On the other hand, the problem of finding stable lower order approximations has been studied by Q in [328], where interesting remarks are made on this scheme and where the possibility of filtering out the spurious pressure modes is considered.

In order to state in a precise way the restrictions that have to be made on the triangulation for higher order approximations, we assume that Ω is a polygon, and that its boundary $\partial\Omega$ has no double points. In other words, there exist two continuous piecewise linear maps $x(t)$, $y(t)$ from $[0,1[$ into $\mathbb{R}$ such that

$$\begin{cases} (x(t_1) = x(t_2) \text{ and } y(t_1) = y(t_2)) \text{ implies } t_1 = t_2, \\ \partial\Omega = \{(x,y) \,|\, x = x(t),\ y = y(t) \text{ for some } t \in [0,1[\}. \end{cases} \tag{8.8.1}$$

Clearly, we will have $\lim_{t\to 1} x(t) = x(0)$ and $\lim_{t\to 1} y(t) = y(0)$. We remark that we are considering a less general case than the one treated by Scott and Vogelius [347]. We shall make further restrictions in what follows, so that we are actually going to present a particular case of their results.

Now let V be a vertex of a triangulation $\mathcal{T}_h$ of Ω and let $\theta_1,\ldots,\theta_n$, be the angles, at V, of all the triangles meeting at V, ordered, for instance, in the anticlockwise sense. If V is an internal vertex, we also set $\theta_{n+1} := \theta_1$. Now we define $S(V)$ according to the following rules.

$$n = 1 \quad \Rightarrow \quad S(V) = 0 \tag{8.8.2}$$

$$n > 1,\ V \in \partial\Omega \quad \Rightarrow \quad S(V) = \max_{i=1,n-1} (\pi - \theta_1 - \theta_{i+1}) \tag{8.8.3}$$

$$V \notin \partial\Omega \quad \Rightarrow \quad S(V) = \max_{i=1,n} (\pi - \theta_i - \theta_{i+1}). \tag{8.8.4}$$

It is easy to check that $S(V) = 0$ if and only if all the edges of $\mathcal{T}_h$ meeting at V fall on two straight lines. In this case, V is said to be singular [347]. If $S(V)$ is positive but very small, then V will be "almost singular". Thus, $S(V)$ measures how close V is to be singular.

We are now able to state the following result.

Proposition 8.8.1 ([347]). *Assume that there exist two positive constants c and δ such that*

$$ch \le h_K \ \forall\, K \in \mathcal{T}_h \tag{8.8.5}$$

and

$$S(V) \ge \delta \textit{ for all } V \textit{ vertex of } \mathcal{T}_h. \tag{8.8.6}$$

Then, the choice $V_h = (\mathcal{L}_k^1)^2 \cap V$, $Q_h = \mathcal{L}_{k-1}^0 \cap Q$, $k \ge 4$, satisfies the inf-sup condition with a constant depending on c and δ but not on h. □

Remark 8.8.1. Condition (8.8.6) is worth a few comments. The trouble is that $S(V) = 0$ makes the linear constraints on $\underline{u}_h$, arising from the divergence-free condition, linearly dependent (see, also, Examples 8.10.2 and 8.10.3). When this linear dependence appears, some part of the pressure becomes unstable. However, we have met this situation in Example 8.10.3 and this was in fact the key to convergence, provided a condition on data was fulfilled. The same analysis would hold here and the unstable part of pressure could be filtered out. □

Remark 8.8.2. The $\underline{P}_k - P_{k-1}$ element can obviously be stabilised by adding bubbles to the velocity space in the spirit of Sect. 8.5.5 (see Proposition 8.5.9). For a less expensive stabilisation, consisting in adding bubbles only in few elements, see [72]. □

8.8.2 Generalised Hood–Taylor Elements

In this subsection, we recall the results proved in [71, 73] concerning the stability of the generalised Hood–Taylor schemes. On triangles or tetrahedra, velocities are approximated by a standard $\underline{P}_k$ element and pressures by a standard *continuous* P_{k-1}, that is, $\underline{v}_h \in (\mathcal{L}_k^1)^n \cap V$ $(n = 2, 3)$, $p \in \mathcal{L}_{k-1}^1 \cap Q$. This choice has an analogue on rectangles or cubes using a $\underline{Q}_k$ element for velocities and a Q_{k-1} element for pressures. The lowest order triangular element (i.e., $k = 2$) has been introduced by Hood and Taylor in [249]. Several papers are devoted to the analysis of this popular element.

The degrees of freedom of some elements belonging to this family are reported in Fig. 8.8.

Remark 8.8.3. Another element that has been used because of the simplicity of its shape functions is the so-called $(\underline{P}_1 - iso - \underline{P}_2) - P_1$ element, It is sketched in Fig. 8.9. It is a composite element assembled from four piece-wise linear elements for velocity while pressure remains linear on the macro-element. The same technique of proof that yields stabiliy of the classical Hood-Taylor element could be used to show the inf-sup condition for this composite element. □

The first proof of convergence was given for the two-dimensional case in [61] where a weaker form of the *inf-sup* condition was used. The analysis was subsequently improved in [375], who showed that the classical *inf-sup* condition is indeed satisfied (see Verfürth's trick in Sect. 8.5.2). The macro-element technique can easily be used for the stability proof of the rectangular and cubic element (of any order) as well as for the tetrahedral case when $k = 2$ (see [351]). In [122], an alternative technique of proof has been presented for the triangular and tetrahedral cases when $k = 2$. This proof generalises to the triangular case when $k = 3$ (see [121]). Finally, a general proof of convergence can be found in [71] and [73] for the triangular and tetrahedral case, respectively.

We now state and prove the theorem concerning the two-dimensional triangular case (see [71]).

Theorem 8.8.1. *Let Ω be a polygonal domain and $\mathcal{T}_h$ a regular sequence of triangular decompositions of it. Then, the choice $V_h := (\mathcal{L}_k^1 \cap H_0^1(\Omega))^2$ and*

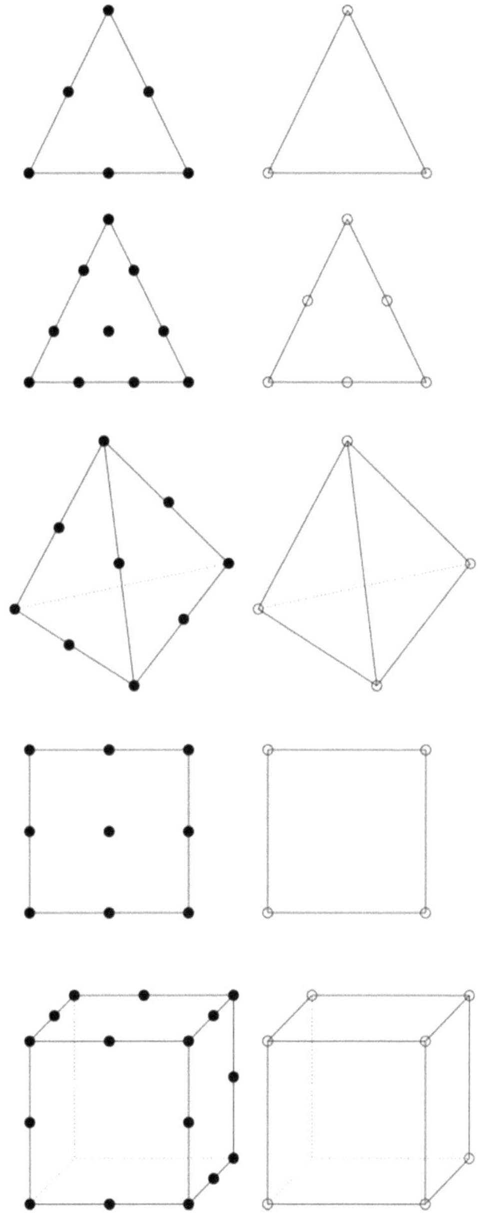

Fig. 8.8 Some stable elements belonging to the Hood–Taylor family

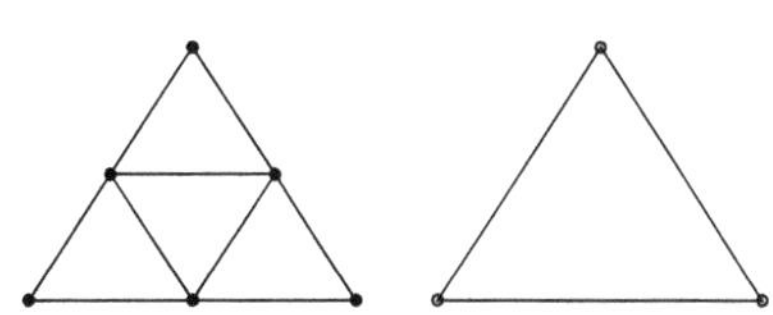

Fig. 8.9 The $(\underline{P}_1 iso \underline{P}_2) - P_1$ element

Fig. 8.10 The reference triangle and its symmetric

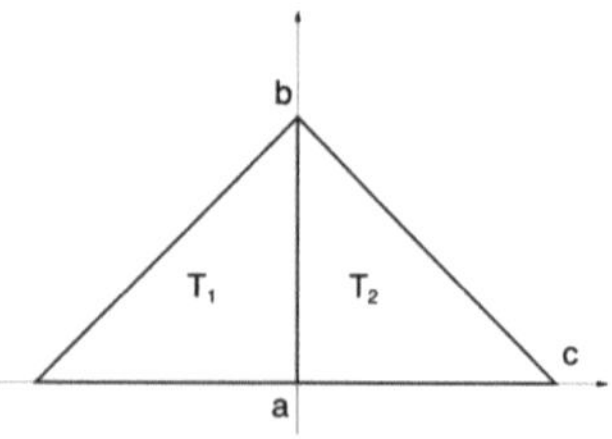

$Q_h := \mathcal{L}^1_{k-1} \cap L^2_0(\Omega)$ satisfies the inf-sup condition (8.2.16) *for any $k \geq 2$ if and only if each triangulation contains at least three triangles.*

Proof (Step 1: necessary part). Let us show first that the hypothesis on the mesh is necessary. If $\mathcal{T}_h$ only contains one element, then it is easy to see that the *inf-sup* constant is zero (otherwise, it should be div $V_h \subset Q_h$, which is not the case since the functions in Q_h are not zero at the vertices). We shall show that if $\mathcal{T}_h$ contains only two triangles T_1 and T_2, then there exists one spurious pressure mode. This implies that, also in this case, the *inf-sup* constant vanishes. We choose the coordinate system (x, y) in such a way that the common edge of T_1 and T_2 lies on the y-axis. Moreover, we suppose that T_2 is the reference triangle and T_1 the symmetric one with respect to the x-axis (see Fig. 8.10). The general case can then be handled by means of suitable affine mappings.

We denote by $\lambda_{i,a}$ and $\lambda_{i,b}$ the barycentric coordinates relative to the vertices a and b, respectively, belonging to the element T_i, $i = 1, 2$. It is easy to check that it holds: $\lambda_{1,a} = 1 + x - y$, $\lambda_{1,b} = y$, $\lambda_{2,a} = 1 - x - y$, and $\lambda_{2,b} = y$. We shall also make use of the function $\lambda_{2,c} = x$. Let $L(x)$ be the Legendre polynomial of degree $k-2$ on the unit interval with respect to the weight $w(x) = x(1-x)^3$ and consider the function $p(x) \in Q_h$ defined as follows:

$$p'(x) = \begin{cases} -L(-x) & \text{for } x < 0, \\ L(x) & \text{for } x > 0. \end{cases} \tag{8.8.7}$$

We shall show that $\underline{\text{grad}}\, p$ is orthogonal to any velocity $\underline{v} \in V_h$. Since p does not depend on y, we can consider the first component v_1 of $\underline{v}$ only, which, by virtue of the continuity at $x = 0$ and of the boundary conditions, has the following general form:

$$v_1 = \begin{cases} \lambda_{1,a}\lambda_{1,b}(C_{k-2}(y) + xA_{k-3}(x, y)) & \text{in } T_1, \\ \lambda_{2,a}\lambda_{2,b}(C_{k-2}(y) + xB_{k-3}(x, y)) & \text{in } T_2, \end{cases} \tag{8.8.8}$$

where the subscripts denote the degrees of the polynomials A, B and C. We then have

$$\begin{aligned}\int_{T_1\cup T_2} \underline{v}\cdot \underline{\text{grad}}\, p\, dx\, dy &= \int_{T_1} v_1 p'\, dx\, dy + \int_{T_2} v_1 p'\, dx\, dy\\ &= \int_{T_2} \lambda_{2,a}\lambda_{2,b} L(x) x (B_{k-3}(x,y) - A_{k-3}(-x,y))\, dx\, dy\\ &= \int_{T_2} \lambda_{2,a}\lambda_{2,b}\lambda_{2,c} L(x) q(x,y)\, dx\, dy,\end{aligned} \tag{8.8.9}$$

where $q(x, y)$ is a polynomial of degree $k - 3$ and where the term involving C disappears by virtue of the symmetries. The last integral reads

$$\int_{T_2} xy(1-x-y) L(x) q(x)\, dx\, dy = \int_0^1 x L(x) Q(x)\, dx \tag{8.8.10}$$

and an explicit calculation shows that $Q(x)$ is of the form

$$Q(x) = (1-x)^3 p_{k-3}(x), \tag{8.8.11}$$

where p_{k-3} is a polynomial of degree $k - 3$. We can now conclude with the final computation

$$\int_{T_1\cup T_2} \underline{v}\cdot \underline{\text{grad}}\, p\, dx\, dy = \int_0^1 x(1-x)^3 L(x) p_{k-3}(x)\, dx = 0. \tag{8.8.12}$$

Step 2: sufficient part. The idea of the proof consists in considering, for each h, a partition of the domain Ω in sub-domains containing exactly three adjacent triangles. By making use of Proposition 8.5.4 and the technique presented in Sect. 8.5.1, it will be enough to prove the *inf-sup* condition for a single macro-element, provided we are able to bound the number of intersections between different sub-domains (basically, every time two sub-domains intersect each other, a factor $1/\sqrt{2}$ shows up in front of the final *inf-sup* constant). Indeed, it is possible to prove that, given a generic triangulation of a polygon, it can be presented as the disjoint union of triplets of triangles and of polygons that can be obtained as unions of triplets with at most three intersections.

Given a generic macro-element $a' \cup b' \cup c'$, consider the (x, y) coordinate system shown in Fig. 8.11, so that the vertices are $B' = (0,0)$, $D' = (1,0)$, $E' = (\alpha, \beta)$. By means of the affine mapping $x' = x + \alpha y$, $y' = \beta y$, the Jacobian of which is β, we can consider the macro-element $a \cup b \cup c$ shown in Fig. 8.12, so that b is the unit triangle. Since $\beta \neq 0$, the considered affine mapping is invertible. With an abuse of notation, we shall now denote by Ω the triplet $a \cup b \cup c$ and by V_h and Q_h the finite element spaces built on it.

We denote by λ^a_{AB} the barycentric coordinate of the triangle a vanishing on the edge AB (analogous notation holds for the other cases). Moreover, we denote by

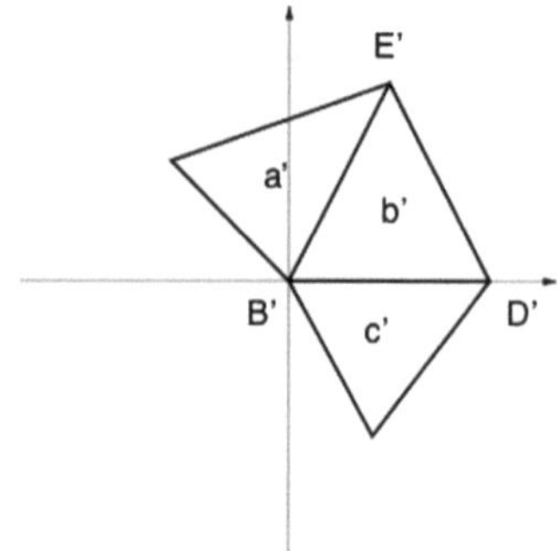

Fig. 8.11 A generic triplet of triangles

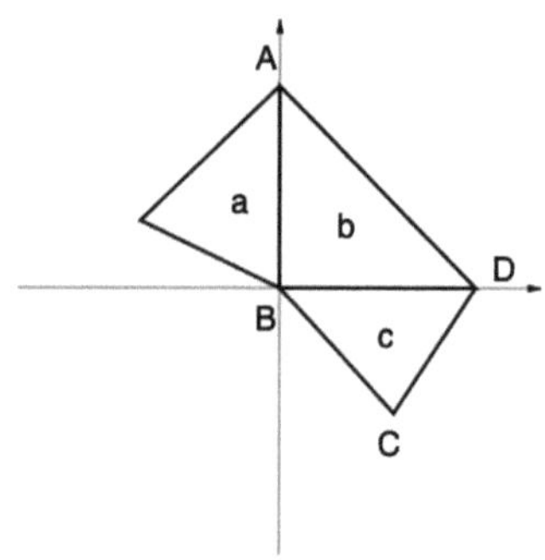

Fig. 8.12 A macro-element where b is the reference triangle

$L^a_{i,x}(x)$ the i-th Legendre polynomial in $[x_A, 0]$, with respect to the measure $\mu_{a,x}$ defined by

$$\int_{x_A}^{0} f(x)\, d\mu_{a,x} = \int_a \lambda^a_{AB}\lambda^a_{AE} f(x)\, dx\, dy \qquad \forall\, f(x) : [x_A, 0] \to \mathbb{R}, \tag{8.8.13}$$

where x_A is the x-coordinate of the vertex A. We shall make use of the following Legendre polynomials, which are defined in a similar way: $L^b_{i,x}$ (its definition involves λ^b_{ED} and λ^b_{BD}), $L^b_{i,y}$ (using λ^b_{BE} and λ^b_{BD}), and $L^c_{i,y}$ (using λ^c_{BC} and λ^c_{CD}).

Standard properties of the Legendre polynomials ensure that we can normalise them, for instance, by requiring that they assume the same value (say 1) at the origin. We now prove by induction with respect to the degree k that a modified *inf-sup* condition holds true (see Verfürth's trick in Sect. 8.5.2). Namely, for any $q_h \in Q_h$, we shall construct $\underline{v}_h \in V_h$ such that

$$\begin{aligned} -\int_{a\cup b\cup c} \underline{v}_h \cdot \underline{\mathrm{grad}}\, q_h\, dx\, dy &\geq c_1 \|\, \underline{\mathrm{grad}}\, q_h \|_0^2, \\ \|\, \underline{v}_h \|_0 &\leq c_2 \|\, \underline{\mathrm{grad}}\, q_h \|_0. \end{aligned} \tag{8.8.14}$$

The case k=2. This is the original Hood–Taylor method. Given $p \in Q_h$, we define $\underline{v}_h = (v_1(x,y), v_2(x,y))$, triangle by triangle, as follows:

$$v_1(x,y)|_a = -\lambda^a_{AB}\lambda^a_{AE}||\underline{\mathrm{grad}}\, p||_0\sigma, \tag{8.8.15}$$

$$v_2(x,y)|_a = -\lambda^a_{AB}\lambda^a_{AE}\frac{\partial p}{\partial y}, \tag{8.8.16}$$

$$v_1(x,y)|_b = -\lambda^b_{ED}\lambda^b_{BD}||\underline{\mathrm{grad}}\, p||_0\sigma - \lambda^b_{ED}\lambda^b_{EB}\frac{\partial p}{\partial x}, \tag{8.8.17}$$

$$v_2(x,y)|_b = -\lambda^b_{ED}\lambda^b_{BD}\frac{\partial p}{\partial y} - \lambda^b_{ED}\lambda^b_{EB}||\underline{\mathrm{grad}}\, p||_0\tau, \tag{8.8.18}$$

$$v_1(x,y)|_c = -\lambda^c_{BC}\lambda^c_{CD}\frac{\partial p}{\partial x}, \tag{8.8.19}$$

$$v_2(x,y)|_c = -\lambda^c_{BC}\lambda^c_{CD}||\underline{\mathrm{grad}}\, p||_0\tau, \tag{8.8.20}$$

where the quantities σ and τ are equal to ± 1 so that the expressions

$$H = \sigma||\underline{\mathrm{grad}}\, p||_0\left(\int_a \lambda^a_{AB}\lambda^a_{AE}\frac{\partial p}{\partial x} + \int_b \lambda^b_{ED}\lambda^b_{BD}\frac{\partial p}{\partial x}\right), \tag{8.8.21}$$

$$K = \tau||\underline{\mathrm{grad}}\, p||_0\left(\int_b \lambda^b_{EB}\lambda^b_{ED}\frac{\partial p}{\partial y} + \int_c \lambda^c_{BC}\lambda^c_{CD}\frac{\partial p}{\partial y}\right) \tag{8.8.22}$$

are non-negative. First of all, we observe that $\underline{v}_h$ is an element of V_h: its degree is at most two in each triangle, it vanishes on the boundary and it is continuous across the internal edges because so is the tangential derivative of p.

It is easy to check that $||\underline{v}_h||_0 \le c_1||\ \underline{\mathrm{grad}}\, p||_0$. In order to prove the first equation in (8.8.14), we shall show that the quantity $|||\underline{\mathrm{grad}}\, p||| = -\int_\Omega \underline{v}_h \cdot \underline{\mathrm{grad}}\, p$ vanishes only when $\underline{\mathrm{grad}}\, p$ is zero. From the equality

$$\begin{aligned} 0 = |||\underline{\mathrm{grad}}\, p||| = &\int_a \lambda^a_{AB}\lambda^a_{AE}\left(\frac{\partial p}{\partial y}\right)^2 + H \\ &+ \int_b \left(\lambda^b_{ED}\lambda^b_{EB}\left(\frac{\partial p}{\partial x}\right)^2 + \lambda^b_{ED}\lambda^b_{BD}\left(\frac{\partial p}{\partial y}\right)^2\right) \\ &+ K + \int_c \lambda^c_{BC}\lambda^c_{CD}\left(\frac{\partial p}{\partial x}\right)^2, \end{aligned} \tag{8.8.23}$$

it follows that

$$\frac{\partial p}{\partial y} = 0 \quad \text{in } a, \tag{8.8.24}$$

$$\frac{\partial p}{\partial x} = \frac{\partial p}{\partial y} = 0 \quad \text{in } b, \tag{8.8.25}$$

$$\frac{\partial p}{\partial x} = 0 \quad \text{in } c, \tag{8.8.26}$$

$$H = K = 0. \tag{8.8.27}$$

These last equations, together with the fact that each component of $\underline{\text{grad}}\, p$ is constant if $p \in Q_h$, easily imply that

$$\underline{\text{grad}}\, p = (0, 0) \quad \text{in } \Omega. \tag{8.8.28}$$

The case $k > 2$. Given p in Q_h, if p is locally of degree $k - 2$, then the result follows from the induction hypothesis. Otherwise, there exists at least one triangle of Ω in which p is exactly of degree $k - 1$. Like in the previous case, we define $\underline{v}_h = (v_1(x, y), v_2(x, y))$ as follows:

$$v_1(x, y)|_a = -\lambda^a_{AB}\lambda^a_{AE}||\,\underline{\text{grad}}\, p||_0 L^a_{k-2,x}\sigma, \tag{8.8.29}$$

$$v_2(x, y)|_a = -\lambda^a_{AB}\lambda^a_{AE}\frac{\partial p}{\partial y}, \tag{8.8.30}$$

$$v_1(x, y)|_b = -\lambda^b_{ED}\lambda^b_{BD}||\,\underline{\text{grad}}\, p||_0 L^b_{k-2,x}\sigma - \lambda^b_{ED}\lambda^b_{EB}\frac{\partial p}{\partial x}, \tag{8.8.31}$$

$$v_2(x, y)|_b = -\lambda^b_{ED}\lambda^b_{BD}\frac{\partial p}{\partial y} - \lambda^b_{ED}\lambda^b_{EB}||\,\underline{\text{grad}}\, p||_0 L^b_{k-2,y}\tau, \tag{8.8.32}$$

$$v_1(x, y)|_c = -\lambda^c_{BC}\lambda^c_{CD}\frac{\partial p}{\partial x}, \tag{8.8.33}$$

$$v_2(x, y)|_c = -\lambda^c_{BC}\lambda^c_{CD}||\,\underline{\text{grad}}\, p||_0 L^c_{k-2,y}\tau, \tag{8.8.34}$$

with the same assumption on σ and τ, so that the terms

$$H = \sigma||\,\underline{\text{grad}}\, p||_0\left(\int_a \lambda^a_{AB}\lambda^a_{AE}L^a_{k-2,x}\frac{\partial p}{\partial x} + \int_b \lambda^b_{ED}\lambda^b_{BD}L^b_{k-2,x}\frac{\partial p}{\partial x}\right), \tag{8.8.35}$$

$$K = \tau||\,\underline{\text{grad}}\, p||_0\left(\int_b \lambda^b_{EB}\lambda^b_{ED}L^b_{k-2,y}\frac{\partial p}{\partial y} + \int_c \lambda^c_{BC}\lambda^c_{CD}L^c_{k-2,y}\frac{\partial p}{\partial y}\right) \tag{8.8.36}$$

are non-negative. The same arguments as for $k = 2$, together with the described normalisation of the Legendre polynomials, show that $\underline{v}_h$ belongs to V_h.

In order to conclude the proof, we need to show that if $|||\,\underline{\text{grad}}\, p||| = -\int_\Omega \underline{v}_h \cdot \underline{\text{grad}}\, p = 0$, then the degree of $\underline{\text{grad}}\, p$ is strictly less than $k - 2$. As before, $|||\,\underline{\text{grad}}\, p||| = 0$ implies

$$\frac{\partial p}{\partial y} = 0 \quad \text{in } a, \tag{8.8.37}$$

$$\underline{\text{grad}}\, p = 0 \quad \text{in } b, \tag{8.8.38}$$

$$\frac{\partial p}{\partial x} = 0 \quad \text{in } c, \tag{8.8.39}$$

$$H = K = 0. \tag{8.8.40}$$

The last equalities imply

$$\int_a \lambda_{AB}^a \lambda_{AE}^a L_{k-2,x}^a \cdot \frac{\partial p}{\partial x} = 0 \tag{8.8.41}$$

and

$$\int_c \lambda_{BC}^c \lambda_{CD}^c L_{k-2,y}^c \cdot \frac{\partial p}{\partial y} = 0. \tag{8.8.42}$$

It follows that the degree of $\underline{\text{grad}}\, p$ is strictly less than $k-2$ in contrast to our assumption. □

Remark 8.8.4. The proof of the theorem shows that the continuity hypothesis on the pressure space Q_h can be weakened up to require that q_h is only continuous on triplets of elements. □

We conclude this subsection by stating the three-dimensional analogue to the previous theorem and by recalling the main argument of the proof presented in [73].

Theorem 8.8.2. *Let Ω be a polyhedral domain and $\mathcal{T}_h$ a regular sequence of decompositions of it into tetrahedra. Assume that every tetrahedron has at least one internal vertex. Then, the choice $V_h := (\mathcal{L}_k^1 \cap H_0^1(\Omega))^3$ and $Q_h := \mathcal{L}_{k-1}^1 \cap Q$ satisfies the inf-sup condition* (8.2.16) *for any $k \geq 2$.*

Proof. We shall make use of the macro-element technique presented in Sect. 8.5.4. In particular, we shall use Proposition 8.5.7 and the comments included in Remark 8.5.6.

We consider an overlapping macro-element partition of $\mathcal{T}_h$ as follows: for each internal vertex x_0, we define a corresponding macro-element M_{x_0} by collecting all elements which touch x_0. Thanks to the regularity assumptions on the mesh, we only have to show that the macro-element condition (8.5.24) holds true (see, in particular, Remark 8.5.6).

Let us consider an element $K \in M = M_{x_0}$ and an edge e of K which touches x_0. With a suitable choice of the coordinate system, we can suppose that the direction of e coincides with that of the x axis. With the notation of Sect. 8.5.4, we shall show that a function in K_M cannot contain functions which depend on x in K.

Namely, given a function $p \in Q_{0,M}$, we can define a function $\underline{v} \in V_{0,M}$ as follows:

$$\underline{v} := \left(-\lambda_{1,i}\lambda_{2,i}\frac{\partial p}{\partial x}, 0, 0\right) \qquad \text{in } K_i,$$

where K_i is a generic element of M sharing the edge e with K and $\lambda_{j,i}$, $j = 1, 2$, are the barycentric coordinates of K_i associated with the two faces of K_i which do not touch e. On the remaining elements, each component of $\underline{v}$ is set equal to zero. It is clear that $\underline{v}$ is a k-th order polynomial in K_i and, since p is continuous in M, $\partial p/\partial x$ is continuous across the faces which meet at e and the function $\underline{v}$ is continuous as well. Hence, $\underline{v}$ belongs to $V_{0,M}$.

From the definition of $Q_{0,M}$, it turns out that

$$0 = \int_M p \operatorname{div} \underline{u} = -\int_M \underline{\operatorname{grad}}\, p \cdot \underline{v} = \sum_i \int_{K_i} \lambda_{1,i}\lambda_{2,i} \left|\frac{\partial p}{\partial x}\right|^2.$$

The last relation implies that p does not depend on x in K_i for any i and, in particular, in K. On the other hand, we can repeat the same argument using as e the other two edges of K meeting at x_0 and, since the directions of the three used edges are independent, we obtain that p is constant in K. □

Remark 8.8.5. From the previous proof, we can deduce that the hypotheses on the triangulation can be weakened, by assuming that each tetrahedron has at least three edges which do not lie on the boundary of Ω and which are not in the same plane. On the other hand, given a generic mesh of tetrahedra, it is not difficult to add suitable elements in order to meet the requirements of the previous theorem. □

Remark 8.8.6. The main argument in the proof of the previous theorem is the straightforward generalisation of the two-dimensional case. Indeed, the proof of Theorem 8.8.1 could be carried out using the macro-element technique as well. □

8.9 Other Developments for Divergence-Free Stokes Approximation and Mass Conservation

From the discussion presented so far, it is clear that, in general, the incompressibility constraint $\operatorname{div}\underline{u} = 0$ is not satisfied exactly at the discrete level. More precisely, the discrete velocity field $\underline{u}_h$ fulfills the following equation

$$\int_\Omega \operatorname{div}\underline{u}_h\, q_h\, dx = 0 \quad \forall q_h \in Q_h,$$

so that the equality $\operatorname{div}\underline{u}_h = 0$ holds in general only if

$$\operatorname{div}(V_h) \subset Q_h. \tag{8.9.1}$$

Almost all stable elements that we have presented up to now do not satisfy (8.9.1), the only exception being the two dimensional Scott–Vogelius scheme $\underline{P}_k - P_{k-1}$, which however requires severe mesh restrictions (see Sect. 8.8.1). Another example is presented in Example 8.10.3, On the other hand, discrete schemes that fail to satisfy the divergence-free condition (at least locally) can lead to undesired instabilities when used for the resolution of more complex problems. This is the case, for instance, when a Stokes solver is used for the approximation of non linear problems (see [39]), or when the incompressibility condition is related to a physical mass conservation property, like in fluid-structure interaction problems (see [78, 280]).

For this reason, a very active research area concerns investigations trying to develop divergence-free Stokes elements, at least in a local sense.

8.9.1 Exactly Divergence-Free Stokes Elements, Discontinuous Galerkin Methods

The simplest idea in order to satisfy (8.9.1) is to use a C^1 approximation of the velocity field and to take as space of pressures exactly $Q_h = \mathrm{div}(V_h)$. Here we do not follow this approach, but we focus on suitably chosen mixed approximations of the Stokes problem.

Early attempts to develop divergence-free finite elements for the approximation of the Stokes problem made use of particular mesh sequences. Besides the already mentioned Scott–Vogelius family (see Sect. 8.8.1), a two dimensional approximation involving a mesh of *rectangles* has been introduced in [253, 385]. The lowest element of the family is constructed as follows: $V_h = \mathcal{L}^1(P_{2,1} \times P_{1,2}, \mathcal{T}_h)$, $Q_h = \mathrm{div}(V_h) \subset \mathcal{L}^1_{[1]}$. It is clear that the use of rectangular elements imposes limitations to the geometry of the domain Ω, which make the scheme unappealing for practical applications.

8.9.1.1 Discontinuous Galerkin Approximations

A more interesting approach arises from the use of *discontinuous Galerkin approximations*. A first possibility is to use a completely discontinuous finite element space for the approximation of the velocity together with a postprocessing procedure (see [156]), or $H(\mathrm{div})$ conforming elements in order to avoid the postprocessing (see [157]).

A more recent approach is based on the idea of using $H(\mathrm{div})$ conforming elements for the approximation of the velocities and to enrich them in order to obtain the stability (see [238]); the enrichment is performed locally by means of divergence-free polynomials (defined as the curl of suitably chosen bubble functions), so that the scheme remains conservative. The construction holds on simplicial meshes in two and three dimensions and is based on $\mathcal{BDM}$ and $\mathcal{RT}$ spaces. This research is further improved in [237], where a conforming divergence-free element, which can be implemented on two dimensional triangular meshes, is presented. In this case the enrichment is based on *rational* functions which ensure stability and modify the tangential components of the basis functions across the interelements in order to guarantee their continuity. Some a posteriori error estimators (Sect. 7.11) for these methods have been considered in [251].

8.9.2 Stokes Elements Allowing for Element-Wise Mass Conservation

From what we have seen, it is not so easy to obtain a discrete velocity with vanishing divergence *pointwise*. On the other hand, in several applications it might be desirable to have a local (element-wise) conservation of mass. From (8.9.1) it is clear that *discontinuous pressure* schemes enjoy automatically a local conservation property. In particular, if Q_h contains piecewise constants, then $\mathrm{div}\underline{u}_h$ has zero mean value on each element. In this respect, we believe that the $\underline{Q}_2 - P_1$ scheme (see Sect. 8.6.3.1) is one of the best performing method for quadrilateral meshes. For simplicial meshes, the SMALL element of Example 8.7.1 provides what seems to be the simplest element ensuring local mass conservation.

Non conforming schemes can also achieve this goal. In particular the use of non conforming piecewise linear element for the velocity and piecewise constants for the pressures yields a simple locally divergence-free scheme. (see [165] and Sect. 8.4.4 for a discussion about this method). The extension to quadrilateral meshes requires a careful choice of the non conforming space (see [330]).

For *continuous pressure* schemes, however, the situation is more complicate. Relation (8.9.1), in particular, shows that the discrete divergence-free condition has to be considered in a non local sense. For this reason, there have been studies trying to modify standard spaces in order to achieve a more local conservation of mass. The main idea behind this technique consists in adding piecewise constants to the pressure space. It is clear that this modification allows for a local mass conservation (actually, the method is transformed into a scheme with discontinuous pressures), but can work only if it does not affect the validity of the inf-sup condition: a larger pressure space is indeed a potential source of trouble for the stability.

Indeed, it can be shown that generalised Hood–Taylor (see Sect. 8.8) can be modified by adding piecewise constants to the pressure space and preserving the inf-sup condition ($k \geq 2$ in two dimensions and $k \geq 3$ in three dimensions). The same procedure can be applied to the $\underline{P}_1\mathrm{iso}\underline{P}_2 - P_1$ element. For the Hood–Taylor scheme, the idea was suggested in [231, 232, 369], where numerical evidence of the improvement was given (see also [144]). The proof of the stability of the enhanced lowest order Hood–Taylor scheme for triangular and rectangular meshes, can be found in [322, 329, 364]. A more comprehensive discussion, including a general proof of stability, can be found in [83].

8.10 Spurious Pressure Modes

As we stated in the introduction to this chapter, the approximation of the Stokes problem has been developed mostly independently of the theory of mixed methods. This led to the use of some approximations which did not satisfy the *inf-sup* conditions and which generated strange results, specially for the pressure components. This generated the concept of spurious pressure modes.

For the Stokes problem with Dirichlet boundary conditions, pressure is defined up to a constant which is the kernel of the gradient operator. This is a natural pressure mode. However, this mode may not be the only one in discrete problems. For a given choice of V_h and Q_h, we define the space S_h of spurious pressure modes as follows:

$$S_h := \mathrm{Ker} B_h^t \setminus \mathrm{Ker} B^t. \tag{8.10.1}$$

It is clear that a necessary condition for the validity of the *inf-sup* condition (8.2.16) is the absence of spurious modes, that is,

$$S_h = \{0\}. \tag{8.10.2}$$

In particular, if S_h is non-trivial, then the solution p_h to the discrete Stokes problem (8.2.11) can be changed to $p_h + s_h$, which is still a solution when $s_h \in S_h$. Spurious modes correspond to null singular values as discussed in Sect. 5.6.2. The existence of spurious modes is in many cases strongly mesh dependent. They will appear on special regular meshes but will remain if such meshes are slightly distorted.

We shall illustrate how this situation may occur with the following example.

Example 8.10.1 (The $Q_1 - P_0$ element). Among quadrilateral elements, the $\underline{Q}_1 - P_0$ element is the first that comes to mind. It is defined as (see Fig. 8.13):

$$V_h := (\mathcal{L}^1_{[1]})^2 \cap V, \qquad Q_h := \mathcal{L}^0_0 \cap Q. \tag{8.10.3}$$

This element is strongly related, for rectangular meshes, to some finite difference methods [206]. Its first appearance in a finite element context seems to be in [255].

However simple it may look, the $\underline{Q}_1 - P_0$ element is one of the hardest elements to analyse and many questions are still open about its properties. This element does not satisfy the *inf-sup* condition: it strongly depends on the mesh. For a regular mesh, the space of spurious modes is one-dimensional. More precisely, $\underline{\mathrm{grad}}_h\, q_h = 0$ implies that q_h is constant on the red and black cells if the mesh is viewed as a *chequerboard* (Fig. 8.14).

This means that one singular value (cf. Chap. 3.4.3) of the operator $B_h = \mathrm{div}_h$ is zero. Moreover, it has been checked by computation [286] that a large number of positive singular values converge to zero when h becomes small. In [263], it has indeed been proved that the second singular value is $O(h)$ and is not bounded below

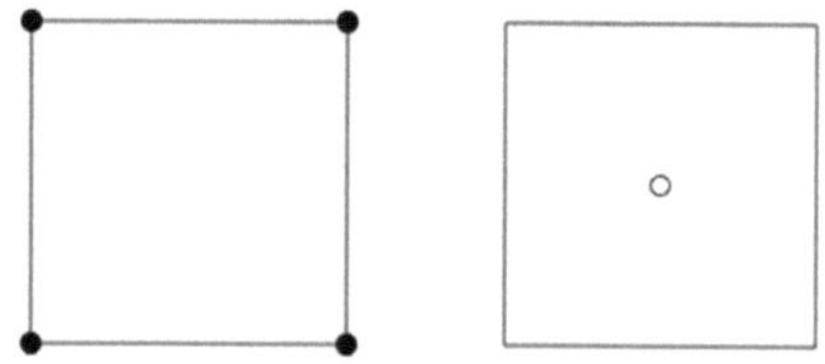

Fig. 8.13 The $\underline{Q}_1 - P_0$ element

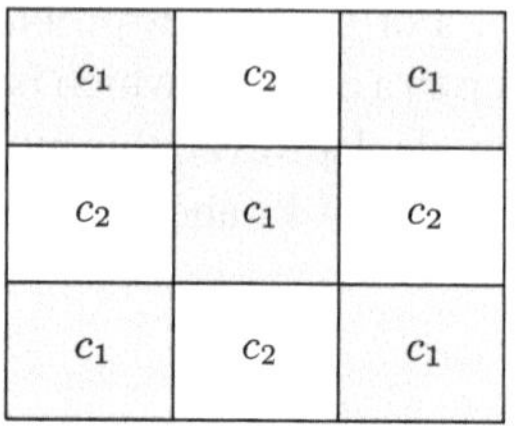

Fig. 8.14 The chequerboard mode

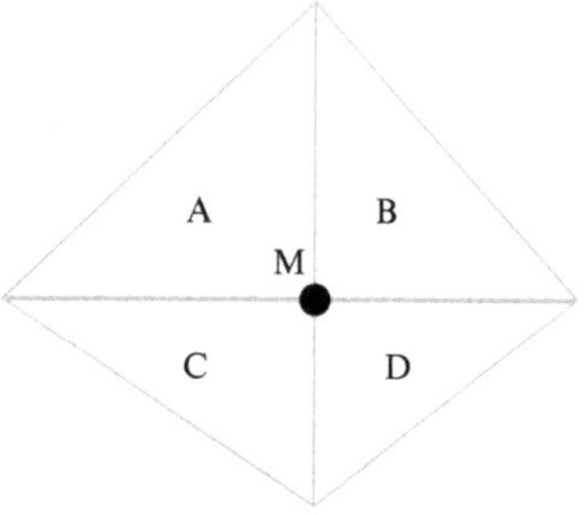

Fig. 8.15 The reference criss-cross

(see also [314]). The $\underline{Q}_1 - P_0$ element has been the subject of a vast literature. We shall come back to it in Sect. 8.10.2. □

We shall now present a few more examples and distinguish between local and global spurious pressure modes.

Example 8.10.2 (The criss-cross $\underline{P}_1 - P_0$ *element).* Let us consider a mesh of quadrilaterals divided into four triangles by their diagonals (Fig. 8.2). We observed, in Example 8.3.2, that the $\underline{P}_1 - P_0$ element, on *general meshes*, is affected by locking, that is, the computed velocity vanishes. On the mesh introduced above, however, it is easy to see that non-zero divergence-free functions can be obtained. The divergence is constant on each triangle. This means that there are four linear relations between the values of the partial derivatives. It is easily seen that one of them can be expressed as a combination of the others, this fact being caused by equality of tangential derivatives along the diagonals. To make things simpler, we consider the case where the diagonals are orthogonal (Fig. 8.15) and we label by A, B, C, D the four triangles. We then have, by taking locally the coordinate axes along the diagonals and by denoting by $\underline{u}^K$ the restriction of a function of V_h to the element K,

$$\frac{\partial u_1^K}{\partial x_1} + \frac{\partial u_2^K}{\partial x_2} = 0, \quad K = A, B, C, D. \tag{8.10.4}$$

On the other hand, one has at the point M

$$\frac{\partial u_1^A}{\partial x_2} = \frac{\partial u_2^B}{\partial x_2}, \quad \frac{\partial u_1^A}{\partial x_1} = \frac{\partial u_1^C}{\partial x_1}, \quad \frac{\partial u_2^C}{\partial x_2} = \frac{\partial u_2^D}{\partial x_2}, \quad \frac{\partial u_1^B}{\partial x_1} = \frac{\partial u_1^D}{\partial x_1}. \tag{8.10.5}$$

It is easy to check that this makes one of the four conditions (8.10.4) redundant. The reader may check the general case by writing the divergence operator in a non orthogonal coordinate system.

The consequence of the above discussion is that on each composite quadrilateral, one of the four constant pressure values will be undetermined. The dimension of $\mathrm{Ker}B_h^t$ will be *at least* as large as the number of quadrilaterals minus one. This is what we shall call *local modes*.

Thus, three constraints remain on each composite quadrilateral element. If we admit that two of them can be controlled, using the methods of Sect. 8.5.5, by the "internal" node M, we obtain an element that is very similar to the $\underline{Q}_1 - P_0$ element with respect to the degrees of freedom. Indeed, it can be checked that on a regular mesh, an additional, *global* chequerboard mode occurs and that the behaviour of this approximation is essentially the same as that of the $\underline{Q}_1 - P_0$ element that will be discussed in details in Sect. 8.10.2. These analogies have been pointed out, for instance, in [82]. □

The above example has shown the existence of two kinds of spurious pressure modes. In the case of the criss-cross $\underline{P}_1 - P_0$ element presented in the previous example, $\dim S_h$ grows as h goes to 0 and there exists a basis of S_h with local support (that is, the support of each basis function can be restricted to one macro-element). We shall refer to these modes as *local spurious modes*. Such pressure modes can be eliminated by considering a composite mesh (in the previous example a mesh of quadrilaterals instead of triangles) and using a smaller space for the pressures by deleting some degrees of freedom from the composite elements. If the original space is to be employed, one must check the extra compatibility conditions. This can often be done by a small change in the data. This will be the case in Example 8.10.3.

If we now consider the $\underline{Q}_1 - P_0$ example (see Example 8.10.1), the dimension of S_h does not grow when h goes to 0 and no basis can be found with a local support. We then have a *global spurious mode* which cannot be eliminated as easily as the local ones. Global modes usually appear on special (regular) meshes and are symptoms that the behaviour of the element at hand is strongly mesh dependent and requires a special care. Some elements may generate both local and global modes as we have seen in the criss-cross $\underline{P}_1 - P_0$ method (see Example 8.10.2).

Example 8.10.3 (The criss-cross $\underline{P}_2 - P_1$ element). Another simple example where a local mode occurs is the straightforward extension of the previous example to the case of a $\underline{P}_2 - P_1$ approximation. This element has an interest because it is the simplest really divergence-free element, that is, $\mathrm{Ker}B_h \subset \mathrm{Ker}B$. Unfortunately, its

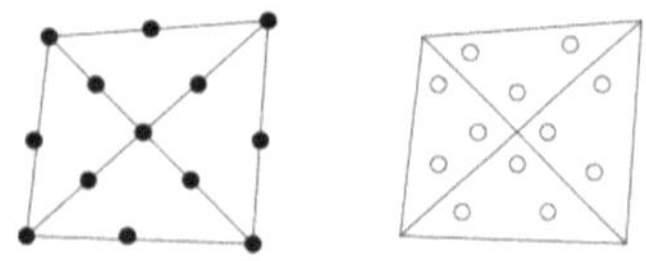

Fig. 8.16 The criss-cross $\underline{P}_2 - P_1$ element

three-dimensional counterpart does not seem to exist. We consider, as in Fig. 8.16, a mesh of quadrilaterals divided into triangles by its diagonal. This means that on each quadrilateral, we have 12 discrete divergence-free constraints, and it is easily seen by the argument of Example 8.10.2, written at the point M, that one of them is redundant. Thus, one spurious mode will appear for each composite quadrilateral. However, in this case, no global mode will appear. The analysis of this element is also related to the work of [153] by considering the stream function associated with a divergence-free function. Considering the space of discrete pressures where the spurious modes are removed, a standard proof using internal degrees of freedom shows that one has a stable approximation. □

8.10.1 *Living with Spurious Pressure Modes: Partial Convergence*

The presence of spurious modes can be interpreted as a signal that the pressure space used is in some sense too rich. We therefore can hope to find a cure by using a strict subspace $\hat{Q}_h$ of Q_h as the space of the discrete pressures, in order to obtain a stable approximation. The question arises whether or not this stability can be used to prove at least a partial result on the original approximation. One can effectively get some results in this direction as discussed in Sect. 5.6.3. In general, we cannot make a direct use of the singular value decomposition but, in some cases, we can identify a guilty subspace.

We suppose, here, that Q and Q_h can be identified to their dual, as it is indeed the case for the Stokes problem.

Following Sect. 5.3.3, we suppose that we know subspaces $\hat{V}_h$ and $\hat{Q}_h$ of V_h and Q_h such that the couple $\hat{V}_h \times \hat{Q}_h$ is stable. We denote $\tilde{Q}_h$ the orthogonal complement of $\hat{Q}_h$ in Q_h. To apply the result of Sect. 5.3.3, we shall need to obtain the following:

$$b(\hat{v}_h, \tilde{q}_h) = 0 \qquad \forall\, \tilde{q}_h \in \tilde{Q}_h,\ \forall\, \hat{v}_h \in \hat{V}_h. \tag{8.10.6}$$

We emphasise that this will be generally possible only on special meshes. We now make the hypothesis that in (8.1.1), g has no component in $\tilde{Q}_h$, that is,

$$(g, \tilde{q}_h) = 0, \quad \tilde{q}_h \in \tilde{Q}_h. \tag{8.10.7}$$

This condition is a restriction on admissible data. In practice, it will imply an extra regularity condition on g which will in turn enable us to obtain (8.10.7) through a *small* modification of g. We mean by small that this modification should not jeopardise the accuracy of the approximation. If we refer to Sect. 5.6.2, by supposing (8.10.7), we have killed the unstable part of $\underline{u}_h$. On the other hand, p_h will have components in $\tilde{Q}_h$. However, the part of p_h in $\hat{Q}_h$ will be stable and will provide a reasonable approximation of the solution. More precisely, under these hypotheses, Proposition 5.3.1 yields the following results:

$$\|u - u_h\|_V \leq c_1\Big(\inf_{\hat{v}_h \in \hat{V}_h} \|u - \hat{v}_h\|_V + \inf_{q_h \in Q_h} \|p - q_h\|_Q\Big) \tag{8.10.8}$$

$$\|p - \hat{p}_h\|_Q \leq c_2\Big(\|u - u_h\|_V + \inf_{q_h \in Q_h} \|p - q_h\|_Q\Big) + \inf_{\hat{q}_h \in \hat{Q}_h} \|p - \hat{p}_h\|_Q. \tag{8.10.9}$$

Example 8.10.4. The simplest example is the case of Example 8.10.3. In this case, we have $\hat{V}_h = V_h$. When the local modes are filtered, pressure will converge, provided g has no component in these modes. This implies a slight restriction of data. □

The most important case is, however, the $\underline{Q}_1 - P_0$ element which we discuss in the next section.

8.10.2 *The Bilinear Velocity-Constant Pressure $\underline{Q}_1 - P_0$ Element*

We now come back to a rapid analysis of what is probably (and unfortunately!) the most popular of all elements for incompressible materials. This is perhaps also the hardest to analyse and as we shall see, only partial results are known. Origins of this element can be traced back to finite difference methods [206] and its peculiar properties were soon recognised. In particular, the chequerboard pressure mode was already a familiar feature long before the scheme used were written in terms of finite elements.

Let us summarise the basic facts. On a regular mesh, for a problem with Dirichlet boundary conditions, two singular values of the matrix (cf. Sect. 5.6.2) vanish instead of one. We thus have a *pure* spurious pressure mode in the terminology of [340, 341]. This spurious mode implies a compatibility condition on the data, which is, in most cases but not always, easily satisfied. When the mesh is slightly distorted, only one singular value is zero, corresponding to constants, but the second one is very small, as the zero has become some value depending on the mesh distortion, thus implying an ill-conditioning of the problem. In many computations, this ill-conditioning is fortunately almost restricted to pressure: we have what [340, 341] call an *impure* pressure mode which can be eventually filtered but often does not

Fig. 8.17 Macro-element and degrees of freedom

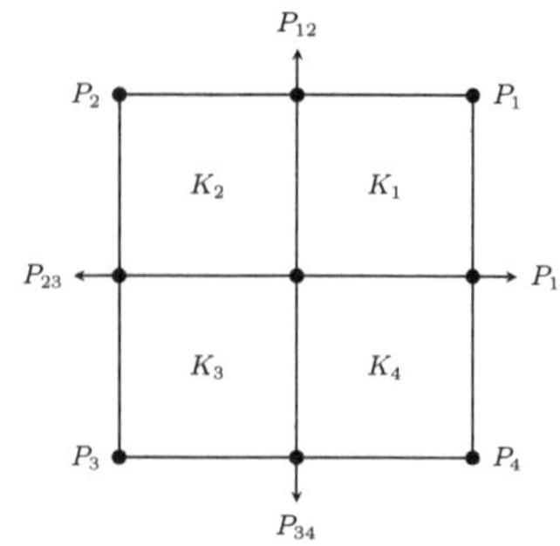

seem to affect (at least substantially) the computation of velocity. This is still not, however, the whole story. One could indeed hope from all this that an *inf-sup* stability condition could hold for the third singular value instead of the second and that we could have stability in a simple quotient space. Experimental evidence showed this hope to be false: on a regular mesh, a large number of eigenvalues converge to zero at order h [286]. Johnson and Pitkäranta [263] indeed proved the constant k_h to be $O(h)$ (see also [99, 100, 288, 314]). The standard estimates would then lead to the conclusion that no convergence will occur, in complete contradiction with experience. The paper of Johnson and Pitkäranta provided a first result by showing, on a regular mesh, that under stricter regularity assumptions than usual on the solution, convergence could take place.

Pitkäranta and Stenberg [324] proved a convergence result, without special regularity assumptions for a special type of mesh. We have already discussed, in Sect. 8.10.1, following Sect. 5.6.2, the underlying algebraical issues involved. If there is a "stable part", the data corresponding to the unstable modes should be null or small. In this case, the velocity can indeed be expected to behave well but the pressure part is doomed. We shall now consider these results for our particular case. To make things simpler, we shall first consider the case of a regular rectangular mesh. On such a mesh, we consider a macro-element (Fig. 8.17) M formed of four quadrilaterals.

On this macro-element, a piecewise constant pressure has four degrees of freedom. We introduce a local basis on M, ϕ_1, ϕ_2, ϕ_3, ϕ_4 described symbolically on the Fig. 8.18.

A chequerboard mode will obviously take its roots in ϕ_4. We therefore introduce quite naturally the space

$$\hat{Q}_h := \sum_M \Big(\sum_{i=1}^{3} \alpha_{iM}\, \phi_{i,M} \Big) \tag{8.10.10}$$

which will be the stable part and

$$\tilde{Q}_h := \sum_M \alpha_{4M}\, \phi_{4,M} \tag{8.10.11}$$

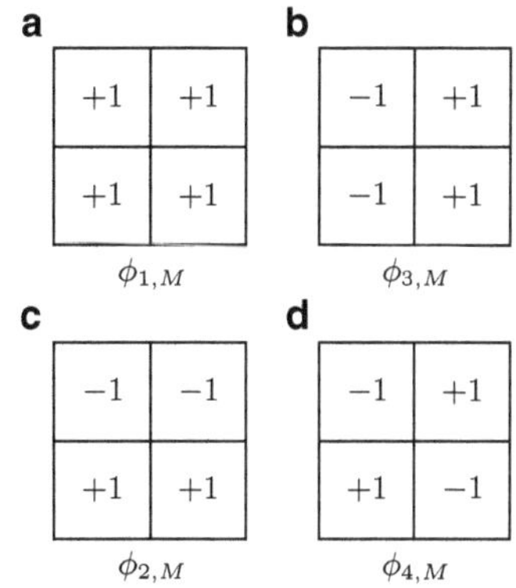

Fig. 8.18 Pressure basis functions on M

the unstable part. Stability of $(V_h, \hat{Q}_h)$ is thus immediate from the standard techniques, building a B-compatible operator. In fact, we shall build it for a subspace $\hat{V}_h$ of V_h which will make it, *a fortiori*, valid for V_h. The choice of $\hat{V}_h$ can be inferred from other well-known elements. We now use, as degrees of freedom, the two values of velocity at the vertices of M and at its barycentre and the *normal value* (rather a correction to this value) at mid-side nodes. The tangential component is thus linear on each edge and determined by the values at the vertices (Fig. 8.17). To build a B-compatible operator, we set

$$\underline{\hat{u}}_h(P_i) = \underline{u}(P_i),\ i = 1, 2, 3, 4. \tag{8.10.12}$$

To determine the normal node on the edges of M, we take

$$\int_e (\underline{u} - \underline{\hat{u}}_h) \cdot \underline{n}_e \, ds = 0, \tag{8.10.13}$$

where $\underline{\underline{n}}_e$ is the normal to e. This is enough to control the flux at interfaces and the part of pressure $(\phi_{1,M})$ which is constant on M is controlled. The $\phi_{2.M}$ and $\phi_{3,M}$ components are controlled by

$$\int_K \operatorname{div}(\underline{u} - \underline{\hat{u}}_h)\phi_{i,M} \, dx = 0,\ i = 2, 3. \tag{8.10.14}$$

It is not difficult to check that $\Pi_h \underline{u} = \underline{\hat{u}}_h$ is a B-compatible operator for $\hat{V}_h \times \hat{Q}_h$.

We would now want to apply Proposition 5.3.1. First, this implies a condition on data.

Remark 8.10.1. Following Sect. 5.3.3, to get reasonable results, the data should satisfy

$$(g, \tilde{q}_h) = \int_\Omega \tilde{q}_h \operatorname{div} \underline{u}_h \, dx = 0 \tag{8.10.15}$$

which means, here,

$$(g, \phi_{4,M}) = \int_{\Omega} \phi_{4,M} \operatorname{div} \underline{u}_h \, dx = 0. \tag{8.10.16}$$

This in fact corresponds to a regularity condition. It is easy to check that on our macro-element, we have

$$\int_M \phi_{4,M} \operatorname{div} \underline{u}_h \, dx = O(h^4) sup|div \frac{\partial^2 \underline{u}}{\partial x \partial y}|. \tag{8.10.17}$$

This is enough to show that this integral can be made null through a small perturbation of data. □

In order to apply Proposition 5.3.1, we now need to check (5.3.19) that is now

$$\int_M \phi_{4M} \ \operatorname{div} \ \underline{\hat{v}}_h \, dx = 0 \ \forall M, \ \forall \underline{\hat{v}}_h. \tag{8.10.18}$$

It should be seen, in order to check this, that the shape function $\underline{w}_1$ associated to vertex P_1, for instance, is a function of $Q_1(M)$ having the whole of M as its support. A straightforward computation then shows that one has

$$\begin{aligned} \int_M \phi_{4M} \ \operatorname{div} \underline{w}_1 \, dx = & \int_{\partial K_1} \underline{w}_1 \cdot \underline{n} \, d\sigma - \int_{\partial K_2} \underline{w}_1 \cdot \underline{n} \, d\sigma \\ & + \int_{\partial K_3} \underline{w}_1 \cdot \underline{n} \, d\sigma - \int_{\partial K_4} \underline{w}_1 \cdot \underline{n} \, d\sigma = 0. \end{aligned} \tag{8.10.19}$$

In the same way, the shape function $\underline{w}_{12}$ associated with node P_{12} satisfies

$$\int_M \phi_{4M} \ \operatorname{div} \underline{w}_{12} \, dx = \int_{\partial K_1} \underline{w}_{12} \cdot \underline{n} \, d\sigma + \int_{\partial K_2} \underline{w}_{12} \cdot \underline{n} \, d\sigma = 0 \tag{8.10.20}$$

and this is also true in the adjacent element because the mesh is aligned. The shape function associated with the barycentre trivially satisfies the condition. Condition (5.3.19) therefore holds and we have, by Proposition 5.3.1,

$$\|\underline{u} - \underline{u}_h\|_V \leq \Big(\inf_{\underline{\hat{v}}_h \in \hat{V}_h} \|\underline{u} - \underline{\hat{v}}_h\|_V + \inf_{q_h \in Q_h} \|q - q_h\|_Q \Big). \tag{8.10.21}$$

In the present case, it is clear that an error estimate in $\hat{V}_h$ has the same order as an estimate in V_h and the result is therefore almost optimal. We also have convergence of (filtered) pressure in $\hat{Q}_h$ by estimate (5.3.25). Following [324], we can now extend this result to the case where the mesh is made from super macro-elements like in Fig. 8.19.

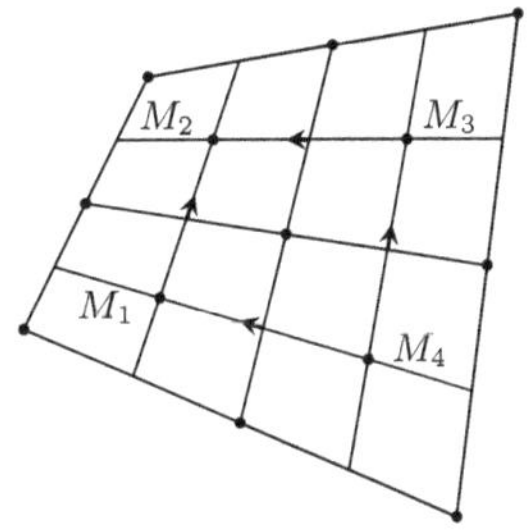

Fig. 8.19 A super-macro SM and its sub-macros

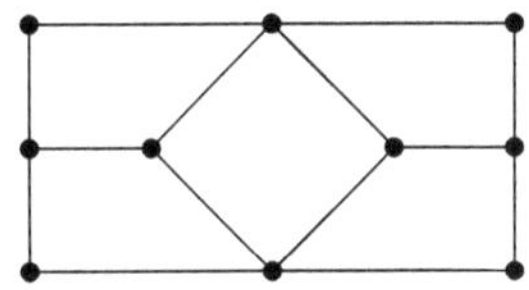

Fig. 8.20 A special macro-element

A general quadrilateral is divided in a regular way into 16 quadrilaterals. It is well-known [202] that on a non rectangular mesh, at least a four by four patch of elements is needed to generate a non-trivial discrete divergence-free function. We thus have four "sub-macros" like in the previous case. The space of filtered pressures $\hat{Q}_h$ is taken exactly as on the regular mesh and is still defined by (8.10.10). The space $\tilde{V}_h$ is defined by the following degrees of freedom: the values of velocity at the vertices of the M_i, the values at the barycentres of the M_i and a correction of the component of velocity parallel to the mesh at the mid-side nodes of the M_i internal to SM. One can directly build an interpolation operator enabling us to check the *inf-sup* condition. Mid-side nodes of SM control the part of pressure constant on the whole of SM. Internal mid-side nodes ensure mass-balance on each M_i and the nodes at the barycentres of the M_i end the job. It must be remarked that the alignment of mid-side velocities along the mesh is an essential feature of the construction.

In order to prove condition (5.3.18), the only hard point is to check that (8.10.18) still holds on every M_i. We refer the reader to [324] for this proof. It is then possible to use Proposition 5.3.1 and to get optimal error estimates.

This is still not the whole story about this peculiar element. It is also possible to prove stability [276, 354] on meshes built from macro-elements like in Fig. 8.20 without filtering or using another subterfuge.

This is coherent with the known experimental fact that on a general distorted mesh, pressure modes disappear and the *inf-sup* constant is independent of h. This last fact is still resisting analysis. It is our hope that the above technique could be generalised to yield the complete result.

The above discussion can be extended to the three-dimensional case. Things are made still more complicated by the fact that on a regular mesh (let say a $n \times n \times n$ assembly of elements to fix ideas), we do not have one spurious pressure mode but $3n-2$ of them. This will also mean the same number of compatibility conditions on

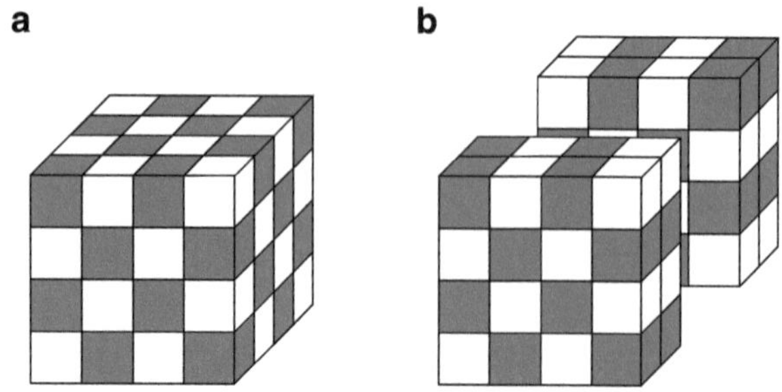

Fig. 8.21 Pressure modes

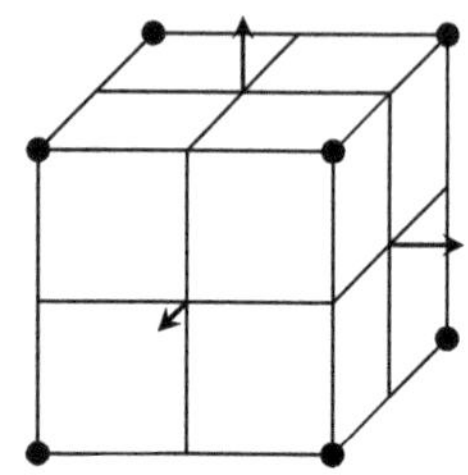

Fig. 8.22 Degrees of freedom for $\hat{V}_h$

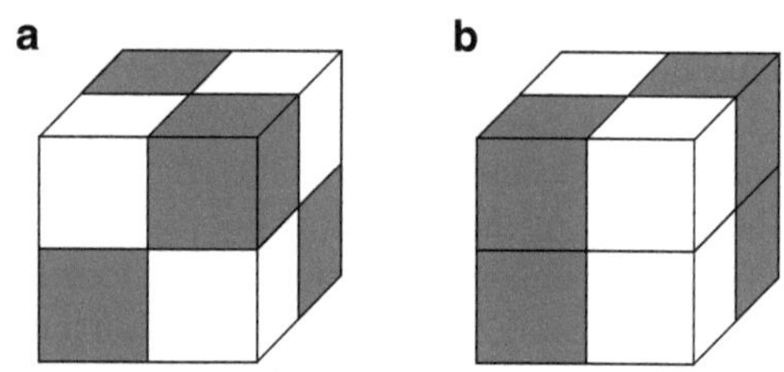

Fig. 8.23 Local spurious modes

data so that trouble should be expected when using apparently reasonable boundary conditions. These spurious modes are depicted in the following Fig. 8.21. One of them is the genuine 3-D chequerboard mode (Fig. 8.21a). The other ones are built from an assembly of 2-D nodes. In Fig. 8.21b, we have sliced the mesh in order to make apparent the internal structure of this mode. There are $3(n-1)$ possible such slices so that we find the number of modes stated above.

We now sketch the extension of the above proof to the 3-D case. We shall only present the rectangular case to avoid lengthening unduly this exposition. We thus suppose that the mesh is built from $2 \times 2 \times 2$ macro-elements (Fig. 8.22).

Our pressure space $\hat{Q}_h$ will be built from Q_h by deleting on each macro-element four $= (3 \times 2 - 2)$ spurious modes sketched in Fig. 8.23.

The mode depicted in Fig. 8.23b has obviously two other symmetrical counterparts. On each macro-element, we thus keep the 3-D analogues of the basis functions $\phi_{1,M}$, $\phi_{2,M}$, $\phi_{3,M}$ of Fig. 8.18. We must now introduce $\hat{V}_h$. This is done again by taking off some degrees of freedom from V_h. The remaining ones are sketched in Fig. 8.22 The internal node at the barycentre of the element is also used. It is now clear that $(\hat{V}_h,\ \hat{Q}_h)$ is a stable pair that provides $O(h)$ convergence. There remains to check condition (5.3.19) that is that $\hat{V}_h$ is transparent with respect to $\tilde{Q}_h$. This is done exactly as in the 2-D case by a simple check of flow balance at the surface of elements. Proposition 5.3.1 then applies and we get $O(h)$ convergence for velocities and filtered pressures.

It could be hoped that the same kind of analysis could be done for equal interpolation continuous pressure methods such as the $\underline{Q}_1 - Q_1$ approximation. Unfortunately, we know of no way in which condition (5.3.19) could be made to hold and an analysis of the convergence properties of these approximations remains an open question. We can however introduce an alternate way of stabilising such approximations and this is done in the following section.

Remark 8.10.2 (However, this is a dangerous element). As we stated at the beginning of this Section, the $\underline{Q}_1 - P_0$ element is widely employed. We presented the above results to provide the reader with enough information about this unduly popular element. It remains that using an unstable element is a dangerous option and that the price to pay for an apparent simplicity may be inaccurate results. □

Remark 8.10.3 (The worst drawback). An important draw-back for the $\underline{Q}_1 - P_0$ element is that the condition number of the dual problem in p is mesh dependent while it is not for stable Stokes elements. When an iterative solution method is used, this leads to a strong slowdown of the convergence. This is especially disastrous for 3-D problems where iterative methods are likely to be necessary. □

8.11 Eigenvalue Problems

We shall briefly consider, here, the application of the results of Chap. 6 to the approximation of eigenvalues for the Stokes problem. The results will also be applicable to incompressible elasticity. They have some importance in this case because of the popular *modal* method in which a problem is approximated using a few eigenvectors as a basis for a Galerkin's method.

We thus consider the eigenvalue problem introduced in (1.3.84), which we recall for simplicity. We now take $V = (H_0^1(\Omega))^2$ and $Q = L^2(\Omega)/\mathbb{R}$ and we look for $\underline{u} \in V$ and $q \in Q$ satisfying

$$\begin{cases} 2\mu \displaystyle\int_\Omega \underline{\underline{\varepsilon}}(\underline{u}) : \underline{\underline{\varepsilon}}(\underline{v})\, dx + \int_\Omega p \,\operatorname{div} \underline{v}\, dx = \lambda \int_\Omega u \cdot v\, dx \quad \forall \underline{v} \in V, \\ \displaystyle\int_\Omega q \operatorname{div} \underline{u} = 0, \quad \forall q \in L^2(\Omega). \end{cases} \tag{8.11.1}$$

The Lagrange multiplier p ensures the incompressibility of the eigenmodes. This is a problem of the type $(f, 0)$. It is easy to see, in the notation of Sect. 1.2.1, that if Ω is, for instance, a convex polygon, Q_0^H is $H^1(\Omega)/\mathbb{R}$ and V_0^H is the subspace of

$(H^2(\Omega) \cap H_0^1(\Omega))^2$ made of divergence-free functions (see [267]). In particular, we can check that $||\underline{u}||_{V_0^H} = ||\Delta \underline{u}||_0 \sim ||\underline{u}||_2$ and $||p||_{Q_0^H} = ||\,\underline{\text{grad}}\, p||_0$.

Let V_h and Q_h be finite dimensional subspaces of V and Q respectively. We then consider the discrete version of (8.11.1),

$$\begin{cases} 2\mu \displaystyle\int_\Omega \underline{\underline{\varepsilon}}(\underline{u}_h) : \underline{\underline{\varepsilon}}(\underline{v}_h)\, dx + \int_\Omega p_h \operatorname{div} \underline{v}_h \, dx = \lambda \int_\Omega u_h \cdot v_h \, dx \quad \forall \underline{v}_h \in V_h, \\ \displaystyle\int_\Omega q_h \operatorname{div} \underline{u}_h = 0, \quad \forall q_h \in Q_h. \end{cases} \tag{8.11.2}$$

In order to apply the theory of Sect. 1.2.1, we must check a few conditions. With respect to ellipticity, we have no problem with conforming approximations. The weak approximability (6.5.41) of Q_0^H will surely hold if

$$\inf_{q_h \in Q_h} ||p - q_h||_0 \le \omega_1(h) ||p||_1 \quad \text{for all } p \in H^1(\Omega)/\mathbb{R},$$

which is satisfied by all choices of finite element spaces that one may seriously think to use in practice.

The strong approximability (6.5.42) of V_0^H, which now reads

$$||\underline{u} - \underline{u}^I||_1 \le \omega_2(h) ||\underline{u}||_2 \quad \text{for all } \underline{u} \in V_0^H, \tag{8.11.3}$$

is more delicate as $\underline{u}^I$ has to be chosen in $\operatorname{Ker} B_h$. If the pair (V_h, Q_h) satisfies the *inf-sup* condition, then the property trivially holds.

Remark, however, that the typical way of proving the *inf-sup* condition, using a B-compatible operator (Sect. 8.4.1) *for every* $\underline{u}$, is more difficult than proving (8.11.3) directly. Moreover, there are choices of elements that fail to satisfy the *inf-sup* condition, for which (8.11.3) holds true. For instance, we may think of the $\underline{Q}_1 - P_0$ element of Sect. 8.10.2.

Let us assume, for simplicity, that Ω is a square and that the decomposition $\mathcal{T}_h$ is made by $N \times N$ macro-elements M as in Fig. 8.17. We have seen that this choice of elements does not satisfy the *inf-sup* condition: the operator B_h^t has a non-trivial kernel (the chequerboard mode), and by discarding it, we still have at best a discrete *inf-sup* condition with $\beta_h \sim h$ (see [101,264,314]). Nevertheless, for $u \in V_0^H \subset \operatorname{Ker} B$, we can construct $\underline{u}^I$ as in the construction of $\hat{V}_h$ in Sect. 8.10.2: let $\underline{\hat{u}}_h$ be the vector in $\hat{V}_h$ satisfying (8.10.12)–(8.10.14). It is not difficult to check that $\underline{u}^I = \underline{\hat{u}}_h$ satisfies (8.11.3) with $\omega_2(h) = O(h)$. We have here an example where the eigenvalues are approximated correctly even though the global matrix associated to (6.5.8) is singular. The same kind of construction could be extended to a mesh of macro-elements as in Fig. 8.19.

Remark 8.11.1. We have thus another instance in which the $\underline{Q}_1 - P_0$ element, very popular for incompressible elasticity problems, manages to give an impression of rectitude. □

8.12 Nearly Incompressible Elasticity, Reduced Integration Methods and Relation with Penalty Methods

8.12.1 Variational Formulations and Admissible Discretisations

We have already seen in Chap. 1 and in Remark 8.1.2 that there are difficulties associated to approximations of nearly incompressible materials when using the standard variational principle. This section will be devoted to showing how these problems arise and how they can be cured from a proper mixed formulation. Consider, to make things simpler, a problem with homogeneous Dirichlet conditions,

$$\inf_{\underline{v}\in(H_0^1(\Omega))^n} \mu \int_\Omega |\underline{\underline{\varepsilon}}(\underline{v})|^2\,dx + \frac{\lambda}{2}\int_\Omega |\operatorname{div}\underline{v}|^2\,dx - \int_\Omega \underline{f}\cdot\underline{v}\,dx. \tag{8.12.1}$$

We already noted in Sect. 8.1 that this problem is closely related to the penalty method used to solve the Stokes problem.

It was soon recognised in practice that a brute force use of (8.12.1) could lead, for large values of λ, to bad results, the limiting case being the locking phenomenon that is an identically zero solution.

Example 8.12.1. The simplest case of such a bad situation would be to employ piecewise linear elements. Then, for λ large, (8.12.1) forces the piecewise constant divergence to be almost null on each element, that is, implicitly using the $\underline{P}_1 - P_0$ element of Example 8.3.2. This led to the still persistent idea that triangular or tetrahedral meshes could not be used for elasticity problems. □

A cure was found in using a reduced (that is, inexact) numerical quadrature when evaluating the term $\lambda\int_\Omega |\operatorname{div}\underline{v}|^2\,dx$ associated with compressibility effects. We refer the reader to the papers of [287] and [60] for a discussion of the long history of this idea. We shall rather develop in detail in this example the relations of reduced integrations and mixed methods and try to make clear to what extent they may be claimed to be equivalent. For this, we first recall from Chap. 1 that problem (8.12.1) can be transformed by a straightforward application of duality techniques into a saddle point problem

$$\inf_{\underline{v}}\sup_{q} \mu \int_\Omega |\underline{\underline{\varepsilon}}(\underline{v})|^2\,dx - \frac{2}{2\lambda}\int_\Omega |q|^2\,dx + \int_\Omega q\,\operatorname{div}\underline{v}\,dx - \int_\Omega \underline{f}\cdot\underline{v}\,dx \tag{8.12.2}$$

for which optimality conditions are, denoting $(\underline{u}, p)$ the saddle point,

$$\mu\int_\Omega \underline{\underline{\varepsilon}}(\underline{u}) : \underline{\underline{\varepsilon}}(\underline{v})\,dx + \int_\Omega p\,\operatorname{div}\underline{v}\,dx = \int_\Omega \underline{f}\cdot\underline{v}\,dx \quad \forall\,\underline{v}\in(H_0^1(\Omega))^2, \tag{8.12.3}$$

$$\int_{\Omega} \operatorname{div} \underline{u}\, q\, dx = \frac{1}{\lambda} \int_{\Omega} pq\, dx \quad \forall\, q \in L^2(\Omega). \tag{8.12.4}$$

This is obviously very close to a Stokes problem and is also an example of the perturbed problem studied in Chap. 4, that is: *find $u \in V$ and $q \in Q$ such that*

$$a(u, v) + b(v, p) = (f, v), \quad \forall\, v \in V, \tag{8.12.5}$$

$$b(u.q) - c(p, q) = (g, q), \quad \forall\, q \in Q. \tag{8.12.6}$$

We then know from Chap. 5, Sect. 5.5.2, that an approximation of (8.12.3) and (8.12.4) (that is, a choice of an approximation for both $\underline{u}$ and p) which leads to error estimates independent of λ must be a good approximation for the limiting case $\lambda = 0$.

Remark 8.12.1. The preceding sections of this chapter therefore give us a good idea of what should (or should not) be used as an approximation. All stable elements of Sects. 8.6 and 8.7 can be employed and the choice depends on the choice of solver and the mesh generation algorithm. □

What we shall now see is that reduced integration methods correspond to *an implicit choice* of a mixed approximation with a discontinuous pressure approximation. The success of the reduced integration method will thus rely on the qualities of this underlying mixed method. We have seen in Sect. 8.8.1 that discontinuous pressure imposing exactly the divergence-free condition requires high degree polynomials and special meshes. Reduced integration is then a way of reducing the degree of the underlying pressure in order to hopefully obtain a stable approximation.

8.12.2 Reduced Integration Methods

Let us consider a (more or less) standard approximation of the original problem (8.12.1). An exact evaluation of the "penalty term" $\lambda \int_{\Omega} |\operatorname{div} \underline{v}|^2 dx$ means that for λ large, one tries to get an approximation of $\underline{u}$ which is *exactly* divergence-free. However, as we have already seen, few finite elements can stand such a condition that will in most cases lead to a locking phenomenon due to over-constraining. In a mixed formulation, one relaxes the incompressibility condition by the choice of the approximation for p. Let us now see how this will be translated as a reduced integration method at least in some cases. Let us then consider $V_h \subset V := (H_0^1(\Omega))^n$, $Q_h \subset Q := L_0^2(\Omega)$, these approximation spaces being built from finite elements defined on a partition of Ω. On each element K, let there be given a set of k points x_i and weights ω_i defining a numerical quadrature formula

$$\int_K f(x)\, dx = \sum_{i=1}^{k} \omega_i\, f(x_i). \tag{8.12.7}$$

Remark 8.12.2. It will be convenient to define the numerical quadrature on a reference element K and to evaluate integrals by a change of variables,

$$\int_K f(x)\,dx = \int_{\hat{K}} f(\hat{x})\,J(\hat{x})\,d\hat{x} = \sum_{i=1}^{k} \omega_1\,f(\hat{x}_i)\,J(\hat{x}_i). \tag{8.12.8}$$

The presence of the Jacobian $J(x)$ should be taken into account when discussing the precision of the quadrature rule on K. □

Let us now make the hypothesis that for $\underline{v}_h \in V_h$ and $p_h,\ q_h \in Q_h$, one has exactly

$$\int_K q_h\ \operatorname{div}\underline{v}_h\,dx = \sum_{i=1}^{k} \omega_i\ \hat{q}_h(\hat{x}_i)\widehat{\operatorname{div}\underline{v}_h}(\hat{x}_i)\ J(\hat{x}_i) \tag{8.12.9}$$

and

$$\int_K p_h\,q_h\,dx = \sum_{k=1}^{k} \omega_i\ \hat{p}_h(\hat{x}_i)\ \hat{q}_h(\hat{x}_i)\ J(\hat{x}_i). \tag{8.12.10}$$

Let us now consider the discrete form of (8.12.4),

$$\int_\Omega \operatorname{div}\underline{u}_h\ q_h\,dx = \frac{1}{\lambda}\int_\Omega p_h\,q_h\,dx,\ \forall\, q_h \in Q_h. \tag{8.12.11}$$

When the space Q_h is built from discontinuous functions, this can be read element by element as

$$\int_K q_h\ \operatorname{div}\underline{u}_h\,dx = \frac{1}{\lambda}\int_K p_h\,q_h\,dx\ \forall\, q_h \in Q_h, \tag{8.12.12}$$

so that using (8.12.9) and (8.12.10), one gets

$$\hat{p}_h(\ddot{x}_i) = \lambda\widehat{\operatorname{div}\underline{u}_h}(\hat{x}_l)\ \text{or}\ p_h(x_i) = \lambda\operatorname{div}\underline{u}_h(x_i). \tag{8.12.13}$$

Formula (8.12.8) can in turn be used in the discrete form of (8.12.3) which now gives

$$2\mu\int_\Omega \underline{\underline{\varepsilon}}(\underline{u}_h) : \underline{\underline{\varepsilon}}(\underline{v}_h)\,dx + \lambda\sum_K\Big(\sum_{i=1}^{k}\omega_i\ J(\hat{x}_i)(\widehat{\operatorname{div}\underline{u}_h}(\hat{x}_i))(\widehat{\operatorname{div}\underline{v}_h}(\hat{x}_i))\Big)$$
$$= \int_\Omega \underline{f}\cdot\underline{v}_h\,dx. \tag{8.12.14}$$

In general, the term $\sum_K \left(\sum_{i=1}^{k} \omega_i \; J(\hat{x}_i)(\widehat{\mathrm{div}\underline{u}_h}(\hat{x}_i))(\widehat{\mathrm{div}\underline{v}_h}(\hat{x}_i)) \right)$ *is not* an exact evaluation of $\int_\Omega \mathrm{div}\underline{u}_h \; \mathrm{div} \; \underline{v}_h \; dx$ and reduced integration is effectively introduced. In the case where (8.12.9) and (8.12.10) hold, there is a perfect equivalence between the mixed method and the use of reduced integration. Whatever will come from one can be reduced to the other one. It will however not be possible, in general, to get equalities (8.12.9) and (8.12.10) and therefore, a further analysis will be needed. However, we shall first consider some examples of this complete equivalence case.

Example 8.12.2. Let us consider the $\underline{Q}_1 - P_0$ approximation on a *rectangle* and a one-point quadrature rule. It is clear that $\mathrm{div}\, \underline{u}_h \in P_1(K)$ and is integrated exactly. In the same way, a one-point rule is exact for $\int_\Omega p_h \; q_h \; dx$ whenever $p_h, \; q_h \in P_0(K)$. There is thus a perfect equivalence between reduced integration and the exact penalty method defined by (8.12.11). □

Example 8.12.3. We now consider again the same $\underline{Q}_1 - P_0$ element on a general quadrilateral. To show that we still have equivalence requires a somewhat more delicate analysis. Indeed, at first sight, the quadrature rule is not exact for $\int_{\hat{K}} \widehat{\mathrm{div}\, \underline{u}_h} \; J_K(\hat{x}) \; d\hat{x}$. Let us however consider in detail the term $\widehat{\mathrm{div}\, \underline{u}_h} = \widehat{\frac{\partial u_1}{\partial x_1}} + \widehat{\frac{\partial u_2}{\partial x_2}}$. Let $B = DF$ be the Jacobian matrix of the transformation F from $\hat{K}$ into K. Writing explicitly

$$F = \begin{cases} a_0 + a_1\hat{x} + a_2\hat{y} + a_3\hat{x}\hat{y} \\ b_0 + b_1\hat{x} + b_2\hat{y} + b_3\hat{x}\hat{y}, \end{cases} \tag{8.12.15}$$

one has

$$B = \begin{pmatrix} a_1 + a_3\hat{y} & b_1 + b_3\hat{y} \\ a_2 + a_3\hat{x} & b_2 + b_3\hat{y} \end{pmatrix} \tag{8.12.16}$$

so that we get

$$B^{-1} = \frac{1}{J(\hat{x})} \begin{pmatrix} b_2 + b_3\hat{x} & -b_1 - b_3\hat{y} \\ -a_2 - a_3\hat{x} & a_1 + a_3\hat{y} \end{pmatrix}. \tag{8.12.17}$$

However,

$$\widehat{\frac{\partial u_1}{\partial x_1}} = \left(\frac{\partial \hat{u}_1}{\partial \hat{x}_1}(b_2 + b_3\hat{x}) - \frac{\partial \hat{u}_1}{\partial \hat{x}_2}(b_1 - b_3\hat{y}) \right) \frac{1}{J(\hat{x})}, \tag{8.12.18}$$

$$\widehat{\frac{\partial u_2}{\partial x_2}} = \left(\frac{\partial \hat{u}_2}{\partial \hat{x}_1}(-a_2 - a_3\hat{x}) + \frac{\partial \hat{u}_2}{\partial \hat{x}_2}(a_1 + a_3\hat{y}) \right) \frac{1}{J(\hat{x})}. \tag{8.12.19}$$

When computing $\int_{\hat{K}} \widehat{\mathrm{div}\, \underline{u}_h} \; J(\hat{x}) \; d\hat{x}$, the Jacobians cancel and one is left with the integral of a function which is linear in each variable and which can be computed exactly by a one-point formula. □

Example 8.12.4. Using a 4-point integration formula on a straight-sided quadrilateral can be seen, as in the previous example, to be exactly equivalent to a $\underline{Q}_2 - Q_1$ approximation [59, 60]. □

The above equivalence is, however, not the general rule. Consider the following examples.

Example 8.12.5. We want to use a reduced integration procedure to emulate the Crouzeix-Raviart element (cf. Sect. 8.6.2). To define a P_1 pressure, we need three integration points which can generate a formula that will be exact for second degree polynomials (but not more). The bubble function included in velocity, however, makes that $\operatorname{div} \underline{u}_h \in P_2(K)$ and $\int_K \operatorname{div} \underline{u}_h\, q_h dx$ will not be evaluated exactly. □

Example 8.12.6. A full isoparametric $\underline{Q}_2 - Q_1$ element is not equivalent to its four-point reduced integration analogue. □

Example 8.12.7. A $\underline{Q}_2 - P_0$ approximation is not, even on rectangles, equivalent to a one-point reduced integration method, for $\operatorname{div} \underline{u}_h$ contains second order terms which are not taken into account by a one-point quadrature. □

8.12.3 *Effects of Inexact Integration*

If we now consider into more details the cases where a perfect equivalence does not hold between the mixed method and some reduced integration procedure, we find ourselves in the setting of Sect. 5.5.4. In particular, $b(\underline{v}_h, q_h)$ is replaced by an approximate bilinear form $b_h(\underline{v}_h, q_h)$. We shall suppose, to simplify, that the scalar product on Q_h is exactly evaluated. Two questions must then be answered.

- Does $b_h(.,.)$ satisfy the *inf-sup* condition?
- Do error estimates still hold without loss of accuracy?

We have already introduced in Sect. 5.5.4 a general setting in which this situation can be analysed. We shall first apply Proposition 5.5.8 in order to check the *inf-sup* condition for two examples and we shall give an example where an inexact integral changes the nature of the problem. We shall then consider consistency error for those three examples.

Example 8.12.8. We in fact come back to Example 8.12.7 and study on a rectangular mesh the $\underline{Q}_2 - P_0$ approximation (see Sect. 8.6.3) with a one-point quadrature rule. This is not, as we have said, equivalent to the standard $\underline{Q}_2 - P_0$ approximation. We now want to check, using Proposition 8.4.1, that it satisfies the *inf-sup* condition. We thus have to build a continuous operator (in $H^1(\Omega)$-norm) such that

$$\int_\Omega \operatorname{div} \underline{u}_h\, q_h\, dx = \sum_K [(\operatorname{div} \Pi_h \underline{u}_h)(M_{0,K}) q_K] \operatorname{area}(K) \qquad (8.12.20)$$

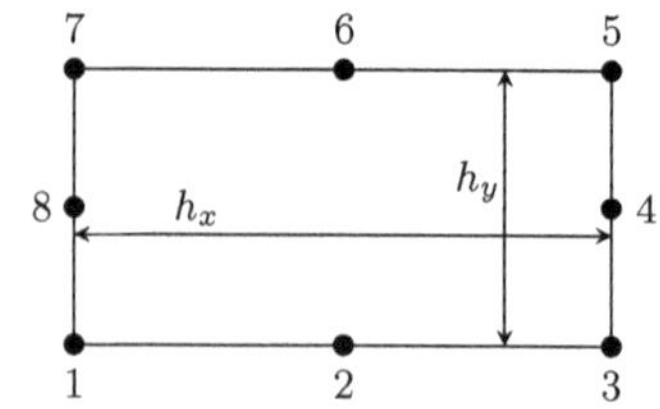

Fig. 8.24 Numbering for the Q_2 element

where $M_{0,K}$ is the barycentre of K and q_K the restriction of q_h to K. As q_h is discontinuous, we can restrict our analysis to one element and we study both sides of equality (8.12.20). We have of course, taking $q_K = 1$,

$$\int_K \operatorname{div} \underline{u}_h \, dx = \int_{\partial K} \underline{u}_h \cdot \underline{n} \, d\sigma. \tag{8.12.21}$$

Using the numbering of Fig. 8.24 and denoting by u_i, v_i the horizontal and vertical components of velocity at node i, we can write (8.12.21) by Simpson's quadrature rule in the form

$$\begin{aligned}\int_K \operatorname{div} \underline{u}_h \, dx = {} & \frac{h_y}{6}[u_5 + 4u_4 + u_3] - \frac{h_y}{6}[u_1 + 4u_8 + u_7] \\ & + \frac{h_x}{6}[v_7 + 4v_6 + v_5] - \frac{h_x}{6}[v_1 + 4v_2 + v_3].\end{aligned} \tag{8.12.22}$$

If we write

$$u_4 = \frac{u_5 + u_3}{2} + \hat{u}_4, \; u_8 = \frac{u_1 + u_7}{2} + \hat{u}_8 \tag{8.12.23}$$

$$v_0 = \frac{v_5 + v_7}{2} + \hat{v}_6, \; v_2 = \frac{v_1 + v_3}{2} + \hat{v}_2, \tag{8.12.24}$$

where $\hat{u}_4, \hat{u}_6, \hat{v}_6$ and $\hat{v}_2$ are corrections with respect to a bilinear interpolation, we may rewrite (8.12.22) as

$$\begin{aligned}\int_K \operatorname{div} \underline{u}_h \, dx = {} & \frac{h_y}{2}[u_5 + u_3 + \frac{4}{3}\hat{u}_4] - \frac{h_y}{2}[u_1 + u_7 + \frac{4}{3}\hat{u}_8] \\ & + \frac{h_y}{2}[v_7 + v_5 + \frac{4}{3}\hat{v}_6] - \frac{h_x}{2}[v_1 + v_3 + \frac{4}{3}\hat{v}_2].\end{aligned} \tag{8.12.25}$$

On the other hand, area (K) $\operatorname{div} \underline{u}_h(M_{0,K})$ can be seen to be equal to

$$\begin{aligned}& \frac{h_y}{2}[u_5 + u_3 + 2\hat{u}_4] - \frac{h_y}{2}[u_1 + u_7 + 2\hat{u}_8] \\ & -\frac{h_x}{2}[u_7 + v_5 + 2\hat{v}_6] - \frac{h_x}{2}[v_1 + v_3 + 2\hat{u}_2].\end{aligned} \tag{8.12.26}$$

If we thus split $\underline{u}_h$ into a bilinear part $\underline{u}_h^0$ and a mid-point correction part $\hat{\underline{u}}_h$, one can define $\Pi_h\underline{u}_h$ by setting

$$\begin{cases} (\Pi_h\underline{u}_h)^0 = \underline{u}_h^0, \\ (\widehat{\Pi_h\underline{u}_h}) = \dfrac{2}{3}\hat{\underline{u}}_h. \end{cases} \tag{8.12.27}$$

Equality (8.12.21) will then hold and (8.12.27) is clearly continuous with a continuity constant independent of h. □

Example 8.12.9. We come back to Example 8.12.5 that is a three-point quadrature rule used in conjunction with the Crouzeix-Raviart element. We shall not give the analysis in detail but only sketch the ideas. The problem is again to check that the *inf-sup* condition holds through Proposition 8.4.1. As the quadrature rule *is exact* when q_h is *piecewise constant*, the obvious idea is to build $\Pi_h\underline{u}_h$ by leaving invariant the trace of $\underline{u}_h$ on ∂K and *only modifying the coefficients of the bubble functions*. This can clearly be done. Continuity is now to be checked and the proof is essentially the same as the standard proof of the *inf-sup* condition (Sect. 8.7.2). □

Example 8.12.10 (A modified $\underline{Q}_1 - P_0$ element). We now present a puzzling example [127] of an element which is stable but for which convergence is tricky due to a consistency error term. We have here a case where using a one-point quadrature rule will change the situation with respect to the *inf-sup* condition. In fact, it will make a stable element from an unstable one but will also introduce an essential change in the problem. The departure point is thus the standard $Q_1 - P_0$ element which was studied in Sect. 8.10.2 and which, as we know, does not satisfy the *inf-sup* condition. We now make it richer by adding to velocity $\underline{u}_h|_K = \{u_1, u_2\}$ what we shall call wave functions. On the reference element $\hat{K} =]-1,1[\times]-1,1[$, those functions are defined by

$$\begin{cases} w_1 = \hat{x}\, b_2(\hat{x},\hat{y}), \\ w_2 = \hat{y}\, b_2(\hat{x},\hat{y}), \end{cases} \tag{8.12.28}$$

where $b_2(\hat{x},\hat{y}) = (1-\hat{x}^2)(1-\hat{y}^2)$ is the Q_2 bubble function. If we now consider

$$\hat{\underline{u}}_h|_K = \{u_1 + \alpha_K w_1, u_2 + \alpha_K w_2\} = \underline{u}_h|_K + \alpha_k \underline{w}_k, \tag{8.12.29}$$

we obtain a new element with an internal degree of freedom. The wave functions that we added vanish on the boundary and nothing is changed for the stability of the mixed method with exact integration. If we rather use a one-point quadrature rule, things become different. We shall indeed check that the modified bilinear form $b_h(\hat{\underline{v}}_h, q_h)$ satisfies the *inf-sup* condition. We thus have to show that

$$\sup_{\hat{\underline{u}}_h} \frac{\sum_K \operatorname{div}\hat{\underline{u}}_h(M_{0,K})p_K\, h_K^2}{\|\hat{\underline{u}}_h\|_1} \geq k_0\, \|p_h\|_0. \tag{8.12.30}$$

This is easily checked by posing on K (we suppose that we have a rectangular mesh to make things simpler)

$$\hat{\underline{u}}_h|_K = h_K\ p_K \underline{w}_K. \tag{8.12.31}$$

We then have $\operatorname{div} \hat{\underline{u}}_h = p_h$ and

$$\|\hat{\underline{u}}_h\|_{1,K} = h\ p_K\ \|\underline{w}_K\|_{1,K}, \tag{8.12.32}$$

which implies

$$\|\underline{u}_h\|_1 \leq c\ \|p_h\|_0, \tag{8.12.33}$$

and (8.12.30) follows. A remarkable point is, now, that even the hydrostatic mode has disappeared. This is an indication that something incorrect has been introduced in the approximation. An analysis of *consistency error* indeed shows that usual error estimates fail and that we are actually approximating a continuous problem in which the incompressibility condition has been replaced by $\operatorname{div} \underline{u} + kp = 0$ where $k = a/b$. We then see that if, in general for the Stokes problem, making the space of velocities richer improves (at least does not reduce) the quality of the method, this fact can become false when numerical integration is used. □

Let us now turn our attention to the problem of error estimation. From Proposition 5.5.6 and Remark 5.5.9, all we have to do is to estimate the consistency terms,

$$\sup_{\underline{v}_h} \frac{|b(\underline{v}_h, p) - b_h(\underline{v}_h, p)|}{\|\underline{v}_h\|_V} \tag{8.12.34}$$

and

$$\sup_{q_h} \frac{|b(\underline{u}, q_h) - b_h(\underline{u}, q_h)|}{\|q_h\|_0}. \tag{8.12.35}$$

We thus have to estimate quadrature errors. It would be out of purpose to enter into details, and we refer the reader to [147, 148] where examples of such analysis are presented exhaustively. The first step is to transform (8.12.34) into a form which is sometimes more tractable. We may indeed write

$$\begin{aligned} b(\underline{v}_h, p) - b_h(\underline{v}_h, p) = {} & (b_h(\underline{v}_h, p - q_h) - b_h(\underline{v}_h, p - q_h)) \\ & + (b(\underline{v}_h, q_h) - b_h(\underline{v}_h, q_h)) \end{aligned} \tag{8.12.36}$$

and

$$\begin{aligned} b(\underline{u}, q_h) - b_h(\underline{u}, q_h) = {} & (b(\underline{u} - \underline{v}_h, q_h) - b_h(\underline{u} - \underline{v}_h, q_h)) \\ & + (b(\underline{v}_h, q_h) - b_h(\underline{v}_h, q_h)). \end{aligned} \tag{8.12.37}$$

The first parenthesis in the right-hand side of (8.12.36) and (8.12.37) can be reduced to an approximation error. The second parenthesis implies only polynomials.

Let us therefore consider (8.12.37) for the three approximations introduced above. For the Crouzeix-Raviart triangle, taking $\underline{v}_h$ the standard interpolate of $\underline{u}$ makes the second parenthesis vanish while the first yields an $O(h)$ estimate. For the two other approximations, taking $\underline{v}_h$ to be a standard bilinear approximation of $\underline{u}$ makes the second parenthesis vanish while the first yields an $O(h)$ estimate, which is the best that we can hope for anyway. The real trouble is therefore with (8.12.34), with or without (8.12.36). In the case of the Crouzeix-Raviart triangle, we can use directly (8.12.34) and the following result of [147, 148].

Proposition 8.12.1. *Let $f \in W_{k,q}(\Omega)$, $p_k \in P_k(K)$ and denote $E_k(fp_k)$ the quadrature error on element K when numerical integration is applied to fp_k. Let us suppose that $E_K(\hat{\phi}) = 0\ \forall\, \hat{\phi} \in P_{2k-2}(K)$. Then, one has*

$$|E_K(fp_k)| \leq c h_K^k\ (meas(K))^{\frac{1}{2}-\frac{1}{9}} |f|_{k,q,K} |p_k|_1. \tag{8.12.38}$$

□

Taking $k = 2$, $q = \infty$ and using the inverse inequality to go from $|p_k|_1$ to $|p_k|_0$, one gets an $O(h^2)$ estimate for (8.12.34).

The two other approximations cannot be reduced to Proposition 8.12.1 and must be studied through (8.12.36). We must study a term like

$$\sup_{\underline{v}_h} \frac{|b(\underline{v}_h, q_h) - b_h(\underline{v}_h, q_h)|}{\|\underline{v}_h\|_1}. \tag{8.12.39}$$

This can at best be *bounded*. For instance in the case of the $\underline{Q}_2 - P_0$ approximation, we can check by hand that the quadrature error on K reduces to $h_K^3\ |\operatorname{div} \underline{v}_h|_{2,K}\ p_k$.

8.13 Other Stabilisation Procedures

We shall now consider, for the Stokes problem, stabilised formulations presented in Sect. 6.1.1 of Chap. 6. It is clear or should be clear from the results presented in the present chapter that the key of success in stabilising incompressible elements is in weakening the discrete divergence-free condition. This was done, up to now, by reducing the space Q_h of pressures or by enriching the space V_h of velocity field. We now consider the other possibility of a modified variational formulation. In many cases, this will amount to explicitly weaken the condition $\operatorname{div}_h\ \underline{u}_h = 0$ by changing it to

$$\operatorname{div}_h\ \underline{u}_h = g_h, \tag{8.13.1}$$

where g_h is a (well chosen) "small" function. One step in this direction had been done in the work of [131] who considered the relaxed condition

$$\int_{\Omega} \operatorname{div} \underline{u}_h \, q_h \, dx = \beta \sum_K h_K^2 \int_K \underline{\mathrm{grad}} \; p_h \cdot \underline{\mathrm{grad}} \; q_h \, dx \tag{8.13.2}$$

in the case of a continuous pressure approximation (that is $Q_h \subset H^1(\Omega)$). In (8.13.2), β is any positive real number. On a regular mesh, this is a discrete form of

$$\operatorname{div} \underline{u} = -\beta h^2 \, \Delta p. \tag{8.13.3}$$

It is easy to understand that appearance of oscillations due to spurious pressure modes will make $\Delta_h \; p_h$ large. This will relax the divergence-free condition, thus preventing the growth of such oscillations. The practical use of (8.13.2) indeed requires a delicate balance between two conflicting phenomena. When α is chosen too small, stabilisation is poor and spurious pressure modes persist. On the other hand, taking α too large spoils the value of p_h near the boundary because of the parasitic Neumann condition $\frac{\partial p_h}{\partial n} = 0$ which is implicit in (8.13.2).

This procedure was later generalised by Hughes and Franca [256] and Hughes et al. [257] in order to improve its consistency, in a way that we present below.

We shall first try to give a unified presentation of this kind of methods using the general theory of stabilisation procedures developed in Chap. 5. We shall first consider augmented methods.

8.13.1 Augmented Method for the Stokes Problem

We consider the stabilised formulation (6.1.50) in the special context of the Stokes problem (1.5.23). We now have

$$V := (H_0^1(\Omega))^n, V' := (H^{-1}(\Omega))^n, H := (L^2(\Omega))^n, Q = Q' := L^2(\Omega)$$

and the dense inclusions $V \subset H \subset V'$. We also have, for $\underline{u} \in V$ and $p \in Q$,

$$\begin{aligned} b(\underline{v}, q) &= \textstyle\int_{\Omega} q \, \operatorname{div} \underline{v} \, dx, \\ B\underline{u} &= -\operatorname{div} \underline{u} \in Q, \; B^t p = \underline{\mathrm{grad}} \, p \in V'. \end{aligned} \tag{8.13.4}$$

As to the operator A, it is defined by

$$\langle A\underline{u}, \underline{v} \rangle = a(\underline{u}, \underline{v}) = 2\mu \int_{\Omega} \underline{\underline{\varepsilon}}(u) : \underline{\underline{\varepsilon}}(v) dx. \tag{8.13.5}$$

The case $t = 1$ of Sect. 6.1.2 (with $\beta_2 = 0$) which corresponds to a stabilisation method introduced by Douglas and Wang [179] now reads as

$$\begin{cases} \langle A\underline{u}_h + \underline{\mathrm{grad}}\, p_h - \underline{f}, \underline{v}_h \rangle_{V' \times V} \\ \qquad\qquad + \beta(A\underline{u}_h + \underline{\mathrm{grad}}\, p_h - \underline{f}, A\underline{v}_h)_{V' \times V'} = 0, \quad \forall\, \underline{v}_h \in V_h, \\ \langle \mathrm{div}\, \underline{u}_h + g, q \rangle_{Q' \times Q} \\ \qquad + \beta(A\underline{u}_h + \underline{\mathrm{grad}}\, p_h - \underline{f}, \underline{\mathrm{grad}}\, q_h)_{V' \times V'} = 0, \quad \forall\, q_h \in Q_h, \end{cases} \tag{8.13.6}$$

while in the case $t = 0$, the extra term is only written in the second equation

$$\begin{cases} \langle A\underline{u}_h + \underline{\mathrm{grad}}\, p_h - \underline{f}, \underline{v}_h \rangle_{V' \times V} = 0, \quad \forall\, \underline{v}_h \in V_h, \\ \langle \mathrm{div}\, \underline{u}_h + g, q \rangle_{Q' \times Q} \\ \qquad + \beta(A\underline{u}_h + \underline{\mathrm{grad}}\, p_h - \underline{f}, \underline{\mathrm{grad}}\, q_h)_{V' \times V'} = 0, \quad \forall\, q_h \in Q_h. \end{cases} \tag{8.13.7}$$

We must now build a computable implementation of these formulations. Indeed, the scalar product in V' is not directly handable. In the present case, as the operator A is an isomorphism from V onto V', we can define the scalar product as

$$(\underline{u}', \underline{v}')_{V' \times V'} := \langle A^{-1}\underline{u}', \underline{v}' \rangle_{V \times V'}. \tag{8.13.8}$$

However, this means being able to compute the exact inverse of A. We now have to introduce an approximation and many options are open.

Example 8.13.1 (Defining a scalar product by an approximation of $\mathbf{A}^{-1}$*).* The first idea that comes to mind is to use some approximate operator S_h^{-1} instead of A^{-1}. This could be done by solving an auxiliary problem in a space richer than V_h. Our problem (8.13.7) would now be changed into

$$\begin{cases} \langle A\underline{u}_h + \underline{\mathrm{grad}}\, p_h - \underline{f}, \underline{v}_h \rangle_{V' \times V} = 0, \quad \forall\, \underline{v}_h \in V_h, \\ \langle \mathrm{div}\, \underline{u}_h + g, q \rangle_{Q' \times Q} \\ \qquad + \beta\langle S_h^{-1}(A\underline{u}_h + \underline{\mathrm{grad}}\, p_h - \underline{f}), \underline{\mathrm{grad}}\, q_h \rangle_{V \times V'} = 0, \quad \forall\, q_h \in Q_h, \end{cases} \tag{8.13.9}$$

with a similar expression for (8.13.6). □

Example 8.13.2 (Defining a scalar product by a change of space). Another way of defining a discrete formulation, introduced in [256] and [257], is to replace

$$\beta(A\underline{u}_h + \underline{\mathrm{grad}}\, p_h - \underline{f}, \underline{\mathrm{grad}}\, q_h)_{V' \times V'} \tag{8.13.10}$$

by an expression of the form

$$\beta \sum_K h_K^2 \int_K (A\underline{u}_h|_K + \underline{\mathrm{grad}}\ p_h - \underline{f}) \cdot \underline{\mathrm{grad}}\ q_h\, dx \tag{8.13.11}$$

and by a similar change on the term appearing in the first equation of (8.13.6) if we want to use $t = 1$. This can be seen as another way (through an "inverse inequality") of defining a discrete scalar product corresponding to the $H^{-1}(\Omega)$ scalar product by using a scalar product in H. If the exact solution $\underline{u}$ is regular enough, this expression written for $\underline{u}$ vanishes and we have what has been termed as "strong consistency". Note that for piecewise linear $\underline{u}_h$, $A\underline{u}_h|_K$ vanishes, leaving only a corrective term of the form

$$\beta \sum_K h_K^2 \int_k (\underline{\text{grad}}\ p_h - \underline{f}) \cdot \underline{\text{grad}}\ q_h\, dx, \tag{8.13.12}$$

which is a variant of (8.13.2). The aim of (8.13.11) is largely to replace the Neumann condition $\frac{\partial p_h}{\partial n} = 0$, which is implicit in (8.13.2), by a more correct one, hopefully making the choice of β easier. □

We now come back to our discrete stabilised problems (8.13.6) or (8.13.9) and we first consider the option of using a discrete operator S_h^{-1}. We shall see that in one important case, both options are equivalent, and that in other cases, we fall back onto known methods.

8.13.2 Defining an Approximate Inverse S_h^{-1}

Let V_h be the finite element space in which we compute $\underline{u}_h$. We introduce a space V_h^+ of new degrees of freedom and the space $W_h := V_h \oplus V_h^+$. We can now define $\underline{s}_h^+ := S^{-1}(A\underline{u}_h + \underline{\text{grad}}\, p_h - \underline{f}) \in V_h^+$ by a "hierarchical" computation in V_h^+

$$a(\underline{s}_h^+, \underline{v}_h^+) = \langle A\underline{u}_h + \underline{\text{grad}}\, p_h - \underline{f}, \underline{v}_h^+ \rangle, \quad \forall\, \underline{v}_h^+ \in V_h^+. \tag{8.13.13}$$

In many cases, V_h^+ will be a space of "bubbles" and this problem will be solvable element by element. For the case $t = 0$, our stabilised problem (8.13.7) would now be read, $\underline{s}_h^+$ being defined from (8.13.13), as

$$\begin{cases} \langle (A\underline{u}_h + \underline{\text{grad}}\, p_h - \underline{f}), \underline{v}_h \rangle_{V' \times V} = 0, & \forall\, \underline{v}_h \in V_h, \\ \langle \text{div}\,(\underline{u}_h + \beta \underline{s}_h^+) + g, q \rangle_{Q' \times Q} = 0, & \forall\, q_h \in Q_h, \end{cases} \tag{8.13.14}$$

which indeed contains a weakened condition of the form (8.13.1). The case $t = 1$ of (8.13.6) would yield

$$\begin{cases} \langle A(\underline{u}_h + \beta_1 \underline{s}_h^+) + \underline{\text{grad}}\, p_h - \underline{f}, \underline{v}_h \rangle_{V' \times V} = 0, & \forall\, \underline{v}_h \in V_h, \\ \langle \text{div}\,(\underline{u}_h + \beta_1 \underline{s}_h^+) + g, q \rangle_{Q' \times Q} = 0, & \forall\, q_h \in Q_h. \end{cases} \tag{8.13.15}$$

Taking into account equation (8.13.13), one can see that for $\beta = 1$, this is nothing but the solution of the Stokes problem using $W_h \times Q_h$ as the finite element space. This will obviously work if this choice of spaces is stable and we have rediscovered that enriching the space V_h is a good way of getting a stable method. As for the case (8.13.14), it would be an approximation of this case, which could hardly be considered as simpler as the matrix of the problem is not symmetric. Proving its stability would require some "quasi-orthogonality" between V_h and V_h^+ of the same type as what is used to study hierarchical error estimators [2, 47]. We shall not try to get much further in this direction. There is however a simple case where both (8.13.14) and (8.13.15) coincide and in fact are equivalent to (8.13.12).

Example 8.13.3 (Defining $\mathbf{S_h^{-1}}$ *with bubbles, the MINI element).* Let us consider the case of a piecewise linear approximation for both V_h and Q_h, which is well known to be unstable. To fix ideas, we shall use for V_h^+ the space B_3 of conforming cubic bubbles, although we might also use other shapes of bubbles or nonconforming quadratic bubbles. Let then b_K be the bubble associated to element K. We can write any function of V_h^+ in the form

$$\underline{v}_h^+ = \sum_K \underline{\beta}_K b_K.$$

The key of what follows is the fact that we have orthogonality between the space of bubbles and V_h in the sense that

$$a(\underline{s}_h^+, v_h) = a(v_h, \underline{v}_h^+) = 0 \quad \forall\, v_h \in V_h, \forall\, \underline{v}_h^+ \in V_h^+. \tag{8.13.16}$$

As we use bubbles, our Eq. (8.13.13) can be solved element by element and we have on every K

$$\delta_K \underline{\beta}_K = \int_K (\underline{f} - A\underline{u}_h - \underline{\mathrm{grad}}\, p_h) \cdot \underline{b}_K\, dx = \int_K (\underline{f} - \underline{\mathrm{grad}}\, p_h) \cdot \underline{b}_K\, dx \tag{8.13.17}$$

where

$$\delta_K = \mu \int_K |\underline{\underline{\varepsilon}}(\underline{b}_K)|^2\, dx. \tag{8.13.18}$$

To make things easier, suppose that f is piecewise constant so that we can rewrite (8.13.17) as

$$\delta_K \underline{\beta}_K = \gamma_K(\underline{f} - \underline{\mathrm{grad}}\, p_h) \tag{8.13.19}$$

where we denote

$$\gamma_K = \int_\Omega b_K\, dx. \tag{8.13.20}$$

We thus obtain

$$S^{-1}(A\underline{u}_h + \underline{\mathrm{grad}}\, p_h - \underline{f}) = \sum_K (\gamma_K/\delta_K)(\underline{f} - \underline{\mathrm{grad}}\, p_h) b_K. \tag{8.13.21}$$

Using (8.13.21), (8.13.14) becomes

$$\begin{cases} \langle A\underline{u}_h + \underline{\mathrm{grad}}\, p_h + \underline{f}, \underline{v}_h\rangle_{V'\times V} = 0, \quad \forall\, \underline{v}_h \in V_h, \\ \langle \mathrm{div}\, \underline{u}_h + g, q\rangle_{Q'\times Q} \\ \qquad - \beta \sum_K (\gamma_K^2/\delta_K)(\underline{\mathrm{grad}}\, p_h - \underline{f}, \underline{\mathrm{grad}}\, q_h)_{L^2(K)\times L^2(K)} = 0, \ \forall\, q_h \in Q_h, \end{cases} \tag{8.13.22}$$

and this is nothing but a slight variant of the stabilisation obtained from (8.13.11), as we can check that

$$\gamma_K^2/\delta_K = c_K h_K^2 \tag{8.13.23}$$

with a constant c_K depending on the shape of the element. The same reasoning can be done with other choices for the space of bubbles.

Remark 8.13.1 (The MINI element). Given the orthogonality of (8.13.16), it is easy to see that the formulations of (8.13.6) and (8.13.7) obtained from (8.13.17) coincide and that for $\beta = 1$, they are nothing but the solution of the Stokes problem with the MINI element. □

We thus see that the technique of Example 8.13.2 can be obtained in different ways. We now proceed to develop an error analysis of these methods.

Example 8.13.4 (Error estimates for the Hughes-Franca stabilisation). We place ourselves in the case of "equal interpolation", that is, using polynomials in $\mathcal{L}_k^1$ for V_h and Q_h. Note however that the space V_h will satisfy boundary conditions while Q_h will not. We present the result in the two-dimensional case but it can easily be extended to the three-dimensional case. We have a space of continuous pressures and we have thus $\underline{\mathrm{grad}}\, p_h \in H = (L^2(\Omega))^2$. To simplify the presentation, we define on H

$$\langle \underline{u}, \underline{v}\rangle_H = \sum_K h_K^2 \int_K \underline{u}\cdot\underline{v}\, dx, \quad [\underline{v}]_H^2 = \langle \underline{v}, \underline{v}\rangle_H. \tag{8.13.24}$$

For any $\underline{u}_h \in V_h$, we also define $\underline{Au_h} \in H$ by

$$\underline{Au_h}|_K = A\underline{u}|_K \tag{8.13.25}$$

and we write a stabilised formulation

$$\begin{cases} \langle A\underline{u}_h + \underline{\mathrm{grad}}\, p_h - \underline{f}, \underline{v}_h\rangle_{V'\times V} = 0, \quad \forall\, \underline{v}_h \in V_h, \\ \langle \mathrm{div}\, \underline{u}_h + g, q\rangle_{Q'\times Q} + \langle (\underline{Au_h}|_K + \underline{\mathrm{grad}}\ p_h - \underline{f}), \underline{\mathrm{grad}}\ q_h\rangle_H = 0, \ \forall\, q_h \in Q_h, \end{cases} \tag{8.13.26}$$

which is the method of [256].

Following the procedure of Chap. 5, Sect. 6.1.2, we must obtain a stability result and estimate the consistency term. For stability, we define the bilinear form

$$\begin{aligned}\mathcal{A}\langle(\underline{u}_h, p_h), (\underline{v}_h, q_h)\rangle &:= a(\underline{u}_h, \underline{v}_h) - b(\underline{v}_h, p_h) + b(\underline{u}_h, q_h) \\ &\quad + \beta\langle(\underline{Au}_h + \underline{\text{grad}}\ p_h), \underline{\text{grad}}\ q_h\rangle_H.\end{aligned} \tag{8.13.27}$$

For $\underline{u}_h = \underline{v}_h$ and $p_h = q_h$, we obtain

$$\mathcal{A}\langle(\underline{v}_h, q_h), (\underline{v}_h, q_h)\rangle = \alpha\|\underline{v}_h\|_V^2 + \beta[\underline{\text{grad}}\, q_h]_H^2 + \beta\langle \underline{Au}_h|_K, \underline{\text{grad}}\, q_H\rangle_H. \tag{8.13.28}$$

To prove stability, we recall that from Verfürth's trick (cf. Sect. 8.5.2 and also Chap. 6.3), we have

$$[\underline{\text{grad}}\, q_h]_h \geq k\|q_h\|_Q. \tag{8.13.29}$$

On the other hand, we bound the last term by

$$\beta\langle A\underline{u}_h|_K, \underline{\text{grad}}\, q_H\rangle_H \leq \beta[A\underline{u}_h]_h[\,\underline{\text{grad}}\, q_h]_H \leq \frac{\beta}{2}[A\underline{u}_h]_h^2 + \frac{\beta}{2}[\underline{\text{grad}}\, q_h]_h^2. \tag{8.13.30}$$

However, using an inverse inequality, we have

$$\int_K |A\underline{u}_h|^2\, dx \leq M\frac{1}{ch_K^2}\|\underline{u}_h\|_{1,K}^2 \tag{8.13.31}$$

and thus

$$\sum_K h_K^2 \int_K |A\underline{u}_h|^2\, dx \leq \frac{M}{c}\|\underline{u}_h\|_V^2. \tag{8.13.32}$$

Using this last result in (8.13.28), we have

$$\mathcal{A}\langle(\underline{v}_h, q_h), (\underline{v}_h, q_h)\rangle(\alpha - \frac{\beta M}{c})\|\underline{v}_h\|_V^2 + \beta[\,\underline{\text{grad}}\, q_h]_H^2, \tag{8.13.33}$$

which implies stability for β small enough. It should be remarked that for the degree of the approximation $k = 1$, $A\underline{u}_h|_K = 0$ and that we then have stability for any value of β.

Following Sect. 6.1.1 of Chap. 6, we now have to bound, $(\underline{u}_I, p_I)$ being an interpolate of $(\underline{u}, p)$, a term of the form

$$\beta\sum_K h_K^2 \int_K \left((|(A\underline{u}_I - A\underline{u})|_K|^2) + |p - p_I|^2\right) dx. \tag{8.13.34}$$

The crucial term is the first one. If polynomials of degree k are employed, we have, by classical interpolation results, an estimate on K,

$$\int_K |(A\underline{u}_I - A\underline{u})|_K|^2\, dx = O(h_K^{2k-2})$$

and the loss of precision is exactly compensated by the choice of the stabilising parameter βh_K^2.

The method does work for any degree of polynomial. However, it should be noted that for $k > 1$, it leads to a non-symmetric system which makes it less appealing.

□

8.13.3 Minimal Stabilisations for Stokes

We now consider another class of stabilisations which contains, as a special case, the method of (8.13.2). Although this could be written in general, we shall restrict ourselves to the case of a first order approximation, $V_h \subset (\mathcal{L}_1^1)^n \cap V$, $Q_h \subset \mathcal{L}_k^1 \cap Q$. The method will be a direct adaptation of Sect. 6.3.1 of Chap. 6. We introduce another space $\tilde{V}_h \in (\mathcal{L}_1^1)^2$, denoting $\tilde{P}$ the projection over $\tilde{V}_h$ in the norm of $H := (L^2(\Omega))^n$ and we consider the following problem:

$$\begin{cases} \langle A\underline{u}_h + \underline{\mathrm{grad}}\, p_h - \underline{f}, \underline{v}_h\rangle_{V'\times V} = 0, \quad \forall\, \underline{v}_h \in V_h, \\ \langle \mathrm{div}\, \underline{u}_h + g, q\rangle_{Q'\times Q} \\ \qquad + r(\underline{\mathrm{grad}}\, p_h - \tilde{P}\ \underline{\mathrm{grad}}\, p_h, \underline{\mathrm{grad}}\, q_h)_H = 0, \quad \forall\, q_h \in Q_h. \end{cases} \tag{8.13.35}$$

This fits entirely into the theory of Chap. 6 and we have the error bound

$$\begin{aligned} &\|\underline{u} - \underline{u}_h\|_V^2 + \|p - p_h\|_Q^2 \\ &\quad \le C\,(\frac{\omega^2(h) + r}{r})\left(\inf_{\underline{v}_h \in V_H} \|\underline{u} - \underline{v}_h\|_V^2\right) \\ &\quad + (1 + \frac{r}{\omega(h)^2})\inf_{q_h \in Q_h} \|p - q_h\|_Q^2 + r\|(I - \tilde{P})(\underline{\mathrm{grad}}\, p)\|_H^2, \end{aligned} \tag{8.13.36}$$

provided that the following assumption holds:

$$\mathbf{A}\begin{cases} \text{There exists a positive constant } \gamma, \text{ independent of } h\text{ , such that} \\ \|P_{V_h}\,\underline{\mathrm{grad}}\, q_h\|^2 \\ \qquad + \|\,\underline{\mathrm{grad}}\, q_h - \tilde{P}\ \underline{\mathrm{grad}}\, q_h\|^2 \ge \gamma\|\,\underline{\mathrm{grad}}\, q_h\|^2 \quad \forall\, q_h \in Q_h. \end{cases} \tag{8.13.37}$$

Now, as we use an approximation of degree one, we would like all terms on the right-hand side of inequality (8.13.36) to be $O(h^2)$ and we consider three cases.

(i) **The Brezzi-Pitkäranta formulation**. If we take $\tilde{V}_h = \{0\}$ and thus $\tilde{P} = I$, assumption **A** evidently holds. The last term in (8.13.36) reduces to $r\|\,\underline{\mathrm{grad}}\,p\|_H^2$ and we need to make $r = O(h^2)$ to get the error bound. The drawback of the method is the boundary layer on $\underline{\mathrm{grad}}\,p_h$.

(ii) **Projection on** V_h. Assumption **A** is again immediate. We must again consider the last consistency term. Some trouble arises because V_h satisfies boundary conditions (e.g., $\underline{v}_h = 0$ on the boundary). We can then expect $\|(I-\tilde{P})(\underline{\mathrm{grad}}\,p)\|_H^2$ to be no better than $O(h)$. Taking $r = O(h)$ would restore the optimal order but the problem with the boundary layer is not cured.

(iii) **Optimal projection**. From the previous discussion, one sees that taking $\tilde{V}_h = (\mathcal{L}_1^1)^2$, that is, suppressing boundary conditions, will work with $r = O(1)$ and will eliminate the spurious boundary layer. The trouble is now with assumption **A**. It was proved in [123] that it indeed holds. This method had been used in [51].

In order to prove our assumption **A**, we first consider the following result.

Proposition 8.13.1. *Let Q_h and V_h be the space of piecewise linear pressures and velocities as above, and let $\tilde{V}_h$ be the space of piecewise linear continuous vectors on $\mathcal{T}_h$ (without boundary conditions.) There exists a constant $\beta^* > 0$, independent of h, such that, for every $q_h \in Q_h$ and for every $\underline{w}_h \in \tilde{V}_h$, there exists a $\underline{v}_h^0 \in V_h$ verifying*

$$\|\underline{v}_h^0\|_0 \leq \|\,\underline{\mathrm{grad}}\,q_h\|_0 \tag{8.13.38}$$

and

$$(\underline{v}_h^0, \underline{\mathrm{grad}}\,q_h)_0 + \|\,\underline{\mathrm{grad}}\,q_h - \underline{w}_h\|_0^2 \geq \beta^* \|\,\underline{\mathrm{grad}}\,q_h\|_0^2. \tag{8.13.39}$$

Proof. Let us consider first a macro-element K made by the collection of triangles having one vertex P of $\mathcal{T}_h$ in common. Split $q_h = q_0 + q_\ell$, where q_0 is such that $\underline{\mathrm{grad}}\,q_0$ has zero mean value in K and q_ℓ is linear on K (hence $\underline{\mathrm{grad}}\,q_\ell$ = constant in K.) It is clear that $(\underline{\mathrm{grad}}\,q_0, \underline{\mathrm{grad}}\,q_\ell)_K = 0$. We now take $\underline{v}_h^0$, piecewise linear, continuous, vanishing on the boundary of K and having value $\sqrt{6}\,\underline{\mathrm{grad}}\,q_\ell$ at the internal vertex P. An easy computation shows that:

$$\|\underline{v}_h^0\|_{0,K} = \|\,\underline{\mathrm{grad}}\,q_\ell\|_{0,K} \tag{8.13.40}$$

and

$$(\underline{v}_h^0, \underline{\mathrm{grad}}\,q_\ell)_K = \sqrt{\frac{2}{3}}\,\|\,\underline{\mathrm{grad}}\,q_\ell\|_{0,K}^2. \tag{8.13.41}$$

On the other hand, $\underline{\text{grad}}\, q_0$ belongs to a space (piecewise constant vectors on K, with continuous tangential components, and zero mean on K) whose intersection with piecewise linear continuous vectors on K is reduced to the zero vector. As we are in finite dimension, there exists a positive constant δ_K such that, for every $\underline{\text{grad}}\, q_0$ and for every $\underline{w}_h$,

$$\| \underline{\text{grad}}\, q_0 - \underline{w}_h \|_0^2 \geq \delta_K \| \underline{\text{grad}}\, q_0 \|_{0,K}^2. \tag{8.13.42}$$

As $\underline{\text{grad}}\, q_\ell$ is clearly continuous and piecewise linear, (8.13.42) easily implies that

$$\begin{aligned} \| \underline{\text{grad}}\, q_h - \underline{w}_h \|_0^2 &= \| \underline{\text{grad}}\, q_0 + \underline{\text{grad}}\, q_\ell - \underline{w}_h \|_0^2 \\ &= \| \underline{\text{grad}}\, q_0 - \underline{\tilde{w}}_h \|_0^2 \geq \delta_K \| \underline{\text{grad}}\, q_0 \|_{0,K}^2, \end{aligned} \tag{8.13.43}$$

and a simple scaling argument shows immediately that δ_K is independent of the *size* of K (notice that (8.13.43) holds *for every* $\underline{w}_h$).

Finally, we explicitly point out that

$$(\underline{v}_h^0, \underline{\text{grad}}\, q_0)_K = \frac{\underline{v}_h^0(P)}{3} \int_K \underline{\text{grad}}\, q_0 \text{d}x = 0, \tag{8.13.44}$$

where P is the only vertex internal to K. From (8.13.41) to (8.13.44), one then gets that, for every q_h and for every $\underline{w}_h$, there is a $\underline{v}_h^0$, piecewise linear, continuous, and vanishing on the boundary of K, such that (8.13.40) holds and

$$(\underline{v}_h^0, \underline{\text{grad}}\, q_h)_K + \| \underline{\text{grad}}\, q_h - \underline{w}_h \|_{0,K}^2 \geq \beta_K \| \underline{\text{grad}}\, q_h \|_{0,K}^2, \tag{8.13.45}$$

for some positive constant β_K independent of q_h and $\underline{w}_h$. The result (8.13.38) and (8.13.39) then follows easily from (8.13.45) by typical instruments (continuity of β_K, splitting of Ω into macro-elements such that each triangle belongs at most to three different macro-elements, and so on). □

With the aid of Proposition 8.13.1, we can now prove Assumption **A**.

Proposition 8.13.2. *Let Q_h, V_h and $\tilde{V}_h$ be as in Proposition 8.13.1. Then, there exists a constant $\tilde{\beta} > 0$ such that*

$$\| P_{V_h} \underline{\text{grad}}\, q_h \|^2 + \| \underline{\text{grad}}\, q_h - P_{\tilde{V}_h} \underline{\text{grad}}\, q_h \|^2 \geq \tilde{\beta} \, \| \underline{\text{grad}}\, q_h \|^2 \qquad \forall\, q_h \in Q_h, \tag{8.13.46}$$

where all the norms are in L^2.

Proof. We start by observing that, for every $\underline{v}_h^0$ and q_h, we have

$$\begin{aligned} (\underline{v}_h^0, \underline{\text{grad}}\, q_h) &= (\underline{v}_h^0, P_{V_h} \underline{\text{grad}}\, q_h) \leq \|\underline{v}_h^0\| \, \| P_{V_h} \underline{\text{grad}}\, q_h \| \\ &\leq \frac{\beta^*}{2} \|\underline{v}_h^0\|^2 + \frac{1}{2\beta^*} \| P_{V_h} \underline{\text{grad}}\, q_h \|^2, \end{aligned} \tag{8.13.47}$$

where the last inequality clearly holds for every positive β^*, but we shall use it for the value of β^* given in (8.13.39). For every q_h, we now take $\underline{v}_h^0$ as given by Proposition 8.13.1, and using (8.13.38), we have

$$(\underline{v}_h^0, \underline{\text{grad}}\, q_h) \leq \frac{\beta^*}{2} \| \underline{\text{grad}}\, q_h \|^2 + \frac{1}{2\beta^*} \| P_{V_h}\ \underline{\text{grad}}\, q_h \|^2, \tag{8.13.48}$$

which, inserted in (8.13.39) with $\underline{w}_h = P_{\tilde{V}_h}\ \underline{\text{grad}}\, q_h$, gives

$$\frac{\beta^*}{2} \| \underline{\text{grad}}\, q_h \|^2 + \frac{1}{2\beta^*} \| P_{V_h}\ \underline{\text{grad}}\, q_h \|^2 + \| \underline{\text{grad}}\, q_h - P_{\tilde{V}_h}\ \underline{\text{grad}}\, q_h \|^2 \geq \beta^*\ \| \underline{\text{grad}}\, q_h \|^2, \tag{8.13.49}$$

and (8.13.46) follows immediately. □

Remark 8.13.2 (Enhanced strain methods). Finally, to conclude this section, we would like to note that another example of stabilisation of the Stokes problem by an enhanced method can be found in the work of [283]. □

8.14 Concluding Remarks: Choice of Elements

We would first like to emphasise that the results of this chapter can be applied as well to flow problems as well as to linear (or linearised) elasticity problems. In this last case, displacement methods also need to be considered from a mixed point of view. Indeed, we have already seen in Sect. 8.12 that there is a close relation between the Stokes problem and linear elasticity problems. However, things are not so simple: fluid people and solid mechanics people form two different communities and information was long to cross the border.

8.14.1 Choice of Elements

We have presented discontinuous pressure and continuous pressure elements. They both have advantages, even though discontinuous pressure is appealing as it enforces an element wise conservation of mass. They can also be implemented by a penalty procedure.

In this respect, the reader should have noticed an important difference between the two-dimensional and the three-dimensional elements presented in this chapter.

- In the 2-D case, we have a choice of discontinuous pressure elements which can be used with a penalty method. Direct solvers are not too sensitive to the ill-conditioning of the resulting system and we thus obtain a good resolution strategy. We can thus recommend the Crouzeix-Raviart element of Example 8.6.1 or its higher order variants.

- In the 3-D case, discontinuous pressure elements satisfying the *inf-sup* condition are expensive as one needs degrees of freedom on the faces. The equivalent of the Crouzeix-Raviart element also uses bubbles of degree four. The ubiquitous $\underline{Q}_1 - P_0$, for which the condition number of the dual problem depends on h, behaves badly with iterative methods which are essential for large-scale simulations. We therefore recommend the Hood-Taylor continuous pressure element.

Remark 8.14.1 (Solvers). The choice of elements is also dictated by the choice of solvers. In the two-dimensional case, where direct solvers are almost always employed, discontinuous pressure elements are desirable as they are compatible with penalty methods. In the three-dimensional case, where iterative solvers are the rule for large problems, penalty methods are to be avoided as they destroy the condition number of the problem. Recent progress in the construction of solvers for indefinite systems [58, 185, 186] however make the use of a continuous pressure element, such as the Taylor-Hood element, possible and efficient. □

Remark 8.14.2 (Meshes). Another consideration is the choice of affine (triangular or tetrahedral) or quadrilateral hexahedral elements. As we already noted, there was a widespread legend that tetrahedra were not suitable for incompressible solid mechanics problems. This was based on a lack of analysis and we advocate the choice of affine elements for two reasons.

- Mesh generation is much easier with tetrahedra than with hexahedra. Indeed, it can, most of times, be done automatically. This is important in complex engineering problems where the domains may be of a complex shape.
- The second reason is mesh adaptation, which is also much easier for tetrahedra. There exist algorithms which can make a mesh optimal to represent a given solution. □

Finally, let us recall that the approximation of incompressible materials is a central issue in many industrial applications. It has therefore been the subject of a vast literature. We believe to have presented the essential points but we also neglected many aspects. Among those, we did not describe finite volume methods, which are mostly amenable to an analysis by the theory of mixed methods. One can find references in [194] and [195].

Chapter 9
Complements on Elasticity Problems

9.1 Introduction

Elasticity problems are probably the most common use of the finite element method. Historically, they were indeed at the origin of the method. We have already considered in Chap. 8, in particular in Sect. 8.12, standard formulations of elasticity problems based on displacement variables. Considerations on the choice of elements have also be presented in Sect. 8.14.1. Our main concern will now be to present mixed methods using explicitly an approximation of the stress tensor, in which the equilibrium condition is strongly imposed on each element.

Mixed methods are indeed an appealing technique for the numerical solution of elasticity problems. They ensure the equilibrium condition (a basic property in solid mechanics) and they make the constitutive law more explicit. The stress tensor becomes the main variable but the symmetry of this tensor makes the construction of suitable elements much more complicated than what can be done, for instance, in thermal problems where families of elements such as the $\mathcal{RT}_k$ and $\mathcal{BDM}_k$ are now classical.

In fact, one should recall that the symmetry of the stress tensor expresses the conservation of angular momentum and that representing *exactly* a conservation law is a difficult task. The idea of using stress tensors having only a *reduced* symmetry goes back to Fraeijs de Veubeke [210], but the introduction and the analysis of specific elements having symmetry only *in average* was done first in [7], while an even weaker form of symmetry (namely, orthogonality to piecewise linear continuous functions) was proposed and studied in [24]. Since then, their use underwent alternate periods of popularity and oblivion. For more information, see e.g. [120, 204, 305, 356, 357] and many others. See also [79, 80] and the references therein.

Recently, a general construction of elements with reduced symmetry was presented in [33] and [30]. Their construction relies on a very elegant but quite abstract procedure, requiring rather sophisticated instruments. In [80], we presented a different analysis of these elements and related ones, using much more elementary

D. Boffi et al., *Mixed Finite Element Methods and Applications*, Springer Series in Computational Mathematics 44, DOI 10.1007/978-3-642-36519-5_9,

and classical techniques. It is clear to us that the construction in [30] still has the merit of having inspired the choice of these elements (and having provided the first proof of their convergence). However, its presentation requires the introduction of a rather heavy theoretical machinery, and therefore we will follow essentially the approach of [80] or, actually, an even simpler one. In particular, the present approach provides a particularly simple analysis of virtually all reduced-symmetry elements present in the literature, in addition to some new ones that generalise the original element of Amara-Thomas [7], whence the name of Generalised Amara-Thomas (GAT).

In this presentation, we shall restrict ourselves to *linear isotropic problems*. As we are interested mainly in discretisation methods, this is not a real restriction: the choice of elements will have in all cases to follow the same rules. Moreover, always to simplify the discussion, we will concentrate on the (rather unrealistic) case of *homogeneous Dirichlet boundary conditions all over the boundary*, and we shall assume that *the domain is convex*.

9.1.1 Continuous Formulation of Stress Methods

We consider a *mixed* approach to linear elasticity problems, that is, we use as main variable a symmetric stress tensor, chosen in a suitable space.

Remark 9.1.1. Throughout the chapter, we will often use 3×3 tensors, say $\underline{\underline{\Phi}}$, that are obtained joining three different vectors $\underline{\phi}^{(1)}$, $\underline{\phi}^{(2)}$, and $\underline{\phi}^{(3)}$ of three components each. In general, we will not distinguish between *row vectors* and *column vectors*. However, in collecting three of them, we would have to distinguish between $\Phi_{ij} = \phi_j^{(i)}$ (patching row-vectors) and $\Phi_{ij} = \phi_i^{(j)}$ (patching column-vectors).

Accordingly, given a space Ψ of three-vectors, we will have to distinguish between the space $(\Psi)^{3r}$ obtained patching row-vectors and $(\Psi)^{3c}$ obtained patching column-vectors. If the space Ψ is itself made as the cube of another space of scalars V (that is, $\Psi = V^3$), then $(\Psi)^{3r} = (\Psi)^{3c} = V^{3\times 3}$. □

We therefore define, in n dimensions,

$$\underline{\underline{H}}(\underline{\mathrm{div}};\Omega) := (H(\mathrm{div};\Omega))^{nr} \equiv \{\underline{\underline{\tau}} \,|\, \underline{\underline{\tau}} \in (L^2(\Omega))^{n\times n},\ \underline{\mathrm{div}}\,\underline{\underline{\tau}} \in (L^2(\Omega))^n\}, \tag{9.1.1}$$

$$\underline{\underline{H}}(\underline{\mathrm{div}};\Omega)_S := \{\underline{\underline{\tau}} \,|\, \underline{\underline{\tau}} \in \underline{\underline{H}}(\underline{\mathrm{div}};\Omega),\ \tau_{i,j} = \tau_{j,i} \quad \forall\, i,\, j = 1,..,n\}, \tag{9.1.2}$$

$$\Sigma := \underline{\underline{H}}(\underline{\mathrm{div}};\Omega),\ \ \Sigma_S := \underline{\underline{H}}(\underline{\mathrm{div}};\Omega)_S,\ \ U := (L^2(\Omega))^n. \tag{9.1.3}$$

We recall the definition of the *trace* of a tensor

$$tr(\underline{\underline{\tau}}) := \sum_{i=1}^{n} \underline{\underline{\tau}}_{ii} \tag{9.1.4}$$

and of the *deviatoric*

$$\underline{\underline{\tau}}^D := \underline{\underline{\tau}} - \frac{1}{n} tr(\underline{\underline{\tau}})\underline{\underline{I}}, \tag{9.1.5}$$

where $\underline{\underline{I}}$ is the *identity* tensor. Note that $tr(\underline{\underline{I}}) = n$ so that in (9.1.5) we have $tr(\underline{\underline{\tau}}^D) = 0$. Note as well that (9.1.5) can equally be written as

$$tr(\underline{\underline{\tau}})\underline{\underline{I}} = n\,(\underline{\underline{\tau}} - \underline{\underline{\tau}}^D), \tag{9.1.6}$$

which, applied to the case of a tensor $\underline{\underline{\tau}} = \underline{\underline{\mathrm{grad}}}\,\underline{v}$ (for some $\underline{v}$), gives

$$(\mathrm{div}\,\underline{v})\,\underline{\underline{I}} \equiv tr(\underline{\underline{\mathrm{grad}}}\,\underline{v})\,\underline{\underline{I}} = n\,(\underline{\underline{\mathrm{grad}}}\,\underline{v} - \underline{\underline{\mathrm{grad}}}\,\underline{v}^D). \tag{9.1.7}$$

At this point, we can set

$$a(\underline{\underline{\sigma}}, \underline{\underline{\tau}}) := \int_\Omega \Big[\frac{1}{2\mu}\underline{\underline{\sigma}}^D : \underline{\underline{\tau}}^D + \frac{1}{n(n\lambda + 2\mu)} tr(\underline{\underline{\sigma}}) tr(\underline{\underline{\tau}})\Big]\,dx, \tag{9.1.8}$$

$$b(\underline{\underline{\tau}}, \underline{v}) := \int_\Omega \underline{\mathrm{div}}(\underline{\underline{\tau}}) \cdot \underline{v}\,dx \tag{9.1.9}$$

and we can write our simple linear elasticity problem as: *find* $(\underline{\underline{\sigma}}, \underline{u}) \in \Sigma_S \times U$ *such that*

$$\begin{cases} a(\underline{\underline{\sigma}}, \underline{\underline{\tau}}) + b(\underline{\underline{\tau}}, \underline{u}) = 0, & \forall\, \underline{\underline{\tau}} \in \Sigma_S, \\ b(\underline{\underline{\sigma}}, \underline{v}) + (\underline{f}, \underline{v}) = 0, & \forall\, \underline{v} \in U. \end{cases} \tag{9.1.10}$$

Remark 9.1.2. The first equation represents the constitutive law and the second one the equilibrium condition. It must be clear that although we consider a linear model, the results can be transposed to more realistic non linear models. □

We thus have to consider the standard conditions for existence and uniqueness of the solution to this problem. It is very easy to check that there exists $C > 0$ such that

$$\inf_{\underline{v}\in U} \sup_{\underline{\underline{\tau}}\in \Sigma_S} \frac{b(\underline{\underline{\tau}}, v)}{\|\underline{\underline{\tau}}\|_1 \|\underline{v}\|_0} \geq c \tag{9.1.11}$$

and that

$$a(\underline{\underline{\tau}}, \underline{\underline{\tau}}) \geq \frac{1}{n(n\lambda + 2\mu)} \|\underline{\underline{\tau}}\|_0^2, \quad \forall\, \underline{\underline{\tau}} \in \Sigma. \tag{9.1.12}$$

We thus have an *inf-sup* condition and coercivity so that our problem is well posed. However, trouble arises when we have to deal with a very large λ (nearly incompressible materials). In fact, it is clear that the coercivity constant which appears in (9.1.12) goes to zero like $1/\lambda$ when $\lambda \to +\infty$ so that the stability properties of problem (9.1.10) seem to deteriorate for large values of λ. Actually, the situation is not as bad as it seems because we do not need coercivity to hold for every $\underline{\underline{\tau}} \in \Sigma$ (or Σ_h) but only for $\underline{\underline{\tau}} \in \mathrm{Ker}B$ (respectively, $\mathrm{Ker}B_h$ for discrete problems). In particular, the continuous formulation (9.1.10) does not break down when $\lambda \to \infty$, because of the following proposition.

Proposition 9.1.1. *There exists a constant* $C > 0$ *such that, for every* $\underline{\underline{\tau}} \in \Sigma$ *satisfying*

$$\int_\Omega tr(\underline{\underline{\tau}})\, dx = 0, \tag{9.1.13}$$

we have

$$\|\underline{\underline{\tau}}\|_0 \leq C(\|\underline{\underline{\tau}}^D\|_0 + \| \underline{div}\, \underline{\underline{\tau}}\|_0). \tag{9.1.14}$$

□

Proof. It is obvious that

$$\|\underline{\underline{\tau}}\|_0 \leq \|\underline{\underline{\tau}}^D\|_0 + \frac{1}{n}\| \operatorname{tr}(\underline{\underline{\tau}})\underline{\underline{I}}\|_0 \tag{9.1.15}$$

and hence it is enough to show that

$$\| \operatorname{tr}(\underline{\underline{\tau}})\|_0 \leq C(\|\underline{\underline{\tau}}^D\|_0 + \| \underline{\operatorname{div}}\, \underline{\underline{\tau}}\|_0) \tag{9.1.16}$$

for some constant C. For this, note that (9.1.13) implies the existence of a $\underline{v} \in (H_0^1)^n$ such that

$$\operatorname{div} \underline{v} = \operatorname{tr}(\underline{\underline{\tau}}), \tag{9.1.17}$$

$$\|\underline{v}\|_1 \leq C\, \| \operatorname{tr}(\underline{\underline{\tau}})\|_0. \tag{9.1.18}$$

Now, from (9.1.17) and (9.1.7), we have:

$$\begin{cases} \|tr(\underline{\underline{\tau}})\|_0^2 = \displaystyle\int_\Omega tr(\underline{\underline{\tau}}) \operatorname{div} \underline{v}\, dx \\ \quad = \displaystyle\int_\Omega \underline{\underline{\tau}} : \underline{\underline{I}} \operatorname{div} \underline{v}\, dx \\ \quad = \displaystyle\int_\Omega \underline{\underline{\tau}} : (\underline{\underline{\operatorname{grad}}}\, \underline{v} - (\underline{\underline{\operatorname{grad}}}\, \underline{v})^D)\, dx \\ \quad = -n \displaystyle\int_\Omega \underline{\operatorname{div}}\, \underline{\underline{\tau}} \cdot \underline{v}\, dx - n \int_\Omega \underline{\underline{\tau}}^D : \underline{\underline{\operatorname{grad}}}\, \underline{v}\, dx \\ \quad \leq n\|\underline{\underline{\tau}}^D\|\, \|\underline{v}\|_1 + n\| \underline{\operatorname{div}}\, \underline{\underline{\tau}}\|_0\|\underline{v}\|_0 \end{cases} \tag{9.1.19}$$

and from (9.1.18) and (9.1.19), we get (9.1.16). □

If we work in the subspace

$$\tilde{\Sigma} := \Big\{\underline{\underline{\tau}} \mid \underline{\underline{\tau}} \in \Sigma,\ \int_{\Omega} tr(\underline{\underline{\tau}})\, dx = 0\Big\}, \tag{9.1.20}$$

we know that the set

$$\mathrm{Ker}B = \big\{\underline{\underline{\tau}} \mid \underline{\underline{\tau}} \in \tilde{\Sigma} \text{ such that } b(\underline{\underline{\tau}}, \underline{v}) = 0\ \forall\, \underline{v} \in U\big\} \tag{9.1.21}$$

is precisely made of tensors satisfying (9.1.13) and

$$\underline{\mathrm{div}}\, \underline{\underline{\tau}} = 0. \tag{9.1.22}$$

Hence, from Proposition 9.1.1, we have

$$a(\underline{\underline{\tau}}, \underline{\underline{\tau}}) \geq \frac{1}{2\mu} \|\underline{\underline{\tau}}^D\|_0^2 \geq C(\mu)\, \|\underline{\underline{\tau}}\|_0^2 = C(\mu) \|\underline{\underline{\tau}}\|^2_{\underline{\underline{H}}(\underline{\mathrm{div}};\Omega)_s}, \quad \forall\, \underline{\underline{\tau}} \in \mathrm{Ker}B. \tag{9.1.23}$$

The stability constant of our problem is therefore independent of λ.

Remark 9.1.3. It must be noted that condition (9.1.13) refers to the fact that with Dirichlet boundary conditions, in incompressible problems, pressure is defined only up to an additive constant. The condition can then be applied a posteriori. It disappears whenever Neumann boundary conditions are imposed on a part of the boundary. From the mathematical point of view, we can also remark that, taking $\underline{\underline{\tau}} = \underline{\underline{I}}$ in the first equation of (9.1.10), we immediately have that the solution $\underline{\underline{\sigma}}$ belongs to $\tilde{\Sigma}$.

To avoid unnecessary complications, we shall often use, in what follows, the spaces Σ and Σ_S instead of $\tilde{\Sigma}$ and $\tilde{\Sigma}_S$. □

9.1.2 Numerical Approximations of Stress Formulations

If we now choose some finite-dimensional subspaces Σ_{Sh} of Σ_S and U_h of U, we must be careful to have the discrete analogues of (9.1.11) and (9.1.23) verified. However, we have to face a delicate point. In order to prove an inequality of type (9.1.23), we *needed*, in Proposition 9.1.1, to have $\underline{\mathrm{div}}\, \underline{\underline{\tau}} = 0$. Hence, our life would be a lot easier if we had the "inclusion of the kernels property": $\mathrm{Ker}B_h \subset \mathrm{Ker}B$. In other words, we would like our spaces Σ_{Sh} and U_h to satisfy the following property:

$$\begin{aligned}\mathrm{Ker}B_h = \{&\underline{\underline{\tau}}_h \in \Sigma_{Sh} : b(\underline{\underline{\tau}}_h, \underline{v}_h) = 0\ \forall\, \underline{v}_h \in U_h\} \\ &\subset \mathrm{Ker}B = \{\underline{\underline{\tau}} \in \Sigma|_S : \underline{\mathrm{div}}\, \underline{\underline{\tau}} = 0\}.\end{aligned} \tag{9.1.24}$$

At the same time, the *inf-sup* condition (9.1.11) is related to the existence of a B-compatible operator $\Pi_h : \Sigma_S \to \Sigma_{Sh}$ such that

$$b(\underline{\underline{\tau}} - \Pi_h \underline{\underline{\tau}}, \underline{v}_h) = 0, \quad \forall \, \underline{v}_h \in U_h, \tag{9.1.25}$$

$$\|\Pi_h \underline{\underline{\tau}}\|_\Sigma \leq c \, \|\underline{\underline{\tau}}\|_\Sigma, \quad \forall \, \underline{\underline{\tau}} \in \Sigma. \tag{9.1.26}$$

We have seen in Chap. 2 many possibilities to approximate $H(\text{div};\Omega)$ and it seems, at first sight, that building a tensor with rows in $H(\text{div};\Omega)$ would be suitable to approximate Σ, but we should not forget the symmetry of the tensors in Σ_S. The problem of finding subspaces of Σ_S and U satisfying (9.1.24)–(9.1.26) is actually very difficult. One of the (nowadays) classical remedies is to give up the symmetry of $\underline{\underline{\tau}}$ and enforce it back in a weaker form by some Lagrange multiplier. This is what we are going to do in the next section.

9.2 Relaxed Symmetry

The idea of relaxing symmetry was, to our knowledge, first used by Fraeijs de Veubeke [212] and his school; it was then used by Amara and Thomas [7] and then by Arnold et al. [24]. Other recent results can be found in [120, 305] and [356, 357].

It is worth recalling that the symmetry of the stress tensor is, in fact, a simplified way of expressing a conservation law, namely the conservation of angular momentum. This should make it easier to understand why symmetry is difficult to enforce. Conservation laws are not easily exactly imposed. In fact, the point of using spaces like $H(\underline{\text{div}};\Omega)$ and its discrete counterparts is to get a strong form for conservation of momentum. Imposing strongly a second conservation law is likely to be difficult.

Before considering a suitable mixed formulation of elasticity problems, we shall first present some results on tensors which will be used throughout the chapter.

9.3 Tensors, Tensorial Notation and Results on Symmetry

Given a second order tensor $\underline{\underline{\tau}}$, we define its skew-symmetric part as

$$\underline{\underline{as}}(\underline{\underline{\tau}}) := \frac{1}{2}\left\{\underline{\underline{\tau}} - \underline{\underline{\tau}}^t\right\}. \tag{9.3.1}$$

The tensor $\underline{\underline{as}}(\underline{\underline{\tau}})$ and in fact all the tensors in X can be identified with a vector in the three-dimensional case and a scalar in the two-dimensional one. Indeed, in two dimensions, for every scalar q, we can define the corresponding skew-symmetric tensor $\underline{\underline{S}}^2(q)$ by

$$\underline{\underline{S}}^2(q) := \begin{pmatrix} 0 & q \\ -q & 0 \end{pmatrix}. \tag{9.3.2}$$

Denoting by $\underline{asp}(\underline{\underline{\tau}}) = \tau_{1,2} - \tau_{2,1}$ the *asymmetric part* of $\underline{\underline{\tau}}$, we have $2\,\underline{\underline{as}}(\underline{\underline{\tau}}) = \underline{\underline{S}}^2(asp(\underline{\underline{\tau}}))$. Similarly in three dimensions, we define, for every vector $\underline{q}$, the tensor $\underline{\underline{S}}^3(\underline{q})$ given by

$$\underline{\underline{S}}^3(\underline{q}) := \begin{pmatrix} 0 & q_3 & -q_2 \\ -q_3 & 0 & q_1 \\ q_2 & -q_1 & 0 \end{pmatrix}. \tag{9.3.3}$$

Then denoting

$$\underline{asp}(\underline{\underline{\tau}}) = \begin{pmatrix} \tau_{23} - \tau_{32} \\ \tau_{31} - \tau_{13} \\ \tau_{12} - \tau_{21} \end{pmatrix}, \tag{9.3.4}$$

we have $2\,\underline{\underline{as}}(\underline{\underline{\tau}}) = \underline{\underline{S}}^3(\underline{asp}(\underline{\underline{\tau}}))$.

As we shall deal mostly with the three-dimensional case, we will concentrate on this case and we will often use just some remarks for the corresponding two-dimensional results.

We first recall the definition of the *permutation tensor* (or pseudo-tensor): for $n = 3$ the triple tensor $\underline{\underline{\underline{\mathbb{P}}}}$ is given by

$$\mathbb{P}_{ijk} := \begin{cases} 1 & \text{if } \{i,j,k\} = \{1,2,3\} \text{ or } \{3,1,2\} \text{ or } \{2,3,1\} \\ -1 & \text{if } \{i,j,k\} = \{3,2,1\} \text{ or } \{1,3,2\} \text{ or } \{2,1,3\} \\ 0 & \text{otherwise.} \end{cases} \tag{9.3.5}$$

For a tensor $\underline{\underline{\tau}}$, we easily check that the vector $\underline{asp}(\underline{\underline{\tau}})$ defined in (9.3.4) can be written as

$$\underline{asp}(\underline{\underline{\tau}}) = \underline{\underline{\tau}} : \underline{\underline{\underline{\mathbb{P}}}} \tag{9.3.6}$$

and the tensor $\underline{\underline{as}}(\underline{\underline{\tau}})$ defined in (9.3.1) becomes

$$2\underline{\underline{as}}(\underline{\underline{\tau}}) \equiv \underline{\underline{S}}^3(\underline{\underline{\tau}} : \underline{\underline{\underline{\mathbb{P}}}}). \tag{9.3.7}$$

Similarly, for each vector $\underline{s}$, we have

$$\underline{\underline{S}}^3(\underline{s}) = \underline{\underline{\underline{\mathbb{P}}}} \cdot \underline{s}. \tag{9.3.8}$$

Then we denote, as usual, by $\underline{x} \wedge \underline{y}$ the external (or *wedge*) product of two vectors, given by

$$(\underline{x} \wedge \underline{y})_i = \mathbb{P}_{ijk} x_j y_k, \tag{9.3.9}$$

where in (9.3.9) (and in all the rest of the section) the Einstein convention of summation of repeated indices is employed.

In a similar way, we can define the left and the right wedge product of a vector $\underline{v}$ with a double tensor $\underline{\underline{\tau}}$ as

$$(\underline{v} \wedge \underline{\underline{\tau}})_{ir} := \mathbb{P}_{ijk} v_j \tau_{kr} \qquad \text{and} \qquad (\underline{\underline{\tau}} \wedge \underline{v})_{ri} := \mathbb{P}_{ijk} \tau_{rj} v_k. \tag{9.3.10}$$

Remark 9.3.1. In the left product, we take the wedge product of $\underline{v}$ with the columns of $\underline{\underline{\tau}}$. Instead, in the right product, we take the wedge product of the rows of $\underline{\underline{\tau}}$ with $\underline{v}$. □

We recall now some useful properties of tensor calculus. We denote by $\underline{x} \equiv (x_1, x_2, x_3)$ the vector containing the independent variables and by ∂_i the partial derivative with respect to x_i. We shall also write, in a classical way, $\underline{\nabla} \equiv (\partial_1, \partial_2, \partial_3)$. In this notation, the (row-wise) $\underline{\underline{\text{curl}}}$ of a tensor $\underline{\underline{\tau}}$ is defined by

$$(\underline{\underline{\text{curl}}}\, \underline{\underline{\tau}})_{ri} = (\underline{\underline{\tau}} \wedge \underline{\nabla})_{ri} = \mathbb{P}_{ijk} \partial_k \tau_{rj},$$

while in the two-dimensional case, we have instead, for every *vector* $\underline{\psi}$,

$$\underline{\underline{\text{curl}}}\, \underline{\psi} \equiv \begin{pmatrix} -\partial_2 \psi_1 & \partial_1 \psi_1 \\ -\partial_2 \psi_2 & \partial_1 \psi_2 \end{pmatrix}. \tag{9.3.11}$$

Coming back to the three-dimensional case, we then introduce the operator

$$\mathcal{A}\underline{\underline{\tau}} = tr(\underline{\underline{\tau}})\underline{\underline{I}} - \underline{\underline{\tau}}^t \tag{9.3.12}$$

and we note that it could also be written as

$$(\mathcal{A}\underline{\underline{\tau}})_{\alpha\beta} = \mathbb{P}_{\alpha ik}\, \mathbb{P}_{jr\beta}\, \delta_{ri}\, \tau_{kj}. \tag{9.3.13}$$

Remark 9.3.2. We point out that, apart from the presence of the trace operator on the main diagonal, the operator $\mathcal{A}$ transfers to the rows the information stored in the columns, and vice-versa. In particular, for a skew-symmetric tensor $\underline{\underline{\gamma}}$, we have $\mathcal{A}\underline{\underline{\gamma}} = -\underline{\underline{\gamma}}^t$. □

We can now recall some useful identities in tensor calculus that will be used in a while. They can be checked by boring but elementary computations.

We thus have for every $\underline{\underline{\Psi}}$ in $(H(\text{curl}))^{3r}$

$$\underline{\underline{\text{curl}}}(\underline{x} \wedge \underline{\underline{\Psi}}) = \mathcal{A}\underline{\underline{\Psi}} + \underline{x} \wedge \underline{\underline{\text{curl}}}\,\underline{\underline{\Psi}} \tag{9.3.14}$$

and

$$\underline{asp}(\underline{\underline{\text{curl}}}\,\underline{\underline{\Psi}}) = \underline{\text{div}}\,\mathcal{A}\underline{\underline{\Psi}}. \tag{9.3.15}$$

We also note that for every $\underline{\underline{\Psi}}$ and every vector $\underline{n}$

$$(\mathcal{A}\underline{\underline{\tau}}) \cdot \underline{n} = \underline{\underline{\underline{P}}} : (\underline{\underline{\tau}} \wedge \underline{n}), \tag{9.3.16}$$

which shows that

$$\underline{\underline{\Psi}} \wedge \underline{n} = 0 \Rightarrow (\mathcal{A}\underline{\underline{\Psi}}) \cdot \underline{n} = 0. \tag{9.3.17}$$

Moreover, for every $\underline{\underline{\tau}}$ given in $\underline{\underline{H}}(\underline{\text{div}}; \Omega)$, we have

$$\underline{\text{div}}(\underline{x} \wedge \underline{\underline{\tau}}) = \underline{x} \wedge (\underline{\text{div}}\underline{\underline{\tau}}) - \underline{asp}(\underline{\underline{\tau}}). \tag{9.3.18}$$

In particular, from (9.3.18), we deduce

$$\underline{\text{div}}(\underline{x} \wedge \underline{\underline{\tau}}) = -\underline{asp}(\underline{\underline{\tau}}) \equiv -\underline{\underline{\tau}} : \underline{\underline{\underline{\mathbb{P}}}} \qquad \text{whenever } \underline{\text{div}}\,\underline{\underline{\tau}} = \underline{0}. \tag{9.3.19}$$

Remark 9.3.3. The reader familiar with the theory of continuum mechanics will recognise the similarity with the classical result that conservation of angular momentum is equivalent to the symmetry of the stress tensor if one already has conservation of momentum. □

Multiplying (9.3.19) times a vector $\underline{p} \in (H^1(K))^3$ and integrating over a domain K, we get

$$\int_K \underline{\text{div}}(\underline{x} \wedge \underline{\underline{\tau}}) \cdot \underline{p}\, d\mathbf{x} = -\int_K \underline{\underline{\tau}} : \underline{\underline{\underline{\mathbb{P}}}} \cdot \underline{p}\, d\mathbf{x} \qquad \text{whenever } \underline{\text{div}}\,\underline{\underline{\tau}} = \underline{0}, \tag{9.3.20}$$

which, integrated by parts, reads

$$\begin{aligned} \int_K \underline{\underline{\tau}} : \underline{\underline{\underline{\mathbb{P}}}} \cdot \underline{p}\, d\mathbf{x} &= -\int_K \underline{\text{div}}(\underline{x} \wedge \underline{\underline{\tau}}) \cdot \underline{p}\, d\mathbf{x} \\ &= -\int_{\partial K} \underline{p} \cdot (\underline{x} \wedge \underline{\underline{\tau}}) \cdot \underline{n}_K ds + \int_K (\underline{x} \wedge \underline{\underline{\tau}}) : \underline{\underline{\text{grad}}}\,(\underline{p})\, d\mathbf{x} \;\; \text{whenever } \underline{\text{div}}\,\underline{\underline{\tau}} = \underline{0}, \end{aligned} \tag{9.3.21}$$

where $\underline{n}_K$ is the outward unit normal vector to ∂K.

The above expressions (9.3.20) and (9.3.21) can also be written in a more suitable form, as we do in the following lemma.

Lemma 9.3.1. *On any three-dimensional domain K, for all divergence-free tensors $\underline{\underline{\tau}} \in \Sigma$ and for all $\underline{p} \in (H^1(K))^3$, we have*

$$\int_K \underline{asp}(\underline{\underline{\tau}}) \cdot \underline{p}\ \mathbf{dx} \equiv \int_K \underline{\underline{\tau}} : \underline{\underline{\underline{\mathbb{P}}}} \cdot \underline{p}\ \mathbf{dx} = \int_{\partial K} (\underline{x} \wedge \underline{p}) \cdot \underline{\underline{\tau}} \cdot \underline{n}_K\, ds - \int_K \underline{\underline{\tau}} : (\underline{x} \wedge \underline{\underline{grad}}\, \underline{p}). \tag{9.3.22}$$

Proof. The proof follows easily using (9.3.6), and then (9.3.21) and the identities

$$\underline{p} \cdot (\underline{x} \wedge \underline{\underline{\tau}}) \cdot \underline{n}_K = -(\underline{x} \wedge \underline{p}) \cdot \underline{\underline{\tau}} \cdot \underline{n}_K \tag{9.3.23}$$

$$(\underline{x} \wedge \underline{\underline{\tau}}) : \underline{\underline{\mathrm{grad}}}\,(\underline{p}) = -\underline{\underline{\tau}} : (\underline{x} \wedge \underline{\underline{\mathrm{grad}}}\, \underline{p}). \tag{9.3.24}$$

□

We can now combine (9.3.22) with the obvious fact that $\underline{asp}(\underline{\underline{\tau}}) = 0$ if and only if $\int_K \underline{asp}(\underline{\underline{\tau}}) \cdot \underline{p}\ \mathbf{dx}$ vanishes for all $\underline{p} \in (H^1(K))^3$ to get the following corollary

Corollary 9.3.1. *On any three-dimensional domain K, for all divergence-free tensors $\underline{\underline{\tau}} \in \Sigma$, the symmetry condition is equivalent to*

$$\int_{\partial K} (\underline{x} \wedge \underline{p}) \cdot \underline{\underline{\tau}} \cdot \underline{n}_K\, ds - \int_K \underline{\underline{\tau}} : (\underline{x} \wedge \underline{\underline{grad}}\, \underline{p})\, dx = 0 \quad \forall \underline{p} \in (H^1(K))^3. \tag{9.3.25}$$

Remark 9.3.4. The formulation (9.3.25) of the symmetry property is particularly suited for imposing *reduced symmetry* conditions. Indeed, requiring (9.3.25) to hold only for certain subspaces of $(H^1(K))^3$ and using (9.3.22), we will obtain weaker symmetry conditions which, however, might still provide discrete solutions with enough accuracy. In particular, these formulations will be equivalent to require that the conservation of angular momentum is satisfied in an element-wise-averaged way (as well as the conservation of momentum). □

9.3.1 Continuous Formulation of the Relaxed Symmetry Approach

We can now define a variational formulation suitable for our purpose. To do so, we first introduce a space of skew-symmetric tensors,

$$X := \{\underline{\underline{\gamma}} \mid \underline{\underline{\gamma}} \in L^2(\Omega)^{n \times n}, \underline{\underline{as}}\,(\underline{\underline{\gamma}}) = \underline{\underline{\gamma}}\} \tag{9.3.26}$$

and we introduce a new bilinear form on $\Sigma \times X$:

$$c(\underline{\underline{\tau}}, \underline{\underline{\gamma}}) := \int_\Omega \underline{\underline{as}}(\underline{\underline{\tau}}) : \underline{\underline{\gamma}}\, dx \equiv \int_\Omega \underline{\underline{as}}(\underline{\underline{\gamma}}) : \underline{\underline{\tau}}\, dx. \tag{9.3.27}$$

We can then consider the following continuous problem: *find* $(\underline{\underline{\sigma}}, \underline{u}, \underline{\underline{\omega}}) \in \tilde{\Sigma} \times U \times X$ *such that*

$$\begin{cases} a(\underline{\underline{\sigma}}, \underline{\underline{\tau}}) + b(\underline{\underline{\tau}}, \underline{u}) + c(\underline{\underline{\tau}}, \underline{\underline{\omega}}) = 0, & \forall\, \underline{\underline{\tau}} \in \tilde{\Sigma}, \\ b(\underline{\underline{\sigma}}, \underline{v}) = (\underline{f}, \underline{v}), & \forall\, \underline{v} \in U, \\ c(\underline{\underline{\sigma}}, \underline{\underline{\gamma}}) = 0, & \forall\, \underline{\underline{\gamma}} \in X. \end{cases} \tag{9.3.28}$$

First of all, we have to prove an existence result for this problem. With respect to the general theory, we first remark that the problem can be written in the form, considered in Sect. 3.2.5,

$$\begin{pmatrix} \mathbb{A} & \mathbb{B}^t \\ \mathbb{B} & 0 \end{pmatrix} = \begin{pmatrix} \mathbb{A} & B^t & C^t \\ B & 0 & 0 \\ C & 0 & 0 \end{pmatrix} \tag{9.3.29}$$

where $\mathbb{A}$ is associated with the bilinear form $a(\cdot,\cdot)$ (now operating in Σ instead of Σ_S) while the operator $\mathbb{B} : \Sigma \to (U \times X)'$ is associated with the bilinear form

$$(\underline{\underline{\tau}}, (\underline{v}, \underline{\underline{\gamma}})) \to b(\underline{\underline{\tau}}, \underline{v}) + c(\underline{\underline{\tau}}, \underline{\underline{\gamma}}).$$

In particular, the kernel of the operator $\mathbb{B}$ is given by

$$\begin{aligned} ker(\mathbb{B}) &= \{\underline{\underline{\tau}} \in \Sigma \text{ s.t. } b(\underline{\underline{\tau}}, \underline{v}) + c(\underline{\underline{\tau}}, \underline{\underline{\gamma}}) = 0 \quad \forall\, \underline{v} \in U, \forall\, \underline{\underline{\gamma}} \in X\} = \\ &= \{\underline{\underline{\tau}} \in \Sigma \text{ s.t. } \underline{\mathrm{div}}\, \underline{\underline{\tau}} = \underline{0} \textbf{ and } \underline{\underline{as}}(\underline{\underline{\tau}}) = \underline{\underline{0}}\}. \end{aligned} \tag{9.3.30}$$

In view of (9.3.30) and (9.1.23), we immediately have the following result.

Proposition 9.3.1. *There exists a constant* $C > 0$*, independent of* λ*, such that for every* $\underline{\underline{\tau}}$ *in* $\tilde{\Sigma} \cap ker(\mathbb{B})$*, we have*

$$a(\underline{\underline{\tau}}, \underline{\underline{\tau}}) \geq C \|\underline{\underline{\tau}}\|^2_{\underline{\underline{H}}(\underline{div};\Omega)}. \tag{9.3.31}$$

□

□

Hence, in order to see (9.3.28) is well posed, we just need an *inf-sup* condition of the form

$$\exists\, C > 0 \text{ such that } \inf_{\underline{v} \in U, \underline{\underline{\gamma}} \in X} \ \sup_{\underline{\underline{\tau}} \in \Sigma} \frac{b(\underline{\underline{\tau}}, \underline{v}) + c(\underline{\underline{\tau}}, \underline{\underline{\gamma}})}{\|\underline{\underline{\tau}}\|_\Sigma \left(\|\underline{v}\|_U + \|\underline{\underline{\gamma}}\|_X\right)} \geq C. \tag{9.3.32}$$

The above condition is indeed satisfied, as we can see in the following proposition.

Proposition 9.3.2. *There exists a constant C such that for any $\underline{v} \in U$ and $\underline{\underline{\gamma}} \in X$, there exists $\underline{\underline{\tau}} \in \Sigma$ such that*

$$b(\underline{\underline{\tau}}, \underline{v}) + c(\underline{\underline{\tau}}, \underline{\underline{\gamma}}) = \|\underline{v}\|_U^2 + \|\underline{\underline{\gamma}}\|_X^2 \tag{9.3.33}$$

and

$$\|\underline{\underline{\tau}}\|_\Sigma \leq C\,(\|\underline{v}\|_U + \|\underline{\underline{\gamma}}\|_X). \tag{9.3.34}$$

Proof. We first give the proof for the two-dimensional case and afterwards for the three-dimensional one. We give it in detail because the technique will be relevant for the construction of the discrete approximations.

The two-dimensional case. The construction of $\underline{\underline{\tau}}$ will be done in two steps. The first one is to build a tensor $\underline{\underline{\tau}}^1 \in \Sigma$ such that

$$\begin{cases} b(\underline{\underline{\tau}}^1, \underline{w}) = (\underline{v}, \underline{w}), & \forall\, \underline{w} \in U, \\ \|\underline{\underline{\tau}}^1\|_\Sigma \leq C\, \|\underline{v}\|_U. \end{cases} \tag{9.3.35}$$

This is easily done, even with a symmetric $\underline{\underline{\tau}}^1$. One could, for instance, solve a classical elasticity problem and take the associated stress field. The second step is to correct this tensor by a divergence-free tensor $\underline{\underline{\tau}}^2$ such that $\underline{\underline{as}}(\underline{\underline{\tau}}^2) = \underline{\underline{\gamma}} - \underline{\underline{as}}(\underline{\underline{\tau}}^1)$. In the two-dimensional case, this divergence-free tensor is obtained by taking the i-th row ($i = 1, 2$) made by the $\underline{\mathrm{curl}}$ (Remark 2.1.5) of the i-th component of a vector $\underline{\Psi} \equiv (\psi_1, \psi_2)$, that is, as we have seen in (9.3.11),

$$\underline{\underline{\tau}}^2 = \begin{pmatrix} -\partial_2\psi_1 & \partial_1\psi_1 \\ -\partial_2\psi_2 & \partial_1\psi_2 \end{pmatrix}. \tag{9.3.36}$$

One sees immediately that the condition $\underline{\underline{as}}(\underline{\underline{\tau}}^2) = \underline{\underline{\gamma}} - \underline{\underline{as}}(\underline{\underline{\tau}}^1)$ is equivalent to

$$\underline{\underline{S}}^2(\partial_1\psi_1 + \partial_2\psi_2) = \underline{\underline{S}}^2(\mathrm{div}\,\underline{\Psi}) = \underline{\underline{as}}(\underline{\underline{\tau}}^2) = \underline{\underline{\gamma}} - \underline{\underline{as}}(\underline{\underline{\tau}}^1) \tag{9.3.37}$$

where $\underline{\underline{S}}^2(q)$ is defined by (9.3.2). In order to satisfy equation (9.3.37) with the required continuity condition, it is then sufficient to solve a Stokes problem for $\underline{\Psi}$.

The three-dimensional case. In the three-dimensional case, the situation is slightly more complex. The first step (9.3.35) can still be carried out easily, but the divergence-free tensor $\underline{\underline{\tau}}^2$ will now be the $\underline{\underline{\mathrm{curl}}}$ of another tensor $\underline{\underline{\Psi}}$. This means that we will look for a $\underline{\underline{\tau}}^2$ of the form:

$$\underline{\underline{\tau}}^2 = \underline{\underline{\mathrm{curl}}}\,\underline{\underline{\Psi}}. \tag{9.3.38}$$

It is convenient to introduce another tensor $\underline{\underline{\Phi}}$ given by

$$\underline{\underline{\Phi}} := \mathcal{A}\underline{\underline{\Psi}}. \tag{9.3.39}$$

According to (9.3.38), it is easy to check that we would always have

$$\underline{\operatorname{div}}\, \underline{\underline{\tau}}^2 = 0 \tag{9.3.40}$$

independently of the choice of $\underline{\underline{\Psi}}$. On the other hand, in order to have $\underline{\underline{as}}(\underline{\underline{\tau}}^2) = \underline{\underline{\gamma}} - \underline{\underline{as}}(\underline{\underline{\tau}}^1)$ (or, rather, $\underline{asp}(\underline{\underline{\tau}}^2) = \underline{asp}(\underline{\underline{\gamma}}) - \underline{asp}(\underline{\underline{\tau}}^1)$), we have to require, as in (9.3.37),

$$\underline{asp}(\underline{\underline{\tau}}^2) \equiv \underline{asp}(\underline{\underline{\operatorname{curl}}}\, \underline{\underline{\Psi}}) \equiv \underline{\operatorname{div}}\, \mathcal{A}\underline{\underline{\Psi}} \equiv \underline{\operatorname{div}}\, \underline{\underline{\Phi}} = \underline{asp}(\underline{\underline{\gamma}}) - \underline{asp}(\underline{\underline{\tau}}^1). \tag{9.3.41}$$

Note that in (9.3.41), we used (9.3.15) and (9.3.39). Now, (9.3.41) can be easily obtained by requiring each line of $\underline{\underline{\Phi}}$ to be the solution of a suitable Stokes problem, that is,

$$\begin{cases} -\Delta \underline{\underline{\Phi}} + \underline{\underline{\operatorname{grad}}}\, \underline{p} = 0 \\ \underline{\operatorname{div}}\, \underline{\underline{\Phi}} = \underline{asp}(\underline{\underline{\gamma}} - \underline{\underline{\tau}}^1). \end{cases} \tag{9.3.42}$$

It is also immediate to verify that the above construction, which relies on the solution of well posed Stokes problems, satisfies the required continuity conditions as well. □

Remark 9.3.5. The above result shows that symmetry or average symmetry of $\underline{\underline{\operatorname{curl}}}\, \underline{\underline{\Psi}}$ is a condition related to the rows of the tensor $\underline{\underline{\Phi}}$ and hence, in view of Remark 9.3.2, to the columns of the tensor $\underline{\underline{\Psi}}$. □

Remark 9.3.6. The above construction is not the most general in the three-dimensional case. It is not necessary to get $\underline{\underline{\Psi}}$ with all the regularity implied by our procedure, that is, for a convex Ω, $\underline{\underline{\Psi}} \in (H^1(\Omega))^{3\times 3}$. In fact, it is sufficient to build $\underline{\underline{\Psi}} \in (H(\operatorname{curl}, \Omega))^{3r}$. This will have as a consequence that some approximations cannot be generated with the discrete solution of Stokes problems. □

9.3.2 Numerical Approximation of Relaxed-Symmetry Formulations

We can now start considering the approximation of the variational formulation (9.3.28). We want to choose subspaces Σ_h, U_h, X_h of Σ, U, X and to solve the problem: find $(\underline{\underline{\sigma}}_h, \underline{u}_h, \underline{\underline{\omega}}_h) \in \tilde{\Sigma}_h \times U_h \times X_h$ such that

$$\begin{cases} a(\underline{\underline{\sigma}}_h, \underline{\underline{\tau}}_h) + b(\underline{\underline{\tau}}_h, \underline{u}_h) + c(\underline{\underline{\tau}}_h, \underline{\underline{\omega}}_h) = 0, & \forall\, \underline{\underline{\tau}}_h \in \tilde{\Sigma}_h, \\ b(\underline{\underline{\sigma}}_h, \underline{v}_h) = (\underline{f}, \underline{v}_h), & \forall\, \underline{v}_h \in U_h, \\ c(\underline{\underline{\sigma}}_h, \underline{\underline{\gamma}}_h) = 0, & \forall\, \underline{\underline{\gamma}}_h \in X_h. \end{cases} \tag{9.3.43}$$

Before starting to look for sufficient conditions on the spaces Σ_h, U_h, X_h that might ensure existence and uniqueness of the solution of (9.3.43), possibly with optimal error bounds, we will need the following slightly refined form of Proposition 9.3.2.

Proposition 9.3.3. *Assume that Ω is convex. Assume that U_h is a subspace of U and that $X_h \equiv \underline{\underline{S}}^3(W_h^3)$ is a finite dimensional subspace of X. Assume further that Ξ_h is a piecewise polynomial space such that (Ξ_h^3, W_h) is a stable element for Stokes. Then, for every $\underline{v}_h \in U_h$ and for every $\underline{\underline{\gamma}}_h \in X_h$, we can find $\underline{\underline{\tau}}(h)$, $\underline{\underline{\tau}}^1(h)$, and $\underline{\underline{\tau}}^2(h)$ such that*

$$\begin{cases} \underline{\underline{\tau}}(h) = \underline{\underline{\tau}}^1(h) + \underline{\underline{\tau}}^2(h) \\ \underline{\underline{\tau}}^1(h) \in (H^1(\Omega))^{3\times 3}, \ \underline{\underline{\tau}}^2(h) \in \underline{\underline{curl}}\,(\Xi_h^{3\times 3}), \ \underline{div}\,\underline{\underline{\tau}}(h) \in U_h \end{cases} \tag{9.3.44}$$

satisfying

$$\underline{div}\,\underline{\underline{\tau}} = \underline{v}_h, \qquad c(\underline{\underline{\tau}}, \underline{\underline{\delta}}_h) = c(\underline{\underline{\gamma}}_h, \underline{\underline{\delta}}_h)\ \forall\, \underline{\underline{\delta}}_h \in X_h, \tag{9.3.45}$$

and

$$\|\underline{\underline{\tau}}^1(h)\|_{H^1} + \|\underline{\underline{\tau}}^2(h)\|_{L^2} \le C\big(\|\underline{v}_h\|_U + \|\underline{\underline{\gamma}}_h\|_X\big), \tag{9.3.46}$$

with C independent of $\underline{v}_h$, $\underline{\underline{\gamma}}_h$, and h.

Proof. The proof follows immediately, going back to the proof of Proposition 9.3.2 and checking that everything fits in place. □

In what follows, whenever the assumptions of Proposition 9.3.3 are satisfied, we will denote by $\Sigma(h)$ the space

$$\Sigma(h) := \{\underline{\underline{\tau}}(h) \in \Sigma \text{ such that } \underline{\underline{\tau}}(h) \text{ has the form (9.3.44)}\}. \tag{9.3.47}$$

We now start choosing suitable assumptions on our discretisations. First of all, in view of Proposition 9.1.1 and of (9.3.30), in order to have ellipticity in the kernel with a constant independent of λ, we shall try to build approximations satisfying the "inclusion of kernels property" that we now have to require in Σ_h (rather than Σ_{Sh} as in (9.1.24)):

$$\mathrm{Ker}B_h \equiv \{\underline{\underline{\tau}}_h \in \Sigma_h |\; b(\underline{\underline{\tau}}_h, \underline{v}_h) = 0 \;\forall\, \underline{v}_h \in U_h\} \subseteq \mathrm{Ker}B \equiv \{\underline{\underline{\tau}} \in \Sigma |\; \underline{\mathrm{div}}\,\underline{\underline{\tau}} = 0\}. \tag{9.3.48}$$

To do so, we can use n copies (one for each line) of some of the finite element discretisations of $H(\mathrm{div}, \Omega)$ available in the literature. In particular, for all our methods, we will assume that the following properties are satisfied. We shall assume that Σ_h and U_h are such that

$$\underline{\mathrm{div}}(\Sigma_h) \subseteq U_h, \tag{9.3.49}$$

and we shall also assume that we can construct an operator Π_h^1 from $\Sigma(h)$ to Σ_h such that for each element K and for each $\underline{\underline{\tau}}(h) \in \Sigma(h)$:

$$\|\Pi_h^1 \underline{\underline{\tau}}^1(h)\|_{\underline{\underline{H}}(\underline{\mathrm{div}};K)} \le C\, \|\underline{\underline{\tau}}^1(h)\|_{(H^1(K))^{n\times n}} \tag{9.3.50}$$

with C a constant independent of $\underline{\underline{\tau}}(h)$ and h, together with

$$\Pi_h^1 \underline{\underline{\tau}} = \underline{\mathrm{div}}\,\underline{\underline{\tau}} \qquad \forall\, \underline{\underline{\tau}} \in (H^1(K))^{n\times n}. \tag{9.3.51}$$

Assuming that we made a choice that takes care of that, we still must check the discrete *inf-sup* condition:

$$\inf_{\underline{v}_h \in U_h, \underline{\underline{\gamma}}_h \in X_h} \; \sup_{\underline{\underline{\tau}}_h \in \Sigma_h} \frac{b(\underline{\underline{\tau}}_h, \underline{v}_h) + c(\underline{\underline{\tau}}_h, \underline{\underline{\gamma}}_h)}{\|\underline{\underline{\tau}}_h\|_\Sigma \,(\|\underline{v}_h\|_U + \|\underline{\underline{\gamma}}_h\|_X)} \ge C > 0. \tag{9.3.52}$$

In order to apply the general theory, we will want to build, following Proposition 5.1.2, a B-compatible interpolation operator $\Pi_h : \Sigma \to \Sigma_h$ satisfying

$$\begin{aligned} &b(\underline{\underline{\tau}} - \Pi_h \underline{\underline{\tau}}, \underline{v}_h) + c(\underline{\underline{\tau}} - \Pi_h \underline{\underline{\tau}}, \underline{\underline{\gamma}}_h) = 0, \forall\, \underline{v}_h \in U_h, \forall\, \underline{\underline{\gamma}}_h \in X_h, \\ &\|\Pi_h \underline{\underline{\tau}}\|_\Sigma \le C\, \|\underline{\underline{\tau}}\|_\Sigma. \end{aligned} \tag{9.3.53}$$

To do this, we shall try to proceed in the same way that we used to prove the continuous *inf-sup* condition: we shall first build $\underline{\underline{\tau}}_h^1$ so that its divergence satisfies the first requirement

$$\begin{aligned} &b(\underline{\underline{\tau}} - \underline{\underline{\tau}}_h^1, \underline{v}_h) = 0, \quad \forall\, \underline{v}_h \in U_h, \\ &\|\underline{\underline{\tau}}_h^1\|_\Sigma \le C\, \|\underline{\underline{\tau}}\|_\Sigma, \end{aligned} \tag{9.3.54}$$

and then we correct this tensor by a divergence-free tensor $\underline{\underline{\tau}}_h^2$ to obtain the required asymmetry:

$$\begin{aligned} &c(\underline{\underline{\tau}} - \underline{\underline{\tau}}_h^2, \underline{\underline{\gamma}}_h) = c(\underline{\underline{\tau}}_h^1, \underline{\underline{\gamma}}_h), \quad \forall \underline{\underline{\gamma}}_h \in X_h, \\ &\|\underline{\underline{\tau}}_h^2\|_\Sigma \le C\,\|\underline{\underline{\tau}} - \underline{\underline{\tau}}_h^1\|_\Sigma. \end{aligned} \tag{9.3.55}$$

Remark 9.3.7. Referring to the continuous case, we can try to build $\underline{\underline{\tau}}_h^2$ by solving (discrete) Stokes problems (one in two dimensions and three in three dimensions), returning a tensor $\underline{\underline{\Psi}}_h$, and then by taking $\underline{\underline{\tau}}_h^2 = \underline{\underline{\mathrm{curl}}}(\underline{\underline{\Psi}}_h)$. It follows that one possible key to our constructions will be stable elements for the Stokes problems, together with the *inclusion of the kernels* property (9.3.48). However, the approach through Stokes problems is less effective for the three-dimensional case, and we shall present an alternative one in Sect. 9.3. □

Before considering specific constructions, let us state the error estimate that one can expect for the discrete problem (9.3.43).

Theorem 9.3.1. *Let us suppose that the spaces $\Sigma_h \times U_h \times X_h$ are such that*

- $\Sigma_h \times U_h$ *satisfies* (9.3.48)
- $\Sigma_h \times U_h \times X_h$ *satisfies* (9.3.52).

Then, (9.3.43) *has a unique solution. Moreover, if* $(\underline{\underline{\sigma}}, \underline{u}, \underline{\underline{\omega}}) \in \Sigma \times U \times X$ *is the solution of* (9.3.28) *and* $(\underline{\underline{\sigma}}_h, \underline{u}_h, \underline{\underline{\omega}}_h) \in \Sigma_h \times U_h \times X_h$ *is the solution of* (9.3.43), *then we have*

$$\begin{aligned} &\|\underline{\underline{\sigma}}_h - \underline{\underline{\sigma}}\|_0 + \|\underline{u}_h - \underline{u}\|_0 + \|\underline{\underline{\omega}}_h - \underline{\underline{\omega}}\|_0 \\ &\quad \le C\Big(\inf_{\underline{\underline{\tau}} \in \Sigma_h} \|\underline{\underline{\tau}}_h - \underline{\underline{\sigma}}\|_0 + \inf_{\underline{v}_h \in U_h} \|\underline{v}_h - \underline{u}\|_0 + \inf_{\underline{\underline{\phi}}_h \in X_h} \|\underline{\underline{\phi}}_h - \underline{\underline{\omega}}\|_0\Big). \end{aligned} \tag{9.3.56}$$

The proof is an easy consequence of the general theory on mixed formulations, for example Theorem 5.2.2. One can see from (9.3.56) that it is important to balance the quality of the approximation for the three components of the solution. In particular, symmetry must be imposed at least to the same precision as the approximation properties for the other variables.

Remark 9.3.8. It is clear that the inclusion of kernels (9.3.48) is not necessary, but it makes the theory easier. We shall discuss later how it can be replaced by a suitable stabilising term. □

Remark 9.3.9. It is easy to see that if the space Σ_h contains (as we implicitly assume) the constant identity tensor $\underline{\underline{I}}$, then solving the problem in $\tilde{\Sigma}_h \times U_h \times X_h$ or in $\Sigma_h \times U_h \times X_h$ gives exactly the same result. □

Remark 9.3.10. In a few cases, we shall be able to build explicitly a basis for the space Σ_{Sh} of discrete symmetric tensors. We shall then be able to consider the problem

$$\begin{cases} a(\underline{\underline{\sigma}}_h, \underline{\underline{\tau}}_h) + b(\underline{\underline{\tau}}_h, \underline{u}_h) = 0, & \forall \underline{\underline{\tau}}_h \in \Sigma_{Sh}, \\ b(\underline{\underline{\sigma}}_h, \underline{v}_h) = (\underline{f}, \underline{v}_h), & \forall \underline{v}_h \in U_h. \end{cases} \tag{9.3.57}$$

□

Remark 9.3.11 (Hybrid methods). We have seen in Chap. 7, in particular in Sect. 7.2.2, that inter-element continuity can be efficiently treated through the introduction of Lagrange multipliers. This technique can evidently be employed here. This was indeed how things were done in [7]. The multipliers can be assimilated to values of the displacement on the boundary of the element and elimination of internal degrees of freedom yields a problem with only displacement unknowns. For incompressible materials, pressure also remains as a variable, as it should be. □

9.4 Some Families of Methods with Reduced Symmetry

Our general strategy will be to show, mimicking essentially Propositions 9.3.2 and 9.3.3, that for every $\underline{v}_h \in U_h$ and for every $\underline{\underline{\gamma}}_h \in X_h$, we can find a tensor $\underline{\underline{\tau}}_h \in \Sigma_h$ of the form

$$\underline{\underline{\tau}}_h = \underline{\underline{\tau}}_h^1 + \underline{\underline{\tau}}_h^2 \tag{9.4.1}$$

such that

$$\underline{\mathrm{div}}\, \underline{\underline{\tau}}_h^1 = \underline{v}_h, \quad \underline{\mathrm{div}}\, \underline{\underline{\tau}}_h^2 = 0, \quad c(\underline{\underline{\tau}}_h^2, \underline{\underline{\delta}}_h) = c(\underline{\underline{\gamma}}_h - \underline{\underline{\tau}}_h^1, \underline{\underline{\delta}}_h)\ \forall \underline{\underline{\delta}}_h \in X_h, \tag{9.4.2}$$

and

$$\|\underline{\underline{\tau}}_h^1\|_\Sigma + \|\underline{\underline{\tau}}_h^2\|_\Sigma \le C\big(\|\underline{v}_h\|_U + \|\underline{\underline{\gamma}}_h\|_X\big) \tag{9.4.3}$$

with C independent of $\underline{v}_h$, $\underline{\underline{\gamma}}_h$, and h.

It is clear that this will imply (9.3.52). In the sequel, we will present some families of choices for the spaces Σ_h, U_h and X_h that will satisfy the above property, together with (9.3.48). For these choices, the assumptions of Theorem 9.3.1 will therefore be satisfied, and consequently, convergence and the optimal error bounds for (9.3.43) will be ensured.

9.4.1 Methods Based on Stokes Elements

The following methods will be based on the availability of stable finite element pairs for Stokes. As already pointed out in [80], this approach will easily allow the introduction and the analysis of all the most interesting reduced elasticity methods in the two-dimensional case, but only a few methods with some interest in the three-dimensional one. Hence, for this class of methods, we will not neglect the two-dimensional case, and we will give a detailed statement and proof for both the three-dimensional and the two-dimensional cases. We start with the three-dimensional one.

Theorem 9.4.1. *Assume that the assumptions of Proposition 9.3.3 are satisfied, together with* (9.3.49), *and that the space* Ξ_h *of Proposition 9.3.3 satisfies*

$$\underline{\underline{curl}}(\Xi_h^{3\times 3}) \subseteq \Sigma_h. \tag{9.4.4}$$

Suppose moreover that the mapping Π_h^1 *from* $\Sigma(h)$ *into* Σ_h *satisfies* (9.3.50) *and* (9.3.51). *Then, the triplet*

$$\Sigma_h \times U_h \times X_h \quad with\ X_h = \underline{\underline{S}}^3(W_h^3) \tag{9.4.5}$$

satisfies the conditions of Theorem 9.3.1.

Proof. For every pair $(\underline{v}_h, \underline{\underline{\gamma}}_h)$ in $U_h \times X_h$, we construct $\underline{\underline{\tau}}(h)$ following Proposition 9.3.3, and then $\underline{\underline{\tau}}_h^1 := \Pi_h^1 \underline{\underline{\tau}}^1(h)$. Using the fact that (in the assumptions of Proposition 9.3.3) the pair (Ξ_h^3, W_h) is a stable Stokes element, we proceed as in the proof of Proposition 9.3.2 and choose $\underline{\underline{\phi}}_h \in \Xi_h^{3\times 3}$ such that

$$\int_\Omega \underline{\mathrm{div}}\, \underline{\underline{\phi}}_h \cdot \underline{\delta}_h \, dx = \int_\Omega \underline{asp}(\underline{\underline{\tau}}(h) - \underline{\underline{\tau}}_h^1) \cdot \underline{\delta}_h \, dx \qquad \forall \underline{\delta}_h \in W_h^3. \tag{9.4.6}$$

Then, we take $\underline{\underline{\tau}}_h^2 := \underline{\underline{\mathrm{curl}}}(\mathcal{A}^{-1} \underline{\underline{\phi}}_h)$, that belongs to Σ_h due to (9.4.4), and set $\underline{\underline{\tau}}_h := \underline{\underline{\tau}}_h^1 + \underline{\underline{\tau}}_h^2$. Using (9.3.15), it is then an easy matter to see that (9.3.53) is satisfied. □

The theorem describing the two-dimensional case would instead be the following one.

Theorem 9.4.2. *Assume that* Ω *is convex. Assume that* U_h *is a subspace of* U *and that* $X_h \equiv \underline{\underline{S}}^2(W_h)$ *is a finite dimensional subspace of* X. *Assume further that* Ξ_h *is a piecewise polynomial space such that* (Ξ_h^2, W_h) *is a stable element for Stokes and that, with the notation of* (9.3.11),

$$\underline{\underline{curl}}(\Xi_h^2) \subseteq \Sigma_h. \tag{9.4.7}$$

Assume further that (9.3.49) *holds and that there exists a mapping* Π_h^1 *from* $\Sigma(h)$ *into* Σ_h *satisfying* (9.3.50) *and* (9.3.51). *Then, the triplet*

$$\Sigma_h \times U_h \times X_h \quad with\ X_h = \underline{\underline{S}}^2(W_h) \tag{9.4.8}$$

satisfies the conditions of Theorem 9.3.1.

Proof. The proof follows closely the one of Theorem 9.4.1. For every pair $(\underline{v}_h, \underline{\underline{\gamma}}_h)$ in $U_h \times X_h$, we construct $\underline{\underline{\tau}}(h)$ following Proposition 9.3.3, and then $\underline{\underline{\tau}}_h^1 := \Pi_h^1 \underline{\underline{\tau}}^1(h)$. Using the fact that (Ξ_h^2, W_h) is a stable Stokes element, we proceed as in the proof of Proposition 9.3.2 and choose a $\underline{\phi}_h \in \Xi_h^2$ such that

$$\int_\Omega \operatorname{div}\underline{\phi}_h \, \delta_h \, dx = \int_\Omega asp((\underline{\underline{\tau}}(h) - \underline{\underline{\tau}}_h^1)) \, \delta_h \, dx \qquad \forall \delta_h \in W_h. \tag{9.4.9}$$

Then we take $\underline{\underline{\tau}}_h^2 := \underline{\underline{\operatorname{curl}}}\, \underline{\phi}_h$, that belongs to Σ_h due to (9.4.4), and set $\underline{\underline{\tau}}_h := \underline{\underline{\tau}}_h^1 + \underline{\underline{\tau}}_h^2$. We note that $asp(\underline{\underline{\tau}}_h) = \operatorname{div}\underline{\phi}_h$ as in (9.3.37). It is then an easy matter to see that (9.3.53) is satisfied. □

Remark 9.4.1. The analysis through a Stokes problem was first introduced for the two-dimensional case in [204] but it was implicit in [7]. There are several examples in the literature of approximations that could be inserted in the above theory.

- The PEERS element of Arnold-Brezzi-Douglas [24], or its variant by Brezzi-Douglas-Marini [120] is obtained through the stability of the MINI element.
- The element of Amara-Thomas [7] can be analysed through the Crouzeix-Raviart element. We shall introduce below a generalisation of this element.
- The elements of Arnold, Falk and Winther recalled below can, as we shall see, be amenable to the Stokes approach in the two-dimensional case. □

We already noted that using a Stokes problem was not the most general way of obtaining the continuous inf-sup condition. In the same way, the above result will enable us to obtain some useful constructions of relaxed symmetry tensors but it does not yield all constructions. However, there exists a good number of elements which are stable for Stokes which can yield useful constructions.

Example 9.4.1 (A three-dimensional family using generalised Taylor-Hood elements). On tetrahedra, we define $\Sigma_h := (RT_k)^{3r}$ and $U_h := (\mathcal{L}_k^0)^3$; it is immediate that we have (9.3.48) and (9.1.11). For the Stokes problem, we consider the generalised Taylor–Hood elements in which velocity is approximated by a space of polynomial elements $(\mathcal{L}_k^1)^3$ and the pressure is continuous piecewise linear (i.e. it belongs to $\mathcal{L}_{k-1}^1$). In particular, this immediately yields a second order elasticity element in which we have

$$\begin{aligned} \Sigma_h &:= (RT_1)^3, \\ U_h &:= (\mathcal{L}_1^0)^3, \\ X_h &:= \underline{\underline{S}}^3((\mathcal{L}_1^1)^3). \end{aligned} \tag{9.4.10}$$

Symmetry is enforced as in the PEERS element of Example 9.4.2 but we now have second order (or higher if wanted) accuracy. Using degrees of freedom on vertices for X_h is a very economical option. It should also be noted that using a continuous X_h is not a serious drawback. Recent developments in iterative methods [185, 186] for saddle-point problems provide an efficient solution method. □

At this point, we recall (see Proposition 8.5.8) that *every Stokes pair* (V_h, Q_h), *provided that* Q_h *is made of continuous functions and* V_h *contains piecewise linear vectors, can be stabilised by bubbles*, meaning that there exists an extended space $V_h^e \supseteq V_h$ (obtained adding suitable bubble functions to V_h) such that (V_h^e, Q_h) is a stable Stokes pair. As a corollary, we have the following result.

Corollary 9.4.1. *Assume that the assumptions of Proposition 9.3.3 are satisfied, together with* (9.3.49)*. Suppose moreover that the mapping* Π_h^1 *from* $\Sigma(h)$ *into* Σ_h *satisfies* (9.3.50)*. Assume finally that* $W_h \subseteq C^0(\overline{\Omega})$*. Then, there exists a space* $\Sigma_h^e \supseteq \Sigma_h$ *(obtained from* Σ_h *through the addition of suitable* $H(div)$*-bubbles) such that the triplet*

$$\Sigma_h^e \times U_h \times X_h \quad \text{with } X_h = \underline{\underline{S}}^3(W_h^3) \tag{9.4.11}$$

satisfies the conditions of Theorem 9.3.1.

Indeed, it will be enough to set

$$\Sigma_h^e := \Sigma_h + \underline{\underline{\mathrm{curl}}}(\mathcal{A}^{-1}(V_h^e)^3) \tag{9.4.12}$$

and note that, as V_h^e is made of bubbles, the normal component of each element of $\underline{\underline{\mathrm{curl}}}(\mathcal{A}^{-1}(V_h^e)^3)$ will vanish at the boundary of each element. On the other hand, $\underline{\mathrm{div}}(\Sigma_h)$ will not be changed, as we add only divergence-free tensors.

Example 9.4.2. As examples of applications of the above strategy to the three-dimensional case, we could consider the three-dimensional version of the PEERS element. We define Σ_h as the space of tensors (in three dimensions) where each line is an element of the lowest order Raviart-Thomas space on tetrahedra. Using for U_h a space of piecewise constant vectors, it is immediate that we have (9.3.48) and (9.1.11). For the Stokes problem, we use the three-dimensional MINI element:

$$\begin{aligned} V_h &:= (\mathcal{L}_1^1)^3 \oplus (B_4)^3, \\ Q_h &:= \mathcal{L}_1^1, \end{aligned} \tag{9.4.13}$$

where B_4 is the space generated by the element-wise *quartic bubbles* b_4 defined in Remark 2.2.4. We can then augment the space Σ_h by adding, in each element, $\underline{\underline{\mathrm{curl}}}(\mathcal{A}(B_4)^{3\times 3})$. This will leave (9.3.48) and (9.1.11) still holding true, and (9.4.7) will now hold as well. All the assumptions of Theorem 9.4.1 will then be satisfied.

□

9.4.2 *Stabilisation by* $H(\underline{\mathrm{curl}})$ *Bubbles*

We have already stated that constructions based on Stokes elements are not a general procedure in the three-dimensional case. A second variant can be found, assuming some additional properties for the operator Π_h^1 and less on the space Ξ_h.

Theorem 9.4.3. *Assume again that the assumptions of Proposition 9.3.3 are satisfied, together with* (9.3.49)*. Suppose moreover that the mapping Π_h^1 from $\Sigma(h)$ into Σ_h satisfies* (9.3.50), (9.3.51), *and*

$$\int_\Omega \underline{asp}(\underline{\underline{\tau}} - \Pi_h^1 \underline{\underline{\tau}})\, dx = \underline{0}. \tag{9.4.14}$$

Let the operator $\underline{\underline{grad}}_h$ be the element-wise gradient and assume finally that we have a space $\mathbb{H}_h$ made of $H(curl)$-bubbles such that there exists $C > 0$ such that

$$\inf_{\underline{w}_h \in W_h^3} \sup_{\underline{\underline{\psi}}_h \in (\mathbb{H}_h)^{3r}} \frac{\int_\Omega \underline{\underline{\psi}} : \mathcal{A}\, \underline{\underline{grad}}_h\, \underline{w}_h\, dx}{\|\underline{\underline{\psi}}_h\|_{(H(curl;\Omega))^{3r}} \|\underline{w}_h\|_{(L^2(\Omega))3}} \geq C \tag{9.4.15}$$

and

$$\underline{\underline{curl}}((\mathbb{H}_h)^{3r}) \subseteq \Sigma_h. \tag{9.4.16}$$

Then, the triplet

$$\Sigma_h \times U_h \times X_h \quad with\ X_h = \underline{\underline{S}}^3(W_h^3) \tag{9.4.17}$$

satisfies the conditions of Theorem 9.3.1.

Remark 9.4.2. Before presenting the proof, we recall that $\underline{\psi}$ is an $H(curl)$-bubble if it belongs to $H(\text{curl}, \Omega)$ and its tangential components vanish at the boundary of each element. We have presented examples of such bubbles in Sects. 2.6.3 and 2.6.4. □

Proof. For every pair $(\underline{v}_h, \underline{\underline{\gamma}}_h)$ in $U_h \times X_h$, we construct $\underline{\underline{\tau}}(h)$ following Proposition 9.3.3, and then $\underline{\underline{\tau}}_h^1 := \Pi_h^1 \underline{\underline{\tau}}(h)$. We now note that (9.4.15) implies that we can choose an element $\underline{\underline{\psi}}_h \in \mathbb{H}_h^{3r}$ such that

$$\int_\Omega \underline{\underline{\psi}}_h : \mathcal{A}\, \underline{\underline{\text{grad}}}\, \underline{w}_h\, dx = -\int_\Omega \underline{asp}(\underline{\underline{\tau}}(h) - \Pi_h^1 \underline{\underline{\tau}}(h)) \cdot \underline{w}_h\, dx \qquad \forall \underline{w}_h \in W_h^3 \tag{9.4.18}$$

with

$$\|\underline{\underline{\psi}}_h\|_{(H(\text{curl};\Omega))^{3r}} \leq (1/C) \|\underline{w}_h\|_{(L^2(\Omega))^3}. \tag{9.4.19}$$

Note that (9.4.18) is compatible, thanks to (9.4.14). We then choose

$$\Pi_h^2\underline{\underline{\tau}}(h) := \underline{\underline{\mathrm{curl}}}\,\underline{\underline{\psi}}_h. \tag{9.4.20}$$

Now, for all $\underline{w}_h \in W_h^3$, we use successively (9.4.20) and (9.3.15), then we integrate by parts using the fact that the lines of $\underline{\underline{\psi}}_h$ are $H(curl)$-bubbles and (9.3.16). From the symmetry of $\mathcal{A}$ and finally (9.4.18), we have

$$\begin{aligned}\int_\Omega \underline{asp}(\Pi_h^2\underline{\underline{\tau}}(h))\cdot\underline{w}_h\,dx &= \int_\Omega \underline{asp}(\underline{\underline{\mathrm{curl}}}\,\underline{\underline{\psi}}_h)\cdot\underline{w}_h\,dx\\ &= \int_\Omega(\underline{\mathrm{div}}\,\mathcal{A}\underline{\underline{\psi}}_h)\cdot\underline{w}_h\,dx = -\int_\Omega(\mathcal{A}\underline{\underline{\psi}}_h):\underline{\underline{\mathrm{grad}}}\,\underline{w}_h\,dx\\ &= -\int_\Omega \underline{\underline{\psi}}_h:\mathcal{A}\,\underline{\underline{\mathrm{grad}}}\,\underline{w}_h\,dx = \int_\Omega \underline{asp}(\underline{\underline{\tau}}-\Pi_h^1\underline{\underline{\tau}}(h))\cdot\underline{w}_h\,dx.\end{aligned} \tag{9.4.21}$$

From (9.4.20) and the first part of (9.4.2), we then easily get

$$\underline{\mathrm{div}}(\Pi_h^1\underline{\underline{\tau}}(h)+\Pi_h^2\underline{\underline{\tau}}(h)) = \underline{\mathrm{div}}(\Pi_h^1\underline{\underline{\tau}}(h)) = \underline{\mathrm{div}}(\underline{\underline{\tau}}(h)) = \underline{v}_h, \tag{9.4.22}$$

while from (9.4.21) and the second part of (9.4.2) we have for all $\underline{\underline{\delta}}_h = \underline{\underline{S}}^3(\underline{w}_h)$ with $\underline{w}_h \in (W_h)^3$

$$c(\Pi_h^1\underline{\underline{\tau}}(h)+\Pi_h^2\underline{\underline{\tau}}(h),\underline{\underline{\delta}}_h) = c(\underline{\underline{\tau}}(h),\underline{\underline{\delta}}_h) = c(\underline{\underline{\gamma}}_h,\underline{\underline{\delta}}_h). \tag{9.4.23}$$

The bounds on $\Pi_h^1\underline{\underline{\tau}}(h) + \Pi_h^2\underline{\underline{\tau}}(h)$ then follow easily from (9.3.50), (9.4.19) and (9.4.3). Consequently, setting $\underline{\underline{\tau}}_h := \Pi_h^1\underline{\underline{\tau}}(h) + \Pi_h^2\underline{\underline{\tau}}(h)$, we have that (9.3.53) is satisfied. □

In a way identical to that of Corollary 9.4.1, we have moreover the following result.

Corollary 9.4.2. *Assume that the assumptions of Proposition 9.3.3 are satisfied, together with* (9.3.49)*. Suppose moreover that the mapping Π_h^1 from $\Sigma(h)$ into Σ_h satisfies* (9.3.50), (9.3.51) *and* (9.4.14)*. Then there exists a space $\Sigma_h^e \supseteq \Sigma_h$ (obtained from Σ_h by adding suitable $H(div)$-bubbles) such that the triplet*

$$\Sigma_h^e \times U_h \times X_h \quad \textit{with}\ X_h = \underline{\underline{S}}^3(W_h^3) \tag{9.4.24}$$

satisfies the conditions of Theorem 9.3.1. □

Indeed, we just have to find a space of H(curl)-bubbles satisfying (9.4.15) and then enlarge Σ_h to enforce (9.4.16).

9.4.3 Two Examples

We now consider two families of elements which can fit into the present framework. The first one has been introduced in [34].

Example 9.4.3 (The Arnold-Falk-Winther family). For $k \geq 1$, we denote by $\mathcal{AFW}_k$ the following choice of finite element spaces.

$$\begin{aligned}&\Sigma_h := (\mathcal{BDM}_k)^{3r},\\&U_h := (\mathcal{L}^0_{k-1})^3,\\&W_h := (\mathcal{L}^{0\,\prime}_{k-1})^3, \qquad X_h := \underline{\underline{S}}^3(W_h) \equiv \underline{\underline{S}}^3((\mathcal{L}^0_{k-1})^3).\end{aligned} \tag{9.4.25}$$

The choice of degrees of freedom made in Chap. 2, Proposition (2.3.2) and Definition 2.3.37, implies that the (standard) interpolation operator Π_h into Σ_h can be defined on each K by

$$\int_f (\underline{\underline{\tau}} - \Pi_h\underline{\underline{\tau}}) \cdot \underline{n}^f_K \cdot \underline{p}_k = 0, \quad \forall \text{ face } f,\ \forall\, \underline{p}_k \in (P_k(f))^3, \tag{9.4.26}$$

and for $k \geq 2$,

$$\int_K (\underline{\underline{\tau}} - \Pi_h\underline{\underline{\tau}}) : \underline{\underline{\eta}}_h \,\mathrm{d}\mathbf{x} \quad \forall \underline{\underline{\eta}}_h \in \mathcal{N}_{k-2}(K) \tag{9.4.27}$$

or equivalently

$$\int_K (\underline{\underline{\tau}} - \Pi_h\underline{\underline{\tau}}) : (\underline{\underline{v}}_h + \underline{\underline{w}}_h \wedge \underline{x})\,\mathrm{d}\mathbf{x} = 0,$$

$$\forall\, \underline{\underline{v}}_h \in (P_{k-2})^{3\times 3}, \forall\, \underline{\underline{w}}_h \in (\hat{P}_{k-2})^{3\times 3}. \qquad \square$$

The choice of $(\mathcal{L}^0_{k-1})^3$ to enforce symmetry yields an unbalanced element. We can obtain the same order of accuracy using fewer degrees of freedom from the following example, which has also been introduced in a different way in [155, 230] and [236].

Example 9.4.4 (The Generalised Amara-Thomas elements family). For $k \geq 2$, we denote by $\mathcal{GAT}_k$ the following approximation

$$\begin{aligned}&\Sigma_h := (\mathcal{BDFM}_k)^{3r},\\&U_h := (\mathcal{L}^0_{k-1})^3,\\&X_h := \underline{\underline{S}}^3((\mathcal{L}^0_{k-1})^3).\end{aligned} \tag{9.4.28}$$

Using again Proposition 2.3.2, the standard interpolation operator Π_h on $\Sigma_h = (\mathcal{BDFM}_k)^{3r}$ will be defined on each K by

$$\int_f (\underline{\underline{\tau}} - \Pi_h \underline{\underline{\tau}}) \cdot \underline{n}_K \cdot \underline{\phi}_h \, ds = 0, \quad \forall \text{ face } f, \forall \underline{\phi}_h \in (P_{k-1})^3 \tag{9.4.29}$$

and as in (9.4.27),

$$\int_K (\underline{\underline{\tau}} - \Pi_h \underline{\underline{\tau}}) : (\underline{\underline{v}}_h + \underline{\underline{w}}_h \wedge \underline{x}) \, d\mathbf{x} = 0$$

$$\forall \underline{\underline{v}}_h \in (P_{k-2})^{3\times 3}, \ \forall \underline{\underline{w}}_h \in (\hat{P}_{k-2})^{3\times 3}. \tag{9.4.30}$$

□

For $k = 2$, one sees that this is an analogue of the two-dimensional element of Amara and Thomas [7] where divergence-free bubbles enable to enforce symmetry to the right order. □

Remark 9.4.3. It must be noted that we did not add bubbles to our spaces. We shall use existing bubbles to get symmetry. One could also consider $\mathcal{GAT}_k$ as an enrichment of $\mathcal{AFW}_{k-1}$ made using "$H(div)$-bubbles". □

It is not difficult to check that taking for Π_h^1 the standard interpolation operator defined for $\mathcal{AFW}_k$ by (9.4.26) and (9.4.27) and for $\mathcal{GAT}_K$ by (9.4.29) and (9.4.30), we have that (9.3.50)–(9.4.14) are satisfied. This last point can easily be deduced from Corollary 9.3.1, taking $p \in \underline{P}_0$ and replacing $\underline{\underline{\tau}}$ by $\underline{\underline{\tau}} - \Pi_h(\underline{\underline{\tau}})$.

It must be noted that $\mathcal{AFW}_k(K)$ and $\mathcal{GAT}_k(K)$ contain the same divergence-free bubbles generated by the Curl of the bubbles of $\mathcal{N}_k(K)$ which are the same as those of $\mathcal{N}_k^r(K)$. Indeed, if we define $\mathbb{H}_k$ as

$$\mathbb{H}_k = \left\{ \underline{\underline{\psi}}_k \,|\, \underline{\underline{\psi}}_k \in (\mathcal{N}_k(K))^{3r}, \underline{\underline{\psi}}_k \wedge \underline{n}_f = 0, \ \forall \text{ face } f \right\}, \tag{9.4.31}$$

it is immediate to check that (9.4.16) is satisfied. Moreover, from the degrees of freedom in $\mathcal{N}_k(K)$ presented in Sect. 2.3.2, we easily deduce that

$$\int_K \underline{\underline{\psi}}_k : \underline{\underline{p}}_{k-2} \, d\mathbf{x}, \quad \forall \underline{\underline{p}}_{k-2} \tag{9.4.32}$$

can be used to define degrees of freedom on $\mathbb{H}_k$. This easily implies that (9.4.15) is also satisfied.

Remark 9.4.4 (The two-dimensional case). In the two-dimensional case, the stability result for the $\mathcal{AFW}$ elements could be easily obtained from Theorem 9.4.2. Indeed, we can use the fact (Remark 8.6.2) that the couple $\underline{P}_{k+1} - P_{k-1}$, $k \geq 1$, is stable for Stokes and note that, with the notation of (9.3.11), $\underline{\underline{\mathrm{curl}}}(\underline{\mathcal{L}}_{k+1}^1) \subseteq (BDM_k)^{2r}$. On the other hand, the three-dimensional case is not amenable to an analysis by the Stokes-based technique as, for instance, the $P_2 - P_0$ element is not stable in the three-dimensional case.

Similarly, the stability result of $\mathcal{GAT}_k$ elements could have easily been obtained from Theorem 9.4.2, using the stability (Proposition 8.6.2) of the $\underline{P}_k^+ - P_{k-1}$ element for Stokes (where P_k^+ is the space of polynomials of degree k enriched by the bubbles of degree $k+1$). Indeed, it is immediate to see that, with the notation of (9.3.11), $\underline{\underline{\text{curl}}}(\underline{P}_k^+) \subseteq (BDFM_k)^{2r}$. □

9.4.4 Methods Based on the Properties of Π_h^1

The following result instead assumes in some cases that *the operator* Π_h^1 *itself* takes care of the asymmetry condition.

Theorem 9.4.4. *Suppose that the assumptions of Proposition 9.3.3 are satisfied, together with* (9.3.49)*. Suppose moreover that the mapping* Π_h^1 *from* $\Sigma(h)$ *into* Σ_h *satisfies* (9.3.50) *and for all* $\underline{\underline{\tau}} \in \Sigma(h)$*, for all* $\underline{v}_h \in U_h$*, and for all* $\underline{w}_h \in W_h$*:*

$$\int_K (\underline{\underline{\tau}} - \Pi_h^1 \underline{\underline{\tau}}) : (\underline{\underline{grad}}\ \underline{v}_h + \underline{x} \wedge \underline{\underline{grad}}\ \underline{w}_h) dx = 0 \qquad \forall \text{ element } K, \tag{9.4.33}$$

$$\int_f (\underline{\underline{\tau}} - \Pi_h^1 \underline{\underline{\tau}}) \cdot \underline{n}_K \cdot (\underline{v}_h + \underline{x} \wedge \underline{w}_h)\, ds = 0 \quad \forall \text{ face } f. \tag{9.4.34}$$

Then, the triplet

$$\Sigma_h \times U_h \times X_h \quad \text{with } X_h = \underline{\underline{S}}^3(W_h) \tag{9.4.35}$$

satisfies the conditions of Theorem 9.3.1.

Proof. The inclusion of kernels (9.3.48) follows easily from (9.3.49). Then, for every $\underline{v}_h \in U_h$ and for every $\underline{\underline{\gamma}}_h \in X_h$, we use Proposition 9.3.3 to construct $\underline{\underline{\tau}} = \underline{\underline{\tau}}(h)$ of the form (9.4.1) satisfying (9.4.2) and (9.4.3). Then, using (9.4.33) and (9.4.34) with $\underline{w}_h = 0$, we have

$$\int_K \underline{\text{div}}(\underline{\underline{\tau}} - \Pi_h \underline{\underline{\tau}}) \cdot \underline{v}_h = 0 \qquad \forall \underline{v}_h \in U_h, \tag{9.4.36}$$

which, joined with (9.3.49), easily gives (9.3.51). Hence, we can apply formula (9.3.22). Using now (9.4.33) and (9.4.34) with $\underline{v}_h = 0$ in (9.3.22), we get

$$\int_K \underline{asp}(\underline{\underline{\tau}} - \Pi_h^1 \underline{\underline{\tau}}) \cdot \underline{w}_h \,\mathbf{dx} = 0 \qquad \forall \underline{w}_h \in W_h^3 \tag{9.4.37}$$

which, using (9.3.6)–(9.3.7) and (9.3.27), gives for all $\underline{\underline{\phi}}_h \equiv \underline{\underline{S}}^3(\underline{w}_h) \in X_h$

$$c(\underline{\underline{\tau}} - \Pi_h^1 \underline{\underline{\tau}}, \underline{\underline{\phi}}_h) \equiv \int_K \underline{\underline{\text{as}}}(\underline{\underline{\tau}} - \Pi_h^1 \underline{\underline{\tau}}) : \underline{\underline{S}}^3(\underline{w}_h) \,\mathbf{dx} = 0.$$

This result, together with (9.3.50) and (9.4.36), gives (9.3.53), which concludes the proof. □

Remark 9.4.5. It is important to point out that, in general, conditions (9.4.33) and (9.4.34) do not **define** the interpolation operator Π_h^1, meaning that the number of degrees of freedom (9.4.33) and (9.4.34) (always in general) is smaller than the dimension of Σ_h, so that other degrees of freedom need to be added in order to define Π_h^1 in a unique way. However, we shall thereafter show that some otherwise defined operator also satisfies the conditions of Theorem 9.4.4. □

Remark 9.4.6. Taking for W_h a space of continuous functions, condition (9.4.34) would reduce to

$$\int_{\partial K} (\underline{\underline{\tau}} - \Pi_h^1 \underline{\underline{\tau}}) \cdot \underline{n}_K \cdot \underline{v}_h = 0 \qquad \forall\, \underline{v}_h \in U_h, \tag{9.4.38}$$

which is satisfied by most reasonable choices of spaces. Hence, somehow, the reduced symmetry would then rely only on internal degrees of freedom. One could thus get an alternate proof for the three-dimensional PEERS element. □

Remark 9.4.7. It is immediate to see that (9.4.26) implies (9.4.34). On the other hand, we underline the difference between $\underline{x} \wedge \underline{\underline{\mathrm{grad}}}\, \underline{w}_h$ in (9.4.33), a condition on columns, and $\underline{\underline{w}}_h \wedge \underline{x}$, a condition on the rows, in (9.4.27). Such a difference forbids the direct use of the interpolator (9.4.26) and (9.4.27) in Theorem 9.4.4 for $k > 1$. However we will be able to prove, for $\mathcal{AFW}_k$, that this interpolator indeed satisfies (9.4.33), while in [80] we had to build explicitly some alternative interpolation operator. On the other hand, the original proof by Arnold-Falk-Winther was based on totally different techniques, related to exterior calculus [30]. □

We shall now show that the standard operator Π_h defined in (9.4.26) and (9.4.27) for the $\mathcal{AFW}_k$ family in three dimensions satisfies the assumptions of Theorem 9.4.4. We shall first need a technical Lemma.

Lemma 9.4.1. *For any $\underline{p}_{k-1} \in \underline{P}_{k-1}$, there exists a $\underline{q}_k \in \underline{P}_k$ and a tensor $\underline{\underline{\eta}}_{k-2} \in (\mathcal{N}_{k-2})^{3r}$ such that*

$$\underline{x} \wedge \underline{\underline{grad}}\, \underline{p}_{k-1} = \underline{\underline{grad}}\, \underline{q}_k + \underline{\underline{\eta}}_{k-2}. \tag{9.4.39}$$

Proof. Using (9.3.14), we see that $\underline{\underline{\mathrm{curl}}}(\underline{x} \wedge \underline{\underline{\mathrm{grad}}}\, \underline{p}_{k-1}) = \mathcal{A}\, \underline{\underline{\mathrm{grad}}}\, \underline{p}_{k-1}$ and therefore belongs to $\underline{\underline{P}}_{k-2}$ (and, actually, to the subspace of $(\mathcal{BDM}_{k-2})^{3r}$ having zero divergence). At this point, we can use the properties of the spaces $\mathcal{N}_{k-2}$ (see e.g. [79]): every vector-valued polynomial of degree $k-2$ with zero divergence can be seen as the curl of an element of $\mathcal{N}_{k-2}$. Applying this property to each row of $\underline{\underline{\mathrm{curl}}}(\underline{x} \wedge \underline{\underline{\mathrm{grad}}}\, \underline{p}_{k-1}) = \mathcal{A}\, \underline{\underline{\mathrm{grad}}}\, \underline{p}_{k-1}$, we construct the rows of $\underline{\underline{\eta}}_{k-2}$, that is,

$$\underline{\underline{\mathrm{curl}}}(\underline{x} \wedge \underline{\underline{\mathrm{grad}}}\, \underline{p}_{k-1}) = \mathcal{A}\, \underline{\underline{\mathrm{grad}}}\, \underline{p}_{k-1} = \underline{\underline{\mathrm{curl}}}\, \underline{\underline{\eta}}_{k-2}. \tag{9.4.40}$$

Hence, $\underline{x} \wedge \underline{\underline{\mathrm{grad}}}\, \underline{p}_{k-1}$ and $\underline{\underline{\eta}}_{k-2}$ have the same curl, and therefore the difference from each row of $\underline{x} \wedge \underline{\underline{\mathrm{grad}}}\, \underline{p}_{k-1}$ and the corresponding row of $\underline{\underline{\eta}}_{k-2}$ is an element in $\underline{P}_{k-1}$ with zero $\underline{\mathrm{curl}}$ and is therefore the gradient of an element in P_k. □

Proposition 9.4.1. *For $\mathcal{AFW}_k$, the (standard) interpolation operator defined by* (9.4.26)–(9.4.27) *satisfies* (9.3.50) *and* (9.4.33)–(9.4.34). *Hence, the conditions of Theorem 9.4.4 are fulfilled.*

Proof. We already noted that (9.4.26) implies (9.4.34). Moreover, we know that $\mathcal{N}_{k-2}$ contains $\underline{\mathrm{grad}}\, P_{k-1}$ so that (9.4.33) holds for $\underline{\underline{w}}_h = 0$. Then, for every $\underline{\underline{\tau}} \in \Sigma(h)$, we have that $\underline{\underline{\tau}} - \Pi_h \underline{\underline{\tau}}$ is divergence-free and therefore for every $\underline{q}_k \in \underline{P}_k$ we have, integrating by parts and using (9.4.26):

$$\int_K (\underline{\underline{\tau}} - \Pi_h \underline{\underline{\tau}})\, \underline{\underline{\mathrm{grad}}}\, \underline{q}_k \, dx = -\int_K \underline{\mathrm{div}}(\underline{\underline{\tau}} - \Pi_h \underline{\underline{\tau}}) \underline{q}_k \, dx - \int_{\partial K} (\underline{\underline{\tau}} - \Pi_h \underline{\underline{\tau}}) \underline{q}_k \, ds = 0. \tag{9.4.41}$$

Then we can consider the second part of (9.4.33), that is, with $\underline{v}_h = 0$. Using (9.4.39), then (9.4.41), and then (9.4.27), we have

$$\begin{aligned} \int_K (\underline{\underline{\tau}} - \Pi_h \underline{\underline{\tau}}) : (\underline{x} \wedge \underline{\underline{\mathrm{grad}}}\, \underline{w}_h)\, dx \\ &= \int_K (\underline{\underline{\tau}} - \Pi_h \underline{\underline{\tau}}) : (\underline{\underline{\mathrm{grad}}}\, \underline{q}_k + \underline{\underline{n}}_{k-2})\, dx \\ &= \int_K (\underline{\underline{\tau}} - \Pi_h \underline{\underline{\tau}}) : \underline{\underline{n}}_{k-2}\, dx = 0, \end{aligned} \tag{9.4.42}$$

showing that (9.4.33) is satisfied, and concluding the proof. □

Remark 9.4.8. It must be noted that on the inter-element boundaries, $\mathcal{AFW}_k$ has the same number of degrees of freedom as $\mathcal{GAT}_{k+1}$ but has the same order of convergence as $\mathcal{GAT}_k$. The elements of $\mathcal{AFW}_k$ are indeed unbalanced with respect to the order of approximation of the different components. The $\mathcal{GAT}_k$ family reaches the same order with much less degrees of freedom on the boundary of the elements. For $k = 2$ in three dimensions, for example, the number of degrees of freedom per face is reduced from 18 to 9, nevertheless permitting to get the same order of convergence. It must also be noted that internal degrees of freedom can be efficiently eliminated by the static condensation process, as we observed in Remark 9.3.11, so that the actual cost is mainly depending on the number of degrees of freedom on interfaces. □

Remark 9.4.9 (Reduced elements). In the case $k = 1$, to specify the tensor $\underline{\underline{\tau}}_h$ we need 9 d.o.f. on each face. This is obviously not optimal in regards to (9.4.33) and (9.4.34) where only six conditions are generated by $\underline{v}_h$ and $\underline{w}_h$ in $U_h = (\mathcal{L}_0^0)^3$.

It is indeed stated in [30] that it is possible to reduce the number of degrees of freedom on each face from nine to six. This is not as good as in the two-dimensional case where the tangential part can be taken constant. The proper choice of space, using the right number of degrees of freedom, will be a consequence of the following general result. □

Let f be a face of some tetrahedron of a mesh. Let $P_k(f)$ and $\hat{P}_k(f)$ denote respectively the set of polynomials and homogeneous polynomials of degree k on f. Let $\underline{n}$ be the normal to the face and denote τ_{nn} and $\underline{\tau}_{nT} = \underline{\tau}_n - \tau_{nn}\underline{n}$ the normal and tangential part of the vector $\underline{\tau}_n$. We also denote $\underline{x}_T^{\perp} := \underline{x} \wedge \underline{n}$. With an abuse of language, we shall consider $\underline{\tau}_{nT}$ and $\underline{x}_T^{\perp}$ as two-dimensional vectors, and more generally, we will identify, whenever convenient, all vectors tangential to the face f with their two-dimensional projection on f.

Theorem 9.4.5. *In order to satisfy* (9.4.34), *we need on each face* f

$$\tau_{nn} \in P_k(f) \tag{9.4.43}$$

and, for the tangential part,

$$\underline{\tau}_{nT} \in N_{k-1}(f) := \underline{P}_{k-1}(f) + \underline{x}_T^{\perp}\hat{\underline{P}}_{k-1}(f). \tag{9.4.44}$$

Proof. Let $U_h := (\mathcal{L}^0_{k-1})^3$ and consider the space $U_h + \underline{x} \wedge U_h$. It is easy to see that in order to generate this space, it is sufficient to consider functions of the form

$$\underline{p} + \underline{x} \wedge \underline{\hat{q}}, \tag{9.4.45}$$

where $\underline{p} \in \underline{P}_{k-1}$ is a *general* vector-valued polynomial of degree $k-1$, but $\underline{\hat{q}} \in \hat{\underline{P}}_{k-1}$ is a vector-valued *homogeneous* polynomial of degree $k-1$. Now, we want to evaluate on a face f of some tetrahedron

$$(\underline{p} + (\underline{x} \wedge \underline{\hat{q}}_h)) \cdot \underline{\underline{\tau}} \cdot \underline{n}. \tag{9.4.46}$$

To do so, we use, on the face, a set of orthogonal co-ordinates defined by the normal $\underline{n}$ and two tangential vectors, $\underline{s}$ and $\underline{t}$. We then write

$$\begin{aligned} \underline{x} &= x_n\underline{n} + x_s\underline{s} + x_t\underline{t}, \\ \underline{p} &= p_n\underline{n} + p_s\underline{s} + p_t\underline{t}, \\ \underline{\hat{q}} &= \hat{q}_n\underline{n} + \hat{q}_s\underline{s} + \hat{q}_t\underline{t}, \\ \underline{\underline{\tau}} \cdot \underline{n} &= \tau_{nn}\underline{n} + \tau_{ns}\underline{s} + \tau_{nt}\underline{t}. \end{aligned} \tag{9.4.47}$$

An elementary computation then yields

$$\begin{aligned} (\underline{p} + (\underline{x} \wedge \underline{\hat{q}})) \cdot \underline{\underline{\tau}} \cdot \underline{n} = (p_n - x_t\hat{q}_s + x_s\hat{q}_t)\tau_{nn} + (p_s - x_n\hat{q}_t + x_t\hat{q}_n)\tau_{ns} \\ + (p_t + x_n\hat{q}_s - x_s\hat{q}_n)\tau_{nt}. \end{aligned} \tag{9.4.48}$$

We now see that τ_{nn} is multiplied by a full polynomial of degree k and we need it to be of the same order. For the tangential terms, we recall that x_n is constant on the face so that the terms $x_n\hat{q}_t$ and $x_n\hat{q}_s$ can be absorbed by p_s and p_t respectively. This leaves us with something of the form

$$\begin{pmatrix} \tau_{ns} \\ \tau_{nt} \end{pmatrix} \cdot \begin{pmatrix} p_s + x_t\hat{q}_n \\ p_t - x_s\hat{q}_n \end{pmatrix} \tag{9.4.49}$$

which shows that the tangential part of $\underline{\tau}_n$ needs only to be in $N_{k-1}(f)$, which is a space smaller than $\underline{P}_k(f)$. □

Remark 9.4.10. A simple count shows that the number of degrees of freedom on each face is then

$$(k+1)(k+2)/2 + k(k+1) + k$$

while for the whole $(\mathcal{BDM}_k)^3$ we would have

$$3(k+1)(k+2)/2.$$

For $k = 1$, we have 6 instead of 9 and for $k = 2$, 14 instead of 18. We refer to [80] for the actual construction of the reduced spaces. □

9.5 Loosing the Inclusion of Kernel: Stabilised Methods

Although the inclusion of kernel condition (9.3.48) is a useful property, it imposes severe restrictions on the construction of approximations. We now present two examples where a simple stabilising term enables to bypass this restriction. Indeed, we have introduced, in Chap. 6, procedures to compensate for the loss of coercivity on the kernel which is the direct consequence of this loss.

We first present an example where the inclusion of the kernel is totally lost so that coercivity on the kernel also is. We only sketch the results and we refer to [125] for more details.

Example 9.5.1. We build a space Σ_h following the idea employed in Sect. 8.4.2. We define, as we did in (8.4.9) to approximate $H(\mathrm{div};\Omega)$,

$$\Sigma_h = (\mathcal{L}_1^1 \oplus B_3)_s^4, \tag{9.5.1}$$

$$U_h = (\mathcal{L}_1^1)^2. \tag{9.5.2}$$

With this choice of space, we only have, as in (9.1.12),

$$a(\underline{\underline{\tau}}, \underline{\underline{\tau}}) \geq \frac{1}{2(\lambda+\mu)} \|\underline{\underline{\tau}}\|_0^2 \tag{9.5.3}$$

and the coercivity constant depends on λ. We cannot obtain something similar to (9.1.9) as inclusion of kernels does not hold. However, using the results of Sects. 1.5 and 6.1.2, that is modifying the variational formulation, we can obtain an approximate solution with error bounds independent of λ. Let us indeed consider the modified variational problem

$$\int_\Omega \frac{1}{2\mu}\underline{\underline{\sigma}}^D : \underline{\underline{\tau}}^D\,dx + \frac{1}{n(n\lambda+2\mu)}\int_\Omega \operatorname{tr}\underline{\underline{\sigma}}\operatorname{tr}\underline{\underline{\tau}}\,dx + \alpha\int_\Omega (\underline{\operatorname{div}}\,\underline{\underline{\sigma}} + \underline{f})\cdot(\underline{\operatorname{div}}\,\underline{\underline{\tau}})\,dx$$
$$+\int_\Omega (\underline{\operatorname{div}}\,\underline{\underline{\tau}})\cdot\underline{u}\,dx = 0 \quad \forall\underline{\underline{\tau}} \in \underline{\underline{H}}(\underline{\operatorname{div}};\Omega)_s, \quad (9.5.4)$$

$$\int_\Omega (\underline{\operatorname{div}}\,\underline{\underline{\sigma}} + \underline{f})\cdot\underline{v}\,dx = 0 \quad \forall\,\underline{v} \in (L^2(\Omega))^2. \quad (9.5.5)$$

This problem is clearly equivalent to the original formulation. However, we now have, instead of (9.1.8),

$$a(\underline{\underline{\sigma}},\underline{\underline{\tau}}) = \frac{1}{2\mu}\int_\Omega \underline{\underline{\sigma}}^D : \underline{\underline{\tau}}^D\,dx + \frac{1}{n(n\lambda+2\mu)}\int_\Omega \operatorname{tr}\underline{\underline{\sigma}}\operatorname{tr}\underline{\underline{\tau}}\,dx + \alpha\int_\Omega \underline{\operatorname{div}}\,\underline{\underline{\sigma}}\cdot\underline{\operatorname{div}}\,\underline{\underline{\tau}}\,dx \quad (9.5.6)$$

and from Proposition 9.1.1 we get the coerciveness property

$$a(\underline{\underline{\sigma}},\underline{\underline{\sigma}}) \geq \alpha_0\|\underline{\underline{\sigma}}\|^2_{\underline{\underline{H}}(\underline{\operatorname{div}};\Omega)_s}, \quad (9.5.7)$$

where α_0 depends on α, μ, and c but is independent of λ.

The bilinear form $b(\underline{\underline{\tau}},\underline{v})$ defined by (9.1.9) is unchanged, but the proof needs the inf-sup condition (cf. [27])

$$\inf_{\underline{v}\in(L^2(\Omega))^2}\ \sup_{\underline{\underline{\tau}}\in(H^1(\Omega))^4_s} \frac{b(\underline{\underline{\tau}},\underline{v})}{\|\underline{\underline{\tau}}\|_1\|\underline{v}\|_0} \geq k > 0. \quad (9.5.8)$$

We now introduce the discretisation already defined by (9.5.1) and (9.5.2). As we have the coercivity property on the whole space Σ, the only delicate point is to obtain a discrete inf-sup condition. We use Proposition 5.4.3 and an operator Π_h defined as in Sect. (8.4.2) to deduce that

$$\|\Pi_h\underline{\underline{\tau}}\|_\Sigma \leq c\|\underline{\underline{\tau}}\|_1. \quad (9.5.9)$$

However, (9.5.8) and (9.5.9) imply, by Proposition 5.4.3, that we have

$$\inf_{\underline{v}_h\in V_h}\ \sup_{\underline{\underline{\tau}}\in\Sigma_h} \frac{b(\underline{\underline{\tau}},\underline{v}_h)}{\|\underline{\underline{\tau}}\|_\Sigma\|\underline{v}_h\|_0} \geq k_0 > 0 \quad (9.5.10)$$

with k_0 independent of h. From the standard theory, we therefore obtain the error estimate

$$\|\underline{\underline{\sigma}} - \underline{\underline{\sigma}}_h\|_\Sigma + \|\underline{u} - \underline{u}_h\|_0 \leq C \left\{ \inf_{\tau_h \in \Sigma_h} \|\underline{\underline{\sigma}} - \tau_h\|_\Sigma + \inf_{v_h \in V_h} \|\underline{u} - v_h\|_0 \right\} \leq Ch(\|\underline{u}\|_2 + \|\underline{\underline{\sigma}}\|_2). \tag{9.5.11}$$

□

We now turn to a construction closer to the approximations studied in the previous Sections. For example, in the construction of the $\mathcal{GAT}_k$ family, we employed a technique which can be seen in two different ways: reducing the number of degrees of freedom at the boundary of elements or adding internal bubbles. In the $\mathcal{GAT}_k$ case, using a reduced version of $\mathcal{BDM}_k$ provided enough internal degrees of freedom to control symmetry.

We shall consider here the reduction of $\mathcal{RT}_k$ or rather an enrichment of $\mathcal{RT}_{k-1}$. This will allow a simple control of symmetry at the price of the loss of coercivity. We shall overcome this loss using the results of Sect. 6.3. The main interest of this construction is that it requires, for a given order of accuracy, fewer degrees of freedom at interfaces than what we had in the $\mathcal{AFW}$, even in the reduced version, or the $\mathcal{GAT}$ families. We shall also consider in detail the lowest order case which provides a stabilised form of the non-conforming method already considered in Sect. 8.4.4 for the Stokes problem.

The construction will rely on internal $H(\text{div})$-bubbles. We thus consider on a simplicial element K, for $k \geq 1$, the space $B_{k+1}^{\mathcal{RT}}$ of vectors $\underline{b}_{k+1}$ in the space $\mathcal{RT}_k$ which have a null normal component on the boundary of the element. They are polynomial vectors of degree $k + 1$ and are bubbles of $H(\text{div}, K)$ in the sense introduced in Sect. 2.6. The elements of $B_{k+1}^{\mathcal{RT}}$ do not include divergence-free bubbles.

Example 9.5.2. In the two-dimensional case, $\mathcal{RT}_1(K)$ contains two bubbles. They are also bubbles of $\mathcal{BDM}_2(K)$ but this space also contains one divergence-free bubble. In the three-dimensional case, we have three bubbles in $\mathcal{RT}_1(K)$ while $\mathcal{BDM}_2(K)$ has six, three of them being divergence-free. □

From (2.3.32), we can specify the elements of $B_{k+1}^{\mathcal{RT}}$ through the degrees of freedom

$$\int_K \underline{b}_{k+1} \cdot \underline{p}_{k-1} dx. \tag{9.5.12}$$

From this, using the notation introduced in Sect. 9.1.1, we can build a space of tensors

$$\underline{\underline{B}}_{k+1}^{\mathcal{RT}} = (B_{k+1}^{\mathcal{RT}})^{3r}. \tag{9.5.13}$$

The degrees of freedom will then be defined by

$$\int_K \underline{\underline{b}}_{k+1} \cdot \underline{\underline{p}}_{k-1} \quad \forall \underline{p}_{k-1} \in \underline{\underline{P}}_{k-1}(K). \tag{9.5.14}$$

We could now add this space of bubble tensors to $\mathcal{RT}_{k-1}^{3n}$ to obtain a space suitable for our purpose. This is equivalent to using a reduced version of $\mathcal{RT}_k$ in which the normal component is of degree $k-1$.

In the sequel, we shall use only the elements of $\underline{\underline{B}}_{k+1}^{\mathcal{RT}}$ which are, loosely speaking, antisymmetric, that is,

$$\int_K \underline{\underline{b}}_{k+1} \cdot \underline{\underline{p}}_{k-1}^S \, dx = 0 \quad \forall \underline{p}_{k-1}^S \in \underline{\underline{P}}_{k-1}^S(K), \tag{9.5.15}$$

where $\underline{\underline{P}}_{k-1}^S(K)$ is the symmetric part of $\underline{\underline{P}}_{k-1}(K)$. We denote this space by $\underline{\underline{B}}_{k+1}^{\mathcal{RT}as}$. Its elements are specified by the degrees of freedom

$$\int_K \underline{\underline{b}}_{k+1}^{as} \cdot \underline{\underline{S}}^n(\underline{p}_{k-1})dx. \tag{9.5.16}$$

Our enriched space will then be

$$\Sigma_h = (\mathcal{RT}_k)^{nr} \oplus \underline{\underline{B}}_{k+2}^{\mathcal{RT}as}. \tag{9.5.17}$$

Example 9.5.3. In the two-dimensional case, we enrich $\mathcal{RT}_0$ by one bubble and by three in the three-dimensional case. This means that we use the minimum to get symmetry. □

To complete our choice of spaces, we take

$$\begin{aligned} &U_h := (\mathcal{L}_k^0)^3, \\ &W_h := (\mathcal{L}_k^0)^3, \qquad X_h := \underline{\underline{S}}^3(W_h) \equiv \underline{\underline{S}}^3((\mathcal{L}_k^0)^3). \end{aligned} \tag{9.5.18}$$

Following Sect. 2.3.1, we then define on Σ_h the interpolation operator Π_h by

$$\int_{\partial K} (\underline{\underline{\tau}} - \Pi_h \underline{\underline{\tau}}) \cdot \underline{n}_K \cdot \underline{p}_k = 0, \quad \forall\, \underline{p}_k \in R_k(\partial K), \tag{9.5.19}$$

$$\int_K (\underline{\underline{\tau}} - \Pi_h \underline{\underline{\tau}}) : \underline{\underline{p}}_{k-1} \, \mathbf{dx} = 0, \quad \forall\, \underline{\underline{p}}_{k-1} \in \underline{\underline{P}}_{k-1}. \tag{9.5.20}$$

Moreover, we determine the extra bubbles by

$$\int_K (\underline{\underline{\tau}} - \Pi_h \underline{\underline{\tau}}) : \underline{\underline{S}}^n(\underline{p}_k)\, d\mathbf{x} = 0, \quad \forall\, \underline{p}_k \in \underline{P}_k. \tag{9.5.21}$$

This implies

$$\int_K \underline{\mathrm{div}}(\underline{\underline{\tau}} - \Pi_h \underline{\underline{\tau}}) : \underline{\underline{p_k}}\, d\mathbf{x} = 0 \quad \forall \underline{\underline{p_k}} \in \underline{\underline{P}}(K). \tag{9.5.22}$$

We could now consider the non stabilised problem

$$\begin{cases} a(\underline{\underline{\sigma}}_h, \underline{\underline{\tau}}_h) + b(\underline{\underline{\tau}}_h, \underline{u}_h) + c(\underline{\underline{\tau}}_h, \underline{\underline{\omega}}_h) = 0, & \forall\, \underline{\underline{\tau}}_h \in \Sigma_h, \\ b(\underline{\underline{\sigma}}_h, \underline{v}_h) = (\underline{f}, \underline{v}_h), \quad \forall\, \underline{v}_h \in U_h, \\ c(\underline{\underline{\sigma}}_h, \underline{\underline{\gamma}}_h) = 0, \quad \forall\, \underline{\underline{\gamma}}_h \in X_h. \end{cases} \tag{9.5.23}$$

The discrete kernel is thus defined by

$$Ker B_h = \{\underline{\underline{\tau}}_h | \int_K \underline{\mathrm{div}}\, \underline{\underline{\tau}} \cdot \underline{p}_k\, d\mathbf{x} = 0 \,\text{and} \int_K (\underline{\underline{\tau}} - \Pi_h \underline{\underline{\tau}}) : \underline{\underline{S}}^3(\underline{q}_k)\, d\mathbf{x} = 0\} \tag{9.5.24}$$

for any $\underline{p}_k \in \underline{P}_k, \underline{q}_{k-1} \in \underline{P}_k$. It should be clear that the interpolation operator on Σ_h is B-compatible and we thus have an *inf-sup* condition.

However, we do not have the inclusion of kernels as (9.5.22) is valid for p_k and not for p_{k+1}. We therefore cannot employ directly this construction in the normal framework, where this condition is a basic assumption. We can, however, employ, instead of the discrete formulation (9.3.43), a stabilised form. To do so, we define

$$R(\underline{\underline{\sigma}}_h, \underline{\underline{\tau}}_h) := \sum_K \int_K ((\underline{\mathrm{div}}\, \underline{\underline{\sigma}}_h - \mathrm{Proj}_k\, \underline{\mathrm{div}}\, \underline{\underline{\sigma}}_h) \cdot \mathrm{div}\, \underline{\underline{\tau}}_h)\, d\mathbf{x}, \tag{9.5.25}$$

where Proj_k is the $L^2(K)$ projection on $\underline{P}_k(K)$.

We then introduce a stabilised version of (9.3.28),

$$\begin{cases} a(\underline{\underline{\sigma}}_h, \underline{\underline{\tau}}_h) + r\, s(\underline{\underline{\sigma}}_h, \underline{\underline{\tau}}_h) + b(\underline{\underline{\tau}}_h, \underline{u}_h) + c(\underline{\underline{\tau}}_h, \underline{\underline{\omega}}_h) = 0, & \forall\, \underline{\tau}_h \in \Sigma_h, \\ b(\underline{\underline{\sigma}}_h, \underline{v}_h) = (\underline{f}, \underline{v}_h), \quad \forall\, \underline{v}_h \in U_h, \\ c(\underline{\underline{\sigma}}_h, \underline{\underline{\gamma}}_h) = 0, \quad \forall\, \underline{\underline{\gamma}}_h \in X_h. \end{cases} \tag{9.5.26}$$

To study the convergence of this formulation, we shall employ the results of Sect. 6.3. We thus have to verify the three hypotheses of this section. We first consider **H.1** and we define on Σ_h the norm

$$[\underline{\underline{\tau}}_h]_h^2 = \|\underline{\underline{\tau}}_h\|_0^2 + \| P_k\, \underline{\mathrm{div}}\, \underline{\underline{\tau}}_h\|_0^2. \tag{9.5.27}$$

From the definition of $KerB_h$ in (9.5.24), we then have

$$a(\underline{\underline{\tau}}_h, \underline{\underline{\tau}}_h) \geq [\underline{\underline{\tau}}_h]^2 \quad \forall \underline{\underline{\tau}}_h \in KerB_h. \tag{9.5.28}$$

Coming back to the non stabilised problem (9.5.23), as we have an inf-sup condition, we deduce that this problem has a unique solution, which is stable in the norm (9.5.27).

Remark 9.5.1. We could try to employ Theorem 5.2.6 to get an error estimate in the norm (9.5.27). As discussed in Sect. 5.2.3, the trouble arises with the continuity constant of the bilinear form $b(\cdot,\cdot)$ and we cannot obtain the right order of convergence. We have to stabilise the formulation. □

Using the construction of Example 8.2.6 to define the operator Φ_h, we get **H.1** of Sect. 6.3. It is clear that the stabilising term (9.5.25) has been tailored to provide **H.2** and **H.3**. We also remark that $R(\underline{\underline{\sigma}}, \underline{\underline{\tau}}_h)$ is well defined for any tensor $\underline{\underline{\sigma}}$ in Σ. From Theorem 6.3.2, we thus get the following estimate for any $\underline{\underline{\sigma}}_I \in \Sigma_h$.

Theorem 9.5.1. *If* $(\underline{\underline{\sigma}}, \underline{u}, \underline{\underline{\omega}}) \in \Sigma \times U \times X$ *is the solution of* (9.3.28) *and* $(\underline{\underline{\sigma}}_h, \underline{u}_h, \underline{\underline{\omega}}_h) \in \Sigma_h \times \underline{U}_h \times \underline{\underline{X}}_h$ *is the solution of* (9.5.26)*, we then have*

$$\|\underline{\underline{\sigma}}_h - \underline{\underline{\sigma}}\|_0 + \|\underline{u}_h - \underline{u}\|_0 + \|\underline{\underline{\omega}}_h - \underline{\underline{\omega}}\|_0$$
$$\leq C\Big(\|\underline{\underline{\sigma}}_I - \underline{\underline{\sigma}}\|_0 + \|\underline{u}_I - \underline{u}\|_0 + \|\underline{\underline{\phi}}_I - \underline{\underline{\omega}}\|_0 + rR(\underline{\underline{\sigma}}_I)\Big). \tag{9.5.29}$$

If we take $\underline{\underline{\sigma}} \in (\mathcal{RT}_k)^3$, the consistency term disappears and we get an $O(h^k)$ estimate, which is the best that we can hope for.

Example 9.5.4. For $k = 0$, we enrich the $\mathcal{RT}_0$ element by three bubbles in the three-dimensional case. On faces, we only have one value of $\underline{\underline{\sigma}}_h \cdot \underline{n}$ to specify. This is really the smallest possible choice... □

Augmented formulations therefore appear as a powerful tool to overcome difficulties associated with problems of coerciveness and enable us to bypass the inclusion of kernel property which is very difficult to obtain in practice. We refer to Chap. 7 for examples where one can employ similar arguments to avoid the inf-sup condition. Examples of applications to elasticity problems can be found in [215].

9.6 Concluding Remarks

We were not able, in this chapter, to present all the possible avenues that have been explored for mixed methods applied to elasticity problems. In particular, we did nor consider some constructions using composite elements which have been developed in [262] and more recently in [27]. The main reason to do this is that we are not aware of any extension of these ideas to the three-dimensional case.

We also did not develop the Hellan-Hermann-Johnson formulation of Example 1.4.5. We refer to the recent results of [318] and [319] for a study of this formulation.

Discontinuous Galerkin methods have been considered in [229].

We also refer to [190], where other possibilities are presented along with many references.

Chapter 10
Complements on Plate Problems

In this chapter, we shall present a few among many applications of mixed methods to plate problems. In the first section, we shall describe a mixed method for the linear thin plates theory and in the second, a dual hybrid method. In the last section, we shall report some recent results on the discretisation of the Mindlin-Reissner formulation for moderately thick plates.

10.1 A Mixed Fourth-Order Problem

10.1.1 The $\psi - \omega$ Biharmonic Problem

Let us now see, as a new example of application of the abstract results of Chaps. 4 and 5, some simple cases of fourth-order problems. We shall start with formulation (1.3.65) which we may now rewrite in the form (4.2.6) by setting

$$V := H^1(\Omega),\ Q := H_0^1(\Omega), \tag{10.1.1}$$

$$a(\omega, \phi) := \int_\Omega \omega\phi\, dx \qquad \forall \omega, \phi \in V, \tag{10.1.2}$$

$$b(\mu, \phi) := \int_\Omega \underline{\operatorname{grad}}\, \mu \cdot \underline{\operatorname{grad}}\, \phi\, dx \qquad \forall \mu \in Q,\ \phi \in V. \tag{10.1.3}$$

We shall denote by (ω, ψ) instead of (u, p) the solution of the problem in order to be consistent with the usual physical notations. It is easy to see that we are now in the situation of Sect. 3.6: the bilinear form $a(\omega, \phi)$ is not coercive on V (nor is it on $\mathrm{Ker}B$ but only on $H := L^2(\Omega)$). A loss of accuracy is therefore to be expected. Another pitfall is that we cannot use the abstract existence results of Chap. 4 for the continuous problem and that we must deduce the existence of a

D. Boffi et al., *Mixed Finite Element Methods and Applications*, Springer Series in Computational Mathematics 44, DOI 10.1007/978-3-642-36519-5_10,

solution through another channel. In the present case, we know that the solution of our mixed problem: *find* $\psi \in H_0^1(\Omega)$ *and* $\omega \in H^1(\Omega)$ *such that*

$$\begin{cases} \int_\Omega \omega\phi\, dx + \int_\Omega \underline{\mathrm{grad}}\,\psi \cdot \underline{\mathrm{grad}}\,\phi\, dx = 0 \quad \forall \phi \in H^1(\Omega), \\ \int_\Omega \underline{\mathrm{grad}}\,\omega \cdot \underline{\mathrm{grad}}\,\mu\, dx = \int_\Omega f\mu\, dx \quad \forall \mu \in H_0^1(\Omega), \end{cases} \tag{10.1.4}$$

should be a solution of a biharmonic problem

$$\Delta^2\psi = f,\ \psi \in H_0^2(\Omega). \tag{10.1.5}$$

From a regularity result on the biharmonic problem, we know, for instance, that if Ω is a convex polygon [234, 281, 362], for $f \in H^{-1}(\Omega)$, the solution of (10.1.5) belongs to $H^3(\Omega)$ so that $\omega = -\Delta\psi$ belongs to $H^1(\Omega)$. It is then direct to verify that we have thus obtained a solution of (10.1.4). This is an example of an "ill-posed" mixed problem. It should be remarked that the discussion of existence made above does not apply when the right-hand side of the first equation of (10.1.4) is not equal to zero.

To get a discrete problem, we take, following the notations of Chap. 2,

$$V_h := \mathcal{L}_k^1,\ Q_h := \mathcal{L}_k^1 \cap H_0^1(\Omega),\ k \geq 2. \tag{10.1.6}$$

The case $k = 1$ requires a more special analysis [197, 226, 344]. We then have that the constant $S(h)$, appearing in (5.2.40), can now be bounded by $S(h) \leq ch^{-1}$ so that a direct application of Proposition 5.2.6 gives

$$\|\omega - \omega_h\|_0 + \|\psi - \psi_h\|_1 \leq ch^{k-1}. \tag{10.1.7}$$

Indeed, the *inf-sup* condition is quite straightforward. The operator B is nothing here but the Laplace operator from $H^1(\Omega)$ to $H^{-1}(\Omega)$, which is obviously surjective. To check the discrete condition, we use the criterion of Proposition 5.4.3: *given* $\omega \in H^1(\Omega)$, *we want to build* $\omega_h \in V_h$ *such that*

$$\int_\Omega \underline{\mathrm{grad}}\,\omega_h \cdot \underline{\mathrm{grad}}\,\mu_h\, dx = \int_\Omega \underline{\mathrm{grad}}\,\omega \cdot \underline{\mathrm{grad}}\,\mu_h\, dx, \quad \forall \mu_h \in Q_h. \tag{10.1.8}$$

We recall, however, that we have chosen $Q_h \subset V_h$ so that (10.1.8) will, a fortiori, hold if we take $\mu_h \in V_h$. However, (10.1.8) is then nothing but a discrete Neumann problem for which a solution exists and can be chosen (it is defined up to an additive constant) so that

$$\|\omega_h\|_1 \leq c\, \|\omega\|_1. \tag{10.1.9}$$

It must be noted that the condition $Q_h \subset V_h$ is essential to the above result. In practice, this is not a restriction as (10.1.6) is a natural and efficient choice.

Result (10.1.7) is far from optimal and may suggest at first sight that the method is not worth being used. It can however be sharpened in two ways. First it is possible to raise the estimate on $|\omega - \omega_h|_0$ by half an order [197, 345] by a quite intricate analysis using L^∞-error estimates. The second way is a more direct variant of the duality method of Sect. 5.5.5 and shows that the expected accuracy can be obtained for $\psi \in H^3(\Omega)$, that is,

$$\|\psi - \psi_h\|_1 \leq ch^k, \tag{10.1.10}$$

and under a supplementary regularity assumption

$$\|\psi - \psi_h\|_0 \leq ch^{k+1}. \tag{10.1.11}$$

We refer the reader to [107, 189, 342, 345] and [192] for this analysis.

On the other hand, the particular structure of problem (10.1.4) allows the use of sophisticated but effective techniques for the numerical solution [150, 225, 227], so that this method and its variants have a considerable practical interest. In fact, it provides a correct setting for the widely used $\psi - \omega$ approximations in numerical fluid dynamics. We refer to [222] for more informations on this subject. Still in the case of fourth-order problems, we could also consider instead formulation (1.3.70) which is more related to plate bending problems. We now set

$$V := (H^1(\Omega))_s^{2\times 2},\ Q := H_0^1(\Omega), \tag{10.1.12}$$

and we define, following (1.3.70) for $\underline{\underline{\sigma}}$ and $\underline{\underline{\tau}}$ in V,

$$a(\underline{\underline{\sigma}}, \underline{\underline{\tau}}) := \frac{12(1-\nu^2)}{Et^3} \int_\Omega [(1+\nu)\,\underline{\underline{\sigma}} : \underline{\underline{\tau}} - \nu\, tr(\underline{\underline{\sigma}})\, tr(\underline{\underline{\tau}})]\, dx. \tag{10.1.13}$$

In order to consider a weaker form of the saddle point problem (1.3.70), we introduce

$$b(v, \underline{\underline{\tau}}) := \int_\Omega (\underline{\mathrm{div}}\, \underline{\underline{\tau}}) \cdot \underline{\mathrm{grad}}\, v\, dx = \int_\Omega \sum_{i,j} \frac{\partial \tau_{ij}}{\partial x_j} \frac{\partial v}{\partial x_i} dx. \tag{10.1.14}$$

This enables us to look for $w \in H_0^1(\Omega)$ instead of $H_0^2(\Omega)$, the second boundary condition being implied by this variational formulation as a natural condition. This is again an "ill-posed" mixed problem: we must obtain existence of a solution through a regularity result on the standard problem. Two approaches have been followed in the approximation of this mixed problem. One of them consists in taking (see [300])

$$V_h := (\mathcal{L}_k^1)_s^{2\times 2},\ Q_h := \mathcal{L}_k^1 \cap H_0^1(\Omega). \tag{10.1.15}$$

With respect to (10.1.14), it is, however, possible to use a second approach and to work not in $V = (H^1(\Omega))_s^{2\times 2}$ but in the weaker space

$$\underline{\underline{H}}(\underline{\mathrm{div}}; \Omega)_s := \{\underline{\underline{\tau}} \mid \tau_{ij} = \tau_{ij},\ \tau_{ij} \in L^2(\Omega),\ \underline{\mathrm{div}}\, \underline{\underline{\tau}} \in (L^2(\Omega))^2\}. \tag{10.1.16}$$

Discretisations of this space can be built through composite elements. We refer to [262] and [27] for the analysis of this case.

In the first case, the results are the same as for the $\psi-\omega$ approximation discussed above. We get, by Proposition 5.2.6, an error estimate which is $O(h^{k-1})$. Duality methods (see [192]) would enable us to lift the estimate on ψ at the right level. For the second case, we can have optimal error estimates (see the above references).

10.1.2 Eigenvalues of the Biharmonic Problem

We now briefly consider the possibility of computing eigenvalues of the biharmonic problem using the elements introduced above. If we refer to Sect. 1.2.1 of Chap. 6, we are considering a $(0, g)$ situation. This means that, fortunately for us, we do not need a coercivity condition. Our eigenvalue problem can indeed be written as: *find $\psi \in H_0^1(\Omega)$ and $\omega \in H^1(\Omega)$ such that*

$$\begin{cases} \displaystyle\int_\Omega \omega\phi\, dx + \int_\Omega \underline{\text{grad}}\,\psi \cdot \underline{\text{grad}}\,\phi\, dx = 0\ \forall \phi \in H^1(\Omega), \\ \displaystyle\int_\Omega \underline{\text{grad}}\,\omega \cdot \underline{\text{grad}}\,\mu\, dx = \lambda \int_\Omega \psi\mu\, dx\ \forall \mu \in H_0^1(\Omega). \end{cases} \tag{10.1.17}$$

In the notation of Sect. 6.5.5, we have $V = H^1(\Omega)$ and $Q = H_0^1(\Omega)$. We take $H_Q = L^2(\Omega)$ and we assume that Ω is a convex polygon. We then have

$$\begin{aligned} V_{Q'}^0 &= \{z \in H^1(\Omega) :\ \exists v \in H_0^2(\Omega) \text{ with } z = \Delta v\} \\ &= \{z \in H^1(\Omega) :\ (z, \mu) = 0\ \forall \mu \in L^2(\Omega) \text{ with } \Delta\mu = 0\} \end{aligned} \tag{10.1.18}$$

so that with obvious notation

$$V_{Q'}^0 = H^3(\Omega) \cap H_0^2(\Omega). \tag{10.1.19}$$

For any given polygon, V_H^0 and Q_H^0 will be slightly more regular, according to the maximum angle (see e.g. [233]).

For every given regular sequence $\{\mathcal{T}_h\}$ of triangulations of Ω and for every integer $k \geq 2$, we can take as in [152, 224, 298]:

$$\begin{aligned} V_h^k &:= \mathcal{L}_k^1 \\ Q_h^k &:= \mathcal{L}_k^1 \cap H_0^1(\Omega). \end{aligned} \tag{10.1.20}$$

Notice that $Q_h^k = V_h^k \cap H_0^1(\Omega)$. We can now define $\Pi_h w$ in V_h as the solution of:

$$(\Delta \Pi_h w, \Delta v_h) = (\Delta w, \Delta v_h) \quad \forall v_h \in V_h^k. \tag{10.1.21}$$

Clearly, (6.5.54)–(6.5.56) hold. Similarly, (6.5.53) holds by taking p^I (here ψ^I) as the usual interpolant. On the other hand, to check (6.5.52), we have to assume quasi-uniformity of the decomposition and then proceed, as we did for Dirichlet's problem in (7.1.43), using an inverse inequality to obtain: *for* $v_h \in KerB_h$ *and* $q \in H^3(\Omega) \cap H_0^2(\Omega)$,

$$(\Delta v_h, \Delta q) = (\Delta v_h, \Delta q - \Delta q^I) \leq C h^{-1} ||v_h||_a \, C h^2 ||q||_3.$$

This shows the utility of the requirement $k \geq 2$. However, a more sophisticated proof, following the arguments of Scholz [344], shows that (1.2.50) also holds for $k = 1$.

We thus have checked all the hypotheses of Theorem 6.5.3 and our eigenvalue problem is properly posed.

10.2 Dual Hybrid Methods for Plate Bending Problems

We now consider as a final example an application of our general theory to hybrid methods. We go back again to Example 1.3.8 and set, for the sake of simplicity, $\nu = 0$ and $Et^3/12 = 1$. The consideration of the true values would not change the mathematical structure of the problem, but would result in more lengthy formulae. The condition $D_2^*(\underline{\underline{\tau}}) = f$ in (1.3.74) is, in general, difficult to enforce directly. Hence, following [321], we may think of working with stresses satisfying $D_2^*(\underline{\underline{\tau}}) = f$ inside each element of a given decomposition. This will imply that we have to enforce some continuity of the stresses by means of a Lagrangian multiplier; moreover, it will be convenient to assume $f \in L^2(\Omega)$. In order to make the exposition clearer, we need some Green's formulae. We have indeed, on any triangle K of a triangulation $\mathcal{T}_h$ of Ω,

$$\int_K \underline{\underline{\tau}} : \underline{\underline{D}}_2(v)\, dx = \int_K D_2^*(\underline{\underline{\tau}}) v \, dx + \int_{\partial K} [M_{nn}(\underline{\underline{\tau}}) \frac{\partial v}{\partial n} - K_n(\underline{\underline{\tau}}) v] \, ds \tag{10.2.1}$$

for all $\underline{\underline{\tau}} \in (H^2(T))_s^{2\times 2}$ and $v \in H^2(T)$, where

$$M_{nn}(\underline{\underline{\tau}}) := (\underline{\underline{\tau}} \cdot \underline{n}) \cdot \underline{n}, \tag{10.2.2}$$

$$K_n(\underline{\underline{\tau}}) := \frac{\partial}{\partial n} \operatorname{tr}(\underline{\underline{\tau}}) - \frac{\partial}{\partial t} [(\underline{\underline{\tau}} \cdot \underline{n}) \cdot \underline{t}], \; \underline{t} = \text{tangent unit vector.} \tag{10.2.3}$$

It is essential, in the definition of K_n, to consider the derivative $\partial/\partial t$ *in the distributional sense*, that is, to take into account the *jumps* of $(\underline{\underline{\tau}} \cdot \underline{n}) \cdot \underline{t}$ at the corners of K (the so-called *corner forces*).

It is easy to check that the condition $D_2^*(\underline{\underline{\tau}}) = f$ in Ω is equivalent to

$$\begin{cases} D_2^*(\underline{\underline{\tau}}) = f \text{ in each } T, \\ \displaystyle\sum_K \int_{\partial K} [M_{nn}(\underline{\underline{\tau}}) \frac{\partial v}{\partial n} - K_n(\underline{\underline{\tau}}) v] \, ds = 0, \qquad \forall v \in H_0^2(\Omega). \end{cases} \tag{10.2.4}$$

Setting

$$\begin{aligned} b(\underline{\underline{\tau}}, v) &:= \sum_K \int_{\partial K} [M_{nn}(\underline{\underline{\tau}}) \frac{\partial v}{\partial n} - K_n(\underline{\underline{\tau}}) v] \, ds \\ &\equiv \int_\Omega \underline{\underline{\tau}} : \underline{\underline{D}}_2(v) \, dx - \sum_T \int_T D_2^*(\underline{\underline{\tau}}) v \, dx, \end{aligned} \tag{10.2.5}$$

$$V_f(\mathcal{T}_h) := \{\underline{\underline{\tau}} \mid \underline{\underline{\tau}} \in (L^2(\Omega))_s^{2\times 2}, \ D_2^*(\underline{\underline{\tau}}) = f \text{ in each } K\}, \tag{10.2.6}$$

the problem can now be written as

$$\inf_{\underline{\underline{\tau}} \in V_f(\mathcal{T}_h)} \sup_{v \in H_0^2} \frac{1}{2} \|\underline{\underline{\tau}}\|_0^2 - b(\underline{\underline{\tau}}, v). \tag{10.2.7}$$

If now $\underline{\underline{\sigma}}^f$ is a given element of $V_f(\mathcal{T}_h)$, that is, a *particular solution* of $D_2^*(\underline{\underline{\sigma}}) = f$ in each K, we have

$$\begin{cases} (\underline{\underline{\sigma}}^0 + \underline{\underline{\sigma}}^f, \underline{\underline{\tau}}) - b(\underline{\underline{\tau}}, w) = 0 & \forall \underline{\underline{\tau}} \in V_0(\mathcal{T}_h), \\ b(\underline{\underline{\sigma}}^0 + \underline{\underline{\sigma}}^f, v) = 0 & \forall v \in H_0^2(\Omega), \end{cases} \tag{10.2.8}$$

where obviously $\underline{\underline{\sigma}}^0 + \underline{\underline{\sigma}}^f := \underline{\underline{\sigma}}$. Problem (10.2.8) has now the form (4.2.6), where $V = V_0(\mathcal{T}_h)$, $Q = H_0^2$, $a(\underline{\underline{\sigma}}, \underline{\underline{\tau}}) = (\underline{\underline{\sigma}}, \underline{\underline{\tau}})$, and $b(\underline{\underline{\tau}}, v)$ is given by (10.2.5). The right-hand side is obviously $-(\underline{\underline{\sigma}}^f, \underline{\underline{\tau}})$ for the first equation and $-b(\underline{\underline{\sigma}}^f, v)$ for the second equation. It is natural to use in V the L^2-norm, and in Q the norm $\|v\|_Q = \|\underline{\underline{D}}_2 v\|_V = \|\underline{\underline{D}}_2 v\|_0$. It is clear that condition (4.2.12), that is, the ellipticity of $a(\cdot,\cdot)$, is trivially satisfied in the whole V (and not only in $\mathrm{Ker}B$) with $\alpha = 1$. A different value for E, t, ν would obviously yield a different value for α, but the V-ellipticity will still be true. It is clear that $\mathrm{Ker}B^t$ cannot be empty; indeed, any v with support in a single K will satisfy $b(\underline{\underline{\tau}}, v) = 0$ for all $\underline{\underline{\tau}}$, and hence is a zero energy mode. However, it is not difficult to see that $\mathrm{Im}B$ is closed.

Proposition 10.2.1. *The image of B is a closed subset of $Q' := H^{-2}(\Omega)$.*

Proof. We have to show that if a sequence $\chi_n := B\underline{\underline{\tau}}_n$ converges to χ in H^{-2}, then $\chi = B\underline{\underline{\tau}}$ for some $\underline{\underline{\tau}} \in V_0(\mathcal{T}_h) =: V$. We first note that

$$\text{if } \underline{\underline{\tau}} \in V_0(\mathcal{T}_h) \text{ and } \phi \in H_0^2(\Omega), \text{ then } b(\underline{\underline{\tau}}, \phi) \equiv (\underline{\underline{\tau}}, \underline{\underline{D}}_2 \phi), \tag{10.2.9}$$

which is quite obvious from (10.2.5) and (10.2.6). Now let $\phi \in H_0^2(\Omega)$ be such that $\Delta^2\phi = \chi$ and let $\underline{\underline{\tau}} := \underline{\underline{D}}_2\phi$ (so that $D_2^*\underline{\underline{\tau}} = \chi$). For every $\phi \in H_0^2$, we have

$$\langle \chi, \phi \rangle_{H^{-2}\times H_0^2} = \langle D_2^*\underline{\underline{\tau}}, \phi \rangle_{H^{-2}\times H_0^2} = (\underline{\underline{\tau}}, \underline{\underline{D}}_2\phi). \tag{10.2.10}$$

Now, since $\chi_n = B\underline{\underline{\tau}}_n \to \chi$ in H^{-2}, we have

$$(\underline{\underline{\tau}}_n, \underline{\underline{D}}_2\phi) = b(\underline{\underline{\tau}}_n, \phi) = \langle B\underline{\underline{\tau}}_n, \phi \rangle = \langle \chi_n, \phi \rangle \to \langle \chi, \phi \rangle = (\underline{\underline{\tau}}, \underline{\underline{D}}_2\phi), \tag{10.2.11}$$

that is, $(\underline{\underline{\tau}}_n - \underline{\underline{\tau}}, \underline{\underline{D}}_2\phi) \to 0$ for all $\phi \in H_0^2(\Omega)$. This easily implies $D_2^*\underline{\underline{\tau}} = 0$ in each T, so that $\underline{\underline{\tau}} \in V_0(\mathcal{T}_h)$. Hence, $\langle \chi, \phi \rangle = (\underline{\underline{\tau}}, \underline{\underline{D}}_2\phi) = b(\underline{\underline{\tau}}, \phi) = \langle B\underline{\underline{\tau}}, \phi \rangle$, that is, $\chi \in \mathrm{Im}B$. □

Proposition 10.2.2. *We have* $\mathrm{Ker}B^t = \prod_K H_0^2(K)$.

Proof. It is obvious from (10.2.5) that if $\phi|_K \in H_0^2(K)$ for all K, then $b(\underline{\underline{\tau}}, \phi) = 0$ $\forall \underline{\underline{\tau}}$ and hence $\phi \in \mathrm{Ker}B^t$. Therefore, we only need to prove that $\mathrm{Ker}B^t \subset \prod_K H_0^2(K)$. For this, let $\phi \in \mathrm{Ker}B^t$, that is,

$$b(\underline{\underline{\tau}}, \phi) \equiv (\underline{\underline{\tau}}, \underline{\underline{D}}_2\phi) = 0 \qquad \forall \underline{\underline{\tau}} \in V_0(\mathcal{T}_h). \tag{10.2.12}$$

We want to show that $\phi \in \prod_K(H_0^2(K))$, that is,

$$\phi|_K \in H_0^2(K) \text{ for all } K. \tag{10.2.13}$$

Let ψ be defined in each K by

$$\psi \in H_0^2(K) \text{ and } \Delta^2\psi = \Delta^2\phi; \tag{10.2.14}$$

clearly, $(\underline{\underline{\tau}}, \underline{\underline{D}}_2\psi) = 0$ for all $\underline{\underline{\tau}}$ in $V_0(\mathcal{T}_0)$ so that from (10.2.12),

$$b(\underline{\underline{\tau}}, \psi - \phi) = (\underline{\underline{\tau}}, \underline{\underline{D}}_2(\psi - \phi)) = 0 \qquad \forall \underline{\underline{\tau}} \in V_0(\mathcal{T}_h). \tag{10.2.15}$$

However, $D_2^*\underline{\underline{D}}_2(\psi - \phi) = \Delta^2(\psi - \phi) = 0$ in each K, so that we can take $\underline{\underline{\tau}} = \underline{\underline{D}}_2(\psi - \phi)$ in (10.2.15) and obtain $\underline{\underline{D}}_2(\psi - \phi) \equiv 0$. Since both ψ and ϕ are in $H_0^2(\Omega)$, this implies $\psi = \phi$ so that from (10.2.14), we get (10.2.13). □

Proposition 10.2.3. *We have*

$$\|\phi\|_{Q/\mathrm{Ker}B^t} = \|\underline{\underline{D}}_2\bar{\phi}\|_0, \tag{10.2.16}$$

where $\bar{\phi}$ is the function in $H_0^2(\Omega)$ such that

$$\phi - \bar{\phi} \in H_0^2(K) \textit{ for each } K, \tag{10.2.17}$$

$$\Delta^2\bar{\phi} = 0 \qquad \textit{in each } K. \tag{10.2.18}$$

Proof. By definition, we have

$$\|\phi\|_{Q/\mathrm{Ker}B^t} = \inf_{\psi\in\mathrm{Ker}B^t} \|\phi-\psi\|_Q. \tag{10.2.19}$$

Now from Proposition 10.2.2 and the definition of $\|\chi\|_Q := \|\underline{\underline{D}}_2\chi\|_0$, we have

$$\|\phi\|_{Q/\mathrm{Ker}B^t} = \inf_{\psi\in\prod_K H_0^2(K)} \|\underline{\underline{D}}^2(\phi-\psi)\|_{0,K}. \tag{10.2.20}$$

It is now an easy matter to check that, for each K,

$$\inf_{\psi\in H_2^2(K)} \|\underline{\underline{D}}_2(\phi-\psi)\|_{0,K}^2 = \inf_{(\psi-\phi)\in H_0^2(K)} \|\underline{\underline{D}}_2\psi\|_{0,K}^2 = \|\underline{\underline{D}}_2\bar{\phi}\|_{0,K}^2, \tag{10.2.21}$$

for $\bar{\phi}$ defined in (10.2.17) and (10.2.18). Hence, (10.2.21) and (10.2.20) prove (10.2.16). □

We are now able to prove the *inf-sup* condition

$$\sup_{\underline{\underline{\tau}}\in V_0(\mathcal{T}_h)} \frac{b(\underline{\underline{\tau}},\phi)}{\|\underline{\underline{\tau}}\|_0\|\phi\|_{Q/\mathrm{Ker}B^t}} = \sup_{\underline{\underline{\tau}}\in V_0(\mathcal{T}_h)} \frac{(\underline{\underline{\tau}},\underline{\underline{D}}_2\phi)}{\|\underline{\underline{\tau}}\|_0\|\underline{\underline{D}}_2\bar{\phi}\|_0}$$

$$\geq \frac{(\underline{\underline{D}}_2\bar{\phi},\underline{\underline{D}}_2\phi)}{\|\underline{\underline{D}}_2\bar{\phi}\|_0^2} = 1 \tag{10.2.22}$$

because $\phi-\bar{\phi}$ is the projection (in Q) of ϕ onto $\mathrm{Ker}B^t$ so that $\bar{\phi}$ and $\phi-\bar{\phi}$ are orthogonal in Q.

Remark 10.2.1. A way of getting rid of $\mathrm{Ker}B^t$ (which is infinite dimensional) is to consider as a space of Lagrange multipliers the space

$$\tilde{Q} := \{\phi \mid \phi\in H_0^2(\Omega),\ \Delta^2\phi = 0 \text{ in each } T\}. \tag{10.2.23}$$

This is what has been done in [114, 127]. The drawback in the choice (10.2.23) is that the actual transversal displacement w does not belong to $\tilde{Q}$ so that, as a solution, we have the unique function $\bar{w}$ in $\tilde{Q}$ that coincides with w (with its first derivatives) at the inter-element boundaries (as in (10.2.17) and (10.2.18)). □

Let us continue our analysis of problem (10.2.8). We already noted that (4.2.12) is satisfied in our case. Hence, we have to check that the right-hand side of the second equation in (10.2.8) (that is $-b(\underline{\underline{\sigma}}^f, v)$) is in $\mathrm{Im}B$; this means that we have to find a particular solution of (10.2.8), which is obvious by taking $\underline{\underline{\sigma}}^f := \underline{\underline{D}}_2 w - \underline{\underline{\sigma}}^f$.

We can now go to the discretisation of (10.2.8); for this, we have to choose subspaces $V_h \subset V_0(\mathcal{T}_h)$ and $Q_h \subset Q$. For instance, for any triple (m,r,s) of integers, we may choose

$$V_h^m := (\mathcal{L}_m^0(\mathcal{T}_h))_s^{2\times 2} \cap V_0(\mathcal{T}_h), \tag{10.2.24}$$

$$Q_h^{r,s} := \{\phi \mid \phi \in H_0^2(\Omega),\ \phi|_{\partial T} \in T_r(\partial T),\ \frac{\partial \phi}{\partial n}|_{\partial T} \in R_s(\partial T) \quad \forall T \in \mathcal{T}_h\}. \tag{10.2.25}$$

Note that V_h is made of tensor-valued polynomials of degree $\le m$ which are completely discontinuous from one element to another and verify $D_2^*\underline{\underline{\tau}} = 0$ in each T. On the other hand, Q_h is clearly infinite dimensional (which is quite unusual); however, this does not show up in the computations, where only the values of ϕ and $\partial\phi/\partial n$ on $\mathcal{E}_h$ are considered. To get coercivity, we now have to choose (m, r, s) in such a way that $\mathrm{Ker}B_h^t \subset \mathrm{Ker}B^t$. This means, in our case, that we have to show

$$\begin{cases} \text{if } \phi \in Q_h^{r,s} \text{ and } b(\underline{\underline{\tau}}, \phi) = 0 \qquad \forall \underline{\underline{\tau}} \in V_h^m \text{ (that is, if } \phi \in \mathrm{Ker}B_h^t\text{)}, \\ \text{then } \phi = \underline{\mathrm{grad}}\,\phi = 0 \text{ on } \mathcal{E}_h, \text{ (that is, } \phi \in \mathrm{Ker}B^t\text{)}. \end{cases} \tag{10.2.26}$$

The proof of (10.2.26) (or, rather, the finding of sufficient conditions on m for having (10.2.26)) will be easier with the following characterisation of V_h^m.

Lemma 10.2.1. *We have*

$$V_h^m \equiv \underline{\underline{S}}\,[(\mathcal{L}_{m+1}^0(\mathcal{T}_h))^2], \tag{10.2.27}$$

where $\underline{\underline{S}}$ *is defined, for* $\underline{q} = (\alpha, \beta)$,

$$\underline{\underline{S}} : (\underline{q}) \to \begin{pmatrix} \partial\alpha/\partial y & -\frac{1}{2}(\partial\alpha/\partial x + \partial\beta/\partial y) \\ -\frac{1}{2}(\partial\alpha/\partial x + \partial\beta/\partial y) & \partial\beta/\partial x \end{pmatrix}. \tag{10.2.28}$$

Proof. The inclusion $\underline{\underline{S}}\,[(\mathcal{L}_{m+1}^0(\mathcal{T}_h))^2] \subseteq V_h^m$ is trivial; the opposite inclusion is an exercise (see [127] for more details). □

We now notice that if $\underline{\underline{\tau}} = \underline{\underline{S}}(\underline{q})$, then

$$b(\underline{\underline{\tau}}, v) = \sum_K \int_{\partial K} \underline{\mathrm{grad}}\, v \cdot \frac{\partial}{\partial t}\underline{q}\, ds, \tag{10.2.29}$$

where $\underline{t}$ is the tangent to ∂T. We also notice that

$$\begin{cases} \phi \in H_0^2(\Omega) \text{ and } \underline{\mathrm{grad}}\,\phi = \text{ constant on } \mathcal{E}_h \\ \text{imply } \phi = 0 \text{ and } \underline{\mathrm{grad}}\,\phi = 0 \text{ on } \mathcal{E}_h. \end{cases} \tag{10.2.30}$$

We may now use (10.2.27)–(10.2.30) in (10.2.26) which becomes

$$\begin{cases} \text{if } \phi \in Q_h^{r,s} \text{ and } \displaystyle\sum_K \int_{\partial K} \underline{\mathrm{grad}}\,\phi \cdot \frac{\partial}{\partial t}\underline{q}\, ds = 0 \quad \forall \underline{q} \in (\mathcal{L}_{m+1}^0(\mathcal{T}_h))^2, \\ \text{then } \underline{\mathrm{grad}}\,\phi = \text{ constant on } \mathcal{E}_h. \end{cases} \tag{10.2.31}$$

Now, (10.2.31) is implied by

$$\begin{cases} \text{if } \phi \in Q_h^{r,s} \text{ and } \displaystyle\int_{\partial K} \underline{\operatorname{grad}}\,\phi \cdot \frac{\partial}{\partial t}\underline{q}\, ds = 0 \quad \forall \underline{q} \in (P_{m+1}(K))^2 \\ \text{then } \underline{\operatorname{grad}}\,\phi = \text{ constant on } \partial T \end{cases} \tag{10.2.32}$$

(but not vice-versa). Now let k be the degree of $\underline{\operatorname{grad}}\,\phi$ on ∂T, that is,

$$k = \max(s, r-1). \tag{10.2.33}$$

The following technical lemma is proved in [127].

Lemma 10.2.2. *If $\phi \in H^1(K)$ and $\phi|_{e_i} \in P_k(e_i)$ $(i = 1, 2, 3)$, and if*

$$\int_{\partial K} \phi \frac{\partial q}{\partial t} ds = 0 \quad \forall q \in P_k(K), \tag{10.2.34}$$

then

$$\phi|_{e_i} = c\ell_k^i(s) + c_1, \ (i = 1, 2, 3), \tag{10.2.35}$$

where, on each e_i, we define ℓ_k^i as the kth Legendre polynomial (normalised with value 1 in the second endpoint in the anticlockwise order).

Formula (10.2.35), for k odd, directly implies that ϕ is constant on ∂K. We therefore have a first result.

Proposition 10.2.4. *If $m + 1 = k = \max(r - 1, s)$ and k is odd, then* (10.2.32) *holds.*

If $m + 1$ is even, we can apply Lemma 10.2.2 to both $\partial\phi/\partial x$ and $\partial\phi/\partial y$ and get

$$\frac{\partial \phi}{\partial x} = c\ell_k^i + c_1, \ \frac{\partial \phi}{\partial y} = \gamma\ell_k^i + \gamma \tag{10.2.36}$$

on each e_i. If now $r - 1 \neq s$, there must exist a combination of $\partial\phi/\partial x$ and $\partial\phi/\partial y$ on each e_i (to get $\partial\phi/\partial n$) which has degree lower than k. This easily implies that both $\partial\phi/\partial x$ and $\partial\phi/\partial y$ are constants on ∂K. We therefore have the following result:

Proposition 10.2.5. *If $m + 1 = k = \max(r - 1, s)$ and $r - 1 \neq s$, then* (10.2.32) *holds.*

We are finally left with the last and worst case in which $r - 1 = s$ is *even*. We have several escapes. First, brutally, we may take $m + 1 = k + 1$. It is easy to see that, then, (10.2.32) always holds. As a second possibility, we may take $m + 1 = k$ and *enrich* $(\mathcal{L}^0_{m+1}(\mathcal{T}_h))^2$ into $(\mathcal{L}^0_{m+1}(\mathcal{T}_h))^2_{enr}$ by adding, in each K, a pair of functions $\underline{q}$ in $(P_{m+1})^2$ such that $\partial q_j/\partial t|_{e_i} = \ell_k^i$ ($j = 1, 2$ and $= 1, 2, 3$). Again, it is easy to check that (10.2.32) is satisfied if we take the enriched space $(\mathcal{L}^0_{m+1}(\mathcal{T}_h))^2_{enr}$ instead of the original one. Then, of course, we must consider $V^m_{h,enr} = \underline{\underline{S}}\,[(\mathcal{L}^0_{m+1}(\mathcal{T}_h))^2_{enr}]$

instead of V_h. Finally, we might give up (10.2.32) and go directly to (10.2.31). It is easy to check that in (10.2.36), the values of c, c_1, γ, and γ_1 must remain constants from one K to another due to the continuity of $\underline{\mathrm{grad}}\,\phi|_e$ across the edges. Hence, since $\phi \in H_0^2(\Omega)$, we must have $c = c_1 = \gamma = \gamma_1 = 0$ and, *actually*, (10.2.31) *holds for* $m + 1 = k = \max(r - 1, s)$ *in any case*, that is, also for $r - 1 = s =$ even. However, we shall see in a moment that (10.2.32) has other basic advantages over (10.2.31) that we are not very willing to give up. We summarise the results in the following theorem.

Theorem 10.2.1. *The condition* $\mathrm{Ker}B_h^t \subset \mathrm{Ker}B^t$ *holds whenever*

$$m + 1 \geq k = \max(r - 1, s). \tag{10.2.37}$$

Moreover, (10.2.32) *holds when* (10.2.37) *is satisfied, unless* $r - 1 = s =$ *even. In that case,* (10.2.32) *is satisfied by taking* $m + 1 > k$ *or by using an enriched* $V_{h,enr}^{k-1}$ *(between* V_h^{k-1} *and* V_h^k*) as described above.*

The condition $\mathrm{Ker}B_h^t = \mathrm{Ker}B^t$ implies, by Proposition 5.5.2, the existence of an operator Π_h from $V_0(\mathcal{T}_h)$ to V_h^m such that

$$b(\underline{\underline{\tau}} - \Pi_h\underline{\underline{\tau}}, v) = 0 \qquad \forall v \in Q_h^{r,s}. \tag{10.2.38}$$

However, in view of the use of Proposition 5.4.3, we would also like to show that there exists a Π_h which satisfies (10.2.38) and

$$\|\Pi_h\underline{\underline{\tau}}\|_0 \leq c\,\|\underline{\underline{\tau}}\|_0, \qquad \forall \underline{\underline{\tau}} \in V_0(\mathcal{T}_h), \tag{10.2.39}$$

with c *independent of* h. Since V_h^m is finite dimensional, (10.2.39) will always hold, but the constant might depend on h. Now, if (10.2.32) holds, we see that Π_h can be defined *element by element*. Now, the dimension of $V_h^m|_K$ depends only on m, but not on h. A continuous dependence argument on the shape of the element can now prove (10.2.39) without major difficulty (but, to be honest, not quickly); we refer to [127] for a detailed proof of (10.2.39). Once we have (10.2.38) and (10.2.39), we apply Proposition 5.4.3 to prove the discrete *inf-sup* condition. Then, Theorem 5.2.5 immediately gives

$$\begin{aligned} \|\underline{\underline{\sigma}} - \underline{\underline{\sigma}}_h\|_0 &= \|\underline{\underline{D}}_2(w - \tilde{w}_h)\|_0 \\ &\leq c\Big\{\inf_{\underline{\underline{\tau}}\in V_h^m} \|\underline{\underline{\sigma}}^0 - \underline{\underline{\tau}}\|_0 + \inf_{\phi\in Q_h^{r,s}} \|\underline{\underline{D}}_2(w - \phi)\|_0\Big\}, \end{aligned} \tag{10.2.40}$$

where $\tilde{w}_h$ is the (unique) element in $Q_h^{r,s}$ that satisfies $\Delta^2\tilde{w}_h = f$ in each K and belongs to the set of discrete solutions.

Theorem 10.2.2. *If* $m+1 \geq \max(r-1, s)$ *(and* $m+1 > s$ *for* $r-1 = s$ *is even), we have*

$$\|\underline{\underline{\sigma}} - \underline{\underline{\sigma}}_h\|_0 + \|\underline{\underline{D}}_2(w - \tilde{w}_h)\|_0 \leq ch^t \left(\|w\|_{t+2} + \sum_K \|\underline{\underline{\sigma}}^f\|^2_{t,K}\right)^{\frac{1}{2}} \tag{10.2.41}$$

with $t = \min(m+1, r-1, s)$.

Proof. The proof is obvious from (10.2.40) and the standard approximation results. □

We end this section with a few computational remarks. First, we notice that our discretisation of (10.2.8) has obviously the matrix structure

$$\begin{pmatrix} A & B^t \\ B & 0 \end{pmatrix}, \tag{10.2.42}$$

where A, corresponding to the approximation of the identity in V_h^m, is obviously block diagonal because V_h^m is made of discontinuous tensors. Hence, one usually makes an a priori inversion of A, to end with the matrix $BA^{-1}B^t$ which operates on the unknown w_h and is symmetric and positive definite. However, the computation of the right-hand side is, in general, a weak point in the use of dual hybrid methods, unless f is very special (zero, Dirac mass, constant, etc.) and allows the use of a simple $\underline{\underline{\sigma}}^f$. A few computational tricks for dealing with more general cases can be found in [127, 289, 290]. Here, we recall from [115] a simple method that works for low-order approximations (more precisely, when t in Theorem 10.2.2 is ≤ 2). We first define the operator $R :=$ orthogonal projection onto V_h. We then remark that the discretisations (10.2.24) and (10.2.25) of (10.2.8) may be written as

$$\begin{cases} (\underline{\underline{\sigma}}_h^0 + \underline{\underline{\sigma}}^f, \underline{\underline{\tau}}) = (\underline{\underline{D}}_2 w_h, \underline{\underline{\tau}}) & \forall \underline{\underline{\tau}} \in V_h, \\ (\underline{\underline{\sigma}}_h + \underline{\underline{\sigma}}^f, \underline{\underline{D}}_2 \phi) = (f, \phi) & \forall \phi \in Q_h. \end{cases} \tag{10.2.43}$$

Solving a priori in $\underline{\underline{\sigma}}_h^0$ from the first equation and substituting into the second equation, we obtain

$$(R\underline{\underline{D}}_2 w_h, \underline{\underline{D}}_2 \phi) = (f, \phi) - (\underline{\underline{\sigma}}^f - R\underline{\underline{\sigma}}^f, \underline{\underline{D}}_2 \phi) \qquad \forall \phi \in Q_h. \tag{10.2.44}$$

Now, the left-hand side of (10.2.44) corresponds to the matrix $BA^{-1}B^t$ acting on the unknown w_h. The right-hand side is actually *computable* because both $(f, \phi) - (\underline{\underline{\sigma}}^f, \underline{\underline{D}}_2 \phi)$ and $(R\underline{\underline{\sigma}}^f, \underline{\underline{D}}_2 \phi)$ depend (looking carefully) only on the values of ϕ and its gradient at the inter-element boundaries. However, the computation, in general, is not easy. Therefore, in some cases, it can be convenient to use a rough approximation of it, for instance

$$(f, \phi) - (\underline{\underline{\sigma}}^f - R\underline{\underline{\sigma}}^f, \underline{\underline{D}}_2 \phi) \simeq \sum_K \frac{meas(K)}{3} \sum_{j=1}^3 f(V_j)\phi(V_j), \tag{10.2.45}$$

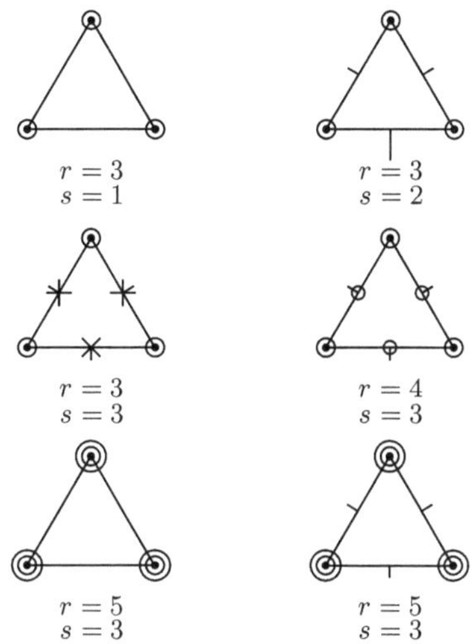

Symbol	Values of
•	ϕ
⊙	$\underline{\text{grad}}\ \phi$
◎	$\partial^2\phi/\partial_n\partial_t$
—	$\underline{\underline{D}}$
×	$\partial\phi/\partial n$
⊖	$\partial\phi/\partial t,\ \partial\phi/\partial t,\ \partial^2\phi/\partial_n\partial_t$

Fig. 10.1 Some common choices for the space $Q_h^{r,s}$

where the V_j are the vertices of K. It can be shown (see [115]) that this involves an additional error of order $O(h^2)$ (essentially because V_h contains *all* piecewise linear stress functions and therefore $\|\underline{\underline{\sigma}}^f - R\underline{\underline{\sigma}}^f\|_0 \leq ch^2$) and hence this procedure is recommended whenever $t \leq 2$ in (10.2.41).

Finally, we provide a few remarks on the choice of the degrees of freedom in V_h^m and $Q_h^{r,s}$. As we have seen, the unknown $\underline{\underline{\sigma}}_h^0$ is usually eliminated a priori at the element level due to the complete discontinuity of V_h^m. As a consequence, the choice of the degrees of freedom in V_h^m is of little relevance. In general, it is more convenient to start from $(\mathcal{L}_{m+1}^0(\mathcal{T}_h))^2$ and to derive V_h through (10.2.27).

When m is "large" (say $m \geq 4$, to fix the ideas), however, the resulting matrix A can be severely ill-conditioned unless the degrees of freedom in V_h^m are chosen in a suitable way. We refer to [289,290] for a discussion of this point. On the other hand, the degrees of freedom in $Q_h^{r,s}$ are the ones that count in the final stiffness matrix, and, besides, they have to take into account the C^{-1} continuity requirements. We sketch in Fig. 10.1 some commonly used choices for different values of r and s.

Remark 10.2.2. It is impossible to say what is, in general, the best choice for r and s. Numerical evidence shows obviously that the accuracy/number of degrees of freedom ratio is improved for large r and s, at least when the solution is smooth.

However, it is clear that the simplest (and most widely used) choice $r = 3$, $s = 1$ allows a much easier implementation. Similar considerations also hold with the choice of m, in particular in the case of an even $r - 1 = s$, for instance for $r = 3$, $s = 2$. The use of the enriched $V^1_{h,enr}$ implies a smaller matrix to be inverted on each element than with the "brutal" choice V^2_h (11×11 instead of 17×17), but the latter may allow some simplification in writing the program. □

Remark 10.2.3. We have used, so far, homogeneous Dirichlet boundary conditions corresponding to a clamped plate. Nothing changes when considering non-homogeneous Dirichlet conditions. If, instead, a part of the plate is simply supported ($w =$ given; $M_{nn} = 0$) or free ($M_{nn} = 0$; $K_n = 0$), then we have two possibilities for dealing with them. Let us discuss a simple case: let $\partial\Omega = \Gamma_D \cup \Gamma_N$ and assume that $w = \partial w/\partial n = 0$ on Γ_D and $M_n = K_n = 0$ on Γ_N. One possibility is to choose $Q^{r,s}_h$ so that its elements vanish only on Γ_D, and to let V^m_h unchanged. In this case, the conditions $M_n = K_n = 0$ on Γ_N will be satisfied only in a *weak* sense. A second possibility is to choose V^m_h in such a way that its elements satisfy, a priori, the boundary condition $M_n = K_n = 0$ on Γ_N. However, care must be taken in this case to enrich conveniently the stress field in the boundary elements so that the *inf-sup* condition still holds. Otherwise, a loss in the order of convergence is likely to occur. □

Remark 10.2.4. One may think to use other discretisations of the dual hybrid formulations than the ones discussed here (see, for instance, the previous remarks). In any case, the *inf-sup* condition should be checked. Although this is not evident from our discussion (because we wanted to deal with many cases at the same time), nevertheless, it is true that to check the *inf-sup* condition in hybrid methods is basically an *easy* task. What is really needed is the following: for any element K, the only displacement modes with zero energy on K, that is, the only modes ϕ such that

$$\int_{\partial K} \left(M_{nn}(\underline{\underline{\tau}})\phi/n - K_n(\underline{\underline{\tau}})\phi \right) ds = 0 \qquad \forall \underline{\underline{\tau}} \in V_h, \tag{10.2.46}$$

must be the *rigid* modes (that is, $\underline{\mathrm{grad}}\,\phi$ = constant on T). If this condition is violated, one can expect trouble (minor or major, depending on the cases). □

10.3 Mixed Methods for Linear Thin Plates

We consider the variational formulation of a problem discussed in Chap. 1 which we recall here for the convenience of the reader. We had

$$L(\underline{\underline{\sigma}}, w) = \inf_{\underline{\underline{\tau}} \in (L^2(\Omega))^{2\times 2}_s} \ \sup_{\phi \in H^2_0(\Omega)} L(\underline{\underline{\tau}}, \phi) \tag{10.3.1}$$

where

$$L(\underline{\underline{\tau}},\phi) := \frac{1}{2}\left(\frac{12}{Et^3}\right)\int_\Omega [(1+\nu)\underline{\underline{\tau}} : \underline{\underline{\tau}} - \nu(\mathrm{tr}(\underline{\underline{\tau}}))^2]\,dx \\ - \int_\Omega \underline{\underline{\tau}} : \underline{\underline{D}}_2\phi\,dx + \int_\Omega f\phi\,dx \tag{10.3.2}$$

$$E = \text{Young's modulus,} \tag{10.3.3}$$

$$t = \text{thickness of the plate,} \tag{10.3.4}$$

$$\nu = \text{Poisson's ratio,} \tag{10.3.5}$$

$$f = \text{transversal load / unit surface,} \tag{10.3.6}$$

$$w = \text{transversal displacement,} \tag{10.3.7}$$

$$\underline{\underline{\sigma}} = \text{stresses (in the Kirchoff assumption).} \tag{10.3.8}$$

In order to use a more compact notation, we set

$$C\underline{\underline{\tau}} := \frac{1}{2}Et^3((1+\nu)\underline{\underline{\tau}} - \nu\,\mathrm{tr}(\underline{\underline{\tau}})\underline{\underline{\delta}}) \tag{10.3.9}$$

and write $L(\underline{\underline{\tau}},\phi)$ as

$$L(\underline{\underline{\tau}},\phi) = \frac{1}{2}(C\underline{\underline{\tau}},\underline{\underline{\tau}}) - (\underline{\underline{\tau}},\underline{\underline{D}}_2\phi) + (f,\phi). \tag{10.3.10}$$

Assume that we are given a triangulation $\mathcal{T}_h$ of Ω and that we are willing to discretise the stress field $\underline{\underline{\sigma}}$ by means of piecewise polynomials for which the normal bending moment

$$M_{nn}(\underline{\underline{\sigma}}) = (\underline{\underline{\sigma}}\cdot\underline{n})\cdot\underline{n} \tag{10.3.11}$$

is continuous from one element to another. We recall the following Green's formulae,

$$\int_K \underline{\underline{\tau}} : \underline{\underline{D}}_2\phi\,dx = -\int_K \underline{\mathrm{div}}\,\underline{\underline{\tau}}\cdot\underline{\mathrm{grad}}\,\phi\,dx + \int_{\partial K} M_{nn}(\underline{\underline{\tau}})\frac{\partial\phi}{\partial n}\,ds \\ + \int_{\partial K} M_{nt}(\underline{\underline{\tau}})\frac{\partial\phi}{\partial t}\,ds, \tag{10.3.12}$$

$$-\int_K \underline{\mathrm{div}}\,\underline{\underline{\tau}}\cdot\underline{\mathrm{grad}}\,\phi\,dx = \int_K D_2^*(\underline{\underline{\tau}})\,\phi\,dx - \int_{\partial K} Q_n(\underline{\underline{\tau}})\,\phi\,ds, \tag{10.3.13}$$

valid for all $\underline{\underline{\tau}}$ and ϕ smooth in K; we recall again that, here, $\underline{t}$ is the unit tangent (anticlockwise) vector and

$$M_{nt}(\underline{\underline{\tau}}) = (\underline{\underline{\tau}}\cdot\underline{n})\cdot\underline{t}, \quad Q_n(\underline{\underline{\tau}}) = \underline{\mathrm{div}}(\underline{\underline{\tau}})\cdot\underline{n}. \tag{10.3.14}$$

If $M_{nn}(\underline{\underline{\tau}})$ is continuous and ϕ is smooth, we can write

$$L(\underline{\underline{\tau}},\phi)=\frac{1}{2}(C\underline{\underline{\tau}},\underline{\underline{\tau}})+\sum_K\Big\{\int_K \underline{\mathrm{div}}(\underline{\underline{\tau}})\cdot\underline{\mathrm{grad}}\,\phi\,dx-\int_{\partial K}M_{nt}(\underline{\underline{\tau}})\frac{\partial\phi}{\partial t}ds\Big\}+(f,\phi). \tag{10.3.15}$$

A little functional analysis shows that every integral in (10.3.15) makes sense (at least as a suitable duality pairing), provided $\underline{\underline{\tau}}$ and ϕ are, respectively, in the following spaces:

$$V:=\{\underline{\underline{\tau}}\mid\underline{\underline{\tau}}|_K\in(H^1(\Omega))_s^{2\times2},\ M_{nn}(\underline{\underline{\tau}})\text{ continuous}\}, \tag{10.3.16}$$

$$Q:=W^{1,p}(\Omega),\ p>2. \tag{10.3.17}$$

Remark 10.3.1 (For mathematicians). We have to choose $p>2$ in (10.3.17) because for $\phi\in H^1(K)$ we have $\partial\phi/\partial t\in H^{-1/2}(\partial K)$ whereas $M_{nt}(\underline{\underline{\tau}})$ is in $\prod_{e_i}H^{1/2}(e_i)$ but *not* in $H^{1/2}(\partial K)$. On the other hand, for $\phi\in W^{1,p}$, we have $\partial\phi/\partial t\in W^{-1/p,p}(\partial K)$. Since $M_{nt}(\underline{\underline{\tau}})$ is in $H^s(\partial K)$ for all $s<1/2$ and since $W^{-1/p,p}(\partial K)\subset H^{-1}(\partial K)$ for $s>1/p$, the boundary integral which appears in (10.3.15) can now be interpreted as a duality pairing between $H^{-s}(\partial K)$ and $H^s(\partial K)$ for $1/p<s<1/2$ (which is possible since $p>2$). □

The Euler equations of (10.3.15) can now be written as:

$$(C\underline{\underline{\sigma}},\underline{\underline{\tau}})+\sum_K\Big\{\int_K \underline{\mathrm{div}}(\underline{\underline{\tau}})\cdot\underline{\mathrm{grad}}\,w\,dx-\int_{\partial K}M_{nt}(\underline{\underline{\tau}})\frac{\partial\phi}{\partial t}\,ds\Big\}=0\quad\forall\underline{\underline{\tau}}\in V, \tag{10.3.18}$$

$$\sum_K\Big\{\int_K \underline{\mathrm{div}}(\underline{\underline{\tau}})\cdot\underline{\mathrm{grad}}\,\phi\,dx-\int_{\partial K}M_{nt}(\underline{\underline{\tau}})\frac{\partial\phi}{\partial t}ds\Big\}=(-f,\phi)\qquad\forall\phi\in Q, \tag{10.3.19}$$

which has the form (5.1.9) if we set

$$a(\underline{\underline{\sigma}},\underline{\underline{\tau}}):=(C\underline{\underline{\sigma}},\underline{\underline{\tau}}), \tag{10.3.20}$$

$$b(\underline{\underline{\sigma}},\phi):=\sum_K\Big\{\int_K \underline{\mathrm{div}}(\underline{\underline{\tau}})\cdot\underline{\mathrm{grad}}\,\phi\,dx-\int_{\partial K}M_{nt}(\underline{\underline{\tau}})\frac{\partial\phi}{\partial t}\,ds\Big\}. \tag{10.3.21}$$

Unfortunately, problem (10.3.18) and (10.3.19), as it stands, does not satisfy any of the conditions given in Chap. 4 in order to have a well posed problem. However, we know that the original problem (1.2.4) has a solution w. If $\underline{\underline{\sigma}}=C^{-1}(\underline{\underline{D}}_2w)$ is in $H^1(\Omega)$, that is if the solution w of (1.2.4) is smooth enough, it is easy to check that the pair $(\underline{\underline{\sigma}},w)$ solves (10.3.18) and (10.3.19). Hence, we only have to prove the *uniqueness* of the solution of (10.3.18) and (10.3.19).

Proposition 10.3.1. *Problem* (10.3.18) *and* (10.3.19) *has a unique solution.*

Proof. It is obvious that

$$a(\underline{\underline{\tau}}, \underline{\underline{\tau}}) \geq \alpha \ \|\underline{\underline{\tau}}\|_0^2, \qquad \forall \underline{\underline{\tau}} \in V. \tag{10.3.22}$$

Let us now check a weaker *inf-sup* condition. For every ϕ in Q, we define $\underline{\underline{\tau}}(\phi)$ by

$$\tau_{11} = \tau_{22} = \phi, \ \tau_{12} = \tau_{21} = 0. \tag{10.3.23}$$

It is immediate to check that $M_{nt}(\underline{\underline{\tau}})$ is continuous across the inter-element boundaries, so that

$$\sum_K \int_{\partial K} M_{nt}(\underline{\underline{\tau}}(\phi)) \frac{\partial \phi}{\partial t} ds = 0 \tag{10.3.24}$$

and therefore

$$b(\underline{\underline{\tau}}(\phi), \phi) = |\phi|_{1,\Omega}^2. \tag{10.3.25}$$

It is also easy to check, using (10.3.23) and the Poincaré's inequality (1.2.14), that

$$\|\underline{\underline{\tau}}(\phi)\|_V \leq c \ |\phi|_{1,\Omega}; \tag{10.3.26}$$

hence, we have from (10.3.25) and (10.3.26) that

$$\begin{aligned} \inf_{\phi \in H_0^1(\Omega)} \sup_{\underline{\underline{\tau}} \in V} \frac{b(\underline{\underline{\tau}}, \phi)}{\|\underline{\underline{\tau}}\|_V |\phi|_{1,\Omega}} &\geq \inf_{\phi \in H_0^1(\Omega)} \frac{b(\underline{\underline{\tau}}(\phi), \phi)}{\|\underline{\underline{\tau}}(\phi)\|_V |\phi|_{1,\Omega}} \\ &\geq \frac{|\phi|_{1,\Omega}}{\|\underline{\underline{\tau}}(\phi)\|_V} \geq \frac{1}{c} > 0. \end{aligned} \tag{10.3.27}$$

Now using (10.3.22) and (10.3.27), we have the desired uniqueness by standard arguments. □

We are now ready to discretise our problem. Following [132] and [261], for any integer $k \geq 0$, we set

$$V_h = (\mathcal{L}_k^0)_s^{2\times 2} \cap V \tag{10.3.28}$$

$$Q_h = \mathcal{L}_{k+1}^1 \tag{10.3.29}$$

with the notation of Chap. 2. Note that the space V_h in (10.3.28) is made of tensors whose normal bending moment is continuous across the inter-element boundaries. The degrees of freedom for Q_h will be the usual ones (see Sect. 2.2). As degrees of freedom for V_h, we may choose, for instance, the following ones:

$$\int_e M_{nn}(\underline{\underline{\tau}}) p(s) \, ds \qquad \forall p \in P_k(e), \quad \forall e \in \mathcal{E}_h, \tag{10.3.30}$$

$$\int_T \underline{\underline{\tau}} : \underline{\underline{p}} \, dx \qquad \forall \underline{\underline{p}} \in (P_{k-1}(T))_s^{2\times 2}, \ \forall K \in \mathcal{T}_h, \ (k \geq 1). \tag{10.3.31}$$

The possibility of choosing (10.3.30) and (10.3.31) as degrees of freedom in V_h is shown by the following lemma and by a standard dimensional count.

Lemma 10.3.1. *Let* $\underline{\underline{\tau}} \in (P_{k-1}(T))^{2\times 2}_s$ *be such that*

$$\int_{e_i} M_{nn}(\underline{\underline{\tau}})p(s)\, ds = 0 \qquad \forall p \in P_k(e_i),\ (i = 1,2,3), \tag{10.3.32}$$

$$\int_K \underline{\underline{\tau}} : \underline{\underline{p}}\, dx = 0 \qquad \forall \underline{\underline{p}} \in (P_{k-1}(T))^{2\times}_s,\ (k \geq 1). \tag{10.3.33}$$

Then, $\underline{\underline{\tau}} \equiv 0$.

Proof. We only give a hint of the proof. From (10.3.32), we get $M_{nn}(\underline{\underline{\tau}}) = 0$. We first show that $D_2^*(\underline{\underline{\tau}}) = 0$. This is trivial for $k \leq 1$; for $k > 1$, take $\underline{\underline{p}} = \underline{\underline{D}}_2 b$ with $b = b_3 D_2^* \underline{\underline{\tau}}$ in (10.3.33) to get $\int_K b_3(D_2^*(\underline{\underline{\tau}}))^2\, dx = 0$ and hence $D_2^*(\underline{\underline{\tau}}) = 0$. Now use the formula (see Sect. 10.2)

$$\int_K \underline{\underline{\tau}} : \underline{\underline{D}}_2 \phi = \int_K D_2^*(\underline{\underline{\tau}})\, \phi + \int_{\partial K} [M_{nn}(\underline{\underline{\tau}}) \frac{\partial \phi}{\partial n} - \mathcal{K}_n(\underline{\underline{\tau}})\phi]\, ds \tag{10.3.34}$$

for $\phi \in P_{k+1}(T)$; thus, we get

$$\int_{\partial K} \mathcal{K}_n(\underline{\underline{\tau}})\, ds = 0 \qquad \forall \phi \in P_{k+1}(T), \tag{10.3.35}$$

and easily obtain that $\mathcal{K}_n(\underline{\underline{\tau}}) = 0$. It is now simple to show that $\underline{\underline{\tau}} = \underline{\underline{S}}(\underline{q})$ (see (10.2.27) for the definition of $\underline{\underline{S}}$) for some $\underline{q} \in (P_{k+1}(K))^2$ with $\underline{q} = \underline{0}$ on ∂K. Therefore, q_1 (for instance) has the form $b_3 z$ with $z \in P_{k-2}(K)$. Let us now choose, in (10.3.33), p_{11} such that $\partial p_{11}/\partial y$ and $p_{12} = p_{22} = 0$. We then get

$$0 = \int_K \tau_{11} p_{11}\, dx = \int_K \frac{\partial q_1}{\partial y} p_{11}\, dx = -\int_K q_1 z\, dx = -\int_K b_3 z^2\, dx \tag{10.3.36}$$

so that $z = 0$ and $q_1 = 0$. Similarly, one proves that $q_2 = 0$. □

We are now able to define the operator Π_h. We set, for $\underline{\underline{\tau}} \in V$,

$$\int_e M_{nn}(\Pi_h \underline{\underline{\tau}} - \underline{\underline{\tau}})\, p(s)\, ds = 0 \qquad \forall p \in P_k(e), \qquad \forall e \in \mathcal{E}_h, \tag{10.3.37}$$

$$\int_K (\Pi_h \underline{\underline{\tau}} - \underline{\underline{\tau}}) : \underline{\underline{p}}\, ds = 0 \qquad \forall \underline{\underline{p}} \in (P_{k-1}(K))^{2\times 2}_s, \qquad \forall K \in \mathcal{T}_h. \tag{10.3.38}$$

Lemma 10.3.2. *Let* Π_h *be defined by* (10.3.37) *and* (10.3.38)*. Then, we have*

$$\|\Pi_h \underline{\underline{\tau}}\|_V \leq c\, \|\underline{\underline{\tau}}\|_V \qquad \forall \underline{\underline{\tau}} \in V \tag{10.3.39}$$

and

$$b(\underline{\underline{\tau}} - \Pi_h \underline{\underline{\tau}}, \phi_h) = 0 \qquad \forall \underline{\underline{\tau}} \in V \qquad \forall \phi_h \in Q_h. \tag{10.3.40}$$

Proof. Formula (10.3.39) is easy to check. Let us prove (10.3.40). From (10.3.12) and (10.3.21), we have

$$b(\underline{\underline{\tau}} - \Pi_h \underline{\underline{\tau}}, \phi) = -\sum_K \Big\{ \int_K (\underline{\underline{\tau}} - \Pi_h \underline{\underline{\tau}}) : \underline{\underline{D}}_2 \phi \, dx - \int_{\partial K} M_{nn}(\underline{\underline{\tau}} - \Pi_h \underline{\underline{\tau}}) \frac{\partial \phi}{\partial n} ds \Big\} \tag{10.3.41}$$

and from (10.3.41), (10.3.37), and (10.3.38), we get (10.3.40). □

Lemma 10.3.3. *If $\underline{\underline{\tau}}_h \in V_h$ is such that*

$$b(\underline{\underline{\tau}}_h, \phi_h) = 0, \qquad \forall \phi_h \in Q_h, \tag{10.3.42}$$

then

$$b(\underline{\underline{\tau}}_h, \phi) = 0, \qquad \forall \phi \in Q. \tag{10.3.43}$$

Proof. We have, from (10.3.13) and (10.3.21),

$$b(\underline{\underline{\tau}}_h, \phi) = -\sum_K \Big\{ \int_K D_2^*(\underline{\underline{\tau}}_h) \phi \, dx + \int_{\partial K} [M_{nt}(\underline{\underline{\tau}}_h) \frac{\partial \phi}{\partial t} - Q_n(\underline{\underline{\tau}}_h) \phi] \, ds \Big\}. \tag{10.3.44}$$

Integrating $\int_{\partial K} M_{nt} \frac{\partial \phi}{\partial t} ds$ by parts and recalling the definition of $\mathcal{K}_n$ in (10.2.3), we then have

$$b(\underline{\underline{\tau}}_h, \phi) = -\sum_K \Big\{ \int_K D_2^*(\underline{\underline{\tau}}_h) \, \phi \, dx - \int_{\partial K} \mathcal{K}_n(\underline{\underline{\tau}}_h) \phi \, ds \Big\}. \tag{10.3.45}$$

Note that (10.3.45) holds for any $\underline{\underline{\tau}}_h$ and ϕ piecewise smooth. If now (10.3.42) holds, we first have $D_2^*(\underline{\underline{\tau}}_h) = 0$ by choosing $\phi|_K = b_3 D_2^*(\underline{\underline{\tau}}_h)$ (for $k \geq 2$, otherwise the property is trivial). Hence, we are left with

$$\sum_K \int_{\partial K} \mathcal{K}_n(\underline{\underline{\tau}}_h) \phi_h \, ds = 0 \qquad \forall \phi \in Q_h. \tag{10.3.46}$$

Since $\mathcal{K}_n$ is made of Dirac measures at the vertices and of polynomials of degree less or equal to $k-1$ on each edge, it is easy to see that (10.3.46) implies $\mathcal{K}_n(\underline{\underline{\tau}}_h) = 0$. Therefore, we have proved that if $\underline{\underline{\tau}}_h \in V_h$ satisfies (10.3.42), then $D_2^*(\underline{\underline{\tau}}_h) = 0$ and $\mathcal{K}_n(\underline{\underline{\tau}}_h) = 0$. We now insert those two equations into (10.3.45) and we get (10.3.43). □

This last property was denoted, in Chap. 5, as $Z_h(0) \subset Z(0)$. We have seen that, together with the existence of the operator Π_h, this property is so important that it can provide optimal error estimates even in desperate situations (no ellipticity, no *inf-sup* condition) like ours.

Actually, we first remark that (10.3.27) and Lemma 10.3.2 provide, through Proposition 5.4.3, the following *inf-sup* type condition:

$$\inf_{\phi_h \in Q_h} \sup_{\underline{\underline{\tau}}_h \in V_h} \frac{b(\underline{\underline{\tau}}_h, \phi_h)}{\|\underline{\underline{\tau}}_h\|_v |\phi_h|_1} \geq c > 0 \text{ (c independent of h)}. \tag{10.3.47}$$

On the other hand, since Q_h and V_h are finite dimensional, (10.3.22) and (10.3.47) ensure that the discrete problem has a unique solution. We are now ready for error estimates.

Proposition 10.3.2. *If* $(\underline{\underline{\sigma}}, w)$ *is the solution of* (10.3.18) *and* (10.3.19) *and* $(\underline{\underline{\sigma}}_h, w_h)$ *is the discrete solution of* (10.3.18) *and* (10.3.19)*, then, through* (10.3.28) *and* (10.3.29)*, we have*

$$\|\underline{\underline{\sigma}} - \underline{\underline{\sigma}}_h\|_0 \leq c\ \|\underline{\underline{\sigma}} - \Pi_h \underline{\underline{\sigma}}\|_0. \tag{10.3.48}$$

□

The proof is immediate from the standard theory of Chap. 5.

From (10.3.48) and standard approximation results, we then have

$$\|\underline{\underline{\sigma}} - \underline{\underline{\sigma}}_h\|_0 \leq ch^{k+1} \|\underline{\underline{\sigma}}\|_{k+1}. \tag{10.3.49}$$

Proposition 10.3.3. *With the notation of Proposition 10.3.2, we have*

$$\|w - w_h\|_1 \leq c\ \{h^{k+1} \|\underline{\underline{\sigma}}\|_{k+1} + h^{k+1} \|w\|_{k+2}\}. \tag{10.3.50}$$

Proof. Let $\phi_h \in Q_h$ to be chosen. From (10.3.47), we have for some $\underline{\underline{\tau}}_h \in V_h$

$$\begin{aligned} c\|\phi_h - w_h\|_1 \|\underline{\underline{\tau}}_h\|_V &\leq b(\underline{\underline{\tau}}_h, \phi_h - w_h) \\ &= b(\underline{\underline{\tau}}_h, \phi_h - w) + b(\underline{\underline{\tau}}_h, w - w_h) \\ &= b(\underline{\underline{\tau}}_h, \phi_h - w) + a(\underline{\underline{\sigma}} - \underline{\underline{\sigma}}_h, \underline{\underline{\tau}}_h). \end{aligned} \tag{10.3.51}$$

It is now elementary to see that ϕ_h can be chosen in such a way that

$$\int_e p \frac{\partial}{\partial t}(w - \phi_h)\, ds = 0 \qquad \forall p \in P_k(e), \quad \forall e \in \mathcal{E}_h, \tag{10.3.52}$$

$$\|w - \phi_h\|_1 \leq ch^{k+1} \|w\|_{k+2}. \tag{10.3.53}$$

With such a choice, we have

$$
\begin{aligned}
b(\underline{\underline{\tau}}_h, w - \phi_h) &= \sum_K \int_K \underline{\mathrm{div}}(\underline{\underline{\tau}}_h) \cdot \underline{\mathrm{grad}}(w - \phi_h) dx \\
&\leq \|\underline{\underline{\tau}}_h\|_V \|w - \phi_h\|_1 \\
&\leq ch^{k+1} \|\underline{\underline{\tau}}_h\|_V \|w\|_{k+2},
\end{aligned}
\tag{10.3.54}
$$

so that from (10.3.51), (10.3.54) and (10.3.49) we get (10.3.50). □

Remark 10.3.2. Result (10.3.50) is not optimal as far as the regularity of w is involved. Actually, it says

$$\|w - w_h\|_1 \leq ch^s \|w\|_{s+2} \; (s \leq k + 1), \tag{10.3.55}$$

while an $(s + 1)$-norm on w should be enough for optimality. Furthermore, a more sophisticated analysis [44, 192] shows that

$$\|w - w_h\|_r \leq ch^{s-r} \|w\|_s \; (s \leq k + 2, \; 0 \leq r \leq 1) \tag{10.3.56}$$

for $k \geq 1$ and

$$\|w - w_h\|_0 \leq ch^2 \|w\|_4 \text{ for } k = 0. \tag{10.3.57}$$

In particular, the approach of [44] has a special interest because, by a suitable use of mesh-dependent norms in V_h and Q_h, they can show that the discretised problem (in the new norms) satisfy the abstract assumptions (5.2.33) and (5.2.34) so that optimal error estimates (in the new norms) can be directly obtained by Theorem 5.2.5. Their approach also works for other fourth-order mixed methods, like those analysed in Sects. 10.1 and 10.2. □

Remark 10.3.3. For the actual solution of the discretised problem, the most convenient method is to disconnect the continuity of $\underline{\underline{\sigma}}_h \cdot \underline{n}$ and to enforce it back via Lagrange multipliers λ_h. Then, one eliminates $\underline{\underline{\sigma}}_h$ at the element level and one solves a symmetric and positive definite system for the unknowns λ_h and w_h. The procedure is identical to the one described in Sect. 7.2 and we refer to it for a detailed description. As far as the error estimates for the Lagrange multipliers λ_h are concerned, recent results have been obtained in [158]. □

Remark 10.3.4. It is interesting to analyse the relationship between the mixed methods described here and some nonconforming methods for fourth-order problems. For instance, the following result is proved in [23]. Let us consider the space built by means of the Morley element $\mathcal{L}_2^{2,NC}$ described in Example 2.2.6 and let us define

$$a_h(\psi_h, \phi_h) := \frac{Et^3}{12(1 - \nu^2)} \sum_K \int_K [(1 - \nu)\underline{\underline{D}}_2 \psi_h : \underline{\underline{D}}_2 \phi_h + \nu \Delta\psi_h \Delta\phi_h] \, dx. \tag{10.3.58}$$

For every $\phi_h \in \mathcal{L}_2^{2,NC}$, let ϕ_h^I be the piecewise linear interpolant of ϕ_h (that is $\phi_h^I \in \mathcal{L}_1^1$ and $\phi_h^I = \phi_h$ at the vertices). Consider now the modified Morley problem: *find* $\psi_h \in \mathcal{L}_2^{2,NC}$ *such that*

$$a_h(\psi_h, \phi_h) = (f, \phi_h^I) \qquad \forall \phi_h \in \mathcal{L}_2^{2,NC}. \tag{10.3.59}$$

Then, we have

$$\underline{\underline{D}}_2 \psi_h = \underline{\underline{\sigma}}_h, \quad \psi_h^I = w^h, \tag{10.3.60}$$

where $(\underline{\underline{\sigma}}_h, w_h)$ is the discrete solution of the mixed problem (10.3.18) and (10.3.19) through (10.3.28) and (10.3.29) for $k = 0$. We note explicitly that, in the case of variable coefficients, the equivalence is more complicated. Also note that $\partial\psi_h/\partial n|_e = \lambda_h|_e$ for all $e \in \mathcal{E}_h$, where λ_h is the Lagrange multiplier introduced in the previous remark. Notice that we have, from [23],

$$\|\psi_h - w\|_{1,h} \leq ch^2 \, \|w\|_3, \tag{10.3.61}$$

which improves (10.3.50) and (10.3.57) since it requires only H^3-regularity on w. This is particularly striking since the cost for computing ψ_h is cheaper (or equal, using λ_h) than the cost for computing $(\underline{\underline{\sigma}}_h, w_h)$. □

10.4 Moderately Thick Plates

10.4.1 Generalities

We end this chapter with a hint on the theory for the so-called "Mindlin–Reissner plates". The corresponding model stands somehow in between the standard three-dimensional linear elasticity and the two-dimensional Kirchhoff theory for thin plates. Let us recall it briefly. Assume that we are given a three-dimensional elastic body that, in absence of forces, occupies the region $\Omega \times]-t, t[$, where $\Omega \subset \mathbb{R}^2$ is a bounded smooth domain and $t > 0$ is "small" (but not "too small") with respect to diam(Ω). This is what we call a "moderately thick" plate. We shall assume, for the sake of simplicity, that the plate is clamped along the entire boundary $\partial\Omega \times]-t, t[$ and that a vertical load $\underline{f} = (0, 0, f_3)$ is imposed.

Here below, we present the "Mindlin-Reissner" model following the classical engineering "derivation". Such derivation is questionable, from the mathematical point of view, at some points, but it has the clear merit of being short and simple. From the mathematical point of view, the derivation of [35] is much more convincing, but it is surely longer and more complicated. As the aim of this book is mainly concentrated on the mathematical properties of models and on their discretisations rather than on the modelling aspects, we decided to stick to the simpler choice.

The Mindlin model assumes that the "in plane" displacements u_1 and u_2 have the form

$$u_1(x,y,z) = -z\,\theta_1(x,y),\ u_2(x,y,z) = -z\,\theta_2(x,y) \tag{10.4.1}$$

and that the "transversal" displacement u_3 has the form

$$u_3(x,y,z) = w(x,y). \tag{10.4.2}$$

The corresponding strain field therefore takes the form:

$$\begin{cases} \varepsilon_{11} = -z\,\partial\theta_1/\partial x;\ \varepsilon_{22} = -z\,\partial\theta_2/\partial y;\ \varepsilon_{33} = 0; \\ 2\varepsilon_{12} = -z(\partial\theta_1/\partial y + \partial\theta_2/\partial x);\ 2\varepsilon_{13} = \partial w/\partial x - \theta_1;\ 2\varepsilon_{23} = \partial w/\partial y - \theta_2; \end{cases} \tag{10.4.3}$$

and assuming a linear elastic material, the stress field is

$$\begin{cases} \sigma_{11} = (\varepsilon_{11} + \nu\varepsilon_{22})\ E/(1-\nu^2);\ \sigma_{22} = (\varepsilon_{22} + \nu_{11})E/(1-\nu^2); \\ \sigma_{ij} = \varepsilon_{ij}E/(1+\nu);\ i,j = 1,2,3,\ i \neq j. \end{cases} \tag{10.4.4}$$

If we now write the total potential energy

$$\Pi = \frac{1}{2}\int_{\Omega\times]-t,t[} (\underline{\underline{\sigma}} : \underline{\underline{\varepsilon}} - 2\,\underline{f}\cdot\underline{u})\,dx\,dy\,dz \tag{10.4.5}$$

in terms of θ and w through (10.4.1)–(10.4.4), we obtain (after some calculations)

$$\Pi = \frac{t^3}{2}(a(\underline{\theta},\underline{\theta}) + \frac{\lambda t}{2}\int_\Omega |\,\underline{\mathrm{grad}}\,w - \underline{\theta}|^2\,dx\,dy - \int_{\Omega\times]-t,t[} f_3\,w\,dx\,dy\,dz, \tag{10.4.6}$$

where the symmetric bilinear form a is identified by

$$\begin{aligned} a(\underline{\theta},\underline{\eta}) := \frac{E}{12(1-\nu^2)}\int_\Omega \Big[\Big(\frac{\partial\theta_1}{\partial x} + \frac{\nu\partial\theta_2}{\partial y}\Big)\frac{\partial\eta_1}{\partial x} + \Big(\frac{\nu\partial\theta_1}{\partial x} + \frac{\partial\theta_2}{\partial y}\Big)\frac{\partial\eta_2}{\partial y} \\ + \frac{(1-\nu)}{2}\Big(\frac{\partial\theta_1}{\partial y} + \frac{\partial\theta_2}{\partial x}\Big)\Big(\frac{\partial\eta_1}{\partial y} + \frac{\partial\eta_2}{\partial x}\Big)\Big]\,dx\,dy, \end{aligned} \tag{10.4.7}$$

where

$$\lambda := \frac{E\ k}{2(1+\nu)} \tag{10.4.8}$$

and k is a correction factor which is often used to account for the "nonconformity" of (10.4.4). Indeed, from (10.4.1)–(10.4.4), we deduce that σ_{13} and σ_{23} are *constants* in z, whereas the physical problem has $\sigma_{13} = \sigma_{23} = 0$ on the upper and lower face of the plate: $\Omega \times \{t\}$ and $\Omega \times \{-t\}$; hence, (10.4.4) is often corrected by assuming that σ_{13} and σ_{23} behave parabolically in z, vanishing for $z = \pm t$ and assuming the value (10.4.4) for $z = 0$. For a mathematically more convincing justification of the classical $5/6$ factor, we refer again to [35]. Actually, for the sake of simplicity, we shall assume, from now on, that

$$\lambda = 1.$$

In fact, as far as we do not expect the true value (10.4.8) to go to zero or to $+\infty$, assuming $\lambda = 1$ will just change the numerical value of the constants appearing in the stability estimates or in the a priori error estimates, but it will not change the behaviour in function of the thickness t or the mesh-size h.

10.4.2 The Mathematical Formulation

The assumed boundary conditions lead to the kinematic constraints

$$\theta_1 = \theta_2 = w = 0 \text{ on } \partial\Omega. \tag{10.4.9}$$

Hence, we define the spaces

$$\Theta := (H_0^1(\Omega))^2;\ Z := H_0^1(\Omega);\ V := \Theta \times Z \tag{10.4.10}$$

with the norm

$$\|(\underline{\eta}, \zeta)\|_V^2 := \|\underline{\eta}\|_1^2 + \|\zeta\|_1^2. \tag{10.4.11}$$

When convenient, the generic element of V will be denoted $\underline{v} = (\underline{\eta}, \zeta)$ with $\underline{\eta} = (\eta_1, \eta_2) \in \Theta$ and $\zeta \in Z$. We finally recall the Korn inequality

$$\exists\, \alpha_{Korn} > 0 \text{ such that } a(\underline{\eta}, \underline{\eta}) \geq \alpha_{Korn}\ \|\underline{\eta}\|_1^2 \quad \forall \underline{\eta} \in \Theta, \tag{10.4.12}$$

where, from now on in this section, the symmetric bilinear form a will be the one given in (10.4.7).

It is easy to check that, for any fixed $t > 0$, functional (10.4.5) has a unique minimiser $(\underline{\theta}, w)$ on V which satisfies

$$t^3\, a(\underline{\theta}, \underline{\eta}) + t \int_\Omega (\underline{\mathrm{grad}}\, w - \underline{\theta}) \cdot \underline{\eta}\, dx\, dy = 0 \quad \forall \underline{\eta} \in \Theta, \tag{10.4.13}$$

$$t \int_\Omega (\underline{\mathrm{grad}}\, w - \underline{\theta}) \cdot \underline{\mathrm{grad}}\, \zeta\, dx\, dy = \int_{\Omega\times]-t,t[} f_3\, \zeta\, dx\, dy\, dz \qquad \forall \zeta \in Z. \tag{10.4.14}$$

In particular, we have

$$\frac{t^3}{2} a(\underline{\eta}, \underline{\eta}) + \frac{t}{2} \int_\Omega |\, \underline{\mathrm{grad}}\, \zeta - \underline{\eta}|^2\, dx\, dy \geq c(t)\ (\|\underline{\eta}\|_1^2 + \|\zeta\|_1^2), \tag{10.4.15}$$

for any $\underline{v} = (\underline{\eta}, \zeta) \in V$. Note that for fixed t, (10.4.15) always guarantees that (10.4.13), (10.4.14) is a nice linear elliptic problem so that, for instance, any reasonable conforming approximation of V will have optimal order of convergence.

The troubles start when we take a *small* t; then, the constant in (10.4.15) deteriorates and so does the constant in front of the optimal error bound. In practice, it is well known that if we use "any reasonable conforming approximation of V", we will get pretty bad answers for small t. Here, we shall make an analysis of the nature of the trouble. We shall also give some sufficient conditions on the discretisation so that it stays good for t smaller and smaller. The one-dimensional case was treated in [15], but the two-dimensional case, as we shall see, is more complicated.

The first thing that we have to do is to construct a *sequence* of physical problems $\mathcal{P}_t$ (for $t > 0$ and, say $t < T_0$) that fulfil the following requirements:

(1) Each $\mathcal{P}_t$ is of type (10.4.13) and (10.4.14) and so has a unique solution $\underline{\theta}(t)$, $w(t)$;
(2) There exists two constants c_1, c_2 with $0 < c_1 < c_2$ such that

$$c_1 \leq \|\underline{\theta}(t)\|_1 + \|w(t)\|_1 \leq c_2 \qquad \forall\, t \in]0, T_0[. \tag{10.4.16}$$

A possible answer is to fix Ω, E, and ν, and to choose, for each $t > 0$, the load $f_3(x, y, z)$ of the form

$$f_3(x, y, z) := \frac{t^2}{2} f(x, y), \tag{10.4.17}$$

with $g(x, y)$ fixed (once and for all) independent of t. It is clear that (10.4.17) implies

$$\int_{\Omega \times]-t,t[} f_3\, w\, dx\, dy\, dz = t^3 \int_\Omega f\, w\, dx\, dy = t^3 (f, w), \tag{10.4.18}$$

where as usual (f, w) denotes the $L^2(\Omega)$ inner product or (with an abuse of notation) whenever f is assumed to be only in $H^{-1}(\Omega)$, the duality pairing between $H^{-1}(\Omega)$ and $H_0^1(\Omega)$. Hence, dividing (10.4.6) by t^3, each problem $\mathcal{P}_t$ will amount to minimise, in V,

$$\Pi_t(\underline{\theta}, w) = \frac{1}{2} a(\underline{\theta}, \underline{\theta}) + \frac{t^{-2}}{2} \int |\,\underline{\mathrm{grad}}\, w - \underline{\theta}|^2\, dx\, dy - (f, w). \tag{10.4.19}$$

Proposition 10.4.1. *Let* $\underline{\theta}(t)$, $w(t)$ *be the minimiser of* (10.4.19) *in* V. *Then,* (10.4.16) *holds with* c_1 *and* c_2 *independent of* t.

Proof. We obviously have

$$a(\underline{\theta}, \underline{\theta}) + t^{-2} \|\,\underline{\mathrm{grad}}\, w - \underline{\theta}\|_0^2 = (f, w). \tag{10.4.20}$$

Using (10.4.12) and a little algebra, we deduce from (10.4.20) that

$$\|\underline{\theta}\|_1^2 + \|w\|_1^2 \leq c(\alpha_{Korn})\|f\|_{-1}\|w\|_1, \tag{10.4.21}$$

which implies the boundedness of $\|\underline{\theta}\|_1 + \|w\|_1$ from above. Then, one observes that the minimum of Π_t over all V is surely smaller than the minimum of Π_t over $V_0 = \{(\underline{\eta}, \zeta) \mid \underline{\eta} = \underline{\mathrm{grad}}\,\zeta\}$ (which is clearly independent of t and negative). Hence,

$$\frac{1}{2}a(\underline{\theta}, \underline{\theta}) + \frac{t^{-2}}{2}\|\underline{\mathrm{grad}}\,w - \underline{\theta}\|_0^2 - (f, w) \leq -c < 0 \tag{10.4.22}$$

for some positive c independent of t, which immediately gives

$$(f, w) \geq c > 0 \tag{10.4.23}$$

which implies that $\|w\|_0$ (and hence $\|\underline{\theta}\|_1 + \|w\|_1$) is bounded from below by a positive constant. This completes the proof. □

According to Proposition 10.4.1, we have now *a sequence* of problems, indexed by the thickness t, whose solutions are bounded uniformly (in t) and also bounded uniformly away from zero.

For the convenience of the reader, we repeat explicitly the general problem of our sequence.

The sequence of minimum problems. Given a bounded domain $\Omega \subset \mathbb{R}^2$ with diameter $T := diam(\Omega)$ and an element $f \in L^2(\Omega)$, for every thickness $t \in]0, T[$, we consider the problem: *find* $(\underline{\theta}(t), w(t))$ *in* $V := (H_0^1(\Omega))^2 \times H_0^1(\Omega)$ *such that*

$$\Pi_t(\underline{\theta}, w) \leq \Pi_t(\underline{\eta}, z) \qquad \forall (\underline{\eta}, z) \in V, \tag{10.4.24}$$

where Π_t is given by (10.4.19).

The sequence (10.4.24) is what we need to analyse the performance of numerical methods. Indeed, we expect a "good and reliable" numerical method to perform *uniformly well* on all the problems of our sequence, regardless of the possible smallness of t. We therefore look for error bounds (in terms of powers of the mesh-size h) which hold uniformly in t.

10.4.3 Mixed Formulation of the Mindlin-Reissner Model

It will be convenient, in order to carry on the analysis, to introduce the auxiliary variable

$$\underline{\gamma}(t) := t^{-2}(\underline{\mathrm{grad}}\,w(t) - \underline{\theta}(t)) \tag{10.4.25}$$

which is related to the shear stresses but does not go to zero with t (and could be considered as a sort of *normalised shear stress*). We can now write the Euler equations for Π_t in the form

$$a(\underline{\theta}, \underline{\eta}) + (\underline{\gamma}, \underline{\text{grad}}\, \zeta - \underline{\eta}) = (f, \zeta), \qquad \forall (\underline{\eta}, \zeta) \in V, \tag{10.4.26}$$

$$\underline{\gamma} = t^{-2}(\underline{\text{grad}}\, w - \underline{\theta}). \tag{10.4.27}$$

This is now taking the form of the abstract problems studied in Chap. 4, especially in Sect. 4.3. In particular, we can define the bilinear forms

$$\mathcal{A}((\underline{\theta}, w), (\underline{\eta}, z)) := a(\underline{\theta}, \underline{\eta}), \tag{10.4.28}$$

where a is defined in (10.4.7), and

$$\mathcal{B}((\underline{\eta}, \zeta), \underline{\delta}) := (\underline{\text{grad}}\, \zeta - \underline{\eta}, \underline{\delta}) \tag{10.4.29}$$

corresponding to the operator

$$\mathbb{B} : (\underline{\eta}, \zeta) \longrightarrow (\underline{\text{grad}}\, \zeta - \underline{\eta}), \tag{10.4.30}$$

and finally the functional

$$(\mathbb{F}, (\underline{\eta}, \zeta)) := (f, \zeta). \tag{10.4.31}$$

With this notation, Eqs. (10.4.26) and (10.4.27) can be written as

$$\mathcal{A}((\underline{\theta}, w), (\underline{\eta}, \zeta)) + \mathcal{B}((\underline{\eta}, \zeta), \underline{\gamma}) = (\mathbb{F}, (\underline{\eta}, \zeta)) \quad \forall\, (\underline{\eta}, \zeta) \in V, \tag{10.4.32}$$

$$\mathcal{B}((\underline{\theta}, w), \underline{\delta}) - t^2(\underline{\gamma}, \underline{\delta}) = 0 \quad \forall\, \underline{\delta}. \tag{10.4.33}$$

As we have already seen on several other examples, it is convenient, from many aspects, to consider (10.4.32) and (10.4.33) as a perturbation of the "limit problem" that we have for $t = 0$, namely

$$\mathcal{A}((\underline{\theta}_0, w_0), (\underline{\eta}, \zeta)) + \mathcal{B}((\underline{\eta}, \zeta), \underline{\gamma}_0) = (\mathbb{F}, (\underline{\eta}, \zeta)) \quad \forall\, (\underline{\eta}, \zeta) \in V, \tag{10.4.34}$$

$$\mathcal{B}((\underline{\theta}_0, w_0), \underline{\delta}) = 0 \quad \forall\, \underline{\delta}. \tag{10.4.35}$$

It is easy to check that the kernel $K := \text{Ker}\mathbb{B}$ is given by

$$K = \{(\underline{\eta}, \zeta) \mid (\underline{\eta}, \zeta) \in V \text{ such that } \underline{\eta} = \underline{\text{grad}}\, \zeta\}. \tag{10.4.36}$$

It is then clear that the Korn inequality (10.4.12) implies that *the bilinear form $\mathcal{A}$, defined in* (10.4.28)*, is elliptic in the kernel K of* $\mathbb{B}$:

$$\mathcal{A}((\underline{\eta}, \zeta), (\underline{\eta}, \zeta)) \geq \alpha_0 \|(\underline{\eta}, \zeta)\|_V^2 \qquad \forall\, (\underline{\eta}, \zeta) \in K, \tag{10.4.37}$$

with α_0 depending only on the Korn constant α_{Korn} appearing in (10.4.12).

On the other hand, we note that we *did not* decide yet what the space Q should be, and hence where $\underline{\delta}$ is allowed to vary in (10.4.33) or in (10.4.35). Recalling the general theory of Chap. 4, we observe that the space Q should be defined in such a way that the operator $\mathbb{B}$, associated with the bilinear form $\mathcal{B}$, is *surjective* from V to Q' (or, at least, that its image is a *closed subspace of* Q'). It is therefore clear that the next, crucial, step has to be the characterisation of the *image of* $\mathbb{B}$, that is $\mathbb{B}(V)$ with V given in (10.4.10).

In what follows, we are going to use the notation introduced in Chap. 2 for the two-dimensional operators

$$\begin{aligned} \underline{\mathrm{curl}} : \phi \longrightarrow \underline{\mathrm{curl}}\, \phi &= \left\{ \frac{\partial \phi}{\partial y}, -\frac{\partial \phi}{\partial x} \right\}, \\ \mathrm{curl} : \underline{\chi} \longrightarrow \mathrm{curl}\, \underline{\chi} &= -\frac{\partial \chi_1}{\partial y} + \frac{\partial \chi_2}{\partial x}. \end{aligned} \tag{10.4.38}$$

Note as well that (for the same reason) we are using here (x, y, z) instead of (x_1, x_2, x_3).

Proposition 10.4.2. *The mapping $\mathbb{B}$ is surjective from V onto the space $\Gamma = H_0(curl, \Omega)$ defined by*

$$H_0(curl; \Omega) = \{\underline{\chi} \mid \underline{\chi} \in (L^2(\Omega))^2,\ curl\, \underline{\chi} \in L^2(\Omega),\ \underline{\chi} \cdot \underline{t} = 0 \text{ on } \partial\Omega\} \tag{10.4.39}$$

$$\|\underline{\chi}\|^2_{H_0(curl;\Omega)} := \|\underline{\chi}\|^2_0 + \|\, curl\, \underline{\chi}\|^2_0 \tag{10.4.40}$$

(where $\underline{t}$ is the unit tangent to $\partial\Omega$) and admits a continuous lifting.

Proof. We shall show that there exists a $\beta_{RM} > 0$ such that: for every $\underline{\chi} \in H_0(\mathrm{rot}; \Omega)$ there exists $(\underline{\eta}, \zeta) \in V$ verifying

$$\underline{\chi} = \underline{\mathrm{grad}}\, \zeta - \underline{\eta} \equiv \mathbb{B}(\underline{\eta}, \zeta), \tag{10.4.41}$$

and

$$\|\zeta\|_1 + \|\underline{\eta}\|_1 \le \frac{1}{\beta_{RM}} \|\underline{\chi}\|_{H_0(\mathrm{curl};\Omega)}. \tag{10.4.42}$$

For this, we first choose $\underline{v} \in (H^1_0)^2$ such that

$$\mathrm{div}\, \underline{v} = -\,\mathrm{curl}\, \underline{\chi}, \tag{10.4.43}$$

$$\|\underline{v}\|_1 \le c \|\,\mathrm{curl}\, \underline{\chi}\|_0; \tag{10.4.44}$$

this is obviously possible because

$$\int_\Omega \mathrm{curl}\, \underline{\chi}\, dx\, dy = \int_{\partial\Omega} \underline{\chi} \cdot \underline{t}\, ds = 0. \tag{10.4.45}$$

Then, we set

$$\underline{\eta} = (\eta_1, \eta_2) := (-v_2, v_1) \tag{10.4.46}$$

so that from (10.4.43) and (10.4.44) we have

$$\operatorname{curl} \underline{\eta} = -\operatorname{curl} \underline{\chi}, \tag{10.4.47}$$

$$\|\underline{\eta}\|_1 \leq \|\operatorname{curl} \underline{\chi}\|_0. \tag{10.4.48}$$

Now choose ζ as the unique solution in $H_0^1(\Omega)$ of

$$\Delta\zeta = \operatorname{div} \underline{\chi} + \operatorname{div} \underline{\eta} \in H^{-1}(\Omega); \tag{10.4.49}$$

we have, using (10.4.48) and (10.4.49),

$$\|\zeta\|_1 \leq c\,(\|\operatorname{div} \underline{\chi}\|_{-1} + \|\operatorname{div} \underline{\eta}\|_{-1}) \leq c\,(\|\underline{\chi}\|_0 + \|\operatorname{curl} \underline{\chi}\|_0). \tag{10.4.50}$$

We now have

$$\begin{cases} \operatorname{div}(\underline{\operatorname{grad}}\,\zeta - \underline{\eta}) = \operatorname{div} \underline{\chi} \text{ in } \Omega, \\ \operatorname{curl}(\underline{\operatorname{grad}}\,\zeta - \underline{\eta}) = \operatorname{curl} \underline{\chi} \text{ in } \Omega, \\ (\underline{\operatorname{grad}}\,\zeta - \underline{\eta}) \cdot \underline{t} = \underline{\chi} \cdot \underline{t} = 0 \text{ on } \partial\Omega, \end{cases} \tag{10.4.51}$$

which easily implies (10.4.41). On the other hand, (10.4.42) follows from (10.4.48) and (10.4.50). □

Proposition 10.4.2 tells us *how to choose* Q in order to have that $\mathbb{B}$ is surjective from V to Q'. Actually, we have little choice: Q' must be equal to the space $\Gamma = H_0(\operatorname{curl}, \Omega)$ defined in (10.4.39). As we are dealing with Hilbert space, this implies that Q has to be the dual space of Γ:

$$Q \equiv \Gamma' := (H_0(\operatorname{curl}; \Omega))'. \tag{10.4.52}$$

On the other hand, a little functional analysis allows us to characterise Γ' as follows:

$$\begin{aligned} \Gamma' &:= (H_0(\operatorname{curl}; \Omega))' \\ &= H^{-1}(\operatorname{div}; \Omega) \\ &= \{\underline{\gamma} \mid \underline{\gamma} \in (H^{-1}(\Omega))^2,\ \operatorname{div}\ \underline{\gamma} \in H^{-1}(\Omega)\} \end{aligned} \tag{10.4.53}$$

with the norm

$$\|\underline{\gamma}\|_Q^2 \equiv \|\underline{\gamma}\|_{\Gamma'}^2 := \|\underline{\gamma}\|_{-1}^2 + \|\operatorname{div}\ \underline{\gamma}\|_{-1}^2\,. \tag{10.4.54}$$

Then, the Closed Range Theorem (see Sect. 4.2.2) tells us that Proposition 10.4.2 can be written in the form of an *inf-sup* condition:

$$\exists \beta_{RM} > 0 \text{ such that } \inf_{\underline{\chi}\in Q} \sup_{(\underline{\eta},\zeta)\in V} \frac{\int_{\Omega}(\underline{\text{grad}}\,\zeta - \underline{\eta})\cdot \underline{\chi}\, dx\, dy}{\|(\underline{\eta},\zeta)\|_V \|\underline{\chi}\|_Q} \geq \beta_{RM}. \tag{10.4.55}$$

Hence, to start with, we can make precise the limit problem (10.4.34) and (10.4.35) as follows

$$\begin{cases} \text{find } (\underline{\theta}_0, w_0) \in V \text{ and } \underline{\gamma}_0 \in Q \text{ such that} \\ \mathcal{A}((\underline{\theta}, w), (\underline{\eta}, \zeta)) + \mathcal{B}((\underline{\eta}, \zeta), \underline{\gamma}_0) = (f, \zeta) \quad \forall\, (\underline{\eta}, \zeta) \in V, \\ \mathcal{B}((\underline{\theta}_0, w_0), \underline{\delta}) = 0 \quad \forall\, \underline{\delta} \in Q. \end{cases} \tag{10.4.56}$$

From (10.4.37) and (10.4.55), using Theorem 4.2.3, we then have the following result on the *limit problem* (10.4.34) and (10.4.35) in the form (10.4.56).

Proposition 10.4.3. *Let* $\mathcal{A}$ *and* $\mathcal{B}$ *be defined as* (10.4.28) *and* (10.4.29)*, respectively. Then, for every* $f \in L^2(\Omega)$*, the limit problem* (10.4.56) *has a unique solution* $(\underline{\theta}_0, w_0, \underline{\gamma_0})$ *and we have*

$$\|\underline{\theta}_0\|_1 + \|w_0\|_1 + \|\underline{\gamma}_0\|_{\Gamma'} \leq c\|f\|_{-1}. \tag{10.4.57}$$

□

Remark 10.4.1. Actually, the abstract theory of Chap. 4 tells us that we could take any framework that is much more general than the one used for problem (10.4.56). For instance, we could have allowed a general $\mathbb{F} \in V'$ (not necessarily of the form (10.4.31)) in the right-hand side of the first equation. Besides, we did not need to assume $f \in L^2(\Omega)$, as $f \in H^{-1}(\Omega)$ would clearly have been sufficient. Moreover, a right-hand side in $Q' = \Gamma$ would also be allowed (instead of zero) in the second equation. We decided, however, to present the result in the framework of our original plate problem. □

Remark 10.4.2. It is not difficult to check that the unique solution of (10.4.56) is related to the solution of the Kirchhoff model: *find* $w_K \in H_0^2(\Omega)$ *such that*

$$\frac{E}{12(1-\nu^2)}\Delta^2 w_K = f \tag{10.4.58}$$

by the relations

$$w_0 = w_K, \qquad \underline{\theta}_0 = \underline{\text{grad}}\, w_K. \tag{10.4.59}$$

□

Remark 10.4.3. In the case of beam problems, the space Γ' is replaced by L^2, which makes things much easier. □

Remark 10.4.4. We now remark that, with our choice, we have $Q' \equiv H_0(\mathrm{curl};\Omega) \hookrightarrow (L^2(\Omega))^2$. As Q' is clearly dense in $(L^2(\Omega))^2$, we also have (identifying, as usual, $(L^2(\Omega))^2$ with its dual space) $(L^2(\Omega))^2 \hookrightarrow Q$. This implies that the perturbation introduced, for positive t, in the full problem (10.4.32) and (10.4.33) has to be regarded as a **singular perturbation** of the limit problem (10.4.34) and (10.4.35). Hence, it has to be dealt with using the instruments of Sect. 4.3.2. □

In view of the previous remark, we introduce the space

$$W := (L^2(\Omega))^2 \tag{10.4.60}$$

and set the mathematical framework for the Mindlin-Reissner problem (10.4.32) and (10.4.33) as follows

$$\begin{cases} \text{find } (\underline{\theta}(t), w(t)) \in V \text{ and } \underline{\gamma}(t) \in W \text{ such that} \\ \mathcal{A}((\underline{\theta}(t), w(t)), (\underline{\eta}, \zeta)) + \mathcal{B}((\underline{\eta}, \zeta), \underline{\gamma}(t)) = (f, \zeta) \quad \forall\, (\underline{\eta}, \zeta) \in V, \\ \mathcal{B}((\underline{\theta}(t), w(t)), \underline{\chi}) = t^2(\underline{\gamma}, \underline{\chi})_W \quad \forall\, \underline{\chi} \in W = (L^2(\Omega))^2. \end{cases} \tag{10.4.61}$$

Having chosen W as well as Q, we can now prove the following result.

Proposition 10.4.4. *Let the spaces V, Q, and W be defined as in* (10.4.10)–(10.4.60)*, respectively, and let the bilinear forms $\mathcal{A}$ and $\mathcal{B}$ and the operator* (10.4.30) *be defined in* (10.4.28)*,* (10.4.29) *and* (10.4.30)*, respectively. Then, there exists an $\tilde{\alpha} > 0$ such that*

$$\tilde{\alpha}\|(\underline{\eta}, \zeta)\|_V^2 \leq \mathcal{A}((\underline{\eta}, \zeta), (\underline{\eta}, \zeta)) + \|\mathbb{B}(\underline{\eta}, \zeta)\|_W^2. \tag{10.4.62}$$

Proof. The result is essentially trivial. Indeed, using (10.4.11), the triangle inequality, and the Poincaré inequality (1.2.14), we have first

$$\|(\underline{\eta}, \zeta)\|_V^2 \leq \|\underline{\eta}\|_1^2 + C_1\|\,\underline{\mathrm{grad}}\,\zeta\|_0^2 \leq C_2(\|\underline{\eta}\|_1^2 + \|\,\underline{\mathrm{grad}}\,\zeta - \underline{\eta}\|_0^2),$$

where C_1 and C_2 depend only on the Poincaré constant. Then, we can use the Korn inequality (10.4.12) and the definition of $\mathcal{A}$ and $\mathbb{B}$ to obtain

$$\|\underline{\eta}\|_1^2 + \|\,\underline{\mathrm{grad}}\,\zeta\|_0^2 \leq \frac{1}{\alpha_{Korn}}\mathcal{A}((\underline{\eta}, \zeta), (\underline{\eta}, \zeta)) + \|\mathbb{B}(\underline{\eta}, \zeta)\|_0^2,$$

and the result follows. □

We can now apply Theorem 4.3.4 (with $g = 0$) and obtain the following result.

Theorem 10.4.1. *With the same assumptions as in Proposition 10.4.4, for every $f \in V'$ and for every $t \in]0,1[$, problem* (10.4.61) *has a unique solution $(\underline{\theta}(t), w(t), \underline{\gamma}(t))$. Moreover, there exists a constant c, depending only on Ω, such that*

$$\|\underline{\theta}(t)\|_1 + \|w(t)\|_1 + \|\underline{\gamma}(t)\|_{\Gamma'} + t\|\underline{\gamma}(t)\|_0 \leq c\|f\|_{V'}. \qquad (10.4.63)$$

□

We can now study the behaviour of the solutions of problem (10.4.61) when $t \to 0$.

Proposition 10.4.5. *With the same assumptions as in Theorem 10.4.1, we have*

$$\begin{aligned} &\underline{\theta}(t) \rightharpoonup \underline{\theta}_0 \text{ in } (H_0^1(\Omega))^2, \\ &w(t) \rightharpoonup w_0 \text{ in } H_0^1(\Omega), \\ &\underline{\gamma}(t) \rightharpoonup \underline{\gamma}_0 \text{ in } \Gamma', \end{aligned} \qquad (10.4.64)$$

where $(\underline{\theta}_0, w_0, \underline{\gamma}_0)$ is the solution of the limit problem (10.4.56).

Proof. The weak convergence (a priori, up to a subsequence) in (10.4.64) just follows from (10.4.16) and (10.4.57). A passage to the limit in (10.4.61) gives (10.4.56). □

Remark 10.4.5. Additional results in this direction can be found in [171]. □

We can now apply the results of Proposition 4.3.5 and of Remarks 4.3.12 and 4.3.14 to estimate the convergence rate as a function of t^2 which plays here the role of λ. This leads us to a convergence rate in $\sqrt{\lambda} = t$. In order to improve this bound and also to enable us later to get sharper error estimates, we now introduce a decomposition principle for (10.4.26) and (10.4.27).

10.4.4 A Decomposition Principle and the Stokes Connection

We shall first prove the following decomposition principle for vector-valued functions in $\Gamma' = H_0(\text{curl}; \Omega)$.

Proposition 10.4.6. *Every element $\underline{\gamma} \in \Gamma'$ can be written in a unique way as*

$$\underline{\gamma} = \underline{grad}\, \psi + \underline{curl}\ p, \qquad (10.4.65)$$

with $\psi \in H_0^1(\Omega)$, $p \in L^2(\Omega)/\mathbb{R}$, and $\underline{curl}\, p = \left\{-\partial p/\partial y, \partial p/\partial x\right\}$. Moreover, we may use

$$\|\underline{\gamma}\|_{\Gamma'}^2 = \|\psi\|_{H_0^1(\Omega)}^2 + \|p\|_{L^2(\Omega)/\mathbb{R}}^2 \qquad (10.4.66)$$

as a norm on Γ'.

Proof. Set $\xi := \operatorname{div} \underline{\gamma} \in H^{-1}(\Omega)$. We define ψ to be the unique solution of $-\Delta\psi = \xi$, $\psi \in H_0^1(\Omega)$ and we set $\underline{\alpha} = \underline{\gamma} - \underline{\text{grad}}\,\psi$. One has $\operatorname{div} \underline{\alpha} = 0$ so that $\underline{\alpha} = \underline{\text{curl}}\, p$ and p is determined up to a constant in $L^2(\Omega)$. Condition (10.4.66) is then immediate. □

Remark 10.4.6. The decomposition introduced in Proposition 10.4.6 also holds for $(L^2(\Omega))^2$ and $H(\text{curl}; \Omega)$. The difference between these spaces lies in the regularity of the p component. Indeed, taking $\underline{\gamma} = \underline{\text{grad}}\,\psi + \underline{\text{curl}}\, p$ with $\psi \in H_0^1(\Omega)$, we have

$$\underline{\gamma} \in (L^2(\Omega))^2 \quad \Leftrightarrow p \in H^1(\Omega)/\mathbb{R}, \tag{10.4.67}$$

$$\underline{\gamma} \in H(\text{rot}; \Omega) \quad \Leftrightarrow p \in H^2(\Omega)/\mathbb{R}, \tag{10.4.68}$$

$$\underline{\gamma} \in H_0(\text{rot}; \Omega) \Leftrightarrow p \in H^2(\Omega)/\mathbb{R} \text{ and } \frac{\partial p}{\partial n} = 0 \text{ on } \partial\Omega. \tag{10.4.69}$$

□

It is now a simple exercise to transform problem (10.4.61) in terms of the new unknowns $\underline{\theta}(t)$, $w(t)$, $\psi(t)$, and $p(t)$. We have indeed the following basic theorem, which is of considerable help in understanding the nature of the Mindlin-Reissner equations.

Theorem 10.4.2. *Any solution of* (10.4.61) *is a solution of the following problem (and conversely) through the change of variables* (10.4.65)*: find* $(\underline{\theta}(t), w(t), \psi(t), p(t))$ *in* $\underline{\Theta} \times Z \times H_0^1(\Omega) \times L^2(\Omega)/\mathbb{R}$ *such that*

$$(\underline{grad}\,\psi, \underline{grad}\,\xi) = (f, \xi) \quad \forall \xi \in H_0^1(\Omega), \tag{10.4.70}$$

$$\begin{cases} a(\underline{\theta}(t), \underline{\eta}) - (\underline{curl}\, p(t), \underline{\eta}) = (\underline{grad}\,\psi, \underline{\eta}) & \forall \underline{\eta} \in (H_0^1(\Omega))^2, \\ -(\underline{\theta}(t), \underline{curl}\, q) - t^2(\underline{curl}\, p(t), \underline{curl}\, q) = 0 & \forall q \in H^1(\Omega)_{/\mathbb{R}}, \end{cases} \tag{10.4.71}$$

$$(\underline{grad}\, w(t), \underline{grad}\,\chi) = (\underline{\theta}(t), \underline{grad}\,\chi) + t^2(\underline{grad}\,\psi, \underline{grad}\,\chi) \quad \forall \chi \in H_0^1(\Omega). \tag{10.4.72}$$

Proof. The proof is immediate: it is enough to make the substitution (10.4.65), and observe that both (10.4.61) and (10.4.70)–(10.4.72) have a unique solution. □

Remark 10.4.7. It must be noted that (10.4.71) implies $\partial p/\partial n|_{\partial\Omega} = 0$ and $p \in H^2(\Omega)$ so that $\underline{\gamma} = \underline{\text{grad}}\,\psi + \underline{\text{curl}}\, p$ is indeed an element of $\Gamma = H_0(\text{curl}; \Omega)$. Note also that $\psi(t)$ is actually *independent* of t. □

Remark 10.4.8. It is important to note that although (10.4.70)–(10.4.72) seems, at first sight, a system of four equations, it actually decomposes immediately into equation (10.4.70) (which allows to compute ψ directly from f), plus equations (10.4.71) (which allow to compute $\underline{\theta}(t)$ and $p(t)$ once we know ψ) plus equation (10.4.72) (which allows to compute $w(t)$ once we know $\underline{\theta}(t)$ and ψ). We have thus reduced, through Theorem 10.4.2, our original problem into the following sequence

- A Dirichlet problem (10.4.70) that is independent of t,
- A "Stokes-like" problem (10.4.71),
- A Dirichlet problem (10.4.72).

□

The decomposition provided by Theorem 10.4.2 shows us that it is the p component of $\underline{\gamma}$ which depends on t. Before coming back to the quantification of this dependency, we rapidly develop the analogy between (10.4.71) and a Stokes problem. Let us set $\underline{\eta}^{\perp} = \{-\eta_2, \eta_1\}$. We can write (10.4.71) in the form

$$\begin{cases} a(\underline{\theta}^{\perp}, \underline{\eta}^{\perp}) + (p, \operatorname{div}\, \underline{\eta}^{\perp}) = \ (\underline{\operatorname{grad}}\, \psi, \underline{\eta}^{\perp}) \qquad \forall \underline{\eta}^{\perp} \in (H_0^1(\Omega))^2, \\ (\operatorname{div}\, \underline{\theta}^{\perp}, q) = t^2\ (\underline{\operatorname{grad}}\, p, \underline{\operatorname{grad}}\, q) \qquad \forall q \in H^1(\Omega)/\mathbb{R}. \end{cases} \tag{10.4.73}$$

The limit problem ($t = 0$) is thus a standard Stokes problem and we shall be able to rely on results of Chap. 8 to build approximations. We shall not analyse here the case $t \neq 0$ in too much detail. However, it is important to see the behaviour of p as $t \to 0$.

Proposition 10.4.7. *Let* $\underline{\theta}(t)$, $w(t)$, $p(t)$, *and* ψ *be the solution of* (10.4.70)–(10.4.72). *We then have*

$$\|\underline{\theta}(t)\|_2 + \|w(t)\|_2 + \|\psi(t)\|_2 + \|p(t)\|_1 + t\ \|p(t)\|_2 \le c\ \|f\|_0 \tag{10.4.74}$$

where the constant c is independent of t. □

We refer to [122] for the proof of this result which is based essentially on the regularity properties of the Dirichlet problem and the Stokes problem.

An important point is that (10.4.74) does not improve too much for a more regular f (even in a smooth domain). It is not possible to bound $\|p(t)\|_2$ uniformly in t. The reason is that the normal derivative of $p(t)$ vanishes although this is not the case for the solution $p(0)$ of the limit problem. We thus have a boundary layer effect which has been studied in [29]. This analysis shows that an analogue of (10.4.74) exists for $\|\underline{\theta}\|_{\frac{5}{2}}$ and $\|p\|_{\frac{3}{2}}$ but not for more regular spaces.

Remark 10.4.9. We can now try to apply Remarks 4.3.12 and 4.3.14 to our problem. Denoting $W_+ := \{p \mid p \in H^2(\Omega)/\mathbb{R},\ \partial p/\partial n|_{\partial\Omega} = 0\}$, it is clear that we have

$$|(\underline{\operatorname{curl}}\, p, \underline{\operatorname{curl}}\, q)| \le c\|p\|_{W_+}\|q\|_{L^2(\Omega)/\mathbb{R}}. \tag{10.4.75}$$

Whenever the solution p_0 of the limit problem is regular enough (this is the case for smooth data and a smooth domain), we shall have

$$p_0 \in [L^2(\Omega), W_+]_\theta \quad \forall\, \theta < \frac{3}{4}. \tag{10.4.76}$$

No improvement is possible because of the fact that $\partial p(0)/\partial n \neq 0$. We can thus apply Remark 4.3.14 to get for $\theta < \frac{3}{4}$

$$\|\underline{\Theta}(t) - \underline{\theta}_0\|_1 + \|p(t) - p_0\|_0 + \|w(t) - w_0\|_1 \leq ct^{2\theta}\, \|p_0\|_\theta, \tag{10.4.77}$$

where $\|p_0\|_\theta$ is the norm of p_0 in $[L^2(\Omega), W_+]_\theta$. We can summarise (10.4.77) by saying that we have an $O(t^{3/2-\varepsilon})$ convergence. This requires, however, a smooth domain. In the case where $\partial\Omega$ is only Lipschitz continuous, the best we can get is $O(t)$. □

10.4.5 Discretisation of the Problem

We now turn our attention to the discretisation of our problem (10.4.26) and (10.4.27). Let us thus assume that we are given finite-dimensional subspaces Θ_h and Z_h of Θ and Z and use $V_h = \Theta_h \times Z_h$ as a subspace of V. We also discretise the space $W = (L^2(\Omega))^2$ by Γ_h and we consider the discretised problem: *find* $(\underline{\theta}^h, w_h, \underline{\gamma}_h)$ *such that*

$$\begin{cases} a(\underline{\theta}_h, \underline{\eta}_h) + (\underline{\gamma}_h, \underline{\mathrm{grad}}\, \zeta_h - \underline{\eta}_h) = (f, \zeta_h) & \forall (\underline{\eta}_h, \zeta_h) \in V_h, \\ (\underline{\mathrm{grad}}\, w_h - \underline{\theta}_h, \underline{\chi}_h) - t^2(\underline{\gamma}_h, \underline{\chi}_h) = 0 & \forall \underline{\chi}_h \in \Gamma_h. \end{cases} \tag{10.4.78}$$

This could also be written with the notation of Sect. 10.4.3, that is, in particular, making use of the bilinear form $\mathcal{A}$ and $\mathcal{B}$ defined in (10.4.28) and (10.4.29). The discrete problem (10.4.78) becomes: *find* $((\underline{\theta}_h, w_h), \underline{\gamma}_h) \in V_h \times Q_h$ *such that*

$$\begin{cases} \mathcal{A}((\underline{\theta}_h, w_h), (\underline{\eta}, \zeta)) + \mathcal{B}((\underline{\eta}, \zeta), \underline{\gamma}_h) = (\mathbb{F}, (\underline{\eta}, \zeta)) & \forall\, (\underline{\eta}, \zeta) \in V_h, \\ \mathcal{B}((\underline{\theta}, w), \underline{\chi}) - t^2(\underline{\gamma}, \underline{\chi}) = 0 & \forall\, \underline{\chi} \in Q_h \equiv \Gamma_h. \end{cases} \tag{10.4.79}$$

In what follows, we shall use either the form (10.4.78) or the form (10.4.79), according to the notational convenience.

Remark 10.4.10. Note that from the second equation of (10.4.78), we *do not have* in general $\underline{\gamma}_h = \lambda t^{-2}(\underline{\mathrm{grad}}\, w_h - \underline{\theta}_h)$ unless we take Θ_h, Z_h, and Γ_H such that $\underline{\mathrm{grad}}\, Z_h - \Theta_h \subseteq \Gamma_h$. This, as we shall see, could be a problem regarding the actual *implementation* of the method. Indeed, in the common engineering practice, one prefers to solve the discrete problems in terms of $\underline{\theta}_h$ and w_h alone. In this case, the use of the mixed formulation (and the introduction of the variable $\underline{\gamma}_h$) should be regarded as a *mathematical artefact* used in order to have *a better understanding of the mathematical structure of the discretised problem.* We will come back several times to this important point. □

It is easy to check that, now, the **discrete kernel** $K_h := \mathrm{Ker}\mathbb{B}_h$ is given by

$$K_h = \{(\underline{\eta}, \zeta) \mid \underline{\eta}, \zeta) \in V_h, (\underline{\eta} - \underline{\mathrm{grad}}\, \zeta, \underline{\delta}) = 0 \quad \forall \underline{\delta} \in \Gamma_h\}, \tag{10.4.80}$$

and we consider the problem of having, for our discrete problem, the *ellipticity in the discrete kernel*;

$$\mathcal{A}((\underline{\eta}, \zeta), (\underline{\eta}, \zeta)) \geq \alpha_0^h \|(\underline{\eta}, \zeta)\|_V^2 \qquad \forall\, (\underline{\eta}, \zeta) \in K_h. \tag{10.4.81}$$

For the continuous case, the Korn inequality (10.4.12) implied that the bilinear form $\mathcal{A}$ is elliptic in the kernel K (see (10.4.37)). As the variable ζ does not appear in the actual expression of $\mathcal{A}((\underline{\eta}, \zeta), (\underline{\eta}, \zeta))$, we deduce that the only possibility in order to have the ellipticity in K_h is that the following property holds

$$\exists \kappa > 0 \text{ s. t. } \{(\underline{\eta}, \zeta) \in K_h\} \Rightarrow \{\|\zeta\|_1 \leq \kappa \|\underline{\eta}\|_1\} \tag{10.4.82}$$

and a simple *necessary* condition for it is that

$$\{(\underline{\mathrm{grad}}\, \zeta, \underline{\delta}) = 0 \quad \forall \underline{\delta} \in \Gamma_h\} \Rightarrow \{\underline{\mathrm{grad}}\, \zeta = \underline{0}\}. \tag{10.4.83}$$

This can easily be satisfied assuming for instance that

$$\underline{\mathrm{grad}}(Z_h) \subseteq \Gamma_h. \tag{10.4.84}$$

As we shall see, the above condition (10.4.84) is not difficult to enforce when choosing the finite element spaces and the vast majority of the good and reliable methods will satisfy it. On the other hand, the discrete *inf-sup* condition

$$\exists \beta_{RM} > 0 \text{ such that } \inf_{\underline{\chi} \in Q_h} \sup_{(\underline{\eta}, \zeta) \in V_h} \frac{\int_\Omega (\underline{\mathrm{grad}}\, \zeta - \underline{\eta}) \cdot \underline{\chi}\, dx\, dy}{\|(\underline{\eta}, \zeta)\|_V \|\underline{\chi}\|_Q} \geq \beta_{RM} \tag{10.4.85}$$

is *a major difficulty*, and most methods will be designed in order to *get around it*. For this, the first methods that we are going to consider are those based on the decomposition principle given in Proposition 10.4.6 and on the re-formulation of the problem given in Theorem 10.4.2.

Remark 10.4.11. It will often be convenient to look as well at the limit problem: *find* $(\underline{\theta}_{0h}, w_{0h}, \underline{\gamma}_{0h}) \in \Theta_h \times Z_h \times \Gamma_h$ *such that*

$$\begin{cases} a(\underline{\theta}_{0h}, \underline{\eta}_h) + (\underline{\gamma}_{0h}, \underline{\mathrm{grad}}\, \zeta_h - \underline{\eta}_h) = (f, \zeta_h) \qquad \forall (\underline{\eta}_h, \zeta_h) \in V_h, \\ (\underline{\chi}_h, \underline{\mathrm{grad}}\, w_{0h} - \underline{\theta}_{0h}) = 0 \qquad \forall \underline{\chi}_h \in \Gamma_h, \end{cases} \tag{10.4.86}$$

that could also be expressed in the form (10.4.79) with $t = 0$. It also comes from the results of Sects. 4.3.2 and 5.5.3 that to get a good approximation of (10.4.61) by (10.4.79) (that is, with convergence properties independent of t), it is necessary for (10.4.86) to be a good approximation of (10.4.34) and (10.4.35). □

We shall first consider the most "naive" case.

Example 10.4.1 (The direct approach). Let us suppose that we are given $\Theta_h \subset \Theta$ and $Z_h \subset Z$, and let us choose

$$\Gamma_h = \underline{\mathrm{grad}}(Z_h) - \Theta_h. \tag{10.4.87}$$

This choice implies that

$$\mathrm{Ker}\mathbb{B}_h = \{(\underline{\eta}_h, \zeta_h) \mid \underline{\eta}_h = \underline{\mathrm{grad}}\,\zeta_h\} \subset \mathrm{Ker}\mathbb{B}, \tag{10.4.88}$$

so that the ellipticity in K_h (10.4.81) evidently holds. It is important to note that the choice (10.4.87) is very easy to use on the computer, as it actually corresponds to minimising the energy functional Π_t given by (10.4.19) on $V_h = \Theta_h \times Z_h$ and that you *do not even see* $\underline{\gamma}_h$ (nor Γ_h). The choice (10.4.87) is then one of the most widely used choices for Γ_h although, in general, one does not realise it.

However, in the limit $t \to 0$, one is lead to minimise

$$\Pi_t \equiv a(\underline{\eta}_h, \underline{\eta}_h) - (f, \zeta_h) \tag{10.4.89}$$

on $\mathrm{Ker}\mathbb{B}_h$. Now, a quick glance to $\mathrm{Ker}B_h$ will make us understand that we have a long way to go. Consider $\underline{\eta}_h^{\perp} = \{-\eta_{2h}, \eta_{1h}\}$, that is, a rotation of $\pi/2$ of $\underline{\eta}_h$. It is clear that if $(\underline{\eta}_h, \zeta_h)$ belongs to $\mathrm{Ker}B_h$, we then have, by (10.4.88),

$$\mathrm{div}\ \underline{\eta}_h^{\perp} = \mathrm{curl}\,\underline{\eta}_h = 0. \tag{10.4.90}$$

Therefore, with choice (10.4.87), we are minimising Π_t in (10.4.89) on a subset of functions $\underline{\eta}_h$ satisfying (10.4.90). However, we have already seen in Chap. 8, for the linear Stokes problem, that it is not recommended to work with velocity fields which are exactly incompressible (because there are too few of them in general). A direct application of (10.4.87) is likely to lead to bad results (e.g. locking) unless a very special choice of Θ_h and Z_h has been made. □

In what follows, we shall mainly concentrate on two groups of finite element approaches: the Methods based on the decomposition principle, and the Methods based on a nonconforming approximation of the original minimisation problem (10.4.24).

10.4.5.1 Methods Based on the Decomposition Principle

The first group of methods that we present is directly guided by the decomposition principle of Propositions 10.4.6 and 10.4.2 in which a Stokes-like problem explicitly appears. For the sake of simplicity, we shall describe *one* possible method in this group, based on the MINI element for Stokes. However, it will be clear that starting from every finite element stable approximation for the Stokes problem using continuous pressures, one can derive a Reissner-Mindlin method belonging to the present group.

The basic idea is to give up a direct approximation of $\underline{\gamma}$ and to approximate instead each component of its decomposition into $\underline{\mathrm{grad}}\,\psi_h + \underline{\mathrm{curl}}\,p_h$. Moreover, as (10.4.71) shows us that θ_h and p_h are analogous to a velocity field and a pressure field in a Stokes problem, we shall try to use some results of Chap. 8 to build a suitable approximation.

We assume that Ω is a convex polygon and that we are given a sequence $\{\mathcal{T}_h\}$ of partitions of Ω into triangles. Let Θ_h be built by employing the MINI element of Chap. 8 , that is, in the notations of Chap. 2,

$$\begin{cases} \Theta_h &= (\mathcal{L}_1^1 \cap H_0^1(\Omega))^2 \oplus B_3, \\ Z_h &= \mathcal{L}_1^1 \cap H_0^1(\Omega). \end{cases} \tag{10.4.91}$$

These are spaces of piecewise linear polynomials enriched by a bubble function in the case of Θ_h. We also introduce

$$\Gamma_h := \underline{\mathrm{grad}}(\mathcal{L}_1^1 \cap H_0^1(\Omega)) \oplus \underline{\mathrm{curl}}\,\mathcal{L}_1^1 \equiv \underline{\mathrm{grad}}\,Z_h \oplus \underline{\mathrm{curl}}\,\mathcal{L}_1^1. \tag{10.4.92}$$

This space is then a strict subspace of piecewise constant vector functions constructed by discretising the ingredients of the decomposition principle of Proposition 10.4.6 and Remark 10.4.6.

It is straightforward to check that $\mathrm{Ker}\mathbb{B}_h$ is made of the pairs $(\underline{\eta}_h, \zeta_h)$ in $\Theta_h \times Z_h$ such that

$$(\underline{\eta}_h, \underline{\mathrm{curl}}\,q_h) = 0 \qquad \forall q_h \in \mathcal{L}_1^1, \tag{10.4.93}$$

$$(\underline{\mathrm{grad}}\,\zeta_h, \underline{\mathrm{grad}}\,\phi_h) = (\underline{\eta}_h, \underline{\mathrm{grad}}\,\phi_h) \qquad \forall \phi_h \in Z_h \equiv \mathcal{L}_1^1 \cap H_0^1(\Omega). \tag{10.4.94}$$

Now, condition (10.4.94) is especially nice as it implies

$$\|\zeta_h\|_1 \le c\,\|\underline{\eta}_h\|_1, \qquad \forall (\underline{\eta}_h, \zeta_h) \in \mathrm{Ker}B_h, \tag{10.4.95}$$

and hence, (10.4.82) holds and we have the ellipticity in the kernel (10.4.81). We still have to check the *inf-sup* condition (10.4.85) and we can do it using Proposition 5.4.3: given $(\underline{\eta}, \zeta)$, we must then be able to build $(\underline{\eta}_h, \zeta_h) = \Pi_h(\underline{\eta}, \zeta)$ such that

$$\mathcal{B}((\underline{\eta},\zeta)-(\underline{\eta}_h,\zeta_h),\underline{\delta}_h)=0 \quad \forall\underline{\delta}_h, \tag{10.4.96}$$

with

$$\|\underline{\eta}_h\|_1+\|\zeta_h\|_1\le c\ (\|\underline{\eta}\|_1+\|\zeta\|_1). \tag{10.4.97}$$

Using the structure $\underline{\delta}=\ \underline{\mathrm{grad}}\,\phi_h+\underline{\mathrm{curl}}\,q_h$, condition (10.4.96) becomes:

$$\begin{cases}(\underline{\mathrm{grad}}\,\phi_h,\underline{\eta}_h-\underline{\mathrm{grad}}\,\zeta_h)-(\underline{\mathrm{grad}}\,\phi_h,\underline{\eta}-\underline{\mathrm{grad}}\,\zeta)=0 \quad \forall\phi_h\in Z_h,\\ (\underline{\mathrm{curl}}\,q_h,\underline{\eta}_h-\underline{\mathrm{grad}}\,\zeta_h)-(\mathrm{c}\underline{\ \mathrm{url}}\,q_h,\underline{\eta}-\underline{\mathrm{grad}}\,\zeta)=0 \quad \forall q_h\in\mathcal{L}^1_1.\end{cases} \tag{10.4.98}$$

In order to construct the operator Π_h, we use the result already obtained in Chap. 8 to deal with the *inf-sup* condition for the MINI element. In particular, we proved that there exists an operator Π_S, from $\Theta=(H^1_0(\Omega))^2$ into Θ_h, such that

$$(\underline{\mathrm{grad}}\,q_h,\underline{\eta}-\Pi_S(\underline{\eta}))=0 \qquad \forall q_h\in\mathcal{L}^1_1, \tag{10.4.99}$$

with $\|\Pi_S(\underline{\eta})\|_1\le C\ \|\underline{\eta}\|_1$ and C independent of h. With the same arguments, we can obviously prove that there exists an operator Π_R from $\Theta=(H^1_0(\Omega))^2$ into Θ_h such that

$$(\mathrm{curl}\,q_h,\underline{\eta}-\Pi_R(\underline{\eta}))=0 \qquad \forall q_h\in\mathcal{L}^1-1, \tag{10.4.100}$$

with

$$\|\Pi_R(\underline{\eta})\|_1\le C\ \|\underline{\eta}\|_1, \tag{10.4.101}$$

with C independent of h. Condition (10.4.100), taking into account the fact that $(\underline{\mathrm{curl}}q_h,\underline{\mathrm{grad}}\,\zeta_h)\ =\ (\underline{\mathrm{curl}}\,q_h,\underline{\mathrm{grad}}\,\zeta)\ \equiv\ 0$ (by Green's formula), tells us that the second equation of (10.4.98) is satisfied if we take $\underline{\eta}_h\ =\ \Pi_r(\underline{\eta})$. We now observe that the first equation of (10.4.98) reduces to

$$(\underline{\mathrm{grad}}\,\phi_h,\ \underline{\mathrm{grad}}\,\zeta_h)=(\ \underline{\mathrm{grad}}\,\phi_h,\ \underline{\mathrm{grad}}\,\zeta-\underline{\eta}+\Pi_R(\underline{\eta})) \qquad \forall\phi_h\in Z_h, \tag{10.4.102}$$

and this is a discrete Dirichlet problem for the Laplace operator for which we have easily $\|\zeta_h\|_1\le c\ (\|\zeta\|_1+\|\underline{\eta}\|_1)$, yielding the second part of (10.4.97).

Remark 10.4.12. It should be clear from our construction that the crucial step is to have an operator Π_R satisfying (10.4.100) and (10.4.101). This, always changing as we did the *div* operator into *rot*, essentially means that we could take, instead of the MINI element, any other finite element pair that is stable for the Stokes problem and which uses continuous pressures. □

Having proved the *inf-sup* condition (10.4.85), we can therefore apply to the limit problem (10.4.86) the basic results of Chap. 5. We can summarise this in the following proposition.

Proposition 10.4.8. *Problem* (10.4.86) *with the choice* (10.4.91) *and* (10.4.92) *has a unique solution. Moreover, if* $(\underline{\theta}_0, w_0, \underline{\gamma}_0)$ *is the solution of* (10.4.34) *and* (10.4.35), *we have*

$$\|\underline{\theta}_0-\underline{\theta}_{0h}\|_1 + \|w_0-w_{0h}\|_1 + \|\underline{\gamma}_0-\underline{\gamma}_{0h}\|_{\Gamma'} \leq ch\,\{\|w_0\|_3 + \|\underline{\gamma}_0\|_{H(\mathrm{div};\Omega)}\}. \tag{10.4.103}$$

□

Remark 10.4.13. The result of Proposition 10.4.2 can be applied to the discrete problem in the present case. Indeed, we built, a priori, Γ_h in order to obtain a decomposition principle. Problem (10.4.78) can be written in the form: *find* $(\underline{\theta}_h(t), w_h(t), \psi_h(t), p_h(t))$ in $\underline{\Theta}_h \times Z_h \times Z_h \times \mathcal{L}_1^1/\mathbb{R}$ *such that*

$$(\underline{\mathrm{grad}}\,\psi_h, \underline{\mathrm{grad}}\,\xi) = (f, \xi) \quad \forall \xi \in Z_h, \tag{10.4.104}$$

$$\begin{cases} a(\underline{\theta}_h, \underline{\eta}) - (\underline{\mathrm{curl}}\,p_h, \underline{\eta}) = (\,\underline{\mathrm{grad}}\,\psi_h, \underline{\eta}) & \forall \underline{\eta} \in \Theta_h, \\ -\,(\underline{\theta}_h, \underline{\mathrm{curl}}\,q) - t^2(\underline{\mathrm{curl}}\,p_h, \underline{\mathrm{curl}}\,q) = 0 & \forall q \in \mathcal{L}^1_{1/\mathbb{R}}, \end{cases} \tag{10.4.105}$$

$$(\underline{\mathrm{grad}}\,w_h,\ \underline{\mathrm{grad}}\,\chi) = (\underline{\theta}_h,\ \underline{\mathrm{grad}}\,\chi) + t^2(\underline{\mathrm{grad}}\,\psi_h,\ \underline{\mathrm{grad}}\,\chi) \quad \forall \chi \in Z_h. \tag{10.4.106}$$

These problems can be solved sequentially and (10.4.105) is a Stokes-like problem using the MINI element of Chap. 8. This approximation has been introduced and studied for $t \neq 0$ in [122]. Using this decomposition and Proposition 10.4.8, recalling that

$$\|\underline{\gamma}\|_{\Gamma'} = \|\psi\|_1 + \|p\|_{0/\mathbb{R}}, \tag{10.4.107}$$

and bringing in the regularity result of Proposition 10.4.7, we have, for $t = 0$, the following estimate:

$$\|\psi_{0h} - \psi_0\|_1 + \|p_0 - p_{0h}\|_{0/\mathbb{R}} \leq ch\ \{\|w_0\|_2 + \|\psi_0\|_2 + \|p_0\|_1\} \leq ch\ \|f\|_0. \tag{10.4.108}$$

From a numerical point of view, (10.4.104)–(10.4.106) can lead to an efficient method, provided one has a Stokes solver available. □

Remark 10.4.14. An easy duality argument would also show that we have the estimate

$$\|\underline{\theta}_0 - \underline{\theta}_{0h}\|_0 + \|w_0 - w_{0h}\|_0 \leq ch^2\{\|w_0\|_3 + \|\underline{\gamma}_0\|_{H(\,\mathrm{div};\Omega)}\}. \tag{10.4.109}$$

□

To end the discussion on this group of methods, we rapidly show how the results of Sect. 5.5 can be applied to the case $t \neq 0$. We consider the error estimate (5.5.52) from Remark 5.5.5, where we denote $V = (H_0^1(\Omega))^2 \times H_0^1(\Omega)$, $Q = \Gamma'$ and $W = (L^2(\Omega))^2$. The parameter λ is, of course, t^2 in the present case. It is easily verified that all conditions are satisfied and that we have, taking into account regularity properties of Remark 10.4.6,

$$\|\underline{\theta}(t)-\underline{\theta}_h(t)\|_1^2 + \|w(t)-w_h(t)\|_1^2 + \|\underline{\gamma}(t)-\underline{\gamma}_h(t)\|_{\Gamma'}^2$$
$$+ t^2 \|\underline{\gamma}(t) - \underline{\gamma}_h(t)\|_0^2 \le C \Big(\inf_{\underline{\eta}_h} \|\underline{\theta}(t) - \underline{\eta}_h\|_1^2 + \inf_{q_h} \|w(t) - q_h\|_1^2$$
$$+ \inf_{\underline{\chi}_h}\{\|\underline{\gamma}(t)-\underline{\chi}_h\|_{\Gamma'}^2 + t^2\|\underline{\gamma}(t)-\underline{\chi}_h\|_0^2\}\Big). \qquad (10.4.110)$$

Using the decomposition principle and the estimate (10.4.74), we can recover the following result of [122].

Theorem 10.4.3. *For every* $t \in]0, T[$, *problem* (10.4.104)–(10.4.106) *with the choices* (10.4.91) *and* (10.4.92) *has a unique solution* $(\underline{\theta}_h(t), w_h(t), \psi_h(t), p_h(t))$. *If moreover* $(\underline{\theta}(t), w(t), \psi(t), p(t))$ *is the solution of* (10.4.70)–(10.4.72), *then we have*

$$\|\underline{\theta}(t) - \underline{\theta}_h(t)\|_1^2 + \|w(t) - w_h(t)\|_1^2$$
$$+ \|\psi(t) - \psi_h(t)\|_1^2 + |p(t) - p_h(t)|_0^2 + t^2 \|p(t) - p_h(t)\|_1^2$$
$$\le c\, h^2\{\|\underline{\theta}(t)\|_2^2 + \|w(t)\|_2^2 + \|\psi(t)\|_2^2 + |p(t)|_1^2 + t^2\|p(t)\|_2^2\}, \qquad (10.4.111)$$

with c independent of h and t.

We therefore have an $O(h)$ convergence uniform in t. This result cannot be (much) improved because of the boundary layer effect already described.

10.4.5.2 Nonconforming Approximations of the Minimum Problem

The previous class of methods is, although interesting, rather remote from the actual engineering practice in which one tries to stick as closely as possible to the original formulation. In particular, as already pointed out in Remark 10.4.10, what is preferred in the engineering practice is to work only in terms of the original unknowns $\underline{\theta}$ and w, and, possibly, having their degrees of freedom at the same nodes (in particular if one wants to extend the methods to *shell problems*).

As we have seen, however, in Example 10.4.1, working directly on the minimisation problem (10.4.24) would require approximations $\underline{\theta}_h(t)$ and $w_h(t)$ that, in the limit for $t \to 0$, satisfy $\underline{\theta}_h(0) = \underline{\text{grad}}\, w_h(0)$, and if we want to use a *conforming approximation* $\Theta_h \subset \Theta$ this would require $w_h \in Z_h$ to belong to $H_0^2(\Omega)$, which is not so easy to obtain, in particular for low degree elements.

The most common escape to the troubles that we are facing is to use some kind of numerical integration (or a nonconforming approximation) for the term

$t^{-2}\| \underline{\text{grad}}\, w - \underline{\theta}\|^2$ which appears in (10.4.19), thus weakening condition (10.4.90). A way of formalising it is the following. We assume that we are given a linear operator r which maps $\Theta_h \times Z_h$ into (for instance) $L^2(\Omega)$. To see an example, consider for instance the possible, but not necessarily recommended, choices:

$$r(\underline{\eta}, \zeta) \in \mathcal{L}_0^0 \text{ and } r(\underline{\eta}, \zeta)|_K = \text{ mean value of } (\underline{\text{grad}}\, \zeta - \underline{\eta}) \text{ on } K \tag{10.4.112}$$

or

$$r(\underline{\eta}, \zeta) \in \mathcal{L}_0^0 \text{ and } r(\underline{\eta}, \zeta)|_K = \text{ value of } (\underline{\text{grad}}\, \zeta - \underline{\eta}) \tag{10.4.113}$$

at the barycentre of K. Then, one minimises, instead of Π_t (as in (10.4.24)), the functional

$$M_t^r := \frac{1}{2} a(\underline{\theta}, \underline{\theta}) + \frac{t^{-2}}{2} \|r(\underline{\theta}, w)\|_0^2 - (f, w) \tag{10.4.114}$$

on $\Theta_h \times Z_h$. This can be regarded as obtained from the problem *find* $(\underline{\theta}_h, w_h, \underline{\gamma}_h) \in \Theta_h \times Z_h \times \Gamma_h$ *such that*

$$\begin{cases} a(\underline{\theta}_h, \underline{\eta}_h) + (\underline{\gamma}_h, r(\underline{\eta}_h, \zeta_h)) - (f, \zeta_h) = 0 & \forall (\underline{\eta}_h, \zeta_h) \in V_h, \\ (\underline{\chi}_h, r(\underline{\theta}_h, w_h)) - t^2(\underline{\gamma_h}, \underline{\chi}_h) = 0 & \forall \underline{\chi}_h \in \Gamma_h, \end{cases} \tag{10.4.115}$$

whenever its second equation is equivalent to

$$\underline{\gamma}_h = t^{-2} r(\underline{\theta}_h, w_h). \tag{10.4.116}$$

This will always be the case for choices of Γ_h that verify

$$r(\Theta_h, Z_h) \subseteq \Gamma_h. \tag{10.4.117}$$

In this case, the limit problem (for $t = 0$) will be: *find* $(\underline{\theta}_h, w_h, \underline{\gamma}_h) \in \Theta_h \times Z_h \times \Gamma_h$ *such that*

$$\begin{cases} a(\underline{\theta}_h, \underline{\eta}_h) + (\underline{\gamma}_h, r(\underline{\eta}_h, \zeta_h)) - (f, \zeta_h) = 0 & \forall (\underline{\eta}_h, \zeta_h) \in V_h, \\ (\underline{\chi}_h, r(\underline{\theta}_h, w_h)) = 0 & \forall \underline{\chi}_h \in \Gamma_h. \end{cases} \tag{10.4.118}$$

With the notation (10.4.28) for $\mathcal{A}$ and setting

$$\tilde{\mathcal{B}}_h((\underline{\eta}, \zeta), \underline{\chi}) = (r(\underline{\eta}, \zeta), \underline{\chi})_{(L^2(\Omega))^2} \qquad \forall (\underline{\eta}, \zeta) \in V_h \quad \forall \underline{\chi} \in \Gamma_h, \tag{10.4.119}$$

we can write the problem (10.4.115) as

$$\begin{cases} \text{find } (\underline{\theta}_h(t), w_h(t)) \in V_h \text{ and } \underline{\gamma}_h(t) \in W_h \equiv \Gamma_h \text{ such that} \\ \mathcal{A}((\underline{\theta}_h(t), w_h(t)), (\underline{\eta}, \zeta)) + \tilde{\mathcal{B}}_h((\underline{\eta}, \zeta), \underline{\gamma}_h(t)) = (f, \zeta) \quad \forall\, (\underline{\eta}, \zeta) \in V_h, \\ \tilde{\mathcal{B}}_h((\underline{\theta}_h(t), w_h(t)), \underline{\chi}) = t^2(\underline{\gamma}, \underline{\chi})_W \quad \forall\, \underline{\chi} \in W = (L^2(\Omega))^2. \end{cases} \tag{10.4.120}$$

The kernel of the operator $\tilde{\mathbb{B}}_h$ associated with $\tilde{\mathcal{B}}$ will then be

$$\text{Ker}\tilde{\mathbb{B}}_h = \{(\underline{\eta}_h, \zeta_h) \in V_h \text{ such that } (r(\underline{\eta}_h, \zeta_h), \underline{\chi}) = 0 \qquad \forall\, \underline{\chi} \in \Gamma_h\}, \tag{10.4.121}$$

which, assuming that (10.4.117) is satisfied, can also be written as

$$\text{Ker}\tilde{\mathbb{B}}_h = \{(\underline{\eta}_h, \zeta_h) \in V_h \text{ such that } r(\underline{\eta}_h, \zeta_h) = 0\}. \tag{10.4.122}$$

All this should be connected to the *ellipticity in the kernel*, or, better, to the following (more powerful) property, strongly related to (5.5.46)

$$\begin{aligned} \exists\, \tilde{\alpha}_{RM} > 0 \text{ such that } & \mathcal{A}((\underline{\eta}, \zeta), (\underline{\eta}, \zeta)) + t^{-2}\|\mathbb{B}_h(\underline{\eta}, \zeta)\|_W^2 \\ & \equiv \mathcal{A}((\underline{\eta}, \zeta), (\underline{\eta}, \zeta)) + t^{-2}\|r(\underline{\eta}, \zeta)\|_0^2 \\ & \geq \tilde{\alpha}_{RM}\|(\underline{\eta}, \zeta)\|_V^2 \quad \forall\, (\underline{\eta}, \zeta) \in V_h \quad \forall t \in]0, T[, \end{aligned} \tag{10.4.123}$$

where T is always the diameter of Ω as in (10.4.19).

We have for this the following result.

Proposition 10.4.9. *Let $\mathcal{A}$ and $\tilde{\mathcal{B}}_h$ be defined as in* (10.4.28) *and* (10.4.119) *for an r that satisfies* (10.4.117)*. If moreover we have*

$$\exists c_r \text{ and } C_r > 0 \text{ such that } \|r(\underline{\eta}, \zeta\|_0^2 \geq C_r\|\, \underline{grad}\, \zeta\|_0^2 - c_r\|\underline{\eta}\|_1^2 \qquad \forall\, (\underline{\eta}, \zeta) \in V_h, \tag{10.4.124}$$

then (10.4.123) *holds.*

Proof. The proof is almost immediate using the Korn inequality (10.4.12). It is sufficient to combine the two inequalities

$$\mathcal{A}((\underline{\eta}, \zeta), (\underline{\eta}, \zeta)) + t^{-2}\|r(\underline{\eta}, \zeta)\|_0^2 \geq \mathcal{A}((\underline{\eta}, \zeta), (\underline{\eta}, \zeta)) \geq \alpha_{Korn}\|\underline{\eta}\|_1^2$$

and

$$\begin{aligned} & \mathcal{A}((\underline{\eta}, \zeta), (\underline{\eta}, \zeta)) + t^{-2}\|r(\underline{\eta}, \zeta)\|_0^2 \\ & \qquad \geq T^{-2}\|r(\underline{\eta}, \zeta)\|_0^2 \geq T^{-2}\Big(C_r\|\, \underline{\text{grad}}\, \zeta\|_0^2 - c_r\|\underline{\eta}\|_1^2\Big). \end{aligned}$$

Condition (10.4.124) might look cumbersome. We have, however, a simple sufficient condition for that.

Proposition 10.4.10. *Assume that $r(\underline{\eta}, \zeta)$ has the form*

$$r(\underline{\eta}, \zeta) := R_h(\underline{\eta}) - \underline{grad}\,\zeta, \tag{10.4.125}$$

where R is a mapping from Θ_h to Γ_h such that

$$\|R_h(\underline{\eta})\|_0 \leq C_R \|\underline{\eta}\|_1, \tag{10.4.126}$$

for some constant C_R independent of h. Then, (10.4.124) *holds.*

The proof is an easy exercise.

We can now use Theorem 5.5.5 and obtain the following abstract error bound.

Theorem 10.4.4. *Assume that R is an operator from Θ_h to Γ_h satisfying* (10.4.126), *and assume that the bilinear form $\tilde{\mathcal{B}}$ is defined through* (10.4.125) *and* (10.4.119). *For every $t \in]0, T[$, let $(\underline{\theta}(t), w(t), \underline{\gamma}(t)$ be the solution of Problem* (10.4.61) *and let $(\underline{\theta}_h(t), w_h(t), \underline{\gamma}_h(t)$ be the solution of* (10.4.120). *Then, for every $(\underline{\theta}_I(t), w_I(t), \underline{\gamma}_I(t)$ in $\Theta_h \times Z_h \times \Gamma_h$ such that*

$$R_h(\underline{\theta}_I) - \underline{grad}\, w_I = t^2 \underline{\gamma}_I, \tag{10.4.127}$$

we have

$$\begin{aligned}
&\|\underline{\theta}_h(t) - \underline{\theta}_I(t)\|_1 + \|w_h(t) - w_I(t)\|_1 + t\|\underline{\gamma}_h(t) - \underline{\gamma}_I(t)\|_0 \\
&\quad \leq C \Big(\|\underline{\theta}(t) - \underline{\theta}_I(t)\|_1 + \|w(t) - w_I(t)\|_1 + \|\underline{\gamma}(t) - \underline{\gamma}_I(t)\|_Q \\
&\qquad + \sup_{\underline{\eta} \in \Theta_h} \frac{(R_h \underline{\eta}, \underline{\gamma}) - (\underline{\eta}, \underline{\gamma}_I)}{\|\underline{\eta}\|_1} \Big)
\end{aligned} \tag{10.4.128}$$

where C is a constant independent of t and h.

Proof. The proof is elementary: using (10.4.126) and Proposition 10.4.10, we obtain (10.4.124). Using Proposition 10.4.9, we obtain (10.4.123), which is the crucial assumption needed to apply Theorem 5.5.5. □

Remark 10.4.15. In many cases, the last term in the right-hand side of (10.4.128) can be better estimated by

$$\sup_{\underline{\eta} \in \Theta_h} \frac{(R_h \underline{\eta}_I - \underline{\eta}, \underline{\gamma})}{\|\underline{\eta}\|_1} + \sup_{\underline{\eta} \in \Theta_h} \frac{(\underline{\eta}, \underline{\gamma} - \underline{\gamma}_I)}{\|\underline{\eta}\|_1} \tag{10.4.129}$$

which, in a sense, separates the errors $\|\underline{\gamma}(t) - \underline{\gamma}_I(t)\|_{-1}$ and $\|R_h - Identity\|$. It has to be pointed out that, in most cases, the difference $R_h \underline{\eta}_I - \underline{\eta}$ will be orthogonal to all (vector-valued) polynomial of a certain degree ℓ so that

$$\sup_{\underline{\eta}\in\Theta_h} \frac{(R_h\underline{\eta}_I - \underline{\eta}, \underline{\gamma})}{\|\underline{\eta}\|_1} \le C\, h\|\underline{\gamma} - \pi_\ell \underline{\gamma}\|_0 \tag{10.4.130}$$

where π_ℓ is the projection operator on polynomials of degree ℓ. □

As we did for the previous class of methods (the ones based on the decomposition), we will not present here a list of all methods of this type available on the market. We will instead present a single method, as an example, in order to show the general guidelines that rule their construction.

We assume again that Ω is a convex polygon and then we are given a sequence $\{\mathcal{T}_h\}$ of partitions of Ω into triangles. We set, with the notation of Chap. 2,

$$\Theta_h := (\mathcal{L}_2^1 + B_3)^2 \cap (H_0^1(\Omega))^2, \qquad Z_h := (\mathcal{L}_2^1 + B_3) \cap H_0^1(\Omega), \tag{10.4.131}$$

$$\Gamma_h := \{\underline{\chi} \in (\mathcal{L}_2^0)^2 \text{ s. t. } \underline{\chi} \mid \underline{\chi}\cdot\underline{t} \in P_1(e)\ \forall \text{ edge } e\} \cap H_0(\text{curl}; \Omega). \tag{10.4.132}$$

Note that this is the *rotated* $\mathcal{BDFM}_2$, following Remark 2.3.2. Together with the spaces (10.4.131), we consider the operator Π_h from, say, $(H^1(\Omega))^2$ into Γ_h defined in each triangle K by

$$\int_e (\Pi_h\underline{\eta} - \underline{\eta})\cdot\underline{t}\mu_1\, ds = 0 \qquad \forall e \in \partial K \qquad \forall \mu_1 \in P_1(e), \tag{10.4.133}$$

$$\int_K (\Pi_h\underline{\eta} - \underline{\eta})\cdot\underline{q}\, dx = 0 \qquad \forall \underline{q} \in \mathcal{RT}_0(K), \tag{10.4.134}$$

where $\mathcal{RT}_0(K)$ is the lowest order Raviart-Thomas space (see Chap. 2).

We can now define the operator r. Following the structure (10.4.125), we set

$$r(\underline{\eta}_h, \zeta_h) = \underline{\text{grad}}\,\zeta_h - \Pi_h\underline{\eta}_h \in \Gamma_h. \tag{10.4.135}$$

The kernel of $\mathbb{B}_h$ as defined in (10.4.121) is now easily characterised as the set of $(\underline{\eta}_h, \zeta_h)$ such that

$$\Pi_h\underline{\eta}_h = \underline{\text{grad}}\,\zeta_h. \tag{10.4.136}$$

Since $\|\Pi_h\underline{\eta}_h\|_0 \le c\|\underline{\eta}_h\|_1$ for some constant c independent of h, we can apply Proposition 10.4.10 and then Proposition (10.4.9) to get

$$\mathcal{A}((\underline{\eta}, \zeta), (\underline{\eta}, \zeta)) + t^{-2}\|r(\underline{\eta}, \zeta)\|_0^2 \ge \tilde{\alpha}_{RM}\|(\underline{\eta}, \zeta)\|_V^2 \tag{10.4.137}$$

that is, more precisely, condition (10.4.137). In order to apply Theorem 10.4.4, we just need to check that condition (10.4.127) holds for suitable $\underline{\theta}_I, w_I, \underline{\gamma_I}$ having optimal approximation properties. For the construction of $\underline{\theta}_I, w_I, \underline{\gamma_I}$, we can use the following lemma.

Lemma 10.4.1. *Assume that*

$$\{\underline{\chi} \in \Gamma_h \text{ such that } curl\,\underline{\chi} = 0\} \subseteq \underline{grad}(Z_h). \tag{10.4.138}$$

Set $\underline{\gamma}_I := \Pi_h \underline{\gamma}$ *and assume that we can find* $\underline{\theta}^i$ *and* w^i *verifying*

$$curl\,\Pi_h(\underline{\theta}^i - \underline{\theta}) = 0 \qquad \Pi_h(\underline{grad}\,w^i - \underline{grad}\,w) = 0. \tag{10.4.139}$$

Then, from (10.4.138) *and* (10.4.139), *one obviously has*

$$\Pi_h(\underline{\theta}^i - \underline{\theta}) = \underline{grad}\,\zeta_h \tag{10.4.140}$$

for some $\zeta_h \in Z_h$. *Then setting*

$$\underline{\theta}_I := \underline{\theta}^i \qquad w_I = w^i - \zeta_h, \tag{10.4.141}$$

one has (10.4.127) *as well as*

$$\|\Pi_h(\underline{\theta}_I - \underline{\theta})\|_1 + \|w_I - w\|_1 \le 2\,\|\Pi_h(\underline{\theta}^i - \underline{\theta})\|_1 + \|w_i - w\|_1. \tag{10.4.142}$$

Note that, in other words, inequality (10.4.142) tells us that we can "arrange" (10.4.127) without losing accuracy. The proof is simple: first we check that

$$\begin{aligned}
\Pi_h\underline{\theta}_I - \underline{\mathrm{grad}}\,w_I & \\
&= \Pi_h\underline{\theta}^i - \underline{\mathrm{grad}}\,w_I = \Pi_h\underline{\theta} + (\Pi_h\underline{\theta}_i - \Pi_h\underline{\theta}) - \underline{\mathrm{grad}}\,w_I \\
&= \Pi_h\underline{\theta} + \underline{\mathrm{grad}}\,\zeta_h - \underline{\mathrm{grad}}\,w_I = \Pi_h\underline{\theta} - \underline{\mathrm{grad}}\,w^i \\
&= \Pi_h\underline{\theta} - \Pi_h\,\underline{\mathrm{grad}}\,w^i = \Pi_h(\underline{\theta} - \underline{\mathrm{grad}}\,w^i) = \Pi_h(t^2\underline{\gamma}) \\
&= t^2\underline{\gamma}_I,
\end{aligned} \tag{10.4.143}$$

giving us (10.4.127). Inequality (10.4.142) then follows immediately from

$$\|w_I - w\|_1 \le \|w_i - w\|_1 + \|\zeta_h\|_1 \le \|w_i - w\|_1 + \|\Pi_h(\underline{\theta}^i - \underline{\theta})\|_1. \tag{10.4.144}$$

□

Then, we just have to construct $\underline{\theta}^i$ and w^i satisfying (10.4.139). The construction of $\underline{\theta}^i$ is easy. Indeed, denoting Π_{CR} the B-compatible operator for the Couzeix-Raviart element for the Stokes problem and by π_1 the projection onto $\mathcal{L}_1^0$, one has from Example 8.6.1

$$\pi_1\,\mathrm{curl}(\underline{\theta} - \mathrm{curl}\,\Pi_{CR}\underline{\theta}) = 0 \tag{10.4.145}$$

and similarly, from the properties of the $\mathcal{BDFM}$ element,

$$\pi_1\,\mathrm{curl}\,\underline{\theta} - \mathrm{curl}\,\Pi_h\underline{\theta} = 0. \tag{10.4.146}$$

We deduce that

$$\operatorname{curl} \Pi_h \underline{\theta} = \pi_1 \operatorname{curl} \underline{\theta} = \pi_1 \operatorname{curl} \Pi_{CR} \underline{\theta} = \operatorname{curl} \Pi_h \Pi_{CR} \underline{\theta}. \tag{10.4.147}$$

This says that the choice

$$\underline{\theta}^i := \Pi_{CR} \underline{\theta} \tag{10.4.148}$$

will satisfy the first condition of (10.4.139). On the other hand, taking

$$w^i(P) = w(P) \qquad \text{for all node } P \text{ in } \mathcal{T}_h, \tag{10.4.149}$$

$$\int_e (w^i - w)\, ds = 0 \qquad \text{for all edge } e \text{ in } \mathcal{T}_h, \tag{10.4.150}$$

and

$$\int_K (w^i - w)\, dx = 0 \qquad \text{for all triangle } K \text{ in } \mathcal{T}_h, \tag{10.4.151}$$

we easily have that

$$\int_e \underline{\operatorname{grad}}(w^i - w) \cdot \underline{t}\, \mu_1, ds = 0 \qquad \forall \text{ edge } e \text{ in } \mathcal{T}_h, \quad \forall\, \mu_1 \in P_1(e) \tag{10.4.152}$$

and

$$\int_T \underline{\operatorname{grad}}(w^i - w) \cdot \underline{q}\, dx = 0 \quad \forall \text{ triangle } T \text{ in } \mathcal{T}_h, \quad \forall \underline{q} \in \mathcal{RT}_{01}(T), \tag{10.4.153}$$

implying the second condition of (10.4.139).

We can therefore use Theorem 10.4.2 and standard interpolation estimates (together with Remark 10.4.15) to obtain the following result.

Theorem 10.4.5. *Consider the discretised problem* (10.4.120) *with the choices* (10.4.131)–(10.4.135)*. Then, for every* $t \in]0, T[$*, it has a unique solution* $(\underline{\theta}_h(t), w_h(t), \underline{\gamma}_h(t)$*. Let moreover* $(\underline{\theta}(t), w(t), \underline{\gamma}(t)$ *be the solution of Problem* (10.4.61)*. Then we have*

$$\begin{aligned} &\|\underline{\theta}_h(t) - \underline{\theta}_I(t)\|_1 + \|w_h(t) - w_I(t)\|_1 + t\, \|\underline{\gamma}_h(t) - \underline{\gamma}_I(t)\|_0 \\ &\qquad \le C\, h^2 \Big(\|\underline{\theta}(t)\|_3 + \|w(t)\|_3 + t\, \|\underline{\gamma}(t)\|_2 + \|\gamma\|_{H^1(\mathrm{div};\Omega)} \Big) \end{aligned} \tag{10.4.154}$$

where C *is a constant independent of* t *and* h.

As we already noted, this estimate is overoptimistic because it ignores the boundary layer effects. From the results of [29], an $O(h^{3/2})$ convergence rate should be expected.

Remark 10.4.16. Similar estimates have been obtained in [117] for the presently discussed element and related ones, including elements defined on quadrilaterals. More refined estimates can be found in [126]. A recent review of different Mindlin-Reissner approximations, including the *Linked interpolation* techniques (that have not been considered here), can be found in [190]. □

Remark 10.4.17. The choice of second-order accuracy has been made only for the sake of simplicity. Higher-order elements are possible and we shall indicate at the end of this chapter a general framework within which they could be built. On the contrary, lower-order elements are more difficult to get; see for instance [54] for the convergence analysis of a similar method, which is only $O(h)$ accurate [55, 258]. We also refer to [28, 54, 122, 126, 181, 182, 323] for other examples. □

Remark 10.4.18. It is possible to use a duality argument to get an $O(h^3)$ estimate for $\|\underline{\theta} - \underline{\theta}_h\|_0$ and $\|w - w_h\|_0$. See [126]. □

Now to end this lengthy section, we are in a position to present general guidelines for the discretisation of Mindlin–Reissner problems.

We must emphasise again that the decomposition principle makes apparent a direct link with the Stokes problem. Indeed, all examples for which a satisfactory analysis could be achieved contained an already proven Stokes element. If we distinguish the case of continuous pressure approximation and the case of discontinuous pressure element, we get two types of strategies.

10.4.6 Continuous Pressure Approximations

- Suppose one knows $\Theta_h \times Q_h$ to be a good approximation of the Stokes problem with $Q_h \subset H^1(\Omega)$.
- Choose Z_h an approximation of $H_0^1(\Omega)$ of the same order of accuracy.
- Write $\Gamma_h = \underline{\mathrm{grad}}\, Z_h + \underline{\mathrm{curl}}\, Q_h$.

In this context, the definition of Γ_h does not lead, in general, to a standard space and the decomposition principle of Theorem 10.4.2 and Remark 10.4.5 is the only way to handle things from a computational point of view. It may, however, happen, for a clever choice of Z_h and Q_h, that Γ_h turns out to be a standard polynomial space. Such a situation has been encountered in [28] where, using for $\Theta_h \times Q_h$ the MINI element, but taking Z_h to be $\mathcal{L}_1^{1,NC}$, that is, a nonconforming P_1 approximation of $H_0^1(\Omega)$, Γ_h comes to be the whole space $(\mathcal{L}_0^0)^2$ and not a proper subspace. For an extension of the Arnold-Falk element to higher degree, see [14, 26].

10.4.7 Discontinuous Pressure Elements

This second class of approximations to the Stokes problem has been the basis for the "reduced integration" method of the last subsection. Here, we shall try to outline the principal features of this strategy in order to provide a guide for possible extensions, some of which can be found in [117].

1. Here again, our starting point is an approximation of the Stokes problem $\Theta_h \times Q_h$, Q_h being a space of discontinuous polynomial functions. This approximation should, of course, satisfy the *inf-sup* condition.
2. We need to match this with an approximation Γ_h of $H_0(\mathrm{curl}, \Omega)$. More precisely, we need a couple of spaces (Γ_h, Q_h) (where Q_h is the same as before) and a uniformly bounded linear operator $\Pi_h \to \Gamma_h$ such that we have the commuting diagram:

$$\begin{array}{ccc} H & \xrightarrow{\mathrm{curl}} & L^2(\Omega) \\ {\scriptstyle \Pi_h}\big\downarrow & & {\scriptstyle P_h}\big\downarrow \\ \Gamma_h & \xrightarrow{\mathrm{curl}} & Q_h \end{array} \tag{10.4.155}$$

   where $\Theta = (H^1(\Omega))^2 \cap H_0\,(\mathrm{curl}, \Omega)$ and P_h is the L^2-projection operator.
3. We finally need a space $Z_h \subset H_0^1(\Omega)$ such that

$$\underline{\mathrm{grad}}\, Z_h = \{\underline{\delta}_h \in \Gamma_h,\ \mathrm{curl}\, \underline{\delta}_h = 0\}. \tag{10.4.156}$$

Ingredients (1), (2), (3) will produce a plate element for which one can essentially repeat the proof of Theorem 10.4.5 and obtain optimal error estimates for $\underline{\theta}$ and w.

Chapter 11
Mixed Finite Elements for Electromagnetic Problems

The finite element approximation of problems arising from electromagnetism has reached a discrete level of maturity and a huge literature is available in different fields of research: in particular in mathematics, engineering, and physics. It is out of the aims of this chapter to give a thorough survey on the topic. We refer the interested reader, for instance, to [248, 302], and to the references therein, for an introduction on computational electromagnetism.

We are interested in showing how the analysis developed in this book can be successfully applied to an active and important research area. For this reason, we are focusing our analysis on the time harmonic formulation of Maxwell's system. The analysis of the finite element approximation of the time harmonic Maxwell system, and of the closely related eigenvalue problem associated with Maxwell's equations, has been a challenging problem for several researchers during at least the last two decades.

It is impossible to understand the approximation of Maxwell's equations without a reasonable knowledge of the properties of the main functional space used for the analysis: $H(\underline{\mathrm{curl}};\Omega)$. We thus start the chapter recalling some important concepts (in particular, about traces and the de Rham complex). We then link the approximation of the time harmonic Maxwell's system to the good approximation of Maxwell eigenvalues and continue our discussion with the state of the art in the approximation of the eigenvalue problem associated with Maxwell's equations.

We conclude the chapter with a short introduction of discrete exterior calculus. It is clear that the de Rham complex and discrete differential forms play an important role in the analysis of the problems we are going to present (indeed, a differential complex and a compatible discretisation of it are recognisable behind all presented results, even when the authors perhaps did not make them explicit).

D. Boffi et al., *Mixed Finite Element Methods and Applications*, Springer Series in Computational Mathematics 44, DOI 10.1007/978-3-642-36519-5_11,

11.1 Useful Results About the Space $H(\underline{\text{curl}};\Omega)$, its Boundary Traces, and the de Rham Complex

Most of the topics of the present section have been already presented in Chap. 2. We summarise them here in a unified setting, in order to recall the main results and the notation. We focus on the three-dimensional setting; some remarks on the two-dimensional case will be given at the end.

$H(\underline{\text{curl}};\Omega)$ is the space of vector fields $\underline{v}$ from Ω to $\mathbb{R}^3$ which satisfy $\underline{\text{curl}}\,\underline{v} \in L^2(\Omega)^3$ with the natural norm $\|\underline{v}\|^2_{\underline{\text{curl}}} = \|\underline{v}\|^2_0 + \|\,\underline{\text{curl}}\,\underline{v}\|^2_0$. The following Green formula, which is valid for smooth functions, is the starting point for the definition of the boundary trace $(\underline{v} \wedge \underline{n})$ for functions $\underline{v} \in H(\underline{\text{curl}};\Omega)$

$$\int_\Omega \underline{v} \cdot \underline{\text{curl}}\,\underline{\phi}\,dx - \int_\Omega \underline{\text{curl}}\,\underline{v} \cdot \underline{\phi}\,dx = \int_{\partial\Omega} (\underline{v} \wedge \underline{n}) \cdot \underline{\phi}\,ds. \tag{11.1.1}$$

We denote by $H_0(\underline{\text{curl}};\Omega)$ the subspace of $H(\underline{\text{curl}};\Omega)$ consisting of vector fields $\underline{v}$ with vanishing trace $\underline{v} \wedge \underline{n}$ along the boundary of Ω.

11.1.1 The de Rham Complex and the Helmholtz Decomposition When Ω Is Simply Connected

Using the terminology of exterior algebra, the space $H(\underline{\text{curl}};\Omega)$ can be identified with an element of the following de Rham complex

$$\mathbb{R} \hookrightarrow H^1(\Omega) \xrightarrow{\ \underline{\text{grad}}\ } H(\underline{\text{curl}};\Omega) \xrightarrow{\ \underline{\text{curl}}\ } H(\text{div};\Omega) \xrightarrow{\ \text{div}\ } L^2(\Omega) \to 0. \tag{11.1.2}$$

If we consider homogeneous boundary conditions, then the de Rham complex takes the following form

$$0 \hookrightarrow H^1_0(\Omega) \xrightarrow{\ \underline{\text{grad}}\ } H_0(\underline{\text{curl}};\Omega) \xrightarrow{\ \underline{\text{curl}}\ } H_0(\text{div};\Omega) \xrightarrow{\ \text{div}\ } L^2(\Omega) \to \mathbb{R}. \tag{11.1.3}$$

If Ω is simply connected, then the two sequences (11.1.2) and (11.1.3) are exact, that is, the range of each operator equals the kernel of the next one. This means, for instance, that every divergence-free vector field of $H(\text{div};\Omega)$ is the $\underline{\text{curl}}$ of an element of $H(\underline{\text{curl}};\Omega)$ or that a curl-free vector field of $H_0(\underline{\text{curl}};\Omega)$ is the gradient of an element of $H^1_0(\Omega)$.

The previous sequences are related to an important tool of vector calculus: the so called Helmholtz decomposition. In the case of a simply connected domain (see, for instance, [223]), the Helmholtz decomposition says that any vector field can be split as the sum of an irrotational and a solenoidal part which refer to a scalar and a vector potential, respectively. More precisely, for any $\underline{v} \in L^2(\Omega)^3$, there exists a unique scalar potential $\varphi \in H^1(\Omega)/\mathbb{R}$ and a unique vector potential $\underline{\psi} \in H(\underline{\text{curl}};\Omega) \cap H_0(\text{div};\Omega)$ with $\text{div}\,\underline{\psi} = 0$ such that

$$\underline{v} = \underline{\text{grad}}\,\varphi + \underline{\text{curl}}\,\underline{\psi}. \tag{11.1.4}$$

If, moreover, $\underline{v}$ belongs to $H_0(\underline{\text{curl}};\Omega)$, then φ can be chosen in $H_0^1(\Omega)$. The Helmholtz decomposition is orthogonal in the sense that $(\underline{\text{grad}}\,\varphi, \underline{\text{curl}}\,\underline{\psi}) = 0$.

11.1.2 The Friedrichs Inequality

The Friedrichs inequality is an important tool when dealing with the space $H_0(\underline{\text{curl}};\Omega)$. It plays the same role as the Poincaré inequality for the space $H_0^1(\Omega)$.

Let us consider the space $K = H_0(\underline{\text{curl}};\Omega) \cap H(\text{div}^0;\Omega)$ consisting of vector fields in $H(\underline{\text{curl}};\Omega)$ with vanishing tangential component along the boundary and with zero divergence in Ω. Then, there exists $\alpha > 0$ such that the following inequality holds true

$$\|\underline{\text{curl}}\,\underline{v}\|_0 \geq \alpha\|\underline{v}\|_0 \quad \forall\, \underline{v} \in K. \tag{11.1.5}$$

It will be clear in the following sections that the Friedrichs constant α is related to the minimum frequency of Maxwell's eigenvalue problem (see (11.2.7)). An immediate consequence of (11.1.5) is that the bilinear form $(\underline{\text{curl}}\,\underline{u}, \underline{\text{curl}}\,\underline{v}) : H_0(\underline{\text{curl}};\Omega) \times H_0(\underline{\text{curl}};\Omega) \to \mathbb{R}$ is coercive in $L^2(\Omega)^3$ when restricted to divergence-free vector fields in K.

11.1.3 Extension to More General Topologies

Electromagnetic devices often involve complicated geometries, so that the simplification made so far (Ω simply connected) is not realistic for several practical applications.

In order to deal with multiply connected domains, a natural mathematical setting consists in assuming the existence of a finite number of mutually disjoint cuts $\Lambda_1, \ldots, \Lambda_N$ such that $\Omega \setminus \cup\Lambda_i$ is simply connected. The mathematical justification of the functional setting required for the description of realistic applications has been the object of a wide investigation. We refer the interested reader, for instance, to [8, 105, 196, 248, 302]. In this case, the sequences (11.1.2) and (11.1.3) are no longer exact. However, the image of each operator is a closed subset of the kernel of the next one. A standard procedure is to consider quotient spaces, which are then called *cohomology* spaces. For instance, looking at the kernel of the $\underline{\text{curl}}$ operator, thought as a subset of $H_0(\underline{\mathrm{curl}};\Omega)$, we can consider the orthogonal complement K of $\underline{\mathrm{grad}}\, H_0^1(\Omega)$. While on simply connected domains $K=\{0\}$, it turns out that on multiply connected domains the dimension of K is exactly equal to N (the number of cuts that we need in order to make Ω simply connected). The space K consists of vector fields $\underline{v}$ which satisfy $\underline{\mathrm{curl}}\,\underline{v}=0$, $\mathrm{div}\,\underline{v}=0$ in Ω and $\underline{v}\wedge\underline{n}=0$ on $\partial\Omega$.

It is clear that the presence of a non-trivial cohomology space K has the effect of changing the construction of Helmholtz's decomposition which now reads

$$\underline{v}=\underline{\mathrm{grad}}\,\varphi+\underline{\mathrm{curl}}\,\underline{\psi}+\underline{\zeta} \tag{11.1.6}$$

with $\underline{\zeta}$ belonging to K.

11.1.4 $H(\underline{\mathrm{curl}};\Omega)$ in Two Space Dimensions

We conclude this section with some remarks concerning two-dimensional domains. As we have already observed in Remark 2.1.5, when Ω belongs to $\mathbb{R}^2$ we have two different curl operators: a *vector* $\underline{\mathrm{curl}}$ operator which acts on scalar functions and returns a vector field and a *scalar* curl operator which acts on vector fields and returns a scalar function.

More precisely, given a vector $\underline{u}(x_1,x_2)=(u_1,u_2)$ we have $\mathrm{curl}\,\underline{u}=\frac{\partial u_2}{\partial x_1}-\frac{\partial u_1}{\partial x_2}$ and, given a scalar $\phi(x_1,x_2)$, we have $\underline{\mathrm{curl}}\,\phi=\left(\frac{\partial\phi}{\partial x_1},-\frac{\partial\phi}{\partial x_2}\right)$.

In this case, the de Rham complex reads

$$\mathbb{R}\hookrightarrow H^1(\Omega)\xrightarrow{\underline{\mathrm{curl}}}H(\underline{\mathrm{curl}};\Omega)\xrightarrow{\mathrm{curl}}L^2(\Omega)\to 0, \tag{11.1.7}$$

or, in presence of boundary conditions,

$$0\hookrightarrow H_0^1(\Omega)\xrightarrow{\underline{\mathrm{curl}}}H_0(\underline{\mathrm{curl}};\Omega)\xrightarrow{\mathrm{curl}}L^2(\Omega)\to\mathbb{R}. \tag{11.1.8}$$

It is clear that the two-dimensional situation is much simpler than the three-dimensional one. In particular, the isomorphism between $\underline{\mathrm{curl}}$ (resp. curl) and $\underline{\mathrm{grad}}$ (resp. div) operators implies that it is enough to study the de Rham complex associated to $H(\mathrm{div};\Omega)$ in order to have all information about $H(\underline{\mathrm{curl}};\Omega)$.

The Helmholtz decomposition takes the same form as in the three-dimensional case, the only difference being that both potentials are scalar. More precisely, if $\Omega\in\mathbb{R}^2$ is simply connected, for any vector field $\underline{v}\in L^2(\Omega)^2$, there exist a unique scalar potential $\varphi\in H^1(\Omega)/\mathbb{R}$ and a unique stream function $\psi\in H_0^1$ such that

$$\underline{v}=\underline{\mathrm{grad}}\,\varphi+\underline{\mathrm{curl}}\,\psi. \tag{11.1.9}$$

11.2 The Time Harmonic Maxwell System

The classical electromagnetic field is described by the four vectors $\mathcal{E}$, $\mathcal{D}$, $\mathcal{H}$, and $\mathcal{B}$ which are functions of the position $\underline{x} \in \mathbb{R}^3$ and of the time $t \in \mathbb{R}$. The vectors $\mathcal{E}$ and $\mathcal{H}$ are referred to as the electric and magnetic field, while $\mathcal{D}$ and $\mathcal{B}$ are the electric and magnetic displacements, respectively.

The time harmonic formulation of Maxwell's equations is based on the following assumptions on the involved quantities:

$$\begin{aligned}
\mathcal{E}(\underline{x}, t) &= \Re\left(e^{-(i\omega t)} E(\underline{x})\right), \\
\mathcal{D}(\underline{x}, t) &= \Re\left(e^{-(i\omega t)} D(\underline{x})\right), \\
\mathcal{H}(\underline{x}, t) &= \Re\left(e^{-(i\omega t)} H(\underline{x})\right), \\
\mathcal{B}(\underline{x}, t) &= \Re\left(e^{-(i\omega t)} B(\underline{x})\right).
\end{aligned} \tag{11.2.1}$$

These assumptions make sense, for instance, when using the Fourier transform in time or when studying the propagation of electromagnetic waves at a given frequency.

In Example 1.2.5 of Chap. 1, we already noticed that the time harmonic assumptions lead to the following time harmonic version of Maxwell's system

$$\begin{aligned}
&- i\omega \underline{\underline{\mu}} H + \underline{\mathrm{curl}}\, E = 0, \\
&\mathrm{div}(\underline{\underline{\varepsilon}} E) = r, \\
&- i\omega \underline{\underline{\varepsilon}} E - \underline{\mathrm{curl}}\, H = -J, \\
&\mathrm{div}(\underline{\underline{\mu}} H) = 0,
\end{aligned} \tag{11.2.2}$$

where $\rho(\underline{x}, t) = \Re\left(e^{-(i\omega t)} r(\underline{x})\right)$ denotes the scalar charge density and $\mathcal{J}(\underline{x}, t) = \Re\left(e^{-(i\omega t)} J(\underline{x})\right)$ is the vector current density.

We can write (11.2.2) as a second order equation by eliminating H as follows

$$\underline{\mathrm{curl}}(\underline{\underline{\mu}}^{-1}\, \underline{\mathrm{curl}}\, E) - \omega^2 \underline{\underline{\varepsilon}} E = F \tag{11.2.3}$$

with $F = i\omega J$ and the divergence condition

$$- \omega^2\, \mathrm{div}(\underline{\underline{\varepsilon}} E) = \mathrm{div}\, F. \tag{11.2.4}$$

Remark 11.2.1. Some of the main mathematical challenges for the approximation of the time harmonic Maxwell system already arise in the case when a unique isotropic material is involved. In such case, one has that the material coefficients $\underline{\underline{\mu}}$ and $\underline{\underline{\varepsilon}}$ are equal to global constants μ and ε times the identity matrix. For this reason and in order to make the presentation more readable, we take in this section

$\mu = \varepsilon = 1$ and denote by $\underline{u}$ the unknown field E. Another assumption that we are going to make in this chapter is that the domain Ω is simply connected. This simplification rules out significant examples of applications but allows us to start with the study of this topic in a simple setting where the mathematical properties of the continuous problem and of its finite element approximation can be presented in a more immediate way.

More general situations can be studied with the help of the considerations made in Sect. 11.1.3. □

In view of Remark 11.2.1 and taking into account perfect conducting boundary conditions, our problem reads: *let Ω be a simply connected domain; given a divergence-free vector field $\underline{f}$ and a positive number ω, find a vector field $\underline{u}$ such that*

$$\begin{aligned} &\underline{\mathrm{curl}}\,\underline{\mathrm{curl}}\,\underline{u} - \omega^2 \underline{u} = \underline{f} && \text{in } \Omega, \\ &\mathrm{div}\,\underline{u} = 0 && \text{in } \Omega, \\ &\underline{u} \wedge \underline{n} = 0 && \text{on } \partial\Omega. \end{aligned} \tag{11.2.5}$$

Remark 11.2.2. It can be easily observed by taking the divergence of the first equation in (11.2.5) that $\mathrm{div}\,\underline{f} = 0$ implies $\mathrm{div}\,\underline{u} = 0$, so that the second equation is indeed redundant. □

From the mathematical point of view, it is clear that the solvability of system (11.2.5) is strictly related to the frequency value ω. In particular, system (11.2.5) is ill-posed if $\omega^2 = \lambda$, where λ satisfies

$$\begin{aligned} &\underline{\mathrm{curl}}\,\underline{\mathrm{curl}}\,\underline{u} = \lambda \underline{u} && \text{in } \Omega, \\ &\mathrm{div}\,\underline{u} = 0 && \text{in } \Omega, \\ &\underline{u} \wedge \underline{n} = 0 && \text{on } \partial\Omega \end{aligned} \tag{11.2.6}$$

for some non-vanishing vector field $\underline{u}$. Problem (11.2.6) is the so called *cavity problem for Maxwell's equations*, which is also known as *interior Maxwell's eigenvalue problem*. We shall refer to the solution of this problem as *interior Maxwell's eigenvalues and eigenfunctions*.

It is clear from this introduction that the study of the Maxwell eigenvalue problem will be of fundamental importance for the understanding of the time harmonic Maxwell's system (11.2.5). This will be done in the next subsection.

11.2.1 Maxwell's Eigenvalue Problem

We are interested in the finite element approximation of problem (11.2.6): *find $\lambda \in \mathbb{R}$ such that, for a non-vanishing $\underline{u}$, it holds*

$$\begin{aligned}
&\underline{\text{curl}}\,\underline{\text{curl}}\,\underline{u} = \lambda \underline{u} && \text{in } \Omega,\\
&\operatorname{div} \underline{u} = 0 && \text{in } \Omega,\\
&\underline{u} \wedge \underline{n} = 0 && \text{on } \partial\Omega.
\end{aligned}$$

The finite element approximation of this problem will be considered in Sect. 11.3.

We shall consider the space $H_0(\underline{\text{curl}}; \Omega)$; some properties related to this space have been summarised at the beginning of this chapter. More information, including finite element spaces approximating it, can be found in Chap. 2. Let V be the subspace of $H_0(\underline{\text{curl}}; \Omega)$ consisting of divergence-free vector fields, i.e., $V = H_0(\underline{\text{curl}}; \Omega) \cap H(\text{div}^0; \Omega)$.

A weak formulation of problem (11.2.6) can be written in a natural way as follows: *find* λ *such that for a non-vanishing* $\underline{u} \in V$ *it holds*

$$(\underline{\text{curl}}\,\underline{u}, \underline{\text{curl}}\,\underline{v}) = \lambda(\underline{u}, \underline{v}) \qquad \forall\, \underline{v} \in V. \tag{11.2.7}$$

Thanks to the fact that the bilinear form $(\underline{\text{curl}}\,\underline{u}, \underline{\text{curl}}\,\underline{v})$ is symmetric, continuous and coercive on the space V and to the compact embedding of V in $L^2(\Omega)^3$, the solution operator associated to problem (11.2.7) is compact and self-adjoint. It follows that problem (11.2.7) admits a countable set of real and positive eigenvalues and that each eigenspace is finite dimensional. Moreover, all the eigenfunctions can be chosen to be real, so that the entire analysis can be performed using real spaces. It is clear that, within this framework, the space V can be decomposed as the direct sum of the eigenspaces.

We observe that $\lambda = 0$ is not an eigenvalue of problem (11.2.7). Indeed, taking $\underline{v} = \underline{u}$, we get $\underline{\text{curl}}\,\underline{u} = 0$, which implies $\underline{u} = 0$ (since $\operatorname{div}\underline{u} = 0$, Ω is simply connected, and $\underline{u} \wedge \underline{n} = 0$ on the boundary). Moreover, putting $\lambda \neq 0$ in the first equation of (11.2.6), easily implies that $\underline{u}$ is divergence-free.

The last remark suggests the introduction of the following *unconstrained* formulation: *find* $\lambda \neq 0$ *such that for a non-vanishing* $\underline{u} \in H_0(\underline{\text{curl}}; \Omega)$ *there holds*

$$(\underline{\text{curl}}\,\underline{u}, \underline{\text{curl}}\,\underline{v}) = \lambda(\underline{u}, \underline{v}) \qquad \forall\, \underline{v} \in H_0(\underline{\text{curl}}; \Omega). \tag{11.2.8}$$

Remark 11.2.3. As a consequence of the previous comments, it turns out that all the eigenvalues of (11.2.8) are positive and coincide with those of (11.2.7). The equivalence is true for the corresponding eigenspaces as well. If the condition $\lambda \neq 0$ is dropped from problem (11.2.8), then the infinite dimensional eigenspace $\underline{\text{grad}}\, H_0^1(\Omega)$ associated with the value $\lambda = 0$ is added to the spectrum. This fact is related to the identity $H_0(\underline{\text{curl}}; \Omega) = V \oplus \underline{\text{grad}}\, H_0^1(\Omega)$. □

In order to study problem (11.2.6), mixed variational formulations are generally used. A first mixed formulation, due to Kikuchi (see [268]), is generally referred to as Kikuchi's formulation and consists in looking for eigenvalues λ and eigenfunctions $\underline{u} \in H_0(\text{curl}; \Omega)$ with $\underline{u} \neq 0$ such that there exists $p \in H_0^1(\Omega)$ satisfying

$$\begin{cases} (\underline{\mathrm{curl}}\,\underline{u}, \underline{\mathrm{curl}}\,\underline{v}) + (\underline{\mathrm{grad}}\,p, \underline{v}) = \lambda(\underline{u}, \underline{v}) & \forall\,\underline{v} \in H_0(\mathrm{curl};\Omega), \\ (\underline{\mathrm{grad}}\,q, \underline{u}) = 0 & \forall\,q \in H_0^1(\Omega). \end{cases} \tag{11.2.9}$$

A second mixed variational formulation has been introduced in [89] and makes use of the space $\mathcal{F} := \underline{\mathrm{curl}}(H_0(\mathrm{curl};\Omega))$. In this case, we are seeking eigenvalues λ and eigenfunctions $\underline{\psi} \in \mathcal{F}$ with $\underline{\psi} \neq 0$ such that there exists $\underline{\sigma} \in H_0(\mathrm{curl};\Omega)$ satisfying

$$\begin{cases} (\underline{\sigma}, \underline{\tau}) + (\underline{\mathrm{curl}}\,\underline{\tau}, \underline{\psi}) = 0 & \forall\,\underline{\tau} \in H_0(\mathrm{curl};\Omega), \\ (\underline{\mathrm{curl}}\,\underline{\sigma}, \underline{\varphi}) = -\lambda(\underline{\psi}, \underline{\varphi}) & \forall\,\underline{\varphi} \in \mathcal{F}. \end{cases} \tag{11.2.10}$$

It is well known that problems (11.2.7), (11.2.9), and (11.2.10) are equivalent in the sense of the following Proposition (see [268] and [89]).

Proposition 11.2.1. *The equivalence of problems* (11.2.7), (11.2.9) *and* (11.2.10) *is expressed by the following three statements.*

(a) Let $(\lambda, \underline{u})$ be an eigenmode of problem (11.2.7)*. Then, the following properties hold:*

- *(i) λ is strictly positive;*
- *(ii) There exists $p \in H_0^1(\Omega)$ such that $(\lambda, \underline{u}, p)$ solves problem* (11.2.9)*;*
- *(iii) There exists $\underline{\psi} \in \mathcal{F}$ such that $(\lambda, \underline{\sigma}, \underline{\psi})$ solves* (11.2.10) *with the choice $\underline{\sigma} = \underline{u}$.*

(b) Let $(\lambda, \underline{u}, p)$ be a solution to (11.2.9)*. Then,*

- *(i) λ is strictly positive;*
- *(ii) $(\lambda, \underline{u})$ solves* (11.2.7)*.*

(c) Let $(\lambda, \underline{\sigma}, \underline{\psi})$ be a solution to (11.2.10)*. Then,*

- *(i) λ is strictly positive;*
- *(ii) $(\lambda, \underline{u})$ solves* (11.2.7) *with $\underline{u} = \underline{\sigma}$.* □

Example 11.2.1. When Ω is a two-dimensional domain, interior Maxwell's eigenproblem (11.2.6) can be reduced to a more standard eigenvalue problem for the Laplace equation with Neumann boundary conditions. Indeed, if (ψ, λ) is a solution to

$$\begin{aligned} -\Delta\psi &= \lambda\psi && \text{in } \Omega, \\ \frac{\partial\psi}{\partial\underline{n}} &= 0 && \text{on } \partial\Omega, \end{aligned} \tag{11.2.11}$$

then it is not difficult to check that $\underline{u} = \underline{\mathrm{curl}}\,\psi$ satisfies $\underline{\mathrm{curl}}\,\mathrm{curl}\,\underline{u} = \lambda\underline{u}$ (since $\Delta\psi = \mathrm{curl}\,\underline{\mathrm{curl}}\,\psi$) and that $\underline{u}$ meets the boundary condition $\underline{u} \cdot \underline{t} = 0$; it can actually be shown that all two-dimensional solutions to (11.2.6) coincide with those to (11.2.11) via the identification $\underline{u} = \underline{\text{curl}}\,\psi$.

For instance, if Ω is the square $(0,\pi)^2$, then the interior Maxwell eigenvalues are given by $\lambda_{mn} = m^2 + n^2$ with $m, n = 0, 1, 2, \ldots$ and $m + n > 0$, and the corresponding eigenfunctions are

$$\underline{u}_{mn} = (-n\cos(mx)\sin(ny), m\sin(mx)\cos(ny)).$$

On the other hand, the situation in three dimensions is quite different. Since this is not the main object of this chapter, we do not stress the reader with the three-dimensional details which can be found with the help of [9, 304]. □

11.2.2 Analysis of the Time Harmonic Maxwell System

A weak formulation of problem (11.2.5) can be naturally obtained with the use of the space $H_0(\underline{\text{curl}};\Omega)$ of vector fields $\underline{v}$ satisfying $\underline{\text{curl}}\,\underline{v} \in L^2(\Omega)^3$ and $\underline{v}\wedge\underline{n} = 0$ on $\partial\Omega$. We refer to the beginning of this chapter and to Chap. 2 for more information on this space and its finite element approximation. After multiplying the first equation of (11.2.5) by a test function $\underline{v} \in H_0(\underline{\text{curl}};\Omega)$, integrating over Ω, and integrating by parts taking into account the boundary conditions, we get in a standard way the following variational formulation of our problem: *given* $\underline{f} \in L^2(\Omega)^3$ *with* $\text{div}\,\underline{f} = 0$, *find* $\underline{u} \in H_0(\underline{curl};\Omega)$ *such that*

$$(\underline{\text{curl}}\,\underline{u}, \underline{\text{curl}}\,\underline{v}) - \omega^2(\underline{u},\underline{v}) = (\underline{f},\underline{v}) \qquad \forall\,\underline{v} \in H_0(\underline{\text{curl}};\Omega). \tag{11.2.12}$$

We did not take into account explicitly the divergence-free condition thanks to Remark 11.2.2. However, it is clear that enforcing the divergence- free condition at the discrete level might be a significant source of trouble. We shall actually observe that suitable mixed formulations will help the understanding of the problem under consideration and its numerical approximation.

One of the most common ways of deriving a mixed formulation for problem (11.2.12) is to consider the first (left) part of the de Rham complex (11.1.2). More precisely, the divergence-free condition can be enforced by requiring $\underline{u}$ to be orthogonal to any element of $\underline{\text{grad}}\,H_0^1(\Omega)$. In this framework, the mixed variational formulation is: *given* $\underline{f} \in L^2(\Omega)^3$ *with* $\text{div}\,\underline{f} = 0$, *find* $\underline{u} \in H_0(\underline{curl};\Omega)$ *and* $p \in H_0^1(\Omega)$ *such that*

$$\begin{cases} (\underline{\text{curl}}\,\underline{u}, \underline{\text{curl}}\,\underline{v}) - \omega^2(\underline{u},\underline{v}) + (\underline{\text{grad}}\,p,\underline{v}) = (\underline{f},\underline{v}) & \forall\,\underline{v} \in H_0(\underline{\text{curl}};\Omega), \\ (\underline{\text{grad}}\,q,\underline{u}) = 0 & \forall\,q \in H_0^1(\Omega). \end{cases} \tag{11.2.13}$$

Let us show that problems (11.2.13) and (11.2.12) are equivalent. It is clear that a solution of (11.2.12) is also solution of (11.2.13) with $p = 0$. Vice versa, we remark that it is admissible to take $\underline{v} = \underline{\text{grad}}\, p$ in (11.2.13) since $\underline{\text{grad}}\, H_0^1(\Omega) \subset H_0(\underline{\text{curl}}; \Omega)$. Hence, since $\underline{\text{curl}}\, \underline{\text{grad}}\, p = 0$, $(\underline{\text{grad}}\, p, \underline{u}) = 0$ and $\text{div}\, \underline{f} = 0$, we get $(\underline{\text{grad}}\, p, \underline{\text{grad}}\, p) = 0$ which, together with the boundary conditions, easily implies $p = 0$.

Remark 11.2.4. We have just observed that formulations (11.2.12) and (11.2.13) are equivalent. This is mainly due to the inclusion

$$\underline{\text{grad}}\, H_0^1(\Omega) \subset H_0(\underline{\text{curl}}; \Omega). \tag{11.2.14}$$

□

It is then natural to define the following bilinear forms

$$\begin{aligned} a(\underline{u}, \underline{v}) &: H_0(\underline{\text{curl}}; \Omega) \times H_0(\underline{\text{curl}}; \Omega) \to \mathbb{R} \\ &:= (\underline{\text{curl}}\, \underline{u}, \underline{\text{curl}}\, \underline{v}) - \omega^2(\underline{u}, \underline{v}) \\ b(q, \underline{v}) &: H_0^1(\Omega) \times H_0(\underline{\text{curl}}; \Omega) \to \mathbb{R} \\ &:= (\underline{\text{grad}}\, q, \underline{v}). \end{aligned} \tag{11.2.15}$$

The following theorem gives the conditions for the well-posedness of problem (11.2.12).

Theorem 11.2.1. *Let us assume that ω^2 is not an interior Maxwell eigenvalue (see problem* (11.2.6)*); then, problem* (11.2.12) *has a unique solution which satisfies the a priori estimate*

$$\|\underline{u}\|_{\underline{curl}} \le C \max\left\{1 + \omega^2, \frac{1 + \lambda_i}{|\lambda_i - \omega^2|}, \; i = 1, 2, \cdots\right\} \|\underline{f}\|_0, \tag{11.2.16}$$

where λ_i, $i = 1, 2, \cdots$, are the interior Maxwell eigenvalues.

Proof. The proof might be carried on by means of the Fredholm alternative theorem. We prefer, however, to make use of the mixed formulation (11.2.13) and to prove appropriate *inf-sup* conditions for the bilinear forms introduced in (11.2.15). This approach will prove very useful when considering the discretisation of our problem.

It is clear that the forms a and b (see (11.2.15)) are linear and continuous, the continuity constant of a being $1 + \omega^2$. In order to show existence and uniqueness for problem (11.2.13), let us prove the *inf-sup* conditions for a and b.

The *inf-sup* condition for the bilinear form b is trivial: given $q \in H_0^1(\Omega)$, we can take $\underline{v} = \underline{\text{grad}}\, q \in H_0(\underline{\text{curl}}; \Omega)$ and get

$$\sup_{\underline{v} \in H_0(\underline{\text{curl}}; \Omega)} \frac{b(q, \underline{v})}{\|\underline{v}\|_1} \ge \frac{\|\, \underline{\text{grad}}\, q \,\|_0^2}{\|q\|_1} \ge C \|q\|_1, \tag{11.2.17}$$

where the constant C is related to the Poincaré inequality.

The *inf-sup* condition for the bilinear form a is less immediate. It is clear that if $\omega^2 = 0$, the bilinear form a is elliptic in the kernel K consisting of divergence-free vector fields in $H_0(\underline{\text{curl}};\Omega)$. This fact is a consequence of the Friedrichs inequality (11.1.5) as it has been already remarked in Sect. 11.1.2.

We need to prove that

$$\sup_{\underline{v}\in K} \frac{a(\underline{u},\underline{v})}{\|\underline{v}\|_{\underline{\text{curl}}}} \geq C\,\|\underline{u}\|_{\underline{\text{curl}}} \quad \forall\, \underline{u} \in K. \tag{11.2.18}$$

We observe that any element of K can be presented as a Fourier series in terms of interior Maxwell's eigenfunctions. More precisely, let $\{\underline{u}_i\}$ and $\{\lambda\}_i$, $i = 1,\dots$ be the set of interior Maxwell's eigenmodes, such that

$$\begin{aligned} &(\underline{\text{curl}}\,\underline{u}_i, \underline{\text{curl}}\,\underline{v}) = \lambda_i(\underline{u}_i,\underline{v}) \qquad \forall\,\underline{v}\in H_0(\underline{\text{curl}};\Omega),\\ &\operatorname{div}\underline{u}_i = 0,\\ &\|\underline{u}_i\|_0 = 1,\\ &(\underline{u}_i,\underline{u}_j) = (\underline{\text{curl}}\,\underline{u}_i, \underline{\text{curl}}\,\underline{u}_j) = 0 \quad \text{for } i\neq j. \end{aligned} \tag{11.2.19}$$

Since $K = \text{span}\{\underline{u}_i\}$, we can write generic elements $\underline{u}$ and $\underline{v}$ in K as

$$\underline{u} = \sum_{i=1}^{\infty} a_i\underline{u}_i, \qquad \underline{v} = \sum_{i=1}^{\infty} b_i\underline{u}_i. \tag{11.2.20}$$

We also observe explicitly that eigenfunctions $\underline{u}_i$ satisfy the Friedrichs equality

$$\|\,\underline{\text{curl}}\,\underline{u}_i\|_0^2 = \frac{\lambda_i}{1+\lambda_i}\|\underline{u}_i\|_{\underline{\text{curl}}}^2. \tag{11.2.21}$$

Given $\underline{u} \subset K$ as in (11.2.20), let us construct $\underline{v}\in K$ in order to prove (11.2.18). Since ω^2 does not coincide with λ_i for any i, we have that

$$(a_i\,\underline{\text{curl}}\,\underline{u}_i, b_i\,\underline{\text{curl}}\,\underline{u}_i) - \omega^2(a_i\underline{u}_i, b_i\underline{u}_i) \neq 0; \tag{11.2.22}$$

we define $b_i = \pm a_i$ with the signs suitably chosen such that the expressions in (11.2.22) are positive for every i. In particular, it turns out that

$$\|\underline{v}\|_{\underline{\text{curl}}} = \|\underline{u}\|_{\underline{\text{curl}}}. \tag{11.2.23}$$

We then have

$$
\begin{aligned}
a(\underline{u},\underline{v}) &= \Big(\sum_i a_i\,\underline{\text{curl}}\,\underline{u}_i, \sum_j b_j\,\underline{\text{curl}}\,\underline{u}_j\Big) - \omega^2\Big(\sum_i a_i\underline{u}_i, \sum_j b_j\underline{u}_j\Big)\\
&= \sum_i a_i b_i(\underline{\text{curl}}\,\underline{u}_i, \underline{\text{curl}}\,\underline{u}_i) - \omega^2\sum_i a_i b_i(\underline{u}_i,\underline{u}_i)\\
&= \sum_i a_i b_i(\underline{\text{curl}}\,\underline{u}_i, \underline{\text{curl}}\,\underline{u}_i) - \omega^2\sum_i \frac{a_i b_i}{\lambda_i}(\underline{\text{curl}}\,\underline{u}_i, \underline{\text{curl}}\,\underline{u}_i)\\
&= \sum_i a_i^2(\underline{\text{curl}}\,\underline{u}_i, \underline{\text{curl}}\,\underline{u}_i)\left|1-\frac{\omega^2}{\lambda_i}\right|\\
&= \sum_i a_i^2\frac{\lambda_i}{1+\lambda_i}\|\underline{u}_i\|^2_{\underline{\text{curl}}}\left|1-\frac{\omega^2}{\lambda_i}\right|\\
&\geq \min_i\frac{|\lambda_i-\omega^2|}{1+\lambda_i}\|\underline{u}\|^2_{\underline{\text{curl}}}.
\end{aligned}
\tag{11.2.24}
$$

Combining (11.2.23) and (11.2.24) gives the *inf-sup* condition for the bilinear form a (11.2.18) with a constant equal to

$$
C = \min_i\frac{|\lambda_i-\omega^2|}{1+\lambda_i}. \tag{11.2.25}
$$

Putting together this estimate for the *inf-sup* constant and the continuity constant of $a(\cdot,\cdot)$ gives the a priori estimate (11.2.16). □

11.2.3 Approximation of the Time Harmonic Maxwell Equations

The finite element approximation of problem (11.2.12) is done as usual by considering a finite dimensional subspace $V_h \subset H_0(\underline{\text{curl}};\Omega)$ and by looking for $\underline{u}_h \in V_h$ such that

$$
(\underline{\text{curl}}\,\underline{u}_h, \underline{\text{curl}}\,\underline{v}) - \omega^2(\underline{u}_h,\underline{v}) = (\underline{f},\underline{v}) \qquad \forall\,\underline{v}\in V_h. \tag{11.2.26}
$$

From the discussion of the previous section, it should be clear that a discrete counterpart of the mixed formulation (11.2.13) will play an important role for the analysis of (11.2.26). Let us consider a finite dimensional approximation of $H_0^1(\Omega)$, that is, $Q_h \subset H_0^1(\Omega)$. We can consider the discretisation of (11.2.13): *find* $(\underline{u}_h, p_h) \in V_h \times Q_h$ *such that*

$$\begin{cases} (\underline{\mathrm{curl}}\,\underline{u}_h, \underline{\mathrm{curl}}\,\underline{v}) - \omega^2(\underline{u}_h, \underline{v}) + (\underline{\mathrm{grad}}\,p_h, \underline{v}) = (\underline{f}, \underline{v}) & \forall\,\underline{v} \in V_h, \\ (\underline{\mathrm{grad}}\,q, \underline{u}_h) = 0 & \forall\,q \in Q_h. \end{cases} \tag{11.2.27}$$

The following proposition states that (11.2.26) and (11.2.27) are equivalent, provided that the inclusion

$$\underline{\mathrm{grad}}\,Q_h \subset V_h \tag{11.2.28}$$

is satisfied (see Remark 11.2.4).

Proposition 11.2.2. *Let $\underline{u}_1 \in V_h$ be a solution to* (11.2.26) *and $(\underline{u}_2, p_h) \in V_h \times Q_h$ be a solution to* (11.2.27) *with the same data ω and $\underline{f}$ with $\mathrm{div}\,\underline{f} = 0$. Let us assume that the inclusion* (11.2.28) *is satisfied. Then, $p_h = 0$ and $\underline{u}_1 = \underline{u}_2$.*

Proof. Taking $\underline{v} = \underline{\mathrm{grad}}\,q$ in (11.2.26), it is clear that we get the second equation of (11.2.27).

It is then enough to show that in (11.2.27) we have $p_h = 0$. Since $\underline{\mathrm{grad}}\,Q_h \subset V_h$, we can take $v = \underline{\mathrm{grad}}\,p_h$ in the first equation of (11.2.27) and, due to $\underline{\mathrm{curl}}\,\underline{\mathrm{grad}}\,p_h = 0$ and $(\underline{u}_2, \underline{\mathrm{grad}}\,p_h) = 0$, we get $(\underline{\mathrm{grad}}\,p_h, \underline{\mathrm{grad}}\,p_h) = (\underline{f}, \underline{\mathrm{grad}}\,p_h) = 0$. The boundary conditions on p_h easily imply $p_h = 0$. □

Remark 11.2.5. It is worth spending a few words about the inclusion (11.2.28). It is clear that (11.2.28) might be satisfied by taking $Q_h = \{0\}$ but this would not be a significant situation. On the other hand, it has been shown in Chap. 2 that edge finite elements satisfy the inclusion with Q_h chosen to be made of standard Lagrangian finite elements. This is part of the de Rham complex which is associated with edge finite elements (see (2.3.64)). In general, a maximal definition of Q_h can be made according to the following argument. Let V_0 be the kernel of the $\underline{\mathrm{curl}}$ operator in the space V_h. Then, any function $\underline{v}$ in V_0 can be written as $\underline{\mathrm{grad}}\,q$ for a suitable q which is uniquely determined up to an additive constant. By virtue of the boundary conditions on $\underline{v}$, the function q has a constant value on the boundary $\partial\Omega$ and we can choose the constant in such a way that q belongs to $H_0^1(\Omega)$. The space Q_h can then be defined as the set of all such q's. □

In order to find the discrete version of Theorem 11.2.1 which will provide us with an a priori error bound for our approximation, we have to consider the discrete interior Maxwell eigenvalues $\lambda_{i,h}$, $i = 1, \cdots, N_h$, which satisfy

$$\underline{u}_h \in V_h, \quad (\underline{\mathrm{curl}}\,\underline{u}_h, \underline{\mathrm{curl}}\,\underline{v}) = \lambda_h(\underline{u}_h, \underline{v}) \qquad \forall\,\underline{v} \in V_h. \tag{11.2.29}$$

Theorem 11.2.2. *Let us assume that ω^2 is not an interior Maxwell eigenvalue (see* (11.2.7)*) and let $\underline{u}$ be the unique solution to* (11.2.12)*. Let us assume, moreover, that ω^2 and h are such that ω^2 does not coincide with any discrete interior Maxwell's eigenvalues $\lambda_{i,h}$, $i = 1, \ldots, N_h$ (see* (11.2.29)*). Then, problem* (11.2.26) *has a unique solution $\underline{u}_h$ which satisfies the error estimate*

$$\|\underline{u} - \underline{u}_h\|_{\underline{curl}} \le C \max\left\{1 + \omega^2, \frac{1 + \lambda_{i,h}}{|\lambda_{i,h} - \omega^2|}, \ i = 1, 2, \dots\right\} \inf_{\underline{v} \in V_h} \|\underline{u} - \underline{v}\|_{\underline{curl}}. \tag{11.2.30}$$

Proof. The proof is analoguous to the one of Theorem 11.2.1. In particular, we shall show the well-posedness of the mixed problem (11.2.27). This consists in showing the discrete *inf-sup* conditions for the bilinear forms a and b introduced in (11.2.15).

Let us choose Q_h such that inclusion (11.2.28) is satisfied. As observed in Remark 11.2.5, this can be done for any choice of V_h.

The *inf-sup* condition for the bilinear form b is immediate (given $q \in Q_h$, take $\underline{v} = \underline{\text{grad}}\, q \in V_h$).

The *inf-sup* condition for the bilinear form a can be proved as in the continuous case by considering a basis of V_h consisting of discrete Maxwell eigenfunctions $\underline{u}_{i,h}$ (see (11.2.29)). The discrete Friedrichs equality in this case reads

$$\| \underline{\text{curl}}\, \underline{u}_{i,h} \|_0^2 = \frac{\lambda_{i,h}}{1 + \lambda_{i,h}} \|\underline{u}_{i,h}\|^2_{\underline{\text{curl}}}. \tag{11.2.31}$$

Arguing as in the proof of Theorem 11.2.1, we get the *inf-sup* constant for the bilinear a

$$C = \min_i \frac{|\lambda_{i,h} - \omega^2|}{1 + \lambda_{i,h}}. \tag{11.2.32}$$

The final estimate (11.2.30) is then obtained by observing that we do not need to estimate the terms $\|p - p_h\|_1$ and $\inf_{q \in Q_h} \|p - q\|_1$ since both p and p_h are zero. □

Remark 11.2.6. The conclusion of Theorem 11.2.2 can be summarised by saying that any choice of V_h which guarantees a good discretisation of $H_0(\underline{\text{curl}}; \Omega)$ provides an optimally convergent scheme for the approximation of the time harmonic equation (11.2.12) under the condition that the distance between the discrete spectrum of Maxwell's eigenproblem and ω^2 remains bounded away from zero. Condition

$$\min_i |\lambda_{i,h} - \omega^2| \ge k > 0 \tag{11.2.33}$$

is achieved, in particular, if V_h provides a good discretisation of Maxwell's eigenvalues. This is the statement of the following corollary □

Corollary 11.2.1. *If V_h is such that the eigenvalues of Maxwell's cavity problem are correctly approximated (see Sect. 11.2.1), then a scheme based on* (11.2.26) *is solvable for h small enough and provides a convergent approximation to the time harmonic Maxwell system* (11.2.12)*. Moreover, the error estimate*

$$\|\underline{u} - \underline{u}_h\|_{\underline{curl}} \le C \inf_{\underline{v} \in V_h} \|\underline{u} - \underline{v}\|_{\underline{curl}}$$

holds true with a constant C which can be computed as in (11.2.26)*.*

Remark 11.2.7. According to the above discussion (see in particular Proposition 11.2.2), it is clear that the solution of (11.2.26) is equivalent to the one of the mixed formulation (11.2.27) in *exact arithmetic*. For this reason, it is recommendable to use the standard formulation (11.2.26) which turns out to be less expensive and easier to solve. It has been however reported that in some cases, one gets more stable results when performing the computation using the mixed formulation (11.2.27). This is particularly visible for very small frequencies (see [371]) or for curvilinear geometries. Computing with the mixed formulation can also be a good debugging tool, since if the Lagrange multiplier is not a machine zero, then this is a good indication that something is wrong. □

11.3 Approximation of the Maxwell Eigenvalue Problem

Remark 11.2.6 raises the question of whether a finite element space sequence $\{V_h\}$ provides a good approximation of interior Maxwell's eigenvalues or not. We recall that $\lambda \in \mathbb{R}$ is an interior Maxwell's eigenvalue and $\underline{u}$ a corresponding eigenfunction if $\underline{u} \not\equiv 0$ and

$$\begin{aligned} \underline{\mathrm{curl}}\,\underline{\mathrm{curl}}\,\underline{u} &= \lambda \underline{u} && \text{in } \Omega, \\ \mathrm{div}\,\underline{u} &= 0 && \text{in } \Omega, \\ \underline{u} \wedge \underline{n} &= 0 && \text{on } \partial\Omega. \end{aligned}$$

This topic has been intensively discussed in the mathematical and engineering literature. The analysis of the two-dimensional case is more standard and can be carried with the help of a mixed formulation which is related to the Neumann problem for Laplace eigenproblem. We refer to [192] for the required estimate and to Sect. 11.3.1 for the details of the two-dimensional analysis. On the other hand, in three space dimensions, the situation is more complicated and has been studied by several authors. We refer, among others, to the pioneer works by Bossavit [102–104] and to the first analysis attempt by Kikuchi [269], where it has been proved that lowest order tetrahedral edge elements satisfy a discrete compactness property. The first rigorous analysis of the three-dimensional case appeared in [89] and it is based on a suitable projection operator which has been constructed in [74]. A proof based on the discrete compactness property can be found in [303] and in [137] where the situation of non-constant coefficients (allowing for heterogeneous and different materials) has been considered. A comprehensive analysis can be found in [76] and we refer the interested reader to [248, 302] and [77] for review contributions.

Here we are adopting the presentation of [76]. We recall that, in Sect. 11.2.1, we introduced a variational formulation of our problem and two equivalent mixed formulations (see, in particular, Proposition 11.2.1). For the reader's convenience, we recall here the three equivalent variational formulations. The first *unconstrained* formulation is: *find $\lambda \in \mathbb{R}$ such that for a non-vanishing $\underline{u} \in H_0(\underline{curl}; \Omega)$ there holds*

$$\begin{cases} (\underline{\mathrm{curl}}\,\underline{u}, \underline{\mathrm{curl}}\,\underline{v}) = \lambda(\underline{u}, \underline{v}) & \forall\, \underline{v} \in H_0(\underline{\mathrm{curl}}; \Omega), \\ \lambda \neq 0. \end{cases} \tag{11.3.1}$$

Notice that the divergence-free constraint has been replaced by the requirement $\lambda \neq 0$ (see also Remark 11.2.3).

The Kikuchi formulation is: *find* $\lambda \in \mathbb{R}$ *such that for* $\underline{u} \in H_0(curl; \Omega)$ *with* $\underline{u} \neq 0$ *there exists* $p \in H_0^1(\Omega)$ *satisfying*

$$\begin{cases} (\underline{\mathrm{curl}}\,\underline{u}, \underline{\mathrm{curl}}\,\underline{v}) + (\underline{\mathrm{grad}}\,p, \underline{v}) = \lambda(\underline{u}, \underline{v}) & \forall\, \underline{v} \in H_0(\mathrm{curl}; \Omega), \\ (\underline{\mathrm{grad}}\,q, \underline{u}) = 0 & \forall\, q \in H_0^1(\Omega). \end{cases} \tag{11.3.2}$$

Finally, the second mixed formulation reads: *find* $\lambda \in \mathbb{R}$ *such that for* $\underline{\psi} \in \mathcal{F}$ *with* $\underline{\psi} \neq 0$ *there exists* $\underline{\sigma} \in H_0(curl; \Omega)$ *satisfying*

$$\begin{cases} (\underline{\sigma}, \underline{\tau}) + (\underline{\mathrm{curl}}\,\underline{\tau}, \underline{\psi}) = 0 & \forall\, \underline{\tau} \in H_0(\mathrm{curl}; \Omega), \\ (\underline{\mathrm{curl}}\,\underline{\sigma}, \underline{\varphi}) = -\lambda(\underline{\psi}, \underline{\varphi}) & \forall\, \underline{\varphi} \in \mathcal{F}. \end{cases} \tag{11.3.3}$$

Let $V_h \subset H_0(\mathrm{curl}; \Omega)$, $Q_h \subset H_0^1(\Omega)$, and $\mathcal{F}_h \subset \mathcal{F}$ be finite element spaces.

The approximation of (11.3.1) consists in looking for eigenvalues λ_h and eigenfunctions $\underline{u}_h \in V_h$ such that $\underline{u}_h \neq 0$ and

$$\begin{cases} (\underline{\mathrm{curl}}\,\underline{u}_h, \underline{\mathrm{curl}}\,\underline{v}) = \lambda_h(\underline{u}_h, \underline{v}) & \forall\, \underline{v} \in V_h, \\ \lambda_h \neq 0. \end{cases} \tag{11.3.4}$$

Notice again that the divergence-free constraint has been replaced by the requirement $\lambda_h \neq 0$ which is common practice in the numerical approximation of (11.2.7).

The approximation of Kikuchi's formulation (11.3.2) consists in looking for eigenvalues λ_h and eigenfunctions $\underline{u}_h \in V_h$ with $\underline{u}_h \neq 0$ such that there exists $p_h \in Q_h$ satisfying

$$\begin{cases} (\underline{\mathrm{curl}}\,\underline{u}_h, \underline{\mathrm{curl}}\,\underline{v}) + (\underline{\mathrm{grad}}\,p_h, \underline{v}) = \lambda_h(\underline{u}_h, \underline{v}) & \forall\, \underline{v} \in V_h, \\ (\underline{\mathrm{grad}}\,q, \underline{u}_h) = 0 & \forall\, q \in Q_h. \end{cases} \tag{11.3.5}$$

Finally, the approximation of problem (11.3.3) consists in looking for eigenvalues λ_h and eigenfunctions $\underline{\psi}_h \in \mathcal{F}_h$ with $\underline{\psi}_h \neq 0$ such that there exists $\underline{\sigma}_h \in V_h$ satisfying

$$\begin{cases} (\underline{\sigma}_h, \underline{\tau}) + (\underline{\mathrm{curl}}\,\underline{\tau}, \underline{\psi}_h) = 0 & \forall\, \underline{\tau} \in V_h, \\ (\underline{\mathrm{curl}}\,\underline{\sigma}_h, \underline{\varphi}) = -\lambda_h(\underline{\psi}_h, \underline{\varphi}) & \forall\, \underline{\varphi} \in \mathcal{F}_h. \end{cases} \tag{11.3.6}$$

In order to state the discrete analogue of Proposition 11.2.1, we need some compatibility assumptions on the discrete spaces Q_h, V_h, and $\mathcal{F}_h$. We shall assume that the following inclusions hold

$$\underline{\text{grad}}\, Q_h \subset V_h, \quad \underline{\text{curl}}\, V_h \subset \mathcal{F}_h. \tag{11.3.7}$$

Remark 11.3.1. Inclusion (11.3.7) extends (11.2.28). In general, a numerical scheme for the approximation of (11.2.6) is given by a suitable definition of V_h. The spaces Q_h and/or $\mathcal{F}_h$ are then constructed in order to consider the mixed formulations (11.3.5) and/or (11.3.6). In Remark (11.2.5), we already described a possible non-trivial construction of Q_h, while $\mathcal{F}_h$ can simply be defined as $\underline{\text{curl}}\, V_h$. □

The following proposition is the discrete analogue of Proposition 11.2.1.

Proposition 11.3.1. *Let us assume that property* (11.3.7) *holds true. Then, the equivalence of problems* (11.3.4), (11.3.5) *and* (11.3.6) *is expressed by the following three statements.*

(a) Let $(\lambda_h, \underline{u}_h)$ *be an eigenmode of problem* (11.3.4)*. Then, the following properties hold:*

- *(i)* λ_h *is strictly positive;*
- *(ii) There exists* $p_h \in Q_h$ *such that* $(\lambda_h, \underline{u}_h, p_h)$ *solves problem* (11.3.5)*;*
- *(iii) There exists* $\underline{\psi}_h \in \mathcal{F}_h$ *such that* $(\lambda_h, \underline{\sigma}_h, \underline{\psi}_h)$ *solves* (11.3.6) *with the choice* $\underline{\sigma}_h = \underline{u}_h$.

(b) Let $(\lambda_h, \underline{u}_h, p_h)$ *be a solution to* (11.3.5)*. Then,*

- *(i)* λ_h *is strictly positive;*
- *(ii)* $(\lambda_h, \underline{u}_h)$ *solves* (11.3.4)*.*

(c) Let $(\lambda_h, \underline{\sigma}_h, \underline{\psi}_h)$ *be a solution to* (11.3.6)*. Then,*

- *(i)* λ_h *is strictly positive;*
- *(ii)* $(\lambda_h, \underline{u}_h)$ *solves* (11.3.4) *for* $\underline{u}_h = \underline{\sigma}_h$. □

11.3.1 Analysis of the Two-Dimensional Case

For the study of the two-dimensional case, we recall that Maxwell's eigenvalue problem can be identified to the more standard Laplace eigenproblem with Neumann boundary conditions (11.2.11) (see Example 11.2.1). Moreover, in this particular situation, the second mixed formulation 11.3.6 reads as follows: *find* $\lambda \in \mathbb{R}$ *such that for* $\psi \in L^2_0(\Omega)$ *with* $\psi \neq 0$ *there exists* $\underline{\sigma} \in H_0(\text{curl}; \Omega)$ *satisfying*

$$\begin{cases} (\underline{\sigma}, \underline{\tau}) + (\operatorname{curl} \underline{\tau}, \psi) = 0 & \forall\, \underline{\tau} \in H_0(\operatorname{curl}; \Omega), \\ (\operatorname{curl} \underline{\sigma}, \varphi) = -\lambda(\psi, \varphi) & \forall\, \varphi \in L^2_0(\Omega). \end{cases} \tag{11.3.8}$$

It is clear that the only formal difference between this formulation and the mixed approximation of the Laplace eigenvalue problem with Neumann boundary conditions is the presence of the curl operator instead of the div one. On the other hand, in two dimensions, we already observed that curl and div operators are isomorphic, and since $\operatorname{curl} \underline{\operatorname{curl}}\, \psi = \operatorname{div} \underline{\operatorname{grad}}\, \psi = \Delta \psi$, we have that (11.3.8) provides indeed the same eigenvalues as the standard mixed approximation of the Laplace eigenproblem. Moreover, the component ψ of the solution is the same as the eigenfunction of the standard Laplace problem, while the component $\underline{\sigma}$ is rotated by an angle $\pi/2$ (it is an approximation of $\underline{\operatorname{curl}}\, \psi$ instead of $\underline{\operatorname{grad}}\, \psi$). This statement is summarised in the next proposition.

Proposition 11.3.2. *When $\Omega \subset \mathbb{R}^2$, the eigensolutions $(\lambda, \underline{u})$ of problem* (11.3.1) *are related to the eigensolutions (μ, ψ) of the following standard Laplace problem*

$$\begin{cases} (\underline{\sigma}, \underline{\tau}) + (\operatorname{div} \underline{\tau}, \psi) = 0 & \forall\, \underline{\tau} \in H_0(\operatorname{div}; \Omega), \\ (\operatorname{div} \underline{\sigma}, \varphi) = -\mu(\psi, \varphi) & \forall\, \varphi \in L^2_0(\Omega), \end{cases}$$

with the identification $\lambda = \mu$ and $\underline{u} = \underline{\operatorname{curl}}\, \psi = \underline{\sigma}^{\perp}$.

Moreover, the discrete eigensolutions $(\lambda_h, \underline{u}_h) \in \mathbb{R} \times V_h$ of problem (11.3.4) *are related to the eigensolutions $(\mu_h, \psi_h) \in \mathbb{R} \times \mathcal{F}$ of the following mixed problem*

$$\begin{cases} (\underline{\sigma}_h, \underline{\tau}) + (\operatorname{div} \underline{\tau}, \psi_h) = 0 & \forall\, \underline{\tau} \in V_h^{\perp} \\ (\operatorname{div} \underline{\sigma}_h, \varphi) = -\mu_h(\psi_h, \varphi) & \forall\, \varphi \in \mathcal{F} \end{cases}$$

with the identification $\lambda_h = \mu_h$, $\underline{u}_h = \underline{\sigma}_h^{\perp}$, and where $\mathcal{F} = \operatorname{div} V_h^{\perp} = \operatorname{curl} V_h$. □

We now come to the choice of finite element space V_h. As already discussed in Chap. 2, if we take as $V_h^{\perp}$ any good approximation of $H_0(\operatorname{div}; \Omega)$, then V_h will provide a good approximation of $H_0(\operatorname{curl}; \Omega)$. Indeed, the next comments apply to any choice of $V_h^{\perp}$ equal to $\mathcal{RT}$, $\mathcal{BDM}$, or $\mathcal{BDFM}$.

The error analysis can be performed by the standard tools already presented in Sect. 1.2.1 and is essentially a consequence of the error estimates of [192]. In particular, according to the theory presented in Sect. 1.2.1 for the eigenvalue problems of the form $(0, g)$, the key condition to be checked is the B-Id-compatibility condition of Definition 6.5.6, we thus need to construct a B-compatible operator satisfying (6.5.56), that is, which converges to the identity in norm. More precisely, we need to construct $\Pi_h : H^1(\Omega)^2 \to V_h^{\perp}$ such that

$$(\operatorname{div}(\underline{\sigma} - \Pi_h \underline{\sigma}), \varphi) = 0 \quad \forall\, \underline{\sigma} \in H^1(\Omega)^2,\ \forall\, \varphi \in \mathcal{F} \tag{11.3.9}$$

and

$$\|\underline{\sigma} - \Pi_h \underline{\sigma}\|_0 \leq \omega(h) \|\underline{\sigma}\|_1, \tag{11.3.10}$$

with $\omega(h)$ tending to zero as h tends to zero.

It is immediate to check that the standard interpolation operator in V_h satisfies (11.3.9), so that (11.3.10) follows by standard interpolation estimates.

The use of Proposition 11.3.1, and in particular of problem (11.3.8), for the analysis of two-dimensional Maxwell's eigenproblem, does not follow the same path when considering quadrilateral finite elements. In this case, we already observed that $\mathcal{RT}$, $\mathcal{BDM}$, or $\mathcal{BDFM}$ do not achieve optimal approximation orders on general quadrilateral meshes. It is clear that poor approximation properties will influence the eigenmode convergence. In particular, lowest order Raviart–Thomas elements do not achieve convergence at all. We refer the interested reader to [64,65].

A possible cure to this bad phenomenon is to use the $\mathcal{ABF}$ element introduced in [21]. In this case, however, problem (11.3.8) does not help for the analysis since it is not immediate to characterise the space $\mathcal{F} = \mathrm{curl}\, V_h$ as a standard finite element space. A careful convergence analysis has been however performed in [216] where it has been shown that the $\mathcal{ABF}$ element is optimally convergent for the approximation of the eigenvalue problem we are interested in.

Another possible cure for the bad behaviour of Raviart–Thomas spaces on general quadrilateral meshes is to use the following reduced integration technique presented in [94]: *find* $\lambda_h \in \mathbb{R}$ *such that for a non-vanishing* $\underline{u}_h \in V_h$ *it holds*

$$\begin{cases} (P\, \underline{\mathrm{curl}}\, \underline{u}_h, P\, \underline{\mathrm{curl}}\, \underline{v}) = \lambda_h (\underline{u}_h, \underline{v}) \quad \forall\, \underline{v} \in V_h, \\ \lambda_h \neq 0, \end{cases} \tag{11.3.11}$$

where V_h is the rotated $\mathcal{RT}$ space of order k and P denotes the L^2 projection onto $\mathcal{L}^1_{[k]}$ (the lowest order is $k = 0$, so that in this case, P turns out to be the L^2 projection onto piecewise constants). In [94], it is shown that the scheme (11.3.11) is optimally convergent and that the projection P can be actually evaluated with the help of a suitable quadrature rule (in the lowest order case, this is the standard midpoint rule). In particular, scheme (11.3.11) turns out to be cheaper with respect to the standard one (which in this case is only suboptimally convergent) and to the scheme based on $\mathcal{ABF}$ element (which is optimally convergent but uses more degrees of freedom).

Another strategy for modifying standard $\mathcal{RT}$–based spaces is presented in [348] and analysed in [93] within the framework of mimetic finite differences. It consists in modifying the standard shape functions of $\mathcal{RT}$ elements by adding an interior bubble which depends on the distortion of the quadrilateral element (no modification on parallelograms).

Remark 11.3.2. The two-dimensional Maxwell eigenvalue problem (together with models related to it) has been considered by many authors; among others, we recall [66], where edge elements are used for the computation of dielectric waveguides, and [169], where a discrete compactness property is proved for two-dimensional edge elements. □

Remark 11.3.3. The extension to three dimensions of the results of the present section is pretty much straightforward for problems involving $H(\mathrm{div};\Omega)$ and has been the object of an intensive research, in particular, for fluid-structure interaction problems [63, 67]. Indeed, also in three space dimensions, the interpolation operator for standard $H(\mathrm{div};\Omega)$ approximations satisfies the B-compatibility condition (11.3.9).

On the other hand, the situation is not so immediate for problems involving $H(\underline{\mathrm{curl}};\Omega)$ in three dimensions. In particular, three-dimensional Maxwell's eigenvalue problem cannot be analysed with the tools of this section, since the interpolation operator for $H(\underline{\mathrm{curl}};\Omega)$ approximations is not a B-compatible operator. □

For the reasons explained above, in the next sections, we will develop a more general theory for the analysis of Maxwell's eigenvalue problem.

11.3.2 Discrete Compactness Property

We have already used the name *discrete compactness*; it has been considered for the approximation of interior Maxwell's eigenvalues for the first time by Kikuchi in [269]. In this section, we give a formal definition of this concept and show how it relates to the variational formulations of the problem we are studying (see also [77, Part 4]).

Definition 11.3.1. We say that the space sequences $\{V_h\}$ and $\{Q_h\}$ satisfy the *Discrete Compactness Property* if any sequence $\{\underline{u}_n\}$ uniformly bounded in $H_0(\mathrm{curl};\Omega)$, with $\underline{u}_n \in V_{h_n}$, such that

$$(\underline{u}_n, \underline{\mathrm{grad}}\, q_n) = 0 \qquad \forall\, q_n \in Q_{h_n},\ \forall\, n, \tag{11.3.12}$$

contains a subsequence strongly converging in $L^2(\Omega)^3$ to a limit $\underline{u}$.

Remark 11.3.4. The definition of discrete compactness is often found in the literature without referring to the arbitrary index choice h_n. This is needed to avoid abstract situations occurring in cases such as when the family $\{V_h\}$ comprises *good* spaces interspersed with an infinite number of *bad* spaces. Without extracting the first arbitrary subsequence associated with h_n, the negative effect of the *bad* spaces might be annihilated by a suitable subsequence choice.

Remark 11.3.5. It is not difficult to see that the limit $\underline{u}$ is actually in $H_0(\mathrm{curl};\Omega)$ (with norm not exceeding the constant that bounds the whole sequence in $H_0(\mathrm{curl};\Omega)$) and that $\mathrm{div}\,\underline{u} = 0$ if the space $\cup_h Q_h$ is dense in $H_0^1(\Omega)$. □

As it should be clear from its name, the DCP mimics, at the discrete level, the compact embedding of $H_0(\mathrm{curl};\Omega)\cap H(\mathrm{div}^0;\Omega)$ into $L^2(\Omega)^3$. The divergence-free constraint is replaced by condition (11.3.12), which is often called the *discrete* divergence-free constraint. We would like to emphasise that the use of a weak divergence condition, like in (11.3.12), is required since functions in V_h are in $H_0(\mathrm{curl};\Omega)$ but need not be (and generally are not) in $H(\mathrm{div};\Omega)$.

The DCP for Maxwell's eigenproblem is a particular situation of a more general picture. We refer the interested reader to the pioneer works by Stummel [359] and Vaĭnikko [373]. Some essential results are summarised in the book of Chatelin [145]. In particular, the *discrete-compact* convergence defined on page 268 of [145] (Sect. 2 of the Appendix) reads exactly the same as Definition 11.3.1 in our setting.

According to Remark 11.3.5, we shall call *Strong Discrete Compactness Property* (SDCP) the following property.

Definition 11.3.2. We say that the space sequences $\{V_h\}$ and $\{Q_h\}$ satisfy the *Strong Discrete Compactness Property* if they satisfy the DCP (see Definition 11.3.1) and the limit $\underline{u}$ has the property $\mathrm{div}\,\underline{u}=0$.

In [76], it has been proved that there is a strong connection between DCP, SDCP and the natural conditions for the approximation of eigenvalues in mixed form that have been reported in Sect. 1.2.1. We recall here the main results; the interested reader can find the proofs in [76].

First of all, let us write the main definitions (already given throughout this book and in particular in Sect. 1.2.1) in the specific case of Maxwell's eigenvalue problem.

The continuous and discrete kernels of the divergence operator are

$$\mathsf{K}^d := \{\underline{v}\mid \underline{v}\in H_0(\mathrm{curl};\Omega),\ \mathrm{div}\,\underline{v}=0 \text{ in } \Omega\}, \tag{11.3.13}$$

$$\mathsf{K}^d_h := \{\underline{v}_h\mid \underline{v}_h\in V_h,\ (\underline{v}_h,\underline{\mathrm{grad}}\,q_h)=0\ \forall\, q_h\in Q_h\}. \tag{11.3.14}$$

Definition 11.3.3. The *ellipticity in the discrete kernel* is satisfied if there exists a positive constant α, independent of h, such that

$$(\underline{\mathrm{curl}}\,\underline{v}_h,\underline{\mathrm{curl}}\,\underline{v}_h)\geq\alpha\|\underline{v}_h\|^2_{L^2}\quad\forall\,\underline{v}_h\in\mathsf{K}^d_h. \tag{11.3.15}$$

Definition 11.3.4. The *weak approximability* of Q is satisfied if there exists $\omega_1(h)$, tending to zero as h goes to zero, such that for every $p\in H^1_0(\Omega)$,

$$\sup_{\underline{v}_h\in\mathsf{K}^d_h}\frac{(\underline{v}_h,\underline{\mathrm{grad}}\,p)}{\|\underline{v}_h\|_{\underline{\mathrm{curl}}}}\leq\omega_1(h)\|p\|_{H^1}. \tag{11.3.16}$$

Let now V_0 be the subspace of $H_0(\mathrm{curl};\Omega)$ containing the solutions $\underline{u}$ to the problem

$$\begin{cases} (\underline{\text{curl}}\,\underline{u}, \underline{\text{curl}}\,\underline{v}) + (\underline{\text{grad}}\,p, \underline{v}) = (\underline{f}, \underline{v}) & \forall\, \underline{v} \in H_0(\text{curl}; \Omega), \\ (\underline{\text{grad}}\,q, \underline{u}) = 0 & \forall\, q \in H_0^1(\Omega), \end{cases} \tag{11.3.17}$$

for all possible $\underline{f} \in L^2(\Omega)^3$. We have $V_0 \subset \mathsf{K}^d$ and, due to the hypotheses on the domain Ω, functions $\underline{v}$ in V_0 satisfy $\underline{v} \in H^{1/2}(\Omega)^3$ and $\underline{\text{curl}}\,\underline{v} \in H^{1/2}(\Omega)^3$. The space V_0 will be endowed with its natural norm.

Definition 11.3.5. The *strong approximability* of V_0 is satisfied if there exists $\omega_2(h)$, tending to zero as h goes to zero, such that for every $\underline{u} \in V_0$, there exists $\underline{u}^I \in \mathsf{K}_h^d$ such that

$$\|\underline{u} - \underline{u}^I\|_{\underline{\text{curl}}} \le \omega_2(h)\|\underline{u}\|_{V_0}. \tag{11.3.18}$$

The continuous and discrete kernels of the $\underline{\text{curl}}$ operator are

$$\mathsf{K}^c := \{\underline{\tau} \mid \underline{\tau} \in H_0(\text{curl}; \Omega), \underline{\text{curl}}\,\underline{\tau} = 0 \text{ in } \Omega\}, \tag{11.3.19}$$

$$\mathsf{K}_h^c := \{\underline{\tau}_h \mid \underline{\tau}_h \in V_h, (\underline{\text{curl}}\,\underline{\tau}_h, \underline{\varphi}_h) = 0\ \forall\, \underline{\varphi} \in \mathcal{F}_h\}. \tag{11.3.20}$$

Let now V^0 be the subspace of $H_0(\text{curl}; \Omega)$ and $\mathcal{F}^0$ the subspace of $\mathcal{F}$ containing the solutions $\underline{\sigma}$ and $\underline{\psi}$ to the source problem

$$\begin{cases} (\underline{\sigma}, \underline{\tau}) + (\underline{\text{curl}}\,\underline{\tau}, \underline{\psi}) = 0 & \forall\, \underline{\tau} \in H_0(\text{curl}; \Omega), \\ (\underline{\text{curl}}\,\underline{\sigma}_h, \underline{\varphi}) = -(\underline{g}, \underline{\varphi}) & \forall\, \underline{\varphi} \in \mathcal{F}. \end{cases} \tag{11.3.21}$$

Definition 11.3.6. The *weak approximability* of $\mathcal{F}^0$ is satisfied if there exists $\omega_3(h)$, tending to zero as h goes to zero, such that, for every $\underline{\varphi} \in V^0$ and every $\underline{\tau}_h \in \mathsf{K}_h^c$,

$$(\underline{\text{curl}}\,\underline{\tau}_h, \underline{\varphi}) \le \omega_3(h)\|\underline{\tau}_h\|_{L^2}\|\underline{\varphi}\|_{\mathcal{F}^0}. \tag{11.3.22}$$

Definition 11.3.7. The *strong approximability* of $\mathcal{F}^0$ is satisfied if there exists $\omega_4(h)$, tending to zero as h goes to zero, such that, for every $\underline{\psi} \in \mathcal{F}^0$, there exists $\underline{\psi}_h \in \mathcal{F}_h$ satisfying

$$\|\underline{\psi} - \underline{\psi}_h\|_{L^2} \le \omega_4(h)\|\underline{\psi}\|_{\mathcal{F}^0}. \tag{11.3.23}$$

Let $\Pi_h : V^0 \to V_h$ be a B-compatible operator that is (see (6.5.55))

$$\begin{cases} (\underline{\text{curl}}(\underline{\sigma} - \Pi_h\underline{\sigma}), \underline{\varphi}_h) = 0 & \forall\, \underline{\sigma} \in V^0,\ \forall\, \underline{\varphi}_h \in \mathcal{F}_h, \\ \|\Pi_h\underline{\sigma}\|_{\underline{\text{curl}}} \le C\,\|\underline{\sigma}\|_{V^0} & \forall\, \underline{\sigma} \in V^0. \end{cases} \tag{11.3.24}$$

Definition 11.3.8. We shall say, following Sect. 6.5.5, that we have a B-Id-compatibility condition if there exists $\omega_5(h)$, tending to zero as h goes to zero, such that

$$\|\underline{\sigma} - \Pi_h \underline{\sigma}\|_{L^2} \leq \omega_5(h) \|\underline{\sigma}\|_{V^0} \quad \forall\, \underline{\sigma} \in V^0. \tag{11.3.25}$$

It is clear that, according to the theory developed in [81] and summarised in Sects. 6.5.4 and 6.5.5, conditions stated in Definitions 11.3.3, 11.3.4, and 11.3.5 are the natural conditions for the good convergence of the eigensolutions of (11.3.5) towards those of (11.2.9) (problem of the form $(f, 0)$), while conditions stated in Definitions 11.3.6, 11.3.7, and 11.3.8 correspond to the good convergence of the eigensolutions of (11.3.6) towards those of (11.2.10) (problem of the form $(0, g)$).

Let us add a final definition which is related to standard approximation properties of V_h.

Definition 11.3.9. V_h is a *good approximation* of V_0 if for any $\underline{v} \in V_0$, there exists a sequence $\{\underline{v}_h^I\} \subset V_h$ such that

$$\|\underline{v} - \underline{v}_h^I\|_{\underline{\text{curl}}} \to 0 \quad \text{as } h \to 0. \tag{11.3.26}$$

The next theorem states the link between DCP and conditions for mixed approximations.

Theorem 11.3.1. *Let us suppose that we are given a finite element space sequence* $\{V_h\}$ *approximating* $H_0(curl; \Omega)$ *and construct* $\{Q_h\}$ *and* $\{\mathcal{F}_h\}$ *according to Remark 11.3.1 so that the inclusions* (11.3.7) *hold true. Then, the following three sets of conditions are equivalent.*

1. *SDCP and approximation property* (11.3.26)*;*
2. *Ellipticity in the kernel, weak approximability of* Q*, and strong approximability of* V_0 *(Definitions 11.3.3, 11.3.4, and 11.3.5, respectively);*
3. *Weak approximability of* $\mathcal{F}^0$*, strong approximability of* $\mathcal{F}^0$*, and B-Id-compatibility (Definitions 11.3.6, 11.3.7, and 11.3.8, respectively).*

Remark 11.3.6. Theorem 11.3.1 basically states that SDCP, together with the natural approximation property (11.3.26), is equivalent to the standard conditions for the approximation of eigenmodes in mixed form. For a more detailed discussion on the differences between DCP and SDCP, we refer the interested reader to [76].

□

11.3.3 Nodal Finite Elements

It is common practice in the approximation of interior Maxwell's equations to call *nodal elements* the finite element spaces which are based on the degrees of freedom

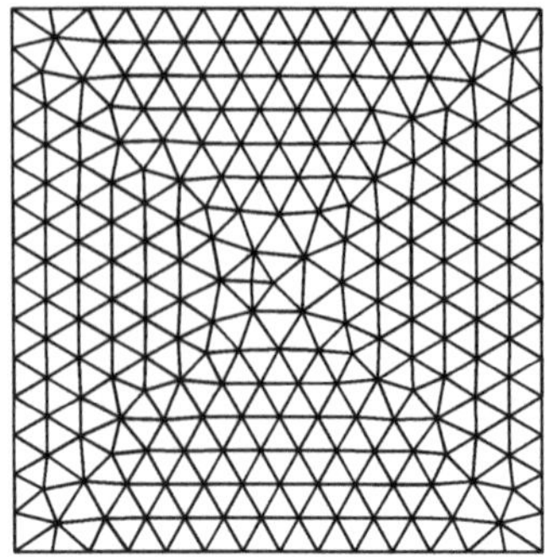

Fig. 11.1 Unstructured mesh of the square

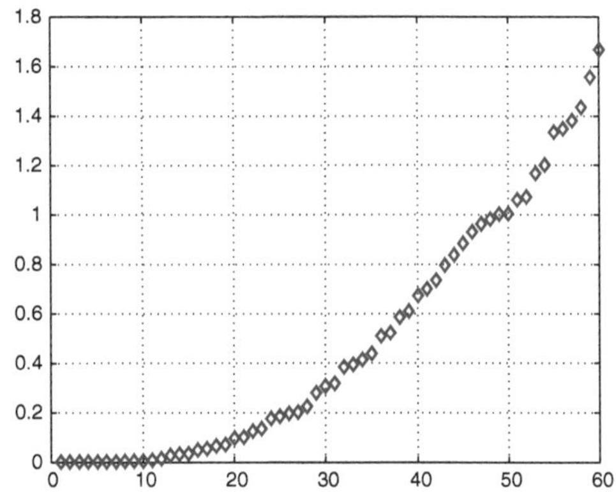

Fig. 11.2 Eigenvalues computed by nodal elements on an unstructured mesh

associated to nodal values. The simplest situation is to consider V_h as the finite element space which approximates each component of a vector of $H_0(\text{curl}; \Omega)$ by means of a continuous piecewise linear function; namely $V_h := (\mathcal{L}_1^1)^n$. In this section, we shall discuss this lowest-order case when Ω is a two-dimensional domain and shall see that, in general, nodal elements do not provide a good scheme for the approximation of Maxwell's eigenvalues. Similar considerations hold for higher-order approximations and for three-dimensional domains. The computations shown in this section have been presented in [89].

We use the two-dimensional version of formulation (11.3.4); that is, we are not enforcing the divergence-free condition, but we are discarding a posteriori the vanishing eigenvalues. Let us take $\Omega =]0, \pi[\times]0, \pi[$ (see Example 11.2.1) so that the exact eigenvalues are given by $\lambda_{mn} = m^2 + n^2, m, n = 0, 1, \ldots, m + n > 0$. The eigenvalues computed on the unstructured mesh of Fig. 11.1 are plotted in Fig. 11.2.

It is clear that the discrete eigenvalues do not have any apparent correlation with the continuous ones. On the other hand, looking at the computed eigenfunctions, it is possible to recognise some reasonable approximations to the correct eigenvalues which are interspersed among a lot of spurious solutions. This is reported in Fig. 11.3 where it is shown that the 49-th and 50-th discrete eigenfunctions provide a good approximation to the first two continuous eigenfunctions, which correspond

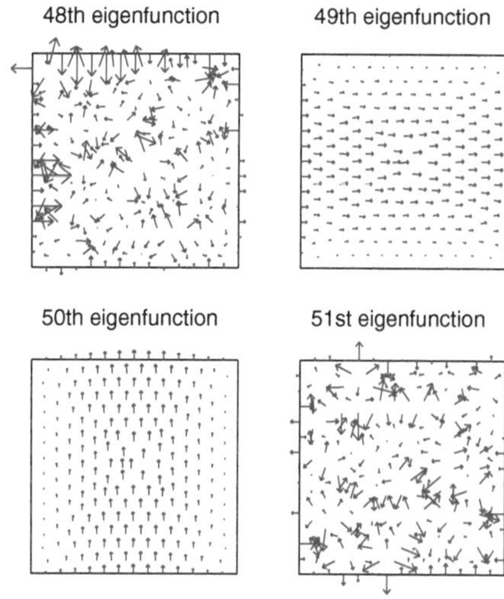

Fig. 11.3 Eigenfunctions computed by nodal elements on an unstructured mesh

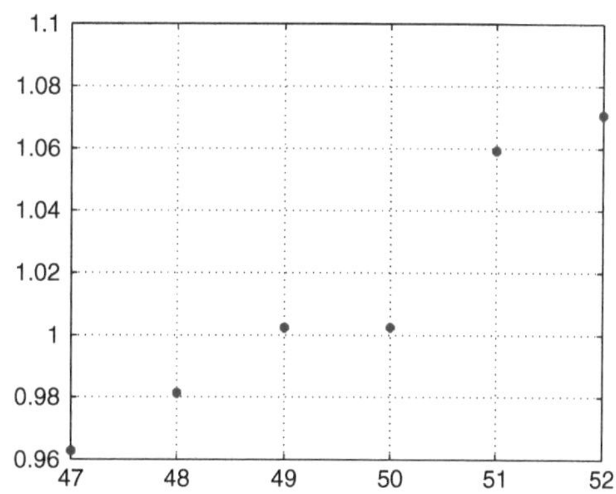

Fig. 11.4 Enlargement of Fig. 11.2 shows that the 49-th and 50-th eigenvalues are good approximations to $\lambda = 1$

to the double eigenvalue $\lambda = 1$. Figure 11.4 shows that the actual value of the 49-th and 50-th eigenvalues is very close to 1.

The behaviour of the plot presented in Fig. 11.2 demonstrates, in particular, that for this discretisation scheme a discrete Friedrichs inequality does not hold with a constant bounded below (see (11.2.31)).

It has been observed that on some particular mesh sequences, the behaviour of the discrete eigenvalues is much better and that the presence of spurious modes is reduced. The mesh presented in [381] and reported in Fig. 11.5, for instance, seems to produce correct results, even if no rigorous proof of convergence has been given so far.

The next example should discourage people from trusting the numerical results when standard nodal elements are used for the approximation of interior Maxwell's eigenvalues. Let us consider the same computational domain $\Omega =]0, \pi[\times]0, \pi[$ and

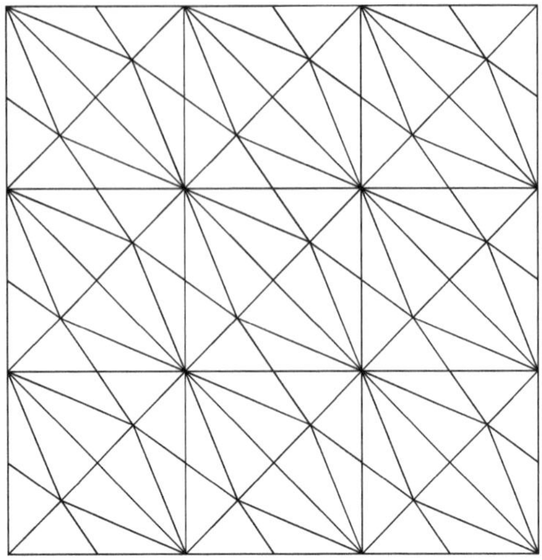

Fig. 11.5 The mesh presented in [381]

Fig. 11.6 A criss-cross triangulation of the square

a sequence of criss-cross meshes like the one presented in Fig. 11.6. The eigenvalues computed with vector-valued piecewise linear elements are reported in Table 11.1.

While the correct eigenvalues are correctly approximated together with the non-physical zero frequency (see last line of the table, where the number of vanishing discrete eigenvalues is reported), it is clear that an additional *spurious* eigenvalue, which seems to converge to the value 6, is present. In Fig. 11.7, the corresponding eigenfunction is plotted. The reader can recognise a chequerboard pattern (typical of criss-cross meshes). This is not the only spurious eigenfunction present on the criss-cross mesh: there are many other spurious solutions with higher frequencies. A theoretical analysis of this phenomenon can be found in [82].

Table 11.1 Nodal approximation on criss-cross mesh

Exact	Computed			
1	1.00428	1.00190	1.00107	1.00068
1	1.00428	1.00190	1.00107	1.00069
2	2.01711	2.00761	2.00428	2.00274
4	4.06804	4.03037	4.01710	4.01095
4	4.06804	4.03037	4.01710	4.01095
5	5.10634	5.04748	5.02674	5.01712
5	5.10634	5.04748	5.02674	5.01712
	5.92293	5.96578	5.98074	5.98767
8	8.27128	8.12151	8.06845	8.04383
9	9.34085	9.15309	9.08640	9.05537
9	9.34085	9.15309	9.08640	9.05537
d.o.f.	254	574	1,022	1,598
# zeros	63	143	255	399

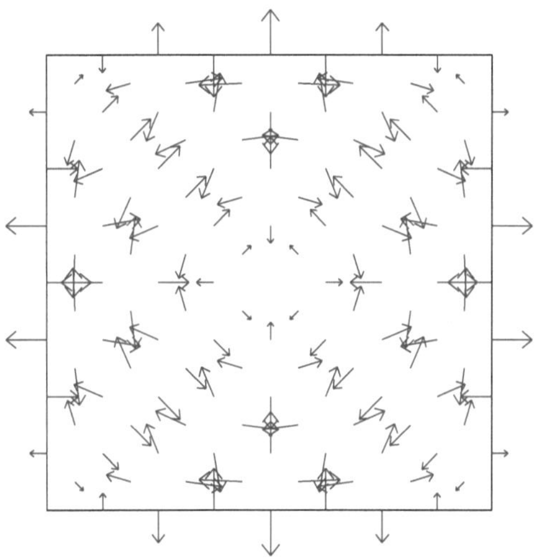

Fig. 11.7 A spurious eigenfunction on the criss-cross mesh

A simple explanation for the appearance of spurious solutions has been given in [88] and [92] for two slightly different situations.

Remark 11.3.7. The quoted references consider the eigenvalue problem corresponding to (11.3.4) for the space $H(\mathrm{div};\Omega)$. This is equivalent to the problem in which we are interested (see discussion of Sect. 11.3.1). □

The analysis of [88] applies to the approximation of the electric field by bilinear elements (component-wise) on a square mesh sequence, where the variational formulation (11.3.4) has been modified by projecting $\underline{\mathrm{curl}}\,\underline{v}$ onto a piecewise constant space. In [92] a modification (P1*) of the standard linear element on the criss-cross mesh has been presented. The P1* element had been introduced in [82] for the analysis of the standard criss-cross element: assuming that a criss-cross mesh of the square is constructed by diving the domain into N^2 sub-squares (macro-elements)

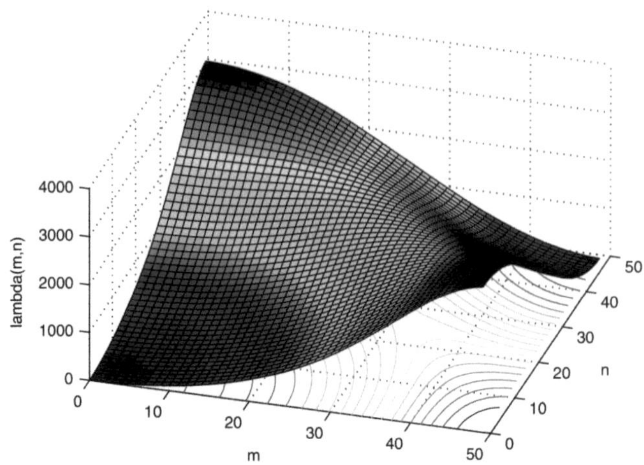

Fig. 11.8 Discrete eigenvalues computed on the criss-cross mesh

which are then partitioned into four sub-triangles by their diagonals, the P1* element is defined by eliminating the degrees of freedom corresponding to the centres of the macro-elements in such a way that the curl of the vectors is constant on each macro-element.

Figure 11.8 represents the surface generated by the discrete eigenvalues (computed on a 50-by-50 criss-cross mesh with the P1* element) as a function of m and n. For this particular mesh and element, it was actually possible to order the discrete eigenvalues in such a way that each of them can be uniquely identified in terms of m and n. While the surface corresponding to the exact eigenvalues should be the paraboloid $m^2 + n^2$, it is clear that the surface reported in Fig. 11.8 does not have the correct behaviour for large m and n. The bad approximation of the highest modes is easily detected with the help of Fig. 11.9, where we represented on the same graph a cut for $m = n$ of the computed surface and of the curve $m^2 + n^2$ corresponding to the exact frequencies. It can be directly computed that, if we fix m and n, then the discrete eigenvalue corresponding to the (m, n) mode converges to $m^2 + n^2$, which is the correct value, as the level of refinement N increases. On the other hand, the $(N - 1, N - 1)$ mode (that is, the value at the right endpoint of Fig. 11.9) converges to six.

From the mathematical point of view, on a criss-cross mesh sequence, piecewise linear elements satisfy a uniform discrete Friedrichs inequality (see (11.2.31)). Ellipticity in the kernel and *inf-sup* properties are satisfied for the mixed formulation (11.3.6), but the eigenvalues are not correctly approximated.

As a final remark, we notice that the number of zero eigenvalues in Table 11.1 is equal to the number of gradients which can be represented by a continuous piecewise linear function on the corresponding mesh; it is equal to the number of C^1 piecewise quadratic functions vanishing on the boundary (in this case the number of criss-crosses minus one, see [326]).

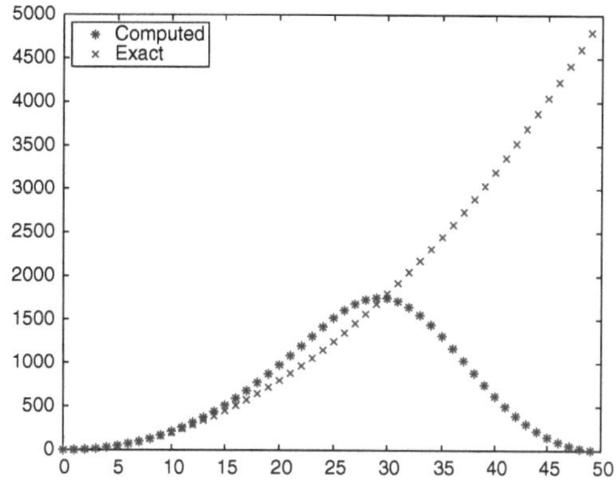

Fig. 11.9 Discrete (⋆) and continuous (x) eigenvalues computed on the criss-cross mesh for $m = n$

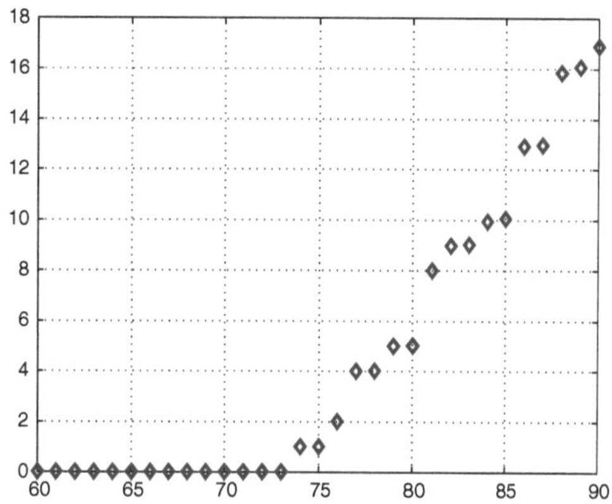

Fig. 11.10 Eigenvalues computed by edge elements on an unstructured mesh

11.3.4 Edge Finite Elements

The use of *edge elements* is universally recognised as the natural discretisation scheme for the approximation of Maxwell's eigenvalue problem. We refer to Sect. 2.5.3 for the description of conforming finite element spaces for the approximation of $H(\underline{\mathrm{curl}}, \Omega)$, also known as edge element spaces. The same two-dimensional version of (11.3.4) as in the previous section is considered; i.e., we are solving the unconstrained formulation and the divergence-free condition is dealt with by discarding the vanishing eigenvalues. We use lowest-order edge elements. The results are presented in Fig. 11.10.

It turns out that, in this case, the eigenvalues are well separated into two families: a number of vanishing (up to machine precision) eigenvalues which correspond to the infinite kernel of the $\underline{\mathrm{curl}}$ operator (see Remark 11.2.3), and positive eigenvalues which are good approximations to the continuous ones. The number of vanishing eigenvalues is equal to the number of gradients which can be represented on the mesh; in the case of lowest order edge elements, this is the number of internal vertices. We refer the interested reader to [89] for more numerical results (including discontinuous materials and non-convex domains) and for a plot of the eigenfunctions.

The convergence analysis for edge element approximations of Maxwell's eigenvalue problem can be carried out in several equivalent ways. With the help of Theorem 11.3.1 and using the results of [81], we can prove directly the Discrete Compactness Property. We could also use the mixed formulation (11.3.5) (thus proving ellipticity in the kernel, weak approximability of Q, and strong approximability of V_0), or use mixed formulation (11.3.6) (which is analysed by means of weak and strong approximabilities of $\mathcal{E}^0$ and of B-Id-compatibility).

It turns out that all these three approaches have been successfully exploited in the literature.

The first convergence proof has been given in [74] where the theoretical setting of [89] is used. The mixed formulation (11.3.6) is considered; while weak and strong approximabilities of $\mathcal{E}^0$ are easy consequences of the space definitions, it is less immediate to see how to construct a B-compatible operator Π_h such that the B-Id-compatibility condition of Definition 11.3.8 is satisfied. This is the main result of [74].

The Discrete Compactness Property (DCP, see Sect. 11.3.2) has been discussed in several papers. Among those, we recall [269] where the DCP has been proved for lowest order edge elements on tetrahedra, even though it is not shown there that DCP implies eigenvalues convergence. DCP is also the main result of [303] where it is given a general result for edge element eigenvalue approximation, and of [137, 138] where a convergence result is given and general domains/materials are considered. In [86, 87], DCP has been used for the analysis of *hp* edge finite elements.

The mixed formulation (11.3.5) has also been used for the analysis of the problem under consideration. For instance, [85] shows the validity of ellipticity in the kernel, weak approximability of Q, and strong approximability of V_0 for the analysis of a modified eigenvalue problem which arises from band gap computation for photonic crystals.

11.4 Enforcing the Divergence-Free Condition by a Penalty Method

We have seen in the previous examples that the divergence-free condition can be (as usual) a source of trouble. In general, we have shown that nodal elements do not deal with such a constraint very well, while edge elements are able to separate in a natural way the solenoidal part of the solution. On the other hand, someone may prefer to deal with nodal elements and for this reason, people have thought of using a penalty method to enforce the constraint in order to be able to use standard nodal elements. The aim of this section is to show that this strategy can be very dangerous: although the method is stable, on general domains, it can converge to

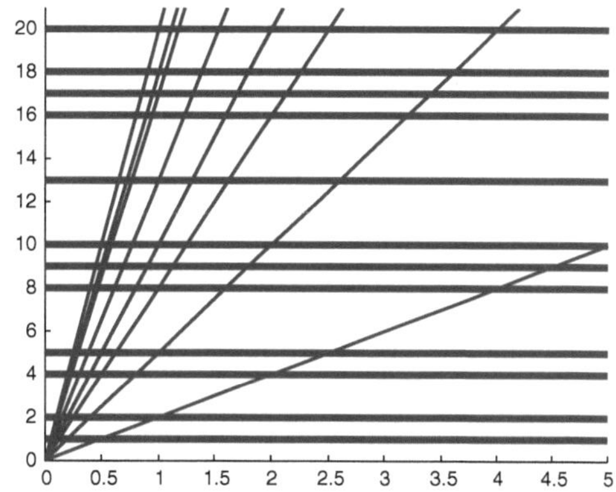

Fig. 11.11 Exact eigenvalues of the penalty formulation 11.4.1

wrong values! This has been observed in [161] (see also [84]) and is a consequence of a coerciveness result by Costabel [160].

To explain this fact, let us consider the penalty formulation of the Maxwell eigenvalue problem (similar remarks apply to the penalty formulation of the time harmonic Maxwell equations): *find* $\lambda \in \mathbb{R}$ *such that for* $\underline{u} \in H_0(\underline{\mathit{curl}};\Omega) \cap H(\mathrm{div};\Omega)$ *with* $\underline{u} \neq 0$ *it holds*

$$(\underline{\mathrm{curl}}\,\underline{u}, \underline{\mathrm{curl}}\,\underline{v}) + \alpha(\mathrm{div}\,\underline{u}, \mathrm{div}\,\underline{v}) = \lambda(\underline{u}, \underline{v}) \quad \forall\, \underline{v} \in H_0(\underline{\mathrm{curl}};\Omega) \cap H(\mathrm{div};\Omega), \tag{11.4.1}$$

where $\alpha > 0$ is the penalty parameter.

Formulation (11.4.1) is very appealing and it can be easily shown that α need not be very large in order to enforce the divergence-free constraint. Actually, since from the Helmholtz decomposition, the vector $\underline{u}$ can be decomposed into a solenoidal and an irrotational part, problem (11.4.1) admits two separate families of solutions: the first one (independent of α) corresponds to the physical solutions we are interested in

$$\begin{aligned} &(\underline{\mathrm{curl}}\,\underline{u}, \underline{\mathrm{curl}}\,\underline{v}) = \lambda(\underline{u}, \underline{v}) \quad \forall\, \underline{v}, \\ &\mathrm{div}\,\underline{u} = 0, \end{aligned} \tag{11.4.2}$$

while the second one depends linearly on α and solves the following problem

$$\begin{aligned} &\alpha(\mathrm{div}\,\underline{u}, \mathrm{div}\,\underline{v}) = \lambda(\underline{u}, \underline{v}) \quad \forall\, \underline{v}, \\ &\underline{\mathrm{curl}}\,\underline{u} = 0. \end{aligned} \tag{11.4.3}$$

Figure 11.11 shows some exact eigenvalues of (11.4.1) as functions of α when Ω is the square of size π. The horizontal lines correspond to the eigenvalues of (11.4.2) (Neumann Laplace eigenproblem), while the oblique lines refer to the eigenvalues of (11.4.3) (Dirichlet Laplace eigenproblem scaled with α). It is clear that, in this case, a penalty parameter $\alpha = 5$ is sufficient in order to have that the first six *distinct* eigenvalues of problem (11.4.1) are interior Maxwell's eigenvalues corresponding to divergence-free eigenfunctions.

One might then think that a conforming discretisation of (11.4.1) can provide a good scheme for the approximation of interior Maxwell's eigenmodes. Moreover, it is not difficult to check numerically if a discrete eigenfunction belongs to family (11.4.2) or (11.4.3) of the spectrum, so that it can be easy to detect whether the penalty parameter α is large enough. While this approach provides good results on convex domains, it is definitely not to be used when the domain presents re-entrant corners. More precisely, any Galerkin approximation of (11.4.1) gives wrong results whenever the space $(H^1(\Omega))^n \cap H_0(\underline{\mathrm{curl}};\Omega)$ is a proper subset of $H_0(\underline{\mathrm{curl}};\Omega) \cap H(\mathrm{div};\Omega)$. The reasons for the troubles come from the following proposition which is a consequence of an ellipticity result proved in [160].

Proposition 11.4.1. *The space $(H^1(\Omega))^n \cap H_0(\underline{curl};\Omega)$ is a closed subspace of $H_0(\underline{curl};\Omega) \cap H(div;\Omega)$ (with respect to the norm of $H_0(\underline{curl};\Omega) \cap H(div;\Omega)$).* □

It turns out that any conforming (piecewise polynomial) finite element space for the approximation of $H_0(\underline{\mathrm{curl}};\Omega) \cap H(\mathrm{div};\Omega)$ must be made of *continuous* functions. Indeed, the conformity in $H(\underline{\mathrm{curl}};\Omega)$ requires the continuity of the tangential component from one element to the other, while the conformity in $H(\mathrm{div};\Omega)$ imposes the continuity of the normal component (cf. Chap. 2). Hence, any conforming approximation of $H_0(\underline{\mathrm{curl}};\Omega) \cap H(\mathrm{div};\Omega)$ is actually a conforming approximation of $(H^1(\Omega))^n \cap H_0(\underline{\mathrm{curl}};\Omega)$ as well. Proposition 11.4.1 implies then that no finite element subspace of $H_0(\underline{\mathrm{curl}};\Omega) \cap H(\mathrm{div};\Omega)$ can approximate a vector field which is in $H_0(\underline{\mathrm{curl}};\Omega) \cap H(\mathrm{div};\Omega)$ and not in $(H^1(\Omega))^n$. This means that if the domain Ω is such that an eigenfunction $\underline{u}$ is not in $(H^1(\Omega))^n$ (and this typically occurs in presence of re-entrant corners, vertices, edges, etc.), then no conforming finite element discretisation of (11.4.1) can approximate it correctly. This phenomenon is shown in Fig. 11.12 where the discrete eigenvalues are shown for different values of α and using continuous finite element spaces (bi-quadratic elements in each component for this test which makes use of a uniform mesh of squares). It is apparent that the two families of solutions (11.4.2) and (11.4.3) do not separate well: there are some values corresponding to horizontal lines (discrete divergence-free eigenvalues) and some values corresponding to oblique lines (discrete irrotational eigenvalues), but also some values lying on curved lines which do not correspond to either family. The eigenvalues of the third set of solutions are originating from exact eigenvalues associated with *singular* eigenfunctions (i.e., not in $(H^1(\Omega))^2$) which cannot be approximated correctly by the finite element space. We refer the interested reader to [161] for more comments on this issue.

The penalty formulation of Maxwell's equations has been investigated by many authors due to the appeal of using nodal elements instead of the (more natural) edge ones and several methods have been proposed in order to overcome the difficulties that we have just described. Some of these methods are based on heuristic facts, others have been analysed rigorously. Due to the already mentioned

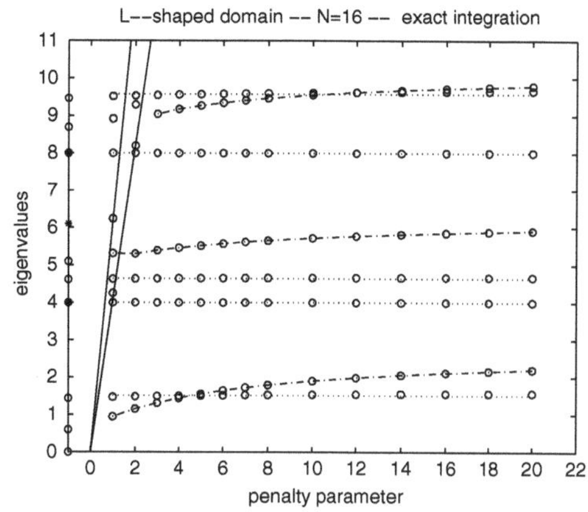

Fig. 11.12 Approximation of (11.4.1) with nodal elements for different values of α

isomorphism between $H(\underline{\text{curl}}; \Omega)$ and $H(\text{div}; \Omega)$ in two space dimensions, we also recall here some two-dimensional schemes originally proposed for a problem analogous to (11.4.1) where an irrotational eigenfunction is sought in $H_0(\text{div}; \Omega)$.

A first possible workaround is to add to the discrete space some suitable functions that can approximate the singular part of the spectrum which is in $H_0(\text{curl}; \Omega) \cap H(\text{div})$ but not in $(H^1(\Omega))^n \cap H_0(\text{curl}; \Omega)$. This procedure has been successfully exploited in two dimensions where the dimension of the singular complement is finite, but it seems difficult to generalise to three-dimensions where the dimension of the singular complement is infinite. We refer the reader interested in this topic to [36, 242] and to the references therein.

Another approach, known as *weighted regularisation*, has been proposed in [162] (see also [163] for a nice review) and consists in weakening the penalisation term in a neighbourhood of non-convex corners. This method has a complete analysis (which in particular shows exponential convergence for its *hp* version [164]) and generalises to three-dimensional non-convex domains. In [327], it has been suggested to drop the penalty term close to re-entrant corners and numerical tests show the good performance of this method for which no analysis is provided.

We conclude this section with some comments on a scheme which has been proposed in [52] for the approximation of a fluid-structure problem. With the natural identifications, the method can be used for the approximation of (11.4.1). It makes use of three finite element space sequences V_h, M_h^1 and M_h^2, and consists in looking for discrete eigenvalues $\lambda_h \in \mathbb{R}$ such that for a non-vanishing $\underline{u}_h \in V_h$ it holds

$$\begin{cases} (P_{M^1} \, \underline{\text{curl}} \, \underline{u}_h, \underline{\text{curl}} \, \underline{v}) + \alpha(P_{M^2} \, \text{div} \, \underline{u}_h, \text{div} \, \underline{v}) = \lambda_h(\underline{u}_h, \underline{v}) \quad \forall \, \underline{v} \in V_h, \\ \lambda_h \neq 0, \end{cases} \tag{11.4.4}$$

where P_{M^i} denotes the L^2 projection onto M_h^i $(i = 1, 2)$. The use of the projections P_{M^i} in (11.4.4) may introduce spurious vanishing frequencies and for this reason, the scheme has to be supplemented with the additional condition $\lambda_h \neq 0$.

The original scheme proposed in [52] uses a mesh of squares and takes V_h as the space of continuous piecewise bi-quadratic functions $(\mathcal{L}^1_{[2]})^2$ with the correct boundary conditions, and M_h^1 and M_h^2 consisting of discontinuous piecewise linear elements $\mathcal{L}^0_1$. The reason of this choice is clearly driven by the analogy with the $\underline{Q}_2$–P_1 Stokes element (see Sect. 8.6.3). We shall refer to this element as the $\underline{Q}_2$–P_1–P_1 element. This element has been shown to be robust for the problem under consideration, even in presence of non-convex domains (see, for instance, [84]), but a complete analysis for it is still lacking. The method performs reasonably well also when using a mesh of general quadrilaterals, provided that the spaces M_h^1 and M_h^2 are constructed using the *unmapped* approach described in Sect. 8.6.3 (see [22]), even though the convergence rates become suboptimal in this case for some eigenvalues.

In [217], an *inf-sup* condition has been shown for the Q_2–P_1–P_0 element and this result has been used for proving the *inf-sup* condition for the Q_2–P_1–P_1 element as well, with the addition of a stabilising term. In [84], it has been actually shown numerically that there is no need for the stabilisation in order to have the *inf-sup* condition.

11.5 Some Remarks on Exterior Calculus

As we have already mentioned several times in this book, the de Rham complex and exterior calculus have become powerful tools for the analysis of mixed finite elements. Even though a rigorous and complete description of this topic is out of the aims of our work, we give here a quick summary on this subject within the framework of edge element approximation of Maxwell's equations. It is not so easy to provide the reader with a comprehensive list of references, since many people worked on this subject using different view points and various approaches: in particular, Douglas and Roberts in [178] identified the commuting diagram property as a key ingredient for the estimates of mixed elements. Bossavit in [102], used the de Rham complex of differential forms for the description of finite element involved with the approximation of Maxwell's equations. Hiptmair [247, 248] used intensively finite element exterior calculus for the approximation of Maxwell's equations; the de Rham complex has been shown as the natural setting for the study of Maxwell's eigenvalues [75] and edge elements [170]. Arnold [17] gave a fundamental contribution to the development of finite element exterior calculus; the state of the art of the theory can be found in [31–33].

The first aim of this section is to provide the reader with suitable tools in order to understand the meaning of the following de Rham complex

$$0 \longrightarrow \Lambda^0(\Omega) \xrightarrow{d} \Lambda^1(\Omega) \xrightarrow{d} \cdots \xrightarrow{d} \Lambda^n(\Omega) \longrightarrow 0 \qquad (11.5.1)$$

when Ω is a domain in $\mathbb{R}^n$.

Given an integer number k with $0 \leq k \leq n$, we consider the space $\text{Alt}^k\mathbb{R}$ of alternating k-forms which has dimension equal to $\binom{n}{k}$ and is spanned by $dx_{\sigma_1} \wedge \cdots \wedge dx_{\sigma_k}$, for $1 \leq \sigma_1 \leq \cdots \leq \sigma_k \leq n$. An element $\omega \in \Lambda^k(\Omega)$ is then a C^∞ function $\omega : \Omega \to \text{Alt}^k\mathbb{R}$; this gives, for $\underline{x} \in \Omega$,

$$\omega_x = \sum_\sigma f_\sigma(\underline{x}) dx_{\sigma_1} \wedge \cdots \wedge dx_{\sigma_k}.$$

In particular, it turns out that $\Lambda^0(\Omega)$ admits a scalar *proxy* $\omega_x = f(\underline{x})$ as well as $\Lambda^n(\Omega)$: $\omega_x = f(\underline{x}) dx_1 \wedge \cdots \wedge dx_n$. This implies that we can identify objects of $\Lambda^0(\Omega)$ and $\Lambda^n(\Omega)$ with scalar functions. Analogously, $\Lambda^k(\Omega)$ admits vector *proxies* for $k = 1$ and $k = n-1$ with the identifications $\omega_x = \sum_{j=1}^n f_j(\underline{x}) dx_j$ for $k = 1$ and $\omega_x = \sum_{j=1}^n f_j(\underline{x})(-1)^{j-1} dx_1 \wedge \cdots \wedge \widehat{dx_j} \wedge \cdots \wedge dx_n$ for $k = n-1$.

Example 11.5.1. When $n = 3$, we have the following four spaces: $\Lambda^0(\Omega)$ can be identified with the space of scalar functions on Ω

$$\omega_x \in \text{Alt}^0\mathbb{R} \leftrightarrow f(\underline{x}) \in \mathbb{R};$$

$\Lambda^1(\Omega)$ is identified with the space of three-dimensional vector fields

$$\omega_x \in \text{Alt}^1\mathbb{R} \leftrightarrow (f_1(\underline{x}), f_2(\underline{x}), f_3(\underline{x})) \in \mathbb{R}^3;$$

$\Lambda^2(\Omega)$ can also be identified with a space of three-dimensional vector fields by $\omega = f_1 dx_2 \wedge dx_3 - f_2 dx_1 \wedge dx_3 + f_3 dx_1 \wedge dx_2$, that is,

$$\omega_x \in \text{Alt}^2\mathbb{R} \leftrightarrow (f_1(\underline{x}), f_2(\underline{x}), f_3(\underline{x})) \in \mathbb{R}^3;$$

in the same way, using the rule $\omega = f dx_1 \wedge dx_2 \wedge dx_3$, the space $\Lambda^3(\Omega)$ can be identified with the space of scalar functions by

$$\omega_x \in \text{Alt}^n\mathbb{R} \leftrightarrow f(\underline{x}) \in \mathbb{R}. \qquad \square$$

The exterior derivative is an operator $d : \Lambda^k\mathbb{R} \to \Lambda^{k+1}\mathbb{R}$ (formally, it should be denoted by d^k, but in general, people use d as this does not generate confusion) defined as follows:

$$d \sum_\sigma f_\sigma dx_{\sigma_1} \wedge \cdots \wedge dx_{\sigma_k} = \sum_\sigma \sum_{j=1}^n \frac{\partial f_\sigma}{\partial x_j} dx_j \wedge dx_{\sigma_1} \wedge \cdots \wedge dx_{\sigma_k}.$$

In particular, we have for 0-forms,

$$df = \sum_{j=1}^n \frac{\partial f_j}{\partial x_j} dx_j.$$

It turns out that d is a differential, that is,

$$d \circ d = 0,$$

and that the following Leibniz rule holds true

$$d(\mu \wedge \nu) = d\mu \wedge \nu + (-1)^k \mu \wedge d\nu, \quad \mu \in \Lambda^k(\Omega),\ \nu \in \Lambda^l(\Omega).$$

Example 11.5.2. Coming back to Example 11.5.1, we can now identify for $n = 3$ the exterior derivatives d^k $(k = 0, 1, 2)$ with standard differential operators. For $k = 0$,we have $df = \frac{\partial f_1}{\partial x_1} dx_1 + \frac{\partial f_2}{\partial x_2} dx_2 + \frac{\partial f_3}{\partial x_3} dx_3$ which gives

$$df \leftrightarrow \left(\frac{\partial f_1}{\partial x_1}, \frac{\partial f_2}{\partial x_2}, \frac{\partial f_3}{\partial x_3}\right),$$

that is,

$$d^0 \leftrightarrow \underline{\mathrm{grad}}\,.$$

For $k = 1$, the definition of exterior derivative gives

$$\begin{aligned} d(f_1 dx_1 + f_2 dx_2 + f_3 dx_3) = \\ &\frac{\partial f_1}{\partial x_1} dx_1 \wedge dx_1 + \frac{\partial f_1}{\partial x_2} dx_2 \wedge dx_1 + \frac{\partial f_1}{\partial x_3} dx_3 \wedge dx_1 \\ &+ \frac{\partial f_2}{\partial x_1} dx_1 \wedge dx_2 + \frac{\partial f_2}{\partial x_2} dx_2 \wedge dx_2 + \frac{\partial f_2}{\partial x_3} dx_3 \wedge dx_2 \\ &+ \frac{\partial f_3}{\partial x_1} dx_1 \wedge dx_3 + \frac{\partial f_3}{\partial x_2} dx_2 \wedge dx_3 + \frac{\partial f_3}{\partial x_3} dx_3 \wedge dx_3 \\ &= \left(\frac{\partial f_3}{\partial x_2} - \frac{\partial f_2}{\partial x_3}\right) dx_2 \wedge dx_3 \\ &+ \left(\frac{\partial f_1}{\partial x_3} - \frac{\partial f_3}{\partial x_1}\right) dx_3 \wedge dx_1 \\ &+ \left(\frac{\partial f_2}{\partial x_1} - \frac{\partial f_1}{\partial x_2}\right) dx_1 \wedge dx_2 \end{aligned}$$

so that we have the identification

$$d(f_1 dx_1 + f_2 dx_2 + f_3 dx_3) \leftrightarrow \left(\frac{\partial f_3}{\partial x_2} - \frac{\partial f_2}{\partial x_3}, \frac{\partial f_1}{\partial x_3} - \frac{\partial f_3}{\partial x_1}, \frac{\partial f_2}{\partial x_1} - \frac{\partial f_1}{\partial x_2}\right)$$

or

$$d^1 \leftrightarrow \underline{\mathrm{curl}}\,.$$

For $k = 2$, we have analogously

$$\begin{aligned} d(f_1 dx_2 \wedge dx_3 - f_2 dx_1 \wedge dx_3 + f_3 dx_1 \wedge dx_2) = & \frac{\partial f_1}{\partial x_1} dx_1 \wedge dx_2 \wedge dx_3 \\ & - \frac{\partial f_2}{\partial x_2} dx_2 \wedge dx_1 \wedge dx_3 \\ & + \frac{\partial f_3}{\partial x_3} dx_3 \wedge dx_1 \wedge dx_2 \end{aligned}$$

which gives

$$d(f_1 dx_2 \wedge dx_3 - f_2 dx_1 \wedge dx_3 + f_3 dx_1 \wedge dx_2) \leftrightarrow \frac{\partial f_1}{\partial x_1} + \frac{\partial f_2}{\partial x_2} + \frac{\partial f_3}{\partial x_3},$$

that is,

$$d^2 \leftrightarrow \operatorname{div}.$$

These identifications allow us to write the de Rham complex (11.5.1) in the more common form

$$0 \longrightarrow C^\infty(\Omega) \xrightarrow{\text{grad}} (C^\infty(\Omega))^3 \xrightarrow{\text{curl}} (C^\infty(\Omega))^3 \xrightarrow{\text{div}} C^\infty(\Omega) \longrightarrow 0.$$

□

Remark 11.5.1. The sequence (11.5.1) is a *complex*, since d is a differential. It is sometimes interesting to see whether a complex is also *exact* (that is the range of each derivative d^k coincides with the kernel of the next one d^{k+1}, meaning that the cohomology of the complex is trivial). This is related to the topology of Ω and/or to the boundary conditions we want to consider. In particular, if $n = 3$, Ω is simply connected and no boundary conditions are considered, then the following complex is exact

$$\mathbb{R} \longrightarrow C^\infty(\Omega) \xrightarrow{\text{grad}} (C^\infty(\Omega))^3 \xrightarrow{\text{curl}} (C^\infty(\Omega))^3 \xrightarrow{\text{div}} C^\infty(\Omega) \longrightarrow 0.$$

Moreover, if standard boundary conditions are considered, the following complex is exact for Ω simply connected:

$$\begin{aligned} 0 \longrightarrow C^\infty(\Omega) \cap H_0^1(\Omega) & \xrightarrow{\text{grad}} (C^\infty(\Omega))^3 \cap H_0(\underline{\text{curl}}; \Omega) \\ & \xrightarrow{\text{curl}} (C^\infty(\Omega))^3 \cap H_0(\text{div}; \Omega) \xrightarrow{\text{div}} C^\infty(\Omega) \longrightarrow \mathbb{R}. \end{aligned}$$

□

The next step in the theory consists in the definition of finite element differential forms. It turns out that basically all conforming spaces presented in Chap. 2 can be reproduced within this framework. Actually, this theory is very powerful and allows the definition of many more finite element spaces (in particular, we observe that we can deal with general n-dimensional spaces and k-forms). The definition starts from the introduction of polynomial differential forms $\mathcal{P}_r\Lambda^k(T)$ on a simplex $T \subset \mathbb{R}^m$ and from the space $\mathcal{P}_r^-\Lambda^k(T)$ which is constructed with the help of the Koszul differential. We refer the interested reader to [33] for the technical details of such a construction, but would like to point out again that the theory includes all basic ingredients of standard mixed finite elements. For instance, we recall the following result which is the translation of the commuting diagram property into the language of differential forms: it is possible to define suitable degrees of freedom, such that we can construct *canonical projection operators* $\Pi : \Lambda^k(T) \to \mathcal{P}_r\Lambda^k(T)$ or $\Pi : \Lambda^k(T) \to \mathcal{P}_r^-\Lambda^k(T)$ which commute with the exterior derivative. For instance, the following diagram commutes

$$\begin{array}{ccc} \Lambda^k(T) & \xrightarrow{d} & \Lambda^{k+1}(T) \\ \Pi\big\downarrow & & \Pi\big\downarrow \\ \mathcal{P}_r\Lambda^k(T) & \xrightarrow{d} & \mathcal{P}_{r-1}\Lambda^{k+1}(T) \end{array}$$

as well as the following one

$$\begin{array}{ccc} \Lambda^k(T) & \xrightarrow{d} & \Lambda^{k+1}(T) \\ \Pi\big\downarrow & & \Pi\big\downarrow \\ \mathcal{P}_r^-\Lambda^k(T) & \xrightarrow{d} & \mathcal{P}_r^-\Lambda^{k+1}(T). \end{array}$$

More details on finite element differential forms can be found, for instance, in [12, 18, 19].

11.6 Concluding remarks

In this chapter we have tried to convince the reader that mixed finite elements can be useful for the approximation of partial differential equations arising from the modelling of electromagnetic problems.

It should be acknowledged that the use of mixed (*edge*) finite element has been widely accepted by the community of people working on the numerical approximation of Maxwell's equation (also because it was soon apparent that the use of standard (*nodal*) finite element was source of trouble.

For a more comprehensive introduction to the finite element approximation of Maxwell's system, more focused references are available. Among those, the reader is referred to [248] and [302]. The finite element approximation of Maxwell's eigenvalue has been discussed in [77], while an extensive discussion on finite element exterior calculus can be found in [34] and the references therein.

References

1. B. Achchab, B. Agouzal, A. Baranger, and J.F. Maitre. Estimateurs d'erreur a posteriori hiérarchiques. Applications aux éléments finis mixtes. *Numer. Math.*, 80(2):159–179, 1998.
2. B. Achchab and J. F. Maître. Estimate of the constant in two strengthened C.B.S. inequalities for F.E.M. systems of 2D elasticity: Application to multilevel methods and a posteriori error estimators. *Numerical Linear Algebra with Applications*, 3:147–159, 1996.
3. D.A. Adams. *Sobolev spaces*. Academic Press, New York, 1975.
4. B. Ahmad, A. Alsedi, F. Brezzi, L.D. Marini, and A. Russo. Equivalent projectors for virtual element methods. Submitted.
5. M. Ainsworth. A posteriori error estimation for lowest order Raviart-Thomas mixed finite elements. *SIAM J. Sci. Comput*, 30(1):189204, 2007.
6. A. Alonso and A. Valli. Some remarks on the characterization of the space of tangential traces of $H(\mathrm{rot};\Omega)$ and the construction of an extension operator. *Manuscripta Math.*, 89(2):159–178, 1996.
7. M. Amara and J.M. Thomas. Equilibrium finite elements for the linear elastic problem. *Numer. Math.*, 33:367–383, 1979.
8. C. Amrouche, C. Bernardi, M. Dauge, and V. Girault. Vector potentials in three-dimensional non-smooth domains. *Math. Methods Appl. Sci.*, 21(9):823–864, 1998.
9. P. Arbenz and R. Geus. Eigenvalue solvers for electromagnetic fields in cavities. Technical report, in cavities, Tech. Rep. 275, Institute of Scientific Computing, ETH Ziirich, Zfirich, 1997.
10. Arbogast, Todd Numerical subgrid upscaling of two-phase flow in porous media. Numerical treatment of multiphase flows in porous media (Beijing, 1999), 35–49, Lecture Notes in Phys., 552, Springer, Berlin, 2000.
11. T. Arbogast, M.F. Wheeler, and I. Yotov. Mixed finite elements for elliptic problems with tensor coefficients as cell-centered finite differences. *SIAM J. Numer. Anal.*, 34(2):828852, 1997.
12. D. N. Arnold. Spaces of finite element differential forms. In U. Gianazza, F. Brezzi, P. Colli Franzone, and G. Gilardi, editors, *Analysis and Numerics of Partial Differential Equations*. Springer, 2013. 19 pages. To appear. arXiv preprint 1204.1351.
13. D. N. Arnold, F. Brezzi, B. Cockburn, and L. D. Marini. Unified analysis of discontinuous Galerkin methods for elliptic problems. *SIAM J. Numer. Anal.*, 39(5):1749–1779, 2001/02.
14. D. N. Arnold, F. Brezzi, and L. D. Marini. Locking-free Reissner-Mindlin elements without reduced integration. *Comput. Methods Appl. Mech. Engrg*, 196:3660–3671, 2007.
15. D.N. Arnold. Discretization by finite elements of a model parameter dependent problem. *Numer. Math.*, 37:405–421, 1981.

D. Boffi et al., *Mixed Finite Element Methods and Applications*, Springer Series in Computational Mathematics 44, DOI 10.1007/978-3-642-36519-5,

16. D.N. Arnold. On nonconforming linear-constant elements for some variants of the Stokes equations. *Istit. Lombardo Accad. Sci. Lett. Rend. A*, 127(1):83–93 (1994), 1993.
17. D.N. Arnold. Differential complexes and numerical stability. In *Proceedings of the International Congress of Mathematicians, Vol. I (Beijing, 2002)*, pages 137–157, Beijing, 2002. Higher Ed. Press.
18. D.N. Arnold and G. Awanou. Finite element differential forms on cubical meshes. *submitted to Mathematics of Computation*, 2012.
19. D.N. Arnold, D. Boffi, and F. Bonizzoni. Approximation by tensor product finite element differential forms. Submitted. arXiv preprint 1212.6559, 2012.
20. D.N. Arnold, D. Boffi, and R.S. Falk. Approximation by quadrilateral finite elements. *Math. Comp.*, 71(239):909–922 (electronic), 2002.
21. D.N. Arnold, D. Boffi, and R.S. Falk. Quadrilateral $H(\mathrm{div})$ finite elements. *SIAM J. Numer. Anal.*, 42(6):2429–2451 (electronic), 2005.
22. D.N. Arnold, D. Boffi, R.S. Falk, and L. Gastaldi. Finite element approximation on quadrilateral meshes. *Comm. Numer. Methods Engrg.*, 17(11):805–812, 2001.
23. D.N. Arnold and F. Brezzi. Mixed and non-conforming finite element methods: implementation, post-processing and error estimates. *Math. Modelling Numer. Anal.*, 19:7–35, 1985.
24. D.N. Arnold, F. Brezzi, and J. Douglas. PEERS: a new mixed finite element for plane elasticity. *Japan J. Appl.Math.*, 1:347–367, 1984.
25. D.N. Arnold, F. Brezzi, and M. Fortin. A stable finite element for the Stokes equations. *Calcolo*, 21:337–344, 1984.
26. D.N. Arnold, F. Brezzi, and L. D. Marini. A family of discontinuous Galerkin finite elements for the Reissner-Mindlin plate. *J. Sci. Comput*, 22/23:25–45, 2005.
27. D.N. Arnold, J. Douglas, and C.P. Gupta. A family of higher order mixed finite element methods for plane elasticity. *Numer. Math.*, 45:1–22, 1984.
28. D.N. Arnold and J.S. Falk. A uniformly accurate finite element method for the Mindlin-Reissner plate. *SIAM. J. Num. Anal.*, 26:1276–1290, 1989.
29. D.N. Arnold and J.S. Falk. The boundary layer for the Reissner-Mindlin plate model. *SIAM J. Math. Anal.*, 21:281–312, 1990.
30. D.N. Arnold, R. Falk, and R. Winther. Mixed finite element methods for linear elasticity with weakly imposed symmetry. *Mathematics of Computation*, 76:1699–1723, 2007.
31. D.N. Arnold, R.S. Falk, and R. Winther. Differential complexes and stability of finite element methods. I. The de Rham complex. In *Compatible spatial discretizations*, volume 142 of *IMA Vol. Math. Appl.*, pages 24–46. Springer, New York, 2006.
32. D.N. Arnold, R.S. Falk, and R. Winther. Differential complexes and stability of finite element methods. II. The elasticity complex. In *Compatible spatial discretizations*, volume 142 of *IMA Vol. Math. Appl.*, pages 47–67. Springer, New York, 2006.
33. D.N. Arnold, R.S. Falk, and R. Winther. Finite element exterior calculus, homological techniques, and applications. *Acta Numerica*, 15:1–155, 2006.
34. D.N. Arnold, R.S. Falk, and R. Winther. Finite element exterior calculus: from Hodge theory to numerical stability. *Bull. Amer. Math. Soc.*, 47:281–354, 2010.
35. D.N. Arnold and A.L. Madureira. Asymptotic estimates of hierarchical modeling. *Math. Models Methods Appl. Sci*, 13:1325–1350, 2003.
36. F. Assous, P. Ciarlet Jr., and E. Sonnendrücker. Resolution of the Maxwell equations in a domain with reentrant corners. *RAIRO Modél. Math. Anal. Numér.*, 32(3):359–389, 1998.
37. J.P. Aubin. *Approximation of elliptic boundary-value problems*. John Wiley and Sons, New York, 1972.
38. J.P. Aubin. *L'analyse non linéaire et ses motivations économiques*. Masson, Paris, 1984.
39. F. Auricchio, L. Beirão da Veiga, C. Lovadina, and A. Reali. The importance of the exact satisfaction of the incompressibility constraint in nonlinear elasticity: mixed FEMs versus NURBS-based approximations. *Comput. Methods Appl. Mech. Engrg.*, 199(5–8):314–323, 2010.
40. I. Babuška. Error-bounds for finite element method. *Numer. Math.*, 16:322–333, 1971.

41. I. Babuška and A.K. Aziz. Survey lectures on the mathematical foundations of the finite element method. In A.K. Aziz, editor, *The Mathematical Foundations of the Finite Element Method with Applications to Partial Differential Equations*. Academic Press, New-York, 1972.
42. I. Babuška and J.E. Osborn. Numerical treatment of eigenvalue problems for differential equations with discontinuous coefficients. *Math. Comp.*, 32:991–1023, 1978.
43. I. Babuška and J.E. Osborn. Eigenvalue problems. In P.G. Ciarlet and J.L. Lions, editors, *Handbook of Numerical Analysis*, volume II, pages 641–788. North–Holland, 1991.
44. I. Babuška, J.E. Osborn, and J. Pitkaranta. Analysis of mixed methods using mesh-dependent norms. *Math. Comp.*, 35:1039–1079, 1980.
45. I. Babuska and W.C. Rheinboldt. Error estimates for adaptive finite element computation. *SIAM Numer. Anal*, 15:736–754, 1978.
46. C. Baiocchi and F. Brezzi. Stabilization of unstable methods. In *Problemi attuali dell'Analisi e della Fisica Matematica, P.E. Ricci Ed.*, pages 59–64. Università "La Sapienza", Roma,, 1993.
47. R. E. Bank and R. K. Smith. A posteriori error estimates based on hierarchical bases. *SIAM Journal on Numerical Analysis*, 30:921–935, 1993.
48. R.E. Bank and A. Weiser. Some a posteriori error estimator for elliptic partial differential equation. *Math. Comp.*, 44:283–301, 1985.
49. J. Baranger, J.F. Maitre, and F. Oudin. Connection between finite volume and mixed finite element methods. *M2AN*, 30:445–465, 1996.
50. V. Barbu and T. Precupanu. *Convexity and optimization in Banach spaces*. Editura Academiei, Bucarest, 1978.
51. G.S. Baruzzi, W.G. Habashi, G. Guèvremont, and M.M. Hafez. A second order finite element method for the solution of the transonic Euler and Navier-Stokes equations. *International Journal for Numerical Methods in Fluids*, 20:671–693, 1995.
52. K.-J. Bathe, C. Nitikitpaiboon, and X. Wang. A mixed displacement-based finite element formulation for acoustic fluid-structure interaction. *Computers & Structures*, 56:225–237, 1995.
53. K.J. Bathe. *Finite Element Procedures in Engineering Analysis*. Prentice Hall, Englewood Cliffs, N.J., 1982.
54. K.J. Bathe and F. Brezzi. The convergence of a four-node plate bending element based on Mindlin-Reissner plate theory and a mixed interpolation. In J.R. Whiteman, editor, *Proceedings of the Conference on Mathematics of Finite Elements and applications V*, pages 491–503. Academic Press, New-York, 1985.
55. K.J. Bathe and E. Dvorkin. A continuum mechanics based four-node shell element for general non-linear analysis. *J. Eng. Comp.*, 1:77–78, 1984.
56. R. Becker, P. Hansbo, and M.G. Larson. Energy norm a posteriori error estimation for discontinuous Galerkin methods. *Comp. Methods Appl. Mech. Engrg.*, 192:2003, 2001.
57. L. Beirão da Veiga, F. Brezzi, A. Cangiani, G. Manzini, L.D. Marini, and A. Russo. Basic principles of virtual element methods. *Math. Models Methods Appl. Sci.*, 23 (2013), no. 1, 199–214.
58. M. Benzi, G.H. Golub, and J. Liesen. Numerical solution of saddle-point problem. *Acta Numerica*, 14:1–137, 2005.
59. M. Bercovier. *Régularisation duale des problèmes variationnels mixtes*. PhD thesis, Université de Rouen, 1976.
60. M. Bercovier. Perturbation of a mixed variational problem, applications to mixed finite element methods. *R.A.I.R.O. Anal. Numer.*, 12:211–236, 1978.
61. M. Bercovier and O.A. Pironneau. Error estimates for finite element method solution of the Stokes problem in the primitive variables. *Numer. Math.*, 33:211–224, 1977.
62. J. Bergh and J. Lofstrom. *Interpolation Spaces: An Introduction*. Springer-Verlag, Berlin, 1976.

63. A. Bermúdez, R. Durán, M. A. Muschietti, R. Rodríguez, and J. Solomin. Finite element vibration analysis of fluid-solid systems without spurious modes. *SIAM J. Numer. Anal.*, 32(4):1280–1295, 1995.
64. A. Bermúdez, P. Gamallo, M. R. Nogueiras, and R. Rodríguez. Approximation of a structural acoustic vibration problem by hexahedral finite elements. *IMA J. Numer. Anal.*, 26(2):391–421, 2006.
65. A. Bermúdez, P. Gamallo, María R. Nogueiras, and R. Rodríguez. Approximation properties of lowest-order hexahedral Raviart-Thomas finite elements. *C. R. Math. Acad. Sci. Paris*, 340(9):687–692, 2005.
66. A. Bermúdez and D.G. Pedreira. Mathematical analysis of a finite element method without spurious solutions for computation od dieletric waveguides. *Numer. Math.*, 61:39–57, 1992.
67. A. Bermúdez and R. Rodríguez. Finite element computation of the vibration modes of a fluid-solid system. *Comput. Methods Appl. Mech. Engrg.*, 119(3–4):355–370, 1994.
68. C. Bernardi, C. Canuto, and Y. Maday. Generalized inf-sup condition for Chebyshev approximation of the Navier-Stokes equations. *SIAM J. Num. Anal.*, 25:1237–1265, 1988.
69. C. Bernardi, Y. Maday, and A. T. Patera. A new nonconforming approach to domain decomposition: the mortar element method. In *Nonlinear partial differential equations and their applications. Collège de France Seminar, Vol. XI (Paris, 1989–1991)*, volume 299 of *Pitman Res. Notes Math. Ser.*, pages 13–51. Longman Sci. Tech., Harlow, 1994.
70. C. Bernardi and G. Raugel. Méthodes d'éléments finis mixtes pour les équations de Stokes et de Navier-Stokes dans un polygone non convexe. *Calcolo*, 18:255–291, 1981.
71. D. Boffi. Stability of higher order triangular Hood–Taylor methods for stationary Stokes equations. *Math. Mod. Meth. Appl. Sci.*, 2(4):223–235, 1994.
72. D. Boffi. Minimal stabilizations of the P_{k+1}-P_k approximation of the stationary Stokes equations. *Math. Models Methods Appl. Sci.*, 5(2):213–224, 1995.
73. D. Boffi. Three-dimensional finite element methods for the Stokes problem. *Siam J. Numer. Anal.*, 34:664–670, 1997.
74. D. Boffi. Fortin operator and discrete compactness for edge elements. *Numer. Math.*, 87(2):229–246, 2000.
75. D. Boffi. A note on the de Rham complex and a discrete compactness property. *Appl. Math. Lett.*, 14(1):33–38, 2001.
76. D. Boffi. Approximation of eigenvalues in mixed form, discrete compactness property, and application to *hp* mixed finite elements. *Comput. Methods Appl. Mech. Engrg.*, 196(37–40):3672–3681, 2007.
77. D. Boffi. Finite element approximation of eigenvalue problems. *Acta Numer.*, 19:1–120, 2010.
78. D. Boffi. The immersed boundary method for fluid-structure interactions: mathematical formulation and numerical approximation. *Bollettino U. M. I.*, (9) V (2012) pp. 711–724.
79. D. Boffi, F. Brezzi, and M. Fortin. *Mixed Finite Element Methods*. Springer-Verlag, Berlin, 2009.
80. D Boffi, F Brezzi, and M. Fortin. Reduced symmetry elements in linear elasticity. *Commun. Pur. Appl. Anal.*, 8:95–121, 2009.
81. D. Boffi, F. Brezzi, and L. Gastaldi. On the convergence of eigenvalues for mixed formulations. *Ann. Scuola Norm. Sup. Pisa Cl. Sci. (4)*, 25(1–2):131–154 (1998), 1997. Dedicated to Ennio De Giorgi.
82. D. Boffi, F. Brezzi, and L. Gastaldi. On the problem of spurious eigenvalues in the approximation of linear elliptic problems in mixed form. *Math. Comp.*, 69(229):121–140, 2000.
83. D. Boffi, N. Cavallini, F. Gardini, and L. Gastaldi. Local mass conservation of Stokes finite elements. *J. Sci. Comput.*, 52(2):383–400, 2012.
84. D. Boffi, C. Chinosi, and L. Gastaldi. Approximation of the grad div operator in nonconvex domains. *CMES Comput. Model. Eng. Sci.*, 1(2):31–43, 2000.
85. D. Boffi, M. Conforti, and L. Gastaldi. Modified edge finite elements for photonic crystals. *Numer. Math.*, 105 (2006), pp. 249–266.

86. D. Boffi, M. Costabel, M. Dauge, and L. Demkowicz. Discrete compactness for the hp version of rectangular edge finite elements. *SIAM J. Numer. Anal.*, 44(3):979–1004 (electronic), 2006.
87. D. Boffi, L. Demkowicz, and M. Costabel. Discrete compactness for p and hp 2D edge finite elements. *Math. Models Methods Appl. Sci.*, 13(11):1673–1687, 2003.
88. D. Boffi, R. Duran, and L. Gastaldi. A remark on spurious eigenvalues in a square. *Appl. Math. Lett.*, Vol. 12, 107–114 (1999).
89. D. Boffi, P. Fernandes, L. Gastaldi, and I. Perugia. Edge approximation of eigenvalue problems arising from electromagnetics. In Désidéri, Le Tallec, Oñate, Périaux, and Stein, editors, *Numerical Methods in Engeneering '96*, pages 551–556. John Wiley & Sons, 1996.
90. D. Boffi and L. Gastaldi. On the quadrilateral $Q_2 - P_1$ element for the Stokes problem. *International Journal for Numerical Methods in Fluids*, 34:664–670, 2002.
91. D. Boffi and L. Gastaldi. On the quadrilateral Q_2-P_1 element for the Stokes problem. *Internat. J. Numer. Methods Fluids*, 39(11):1001–1011, 2002.
92. D. Boffi and L. Gastaldi. Analysis of finite element approximation of evolution problems in mixed form. *SIAM J. Numer. Anal.*, 42(4):1502–1526 (electronic), 2004.
93. D. Boffi and L. Gastaldi. Some remarks on quadrilateral mixed finite elements. *Comput. Struct.*, 87 (2009) 751–757. doi:10.1016/j.compstruc.2008.12.006.
94. D. Boffi, F. Kikuchi, and J. Schöberl. Edge element computation of Maxwell's eigenvalues on general quadrilateral meshes. *Math. Models Methods Appl. Sci.*, 16(2):265–273, 2006.
95. D. Boffi and C. Lovadina. Analysis of new augmented Lagrangian formulations for mixed finite element schemes. *Numer. Math.*, 75(4):405–419, 1997.
96. R. Bois, M. Fortin, and A. Fortin. A fully optimal anisotropic mesh adaptation method based on a hierarchical error estimator. *Computer Methods in Applied Mechanics and Engineering*, 209–212:12–27, 2012.
97. R. Bois, M. Fortin, A. Fortin, and A. Couët. High order optimal anisotropic mesh adaptation using hierarchical elements. *European Journal of Computational Mechanics/Revue Europenne de Mcanique Numrique*, 21(1–2):72–91, 2012.
98. J. M. Boland and R. A. Nicolaides. Stability of finite elements under divergence constraints. *SIAM J. Numer. Anal.*, 20(4):722–731, 1983.
99. J. M. Boland and R. A. Nicolaides. On the stability of bilinear-constant velocity-pressure finite elements. *Numer. Math.*, 44(2):219–222, 1984.
100. J. M. Boland and R. A. Nicolaides. Stable and semistable low order finite elements for viscous flows. *SIAM J. Numer. Anal.*, 22(3):474–492, 1985.
101. J.M. Boland and R. Nicolaides. On the stability of bilinear–constant velocity–pressure finite elements. *Numer. Math.*, 44:219–222, 1984.
102. A. Bossavit. Whitney forms: a class of finite elements for three-dimensional computations in electromagnetism. *IEEE Proc. A*, 135:493–500, 1988.
103. A. Bossavit. Un nouveau point de vue sur les éléments mixtes. *Matapli (Bull. Soc. Math. Appl. Industr.)*, 20:22–35, 1989.
104. A. Bossavit. Solving Maxwell's equations in a closed cavity, and the question of spurious modes. *IEEE Trans. MAG*, 26:702–705, 1990.
105. A. Bossavit. *Computational electromagnetism*. Electromagnetism. Academic Press Inc., San Diego, CA, 1998.
106. D Braess. *Finite Elements: Theory, Fast Solvers and Applications in Solid Mechanics*. Cambridge University Press, Cambridge, 2007.
107. J.H. Bramble and R.S. Falk. Two mixed finite element methods for the simply supported plate problem. *R.A.I.R.O. Anal. Numer.*, 17:337–384, 1983.
108. J.H. Bramble, R.D. Lazarov, and J.E. Pasciak. Least-squares for second order elliptic problems. *Comp. Meth. Appl. Mech. Engng.*, 152:195–210, 1998.
109. J.H. Bramble and J.E. Osborn. Rate of convergence for nonselfadjoint eigenvalue approximations. *Math. Comp.*, 27:525–549, 1973.
110. J.H. Bramble and A.H. Schatz. Estimates for spline projections. *R.A.I.R.O. Anal. Numer.*, 10:5–37, 1976.

111. J.H. Bramble and A.H. Schatz. Higher order local accuracy by averaging in the finite element method. *Math. Comp.*, 31:94–111, 1977.
112. F. Brezzi. On the existence, uniqueness and approximation of saddle–point problems arising from Lagrangian multipliers. *R.A.I.R.O. Anal. Numer.*, 8:129–151, 1974.
113. F. Brezzi. Sur une méthode hybride pour l'approximation du problème de la torsion d'une barre élastique. *Ist. Lombardo (Rend. Sc.*, A108:274–300, 1974.
114. F. Brezzi. Sur la méthode des éléments finis hybrides pour le problème biharmonique. *Numer. Math.*, 24:103–131, 1975.
115. F. Brezzi. Hybrid approximations of nonlinear plate-bending problems. In *Hybrid and mixed finite element methods (Atlanta, Ga., 1981)*, pages 267–280. Wiley, Chichester, 1983.
116. F. Brezzi and K.J. Bathe. A discourse on the stability conditions for mixed finite element formulations. *CMAME*, 82:27–57, 1990.
117. F. Brezzi, K.J. Bathe, and M. Fortin. Mixed-interpolated elements for Reissner-Mindlin plates. *Int. J. Num. Meth. Eng.*, 28:1787–1801, 1989.
118. F. Brezzi, J. Douglas, R. Duran, and M. Fortin. Mixed finite elements for second order elliptic problems in three variables. *Numer. Math.*, 51:237–250, 1987.
119. F. Brezzi, J. Douglas, M. Fortin, and L.D. Marini. Efficient rectangular mixed finite elements in two and three space variables. *Math. Model. Numer. Anal.*, 21:581–604, 1987.
120. F. Brezzi, J. Douglas, and L.D. Marini. Recent results on mixed finite element methods for second order elliptic problems. In Balakrishanan, Dorodnitsyn, and Lions, editors, *Vistas in Applied Math., Numerical Analysis, Atmospheric Sciences, Immunology*. Optimization Software Publications, New York, 1986.
121. F. Brezzi and R.S. Falk. Stability of higher-order Hood-Taylor methods. *SIAM J. Numer. Anal.*, 28(3):581–590, 1991.
122. F. Brezzi and M. Fortin. *Mixed and hybrid finite element methods*. Springer-Verlag, New York, 1991.
123. F. Brezzi and M. Fortin. A minimal stabilisation procedure for mixed finite element methods. *Numer. Math.*, 89:457–492, 2001.
124. F. Brezzi, M. Fortin, and L. D. Marini. Mixed finite element methods with continuous stresses. *Math. Models Methods Appl. Sci.*, 3(2), 1993.
125. F. Brezzi, M. Fortin, and L.D. Marini. Error analysis of piecewise constant approximations of Darcy's law. *Comput. Methods Appl. Mech. Engrg*, 195:1547–1599, 2006.
126. F. Brezzi, M. Fortin, and R. Stenberg. Error analysis of mixed-interpolated elements for Reissner-Mindlin plates. *Math. Models Methods Appl. Sci.*, 1(2):125–151, 1991.
127. F. Brezzi and L.D. Marini. On the numerical solution of plate bending problems by hybrid methods. *R.A.I.R.O. Anal. Numer.*, pages 5–50, 1975.
128. F. Brezzi, L.D. Marini, and P. Pietra. Méthodes d'éléments finis mixtes et schéma de Scharfetter-Gummel. *C.R.A.S. Paris*, 305, I:599–604, 1987.
129. F. Brezzi, L.D. Marini, and P. Pietra. Numerical simulation of semi conductor devices. *Comp. Math. Appl. Mech. Eng.*, 75:493–514, 1989.
130. F. Brezzi, L.D. Marini, and P. Pietra. Two dimensional exponential fitting and application to drift-diffusion models. *SIAM. J. Num. Anal.*, 26:1347–1355, 1989.
131. F. Brezzi and J. Pitkäranta. On the stabilization of finite element approximations of the Stokes equations. In W. Hackbush, editor, *Efficient Solutions of Elliptic Systems*, volume 10 of *Notes on Numerical Fluid Mechanics*. Braunschweig, Wiesbaden, Vieweg, 1984.
132. F. Brezzi and P.A. Raviart. Mixed finite element methods for 4th order elliptic equations. In J. Miller, editor, *Topics in Numerical Analysis III*. Academic Press, New-York, 1978.
133. A. Buffa. Hodge decomposition on the boundary of a polyhedron: the multi-connected case. *Math. Mod. Meth. Appl. Sci.*, 11(9):1491–1504, 2001.
134. A. Buffa and P. Ciarlet Jr. On traces for functional spaces related to Maxwell's equations. I. An integration by parts formula in Lipschitz polyhedra. *Math. Methods Appl. Sci.*, 24(1):9–30, 2001.

135. A. Buffa and P. Ciarlet Jr. On traces for functional spaces related to Maxwell's equations. II. Hodge decompositions on the boundary of Lipschitz polyhedra and applications. *Math. Methods Appl. Sci.*, 24(1):31–48, 2001.
136. A. Buffa, M. Costabel, and D. Sheen. On traces for $H(curl, \Omega)$ in Lipschitz domains. *J. Math. Anal. Appl.*, 276(2):845–867, 2002.
137. S. Caorsi, P. Fernandes, and M. Raffetto. On the spurious modes in deterministic problems. *COMPEL*, 13, Supplement A:317–332, 1994.
138. S. Caorsi, P. Fernandes, and M. Raffetto. Towards a good characterization of spectrally correct finite element methods in electromagnetics. *COMPEL*, 15, 1996.
139. Carsten C. A posteriori error estimate for the mixed finite element method. *Math. Comp*, 66, 303 1997.
140. C. Carstensen and G. Dolzmann. A posteriori error estimates for mixed FEM in elasticity. *Numer. Math.*, 81:187209, 1998.
141. J. Céa. Approximation variationnelle des problèmes aux limites. *Ann. Inst. Fourier*, 14:2, 1964.
142. J. Céa. Approximation variationnelle et convergence des éléments finis; un test. *Journées Eléments Finis*, Université de Rennes, 1976.
143. M. Cessenat. *Mathematical methods in electromagnetism*, volume 41 of *Series on Advances in Mathematics for Applied Sciences*. World Scientific Publishing Co. Inc., River Edge, NJ, 1996.
144. D. Chapelle and K.-J. Bathe. The inf-sup test. *Comput. & Structures*, 47(4–5):537–545, 1993.
145. F. Chatelin. *Spectral approximation of linear operators*. Computer Science and Applied Mathematics. Academic Press Inc. [Harcourt Brace Jovanovich Publishers], New York, 1983. With a foreword by P. Henrici, With solutions to exercises by Mario Ahués.
146. P.G. Ciarlet. *The finite element method for elliptic problems*. North-Holland, Amsterdam, 1978.
147. P.G. Ciarlet. *Mathematical elasticity. Vol. I*. North-Holland Publishing Co., Amsterdam, 1988. Three-dimensional elasticity.
148. P.G. Ciarlet. *Mathematical elasticity. Vol. II*. North-Holland Publishing Co., Amsterdam, 1997. Theory of plates.
149. P.G. Ciarlet and P. Destuynder. A justification of the two-dimensional linear plate model. *J. Mécanique*, 18:315–344, 1979.
150. P.G. Ciarlet and R. Glowinski. Dual iterative techniques for solving a finite element approximation of the biharmonic equation. *Comp. Meth. Appl. Mech. Eng.*, 5:227–295, 1975.
151. P.G. Ciarlet and P.A. Raviart. Interpolation theory over curved elements with applications to finite element methods. *Comp. Meth. Appl. Mech. Eng.*, 1:217–249, 1972.
152. P.G. Ciarlet and P.A. Raviart. A mixed finite element method for the biharmonic equation. In C. de Boor, editor, *Mathematical Aspects of Finite Element in Partial Differential Equations*. Academic Press, New York, 1974.
153. J.F. Ciavaldini and J.C. Nédélec. Sur l'élément de Fraeijs de Veubeke et Sander. *R.A.I.R.O. Anal.Numer.*, 8:29–45, 1974.
154. P. Clément. Approximation by finite element functions using local regularization. *R.A.I.R.O. Anal. Mumer.*, 9:77–84, 1975.
155. B. Cockburn, J. Gopalakrishnan, and J. Guzmán. A new elasticity element made for enforcing weak stress symmetry. *Math. Comp*, 79:1331–1349, 2010.
156. B. Cockburn, G. Kanschat, and D. Schötzau. A locally conservative LDG method for the incompressible Navier-Stokes equations. *Math. Comp.*, 74(251):1067–1095 (electronic), 2005.
157. B. Cockburn, G. Kanschat, and D. Schötzau. A note on discontinuous Galerkin divergence-free solutions of the Navier-Stokes equations. *J. Sci. Comput.*, 31(1–2):61–73, 2007.
158. L. Comodi. The Hellan-Hermann-Johnson method: error estimates for the Lagrange multipliers and post processing. *Math. Comp.*, 52:17–30, 1989.
159. P. Constantin and C. Foias. *Navier-Stokes Equations*. University of Chicago Press, Chicago, 1988.

160. M. Costabel. A remark on the regularity of solutions of Maxwell's equations on Lipschitz domains. *Math. Methods Appl. Sci.*, 12(4):365–368, 1990.
161. M. Costabel and M. Dauge. Maxwell and Lamé eigenvalues on polyhedra. *Math. Methods Appl. Sci.*, 22(3):243–258, 1999.
162. M. Costabel and M. Dauge. Weighted regularization of Maxwell equations in polyhedral domains. A rehabilitation of nodal finite elements. *Numer. Math.*, 93(2):239–277, 2002.
163. M. Costabel and M. Dauge. Computation of resonance frequencies for Maxwell equations in non-smooth domains. In *Topics in computational wave propagation*, volume 31 of *Lect. Notes Comput. Sci. Eng.*, pages 125–161. Springer, Berlin, 2003.
164. M. Costabel, M. Dauge, and C. Schwab. Exponential convergence of *hp*-FEM for Maxwell equations with weighted regularization in polygonal domains. *Math. Models Methods Appl. Sci.*, 15(4):575–622, 2005.
165. M. Crouzeix and P.A. Raviart. Conforming and non-conforming finite element methods for solving the stationary Stokes equations. *R.A.I.R.O. Anal. Numer.*, 7:33–76, 1973.
166. R. Dautray and J.-L. Lions. *Mathematical analysis and numerical methods for science and technology. Vol. 1.* Springer-Verlag, Berlin, 1990.
167. R. Dautray and J.L. Lions. *Analyse mathématique et calcul numérique pour les sciences et les techniques.* Collection Commissariat à l'Energie Atomique. Masson, Paris, 1984.
168. M.C. Delfour and Z.-P. Zolésio. *Shapes and Geometries: Analysis, Differential Calculus, and Optimization.* Advances in Design and Control. SIAM, Philadelphia, 2001.
169. L. Demkowicz, P. Monk, Ch. Schwab, and L. Vardapetyan. Maxwell eigenvalues and discrete compactness in two dimensions. *Comput. Math. Appl.*, 40(4–5):589–605, 2000.
170. L. Demkowicz, P. Monk, L. Vardapetyan, and W. Rachowicz. De Rham diagram for *hp* finite element spaces. *Comput. Math. Appl.*, 39(7–8):29–38, 2000.
171. P. Destuynder. *Une théorie asymptotique des plaques minces en élasticité linéaire.* Masson, Paris, 1986.
172. J. Dieudonné. *Foundation of Modern Analysis.* Institut des Hautes Études Scientifiques, Paris. Academic Press, New York and London, 1960.
173. J. Douglas, R. Duran, and P. Pietra. Alternating-direction iteration for mixed finite element methods. In R. Glowinski and J.L. Lions, editors, *Computing Methods in Applied Sciences and Engineering 7.* North-Holland, Amsterdam, 1986.
174. J. Douglas, R. Duran, and P. Pietra. Formulation of alternating-direction iterative methods for mixed methods in three space. In E.L. Ortiz, editor, *Numerical Approximation of Partial Differential Equations.* North-Holland, Amsterdam, 1987.
175. J. Douglas and F. Milner. Interior and superconvergence estimates for mixed methods for second order elliptic problems. *Math. Modelling Numer. Anal.*, 19:397–428, 1985.
176. J. Douglas and P. Pietra. A description of some alternating-direction iterative techniques for mixed finite element methods. In W.E. Fitzgibbon, editor, *Mathematical andComputational Methods in Seismic Exploration and Reservoir Modeling.* SIAM, Philadelphia, 1986.
177. J. Douglas and J.E. Roberts. Mixed finite element methods for second order elliptic problems. *Math. Applic. Comp.*, 1:91–103, 1982.
178. J. Douglas and J.E. Roberts. Global estimates for mixed methods for second order elliptic equations. *Math. Comp.*, 44:39–52, 1985.
179. J. Douglas and J. Wang. An absolutely stabilized finite element method for the Stokes problem. *Math. Comput.*, 52:495–508, 1989.
180. T. Dupont and L.R. Scott. Polynomial approximation of functions in Sobolev spaces. *Math. of Comp.*, 34:441–463, 1980.
181. R. Duran and E. Liberman. On mixed finite element methods for the Reissner-Mindlin plate model. *Math. Comp*, 58(198):561–573, 1992.
182. R. G. Duran and E. Liberman. On the convergence of a triangular mixed finite element method for Reissner-Mindlin plates. *Math. Models Methods Appl. Sci*, 6:339–352, 1996.
183. G. Duvaut and J.L. Lions. *Les inéquations en mécanique et en physique.* Dunod, Paris, 1972.
184. I. Ekeland and R. Temam. *Analyse convexe et problèmes variationnels.* Dunod Gauthier Villars, Paris, 1974.

185. A. El Maliki, M. Fortin, N. Tardieu, and A. Fortin. Iterative solvers for 3D linear and nonlinear elasticity problems: Displacement and mixed formulations. *Int. J. Numerical Methods in Engineering*, 83, 2010.
186. H.C. Elman, D.J. Silvester, and A.J. Wathen. *Finite elements and fast iterative solvers: with applications in incompressible fluid dynamics*. Oxford University Press, 2005.
187. R. Eymard, Galloü, and R. Herbin. *Finite Volumes Methods*. Handbook of Numerical Analysis. Elsevier, 2000.
188. R.S. Falk and J.E. Osborn. Error estimates for mixed methods. *R.A.I.R.O. Anal. Numer.*, 4:249–277, 1980.
189. R.S. Falk. Approximation of the biharmonic equation by a mixed finite element method. SIAM J. Numer. Anal., 15, 556–567 (1978)
190. R.S. Falk. Finite elements for the Reissner-Mindlin plate. In D. Boffi and L. Gastaldi, editors, *Mixed Finite Elements,Compatibility Conditions and Applications*. Springer-Verlag, Berlin, 2008.
191. R.S. Falk, P. Gatto, and P. Monk. Hexahedral H(div) and H(curl) finite elements. *ESAIM: Mathematical Modelling and Numerical Analysis*, 45:115–143, 2011.
192. R.S. Falk and J.E. Osborn. Error estimates for mixed methods. *RAIRO Anal. Numér.*, 14(3):249–277, 1980.
193. Farhat, Charbel; Macedo, Antonini; Lesoinne, Michel A two-level domain decomposition method for the iterative solution of high frequency exterior Helmholtz problems. Numer. Math. 85 (2000), no. 2, 283–308.
194. M. Farhloul and M. Fortin. A new mixed finite element for the Stokes and elasticity problems. *SIAM Journal on Numerical Analysis*, 30(4), 1993.
195. M. Farhloul and M. Fortin. Review and complements on mixed-hybrid finite element methods for fluid flows. *Journal of Computational and Applied Mathematics*, 149(1–2), 2002.
196. P Fernandes and G. Gilardi. Magnetostatic and electrostatic problems in inhomogeneous anisotropic media with irregular boundary and mixed boundary conditions. *Math. Models Methods Appl. Sci.*, 7:957–991, 1997.
197. G.J. Fix, M.D. Gunzburger, and R.A. Nicolaides. Theory and applications of mixed finite element methods. In C.V. Coffman and G.J. Fix, editors, *Constructive Approaches to Mathematical Models*, pages 375–393. Academic Press, New York, 1979.
198. A. Fortin. *Méthodes d'éléments finis pour les équations de Navier–Stokes*. PhD thesis, Université Laval, 1984.
199. A. Fortin and M. Fortin. Newer and newer elements for incompressible flow. In R.H. Gallagher, G.F. Carey, J.T. Oden, and O.C. Zienkiewicz, editors, *Finite Elements in Fluids*, volume 6. John Wiley, Chichester, 1985.
200. M. Fortin. Utilisation de la méthode des éléments finis en mécanique des fluides. *Calcolo*, 12:405–441, 1975.
201. M. Fortin. An analysis of the convergence of mixed finite element methods. *R.A.I.R.O. Anal. Numer.*, 11:341–354, 1977.
202. M. Fortin. Old and new finite elements for incompressible flows. *Int. J. Num. Meth. in Fluids*, 1:347–364, 1981.
203. M. Fortin. A three-dimensional quadratic non-conforming element. *Mumer. Math.*, 46:269–279, 1985.
204. M. Fortin and M. Farhloul. Dual hybrid methods for the elasticity and the Stokes problems: a unified approach. *Numerische Mathematik*, 76:419–440, 1997.
205. M. Fortin and R. Glowinski. *Augmented Lagrangian methods*. North-Holland, Amsterdam, 1983.
206. M. Fortin, R. Peyret, and R. Temam. Résolution numérique des équations de Navier-Stokes pour un fluide visqueux incompressible. *Journal de mécanique*, 10, 3:357–390, 1971.
207. M. Fortin and R. Pierre. Stability analysis of discrete generalized Stokes problems. *Numer. Methods Partial Differential Equations*, 8(4):303–323, 1992.
208. M. Fortin and Guenette R. A new mixed finite element method for computing viscoelastic flows. *J. Non-Newtonian Fluid Mech.*, 60:27–52, 1995.

209. M. Fortin and M. Soulie. A non-conforming piecewise quadratic finite element on triangles. *Int. J. Num. Meth. Eng.*, 19:505–520, 1983.
210. B. Fraeijs de Veubeke. Displacement and equilibrium models in the finite element method. In O.C. Zienkiewicz and G. Holister, editors, *Stress Analysis*. John Wiley and Sons, New York, 1965.
211. B. Fraeijs de Veubeke. Variational principles and the patch test. *Int. J. Numer. Meth. Eng.*, 8:783–801, 1974.
212. B. Fraeijs de Veubeke. Stress function approach. In *World Congress on the Finite Element Method in Structural Mechanics*, Dorset, England, 1975. Bournemouth.
213. L.P. Franca. *New mixed finite element methods*. PhD thesis, Appl. Mech. Divi., Stanford University, 1987.
214. Franca, Leopoldo P.; Macedo, Antonini P. A two-level finite element method and its application to the Helmholtz equation. Internat. J. Numer. Methods Engrg. 43 (1998), no. 1, 23–32.
215. L.P. Franca and R. Stenberg. Error analysis of some Galerkin least-squares methods of the elasticity equations. Technical report, INRIA, 1989.
216. F. Gardini. Discrete compactness property for quadrilateral finite element spaces. *Numer. Methods Partial Differential Equations*, 21(1):41–56, 2005.
217. L. Gastaldi. Mixed finite element methods in fluid structure systems. *Numer. Math.*, 74(2):153–176, 1996.
218. L. Gastaldi and R.H. Nochetto. Optimal L^{∞}-error estimates for nonconforming and mixed finite element methods of lowest order. *Numer. Math.*, 50(5):587–611, 1987.
219. L. Gastaldi and R.H. Nochetto. Sharp maximum norm error estimates for general mixed finite element approximations to second order elliptic equations. *RAIRO Modél. Math. Anal. Numér.*, 23(1):103–128, 1989.
220. G. N. Gatica and M Maischak. A posteriori error estimates for the mixed finite element method with Lagrange multipliers. *Numer. Methods Partial Differential Eq.*, 21:421450, 2005.
221. P.L. George and H. Bourouchaki. *Delauney Triangulation and Meshing: Application to Finite Elements*. Hermes Science Publications, 1998.
222. V. Girault and P.A. Raviart. *Finite Element Approximation of Navier-Stokes Equations*, volume 749 of *Lectures Notes in Math.* Springer-Verlag, Berlin, 1979.
223. V. Girault and P.A. Raviart. *Finite Element Methods for Navier-Stokes Equations, Theory and Algorithms*. Springer-Verlag, Berlin, 1986.
224. R. Glowinski. Approximations externes par éléments finis de Lagrange d'ordre un et deux, du problème de Dirichlet pour l'opérateur biharmonique. Méthodes itératives de résolution des problèmes approchés. In J. Miller, editor, *Topics in Numerical Analysis*, pages 123–171, New York, 1973. Academic Press.
225. R. Glowinski. Approximations externes par éléments finis d'ordre 1 et 2 du problème de Dirichlet pour l'opérateur biharmonique,methode itérative de résolution des problèmes approchés. In J. Miller, editor, *Topics in Numerical Analysis*. Academic Press, New York, 1973.
226. R. Glowinski. *Numerical Methods for Nonlinear Variational Problems*. Springer-Verlag, Berlin, 1984.
227. R. Glowinski and O. Pironneau. Numerical methods for the first biharmonic equation and for the two-dimensional Stokes problem. *SIAM Review*, 17:167–212, 1979.
228. G.H. Golub and C.F. Van Loan. *Matrix Computations (3rd edition*. Johns Hopkins, Baltimore, 1996.
229. J. Gopalakrishnan and J. Guzmán. Symmetric non-conforming mixed finite elements for linear elasticity. *SIAM Journal on Numerical Analysis*, 49(4):1504–1520, 2011.
230. J. Gopalakrishnan and J. Guzmán. A second elasticity element using the matrix bubble. *IMA J. Numer. Anal.*, 32:352–372, 2012.
231. P. M. Gresho, R. L. Lee, S. T. Chan, and J. M. Leone. A new finite element for Boussinesq fluids. In *Pro. Third Int. Conf. on Finite Elements in Flow Problems*, pages 204–215. Wiley, New York, 1980.

232. D. F. Griffiths. The effect of pressure approximation on finite element calculations of compressible flows. In *Numerical Methods for Fluid Dynamics*, pages 359–374. Academic Press, Morton, K. W. and Baines, M. J. edition, 1982.
233. P. Grisvard. *Elliptic Problems in Non-Smooth Domains*. Pitman, London, 1985.
234. P. Grisvard. *Singularities in Boundary Value Problems*. Masson, Paris, 1992.
235. C. Gruau and T. Coupez. 3D tetrahedral, unstructured and anisotropic mesh generation with adaptation to natural and multidomain metric. *Comp. Meth in Appl. Mech and Engrg*, 194:4951–4976, 2005.
236. J. Guzmán. A unified analysis of several mixed methods for elasticity with weak symmetry. *J. Sci. Comp.*, 44:156–169, 2010.
237. J. Guzmán and M. Neilan. Conforming and divergence-free Stokes elements on general triangular meshes. Math. Comp., to appear
238. J. Guzmán and M. Neilan. A family of non-conforming elements for the Brinkman problem. *IMA J. Num. Anal.*, 32(4):1484–1508, 2012.
239. W.G. Habashi, J. Dompierre, Y. Bourgault, D. Ait Ali Yahia, M. Fortin, and M.G. Vallet. Anisotropic mesh adaptation: towards user-independent mesh-independent and solver-independent CFD. part i: general principles. *International Journal for Numerical Methods in Fluids*, 32:725–744, 2000.
240. F.H. Harlow and R.E. Welsch. Numerical calculation of time dependent viscous incompressible flow. *Phys. Fluids*, 8:–2182, 1965.
241. Y. Haugazeau and P. Lacoste. Condensation de la matrice masse pour les éléments finis mixtes de H(rot). *C. R. Acad. Sci. Paris*, 316, série I:509–512, 1993.
242. C. Hazard. Numerical simulation of corner singularities: a paradox in Maxwell-like problems. *C.R. Mecanique*, 330:57–68, 2002.
243. K. Hellan. *Analysis of elastic plates in flexure by a simplified finite element method*, volume 46 of *Civil Engineering Series*. Acta Polytechnica Scandinavia, Trondheim, 1967.
244. J.P. Hennart, J. Jaffré, and J.E. Roberts. A constructive method for deriving finite elements of nodal type. *Numer. Math.*, 55:701–738, 1988.
245. L.R. Herrmann. Finite element bending analysis for plates. *J. Eng. Mech. Div. ASCE*, 93, EM5:13–26, 1967.
246. M. Hestenes. Multiplier and gradient methods. *J. Opt. Theory and App.*, 4:303–320, 1969.
247. R. Hiptmair. Canonical construction of finite elements. *Math. Comp.*, 68(228):1325–1346, 1999.
248. R. Hiptmair. Finite elements in computational electromagnetism. *Acta Numer.*, 11:237–339, 2002.
249. P. Hood and C. Taylor. Numerical solution of the Navier–Stokes equations using the finite element technique. *Comput. Fluids*, 1:1–28, 1973.
250. Hou, Thomas Y.; Wu, Xiao-Hui; Cai, Zhiqiang Convergence of a multiscale finite element method for elliptic problems with rapidly oscillating coefficients. Math. Comp. 68 (1999), no. 227, 913–943.
251. P. Houston, D. Schötzau, and T.P. Wihler. Energy norm a posteriori error estimation for mixed discontinuous Galerkin approximations of the Stokes problem. *Journal of Scientific Computing*, 22–23(1):347–370, 2005.
252. W. Huang. Metric tensors for anisotropic mesh generation. *J. Comput. Phys.*, 204:633–665, 2005.
253. Y. Huang and S. Zhang. A lowest order divergence-free finite element on rectangular grids. *Front. Math. China*, 6(2):253–270, 2011.
254. T.J.R. Hughes. *The finite element method: linear static and dynamic finite element analysis*. Prentice-Hall, Englewood Cliffs N.J., 1987.
255. T.J.R. Hughes and H. Allik. Finite elements for compressible and incompressible continua. In *Proceedings of the Symposium on Civil Engineering*, pages 27–62, Nashville Tenn., 1969. Vanderbilt University.

256. T.J.R. Hughes and L.P. Franca. A new finite element formulation for computational fluid dynamics: VII. the Stokes problem with various well-posed boundary conditions, symmetric formulations that converge for all velocity-pressure spaces. *Comp. Meth. Appl. Mech. Eng.*, 65:85–96, 1987.
257. T.J.R. Hughes, L.P. Franca, and M. Balestra. A new finite element formulation of computational fluid dynamics: a stable Petrov-Galerkin formulation of the Stokes problem accomodating equal-order interpolations. *Comp. Meth. Appl. Mech.Eng.*, 59:85–99, 1986.
258. T.J.R. Hughes and T.E. Tezduyar. Finite elements based upon Mindlin plate theory with particular reference to the four-node bilinear isoparametric element. *Jour. of App. Mech.*, 48:587–596, 1981.
259. B.M. Irons and A. Razzaque. Experience with the patch-test for convergence of finite elements. In A.K. Aziz, editor, *Mathematics of Finite Element Method with Applications to Partial Differential Equations*. Univ. of Maryland, Baltimore, 1972.
260. P. Jamet. Estimation d'erreur pour des éléments finis droits presque dégénérés. *R.A.I.R.O. Anal. Numer.*, 10, 3:43–62, 1976.
261. C. Johnson. On the convergence of a mixed finite element method for plate bending problems. *Numer. Math.*, 21:43–62, 1973.
262. C. Johnson and B. Mercier. Some equilibrium finite element methods for two-dimensional elasticity problems. *Numer. Math.*, 30:103–116, 1978.
263. C. Johnson and J. Pitkäranta. Analysis of some mixed finite element methods related to reduced integration. *Math. Comp.*, 38:375–400, 1982.
264. C. Johnson and J. Pitkäranta. Analysis of some mixed finite element methods related to reduced integration. *Math. Comp.*, 38:375–400, 1982.
265. C. Johnson and V. Thomée. Error estimates for some mixed finite element methods for parabolic type problems. *R.A.I.R.O. Anal. Numer.*, 15:41–78, 1981.
266. J. E. Jones, Z. Cai, S. F. Mccormick, and T. F. Russell. Control-volume mixed finite element methods. *Comput. Geosci.*, 1:289–315, 1997.
267. R.B. Kellogg and J.E. Osborn. A regularity result for the Stokes problem. *J. Funct. Anal.*, 21:397–431, 1976.
268. F. Kikuchi. Mixed and penalty formulations for finite element analysis of an eigenvalue problem in electromagnetism. In *Proceedings of the first world congress on computational mechanics (Austin, Tex., 1986)*, volume 64 (1–3), pages 509–521, 1987.
269. F. Kikuchi. On a discrete compactness property for the Nédélec finite elements. *J. Fac. Sci. Univ. Tokyo Sect. IA Math.*, 36(3):479–490, 1989.
270. F. Kikuchi, M. Okabe, and H. Fujio. Modification of the 8-node serendipity element. *Comp. Methods Appl. Mech. Engrg.*, 179:91–109, 1999.
271. N. Kikuchi and J.T. Oden. *Contact problems in Elasticity: A Study of Variational Inequalities and Finite Element Methods*. SIAM Studies in Applied mathematics. SIAM, Philadelphia, 1988.
272. O.A. Ladyzhenskaya. *The mathematical theory of viscous incomprehensible flow*. Gordon and Breach Science Publishers, New York, 1969.
273. L. D. Landau and E. M. Lifshitz. *Course of theoretical physics. Vol. 8*. Pergamon International Library of Science, Technology, Engineering and Social Studies. Pergamon Press, Oxford, 1984.
274. M.G. Larson and A. Målqvist. A posteriori error estimates for mixed finite element approximations of elliptic problems. *Numer. Math*, 108:487500, 2008.
275. P. Lascaux and P. Lesaint. Some non-conforming finite elements for the plate bending problem. *R.A.I.R.O. Anal. Numer.*, 9:9–53, 1975.
276. P. Le Tallec and V. Ruas. On the convergence of the bilinear velocity-constant pressure finite method in viscous flow. *Comp. Meth. Appl. Mech. Eng.*, 54:235–243, 1986.
277. R. Leis. *Initial-boundary value problems in mathematical physics*. B. G. Teubner, Stuttgart, 1986.
278. P.G. Lemarié-Rieusset. *Recent Developments in the Navier-Stokes Problem*. CRC Research Notes in Mathematics 431. Chapman and Hall, Boca Raton, 2002.

279. P. Lesaint. Nodal methods for the transport equation. In *The Mathematics of Finite Elements and Applications V*, pages 562–569, Oxbridge, England, 1985. MAFELAP 984, Proc 5th Conf.
280. A. Linke. Collision in a cross-shaped domain—a steady 2d Navier-Stokes example demonstrating the importance of mass conservation in CFD. *Comput. Methods Appl. Mech. Engrg.*, 198(41–44):3278–3286, 2009.
281. J.L. Lions and E. Magenes. *Problèmes aux limites non-homogènes et applications*. Dunod, Paris, 1968.
282. M. Lonsing and Verfürth. A posteriori error estimators for mixed finite element methods in linear elasticity. *Numer. Math*, 97:757778, 2004.
283. C. Lovadina. Analysis of strain-pressure finite element methods for the Stokes problem. *Numer. Methods for PDEs*, 13:717–730, 1997.
284. C. Lovadina and Auricchio F. On the enhanced strain technique for elasticity problems. *Computers and Structures*, 81:777–787, 2003.
285. C. Lovadina and R. Stenberg. Energy norm a posteriori error estimates for mixed finite element methods. *Math.Comp.*, 75:1659–1674, 2006.
286. D.S. Malkus. Eigenproblems associated with the discrete LBB-condition for incompressible finite elements. *Int. J. Eng. Sci.*, 19:1299–1310, 1981.
287. D.S Malkus and T.J.R. Hughes. Mixed finite element methods. reduced and selective integration techniques: a unification of concepts. *Comp. Methods Appl. Mech. Eng.*, 15:63–81, 1978.
288. L. Mansfield. On finite element subspaces on quadrilateral and hexahedral meshes for incompressible viscous flow problems. *Numer. Math.*, 45:165–172, 1984.
289. L.D. Marini. Implementation of hybrid finite element methods and associated numerical problems, part 1. Technical Report 36, IAN-CNR, Pavia, 1976.
290. L.D. Marini. Implementation of hybrid finite element methods and associated numerical problems, part 2. Technical Report 182, IAN-CNR, Pavia, 1978.
291. L.D. Marini. An inexpensive method for the evaluation of the solution of the lower order Raviart-Thomas mixed method. *SIAM J. Numer. Anal.*, 22:493–496, 1985.
292. L.D. Marini and P. Pietra. An abstract theory for mixed approximations of second order elliptic problems. *Mat. Apl. Comput.*, 8(3):219–239, 1989.
293. L.D. Marini and P. Pietra. New mixed finite element schemes for current continuity equations. *COMPEL*, 9(4):257–268, 1990.
294. L.D. Marini and A. Savini. Accurate computation of electric field in reverse biased semiconductor devices: a mixed finite element approach. *Compel*, 3:123–135, 1984.
295. J.E. Marsden and T.J.R. Hughes. *Mathematical Foundations of Elasticity*. Prentice Hall, New York, 1983.
296. G. Matthies. Mapped finite elements on hexahedra. Necessary and sufficient conditions for optimal interpolation errors. *Numer. Algorithms*, 27(4):317–327, 2001.
297. R. H. McNeal and R. L. Harder. Eight nodes or nine? *Int. J. Numer. Methods Engrg.*, 33:1049–1058, 1992.
298. B. Mercier. Numerical solution of the biharmonic problem by mixed finite elements of class C^0. *Boll. U.M.I.*, 10:133–149, 1974.
299. B. Mercier, J. Osborn, J. Rappaz, and P.A. Raviart. Eigenvalue approximation by mixed and hybrid methods. *Math. Comp.*, 36:427–453, 1981.
300. T. Miyoshi. A finite element method for the solution of fourth order partial differential equations. *Kumamoto J. Sci. Math.*, 9:87–116, 1973.
301. B. Mohammadi, P-L. George, F. Hecht, and E. Saltel. 3D mesh adaptation by metric control for CFD. *Revue européenne des éléments finis*, 9:439–449, 2000.
302. P. Monk. *Finite element methods for Maxwell's equations*. Numerical Mathematics and Scientific Computation. Oxford University Press, New York, 2003.
303. P. Monk and L. Demkowicz. Discrete compactness and the approximation of Maxwell's equations in $\mathbb{R}^3$. *Math. Comp.*, 70(234):507–523, 2001.

304. P. Monk, Y. Wang, and B. Szabo. Computing cavity models using the p-version of the finite element method. *IEEE Transactions on Magnetics*, 32:37–46, 1996.
305. M.E. Morley. A family of mixed finite elements for linear elasticity. *Numer. Math.*, 55(6):633–666, 1989.
306. S.E. Mousavi and Sukumar N. Numerical integration of polynomials and discontinuous functions on irregularconvex polygons and polyhedrons. *Comput.Mech.*, 47:535–554, 2011.
307. Russel T.F. Naff R.L. and Wilson J. D. Shape functions for velocity interpolation in general hexahedral cells. *Comput. Geosci.*, 6:285–314, 2002.
308. A. Naga and Z. Zhang. A posteriori error estimates based on the polynomial preserving recovery. *SIAM J. Numer. Anal*, 42:1780–1800, 2005.
309. J. Nečas. *Les méthodes directes en théorie des équations elliptiques*. Masson, Paris, 1967.
310. J.-C. Nédélec. Mixed finite elements in $\mathbb{R}^3$. *Numer. Math.*, 35(3):315–341, 1980.
311. J.-C. Nédélec. A new family of mixed finite elements in $\mathbb{R}^3$. *Numer. Math.*, 50(1):57–81, 1986.
312. R.A. Nicolaides. Existence, uniqueness and approximation for generalized saddle point problems. *SIAM J. Numer. Anal.*, 19:349–357, 1982.
313. J. Nitsche. Ein Kriterium für die Quasi-Optimalität des Ritzchen Verfahrens. *Numer. Math.*, 11:346–348, 1968.
314. J.T. Oden and O. Jacquotte. Stability of some mixed finite element methods for Stokesian flows. *Comp. Meth. App. Mech. Eng.*, 43:231–247, 1984.
315. J.T. Oden and J.N. Reddy. On mixed finite element approximations. *SIAM J. Numer. Anal.*, 13:393–404, 1976.
316. J.E. Osborn. Eigenvalue approximations by mixed methods. In R. Vichnevetsky and R. Stepleman, editors, *Advances in Computer Methods for Partial Differential Equations III*, pages 158–161. New Brunswick, 1979.
317. L. Paquet. Problèmes mixtes pour le système de Maxwell. *Ann. Fac. Sci. Toulouse Math. (5)*, 4(2):103–141, 1982.
318. A. Pechstein and J. Schöberl. Tangential-displacement and normal-normal-stress continuous mixed finite elements for elasticity. *Mathematical Models and Methods in Applied Sciences*, 21(8):1761–1782, 2011.
319. A. Pechstein and J. Schöberl. Anisotropic mixed finite elements for elasticity. *Int. J. Numer. Meth. Engng*, 90:196–217, 2012.
320. T.H.H. Pian. Formulations of finite element methods for solid continua. In R.H. Gallagher, Y. Yamada, and J.T. Oden, editors, *Recent Advances in Matrix Methods Structural Analysis and Design*. The University of Alabama Press, Alabama, 1971.
321. T.H.H. Pian and P. Tong. Basis of finite element methods for solid continua. *Int. J. Num. Meth. Eng.*, 1:3–28, 1969.
322. R. Pierre. Local mass conservation and C^0-discretizations of the Stokes problem. *Houston J. Math.*, 20(1):115–127, 1994.
323. J. Pitkäranta. Analysis of some low-order finite element schemes for Mindlin-Reissner and Kirchhoff plates. *Numer. Math.*, 53:237–254, 1988.
324. J. Pitkäranta and R. Stenberg. Analysis of some mixed finite element methods for plane elasticity equations. *Math. Comp.*, 41:399–423, 1983.
325. M.J.D. Powell. A method for non-linear constraints in minimization problems. In R. Fletcher, editor, *Optimization*. Academic Press, London, 1969.
326. M.J.D. Powell. *Piecewise quadratic surface fitting for contour plotting*, pages 253–271. Software for numerical mathematics. London: Academic, 1974.
327. K. Preis, O. Bíró, and I. Ticar. Gauged current vector potential and reentrant corners in the FEM analysis of 3D eddy currents. *IEEE Transaction on Magnetics*, 36(4):840–843, 2000.
328. J. Qin. *On the convergence of some simple finite elements for incompressible flows*. PhD thesis, Penn State University, 1994.
329. J. Qin and S. Zhang. Stability of the finite elements $9/(4c+1)$ and $9/5c$ for stationary Stokes equations. *Comput. & Structures*, 84(1–2):70–77, 2005.

330. R. Rannacher and S. Turek. Simple nonconforming quadrilateral Stokes element. *Numer. Methods Partial Differential Equations*, 8(2):97–111, 1992.
331. P.A. Raviart and J.M. Thomas. A mixed finite element method for second order elliptic problems. In I. Galligani and E. Magenes, editors, *Mathematical Aspects of the Finite Element Method*, volume 606 of *Lectures Notes in Math.* Springer-Verlag, New York, 1977.
332. P.A. Raviart and J.M. Thomas. Primal hybrid finite element methods for second order elliptic equations. *Math. of Comp.*, 31:391–413, 1977.
333. P.A. Raviart and J.M Thomas. Dual finite element models for second order elliptic problems. In R. Glowinski, E.Y. Rodin, and O.C. Zienkiewicz, editors, *Energy methods in Finite Element Analysis*. John Wiley and Sons, Chichester, 1979.
334. P.A. Raviart and J.M. Thomas. *Introduction à l'analyse numérique des équations aux dérivées partielles*. Masson, Paris, 1983.
335. E. Reisnerr. On a variational theorem for finite elastic deformations. *J. Math. Physics*, 32:129–135, 1953.
336. E. Reisnerr. On the variational theorem in elasticity. *J. Math. Physics*, 29:90–95, 1958.
337. J.E. Roberts and J.M. Thomas. Mixed and hybrid methods. In P.G. Ciarlet and J.L. Lions, editors, *Handbook of Numerical Analysis, Finite Element Methods Part 1*, volume II. North-Holland, Amsterdam, 1989.
338. R.T. Rockafellar. *Convex Analysis*. Princeton University Press, Princeton, N.J., 1970.
339. G. Sangalli, Capturing small scales in elliptic problems using a Residual-Free Bubbles Finite Element Method, Multiscale Modeling and Simulation: A SIAM Interdisciplinary Journal, Vol. 1 (3), pp. 485–503, 2003
340. R. L. Sani, P. M. Gresho, R. L. Lee, D.F. Griffiths, and M. Engelman. The cause and cure (!) of the spurious pressures generated by certain FEM solutions of the incompressible Navier–Stokes equations. II. *Internat. J. Numer. Methods Fluids*, 1(2):171–204, 1981.
341. R.L. Sani, P.M. Gresho, R.L. Lee, and D.F. Griffiths. The cause and cure (?) of the spurious pressures generated by certain FEM solutions of the incompressible Navier–Stokes equations. I. *Internat. J. Numer. Methods Fluids*, 1(1):17–43, 1981.
342. R. Scholz. Approximation von Sattelpunkten mit Finiten Elementen. *Bonner Mathematischen Schriften*, 89:54–66, 1976.
343. R. Scholz. L^∞-convergence of saddle point approximations for second order problems. *R.A.I.R.O. Anal. Numer.*, 11:209–216, 1977.
344. R. Scholz. A mixed method for fourth order problems using linear finite elements. *R.A.I.R.O. Anal. Numer.*, 12:85–90, 1978.
345. R. Scholz. A remark on the rate of convergence for a mixed finite element method for second order problems. *Numer. Funct. Anal. Optim.*, 4:269–277, 1982.
346. R. Scholz. Optimal L^∞-estimates for a mixed finite element method for elliptic and parabolic problems. *Calcolo*, 20:355–377, 1983.
347. L.R. Scott and M. Vogelius. Norm estimates for a maximal right inverse of the divergence operator in spaces of piecewise polynomials. *Math. Modelling Numer. Anal.*, 9:11–43, 1985.
348. J. Shen. Mixed finite element methods on distorted rectangular grids. Technical Report ISC-94-13-MATH, Texas A&M University, 1994.
349. D.J. Silvester and N. Kechkar. Stabilized bilinear-constant velocity-pressure finite elements for the conjugate gradient solution of the Stokes problem. *Comput. Methods Appl. Mech. Engrg.*, 79:7186, 1990.
350. H. Sohr. *The Navier-Stokes Equations*. Birkhauser, Basel, 2008.
351. R. Stenberg. Analysis of mixed finite element methods for the Stokes problem: a unified approach. *Math. of Comp.*, 42:9–23, 1984.
352. R. Stenberg. On the construction of optimal mixed finite element methods for the linear elasticity problem. *Numer. Math.*, 48:447–462, 1986.
353. R. Stenberg. On some three-dimensional finite elements for incompressible media. *Comp. Meth. Appl. Mech. Eng.*, 63:261–269, 1987.
354. R. Stenberg. On the postprocessing of mixed equilibrium finite element methods. In W. Hackbusch and K. Witsch, editors, *Numerical Tehchniques in Continuum Mechanics*. Veiweg, Braunschweig, 1987. Proceedings of the Second GAMM-Seminar, Kiel 1986.

355. R. Stenberg. Error analysis of some finite element methods for the Stokes problem. Technical Report 948, INRIA, Domaine de Voluceau, B.P.105, 78153, Le Chesnay, France, 1988.
356. R. Stenberg. A family of mixed finite elements for the elasticity problem. *Numer. Math.*, 53:513–538, 1988.
357. R. Stenberg. Two low-order mixed methods for the elasticity problem. In J.R. Whiteman, editor, *The mathematics of finite elements and applications*, pages 271–280. Academic Press, London, 1988.
358. G. Strang and G.J. Fix. *An Analysis of the Finite Element Method*. Prentice Hall, New York, 1973. now published by Wellesley-Cambridge Press.
359. F. Stummel. Diskrete Konvergenz linearer Operatoren. I. *Math Ann.*, 190:45–92, 1970/71.
360. F. Stummel. The generalized patch test. *SIAM, J. Numer. Anal.*, 16:449–471, 1979.
361. S. Sun and M. F. Wheeler. Symmetric and non symmetric discontinuous Galerkin methods for reactive transport in porous media. *SIAM J. Numer. Anal.*, 43(1):195–219, 2005.
362. R. Temam. *Navier-Stokes Equations*. North-Holland, Amsterdam, 1977.
363. R. Temam. *Navier-Stokes Equations and Nonlinear Functional Analysis, 2nd Ed.* SIAM, Philadelphia, 1995.
364. R. W. Thatcher. Locally mass-conserving Taylor-Hood elements for two- and three-dimensional flow. *Internat. J. Numer. Methods Fluids*, 11(3):341–353, 1990.
365. J.M Thomas. Méthode des élḿents finis hybrides duaux pour les problèmes du second ordre. *R.A.I.R.O. Anal. Numer.*, 10:51–79, 1976.
366. J.M Thomas. Sur l'analyse numérique des méthodes d'éléments finis hybrides et mixtes. Master's thesis, Université Pierre et Marie Curie, Paris, 1977.
367. T.P. Thomas-Peter Fries and Ted Belytschko T. The extended/generalized finite element method: An overview of the method and its applications. *Int. J. Numer. Meth. Engng.*, 84(3):253–304, 2010.
368. F. Thomasset. *Implementation of finite element methods for Navier–Stokes Equations*. Springer Series in Comp. Physics. Springer-Verlag, Berlin, 1981.
369. D. M. Tidd, R. W. Thatcher, and A. Kaye. The free surface flow of Newtonian and non-Newtonian fluids trapped by surface tension. *Internat. J. Numer. Methods Fluids*, 8(9):1011–1027, 1988.
370. P. Tong and T.H.H. Pian. A variational principle and the convergence of a finite element method based on assumed stress distribution. *Int. J. Solids Struct.*, 5:463–472, 1969.
371. L. Vardapetyan, L. Demkowicz, and D. Neikirk. hp-vector finite element method for eigenmode analysis of waveguides. *Comput. Methods Appl. Mech. Engrg.*, 192(1–2):185–201, 2003.
372. R. Varga. *Matrix Iterative Analysis*. Prentice-Hall, Englewood Cliffs, N.J., 1962.
373. G.M. Vaĭnikko. Discretely compact sequences. (russian). *Ž. Vyčisl. Mat. i Mat. Fiz.*, 14:575–583, 1974.
374. R. Verfürth. A combined conjugate gradient multi-grid algorithm for the numerical solution of the Stokes problem. *IMA J. Numer.Anal.*, 4:441–455, 1984.
375. R. Verfürth. Error estimates for a mixed finite element approximation of the Stokes equation. *R.A.I.R.O. Anal. Numer.*, 18:175–182, 1984.
376. R. Verfürth. *A Review of a Posteriori Error Estimation and Adaptive Mesh-Refinement Techniques*. Advances in numerical mathematics. Wiley-Teubner, 1996.
377. E. L. Wachspress. *A Rational Finite Element Basis*. Academic Press, New York, 1975.
378. X. Wang and K.-J. Bathe. On mixed elements for acoustic fluid-structure interaction. *Math. Models Methods Appl. Sci.*, 7(3):329–343, 1997.
379. H. Whitney. *Geometric integration theory*. Princeton University Press, Princeton, N. J., 1957.
380. B.I. Wohlmuth and R. H. W. Ronald H. W. Hoppe. A comparison of a posteriori error estimators for mixed finite element discretizations by Raviart-Thomas elements. *MATH. COMP*, 68:1347–1378, 1999.
381. S. Wong and Z. Cendes. Combined finite element-modal solution of three-dimensional eddy current problems. *IEEE Transactions on Magnetics*, 24:2685–2687, 1988.
382. K. Yosida. *Functional analysis*. Springer-Verlag, New York, 1966.

383. A. Younès, P. Ackerer, and G. Chavent. From mixed finite elements to finite volumes for elliptic pdes in two and three dimensions. *Internat. J. Numer. Methods Engrg*, 59:365388, 2004.
384. J. Zhang and F. Kikuchi. Interpolation error estimates of a modified 8-node serendipity finite element. *Numer. Math.*, 85(3):503–524, 2000.
385. S. Zhang. A family of $Q_{k+1,k} \times Q_{k,k+1}$ divergence-free finite elements on rectangular grids. *SIAM J. Numer. Anal.*, 47(3):2090–2107, 2009.
386. O. C. Zienkiewicz and R. L. Taylor. *The Finite Element Method, Fourth edition, Volume 1: Basic Formulation and Linear Problems*. McGraw-Hill, London, 1989.
387. O.C. Zienkiewicz. *The finite element method*. McGraw-Hill, London, 1977.
388. O.C. Zienkiewicz, S. Qu, R.L. Taylor, and S. Nakazawa. The patch text for mixed formulations. *Int. J. Num. Meth. Eng.*, 23:1873–1883, 1986.

Index

D. Boffi et al., *Mixed Finite Element Methods and Applications*, Springer Series in Computational Mathematics 44, DOI 10.1007/978-3-642-36519-5,

MIX
Papier aus verantwortungsvollen Quellen
Paper from responsible sources
FSC® C105338

If you have any concerns about our products,
you can contact us on
ProductSafety@springernature.com

In case Publisher is established outside the EU,
the EU authorized representative is:
Springer Nature Customer Service Center GmbH
Europaplatz 3, 69115 Heidelberg, Germany

Printed by Libri Plureos GmbH
in Hamburg, Germany